AF692265

Zebrafish as a Model for Parkinson's Disease

The increasing demand for innovative techniques arises from the lack of safe, effective, and patient-friendly therapies for neurodegenerative disorders. With this objective in mind, the chapters of the book are structured to offer a thorough insight into recent advancements in utilizing the zebrafish (ZF) as a model for studying Parkinson's disease (PD). This book aims to present readers with a comprehensive understanding of the clinical application of the ZF model in treating PD, encompassing the latest developments, challenges, safety considerations, toxicity issues, regulatory aspects, future potential, and limitations. Individuals in academia, the scientific community, business, and education seeking a more effective approach to target the brain stand to benefit from this resource.

Key Features

- Provides a comparative perspective of the zebrafish–Parkinson's disease model
- Highlights the restrictions of available medicines
- Describes biochemical and histopathological characteristics, advantages, and disadvantages of this model
- Emphasizes distinct facets of histopathology
- Presents advances and developments of the future potential perspectives

Zebrafish as a Model for Parkinson's Disease

Edited by Wael Mohamed

CRC Press
Taylor & Francis Group
Boca Raton London New York

CRC Press is an imprint of the
Taylor & Francis Group, an **informa** business

Designed cover image: ShutterStock Images

First edition published 2025

by CRC Press
2385 NW Executive Center Drive, Suite 320, Boca Raton FL 33431

and by CRC Press
4 Park Square, Milton Park, Abingdon, Oxon, OX14 4RN

CRC Press is an imprint of Taylor & Francis Group, LLC

© 2025 Taylor & Francis Group, LLC

Reasonable efforts have been made to publish reliable data and information, but the author and publisher cannot assume responsibility for the validity of all materials or the consequences of their use. The authors and publishers have attempted to trace the copyright holders of all material reproduced in this publication and apologize to copyright holders if permission to publish in this form has not been obtained. If any copyright material has not been acknowledged please write and let us know so we may rectify in any future reprint.

Except as permitted under U.S. Copyright Law, no part of this book may be reprinted, reproduced, transmitted, or utilized in any form by any electronic, mechanical, or other means, now known or hereafter invented, including photocopying, microfilming, and recording, or in any information storage or retrieval system, without written permission from the publishers.

For permission to photocopy or use material electronically from this work, access www.copyright.com or contact the Copyright Clearance Center, Inc. (CCC), 222 Rosewood Drive, Danvers, MA 01923, 978–750–8400. For works that are not available on CCC please contact mpkbookspermissions@tandf.co.uk

Trademark notice: Product or corporate names may be trademarks or registered trademarks and are used only for identification and explanation without intent to infringe.

ISBN: 9781032515779 (hbk)
ISBN: 9781032515786 (pbk)
ISBN: 9781003402893 (ebk)

DOI: 10.1201/9781003402893

Typeset in Times
by Apex CoVantage, LLC

I express heartfelt gratitude and deep appreciation as I dedicate this work to my wife, Dr. Rehab Ismaeil. Your unwavering support and unconditional love have been constant pillars throughout my academic journey and beyond. I am truly grateful for the love and inspiration you bring to my life, and I acknowledge that none of this would be possible without you.

Wael Mohamed
Pahang, Malaysia

Contents

SECTION II Zebrafish as a Model of Parkinson's Disease

Preface

> "To study the phenomenon of disease without books is to sail an uncharted sea, while to study books without patients is not to go to sea at all."
>
> **—William Osler (1849–1919)**

Neurodegenerative disorders, encompassing conditions like Alzheimer's and Parkinson's, play a significant role in the global burden of chronic and debilitating illnesses. In essence, they constitute the leading cause of disability-adjusted life years (DALYs) and rank as the second leading cause of mortality worldwide. The lack of effective treatments and public awareness has exacerbated the severity of this issue. Parkinson's disease (PD), the second most prevalent neurodegenerative ailment after Alzheimer's, is characterized by the loss of midbrain dopaminergic neurons and age-dependent motor impairment. A pathogenic aspect of PD and associated illnesses, such as dementia with Lewy bodies, involves the deposition of neuronal inclusions known as Lewy bodies (LBs) in affected areas. It is believed that α-synuclein aggregation and fibrillation trigger the LB formation process. Experimental investigations are ongoing to make PD risk predictable and manageable, drawing insights from epidemiological and genetic studies. This book focuses on cutting-edge techniques for creating and validating PD Zebrafish as a translational experimental model, exploring characteristics such as the prion-like properties of α-synuclein, mitochondrial functions linked to various pathways, and genetic associations using mammalian models. The book aims to serve as an introduction to fundamental studies on PD pathogenesis.

The primary question in the field of PD research is why there is still no curative treatment for PD and how early diagnosis and screening can be achieved. Many PD cases seek medical advice at an advanced disease stage, posing a clinical challenge. The translational model of PD faces the task of generating a suitable model that captures all the characteristics of PD in humans. The current book aims to address these questions and advance the understanding of experimental models of PD and related movement disorders.

PD is a neurodegenerative condition impacting the motor system, characterized by the presence of Lewy bodies and intracellular protein aggregates, primarily composed of α-synuclein. It is the most prevalent synucleinopathy, marked by the loss of midbrain cells leading to dopamine deficiency and motor symptoms. While the exact cause of PD is uncertain, both genetic and environmental factors are implicated. Various in vitro and in vivo models have been employed to study PD, but differences in findings exist. A continuous observation and experimental modification platform are crucial for studying progressive neurodegenerative diseases. To explore new remedies for PD symptoms, different animal models, including genetic and neurotoxic models, are utilized, each with its own set of advantages and drawbacks. The current book aims to gather contributions from renowned scientists in the PD experimental field, seeking their opinions on improving the Zebrafish model of PD.

The escalating demand for innovative techniques arises from the absence of safe, effective, and patient-friendly therapies for neurodegenerative disorders. With this objective, the book's chapters are organized to offer a comprehensive overview of recent developments using Zebrafish as a model for studying PD. Providing a thorough understanding of the clinical translation of the Zebrafish model for PD treatment, the book covers the latest developments, challenges, safety considerations, toxicity issues, regulatory factors, future potential, and limitations. This resource is intended for individuals in academia, the scientific community, the business world, and the educational community seeking a more effective strategy to target the brain.

Editor
Wael Mohamed, MBBCH, MMSc, MD, PhD
IIUM, Kuantan, Pahang, Malaysia

Acknowledgments

First, I would like to send great appreciation for all the authors who contributed to this timely project. The high level of devotion and dedication between the authors and editors made writing this book an enjoyable journey. In addition, I also extend gratefulness for the authors who are in the fields of neurology and neuropsychiatric research for delivering years of their experience and work in different areas of psychiatric disorders to deliver such an elegant piece of work. The herein discussed topics and applications are of great value in the areas of biology, neurological disorders, and neurodegeneration. Finally, I would like to thank my friends and colleagues for their unconditional love, encouragement, and inspiration throughout the endeavour of the project. Thank you.

Editor
Wael Mohamed, MBBCH, MMSc, MD, PhD
IIUM, Kuantan, Pahang, Malaysia

About the Editor

Wael Mohamed, *BBCH (Egypt), MMSc (Egypt), MD/PhD (USA)*, a physician neuroscientist, earned his PhD from Pennsylvania State University in the USA and currently holds the position of Professor Madya at the International Islamic University Malaysia Medical School. With an extensive background, Dr. Mohamed has been invited to deliver more than 150 lectures both domestically and internationally. His contributions to neuroscience/psychiatry are notable, with the publication of more than 120 peer-reviewed papers, boasting an h-index of 23. Additionally, Dr. Mohamed serves as an editor for *Frontiers in Neurology* and *PLOS-ONE* journals, where he has contributed to various journal special issues on brain disorders. In the realm of books, he is actively involved with Springer, CRC Press, Elsevier and Cambridge in the field of neuroscience. His research endeavors have attracted numerous grants from esteemed national and international organizations, including IBRO, ISN, MJF, STDF, FRGS, and INDO-ASEAN, amounting to a total research funding of half a million US$. Notably, he is the founder and chair of the AfrAbia-PD-Genomic Consortium funded by MJF and GP2.

Contributors

Raihan Abdullah
International Islamic University Malaysia
Kuantan, Pahang

Taoheed Adedeji Abdulsalam
Bioresources Development Centre (BIODEC), National Biotechnology Development Agency
Ogbomoso, Nigeria

Ebele Joan Ajagun
National Biotechnology Development Agency
Ogbomoso, Nigeria

Ayodeji Johnson Ajibare
Lead City University
Ibadan, Nigeria

Towseef Akram
Sher-e-Kashmir University of Agricultural Sciences and Technology of Kashmir
Jammu and Kashmir, India

Wasim Akram
The Neotia University
West Bengal, India

Shaimaa Nasr Amin
The Hashemite University
Zarqa, Jordan

Abraham Olufemi Asuku
National Biotechnology Development Agency
Ogbomoso, Nigeria

Maryam Tayo Ayinla
University of Ilorin
Ilorin, Nigeria

Yashwanth Bomma
Eastman Chemical Company
Kingsport, Tennessee, USA

Fatemeh Delkhosh-Kasmaie
Royan Institute for Stem Cell Biology and Technology, ACECR
Tehran, Iran

Manar A. Eissa
Merit University Egypt (MUE)
New Sohag, Egypt

Dalia Azmy Elberry
Cairo University
Cairo, Egypt

Nicola Facchinello
Neuroscience Institute
Italian Research Council (CNR)
Padova, Italy

Hagar F. Forsan
Agricultural Research Center (ARC)
Dokki, Egypt

Nazir Ahmad Ganai
Sher-e-Kashmir University of Agricultural Sciences and Technology of Kashmir
Jammu and Kashmir, India

Kaosara Oyinola Ganiyu
University of Ilorin
Ilorin, Nigeria

Dalia Hamza
Cairo University
Giza, Egypt

Yusuf Olamilekan Hamza
University of Ilorin
Ilorin, Nigeria

Ahmad Hassan
Washington University in St. Louis School of Medicine
St. Louis, Missouri, USA

Sarah Hassan
The Chicago School of Professional Psychology
Chicago, Illinois, USA

Elshaimaa Ismael
Cairo University
Giza, Egypt

Inain Jaias
Sher-e-Kashmir University of Agricultural Sciences and Technology of Kashmir
Jammu and Kashmir, India

Burhan Kantawala
Yerevan State Medical University
Yerevan, Armenia

Hasiba Karimi
Bezmialem Vakıf University
Istanbul, Turkey

Xiuyun Liu
Tianjin University
Tianjin, P.R. China

Hakim Mudasir Maqsood
The Neotia University
West Bengal, India

Asra Imtiyaz Mattoo
Sher-e-Kashmir University of Agricultural Sciences and Technology of Kashmir
Jammu and Kashmir, India

Mohd Hamzah Mohd Nasir
International Islamic University Malaysia
Kuantan, Pahang

Zahra Morshed
Royan Institute for Stem Cell Biology and Technology, ACECR
Tehran, Iran

Sobanaa Murugesan
Pondicherry University
Puducherry, India

Jabra Mustafa
Loyola University Chicago
Stritch School of Medicine
Chicago, Illinois, USA

Ider Oujamaa
Cadi Ayyad University
Marrakesh, Morocco

Rajesh Pamanji
Pondicherry University
Puducherry, India

Meijun Pang
Tianjin University
Tianjin, P.R. China

Rui Peng
Tianjin University
Tianjin, P.R. China

Vishnu Prasad Nair R U
Pondicherry University
Puducherry, India

Azhwar Raghunath
University of Michigan
School of Medicine
Ann Arbor, Michigan, USA

Mohammad Rezaei
Royan Institute for Stem Cell Biology and Technology, ACECR
Tehran, Iran

Tobiloba Samuel Olajide
University of Medical Sciences
Ondo, Nigeria

Vivek Sanker
Trivandrum Medical College
Kerala, India

Joseph Selvin
Pondicherry University
Puducherry, India

Faez Sharif
International Islamic University Malaysia
Kuantan, Pahang

Sanobar Shariff
Yerevan State Medical University
Yerevan, Armenia

Gisha Sivan
SRM Institute of Science and Technology
Chennai, India

Qamar Sultana
Deccan College of Medical Sciences
Hyderabad, India

Yaser Tahamtani
Royan Institute for Stem Cell Biology and Technology, ACECR
Tehran, Iran

Mahsa Valizadeh
Royan Institute for Stem Cell Biology and Technology, ACECR
Tehran, Iran

Hala Zaher
Cairo University
Giza, Egypt

Section I

From Aquarium to Lab

The Rise of Zebrafish as a Model Organism

1 General Methods for Zebrafish Care

Fatemeh Delkhosh-Kasmaie, Yaser Tahamtani, and Mohammad Rezaei

1.1 ZEBRAFISH FEEDING

Zebrafish (*Danio rerio*) are omnivores, which means they eat plants and animals. These fish feed on a variety of small aquatic organisms, namely zooplankton, insect larvae, and small crustaceans in the wild. Because of the digestive tract of zebrafish, with a large stomach (no true stomach acid) relative to body size and a short intestine, they can consume large foods in moments and digest them quickly. It helps them survive in various bodies of aquatic environment, including labs (Del Vecchio et al., 2022; Sullins, 2023).

From a scientific point of view, feeding efficiency of the fish is more important, which is why different ratios of nutrients were suggested. They determined nutrient requirements to meet essential body needs that provide maximum growth, survival, reproduction, and immune function of various fish from larvae to adults (Frederickson et al., 2021; Harper & Lawrence, 2016). They achieved a practical and balanced diet with five main nutrient classes: protein/amino acids, lipids, carbohydrates, vitamins, and minerals (Del Vecchio et al., 2022; Farias & Certal, 2016).

Biological functions such as growth, reproduction, maintenance, and tissue repair depend on amino acids either nonessential or essential. Zebrafish require a sufficient amount of protein in their diet in order to synthesize it, which can vary based on feeding habits, environmental conditions, and developmental stages; zebrafish requirement typically decreases with growth (Harper & Lawrence, 2016). Uptaking an accurate protein percentage is critical because the excess protein is converted into energy or energy storage products. They do this naturally by controlling growth through reduced food consumption. The minimum protein requirement varies between 20% and 60%, depending on the zebrafish developmental stages; this amount also decreases with growth (Avdesh et al., 2012; Del Vecchio et al., 2022; Farias & Certal, 2016).

Another source of energy required for normal development and growth is lipids. In order to absorb fat-soluble vitamins, an adequate intake of fat is necessary. Zebrafish cannot synthesize some of the polyunsaturated fatty acids (PUFA) such as linoleate and linolenate (Harper & Lawrence, 2016).

Carbohydrates are one of the important sources of energy, which can help reduce the use of proteins and lipids, as well as their binding properties, although fish have no need to uptake carbohydrate. Zebrafish consume insects and phytoplankton, and therefore are able to digest carbohydrate molecules. Carbohydrate-containing nutrition increases the rate of growth, feed conservation, and weight gain in zebrafish too (Harper & Lawrence, 2016).

Fat-soluble vitamins such as A, D, E, and K and water-soluble vitamins including C and B groups vitamins, even though infinitesimally small, are essential to life. Vitamin deficiency leads to a variety of non-specific symptoms. Accordingly, ensuring a sufficient supply to cover the fish requirement is a conspicuous food formulation.

Fish get minerals through their gills or food. Minerals directly affect some biological functions. It should be noted that the amount of dietary minerals changes depending on the concentration in the water.

Another thing that must be considered is the relationship between feeding regimen (frequency and rate) and swimming performance and metabolism (Dametto et al., 2018; Farias & Certal, 2016).

1.1.1 Food Types

Zebrafish feed on two different food classes, namely processed and live food. Both types meet the needs of biological processes. In addition, fish feeding should have balanced nutritional profiles appropriate to mass culture. Live food includes *Artemia*, rotifers, and *Paramecium*, part of zooplankton species that fish consume in the wild. Live food omission from the fish diet seems non-practical. The exact fish requirements are not known, and despite the risk of pathogens and toxins, as well as the cost of this feeding, it is not reasonable that live food be replaced with other foods (Morais et al., 2004; Sullins, 2023).

1.1.1.1 Live Foods

Artemia (brine shrimp) is an aquatic crustacean and is consumed as a quick food source for farmed fish. *Artemia* populations have been found all over the world, particularly in saltwater lakes. The first historical *Artemia* date back to Iran, in Lake Urmia. They produce metabolically inactive eggs surrounded by a hard shell. If these cysts are packed in a vacuum and in cool conditions, they can be stored for years. When they are needed, just use fresh saltwater and wait 18–24 hours at 28–29 degrees Celsius to hatch. Zebrafish are fed with free-swimming *nauplii*. There is a limited time for consumption of *nauplii*, maximum after 48 hours, when the nutritional profile decreases. This life stage is more appropriate for feeding zebrafish, especially larvae that are small and less able to swim because they digest easily and are visible to fish. *Nauplii* are enriched with something like fatty acid emulsification, based on

DOI: 10.1201/9781003402893-2

requirements, management opinion, or special conditions. Brine shrimp have a different nutritional profile, but they can provide everything fish need by the enrichment process. Decapsulating with hypochlorite/sodium hydroxide solution is always worthwhile for two reasons. First, hard shell removal reduces pathogen risks and later, *nauplii* spend shorter time and less energy for hatching, thus containing more energy to transfer (Harper & Lawrence, 2016).

The second live food is *Paramecium*, which is used for initial feeding and in the early larval stage. These small organisms (150–250 µm) are easy to cultivate. When not in use, they keep on living and reproducing in tanks. Feeding with microalgae and *Spirulina* can improve the nutrient profile of paramecia.

Rotifers, the third live food, are the best for zebrafish larvae. Small size (100–250 µm), slow swimming, and swallowing of microalgae, bacteria, *Spirulina*, and fatty acid emulsification are what make them (after *Artemia*) the best choice to feed larvae. Rotifers also help with nutrient recycling and cleaning of water tanks (Allen et al., 2016; Lawrence et al., 2012).

1.1.1.2 Processed Foods

Keep in mind that the processed food should have the appropriate qualities in all aspects, demonstrate good water stability (meaning that nutrients will not leach quickly into water), and remain in suspension for a long period of time to be more easily accessible (Harper & Lawrence, 2016).

Larvae Diet: Achieving satisfactory survival and growth rates is the main challenge in aquaculture breeding. As mentioned earlier, live food is the better option. Balanced nutrition, visual and chemical attraction, digestibility, and suspension in water demonstrate why live food is the best option (Harper & Lawrence, 2016). Although zebrafish larvae can only culture on processed food, compared to live food they have lower survival and growth rates (about 80–90%), possibly due to modern facilities such as automated feeders, this rate increased.

Juvenile/Adult Diets: Beyond the larval stage, fish can feed by processing food alone. Of course, palatability, digestibility, and water stability as important aspects that should be considered. Processed food must be designed to gain satisfactory growth and reproduction rates, although this needs investigation for more information. The effectiveness and quality of pre-prepared food are becoming better and more appropriate than in the past. Currently, utilizing a combination diet—live and processed food—is preferable to using dry food alone (Carvalho et al., 2006; Siccardi et al., 2009).

1.1.2 Life Stage and Feeding

The requirements change with zebrafish growth from initial feeding to adult and reproductive stages; therefore, they need different quality and quantity, which means nutritional profile and density and frequency of feed, respectively. As previously mentioned, it is essential to formulate requirements and match these to the diet to ensure needs are met, that result in keeping and growing a healthy population (Lawrence, 2007).

1.1.2.1 Larvae

Almost 5 days post fertilization (5dpf), larvae have a developed swim bladder and digestive system, opened mouth, and a depleted yolk-sac, so they have demanded more energy, which they will get through food consumed. The diet should be rich in protein, sufficient lipids, and appropriate vitamins and minerals to achieve higher growth and survival rates. Up to 45–60% protein, 6–10% fat, and less than 5% carbohydrate, an appropriate amount of minerals and vitamins, generally meet the need (Harper & Lawrence, 2016). An excellent food source is live food, although processed food can be useful.

The following items are considered for zebrafish processed food:

1. Stability: it would be ideal if this food were as stable in water as live food.
2. Suspension: food must be suspended so that it is available to fish for effective feeding. Processed food with low density can float for adequate time.
3. Size: this item is crucial. As they grow, they can consume food of larger size. At the larvae stage, they can consume 150–200 µm prey, e.g., *Paramecium*, rotifers, newly hatched *Artemia*, and a variety of processed food with appropriate size.
4. Digestibility: highly digestible means that nutrients are easily absorbed. The content of live food proteolytic enzymes in *Artemia* and others ensures a digestible state, but processed foods should be tested and confirmed.
5. Acceptability: visual and chemical stimuli attract the zebrafish; in fact, they hunt food. Free amino acids in both types of food are attractive for fish, although live food is more desirable because of slow movement (swimming).

To ensure the larvae's high encounter rate with food, feed them with higher volume, because of zebrafish's poor swimming ability and also small size. Larvae can consume food between 50–300% of their body weight. *Paramecium* is the best food and can be added to the culture without concern about affecting zebrafish. They aren't consumed totally and survive, can reproduce, and turn into a constant food supply. Frequent small meals of live/processed food cover the high metabolism and growth of the larvae.

1.1.2.2 Juvenile

The body structure of juveniles (21–28 days postfertilization) matures and develops rapidly. They should be fed according to their needs. For these reasons and also sexual maturity, protein and lipid requirements are high in the larvae stage. Despite the enlargement of the gastrointestinal system, carbohydrates have no proven effect on the zebrafish's nutritional requirements. Vitamins and minerals are necessary for growth and metabolism: 45–60% protein, 6–15% lipids, less than 5% carbohydrates, and the required amount of vitamins and minerals have been generally recommended (Harper & Lawrence, 2016).

Some food characteristics that are important to zebrafish include:

1. Stability: especially when they adapt to a new diet.
2. Suspension: fish swim faster; therefore, increasing the suspension time is desirable. Food that settles at the bottom is washed out of the tanks, which is not economical.
3. Size: they can eat food of various sizes from 400–600 µm because the ability to tear and break food is promoted. *Artemia* and pellet/crumble processed food seem suitable for feeding.
4. Digestibility: despite the development of the digestive system, food is better to be highly digestible in order to provide more energy for maintaining growth and maturation.
5. Acceptability: which is almost like the larvae stage, of course, larger mouth size and swimming ability are more effective.

Juveniles eat about 25–50% of their body weight; they feed more efficiently and consume a larger amount of food. It must be taken into account that their swimming skills have improved and they miss less amount of food as sediment. Frequent small meals seem to be the better option, and overfeeding must be avoided. Nutrient absorption correlates with the retention time of the gut. Feeding 4 to 8 times per day seems more appropriate.

1.1.2.3 Adult

The main goal of zebrafish culture is embryo production. The diet of adult zebrafish is designed to support the continued production of numerous embryos. The growth rate is decreased while they consume energy to maintain homeostasis and reproduction. Diet still requires protein but less than in the previous stages. Lipids play a slightly more important role at this stage. In fact, fertilization and egg production are influenced by the amount of lipids in the diet, especially n3 and n6 fatty acids. Minimum carbohydrate content is required. Vitamins and minerals must be in the diet, although vitamin A plays a key role in reproductive function. The diet contains at least 45–55% proteins, 10–15% fats, and less than 5% carbohydrates, vitamins, and minerals. The diet also contains ascorbic acid and vitamin A necessarily (Harper & Lawrence, 2016).

Like other developing stages, some characteristics of food must be considered, including:

1. Stability: it is important to know that food meets all the nutrients needs, especially fatty acids and vitamins.
2. Suspension: adult fish eat at the surface and bottom of the tank or within the water column, but suspension feed at the surface is preferred.
3. Size: the adult can prey on items larger than 600 µm, although 400–600 µm is more recommended.
4. Digestibility: it may reduce energy loss.
5. Acceptability: it is the same standard as other developmental stages.

Typically, adult fish feed about 1–10% of their body weight (food densities). Avoid overfeeding as reproductive rates and water quality will decrease. Depending on the reproduction period, it may be ideal to feed adults 3–5 times per day, even reducing fish density (Avdesh et al., 2012).

1.1.3 Feeding Procedures

1.1.3.1 Artemia Procedure

1.1.3.1.1 Materials and Equipment

Dried artemia cysts (2 mg/liter), bleach (Cl), sea salt (30 g/liter), hatching cone (10-liter) with aeration, heater, white light, nylon mesh (REF352340; 40 µm), graduated cylinders (1-liter), beaker (1-liter), pipet pastors (3/5), and RO water (reverse osmosis water), measuring spoon, lab balance scales (Figures 1.1 and 1.2).

1.1.3.1.2 Cyst Hydration and Decapsulation

1. Preparing 2 grams per liter of dried artemia cysts (Figure 1.3a). In detail, add 8 grams of artemia cysts into 800 ml RO water with an air pump for about 2 hours at room temperature.
2. Check the cyst under stereomicroscope; fully hydrated cysts are utterly spherical in shape (Figure 1.3b).
3. Add 100 ml bleaching, and stay for 10 minutes or more for cyst decapsulations (Figure 1.3c) and asepsis.
4. When the color changing from dark brown to orange is observed, this suspension must be rinsed 2–3 times with RO water (remove cyst capsule and remaining bleach).

1.1.3.1.3 Decapsulated Artemia Hatching Instruction

Tip 1. The optimal conditions for hatching brine shrimp are: 28–29 °C temperature and keeping them in a lighted area with aeration.

Tip 2. Rinse all equipment that is used with RO water and let them dry.

1. Make sea saltwater solution, 30 g/l: add 120 grams of salt to some RO water for starting.
2. Sea salt water and decapsulated cysts are added to the hatchery and adjusted to 4 liters with RO water.
3. Incubation of these suspensions is for 24–48 hours at 28.5 °C.

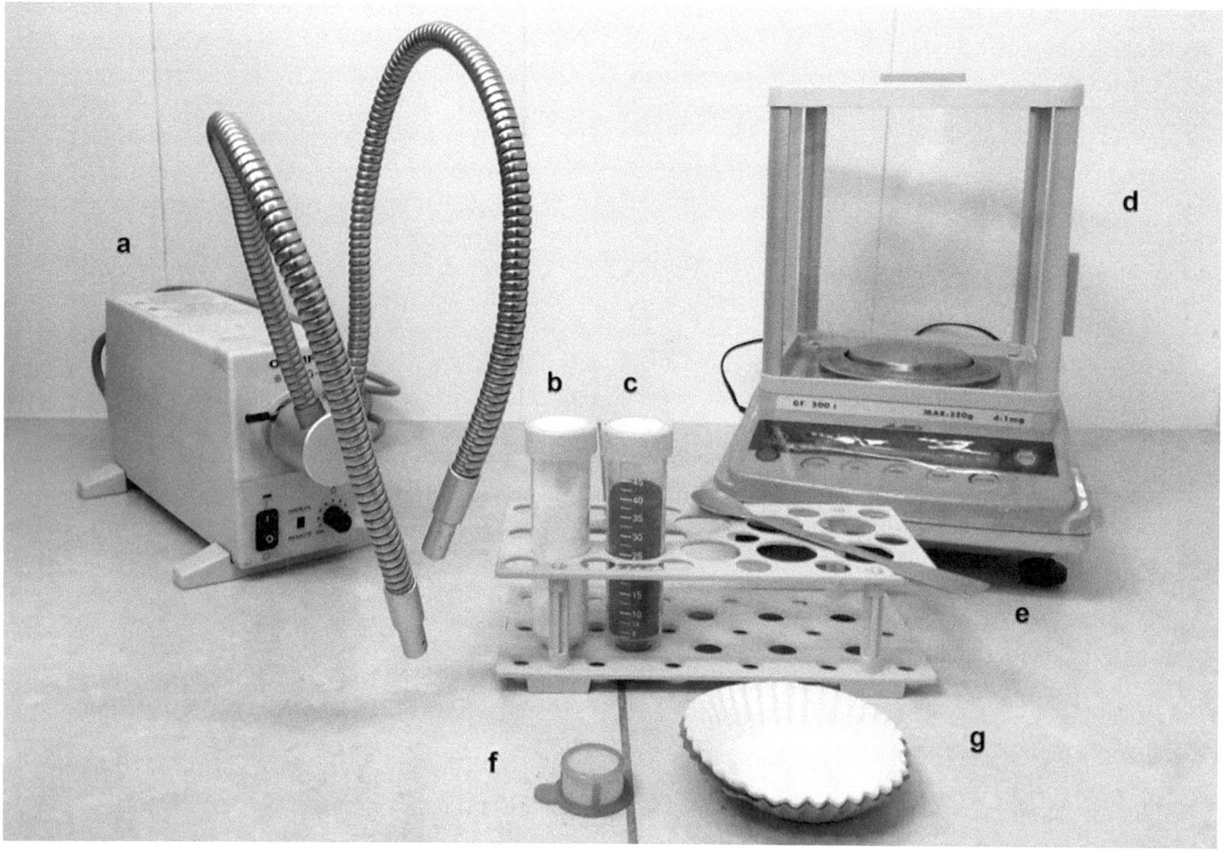

FIGURE 1.1 Materials and equipment for Artemia preparation: (a) laboratory illuminator, (b) sea salt, (c) dried Artemia cyst, (d) lab balance scales, (e) measuring spoon, (f) 40 μm nylon filter, (g) piece of paper.

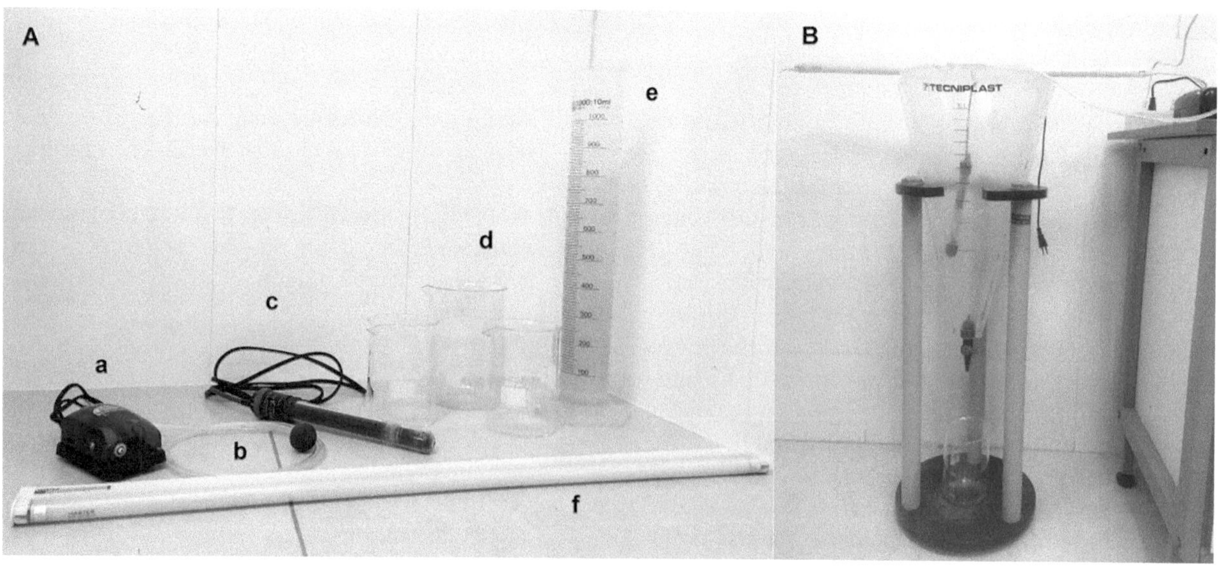

FIGURE 1.2 A: Materials and equipment for Artemia hatching: (a) air pump, (b) air stone, (c) heater, (d) beakers, (e) 1-L graduated cylinder, (f) white light. B: The assemble hatching cone with aeration and other equipment.

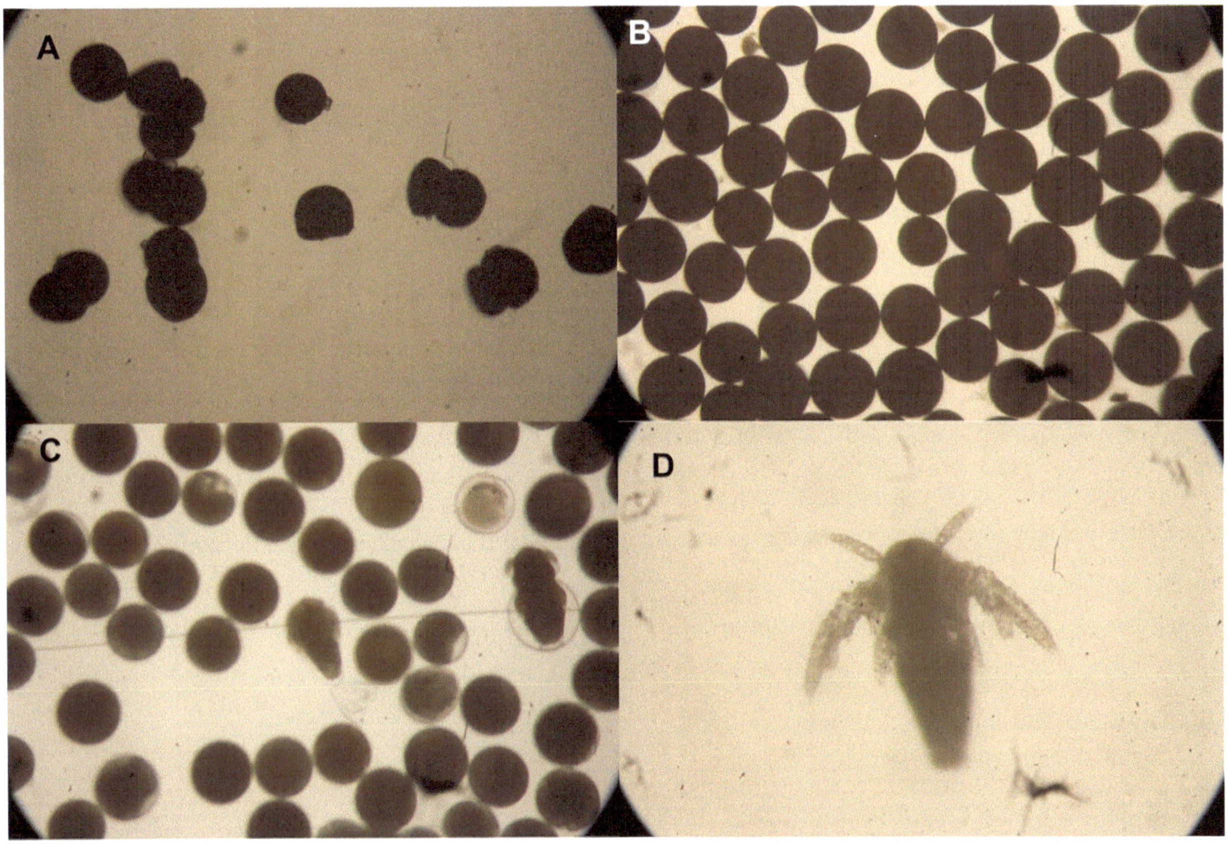

FIGURE 1.3 Different stages of Artemia life cycle: (a) dried cyst, (b) hydrated cyst, Λ c) decapsulated cyst, (d) nauplii.

4. Note: In less than 24 hours, most cysts remain unhatched, and after more than 48 hours the Artemia nauplii (Figure 1.3d) need to feed therefore they are not of adequate quality for feeding zebrafish.
5. After 48 hours, the hatchery consists of artemia nauplii and unhatched cysts.
6. Harvest this suspension to the beaker, leave it for 1–2 minutes until sedimentation, and use the focal lamp to collect nauplii.
   Note: the whole of artemia is not transferred in one part, because of changes in temperature and oxygen, and some live artemia are lost.
7. Utilized pipette to collect them while trying to avoid collecting any unhatched eggs, or dead shrimps.
8. Use a 40 filter (mesh) to remove salt and extra water (rinse the nauplii with RO water).
9. Feed the fish with the appropriate amount of suspension of live food (about 1.5 ml per 10 fish) which can vary depending on age, size, and number of fish in each tank.
10. Place the pipette on top of the tank and dispense a sufficient amount of artemia.
11. After consuming the harvested Artemia, repeat steps 7–10 until all the nauplii have been ate. When artemia is collected several times, this occurs spontaneously.
12. Clean and reset the cone.

1.1.3.2 Paramecia Procedure

1.1.3.2.1 Materials and Equipment

paramecia seeds, RO water, nutritional brewer's yeast, autoclaved dry whole-wheat (or boiled wheat seeds), 15×25 mm petri dishes, 100 ml beaker, measuring spoon, small kitchen strainer, 55–10 μm filter (mesh),

Tip 1. Culture containers with large surfaces should be easy to wash as well as be able to withstand high temperatures.

Tip 2. For dense paramecia culture, it seems to need a large water surface-to-air ratio.

Tip 3. A 1–4 weeks period and 24–26 °C temperature are vital for steady and optimal density.

1.1.3.2.2 Making Culture

1. Start with a clean work surface, and autoclave dry whole- wheat.

2. Boil the 1–2 tbsp of autoclave seeds and RO water for 10 minutes.
3. Remove the wheat from the hot plate and pour off the liquid, leave it to cool.
4. Prepare Petri dishes and label them with date, then fill with 100 ml RO water and add 100 mg of powdered brewer's yeast and 15 wheats to each petri dish.

1.1.3.2.3 Inoculation

1. Split a single concentrated culture of paramecia (150 ml) into 3 new culture dishes (density of at least 100 cells/ml).
2. Remove the old wheat seeds with a strainer (filter) from the starter culture. Use the Ro water for rinse and remove any attached paramecia. Collect the strained solution in a large beaker.
3. Divided strained culture of at least 40–60 ml per Petri-dishes and mix them gently and repeat during the process, because paramecia tend to concentrate on the surface.
4. Cover Petri dishes and label them with a date.
5. Maintain paramecia cultures between 24–26 °C. Avoid high-temperature shock.
6. Leave them for 1–4 weeks to reach optimal density. After 4 weeks the paramecia tend to decline in density therefore add some wheat and yeast to improve culture with a longer than 4 weeks period, the optimal density feeding is between 2–4 weeks.

1.1.3.2.4 Harvesting Paramecia from Culture for Larval Feeding

1. A single petri dish feeds 100 larva fish. Approximately 10 ml of concentrated paramecia (500/1500 paramecia/ml) in each tank with 50 larvae.
2. Start to feed 5 dpf larvae (after transport larvae to the housing system) twice per day for 5 days (ending 8–11 dpf).
3. Use a 55 μm filter to remove the large particle debris (wheat and yeast). Collect them in a beaker.
4. Use a 10–15 μm filter to eliminate ammonia wastewater. Repeat the process 2–3 times to rinse paramecia well. Pour the paramecia into a clean beaker after the final rinse.
5. Re-suspend the paramecia to achieve final concentration (500–1500 paramecia per ml).

1.1.3.3 Rotifer Procedure

1.1.3.3.1 Materials and Equipment

Algae, soda bottles, modified MBL, vitamins, fluorescent, air pump, plastic soda bottles, 53 ml filter

1.1.3.3.2 Algae Culture

1. Utilize Algae to inoculate 1-liter algae cultures in 2-liter soda bottles at room temperature in a modified MBL medium with 18 ppt (part per thousand) vitamins under fluorescent light and constant aeration.
2. Use ready culture in 1–2 weeks in a bright green color.
   Note: Add doubled or tripled the amount of vitamins if they don't reach a bright green color. They reach a shade of yellow or with, culture is crashed.
3. Interchange the modified MBL medium when depleted to around 500 ml.
   Note: Be careful of unidentified protozoan contamination.

1.1.3.3.3 Making Salt Water

1. Add some salt to the bucket, then add RO water and adjust to 30-L.
2. Use the formulation to get the correct salinity:

$$\frac{(41.8 \times ppt)}{35} = Salt\ needed\ (grams\ per\ liter)$$

   15 ppt: 17.91 g/L or 537g in 30 L
   5 ppt: 5.97 g/L
3. Add an air diffuser to mix.

1.1.3.3.4 Assembly of 1.6 L Rotifer Culture Bottles (RCB)

1. Rinse two 2-liter plastic soda bottle with RO water, remove label
2. Cut one of them along the horizontal axis 11 cm from the bottom to create the base.
3. Cut another one along the horizontal axis 24 cm from the top of the bottle to create the top portion.
4. Put upside down into the base and tape the two units.
5. Indicate on the side of the bottle approximately 200 ml up to a volume of 1.6 L.

1.1.3.3.5 Rotifer Culture

1. Develop culture 3000–5000 rotifer cysts in the aerated 1.6-L RCB at the highest salinity (15ppt) at room temperature under constant fluorescent light.
2. Fed rotifer once daily with 25 ml algae per 1-L culture, then adjust down to 15 ppt with RO water until cultures are slightly green and cloudy.
3. Harvest approximately 50% of each culture to restrict sexual reproduction.
4. Transfer the culture to a new RCB to prevent algae accumulation. Wash the old one.
   Note: Construct a simple wooden frame to avoid tipping over the culture bottles (Figure 1.4).

1.1.3.3.6 Zebrafish Feeding

1. Use the 53-ml filter to harvest slightly less than half of two rotifer cultures.

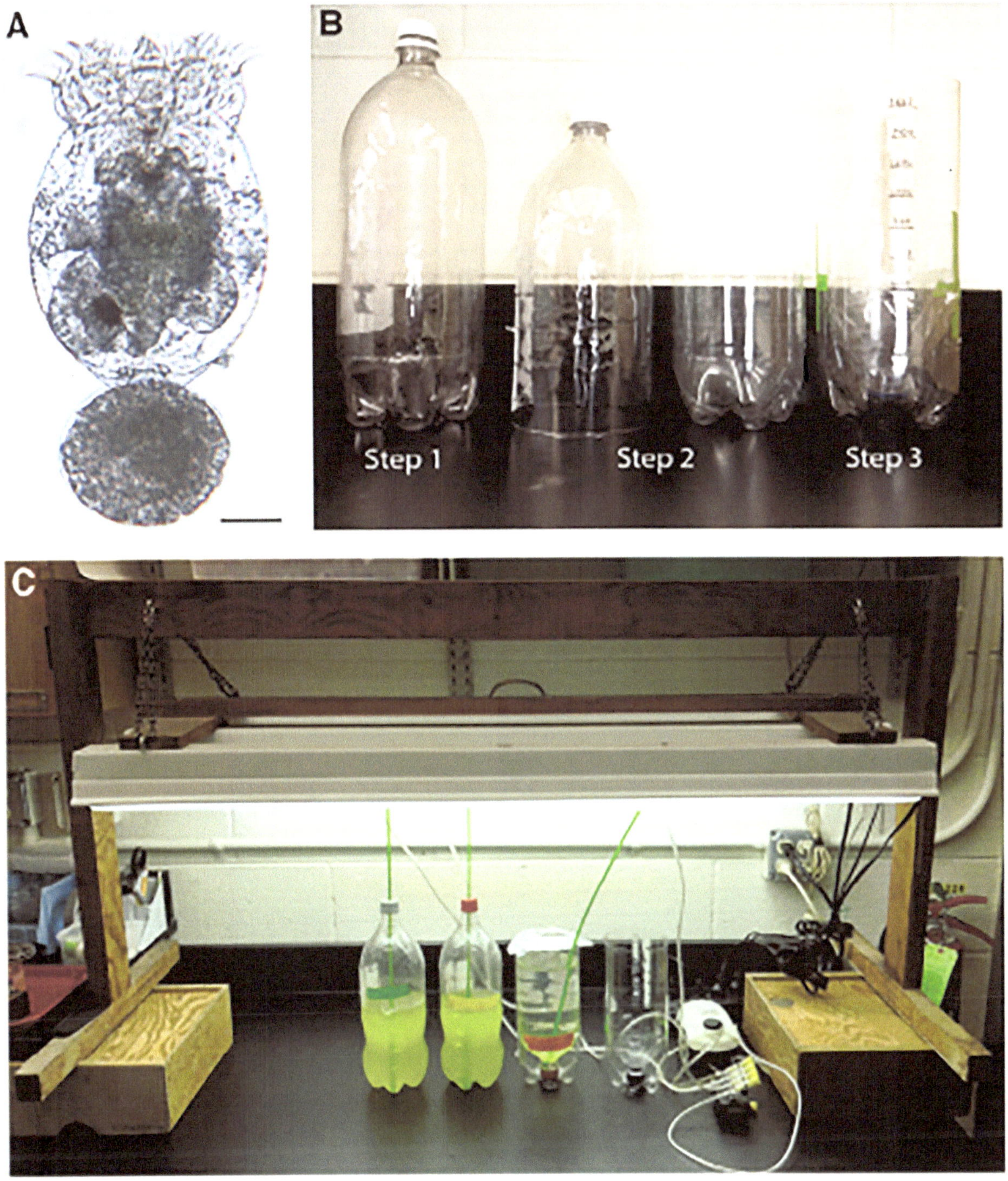

FIGURE 1.4 Rotifer culture: (a) *Brachionus plicatilis* and mictic culture. (b) Rotifer culture apparatus assembly. Step 1: clean and rinse two clear 2-L soda bottles. Step 2: cut one bottle 11 cm from the bottom and the other 24 cm from the top. Step 3: invert the top of the second bottle into the bottom portion of the first bottle. Tape the sections together and mark the appropriate volumes of the bottle. (c) Two aerated cultures of *Nan-nochloropsis* grown in 18 parts per thousand (ppt) salt+MBL media, 15 an aerated culture of *B. plicatilis* (15 ppt salt water), and an empty rotifer culture bottle at room temperature under fluorescent lights (left to right, respectively).

Source: (Allen et al., 2016)

2. Rinse them with 6 ppt salinity and concentrate them into a beaker in 100 ml of 6 ppt.
3. Feed larvae with 25 ml of the solution (250 rotifer/ml) twice daily.
   Note: create rotifer suspension for morning and afternoon consumption.
   Note 2: resuspend the rotifer by hand because it begins to settle to the bottom because of osmotic shock during this process.

1.1.3.4 Processed Food Procedure

Tip 1. Prepare food size from 100 μm for larvae to 300/400 μm for adult fish (Figure 1.5).

Tip 2. Regulate zebrafish feeding-frequency, and food size- based on their life stage.

Tip 3. Advise to feed once or twice per day for adults and 5–6 times for larvae.

Tip 4. Store the remained food at -18 °C.

1.1.3.5 Zebrafish Feeding Schedule

1.1.3.5.1 Definitions

Larvae: 5–21 dpf.

Juvenil fish: 21–90 days old fish

Adult: +91 days old fish

Tip. Larvae fish fed up to 50–300% of their body weight each day, compared to adults that fed about 1–10% body weight per day.

Note: Decrease the numbers of feeds as fish get older.

1.1.3.5.2 Weekday Feeding Schedule

- Feed larvae and juvenile 3–5 times with processed and live foods. (9:00–10:00 A.M./ 2:00–3:00 P.M./ 6:00–7:00 P.M.)
- Feed adults twice daily with live and processed foods.

Tip1. Check the tank for unfed food after 10 minutes to achieve an excellent food quantity.

Tip2. Be careful to not over-feed to prevent fat accumulation and eliminate egg/sperm.

1.1.3.5.3 Weekend Feeding Schedule

- Feed all of them twice for day in the morning and afternoon.

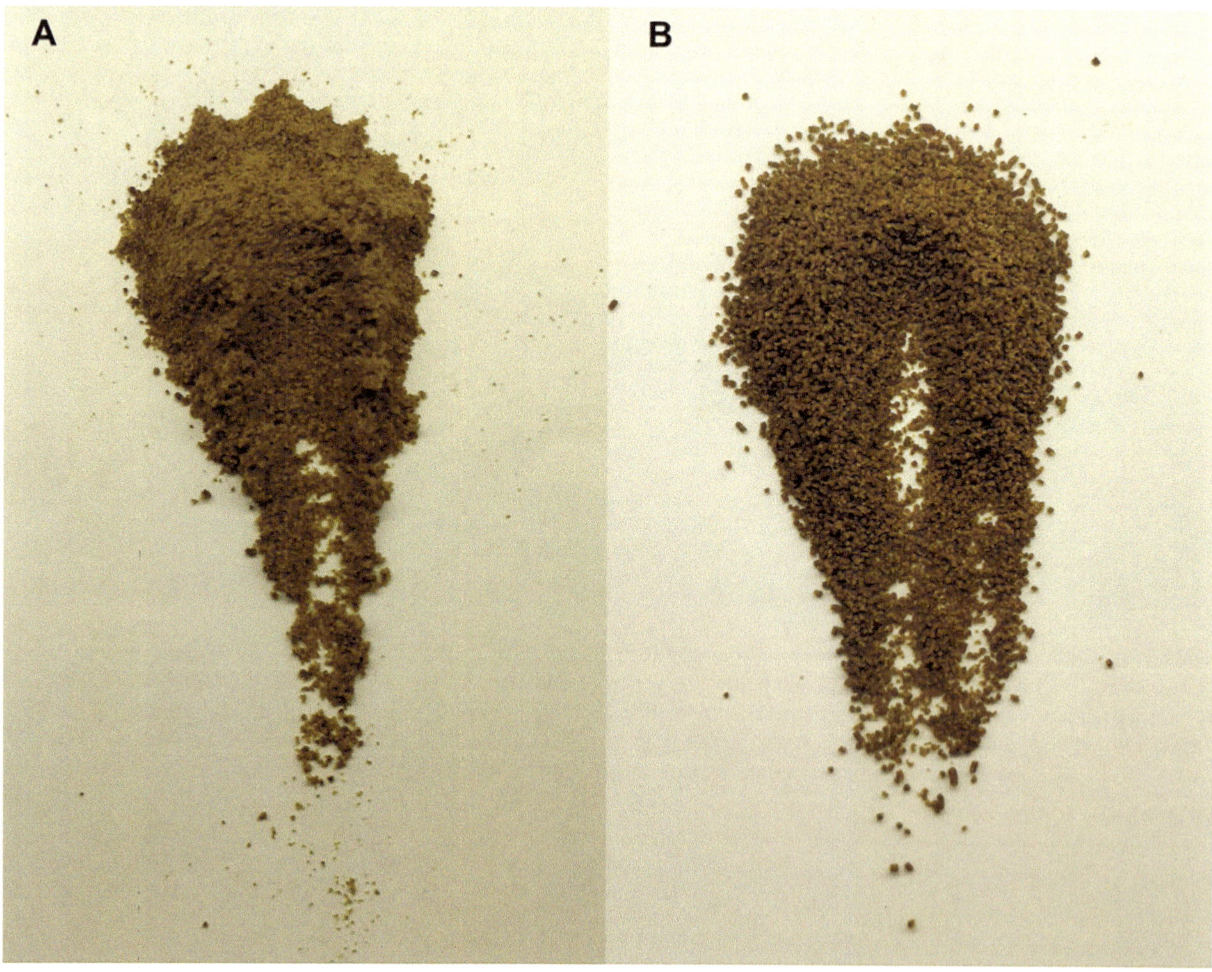

FIGURE 1.5 Processed food for zebrafish feeding: (a) larvae food, (b) adult food.

1.2 WATER

Water is the living medium of fish such as air for humans, so fish are very sensitive to its quality. For osmoregulation, freshwater fish absorb ions across the gills and digestive tract and excrete extra water by the kidneys, but this procedure in marine fish is the opposite, so water is in contact with fish externally and internally (Beyenbach, 2004). For wild fish, water quality is adjusted naturally, but in the culture system because of low input, is deteriorated by respiration and nitrogen metabolism of fish (Tumwesigye et al., 2022). Water quality is classified into three categories, including physical, chemical, and biological parameters, and by any change in these parameters, fish become stressed and have negative effects on survival, growth, reproduction, and management of fish. Quality of water must be retained in physiological requirements for maximum growth and breeding of farmed fish (Mohammad & Haque, 2021), so for any species we need a comprehensive understanding of their environmental biotic and abiotic factors and also their natural behaviors. The factors that must be refined for husbandry and breeding standards of culture fish include salinity, pH, temperature, dissolved gases, nitrogenous wastes (ammonia, nitrite, and nitrate) of fish, turbidity, light cycle, and illumination levels (Tsang & Gerlai, 2022; Tsang et al., 2017).

Zebrafish has a broad geographic range and is a native species to freshwater rivers in India, Bangladesh, and Nepal (Spence et al., 2008). They prefer living in slow-moving or standing water bodies, edge of streams, ponds, and even rice paddies, but zebrafish are found in diverse environments from rivers with clear water to ponds with turbid and dense vegetation water (Engeszer et al., 2007; Sundin et al., 2019). This diversity of place results in a wide range of environmental conditions, including temperature from 24.6°C to 35.3°C, conductivity of 10 μs to 271 μs, pH from 5.9 to just below 10, different bottom substrates and depth from surface to 60 cm (Lawrence, 2007). This is necessary to make a similar condition for the culture of zebrafish in a research laboratory.

For traditional aquaculture, two main sources of water include surface water and ground water or spring water. Water quality in these sources could be different according to season and site (Summerfelt, 2000; Swann, 1993). Ground water is more stable in composition, quality, and quantity and also is free of anthropogenic and pathogen contamination (Arabi et al., 2011). But with regards to the sensitivity of laboratory animals, it is necessary to do a comprehensive chemical and microbial analysis regularly for any water sources used for these animals. Zebrafish facilities are usually located inside or near to cities, so tap water is the likely water source for these facilities. According to the importance of these resources for humans, in addition to availability, they have high quality and are monitored regularly for composition and contaminations. Elimination of pathogens in drinking water is acquired by disinfection protocols, the most abundant one of which is the chlorination method (chlorine or chloramine) (Tsitsifli & Kanakoudis, 2018). Chloramine is a combination of ammonia and chlorine that is longer lasting than chlorine, but the use of chlorine is more common, and its elimination is easier. In water supplies of cities, the normal concentration of chlorine is between 0.5 and 2.0 mg/L, and this concentration is toxic for fish and most other aquatic organisms (El-Bouhy et al., 2006). The presence of these compounds in water can cause severe damage to the gills, skin, and eyes of fish and result in hypoxia, trouble swimming, fading of color, and sudden death (El-Bouhy et al., 2006; Svecevičius et al., 2005). Therefore, removal of these compounds is necessary for water that is used in zebrafish facilities. The most common methods for chlorine removal are vigorous aeration of tap water for at least 24 hours in an open-topped container, the use of sodium thiosulfate with a molar ratio of 3:1 to chlorine, and passing water through granular activated carbon (Giles & Danell, 1983). Chloramine is more stable so aeration is not effective for that; the recommended methods are the use of sodium thiosulfate, which results in the elimination of chlorine and release of ammonia into the water, and also passing water through granular activated carbon (Noga, 2010; Roberts & Palmeiro, 2008). However, these filters could not provide ideal water quality for zebrafish facilities, because any impurities or fluctuation in water quality has big effects on research results. The most recommended approach to water treatment is a combination of a deionization system and reverse osmosis. Two prefilters are used before reverse osmosis, including a mechanical particulate cartridge and activated carbon filter (Lawrence & Mason, 2012). The first one is a mechanical filter that removes debris such as sand, organic matter, and clay, where the size of filtered particles depends on the cartridge type. Carbon filters are used to eliminate chlorine and chloramine, but they are also effective to remove certain chemicals, particularly organic chemicals that are dissolved in water. In the reverse osmosis (RO) stage, water is forced through a semi-permeable membrane in reverse of its natural osmotic flow, and the pores of this membrane just allow the passage of water molecules (Aleström et al., 2020; Suzaimi et al., 2021). Some harmful ions are eliminated in the RO stage, but using the deionization stage is necessary to remove the remaining ionized minerals and salts through the ion exchange process. These treatments result in pure water with near-zero total dissolved solids but lack some beneficial substances (calcium, magnesium) that are necessary for fish. So, some synthetic sea salts are added to this water in the fish housing system (Aleström et al., 2020).

1.2.1 Recirculating Aquaculture Technology

There are two types of rack systems for zebrafish maintenance. In a multi-rack system, a centralized filtration unit supports a group of racks that are linked together, and this system is common in big facilities. The standalone rack (Figure 1.6a) includes an individual rack with a complete filtration unit and is suitable for small facilities and

FIGURE 1.6 Equipment of stand-alone housing systems: (a) stand-alone rack, (b) RO water purifier, (c) biological filters, (d) UV sterilizer.

quarantine rooms. Both of these systems use recirculating aquaculture technology to prepare water for zebrafish maintenance (Teoh et al., 2020). This technology includes three filters (mechanical, biological, and chemical) and a disinfection unit to treat water by eliminating different organic materials, such as metabolic wastes of fish, and reusing this filtered water in zebrafish housing systems. These systems received water from an RO water purifier (Figure 1.6b). The solid wastes are trapped by a mechanical filter, and this filter could be in different forms, including stainless steel floss, foam, or pad. Biological filters (Figure 1.6c) are responsible for providing a high surface area for nitrifying bacteria that conduct the oxidation process of ammonium to nitrite and then to nitrate. Different forms of biological filters could be used, such as a mesh bag of biomedia, fluidized beds, or floating media. The effectiveness of biological filters is related to the size and volume of biofilters, quality, circulation, and retention time of water in these filters (Hammer, 2020). For chemical filters, most zebrafish facilities use activated carbon or zeolite. Activated carbon could remove chlorine and chloramine, dissolved protein, water discoloration, and bad smells, but zeolite binds to positively charged ions like ammonium and removes them from water. Finally, water goes through a UV sterilizer (Figure 1.6d) to decrease the population of microorganisms such as viruses, fungi, or bacteria that could be pathogenic for fish. The practical goal of these systems is to maintain an optimal and stable water quality that provides the health and function of fish and also nitrifying bacteria in housing systems (Lawrence & Mason, 2012).

Water quality is determined by different parameters that interact with fish and other microorganisms such as bacteria. Here we describe some of the most important of these parameters and their effects on fish and also the manner in which they must be monitored.

1.2.2 Temperature

Fish are known as poikilothermic animals, which means their body temperatures are similar to the water temperature they live in, and their inner mechanisms cannot provide body warming (Akhter et al., 2021). In fish, temperature adjusts the rate of all physiological processes, so it is the most important abiotic factor for them. For warm-water fish, metabolic processes increase alongside with increase in temperature, but for cold-water fish, this procedure is the opposite, and high metabolism occurs at lower temperatures (Volkoff & Rønnestad, 2020). The required metabolic rate for life maintenance of fish is increased with temperature, leading to a decrease of the required energy for growth and reproduction, so out of optimal temperatures these factors are decreased significantly (Fekri et al., 2018; Khallaf et al., 2021). Fish are adapted to seasonal changes of water temperature naturally, but abrupt fluctuation would be lethal for them, especially in larval stages (Hernandez et al., 2007), so any change in culture temperature requires immediate attention. Zebrafish are known as a tropical fish, but their environmental temperatures fluctuate both daily and seasonally around 5.6°C and 30°C respectively (Cortemeglia & Beitinger, 2005). In laboratory conditions, the optimal temperature is advised between 26°C and 28.5°C. Incubated embryos in below (<22°C) or above (>34°C) optimal temperature could not complete embryogenesis and lead to 100% mortality of these embryos (Schirone & Gross, 1968). Fluctuation of temperature in the embryonic stage has crucial effects on the cardiac shape and swimming capacity of adult fish (Dimitriadi & Koumoundouros, 2022). The influence of temperature on sex ratio has been confirmed in zebrafish, where exposure of embryos and larvae to high temperature resulted in a high significant rate of masculinization that persists for multigenerations (Ribas et al., 2017; Valdivieso et al., 2020). In addition to the effects of temperature on fish, it also affects the equilibrium of total ammonia nitrogen and dissolved oxygen (DO) solubility (Li et al., 2013); by a 10°C rise in temperature, the ratio of ammonia to ammonium increases two-fold, and DO solubility is decreased significantly (Chapra et al., 2021; Yan et al., 2019), so the temperature must be monitored daily. In a zebrafish laboratory, a consistent water temperature could be achieved by heaters that are placed in a housing system or by control of culture room temperature, and a combination of them would provide stable conditions (Lawrence & Mason, 2012; Linbo, 2009). Because of solid accumulation on heater elements, they should be cleaned regularly. It is recommended that the laboratory hold extra heaters and related fuses for emergency situations, and lab technicians must be familiar with this equipment.

1.2.3 pH

pH is the reciprocal of hydrogen ion concentration in logarithm form, which ranges between 0- and 14-gram atoms per liter. This symbol measures the acidity or alkalinity of aqueous solutions. In a neutral situation, the concentration of hydrogen (H+) and hydroxyl (OH-) ions are equal, and the pH level is 7.0. Solutions with pH less than 7 are defined as acidic, and solutions with pH more than 7 are alkaline (Weiner & Matthews, 2003). Considering that the pH formula is logarithmic, by one unit change the concentration of hydrogen ion is changed by a factor of 10 (Hopkins et al., 2022). pH is effective in total ammonia nitrogen equilibrium in fish culture water, so any change in this factor must be addressed. In the culture system, biological processes such as the conversion of NH_3 into NO_3 decrease the pH, but fish waste breakdown increases that by releasing ammonia (Kutty, 1987). In recirculating systems, water tends to become acidic because of biological processes, but additions of sodium bicarbonate as a buffer compensate for it. The influence of pH on the stress response, growth, reproduction, mortality, and disease resistance of larvae and adult fish has been demonstrated, and the optimal pH for most freshwater fish to be healthy and have desirable growth is in the range of 6.5 to 9.0 (Sahu, 2018; Swain et al., 2020). The recommended range of pH for zebrafish in the culture system is between 7 and 8, which is also optimal for bacteria of biological filters, but in natural habitats, pH fluctuates between 6 to 10 (Aleström et al., 2020). Zebrafish can adapt to pH as low as 4 in chronic exposure, but it has disruptive effects on their reproductive performance. Acute exposure of zebrafish to a low or high level of pH leads to different responses, such as excessive mucus production that covers the gills and skin and disrupts ion and gas exchange, destruction of the respiratory epithelium of gills, and behavioral changes that are followed by death (Kwong et al., 2014). So, the stability of pH is necessary for the maintenance of fish in healthy condition.

1.2.4 Salinity

The ability of water to pass an electric current is measured and defined by conductivity term and expressed in micro-Siemens unit (μS). Electrical current is conducted by dissolved salts, so conductivity is increased when water contains more dissolved ions such as sodium (Na^+), chloride (Cl^-), calcium (Ca^{+2}), and carbonate (co^{3-2}). The range of conductivity for tap water is between 80–800 μS, distilled water 0.5–3.0 μS, and sea water 55.000 μS (Bhateria & Jain, 2016). Temperature is effective on conductivity, so this parameter must be paid attention when conductivity is measured. The maintenance of osmotic homeostasis between body fluid and the aquatic environment in fish is conducted by osmoregulation procedures in the gills, skin, and kidneys (McCormick et al., 2020). For this purpose, zebrafish as freshwater fish actively absorb ions from water and food into body fluid, reabsorb salts in the kidneys, and secrete high amounts of diluted urine (Guh et al., 2015). This procedure is energy consuming, so regulation of conductivity at the optimal level in the culture system increases the use of energy for the growth and reproduction of fish.

1.2.5 Ammonia

Among water quality parameters, ammonia after oxygen could be the most important one for fish health. In an aquaculture system, digestion of protein in fish food produces ammonia that is excreted by diffusion through the gills and feces of fish, so in an intensive system, toxicity of ammonia is more prevalent. Ammonia exists in two forms including ionized ammonia ($NH4^+$), which is less toxic, and the un-ionized form (NH3), with has high toxicity for fish (Sergeant, 2014). These two forms are in equilibrium, but pH and temperature could change their concentration, and the toxic form is increased in alkaline conditions (Table 1.1) (Nagaraju et al., 2023). The toxicity of ammonia increased 10 times by one unit increase of pH. Ammonia mostly has toxic effects on the central nervous system, and also chronic exposure reduces the growth and survival of larvae (Gomułka et al., 2014). The concentration of more than 0.05 mg/L of NH_3 leads to tissue damage in fish, and at 20 mg/l they become sensitive to infectious disease and will die (Mustapha & Akinshola, 2016). So, in the culture system, the ammonia concentration must be retained near 0 mg/L.

1.2.6 Nitrite

In the bacterial oxidation procedure of ammonia, nitrite is generated as an intermediate and toxic product. The accumulation of nitrite in the culture system is a big problem because it can be absorbed actively by the gill epithelium, which leads to the high concentration of nitrite in body fluids. The elevated nitrite concentration in fish bodies results in the oxidization of hemoglobin to met-hemoglobin, which is called brown blood disease (Kroupova et al., 2005). Different factors affect the toxicity of nitrite in fish culture systems, including temperature, salinity, fish species, and fish age. In the following section, we describe them. By increasing temperature, the metabolic rate of fish and demand for oxygen is increased too, but nitrite disrupts the transport of oxygen in the bloodstream, which is lethal for aquatic animals (Boyd, 2014; Ciji & Akhtar, 2020). Fish uptake nitrite by chloride cells of the gills, and in freshwater species the affinity of these cells for nitrite is low, so chloride in a competitive route could inhibit its uptake. Some species are more susceptible to nitrite toxicity; coldwater fish such as salmonids are more sensitive than warm-water species. Studies have indicated that in similar conditions, younger fish are more tolerant to nitrite than older individuals are, and there is higher activity of methemoglobin reductase in younger fish, and different respiratory systems in larvae and adult fish could explain this difference (Ciji & Akhtar, 2020). The 96-hour LC50 for freshwater fish ranges from 0.66 to 200 ppm. In zebrafish embryos and early larval stages, 96-hour LC50 of nitrite was 411 ppm. Exposure to 10 and 300 ppm for 96 hours resulted in craniofacial malformations, immobility, and swim bladder noninflation, but for adult zebrafish 96-hour LC50 at 386 ppm are reported (Simmons et al., 2012).

1.2.7 Nitrate

Nitrification is a process in the oxidation of ammonia to nitrite and subsequently to nitrate, so nitrate is the end product of this process. In a recirculation aquaculture system, nitrate is accumulated rapidly as fish excrete ammonia, and in biological filters converts to nitrate (Gomez Isaza et al., 2020). Nitrate in low concentration is considered a harmless product, but chronic exposure has toxicity effects (Torno et al., 2018). This toxicity is similar to nitrite, which decreases the efficiency of oxygen transport and leads to brown blood disease. Age, body size, water quality, and species of fish could be effective in toxicity of nitrate. Salmonids are the most sensitive species to nitrate toxicity, and in *oncorhynchus mykiss*, accumulation of nitrate with a concentration of 80–100 ppm resulted in health and welfare impacts in juveniles, such as abnormal swimming behavior (Davidson et al., 2014). The 96-hour LC50 for freshwater fish ranges from 0.66 to 200 ppm. Nitrite at a concentration of 200 ppm didn't show any effect on zebrafish larvae, but at 400 ppm survival decreased and some morphological abnormalities were observed too (Learmonth & Carvalho, 2015). However, the recommended level of nitrate for larvae and adult cultures of zebrafish is below 50 ppm.

TABLE 1.1
Percentage of Un-Ionized Ammonia in Water of Fish Culture Systems with Respect to Water Temperature and pH

	Water Temperature						
pH	6°C	10°C	18°C	22°C	26°C	28°C	30°C
6.5	0.04	0.06	0.10	0.14	0.20	0.22	0.25
7	0.13	0.18	0.34	0.45	0.60	0.70	0.80
7.5	0.43	0.59	1.06	1.42	1.88	2.17	2.48
8	1.34	1.83	3.28	5.35	5.74	6.56	7.46
8.5	4.12	5.55	9.68	12.56	16.80	18.2	20.3
9	11.96	15.68	25.30	31.23	37.83	41.2	44.6

1.2.8 Water Quality Procedures

Adult zebrafish could be maintained in treated tap water, but embryo and young larvae are more sensitive to water quality and must be retained in embryo medium (E2 or E3).

1.2.8.1 E2 Medium Procedure

1.2.8.1.1 Materials and Equipment

NaCl, KCl, KH_2PO_4, Na_2HPO_4_$7H_2O$, $CaCl_2$_$2H_2O$, $NaHCO_3$, methylene blue, HCl, distilled water, 1L and 2L beaker, pH meter, magnetic stirrer

The working solution of E2 medium is 0.5X, and we need to prepare three stock solutions (E2A, E2B, and E2C) to prepare this medium.

1.2.8.1.2 E2A 100X Stock

Prepare 100X E2A by combining the following ingredients in a final volume of 2000 ml:

175.0 g Sodium Chloride (NaCl)
7.5 g Potassium Chloride (KCl)
49.3 g Magnesium Sulfate Heptahydrate ($MgSO_4$_$7H_2O$) or 24.1 g of Anhydrous $MgSO_4$
40 ml E2A Buffer Mix
Add 1600 ml distilled water to a 2 L beaker.
Add the above reagents and stir to dissolve them.

Tip. Dissolve each one completely before adding the next reagent.

Add 40 ml E2A Buffer Mix and mix with stirring.
Adjust final volume to 2000 ml by adding distilled water and mix completely.
Sterilize by autoclave or filter.
Store in refrigerator at 4°C.

1.2.8.1.3 E2A Buffer Mix

Prepare E2A Buffer Mix by combining the following ingredients in a final volume of 1000 ml:

102.1 g Potassium Phosphate Monobasic (KH_2PO_4)
67.0 g Sodium Phosphate Dibasic Heptahydrate (Na_2HPO_4_$7H_2O$) or 35.5 g of Anhydrous Na_2HPO_4
Add 750 ml distilled water to a 1 L beaker.
Add the above reagents and stir to dissolve them.

Tip. Dissolve each one completely before adding the next reagent.

Adjust final volume to 1000 ml by adding distilled water and mix completely.
Sterilize by autoclave or filter.
Store in refrigerator at 4°C.

1.2.8.1.4 E2B 500X Stock

Prepare 500X E2B by combining the following ingredients in a final volume of 1000 ml:

73.5 g Calcium Chloride Dihydrate ($CaCl_2$_$2H_2O$)
Add 800 ml distilled water to a 1 L beaker.
Add the above reagent and stir to dissolve that.
Adjust final volume to 1000 ml by adding distilled water and mix completely.
Sterilize by autoclave or filter.
Store in refrigerator at 4°C.

1.2.8.1.5 E2C 500X Stock

Prepare 500X E2C by combining the following ingredients in a final volume of 500 ml:

14.7 g Sodium Bicarbonate ($NaHCO_3$)
Add 350 ml distilled water to a 1 L beaker
Add the above reagent and stir to dissolve that.
Adjust final volume to 500 ml by adding distilled water and mix completely.
Sterilize by autoclave or filter.
Aliquot into 20 ml portions.
Store in freezer at -20°C.

Tip. Thaw the frozen stock just prior to use.

1.2.8.1.6 Methylene Blue 1% Stock

Prepare 1% Methylene Blue by combining the following ingredients in a final volume of 1000 ml:

10 g Methylene Blue
Add 1000 ml distilled water to a 1 L beaker.
Add the above reagent and shake to dissolve that.
Store in refrigerator at 4°C.

1.2.8.1.7 E2 0.5X Working Solution with Methylene Blue

Prepare 0.5X E2 by combining the following ingredients in a final volume of 20 L:

100 ml 100X E2A
20 ml 500X E2B
20 ml 500X E2C
1 ml 1% Methylene Blue
Adjust final volume to 20 L by adding Reverse Osmosis water and mix completely.
Check conductivity and pH of solution.
Use HCL or NaOH to adjust pH to 7.2–7.6.
Store at room temperature.

Tip. Make the fresh working solution weekly.

1.2.8.2 E3 Medium Procedure

1.2.8.2.1 Materials and Equipment

NaCl, KCl, $CaCl_2$_$2H_2O$, $MgSO_4$_$7H_2O$, HCL, NaOH, Methylene Blue, 2L beaker, pH meter

The working solution of E3 medium is 1X.
We prepare the 60X stock solution by combining the following ingredients in a final volume of 2000 ml:

34.8 g Sodium Chloride (NaCl)
1.6 g Potassium Chloride (KCl)
5.8 g Calcium Chloride Dihydrate ($CaCl_2$_$2H_2O$)
9.78 g Magnesium Sulfate Heptahydrate ($MgSO_4$_$7H_2O$)
Add 1600 ml distilled water to a 2 L beaker.
Add the above reagents and stir to dissolve them.

Tip. Dissolve each one completely before adding the next reagent.

Adjust final volume to 2000 ml by adding distilled water and mix completely.
Check conductivity and pH of solution.
Use HCl or NaOH to adjust pH to 7.2–7.6.
Sterilize by autoclave.
Store at room temperature.

1.2.8.2.2 E3 1X Working Solution with Methylene Blue

Prepare 1X E3 by combining the following ingredients in a final volume of 2 L:

16.5 ml 60X E3 solution
0.1 ml 1% Methylene Blue
Store in incubator (28°C).

Tip. Make the fresh working solution weekly.

Adult fish are maintained in a housing system, and these systems must provide the optimal water quality for fish. So, for any water quality parameters, we need level range and also monitoring schedules that are outlined as follows (Table 1.2).

Tip 1. Source water supply must be stable and provide sufficient volume for the needs of the system.
Tip 2. Chlorine and chloramines must be removed from inlet water.

1.2.8.3 Water Quality Monitoring Procedure

1.2.8.3.1 Materials and Equipment

Sea salt, sodium bicarbonate, commercial kits for pH, alkalinity, ammonia and hardness, pH calibration buffer, conductivity calibration buffer

Daily maintenance

1. Touch screen display: the display must be checked several times per day to be sure about its operation and absence of alarm messages. All information about water quality must be recorded in the room status sheet (Figure 1.7).
2. System function: the functionality of the system is verified by checking the display and water levels.
3. Tank inspection: the water level for each tank must be checked visually.
4. Salt dosing system: the proper functionality and level of salt solution must be checked visually, to add solution as needed.
5. Bicarbonate dosing system: the proper functionality and level of bicarbonate solution must be checked visually, to add solution as needed.

Tip. Before adding solutions, double-check that add any solution to its own container.

Weekly maintenance

1. Water quality: a water analysis using commercial kits must be done to determine pH, ammonia, alkalinity, and hardness. All obtained information must be recorded on the room status sheet.
2. Spot clean tanks: once per week, tanks must be checked, and any ones that seem to be dirty should be washed.

Biweekly

1. Checking heating elements: any malfunction will be indicated on the system screen as an alarm.
2. Check UV lamps: the alarm on system screen indicates the malfunction.
3. Check conductivity probe: check it for any damage by bringing it out of the water.
4. Check pH probe: check it for any damage by bringing it out of the water.

TABLE 1.2
Description of Water Quality Parameters, Testing Frequency, and Level Range for Zebrafish Culture System

Parameters	Testing Frequency	Level Range
Temperatures	Daily	26–29°C
pH	Daily	7–8
Conductivity	Daily	200–3000 μS
Total ammonia nitrogen	Weekly	Close to 0 ppm or less than 1
Nitrite	Weekly	Close to 0 ppm or less than 0.5 ppm
Nitrate	Weekly	Less than 50 ppm

Room Status Sheet | Zebrafish Room Sheet
Month: | Year: | Room: | Technician:

	Check Record Daily									Check Once Week						Daily			2 Week
	Room		Feed		Water parameters								Tanks		Observed			
	Temp	Humid	AM	PM	Salin	pH	Temp.	Salt tank	Bicarb tank	NH3	NO3	NO2	Spot clean tank	Tank system	AM	PM	Tech	initial
1																		
2																		
3																		
4																		
5																		
6																		
7																		
8																		
9																		
10																		
11																		
12																		
13																		
14																		
15																		
16																		
17																		
18																		
19																		
20																		
21																		
22																		
23																		
24																		

FIGURE 1.7 Room status sheet. All daily information should be recorded on this sheet.

Monthly Maintenance

1. pH probe: it must be checked for damage.
2. Conductivity probe: it must be checked for damage.
3. Check float switches: in the standard condition, the water level should be 2.5 cm below the overflow outlet. So, when water level is above, the replacement of float switches is necessary.
4. Check drum filter: the washing pressure gauge must show 7 bars.
5. Check carbon filters: these filters should be checked regularly for any damage and replaced if necessary.

Three-month maintenance

1. Check biofilter air distribution: biofilter chips must be moving constantly. Check the air nozzle to be sure that air is flowing.
2. Heating element: the problem in heating of element can be seen on the home screen.
3. Calibrate pH probe: calibration must be done with pH calibration standard buffer solutions pH 4.00 and 7.00.
4. Calibrate conductivity probe: calibration must be done with conductivity calibration standard solutions.

Tip. The metal end of the probe must never be touched or wiped.

Six-month maintenance

1. Check sump level: it must be checked and cleaned of any residue.

Annual maintenance

Change UV lamps: The UV lamp of system must be on permanently (except at replacing time), and it must be changed every 9000 hours. For this purpose, turn off the system, unplug the UV lamp, and replace it with a new one, then return the lamp into the UV unit and turn on the system.

Tip 1. Be sure that all parts are well sealed.
Tip 2. Don't touch lamps with dirty fingers.
Tip 3. UV lamps can cause damage to eyes and skin.

1. Clean biofilter tank: it must be performed by an expert technician.
2. System cleaning: system needs to be cleaned regularly to inhibit the accumulation of waste products and provide healthy condition for fish.
   A: Tanks must be cleaned at a schedule that provides the quality of water and allows monitoring of fish. Accumulation of food particles on

tank lids lead to fungi contamination, so these components need to be cleaned weekly.

Tip. Tanks could be disinfected by bleach solution, but it is necessary to rinse them very well and dry completely before reusing.

B: Clean the racks and other equipment as needed

1.3 ZEBRAFISH HEALTH CARE

Even more important than ethical goals are determining and reporting on the health of animals. Pathogens can increase morbidity and mortality and reduce larvae survival. It should be noted that infectious diseases have an impact on research results; for example, some species of mycobacteria can alter gene expression and upregulate various inflammatory cytokines (Lieggi, 2020). Please refer to Table 1.3 for further information.

It is important to inform researchers about the diseases and their impact on study results. Health status reporting is also one of the important ethical obligations to reduce the risk of infection prevalence and distribution among zebrafish laboratories.

1.3.1 Animal Welfare Indicators for Fish

There are well-established welfare indicators for mammalian species, such as ethics, neuroscience, animal research, etc. However, it is not clarifying to zebrafish. Some indicators based on previous aquaculture industries are suggested as follows.

Change in color: Zebrafish have three types of pigment cells, including melanophores, xanthophores, and iridophores, whose distributions are responsible for the color and patterns of the zebrafish body (Silver et al., 2006). Skin color changes due to some stimuli, such as the environment, social subordination, or physiological stress. Various stimuli, including camouflage, light intensity, and aggressive behavior, cause aggregate or disperse the melanophores based on melanin-concentration hormone (MCH) (Parichy, 2020).

Respiratory rate: The movement and flow of the water through the gills, respiratory rate is measured visually. Several factors increase the respiratory rate. For example, the fish have difficulty breathing because of reduced water oxygen or the functional surface areas of the gills, and even certain physiological requirements (Messerli et al., 2020).

Swimming and behavior patterns: Several factors such as health and environmental conditions, stress, and predators have influenced swimming behavior, pattern, speed, and position in the water column, which can be one of the indicators of welfare. Determination of these items is important for zebrafish behavioral research (Øverli et al., 2005).

Growth rate: Genetics, environmental parameters, density, chronic stress, and nutrition can affect growth rates (Siccardi et al., 2009). Also, population density, body size, and age influence growth rate (Tsai et al., 2007). In

TABLE 1.3
Examples of Clinical Impact and Research Impact of Select Infectious Organisms

Agent	Clinical Impact	Interference with Research
Pseudoloma neurophilia	Subclinical or variable morbidity and mortality; skeletal deformities; reduced growth; emaciation; decreased fecundity; lethargy; decreased larval survival	Histologic alterations of involved tissue; myonecrosis and autofluorescence interferes with muscular development and imaging studies; altered behavioral studies, including startle response and shoaling behavior; increased mortality following gamma irradiation
Pseudocapillaria tomentosa	Variable morbidity and mortality; diffuse granulomatous peritonitis; emaciation; decreased growth rates and fecundity	Histologic alterations of involved tissue; hematologic and gastrointestinal eosinophilia; altered results of carcinogenesis studies
M. marinum	Variable morbidity and mortality (exacerbated by stress); dermal lesions; granuloma formation; zoonotic potential; coelomic edema (dropsy), zoonotic potential	Activation of innate immune response; alterations of gene expression; tumor promoter (Medaka); reactivation of latent infection by gamma irradiation
M. haemophilum	Moderate to significant morbidity/mortality, emaciation, lethargy, dropsy, dermal lesions, granuloma formation, zoonotic potential	Tumor promoter evokes chronic inflammation with alteration if inflammatory mediated pathways, granulomatous nephritis and hepatitis may impact kidney and liver function
M. chelonae	Typically subclinical; aerocystitis; zoonotic potential	Autofluorescence interferes with imaging studies
Pleistophora hyphessobryconis	Lethargy, morbidity, mortality	Severe necrosis and expansion of myofibers; muscle deformity
Edwardsiella ictaluri	High mortality epizootics	Necrotizing inflammation in various visceral organs

Source: (Lieggi, 2020).

order to use growth rate as a welfare indicator, the determination of optimal growth and maintenance should be considered.

Body condition: Body condition score can be easily assessed based on weight gain, weight loss, breeding cycle, non-infectious illness, and infections. Infectious diseases such as *Pseudoloma neurophilia* cause body curvature (Cade et al., 2008).

External morphological abnormalities: Fin, gill, and eye abnormalities are the general indicator of health and well-being that can be influenced by husbandry and environmental conditions (Noga, 1998).

Reproductive performance: Many factors such as chronic stress, water quality, nutritional imbalance, etc. affected the reproductive performance of fish (Harper & Lawrence, 2016).

Biomedical welfare indicators: Physiological parameters change after the stressful state. These parameters include energy use, osmoregulation, respiration, and immunity. When fish face short-term stress, they seem to behave normally but deplete their energy reserves (Floyd, 2002).

Food consumption: One of the important welfare indicators is food consumption, which can be influenced by chronic and acute stress. In response to stress, fish reduce food consumption and mobilize energy reserves (Olsen et al., 2008).

1.3.2 Health Care Procedures

1.3.2.1 Histology Procedure

Fish that are reared in large-scale research facilities reveal novel pathogens and diseases. The primary diagnostic test is usually histopathology, which has several strengths such as documentation of tissue/organ changes without guesses. In fact, all important laboratory zebrafish diseases were recognized with histopathological sections in moribund fish (Kent et al., 2020).

Tip: Buffered formalin 10% or Dietrich's solution and paraformaldehyde are used to preserve diagnostic samples for adult fish and larvae, respectively.

1.3.2.1.1 Preparing Fish and Fixation

1. Euthanize zebrafish with 160 mg/L Tricaine-S in system water or hypothermal shock (rapid cooling). Do not freeze fish.
2. Cut off the tail just forward of the tail fin, except for tail lesions or fish less than 1.5 cm in length.
3. Cut the belly skin with scissors, open the body cavity, and drain internal organs, except skinny fish or less than 1 cm in length.
4. Place fish into a fixator that is at least ten times the volume of the specimen (10 ml of fixative is sufficient for average-size adult zebrafish).
5. Lid the container tightly and label it with information such as fish stock number, clinical signs, and diseases.

1.3.2.1.2 Preparing for Histology

1. Transfer the fixed fish to 5% trichloroacetic acid (TAC) in Dietrich's fixative to soften scales and skin, in addition to decalcifying.
2. Place them on the rotor overnight.
3. Replace TCA-containing Dietrich's with ethanol 70% for 10 minutes on the rotor. Repeat to rinse twice. Replace with the refresh ethanol 70%.
4. Bisect along their length, and cut parallel toward the left side of the spinal cord.

1.3.2.1.3 Fixative

1. Dietrich's fixative: Mix the ethanol 95% (30 ml), formalin (10 ml), glacial acetic acid (2 ml), and distilled water (58 ml): total volume 100 ml.
2. Formalin 10%: formalin 37% (100 ml/1 part), distill water (90 ml/9 part): total volume 100 ml.

1.3.2.2 Molecular Detection Procedure

Detecting pathogen DNA is a key tool to assess the health of zebrafish that for some reasons, including increased sensitivity, quick obtain results, high-throughput sampling, and multiplex ability, turn into a mainstay in the diagnostic laboratory (Kent et al., 2020). Also, it is a cost-effective method compared to others such as microbial culture and histology.

1.3.2.2.1 Tissue Collection

Adult fish fin clip

1. Anesthetize fish with Tricaine-S 0.004%, and wait until gill movement slows.
2. Put the fish on a clean surface and cut a piece of approximately two-thirds of the tail. If a bigger piece of tail is cut, it will not be regenerated again.
3. Bring back the fish to fresh water. Pick up the fin with a sterile pipette and transfer it to the requisite solution.

For embryo tail clip

1. Anesthetize the embryo in 0.004% Tricaine-S.
2. Wait 2 minutes and cut a piece of tail with forceps under a dissecting microscope.
3. Put the tail on the appropriate solution.

1.3.2.3 Microbiology Procedure

A bacterial culture or PCR analysis is performed with histopathology to connect novel microorganisms with diseases that are particularly useful to separate the risk of internal structure bacterial contamination. Histopathology determines what Bactria are infections and associated with pathological changes (Kent et al., 2020).

1.3.2.4 Necropsy Procedure

Necropsy creates an opportunity to assess external and internal observations and measure condition indices, including:

- Hepatosomatic index (liver weight/ total body weight): fitness or energy reserve indicator
- Gonadosomatic index (gonad weight/total body weight): reproductive health indicator
- Quantitative health assessment: prevalence of individual lesion type or percentage of normal individuals
- Age analysis

Depending on the question, necropsy-based assessment augments the grossly visible assessment. Collecting biometric data (length, weight, liver weight, gonad weight), plasma/serum analysis, documenting visible abnormalities, collection of tissue for microscopic/gene expression, and functional immune analyses are the necropsy routine approaches (Blazer et al., 2018; Gupta & Mullins, 2010).

1.3.2.4.1 Dissection

1. Euthanize the zebrafish with Tricaine 0.2% or incubate in ice water for 15 minutes.
2. Dry the fish on a paper towel and place it on a dissection mat.
3. Pin the fleshy part of the fishtail and the ventral part of the eye socket.
4. Cut the skin from the anal fin to the operculum (Figure 1.8).
5. Remove the operculum and pectoral fin.
6. Remove skin and underlying muscle from above the exposed gill alongside the fish to the anal fin.

Now, internal organs are visible.

1.4 ZEBRAFISH ROOM

As mentioned in section 1.2.1, there are two types of housing systems for zebrafish: centralized and packaged systems. Design of the housing room depends on the housing types and also the required techniques for the research program (Lawrence & Mason, 2012). Centralized systems in comparison with packaged types need more space but support a bigger population of fish. The space between maintenance racks is determined according to study types, but a standard space is needed for enough light receiving and also movement of tanks. In addition to active service, the expansion program for the future must be considered when designing a room (Zynda, 2020). Adjusting of temperature, humidity, sound, and lighting in a zebrafish room is necessary to provide ideal conditions for zebrafish maintenance in healthy condition.

1.4.1 Humidity

Humidity is of little importance to zebrafish welfare, although humidity affects room and tank temperatures; tanks are cooler in dry rooms, and the temperature can fluctuate due to evaporation of dew on the tank. It must be taken into account that the main reasons for humidity control are related to mold condensation and growth, as well as protecting the external support system of the zebrafish facility systems.

1.4.2 Light

Lighting devices are one of the most important parts of zebrafish facilities that minimize fish stress and impact successful breeding. In addition, physiological and behavioral processes are influenced by light; therefore, appropriate lighting ensures optimal growth, development, and reproduction. Light and other environmental factors such as temperature can work synergistically, making it difficult to predict the outcome of an inappropriate lighting system (Stickney, 1994; Villamizar et al., 2014).

The standard lighting regimes are 14 hours of light, 10 hours of dark, or a 12:12 hour cycle. The longer lighting or dark period is considered harmful to reproductive success. If artificial light is used for the holding facility, then ensure that all windows are covered to avoid disrupting

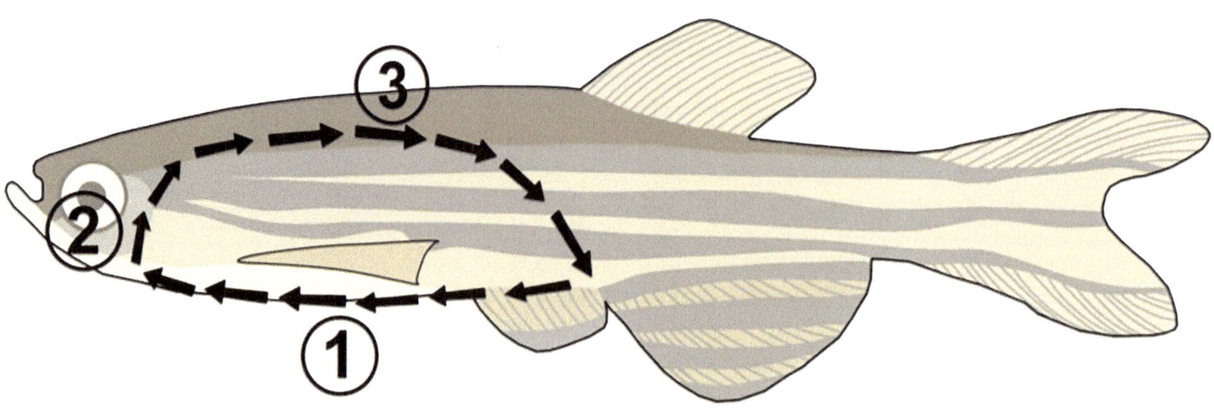

FIGURE 1.8 Zebrafish dissection. Removing of operculum and pectoral fin for accessibility to organs.

the light-dark. Red light should also not be used as it has negative effects. When using fish in experiments, light phases should be chosen. Circadian rhythm must be followed every day and in every facility while mimicking the natural circadian cycle. When the light changes quickly, fish are easily startled; therefore, a gradual brightening and dimming period of about 20 to 30 minutes seems essential (Figure 1.9) (Adatto et al., 2016; Frøland Steindal & Whitmore, 2019; Stephenson et al., 2011).

Larvae react to light changes, and in fact their movement and thus their food intake are affected. It has been reported that visual and behavioral deficits without anatomical abnormalities were found under constant light. Additionally, constant darkness has been shown to delay general embryo development. A normal photoperiod can improve survival rates (Bilotta, 2000; Burgess & Granato, 2007; Villamizar et al., 2014).

Zebrafish development is influenced by the light spectrum. Blue and violet light showed the highest hatching rate, while red light resulted in reduced food intake and poor survival rate. There are no specific requirements for the light spectrum, but further research is needed (Saszik et al., 1999; Villamizar et al., 2014).

1.4.3 Sound and Vibration

Zebrafish become accustomed to environmental noises and vibrations but react to loud noises or novel vibrations. Some sounds such as talking and music have unknown effects (Barcellos et al., 2018; Vandenberg et al., 2012).

1.5 QUARANTINE ROOM

The zebrafish facility must be composed of two physically separated rooms: the main room and the quarantine room. The main room contains the central or stand-alone housing system for the culture and breeding of broodstocks and experimental animals (Borges et al., 2016). The quarantine area is a single room that is equipped with a stand-alone housing system; this room is used for rearing and monitoring newly imported embryos and adult fish to inhibit pathogen agents from disseminating to the main room. Embryos must be surface disinfected and then could be entered into the housing systems of the quarantine; after reaching sexual maturity, the resulting embryos from their breeding are disinfected and housed in the main room. If imported adult fish look unhealthy, they shouldn't be introduced to the quarantine room. Fish that are found to be free from harmful pathogens are bred, and their bleached embryos are moved to the housing facility of the main room. The equipment (nets, tanks, spawning tanks, etc.) of this room should be used only in this unit and shouldn't be mixed with the main room's equipment (Aleström et al., 2020; Varga, 2011).

1.6 TRANSPORTATION

There are several reasons for transporting zebrafish, including demand for wild-type, off-site breeding and the introduction of new transgenic and mutant strains around the world. They have to leave for distant destinations, sometimes to another

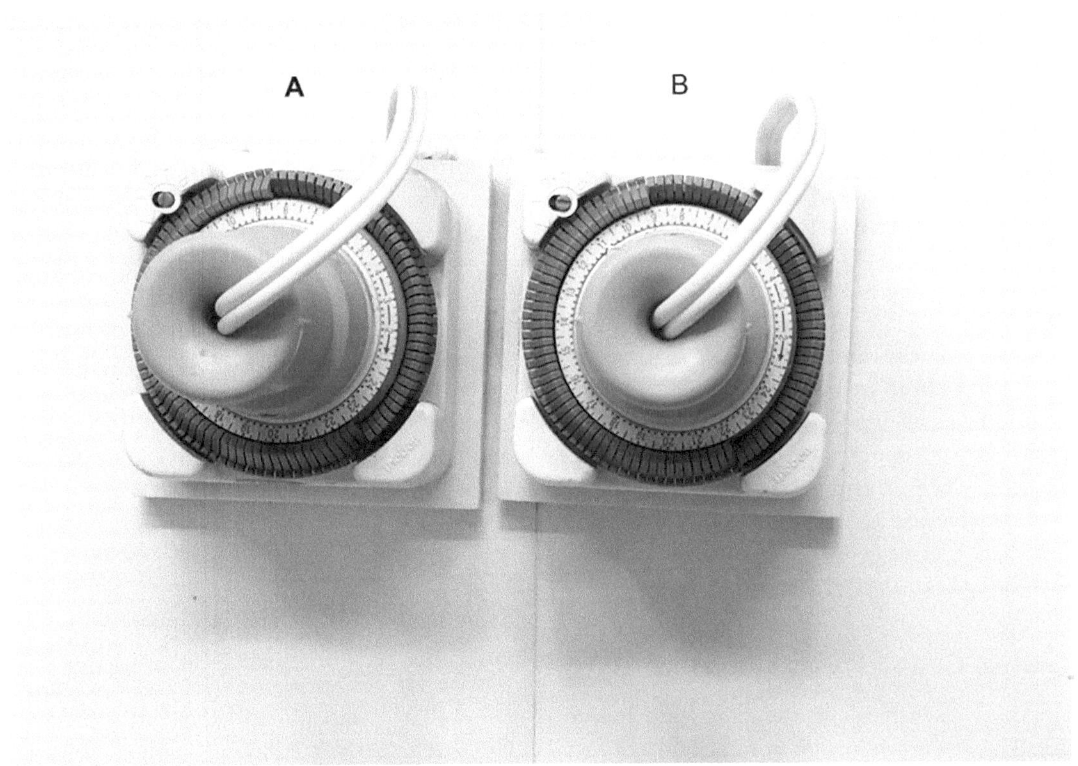

FIGURE 1.9 Light timer: (a) main light timer, (b) side light timer.

country or continent. Shipping between institutions—both within and beyond national borders—presents a number of challenges. The health and welfare of zebrafish are at risk during long journeys. For example, the immune system is decreased due to transport stress. Additionally, there are some regulatory requirements for the transport of live animals across international borders. Therefore, it is preferred to transport fertilized eggs or early larvae that are younger than 5 days. Packing and isolating must be done carefully to minimize the potential of injury or stress (Ostrander, 2000).

After the fish have been health and welfare checked and approved by the Animal Welfare Committee, they are placed in quarantine. To make fish updates easier, it is important to know the water temperature and quality. Depending on the water quality, gradual water changes are recommended. When fish arrive at the facility, manipulate the fish as little as possible to avoid stress and temperature shock. Temperature changes greater than 2–3°C and pH (ammonia content) prior to release must be determined. Finally, the exchange between laboratories also involves the risk of the spread of pathogens. The disease status of fish and zebrafish-relevant pathogens must be determined (Barton et al., 2020).

1.6.1 Shipping General Physical Requirements

The biggest challenges in transportation are related to maintaining appropriate water quality; the crucial water parameters are ammonia and dissolved oxygen during transport (Harper & Lawrence, 2016). The consumption of oxygen and production of ammonia are the result of normal metabolism and can present challenges in the context of a closed, static unit such as a shipping container. The key role in packing fish is to minimize these changes. Fish density has an impact on increasing the rate of both. The best way to deal with the problem is to reduce the density of fish in tanks.

Lim et al. suggested packing animals with a volume of air that is three times the volume of water to maintain a suitable oxygen level (Lim et al., 2003). Pure oxygen is added to the bags. Fasting fish for 24 to 36 hours before packing reduces ammonia production; even on longer trips, ammonia-binding products can be beneficial. Maintaining a stable water temperature is important during shipment. The temperature becomes warm, resulting in reduced dissolved oxygen and increased ammonia toxicity. Fish suffer from hypothermia when the temperature gets too low. Use thick-walled Styrofoam shipping boxes to protect against physical disturbance, or using extra water bags can be the practical solution to keep an adequate temperature.

1.6.2 Receiving General Physical Requirements

A few factors are taken into consideration when receiving zebrafish. The fish are immediately inspected and then transferred from the shipping containers. It is important to recognize the chemical changes that occur during transportation in shipping containers to determine how the fish will be unpacked and transported. The concentration of oxygen and ammonia has changed in the static, closed environment of shipping bags. Temperature and pH have a direct impact on the toxicity of total ammonia nitrogen (TAN), meaning that toxicity increases with rising pH and temperature. Carbon dioxide released by fish reacts with water to form a weak acid that lowers pH and lowers problem levels. The main challenge is opening the bag and exposing it to the environment, which results in the release of CO_2 into the atmosphere and causes a rapid increase in the pH of the water and subsequently an increase in total ammonia nitrogen.

Shipping bags should not be opened until the fish is ready for transfer. Once the bag is opened, the fish should immediately take in new water. This is a crucial step in the receiving protocol when shipping densities and temperatures are high. To avoid thermal shock, it is more important to wait to transfer fish until the transfer water and receiving water are at the same temperature.

1.6.3 Transport Procedures

1.6.3.1 Shipping Procedure

1.6.3.1.1 Materials and Equipment

Styrofoam box, plastic bags, labels, oxygen, 20–250 ml tissue culture flask, methylene blue

1.6.3.1.2 Pack Adult/Eggs and Larvae

Adult fish: A double-layer with the best quality plastic is chosen while it has a capacity of about 10 fish/2 L of water, and it must be just one-third full of water and the rest full of oxygen. Don't forget to tape the bottom point of the bag to prevent fish from getting caught and injured in the corner. It is better that fish are off feed for one day to produce less ammonia and feces. For longer travel, using an ammonia sequester seems reasonable.

Eggs and larvae: Disinfected eggs are packed in a 200–250 ml tissue culture flask, each 1–2 eggs/ml of E3 environment while 50% of the flask is full of embryos and the rest full of air. It is possible to add methylene blue 0.5 mg/ml to prevent fungal growth. The packing box should be isolated.

1.6.3.1.3 Prepare the Following Labels and Documents

Live fish label (genotype, how many of each sex, how old, . . .) (Figure 1.10), address label, health and custom certification for international shipment as well as "keep at room temperature", "handle with care", "do not X-ray", "non-toxic", "nonhazardous", "no commercial value"

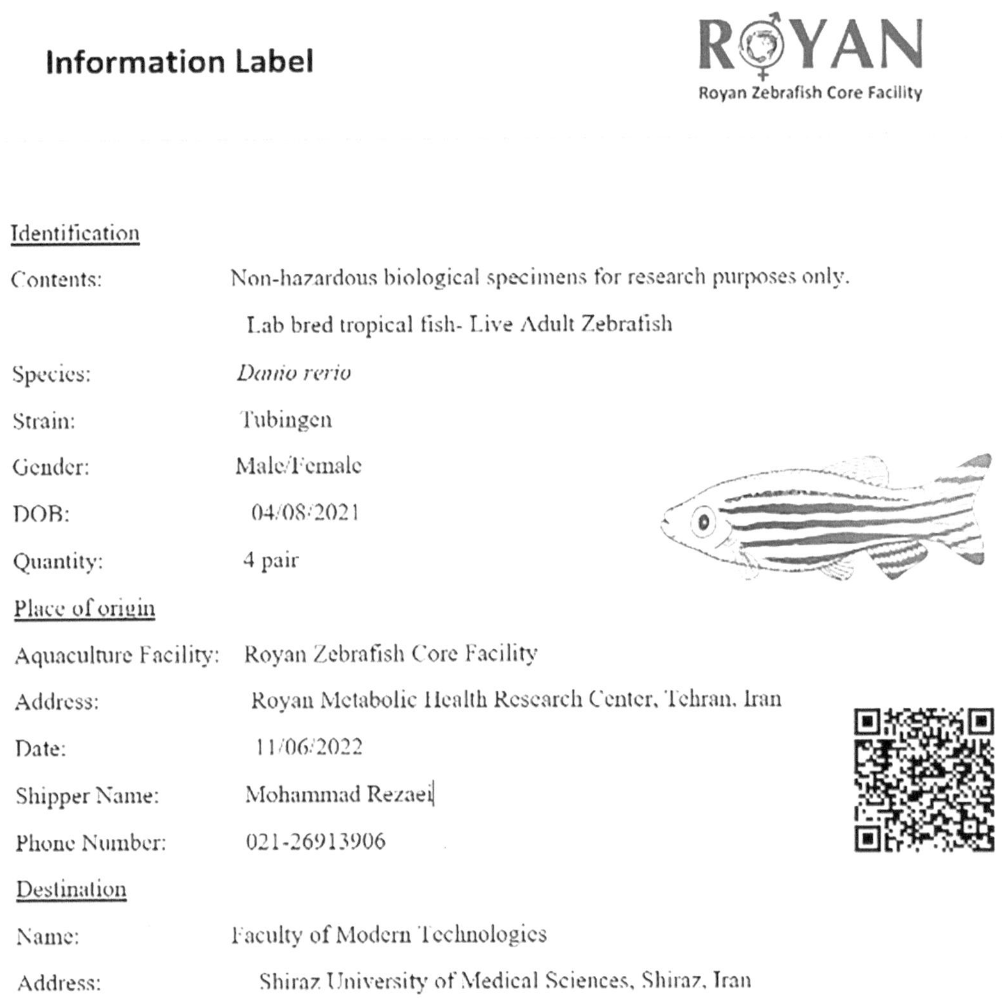

Information Label

ROYAN
Royan Zebrafish Core Facility

<u>Identification</u>

Contents: Non-hazardous biological specimens for research purposes only.
Lab bred tropical fish- Live Adult Zebrafish

Species: *Danio rerio*

Strain: Tubingen

Gender: Male/Female

DOB: 04/08/2021

Quantity: 4 pair

<u>Place of origin</u>

Aquaculture Facility: Royan Zebrafish Core Facility

Address: Royan Metabolic Health Research Center, Tehran, Iran

Date: 11/06/2022

Shipper Name: Mohammad Rezaei

Phone Number: 021-26913906

<u>Destination</u>

Name: Faculty of Modern Technologies

Address: Shiraz University of Medical Sciences, Shiraz, Iran

FIGURE 1.10 Information label. Essential information about shipped fish must be mentioned in this label.

1.6.3.1.4 Prepare the Pack for Sending

- Seal the bags by knotting or twisting the neck and secure rubber bands tightly around the knot.
- Put the sealed pack into the protected case (Styrofoam box) and stabilize the bags; it is important that bags don't move in the box while passing the routes.

1.6.3.2 Receiving Procedure

Tip 1. Specify transport itinerary and condition, breeding history of fish, dietary background, and health status.

Tip 2. Have the health and welfare of the animals checked upon arrival by trained and informed animal care staff.

Tip 3. Be informed of the water temperature and quality conditions during transport to acclimate these conditions in the laboratory.

1. Check the bags for any damage and fish for any sign of stress, trauma, and mortality.
2. Float the unopened bag in water that is the same temperature as the receiving water.
3. Transfer the fish to the receiving water after equilibrating the temperature.
   Note: Some have suggested adding small amounts of the receiving water; however, if the ammonia level is high, fish will be endangered because of chemical changes (toxic ammonia).
4. Approve the source of zebrafish, and put them into the appropriate tank in the quarantine room (they are isolated due to health concerns).
5. Check them for disease and mortality; it is better that they be examined by a veterinarian.
6. After a while, put the tank into the circulatory system until changing the water slowly.

1.7 THE ZEBRAFISH DATABASE SYSTEM AND TANK LABELS

Proper identification of zebrafish is the essential element for zebrafish core facilities management. Labeling and record keeping, as well as preparation of monthly and annual reports, are more beneficial for organizing, protecting the stocks, and managing the facility requirements.

1.7.1 Tank Labels

Effective management of zebrafish breeding and characterization is related to correct data recording. Information is recorded with traditional methods such as bookkeeping or some modern method, such as using Excel or other applications (Figure 1.11). Data to be recorded for each tank includes zebrafish strain, a descriptive name, housing system code, generation number, date of birth/age, type of fertilization, the number of parents, and the total number of larvae or adult fish in each tank (Figure 1.12).

These documents are attached to each tank and also kept in the same room. Tank labels are important for providing quick visual cues, and these records would also be helpful for colony management. Quality of brood stock and the survival rate of larvae are achieved by this data that reveal the need for outbreeding or replacement of fish by a new

Zebrafish Lines Records

Year: Room: Rack: Technician:

Tank Number	Fish line	Date of Birth	Tank Label	Euthanasia Date	Veterinarian initial
1A	mbp:eGFP NTR	23/12/2019	2A natural 23DEC19 mbp eGFP NTR 5pair 36	23/06/2021	
2A	fli:eGFP	01/08/2019	6A natural 01AUG19 fli eGFP 8pair 65	01/02/2021	
3A	AB	10/05/2019	4A natural 10MAY19 AB 15pair 94	10/12/2021	
4A	gata:dsRED	14/08/2019	6A natural 14AUG19 gata dsRED 3pair 81	14/02/2021	
5A	ins:kaede-NTR	13/05/2019	2A natural 13MAY19 ins kaede 6pair 120	13/12/2021	

FIGURE 1.11 Fish lines information. This information is recorded for colony management.

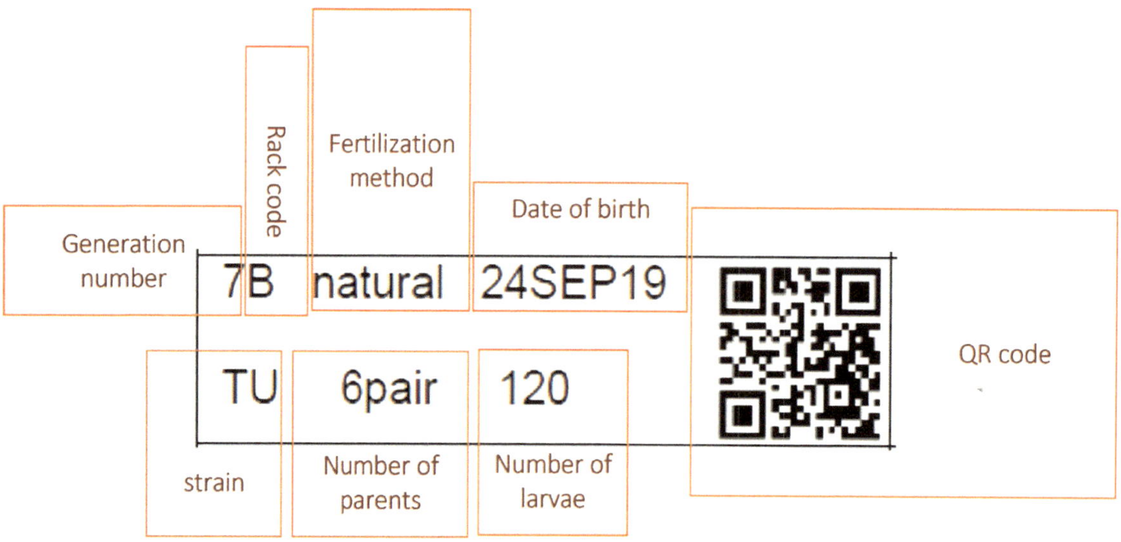

FIGURE 1.12 Tank labels. These labels provide information about living fish in tanks.

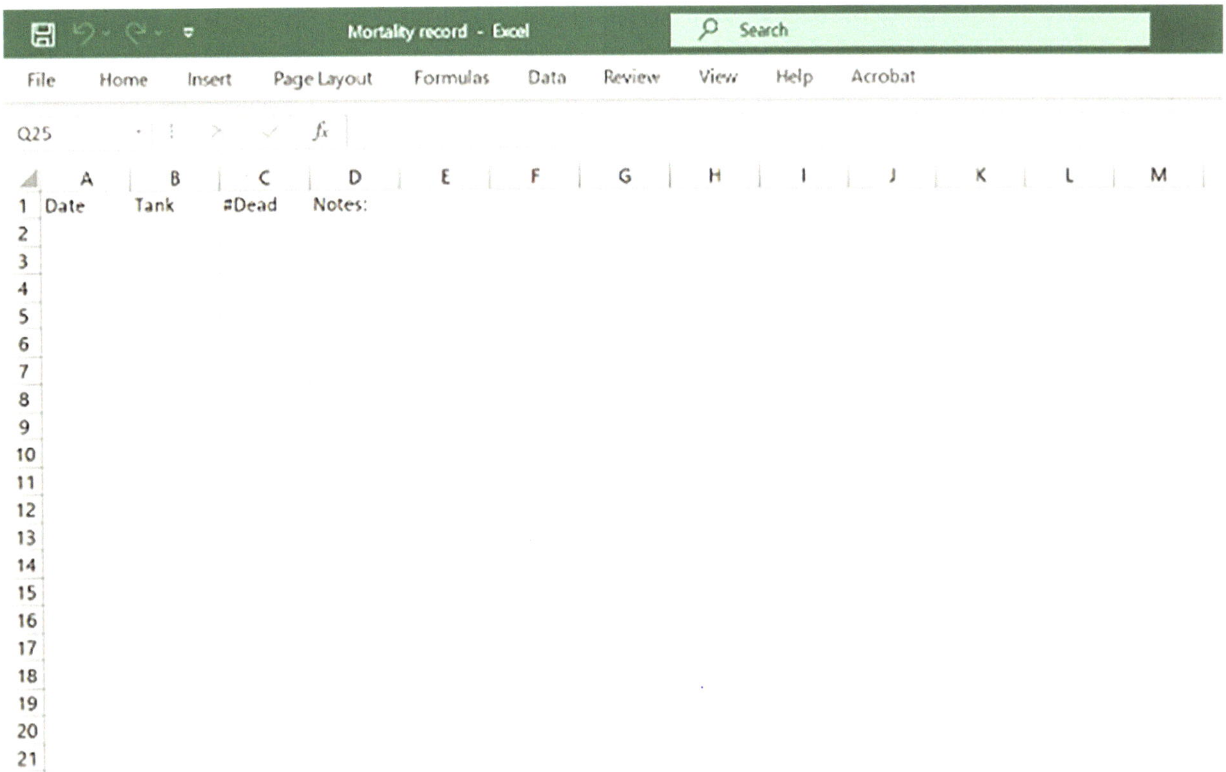

FIGURE 1.13 Mortality rate record. Mortality of fish in any tank should be recorded daily.

batch. In addition, disease detection and spreading as well as reproducibility should be recorded. By preserving these documents, precise history of the laboratory could be determined; for example, which tank has bred successfully without diseases and which tank is not suitable for breeding.

1.7.2 Record Keeping

A complete tracking system (record keeping) covering husbandry, maintenance, reproduction, and experiments provides important information and is also helpful in ensuring ethical rules. This information is reported monthly and annually and documented in the institute and laboratory. Data should be regularly collected and recorded in a simple way that reveals information easily to everyone.

The annual report involves the breeding strategy, reproductive rates, and reproductive conflicts, especially mating of specific strains with wild types, morbidity and mortality rate (Figure 1.13), import/export, and other effective factors related to core facility such as water quality, equipment maintenance and repair, and feeding.

The Zebrafish Information Network (Zfin) has proposed three separate databases (files): the Zebrafish Database, the Tank Status Database, and the User Database. They contain the most information about individual fish stocks, allow easy checking of tank status and their locations, and explain each user code in the main database, respectively (Westerfield, 2007).

1.7.3 Tank Label Procedure

Tip 1. Tank label must be resistant to water and fit to attach them on the front of the tanks.

Tip 2. It is better that technicians store the shorting history as a QR code.

The attached label to each tank should contain this information:

- Strain
- Date and type of fertilization
- Number of used broodstock
- Number of larvae in each tank
- Rack code
- Generation number

REFERENCES

Adatto, I., Krug, L., & Zon, L. I. (2016). The red light district and its effects on Zebrafish reproduction. *Zebrafish*, *13*(3), 226–229.

Akhter, F., Siddiquei, H. R., Alahi, M. E. E., & Mukhopadhyay, S. C. (2021). Recent advancement of the sensors for monitoring the water quality parameters in smart fisheries farming. *Computers*, *10*(3), 26.

Aleström, P., D'Angelo, L., Midtlyng, P. J., Schorderet, D. F., Schulte-Merker, S., Sohm, F., & Warner, S. (2020). Zebrafish: Housing and husbandry recommendations. *Laboratory Animals*, *54*(3), 213–224.

Allen, R. L., Wallace, R. L., & Sisson, B. E. (2016). A rotifer-based technique to rear Zebrafish larvae in small academic settings. *Zebrafish*, *13*(4), 281–286.

Arabi, S., Kwaya, M., Jauro, I., & Jaoji, A. (2011). Assessment of surface and groundwater quality for use in aquaculture in parts of northern Nigeria. *Research Opinions in Animal & Veterinary Sciences*. www.academia.edu/804181/Assessment_of_Surface_and_Groundwater_Quality_for_use_in_Aquaculture_in_Parts_of_northern_Nigeria.pdf.

Avdesh, A., Chen, M., Martin-Iverson, M. T., Mondal, A., Ong, D., Rainey-Smith, S., . . . Verdile, G. (2012). Regular care and maintenance of a Zebrafish (*Danio rerio*) laboratory: An introduction. *Journal of Visualized Experiments: JoVE*, (69), e4196.

Barcellos, H. H. A., Koakoski, G., Chaulet, F., Kirsten, K. S., Kreutz, L. C., Kalueff, A. V., & Barcellos, L. J. G. (2018). The effects of auditory enrichment on Zebrafish behavior and physiology. *PeerJ*, *6*, e5162.

Barton, C. L., Baumann, D. P., & Cox, J. D. (2020). Export and transportation of Zebrafish. In *The Zebrafish in biomedical research* (pp. 443–450). Elsevier.

Beyenbach, K. W. (2004). Kidneys sans glomeruli. *American Journal of Physiology-Renal Physiology*, *286*(5), F811–F827.

Bhateria, R., & Jain, D. (2016). Water quality assessment of lake water: A review. *Sustainable Water Resources Management*, *2*, 161–173.

Bilotta, J. (2000). Effects of abnormal lighting on the development of Zebrafish visual behavior. *Behavioural Brain Research*, *116*(1), 81–87.

Blazer, V. S., Walsh, H. L., Braham, R. P., & Smith, C. (2018). Necropsy-based wild fish health assessment. *Journal of Visualized Experiments*, (139). https://doi.org/10.3791/57946

Borges, A. C., Pereira, N., Franco, M., Vale, L., Pereira, M., Cunha, M. V., . . . Rebelo, M. (2016). Implementation of a Zebrafish health program in a research facility: A 4-year retrospective study. *Zebrafish*, *13*(S1), S-115–S-126.

Boyd, C. E. (2014). Nitrite toxicity affected by species susceptibility, environmental conditions. *Global Aquaculture Advocate*, *17*, 34–36.

Burgess, H. A., & Granato, M. (2007). Modulation of locomotor activity in larval Zebrafish during light adaptation. *Journal of Experimental Biology*, *210*(14), 2526–2539.

Cade, B. S., Terrell, J. W., & Porath, M. T. (2008). Estimating fish body condition with quantile regression. *North American Journal of Fisheries Management*, *28*(2), 349–359. https://doi.org/10.1577/M07-048.1

Carvalho, A. P., Araújo, L., & Santos, M. M. (2006). Rearing Zebrafish (*Danio rerio*) larvae without live food: Evaluation of a commercial, a practical and a purified starter diet on larval performance. *Aquaculture Research*, *37*(11), 1107–1111. https://doi.org/10.1111/j.1365-2109.2006.01534.x

Chapra, S. C., Camacho, L. A., & McBride, G. B. (2021). Impact of global warming on dissolved oxygen and BOD assimilative capacity of the world's rivers: Modeling analysis. *Water*, *13*(17), 2408.

Ciji, A., & Akhtar, M. S. (2020). Nitrite implications and its management strategies in aquaculture: A review. *Reviews in Aquaculture*, *12*(2), 878–908.

Cortemeglia, C., & Beitinger, T. L. (2005). Temperature tolerances of wild-type and red transgenic zebra danios. *Transactions of the American Fisheries Society*, *134*(6), 1431–1437.

Dametto, F. S., Fior, D., Idalencio, R., Rosa, J. G. S., Fagundes, M., Marqueze, A., . . . Barcellos, L. J. G. (2018). Feeding regimen modulates Zebrafish behavior. *PeerJ*, *6*, e5343.

Davidson, J., Good, C., Welsh, C., & Summerfelt, S. T. (2014). Comparing the effects of high vs. low nitrate on the health, performance, and welfare of juvenile rainbow trout Oncorhynchus my kiss within water recirculating aquaculture systems. *Aquacultural Engineering*, *59*, 30–40.

Del Vecchio, G., Mazzei, A., Schiavone, R., Gomes, A. S., Frangelli, G., Sala, T., . . . Rønnestad, I. (2022). Rearing conditions and automated feed distribution systems for Zebrafish (*Danio rerio*). *Applied Sciences*, *12*(21), 10961.

Dimitriadi, A., & Koumoundouros, G. (2022). Elevated embryonic temperature has persistent adverse effects on Zebrafish swimming capacity. *Fishes*, *7*(6), 373.

El-Bouhy, Z., Saleh, G., El-Nobi, G., & Reda, R. (2006). Study on the effect of chlorine on health and growth of some ornamental fishes. *Zagazig Veterinary Journal*, *34*(2), 164–172.

Engeszer, R. E., Patterson, L. B., Rao, A. A., & Parichy, D. M. (2007). Zebrafish in the wild: A review of natural history and new notes from the field. *Zebrafish*, *4*(1), 21–40.

Farias, M., & Certal, A. C. (2016). Different feeds and feeding regimens have an impact on Zebrafish larval rearing and breeding performance. *International Journal of Marine Biology and Research*, *1*(1), 1–8.

Fekri, L., Affandi, R., Rahardjo, M. F., Budiardi, T., Simanjuntak, C. P. H., Fauzan, T., & Indrayani, I. (2018). The effect of temperature on the physiological condition and growth performance of freshwater eel elver *Anguilla bicolor bicolor* (McClelland, 1844). *Jurnal Akuakultur Indonesia*, *17*(2), 181–190.

Floyd, R. F. (2002). Stress, its role in fish disease. In *IFAS Extension: CIR* (p. 919). University of Florida.

Frederickson, S. C., Steinmiller, M. D., Blaylock, T. R., Wisnieski, M. E., Malley, J. D., Pandolfo, L. M., & Castranova, D. (2021). Comparison of juvenile feed protocols on growth and spawning in Zebrafish. *Journal of the American Association for Laboratory Animal Science*, *60*(3), 298–305.

Frøland Steindal, I. A., & Whitmore, D. (2019). Circadian clocks in fish—what have we learned so far? *Biology*, *8*(1), 17.

Giles, M., & Danell, R. (1983). Water dechlorination by activated carbon, ultraviolet radiation and sodium sulphite: A comparison of treatment systems suitable for fish culture. *Water Research*, *17*(6), 667–676.

Gomez Isaza, D. F., Cramp, R. L., & Franklin, C. E. (2020). Simultaneous exposure to nitrate and low pH reduces the blood oxygen-carrying capacity and functional performance of a freshwater fish. *Conservation Physiology*, *8*(1), coz092.

Gomułka, P., Żarski, D., Kupren, K., Krejszeff, S., Targońska, K., & Kucharczyk, D. (2014). Acute ammonia toxicity during early ontogeny of ide Leuciscus idus (Cyprinidae). *Aquaculture International*, *22*, 225–233.

Guh, Y.-J., Lin, C.-H., & Hwang, P.-P. (2015). Osmoregulation in Zebrafish: Ion transport mechanisms and functional regulation. *EXCLI Journal*, *14*, 627.

Gupta, T., & Mullins, M. C. (2010). Dissection of organs from the adult Zebrafish. *Journal of Visualized Experiments*, (37). https://doi.org/10.3791/1717

Hammer, H. S. (2020). Recirculating aquaculture systems (RAS) for Zebrafish culture. In *The Zebrafish in biomedical research* (pp. 337–356). Elsevier.

Harper, C., & Lawrence, C. (2016). *The laboratory Zebrafish*. CRC Press.

Hernandez, J. M., Leon-Santana, M., & Leon, C. J. (2007). The role of the water temperature in the optimal management of marine aquaculture. *European Journal of Operational Research*, *181*(2), 872–886.

Hopkins, E., Sanvictores, T., & Sharma, S. (2022). Physiology, acid base balance. In *StatPearls [Internet]*. StatPearls Publishing.

Kent, M. L., Sanders, J. L., Spagnoli, S., Al-Samarrie, C. E., & Murray, K. N. (2020). Review of diseases and health management in Zebrafish *Danio rerio* (Hamilton 1822) in research facilities. *Journal of Fish Diseases*, *43*(6), 637–650. https://doi.org/10.1111/jfd.13165

Khallaf, E. A., Alne-Na-Ei, A., El-Messady, F., & Hanafy, E. (2021). Effect of temperature rise on growth performance, feed intake, feed conversion ratio and sex ratio of the Nile Tilapia, Oreochromis niloticus. *Egyptian Journal of Aquatic Biology & Fisheries*, *25*(3).

Kroupova, H., Machova, J., & Svobodova, Z. (2005). Nitrite influence on fish: A review. *Veterinarni Medicina-Praha-*, *50*(11), 461.

Kutty, M. (1987). *Fisheries and aquaculture management. Site selection for aquaculture: Chemical features of water.* United Nations Development Programme, African Regional Aquaculture Center, Port Hartcour, RAF/82/009. Retrieved from https://www.fao.org/3/ac175e/ac175e00.htm.

Kwong, R. W., Kumai, Y., & Perry, S. F. (2014). The physiology of fish at low pH: The Zebrafish as a model system. *Journal of Experimental Biology*, *217*(5), 651–662.

Lawrence, C. (2007). The husbandry of Zebrafish (*Danio rerio*): A review. *Aquaculture*, *269*(1–4), 1–20.

Lawrence, C., & Mason, T. (2012). Zebrafish housing systems: A review of basic operating principles and considerations for design and functionality. *ILAR Journal*, *53*(2), 179–191.

Lawrence, C., Sanders, E., & Henry, E. (2012). Methods for culturing saltwater rotifers (Brachionus plicatilis) for rearing larval Zebrafish. *Zebrafish*, *9*(3), 140–146.

Learmonth, C., & Carvalho, A. P. (2015). Acute and chronic toxicity of nitrate to early life stages of Zebrafish—setting nitrate safety levels for Zebrafish rearing. *Zebrafish*, *12*(4), 305–311.

Li, H. Y., Xu, J., & Xu, R. Q. (2013). The effect of temperature on the water quality of lake. *Advanced Materials Research*, *821*, 1001–1004.

Lieggi, C. (2020). Health surveillance programs. *The Zebrafish in Biomedical Research*, 415–429.

Lim, L. C., Dhert, P., & Sorgeloos, P. (2003). Recent developments and improvements in ornamental fish packaging systems for air transport. *Aquaculture Research*, *34*(11), 923–935.

Linbo, T. L. (2009). *Zebrafish (Danio rerio) husbandry and colony maintenance at the Northwest Fisheries Science Center.* NOAA technical memorandum NMFS-NWFSC; 100. Retrieved from https://repository.library.noaa.gov/view/noaa/3645.

McCormick, S. D., Taylor, M. L., & Regish, A. M. (2020). Cortisol is an osmoregulatory and glucose-regulating hormone in Atlantic sturgeon, a basal ray-finned fish. *Journal of Experimental Biology*, *223*(18), jeb220251.

Messerli, M., Aaldijk, D., Haberthür, D., Röss, H., García-Poyatos, C., Sande-Melón, M., . . . Djonov, V. (2020). Adaptation mechanism of the adult Zebrafish respiratory organ to endurance training. *PLoS One*, *15*(2), e0228333. https://doi.org/10.1371/journal.pone.0228333

Mohammad, N., & Haque, M. M. (2021). Impact of water quality on fish production in several ponds of Dinajpur Municipality Area, Bangladesh. *International Journal of Science and Innovative Research*, *2*, 01000421JESIR.

Morais, S., Lacuisse, M., Conceição, L. E. C., Dinis, M. T., & Rønnestad, I. (2004). Ontogeny of the digestive capacity of Senegalese sole (Solea senegalensis), with respect to digestion, absorption and metabolism of amino acids from Artemia. *Marine Biology*, *145*(2), 243–250. https://doi.org/10.1007/s00227-004-1326-6

Mustapha, M., & Akinshola, F. (2016). Ammonia concentrations in different aquaculture holding tanks. *West African Journal of Applied Ecology*, *24*(1), 1–8.

Nagaraju, T. V., Sunil, B., Chaudhary, B., Prasad, C. D., & Gobinath, R. (2023). Prediction of ammonia contaminants in the aquaculture ponds using soft computing coupled with wavelet analysis. *Environmental Pollution*, *331*, 121924.

Noga, E. J. (1998). Toxic algae, fish kills and fish disease. *Fish Pathology*, *33*(4), 337–342.

Noga, E. J. (2010). *Fish disease: Diagnosis and treatment.* John Wiley & Sons.

Olsen, R. E., Sundell, K., Ringø, E., Myklebust, R., Hemre, G.-I., Hansen, T., & Karlsen, Ø. (2008). The acute stress response in fed and food deprived Atlantic cod, Gadus morhua L. *Aquaculture*, *280*(1–4), 232–241.

Ostrander, G. (2000). *The laboratory fish.* Elsevier.

Øverli, Ø., Winberg, S., & Pottinger, T. G. (2005). Behavioral and neuroendocrine correlates of selection for stress responsiveness in rainbow trout—a review. *Integrative and Comparative Biology*, *45*(3), 463–474. https://doi.org/10.1093/icb/45.3.463

Parichy, D. M. (2020). Zebrafish pigmentation. *The Zebrafish in Biomedical Research*, 97–102.

Ribas, L., Liew, W. C., Díaz, N., Sreenivasan, R., Orbán, L., & Piferrer, F. (2017). Heat-induced masculinization in domesticated Zebrafish is family-specific and yields a set of different gonadal transcriptomes. *Proceedings of the National Academy of Sciences*, *114*(6), E941–E950.

Roberts, H., & Palmeiro, B. S. (2008). Toxicology of aquarium fish. *Veterinary Clinics of North America: Exotic Animal Practice*, *11*(2), 359–374.

Sahu, S, D. S. (2018). *Effect of water pH on growth and survival of Trichogaster lalius (Hamilton, 1822) under captivity* (7 Honors Theses. 870), Ouachita Baptist University. Retrieved from https://scholarlycommons.obu.edu/honors_theses/870.

Saszik, S., Bilotta, J., & Givin, C. M. (1999). ERG assessment of Zebrafish retinal development. *Visual Neuroscience*, *16*(5), 881–888.

Schirone, R. C., & Gross, L. (1968). Effect of temperature on early embryological development of the zebra fish, *Brachydanio rerio. Journal of Experimental Zoology*, *169*(1), 43–52.

Sergeant, C. (2014). The management of ammonia levels in an aquaculture environment. *Water/Wastewater*, 2–3.

Siccardi III, A. J., Garris, H. W., Jones, W. T., Moseley, D. B., D'Abramo, L. R., & Watts, S. A. (2009). Growth and survival of Zebrafish (*Danio rerio*) fed different commercial and laboratory diets. *Zebrafish*, *6*(3), 275–280. https://doi.org/10.1089/zeb.2008.0553

Silver, D. L., Hou, L., & Pavan, W. J. (2006). The genetic regulation of pigment cell development. *Neural Crest Induction and Differentiation*, 155–169.

Simmons, A. E., Karimi, I., Talwar, M., & Simmons, T. W. (2012). Effects of nitrite on development of embryos and early larval stages of the Zebrafish (*Danio rerio*). *Zebrafish*, *9*(4), 200–206.

Spence, R., Gerlach, G., Lawrence, C., & Smith, C. (2008). The behaviour and ecology of the Zebrafish, *Danio rerio. Biological Reviews*, *83*(1), 13–34.

Stephenson, J. F., Whitlock, K. E., & Partridge, J. C. (2011). Zebrafish preference for light or dark is dependent on ambient light levels and olfactory stimulation. *Zebrafish*, *8*(1), 17–22.

Stickney, R. R. (1994). *Principles of aquaculture*. John Wiley & Sons.

Sullins, A. (2023). *Zebrafish feeding and breeding: Adapting best practices for Zebrafish husbandry* (Honors Theses. 870), Ouachita Baptist University. Retrieved from https://scholarlycommons.obu.edu/honors_theses/870.

Summerfelt, R. C. (2000). Water quality considerations for aquaculture. *Department of Animal Ecology*, 2–7.

Sundin, J., Morgan, R., Finnøen, M. H., Dey, A., Sarkar, K., & Jutfelt, F. (2019). On the observation of wild Zebrafish (*Danio rerio*) in India. *Zebrafish*, *16*(6), 546–553.

Suzaimi, N. D., Goh, P. S., & Ismail, A. F. (2021). The applications of integrated osmosis processes for desalination and wastewater treatment. In *Osmosis Engineering* (pp. 313–332). Elsevier.

Svecevičius, G., Šyvokienė, J., Stasiūnaitė, P., & Mickėnienė, L. (2005). Acute and chronic toxicity of chlorine dioxide (ClO2) and chlorite (ClO2) to rainbow trout (Oncorhynchus mykiss)(4 pp). *Environmental Science and Pollution Research*, *12*, 302–305.

Swain, S., Sawant, P. B., Chadha, N. K., Chhandaprajnadarsini, E., & Katare, M. B. (2020). Significance of water pH and hardness on fish biological processes: A review. *International Journal of Chemical Studies*, *8*(4), 330–337.

Swann, L. (1993). *Water quality water sources used in aquaculture*. Water Quality Fact Sheet as-486. Aquaculture Extension, Illinois-Indiana Sea Grant Program, Purdue University, West Lafayette, IN, USA.

Teoh, S. L., Sapri, S. R. B., Yusof, M. R. B. M., Yahaya, M. F., & Das, S. (2020). Construction of an affordable open-design recirculating Zebrafish housing system. *Journal of the American Association for Laboratory Animal Science*, *59*(5), 512–518.

Torno, J., Einwächter, V., Schroeder, J. P., & Schulz, C. (2018). Nitrate has a low impact on performance parameters and health status of on-growing European sea bass (Dicentrarchus labrax) reared in RAS. *Aquaculture*, *489*, 21–27.

Tsai, S. B., Tucci, V., Uchiyama, J., Fabian, N. J., Lin, M. C., Bayliss, P. E., . . . Kishi, S. (2007). Differential effects of genotoxic stress on both concurrent body growth and gradual senescence in the adult Zebrafish. *Aging Cell*, *6*(2), 209–224.

Tsang, B., & Gerlai, R. (2022). Breeding and larviculture of Zebrafish (*Danio rerio*). In *Laboratory fish in biomedical research* (pp. 63–80). Elsevier.

Tsang, B., Zahid, H., Ansari, R., Lee, R. C.-Y., Partap, A., & Gerlai, R. (2017). Breeding Zebrafish: A review of different methods and a discussion on standardization. *Zebrafish*, *14*(6), 561–573.

Tsitsifli, S., & Kanakoudis, V. (2018). *Disinfection impacts to drinking water safety—A review*. Paper presented at the proceedings.

Tumwesigye, Z., Tumwesigye, W., Opio, F., Kemigabo, C., & Mujuni, B. (2022). The effect of water quality on aquaculture productivity in Ibanda district, Uganda. *Aquaculture Journal*, *2*(1), 23–36.

Valdivieso, A., Ribas, L., Monleón-Getino, A., Orbán, L., & Piferrer, F. (2020). Exposure of Zebrafish to elevated temperature induces sex ratio shifts and alterations in the testicular epigenome of unexposed offspring. *Environmental Research*, *186*, 109601.

Vandenberg, L. N., Stevenson, C., & Levin, M. (2012). Low frequency vibrations induce malformations in two aquatic species in a frequency-, waveform-, and direction-specific manner. *PLoS One*, *7*(12), e51473.

Varga, Z. M. (2011). Aquaculture and husbandry at the Zebrafish international resource center. *Methods in Cell Biology*, *104*, 453–478.

Villamizar, N., Vera, L. M., Foulkes, N. S., & Sánchez-Vázquez, F. J. (2014). Effect of lighting conditions on Zebrafish growth and development. *Zebrafish*, *11*(2), 173–181.

Volkoff, H., & Rønnestad, I. (2020). Effects of temperature on feeding and digestive processes in fish. *Temperature*, *7*(4), 307–320.

Weiner, R. F., & Matthews, R. A. (Eds.). (2003). Chapter 5—Measurement of water quality. In *Environmental engineering* (4th ed., pp. 81–106). Butterworth-Heinemann.

Westerfield, M. (2007). *The Zebrafish book: A guide for the laboratory use of Zebrafish (Danio Rerio)*. University of Oregon Press.

Yan, Z.-G., Fan, J.-T., Zheng, X., Wang, S.-P., Guo, X.-S., Zhang, T.-X., . . . Zhang, Y.-Z. (2019). Neglect of temperature and pH impact leads to underestimation of seasonal ecological risk of ammonia in Chinese surface freshwaters. *Journal of Chemistry*, *2019*, 1–7.

Zynda, J. R. (2020). Aquatics facility design considerations: Incorporating aquatics into an Animal facility. In *The Zebrafish in biomedical research* (pp. 265–277). Elsevier.

2 Breeding of Zebrafish and Its Life Cycle

Hakim Mudasir Maqsood, Inain Jaias, Asra Imtiyaz Mattoo, Towseef Akram, Wasim Akram, and Nazir Ahmad Ganai

2.1 INTRODUCTION

The zebrafish (*Danio rerio*), native to the freshwater systems of South Asia, has become one of the most widely utilized vertebrate models in scientific investigations. With their small size, low maintenance requirements, and high reproductive capacity, zebrafish offer researchers an exceptional opportunity to study diverse biological phenomena in a cost-effective and time-efficient manner. They serve as an essential model organism for studying various aspects of development, including sexual differentiation, reproductive behavior, and embryonic development. Zebrafish are characterized by their striking blue and silver stripes running horizontally along their slender bodies. Typically, they grow to a length of about 2.5 to 5 centimeters, making them convenient subjects for laboratory experimentation. Their transparency during embryonic stages allows researchers to easily observe and study the development of internal structures and organs, a feature that sets them apart from many other model organisms. Understanding zebrafish breeding and life cycle not only provides essential knowledge for researchers and aquaculturists but also fosters the development of innovative strategies to address critical scientific questions and challenges. By uncovering the secrets of zebrafish reproduction and development, we gain valuable insights into fundamental biological processes, human health, and the impacts of environmental factors on aquatic organisms.

To comprehend the breeding process of zebrafish, we will delve into their reproductive anatomy and behavior, shedding light on the intricacies of mating rituals, courtship displays, and the underlying mechanisms involved in fertilization. Moreover, we will explore the various factors that influence successful zebrafish breeding, including water quality, temperature, lighting, and dietary considerations. We will delve into the intricate details of zebrafish breeding, covering topics such as spawning induction, egg collection and handling, embryo development, and the establishment of breeding colonies. By acquiring a comprehensive understanding of zebrafish reproduction, researchers can harness the full potential of this remarkable model organism and advance our knowledge in various disciplines. The life cycle of zebrafish encompasses distinct stages, each marked by unique developmental milestones and morphological transformations. From the fertilized egg to the adult fish, we will examine the key events and transformations that occur during embryogenesis, larval development, and the maturation of zebrafish into sexually mature adults. We will also discuss the importance of maintaining appropriate housing conditions and managing environmental parameters throughout the different life stages to ensure optimal growth, health, and reproductive success.

2.2 MATING AND EGG PRODUCTION IN ZEBRAFISH

All zebrafish have embryonic ovaries at the beginning of their sexual development. The ovaries of fish that develop into females mature, whereas the ovaries of fish that develop into males retreat and the testicles grow (Liew & Orban, 2014). Zebrafish are thought to turn into males during sexual differentiation due to high amounts of the stress hormone cortisol (Ribas *et al.*, 2017). The presence of spermatogenic cysts, which are groups of Sertoli cells encasing germ cells that develop synchronously rather than a small number of germ cells with various developmental stages as seen in Sertoli cells in higher vertebrates, is one distinctive spermatogenesis pattern found in zebrafish (Schulz *et al.*, 2018). Male and female zebrafish can be distinguished morphologically by several key features. Females typically have a rounder and fuller abdomen, especially when gravid, due to the presence of mature eggs. In contrast, males generally possess a slimmer abdomen. Another distinguishing characteristic is the anal fin; in males, it is often more pointed and elongated with a slight concave shape, while in females, it tends to be shorter and more rounded. Additionally, males may exhibit brighter and more intense coloration, especially on their fins and bodies, as part of their courtship displays, although coloration can vary among strains. These morphological differences provide useful cues for identifying the sex of zebrafish, which is particularly valuable for breeding and research purposes. In females, the ovary develops in four stages: (1) the primary oocyte stage, where relatively small spherical cells can be seen; (2) the cortical-alveolar stage, where oocytes are enlarged and filled with cortical alveoli; (3) the vitellogenic stage, where egg yolk is present in the oocytes; and (4) the maturation stage, where oocytes with an irregular layer can be seen (Çakici & Üçüncü, 2007).

Once mature, zebrafish spawn in natural settings in shallow water with aquatic vegetation (Lawrence, 2007). Zebrafish are known to spawn in groups in nature and are known to shoal (Spence *et al.*, 2008). The photoperiod has

DOI: 10.1201/9781003402893-3

a considerable impact on the endogenous reproduction rhythm of zebrafish, which has been traditionally used for breeding at a cycle of 10 hours of light and 14 hours of darkness (Blanco-Vives & Sanchez-Vazquez, 2009). Spawning normally occurs during morning hours (Darrow & Harris, 2004). Male and female fish swim back and forth in their habitat for around ten minutes during mating. The male eventually starts to pursue the female intently. The male continuously swims between the female and the shallow parts of the mating tank, perhaps leading the female to spawning places. The male oscillates his body towards the female as they swim in circles together. The two occasionally appear to court. When the male wraps the female in his trunk, both sexes release their gametes, and this is known as spawning (Zempo *et al.*, 2021). A single female zebrafish can generate 200–300 eggs in one spawning session after a successful pairing (Nasiadka & Clark, 2012).

On average, 58–78% of released eggs become fertilized, depending on the quality of the eggs and sperm (Riesco *et al.*, 2019). The amount of eggs produced in each spawning did not connect with the male's dominance rank, and it was discovered that females did not respond preferentially to males of varied dominance status (Spence & Smith, 2006). In general, zebrafish eggs are larger than those of other fish, measuring about 0.7 mm in diameter and being optically translucent (Spence *et al.*, 2008).

2.3 EMBRYONIC DEVELOPMENT OF ZEBRAFISH

According to Aral *et al.* (2011), there are five stages in the life cycle of fish: the embryonic phase, the larval phase, the fry phase, the ripe phase, and the senescent phase. The embryonic phase is divided into number of sub-phases, which finally results in the development of larvae. The various larval phases involved in the development of zebrafish are presented in Figure 2.1.

About 40 minutes after fertilization, the freshly fertilized egg is in the zygote period until the first cleavage happens. At the time of fertilization, the zygote has a diameter of around 0.7 mm. The cells, or blastomeres, continue to divide after the initial cleavage, happening around every 15 minutes. The blastomeres, or a certain subset of them depending on the stage, remain connected by cytoplasmic bridges despite the meroblastic cytoplasmic division of the blastodisc (Kimmel & Law, 1985). Various celled stages appear during the cleavage period including two-celled, four-celled, eight-celled, 16-celled and 64-celled stages. At the animal pole of a large yolk cell, the first 12 divisions take place simultaneously and form a mound of cells. This group of cells makes up the blastoderm. The daughter cells are significantly less uniformly structured than they were during the preceding cleavage phase during the early stages of the blastoderm, where near-synchronous or metasynchronous cell division takes place (Webb & Miller, 2006). The mid blastula transition starts in zebrafish during the tenth cell cycle (Newport & Kirschner, 1982). Moreover, blastomere motility and subsequent morphogenetic motions begin with mid blastula transition. Once the embryo has passed the blastula stage, the first coordinated cell movement known as epiboly in zebrafish starts (Warga & Kimmel, 1990). Epiboly, a prominent cell movement during gastrulation in zebrafish, occurs when a squamous epithelium (the enveloping layer), a multi-layer of loosely packed cells (the deep

FIGURE 2.1 Different phases in embryonic development of zebrafish.

Source: (Ribas & Piferrer, 2014).

cells), and a yolk nuclear syncytium (the yolk syncytial layer) expand in concert to engulf the yolk and close the blastopore (Bruce, 2016). Four primary activities make up epiboly: (1) the enveloping epithelial layer sheet expands; (2) the internal yolk syncytial layer expands; (3) the yolk cytoplasmic layer is diminished and eliminated; and (4) the enveloping layer and yolk syncytial layer undergo profound cell rearrangements (Trinkaus, 1984). The activities involved are shown in Figure 2.2.

The major germ layers and the embryonic axis are produced while epiboly continues along with the morphogenetic cell movements of involution, convergence, and extension. As far as we can determine, the start of gastrulation defines the beginning of involution and happens at 50% epiboly. The embryonic germ layers are formed during the morphogenetic phase of gastrulation (Rohde & Heisenberg, 2007). Cells gather at the border zone and stream downward towards the yolk cell to internalize to form the mesoderm and endoderm (mesendoderm), resulting in a thicker infolding known as the "germ ring" (Montero & Heisenberg, 2004). After this, mesendodermal cells move inside to form an interior layer known as the "hypoblast," while the cells on top, known as the "epiblast," form an ectodermal layer that continues to epibolize.

At 50% epiboly, when internalization starts, convergent extension cell movements also start at the same time. The axis narrows mediolaterally as a result of convergence, which is the movement of hypoblast and epiblast cells towards the embryo's future dorsal side. As gastrulation develops, convergent extension motions become constrained, preventing the mixing of distinct mesodermal subtypes (Glickman *et al.*, 2003). Generally, mesoderm cells in more dorsal locations exhibit more prominent convergent extension activities (Myers *et al.*, 2002). Cell adhesion and cytoskeletal reorganization are the main determinants of gastrulation cell motions. Moreover, the enveloping layer and yolk syncytial layer have been described as having tissue interactions that may contribute to gastrulation (Rohde & Heisenberg, 2007). It is known that signalling pathways such as the PDGF-PI3K, Wnt/Calcium, Jak/Stat, and others control gastrulation in zebrafish. Epiblast cells on the dorsal side begin to elongate mediolaterally and extend protrusions after 70% epiboly, a behavior that is most likely related to the mediolateral intercalation seen in the epiblast (D'Amico & Cooper, 2001).

During the sedimentation phase, a tremendous variety of morphogenetic movements take place. The embryo elongates swiftly, the somites form, the beginnings of the primary organs can be seen, the tail bud becomes more noticeable, and body movement begins (Kimmel *et al.*, 1995). The medial somitic epithelium, also known as the "adaxial" zone, is located dorsoventrally in the middle of each somite and contains some of the early cells that elongate into muscle fibers (Thisse *et al.*, 1993). During this stage, the development of the optic, olfactory, and ear tissues begins. It is clear from this that the sensory

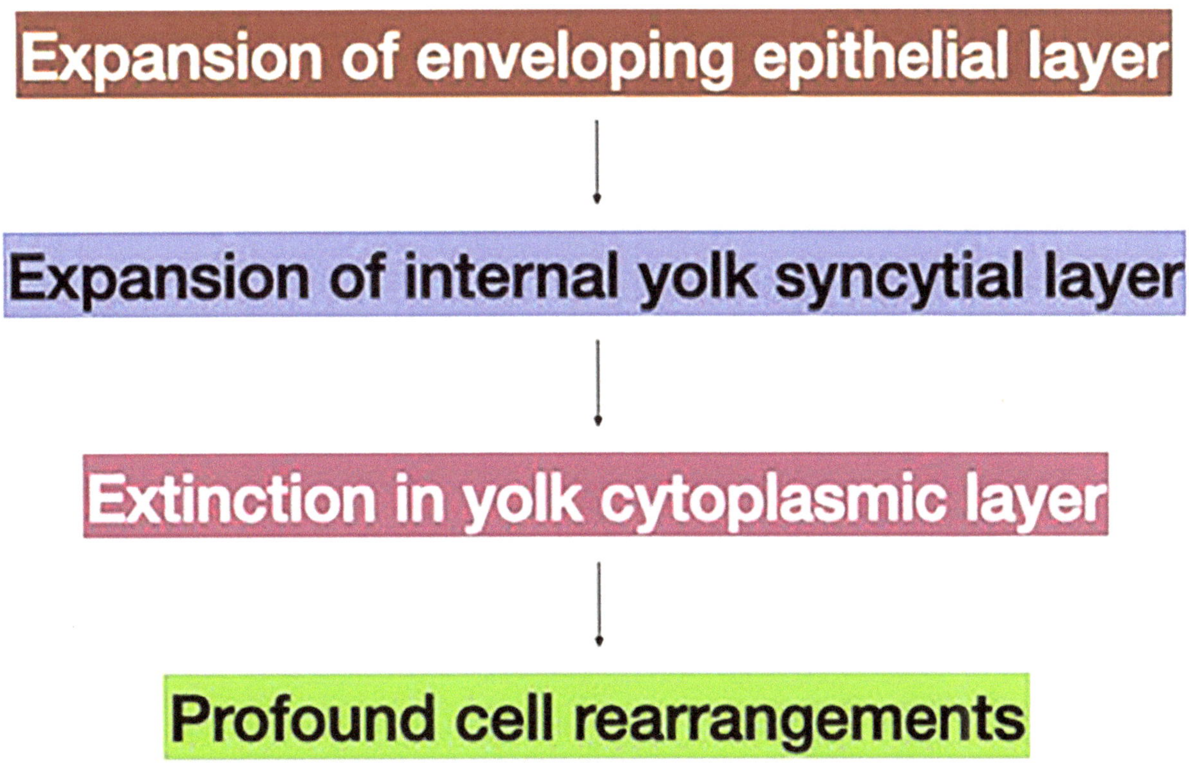

FIGURE 2.2 Steps involved in the process of epiboly in zebrafish.

neurons and the motor neurons also have a significant impact on the zebrafish's behavioral reactivity and motor reaction to light.

In order to develop skin or other organ linings, such as pigments, there occurs a migration from the germ layer during the pharyngeal stage (Zahro *et al.*, 2022). The construction of the pharyngeal arches, which will eventually give rise to the mandibula and gills, occurs during the pharyngula phase. The development of the embryo slows down at this stage, and the head is compressed, the fins begin to form, cells begin to pigment, and last but not least, the circulatory system forms and the heart starts to beat (Reib, 1973).

It is commonly accepted that mechanical hatching and enzymatic hatching are the two mechanisms involved in egg envelope breaking (Ishida, 1985). The movement of embryos and the elongation of the embryo body are the major mechanisms for the mechanical hatching of fish, while the process of enzymatic choriolysis by hatching enzyme, secreted from the gland cell, is known as enzymatic hatching (Fu *et al.*, 2021). In order to lessen the risk of hatchling predation, embryos can either delay time to hatch or accelerate time to hatch (Laurila *et al.*, 2002; Kusch & Chivers, 2004). Additionally, before hatching, embryos can learn to recognize predators by paying attention to exterior stimuli (Horn *et al.*, 2019). Moreover, studies indicate that early-hatching embryos have undeveloped bodies and exhibit relatively poor swimming abilities (Wisenden *et al.*, 2022). The growth of the pectoral fin is still a helpful trait for timing the hatching time. The paired fin rudiments are elongated buds at the beginning of the period, and each one already has centrally positioned mesenchymal condensations that will eventually form the girdle cartilages (Kimmel *et al.*, 1995).

Following the embryonic stage, the fish undergo a 6-week larval development period, during which their length more than triples and they undergo a series of morphological changes that give them the fins, coloration pattern, and overall body morphology of juvenile fish. Earlier research indicates that diet and microRNAs may control fish metamorphosis (Kaushik *et al.*, 2011). Although not being sexually mature, juveniles have morphology that closely matches that of adults since they have shed their larval fin fold and grown scales. Zebrafish reach sexual maturity and are regarded as adults after 3 months at a temperature of 28.5 °C (Parichy *et al.*, 2009).

2.4 LABORATORY REARING AND MANAGEMENT OF ZEBRAFISH

Zebrafish are typically housed in glass or acrylic tanks with a filtration system to maintain water quality. Housing densities have a considerable impact on laboratory zebrafish welfare, so it is recommended to keep them at low densities of no more than 5 fish per liter. Maintaining a clean environment with a properly operating aquarium system is critical for sustaining their health. Zebrafish are typically kept in slightly acidic to slightly alkaline waters (pH 6.5–8.0) with high clarity, under a 14:10 h light:dark cycle, and at a temperature of 28.5 °C (Table 2.1).

Water quality should be monitored often, including parameters such as alkalinity, pH, temperature, hardness, ammonia, nitrate, nitrite, dissolved oxygen, salinity, and conductivity. Nitrate levels should not exceed 50 mg/L^{-1}, and excess nitrate can be lowered by adding fresh system water to the circulating system (Table 2.1). Temperature plays a crucial role in the successful rearing of zebrafish in laboratory settings. Maintaining a stable and appropriate temperature is essential for their overall health and development. Zebrafish are poikilothermic animals, meaning their body temperature is influenced by the surrounding environment. Typically, a temperature of 28.5 °C is considered optimal for zebrafish husbandry (Aleström *et al.*, 2020), as it supports their metabolic rate, growth, and reproductive activities. Deviations from this range can lead to stress, decreased fertility, and compromised immune function. The temperature limits for zebrafish rearing, as highlighted by López-Olmeda and Sanchez-Vazquez (2011), have been presented in Figure 2.3. Precise temperature control is imperative to ensure consistent and reliable results in zebrafish research, making it a fundamental parameter in the successful breeding and maintenance of this model organism in laboratory conditions.

TABLE 2.1
Recommended Parameters Necessary in Zebrafish Rearing

S. No.	Parameter	Recommended Level	Reference
1	Dissolved oxygen	>6 mg l^{-1}	Longkumer *et al.* (2020)
2	Ammonia	0 mg l^{-1}	Murray *et al.* (2020)
3	Nitrite	<0.5 mg l^{-1}	Hammer (2020)
4	Nitrate	<100 mg l^{-1}	Pereira *et al.* (2017)
5	pH	6.8–7. 5	Avdesh *et al.* (2012)
6	Salinity	0.5–2 g l^{-1}	Harper and Lawrence (2016)
7	Temperature	28.5 °C	Aleström *et al.* (2020)

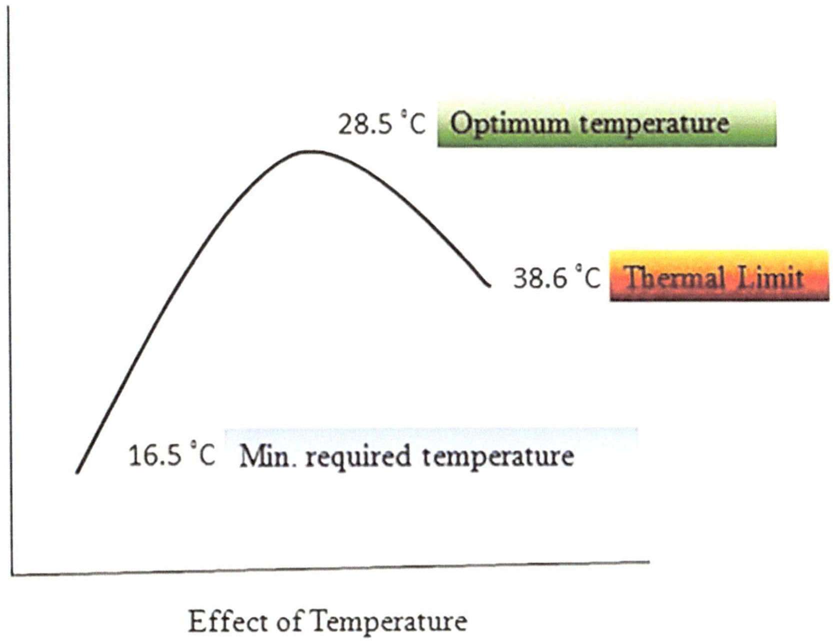

FIGURE 2.3 Effect of temperature on the development of zebrafish.

Zebrafish are omnivorous and can be fed commercially available fish flakes or pellets specifically formulated for them. Live or frozen food like brine shrimp or daphnia can be provided as a supplement to the diet. Feeding frequency and portion size should be adjusted based on the age and size of the fish. However, live feed and associated fish prey-capture activity have an enriching impact; the dry feed diets are typically thought to be nutritionally sufficient (Varga, 2016). It has been observed that the feeding regimen highly influences and modulates fish behavior (Dametto *et al.*, 2018). Zebrafish can reproduce all year round, and a single pair has the capacity to create more than 100 embryos in a single day (Spence *et al.*, 2008). Appropriate sex ratios and suitable breeding substrates, such as marbles or mesh grids, should be provided to encourage reproduction. Eggs can be collected, transferred to a separate tank, and kept under controlled conditions for further development.

In the context of zebrafish research, meticulous sanitization of equipment and adherence to stringent hygiene practices are of paramount importance. Contaminants, such as bacteria, viruses, or chemicals, can severely impact the health of zebrafish and compromise experimental outcomes. Proper sterilization of tanks, water filtration systems, nets, and other equipment used in housing and feeding zebrafish is essential to maintain a controlled and pathogen-free environment. Additionally, strict personal hygiene measures, including the use of gloves and lab coats, are crucial for preventing cross-contamination. Regular monitoring and cleaning of the laboratory facility, including benchtops and floors, help to reduce the risk of disease transmission. By prioritizing equipment sanitization and hygiene protocols, researchers can ensure the well-being of zebrafish colonies, enhance the reliability of experimental results, and maintain the integrity of this valuable model organism in laboratory settings. Measures like water sterilization using UV filters and routine screening for common fish pathogens are often implemented in zebrafish facilities.

2.5 INDUCED BREEDING WITH HORMONES

Aquaculturists must efficiently manage and supervise the full life cycle of the fish they are raising in order to maintain a profitable fish farming business. Monitoring environmental factors such as temperature, light intensity, water quality, and food safety can help achieve this goal. The hypothalamo-hypophyseal-gonadal endocrine axis, which controls gonadal development and reproduction, is usually easier to affect. There are two approaches to hormonal intervention along this axis:

- The first is the hypophyseal approach, or hypophysation. This method induces an artificial surge of gonadotropin (GtH) by injecting a preparation that contains a gonadotropic hormone derived from fish or mammals (such as GtH-containing carp pituitary extract or human chorionic gonadotropin, hCG).

- In the second method, gonadotropin-releasing hormone (GnRH), a hypothalamic peptide, is used, occasionally in combination with a dopamine receptor antagonist. This combination causes the release of endogenous GtH, which in turn causes a comparable surge.

Like many other fish species, zebrafish use induced breeding as a technique to control and promote reproduction in a regulated setting. GnRH is essential to this process. Gonadotropin release is stimulated by the neuropeptide hormone known as GnRH. The reproductive system is regulated by these gonadotropins in turn (Figure 2.4).

The complex process of employing GnRH to stimulate spawning in zebrafish involves regulating the release of luteinizing hormone (LH) and the interaction of several variables. Vasoactive intestinal peptides (VIP) has been found to directly regulate GnRH neurons in the brain of zebrafish, which in turn leads to LH secretion (Mizrahi & Levavi-Sivan, 2023).

Furthermore, GnRH3—the hypophysiotropic form of GnRH—has been linked to the regulation of LH secretion in zebrafish (Tanaka *et al.*, 2022a). However, since zebrafish reproduction was unaffected by GnRH3 gene disruption through knockout, the exact role of GnRH3 in zebrafish reproduction is still unknown (Tanaka *et al.*, 2022b). Furthermore, it has been observed that low temperatures can influence the transcriptional regulation of zebrafish reproduction, including the GnRH signalling pathway (Ge *et al.*, 2022). Overall, the process of inducing spawning in zebrafish through GnRH entails a complex interplay of various factors, including VIP, GnRH3, and the regulation of temperature (Falahatkar *et al.*, 2013).

When zebrafish are bred in captivity, their failure to ovulate and spawn is caused by the GnRH system failing (Mylonas & Zohar, 2001). This also affects commercially important aquaculture species. Understanding the fundamental cause of this failure and trying to overcome it by giving exogenous GnRH and its analogs prompted focused studies of the GnRH system in important aquaculture species such as salmonids, cyprinids, and many perciform species. The verification of zebrafish as a vertebrate model, despite sits recognized benefits, in conjunction with revolutionary developments in molecular platforms like transgenesis, genomics, and loss of function, has enabled the acquisition of new understandings and paradigmatic changes in the understanding of the GnRH system.

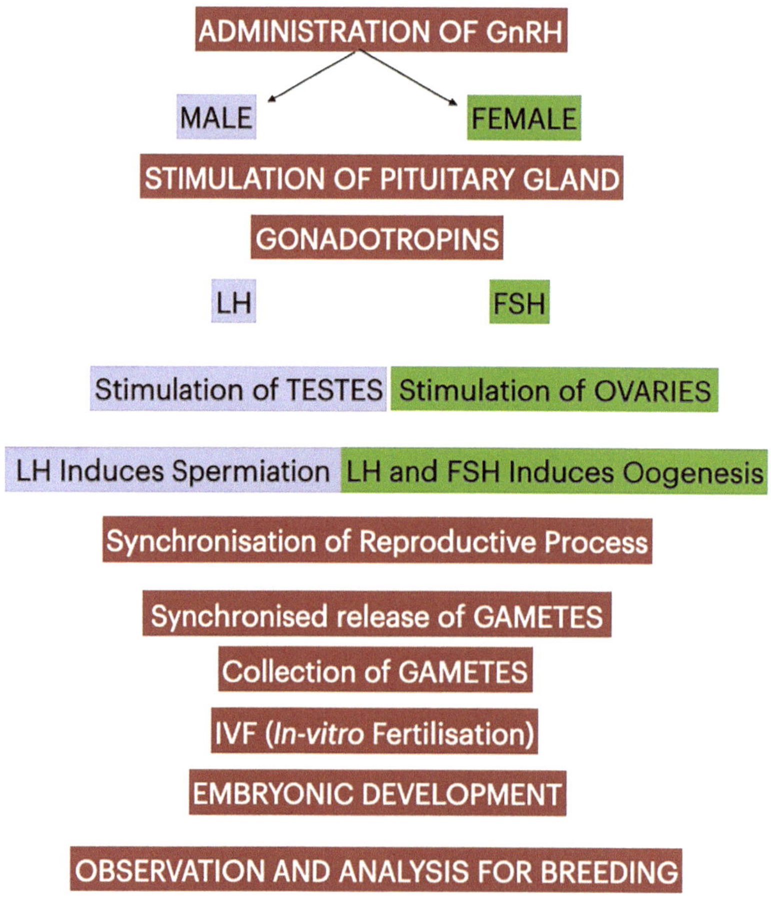

FIGURE 2.4 Key steps involved in GnRH-induced breeding in zebrafish.

In fish, a tiny structure called the neurohypophysis connects the pituitary gland to the hypothalamus. This neurohypophysis is made up of axons that pass through the pituitary gland and come from neurosecretory cells in the brain. Notably distinct from tetrapods, the fish pituitary lacks the median eminence and portal vasculature system that typically facilitates the connection between the hypothalamus and the pituitary in tetrapods. Consequently, neurohormones influencing the pituitary function are discharged from neurosecretory terminals, more or less directly, into the vicinity of the target cells (Blanton & Specker, 2007). A two-layered basal membrane separates the neurosecretory terminals from the adenohypophyseal cells in some fish species, including eels and salmonids. Neurosecretory products are released by this membrane, which then allows them to diffuse through the double basal membrane and eventually move through the intercellular gaps. The neurohormones can efficiently reach and affect their target cells thanks to this complex procedure. Highly relevant neuroanatomical evidence is the presence of terminals that release chemicals such as dopamine, gamma-amino butyric acid (GABA), neuropeptide Y (NPY), gonadotropin-releasing hormone (GnRH), or gonadotropin-inhibitory hormone (GnIH) in close proximity to gonadotropic cells. This existence has been essential in identifying the precise actions of these neurohormones on gonadotropin secretion and their influence on fish reproductive cycles (Paullada-Salmerón *et al.*, 2016). A complex vascular network was recently discovered throughout the entire pituitary using transgenic zebrafish that expressed combination levels of gonadotropin-releasing hormone 3 (GnRH3), luteinizing hormone (Lh), and follicle-stimulating hormone (FSH). It was shown that this complex network actively transmits GnRH3 signaling to the pituitary, which adds fresh insight into the functioning of neuroendocrine interactions in this setting (Golan *et al.*, 2015). In teleosts, the vascular system contributes to neurohormonal signaling even in the absence of a structure resembling the median eminence, aside from the direct innervation of gonadotropes by GnRH neurons. Human chorionic gonadotropin (hCG) triggers spawning in zebrafish by activating the LH signaling pathway. LH, in turn, governs the expression of insulin-like growth factors (IGFs), with a notable emphasis on IGF-3. The pivotal role of IGF-3 lies in mediating the action of LH on oocyte maturation (Li *et al.*, 2015).

Exposure to dydrogesterone in zebrafish induces oocyte maturation and ovulation, potentially via the arachidonic acid (AA) metabolic pathway and purinergic signaling (Jiang *et al.*, 2019). The administration of GnRH, luteinizing hormone-releasing hormone (LHRH), and their analogs through a sustained release delivery system results in a protracted elevation of plasma levels of GtH, ultimately leading to the induction of ovulation and spawning in fish (Alawi *et al.*, 2013).

The administration of GnRH, LHRH, and their analogs into fish through a polymer-based sustained release delivery system has been found to be an effective way to cause a sustained rise in the fish's plasma levels of gonadotropin GtH. This elevated state, which can be induced both inside and outside of the natural spawning season, results in ovulation and spawning. As a result, this administration strategy offers a novel way to control fish reproduction, which is an essential component of the financial management of fish farming.

2.5.1 Induced Spawning Using Manipulation of Light Cycles

As oviparous creatures, zebrafish are distinguished by their absence of parental care and external fertilization. These fish disperse mature, nonadhesive eggs in batches and are asynchronous spawners, meaning they do not synchronize their reproductive activities. This allows them to spawn eggs at different times, contributing to a continuous reproductive cycle (Breder *et al.*, 1966). A single female can produce several hundred eggs during a single spawning session. These eggs are released in high fecundity. Males in the surroundings increase female ovulation and subsequent oviposition (van den Hurk & Resink, 1992).

Zebrafish reproduction is influenced by the length of the photoperiod, the light spectrum and intensity produced by artificial illumination, and spawning, egg hatching, larval development, fry growth rate, and behavior. In contrast to the day-to-day progressive shift that occurs naturally in natural environments, laboratory settings frequently involve an abrupt change in light levels. This change is usually triggered by a timer that turns on all lights at a predetermined time and turns them out at the conclusion of the light phase. That being said, this abrupt change might not be the best. Studies have revealed that zebrafish that are suddenly exposed to different illumination conditions display "frantic behavior," which includes jumping and slamming into the glass tank's walls. These reactions can cause internal bleeding or external damage to the fish. Even if there is no visible damage, the abrupt change in lighting can be an unnecessary stressor and may cause long-lasting behavioral changes in breeding fish and their developing offspring. One way to lessen the negative effects of abrupt changes in lighting is to use dimmers, which allow for a gradual reduction and increase in light intensity. Some researchers contend that zebrafish can adjust to subjective day or night lighting conditions in 20 to 30 minutes, which emulates the natural cycles of dawn and dusk.

In adult zebrafish, the light cycle is essential for regulating a variety of rhythmic biochemical, physiological, neurological, and behavioral changes. It can also have a significant impact on the development of the embryo. The ability of zebrafish's internal body clocks to synchronize and adapt to the photoperiod is essential for spawning success.

Zebrafish in lab settings have a preferred circadian light cycle that consists of roughly 14 hours of light and 10 hours of total darkness. Zebrafish spawn at dawn in their natural habitat, signaling the start of the light cycle. Similar to this, in lab settings, sexes should be kept apart until morning light is introduced for optimal spawning to take place. Fish exposed to constant light have far lower breeding success rates. Sustaining an appropriate light cycle is also essential for healthy embryonic development. Studies have shown that various visible light wavelengths are exposed to fertilized eggs and larvae within a 12-hour light and 12-hour dark cycle until 30 days after hatching. This causes wavelength-specific changes in gene expression, hatching rate, survival rate, and growth rate.

2.5.2 Temperature Control for Spawning Induction

As a poikilothermic species, zebrafish undergo significant temperature changes that closely resemble their natural surroundings. Interestingly, there are noticeable differences in the eating, behavior, and growth patterns of zebrafish depending on the temperature ranges to which they are exposed. Thus, when it comes to the husbandry and reproduction of zebrafish, the careful selection and upkeep of a suitable temperature regimen are critical factors.

Although zebrafish are classified as eurythermal teleosts, meaning they can survive in a wide range of temperatures, studies investigating the effects of temperature on development point to an ideal but very small range of temperatures. Interestingly, raising zebrafish at 22°C produces a primarily male population (87.1% males), whereas raising them at 31°C produces a primarily female population (82.4% females). Apart from the unequal sex ratios caused by temperature fluctuations or temperature cycles, zebrafish kept at temperatures higher than 30°C have a significantly shorter life span. Further research has supported these findings by highlighting the fact that zebrafish survival drops sharply below 50% in situations where temperatures rise above 34°C.

2.5.3 Artificial and Natural Breeding in Zebrafish

In a lab setting, zebrafish are bred using two distinct techniques: natural spawning and artificial fertilization. Male and female zebrafish are paired together in a suitable breeding tank for natural spawning. Environmental cues including variations in temperature, light intensity, or water quality cause spawning to occur. Fertilization occurs when the male and female release their sperm and eggs at the same time. On the other hand, zebrafish are harvested for their eggs and milt, respectively, using a mild extraction technique. In order to aid in fertilization, the retrieved eggs and sperm are then mixed in a regulated setting (such as a Petri dish or tiny container).

Zebrafish can be made to spawn naturally by using a variety of breeding methods and systems. Zebrafish have a new breeding system that consists of a spawning support with a platform raised above the tank floor and a water circulation system to flood the platform (Dijoux *et al.*, 2017). A different type of breeding box has been suggested, which features a wavy fastening plate at the bottom to avoid fish eggs from dropping into the outer box and from being eaten by zebrafish. Similar egg output was achieved by another researcher's approach of rearing zebrafish in a home tank as opposed to a spawning tank (Tsang *et al.*, 2017).

There are several techniques for artificially fertilizing zebrafish. One method is to induce triploidy in zebrafish by administering cold shock and heat shock treatments (Franěk *et al.*, 2019). In some females, heat shock therapy has been found to be more effective, resulting in a 100% triploid generation rate (Dai *et al.*, 2022). Another technique uses melatonin to keep ovulated eggs from aging and adds sunflower pollen extract to male zebrafish to encourage gonadal growth and sperm quality (Jones *et al.*, 2023). Zebrafish have also benefited from intracytoplasmic sperm injection (ICSI), with a success rate of 13.4% in medaka individuals (Otani *et al.*, 2009). ICSI replicates the process of natural fertilization by injecting a demembranated spermatozoon through the micropyle into the egg cytoplasm. These techniques offer useful approaches for zebrafish artificial fertilization.

2.5.4 Future Prospects

Since fish are a well-established model vertebrate, they are a useful tool for studying the complexities of the GnRH system and figuring out how GnRH multiplicity affects function. The first studies, which concentrated primarily on aquaculture species that are economically important, sought to understand and address the GnRH-related deficits that cause confined finfish to not ovulate or spawn. The focus has recently switched to fish models, specifically zebrafish, due to their remarkable traits and the remarkable advances in genome editing, post-genomic, genomic, and molecular technologies that have produced these discoveries.

Future research is well-positioned to benefit from the zebrafish model, which has two different forms of GnRH. This model can be used to investigate possible new functions of GnRHs, such as coordinating feeding, body metabolism, and reproduction. It can also be used to investigate the transduction of environmental cues, like pheromones and photoperiod, which are facilitated by the olfactory and pineal systems, respectively, and how these affect reproductive behaviour. The role of GnRH2, which is common in most vertebrates, is still unclear in all species, from fish to humans. Small fish models present a special chance to clarify the importance of GnRH2 and clarify its widespread distribution in the future. Aditionallly, utilizing zebrafish models allows for the exploration of the intericate interplay between GnRH signaling and various physiological processes, shedding light on evolutionary conservation and divergence in reproductive regulation across species. This research avenue holds promise for uncovering novel insights into the fundamental mechanisms governing reproduction and its integration with broader biological functions, paving the way for potential applications in both basic science and clinical contexts.

2.6 GENOTYPE MAINTENANCE AND LINE DEVELOPMENT IN ZEBRAFISH

The colonies' genetic variety is crucial to the health and viability of zebrafish populations in research settings. Zebrafish are vital models in scientific research because of their genetic similarity to humans and their translucent embryos, which allow for non-invasive monitoring of developmental processes. A large genetic pool enhances the representativeness of trials, allowing researchers to draw conclusions about gene functions, illness causation, and treatment responses with more reliability. Additionally, genetic diversity protects zebrafish colonies from potential threats such as diseases and environmental changes, ensuring their long-term viability. Variations in genetic characteristics within a population allow for the examination of a wide range of biological events, resulting in comprehensive and insightful scientific research that assists advancements in drug discovery and biomedical research. Therefore, to guarantee the model organism's continued efficacy in a variety of experimental contexts, it is imperative to maintain and regulate genetic heterogeneity in zebrafish colonies.

The maintenance and evolution of zebrafish genotypes require a sophisticated approach that involves meticulous record-keeping, controlled breeding programs, advanced genotyping techniques, and a commitment to line stability. Howe *et al.* (2017) talk about how storing and recording genetic data is critical to maintaining traceability and reproducibility in zebrafish lines. These strategies will be refined as zebrafish research develops, maintaining the zebrafish's status as an invaluable model organism in numerous scientific domains. The scientific community's collaborative efforts in sharing verified lines will further encourage improvements in developmental biology, genetics, and drug discovery, reinforcing the zebrafish's critically important role in advancing scientific understanding and improving medical outcomes. The first step in maintaining the integrity of zebrafish genotypes is thorough documentation. Researchers need to carefully document the genetic background, origin, and specific mutations of zebrafish strains. Using outcrossing or inbred lines is a calculated move that strikes a balance between the necessity to preserve particular qualities within a population and the need for controlled breeding to preserve and control particular genotypes. One crucial aspect of maintaining genotypes is controlled breeding (Boudry *et al.*, 2021). Researchers use controlled breeding to ensure that the intended genotype is propagated. Breeding couples must be carefully chosen in order to achieve the desired mutations and genetic background. Genotyping processes are necessary to validate and maintain genotypes in zebrafish colonies. To confirm the presence of targeted modifications and remove undesirable alleles, routine genotyping using technologies such as PCR-based methods is essential (Milbury *et al.*, 2009; Huan *et al.*, 2013). Genotyping accuracy is an essential part of genotype maintenance strategies and directly affects the reliability of study findings. Cryopreservation techniques considerably aid in the long-term preservation of zebrafish genotypes (Hagedorn & Carter, 2011). By preserving sperm and embryos, researchers can prevent genetic drift over time and preserve valuable genetic material. To preserve zebrafish as a model organism, cryopreserved repositories are established to ensure that certain genotypes are available for study in future (Liu *et al.*, 2019).

Selective breeding is one of the most crucial methods for producing zebrafish lines with enhanced phenotypic traits. By focusing on breeding individuals with certain qualities of interest, researchers can gradually enhance the expression of desirable phenotypes within a population. Sibling or paired mating in controlled genetic crosses facilitates the deliberate establishment of stable genetic lineages. Forward and reverse genetic screenings are useful methods for developing zebrafish lines (Fuentes *et al.*, 2018). Scientists can learn more about the role of genes and phenotypic variety by using forward genetic screens to find novel mutations. Shang *et al.* (2017) claim that the advent of genome editing technologies, particularly CRISPR/Cas9, has fundamentally altered reverse genetic screening by enabling accurate modifications to the zebrafish genome to produce lines with desired genetics changes. A review by Geisler *et al.* (2016) discusses techniques such as cryopreservation of gametes and embryos, in vitro fertilization, and the use of molecular tools such as CRISPR/Cas9 for targeted gene editing. These strategies contribute to the precise control of genetic traits in zebrafish lines.

Balance between inbreeding and outbreeding is necessary for evolved lineages to have the maximum amount of genetic diversity and stability. Even though inbreeding can be utilized to fix desired characteristics, excessive inbreeding can lead to inbreeding depression. Researchers need to carefully assess the genetic health of lines and adjust breeding strategies as necessary in order to maintain a delicate balance between fixing traits and preserving genetic variation through outbreeding. Regular cross-validation is necessary to ensure that generated lines are stable and repeatable. Sharing certified lines with the scientific community promotes collaboration and increases the overall reliability of zebrafish research. Because zebrafish researchers openly exchange verified lines, our knowledge of zebrafish biology and genetics as a whole is expanded through collaborative research.

2.7 CHALLENGES IN ZEBRAFISH BREEDING

Breeding and maintaining healthy zebrafish colonies in the laboratory pose challenges (Castranova & Wang, 2020; Tsang *et al.*, 2017). Emerging trends include a focus on understanding breeding complexities, with mate choice, mating behavior, and spawning as key considerations. Standardization of breeding and housing methods is crucial for cross-laboratory comparability. The need for well-managed colonies and maintaining genetic diversity are addressed through strategies such as single-pair mating and group crossing (Nasiadka & Clark, 2012). Refinement of husbandry and management methods is vital for the continued growth of zebrafish research (Lawrence, 2016). Commonly encountered challenges in breeding of zebrafish include the following.

2.7.1 Low Fertility

One challenge in zebrafish breeding is achieving consistent and high fertility rates. A study by Gao *et al.* (2018) suggests that factors such as suboptimal environmental conditions, inadequate nutrition, or genetic factors may contribute to reduced fertility. To address this, researchers recommend optimizing breeding tank conditions, providing a well-balanced diet rich in essential nutrients, and selectively breeding for improved reproductive performance.

2.7.2 Poor Egg Quality

Poor egg quality can impact the success of zebrafish breeding programs. An excellent review by Cheung *et al.* (2019) in *Zebrafish* discusses how factors such as maternal age, nutritional status, and environmental stressors can influence egg quality. The review suggests that dietary supplementation with specific nutrients, such as omega-3 fatty acids and antioxidants, may enhance egg quality. Additionally, maintaining optimal water quality and minimizing stressors in the breeding environment are crucial for improving overall reproductive success.

2.7.3 Disease Outbreaks

Disease incidence poses a significant challenge in zebrafish facilities (Jørgensen, 2020). A study by Kent *et al.* (2020) emphasizes the importance of biosecurity measures to prevent the introduction and spread of pathogens. The research highlights the need for routine health monitoring, quarantine protocols, and the development of specific pathogen-free (SPF) and specific pathogen-tolerant (SFT) zebrafish lines. Implementing strict hygiene practices and optimizing the health management protocols contribute to minimizing the risk of disease outbreaks and maintaining a healthy breeding population.

These challenges underscore the complexity of zebrafish breeding, and addressing them requires a multidisciplinary approach. Researchers continue to investigate new strategies, guided by insights from peer-reviewed studies, to overcome these challenges and improve the efficiency and sustainability of zebrafish breeding programs.

2.8 ETHICAL CONSIDERATIONS IN ZEBRAFISH BREEDING

Ethical considerations in zebrafish breeding revolve around ensuring the well-being of the animals throughout their lifecycle. Research studies, such as those by Lawrence (2007) in "Zebrafish," emphasize the importance of minimizing stress, providing appropriate housing conditions, and using humane euthanasia methods. Zebrafish, used in behavior and neuroscience research, highlight the importance of nutrition for larval development and breeding success (Faria *et al.*, 2015). Ethical breeding practices also involve transparency in reporting methodologies and results.

2.8.1 Compliance with Animal Welfare Guidelines

Researchers working with zebrafish are expected to comply with established animal welfare guidelines. The Guide for the Care and Use of Laboratory Animals (Institute of Laboratory Animal Resources (US), Committee on Care & Use of Laboratory 1986) provides general principles for the ethical treatment of animals in research. Compliance with these guidelines involves proper housing, nutrition, veterinary care, and adherence to protocols that minimize pain and distress during experimental procedures.

2.8.2 Responsible Breeding Practices

Responsible breeding practices in zebrafish research include maintaining genetic diversity, avoiding inbreeding, and employing strategies to reduce the production of surplus animals. A study by Kütter *et al.* (2023) in *Zebrafish* discusses the importance of genetic management to maintain diverse and healthy populations. Researchers often adopt breeding strategies that minimize the number of animals produced while achieving the experimental objectives.

2.8.3 Considerations for Experimental Use and Surplus Animals

Ethical considerations extend to the use of zebrafish in experiments and the management of surplus animals. A review by Canedo *et al.* (2022) discusses the 3Rs (replacement, reduction, refinement) principles as applied to zebrafish research, emphasizing the exploration of alternative methods, reducing the number of animals used, and refining procedures to minimize potential harm. Responsible breeding practices should also address the fate of surplus animals through appropriate housing, adoption, or humane euthanasia.

2.9 FUTURE PERSPECTIVE IN ZEBRAFISH BREEDING

Zebrafish research is relevant in various fields, modeling human diseases and aiding drug discovery (Phillips & Westerfield, 2014). In cancer research, zebrafish models study tumor development, angiogenesis, metastasis, and drug screening (Zhao *et al.*, 2015). Zebrafish's genetic homology to humans and transparent embryos make them valuable in biomedical research, especially for studying vertebrate development and modeling human diseases (Tavares & Lopes, 2013). Zebrafish contribute to vision research, drug discovery, and toxicity testing through live imaging, genetic tools, and automated systems (Cleghorn &

Brockerhoff, 2020). Their use in disease modeling accelerates understanding molecular mechanisms in developmental, mental, and metabolic disorders, providing diagnostic and therapeutic solutions (Choi *et al.*, 2021). Zebrafish's versatility makes them ideal for functional genomics and aquatic biomedicine (Aleström, 2006). Zebrafish breeding has witnessed significant advancements in recent years, fostering progress in various research areas, particularly in fish biotechnology. Notably, researchers have explored refined techniques for controlled reproduction, such as optimizing environmental conditions, diet, and photoperiod manipulation. These efforts aim to enhance reproductive performance, ensuring a consistent and reliable supply of zebrafish for experimentation. Furthermore, advancements in assisted reproductive technologies, including in vitro fertilization and cryopreservation of gametes, offer new avenues for precise genetic manipulations and conservation efforts. In the realm of genomics, the utilization of CRISPR/Cas9 technology has revolutionized targeted gene editing in zebrafish, enabling researchers to model human diseases more accurately.

Looking ahead, the future prospects of zebrafish breeding involve a deeper understanding of their reproductive physiology and the development of innovative tools for genetic manipulation. Emerging technologies, such as single-cell sequencing and advanced imaging, hold promise in unraveling the intricacies of zebrafish development and reproductive processes. The use of reverse genetics techniques, such as gene-targeting strategies, is an important trend (Konantz & Antos 2014). The integration of artificial intelligence and machine learning in data analysis may also expedite the identification of key genes and pathways governing reproduction.

Considering zebrafish welfare in housing is paramount for ethical research practices (Sneddon *et al.*, 2017). Recent studies emphasize the importance of maintaining optimal environmental conditions, including water quality, temperature, and lighting, to ensure the well-being of zebrafish. Adoption of enriched housing environments that mimic their natural habitats contributes to reducing stress and enhancing overall health. Continued research into refining housing conditions and implementing measures to minimize any potential adverse effects on zebrafish welfare will be essential for sustaining high-quality experimental outcomes and maintaining ethical standards in fish biotechnology research.

Recent trends in zebrafish research underscore the versatility of this model organism and its growing significance in various scientific disciplines. According to a review article by Howe *et al.* (2012) published in the journal *Trends in Genetics*, one prominent trend is the increasing use of zebrafish in the study of complex human diseases. The zebrafish's genetic and physiological similarities to humans make it an excellent model for understanding the molecular basis of diseases such as cancer, neurodegenerative disorders, and cardiovascular diseases. Another noteworthy trend, highlighted in a comprehensive review by Goessling and North (2020) in *Disease models and mechanism*, is the expanding role of zebrafish in regenerative medicine. Zebrafish possess remarkable regenerative capabilities, particularly in tissues like the heart and central nervous system. Researchers are exploring these regenerative mechanisms to develop therapeutic strategies for human tissue repair and regeneration.

In the realm of drug discovery and toxicology, a study by Kanungo et al. (2014) in the *Journal of Current Pharmacetual Design* discusses the increasing use of zebrafish in high-throughput drug screening. Zebrafish embryos provide a cost-effective and efficient platform for assessing the safety and efficacy of potential drug candidates. Furthermore, advancements in imaging technologies, as highlighted in a different works by Cutrale *et al.* (2019), Mikut *et al.* (2013), and Høgset *et al.* (2020), have facilitated unprecedented insights into zebrafish development and behavior. High-resolution imaging techniques, coupled with innovative molecular tools, enable researchers to visualize cellular and subcellular processes with remarkable detail, enhancing our understanding of zebrafish biology.

Overall, these trends underscore the dynamic nature of zebrafish research, encompassing diverse areas from disease modeling to regenerative medicine, drug discovery, and advanced imaging techniques. These trends contribute to the continuous evolution of zebrafish as a powerful model organism in biological research.

2.10 CONCLUSION

In conclusion, the study of the breeding and life cycle of zebrafish (*Danio rerio*) is of paramount importance in scientific research. Zebrafish offer numerous advantages as a model organism, from their ease of breeding in a controlled laboratory environment to their well-defined and easily observable life cycle stages. By understanding the intricacies of zebrafish breeding and their life cycle, researchers can harness this model organism to advance our knowledge in fields such as developmental biology, genetics, toxicology, and disease modeling. Zebrafish have proven to be a versatile and powerful tool in scientific investigations, offering the potential for groundbreaking discoveries and the development of innovative treatments. As such, the study of zebrafish breeding and life cycle is pivotal in advancing our understanding of fundamental biological processes and their implications for health and disease.

REFERENCES

Alawi, H., Asiah, N., & Priyatama, A. T. (2013). Induced spawning of selais fish (Ompok hypopthalmus) under different doses of human chorionic gonadotropin hormon (hCG). *Jurnal Perikanan dan Kelautan*, *17*(2).

Aleström, P., D'Angelo, L., Midtlyng, P. J., Schorderet, D. F., Schulte-Merker, S., Sohm, F., & Warner, S. (2020). Zebrafish: Housing and husbandry recommendations. *Laboratory Animals*, *54*(3), 213–224.

Aleström, P., Holter, J. L., & Nourizadeh-Lillabadi, R. (2006). Zebrafish in functional genomics and aquatic biomedicine. *Trends in Biotechnology, 24*(1), 15–21.

Aral, F., Sahinoz, E., & Doğu, Z. (2011). Embryonic and larval development of freshwater fish. *Recent Advances in Fish Farms*, 83–94.

Avdesh, A., Chen, M., Martin-Iverson, M. T., Mondal, A., Ong, D., Rainey-Smith, S., . . . Martins, R. N. (2012). Regular care and maintenance of a Zebrafish (*Danio rerio*) laboratory: An introduction. *Journal of Visualized Experiments: JoVE*, (69), e4196.

Blanco-Vives, B., & Sanchez-Vazquez, F. J. (2009). Synchronisation to light and feeding time of circadian rhythms of spawning and locomotor activity in Zebrafish. *Physiology & Behavior, 98*(3), 268–275.

Blanton, M. L., & Specker, J. L. (2007). The hypothalamic-pituitary-thyroid (HPT) axis in fish and its role in fish development and reproduction. *Critical Reviews in Toxicology, 37*(1–2), 97–115.

Boudry, P., Allal, F., Aslam, M. L., Bargelloni, L., Bean, T. P., Brard-Fudulea, S., . . . Houston, R. D. (2021). Current status and potential of genomic selection to improve selective breeding in the main aquaculture species of International Council for the Exploration of the Sea (ICES) member countries. *Aquaculture Reports, 20*, 100700.

Breder, C. M. & Rosen, D. F. 1966. *Modes of Reproduction in Fishes*. Neptune City, N.J.: T.F.H.

Bruce, A. E. (2016). Zebrafish epiboly: Spreading thin over the yolk. *Developmental Dynamics, 245*(3), 244–258.

Çakici, Ö., & Üçüncü, S. İ. (2007). Oocyte development in the Zebrafish, *Danio rerio* (Teleostei: Cyprinidae). *Ege Journal of Fisheries and Aquatic Sciences, 24*(1), 137–141.

Canedo, A., Saiki, P., Santos, A. L., Carneiro, K. D. S., Souza, A. M. D., Qualhato, G., . . . Rocha, T. L. (2022). Zebrafish (Danio rerio) meets bioethics: The 10Rs ethical principles in research. *Ciência Animal Brasileira, 23*, e-70884.

Castranova, D., & Wang, C. (2020). Zebrafish breeding and colony management. In *The Zebrafish in biomedical research* (pp. 357–364).

Cheung, C. T., Nguyen, T. V., Le Cam, A., Patinote, A., Journot, L., Reynes, C., & Bobe, J. (2019). What makes a bad egg? Egg transcriptome reveals dysregulation of translational machinery and novel fertility genes important for fertilization. *BMC Genomics, 20*, 1–14.

Choi, T. Y., Choi, T. I., Lee, Y. R., Choe, S. K., & Kim, C. H. (2021). Zebrafish as an animal model for biomedical research. *Experimental & Molecular Medicine, 53*(3), 310–317.

Cleghorn, W. M., & Brockerhoff, S. E. (2020). Zebrafish in biomedical research: The retina and vision. In *The Zebrafish in biomedical research* (pp. 237–244).

Cutrale, F., Fraser, S. E., & Trinh, L. A. (2019). Imaging, visualization, and computation in developmental biology. *Annual Review of Biomedical Data Science, 2*, 223–251.

D'Amico, L. A., & Cooper, M. S. (2001). Morphogenetic domains in the yolk syncytial layer of axiating Zebrafish embryos. *Developmental Dynamics: An Official Publication of the American Association of Anatomists, 222*(4), 611–624.

Dai, X., Qiu, L., Zhang, H., Xu, C., Zhao, B., Gao, Y., . . . Chu, Z. (2022). Tea polyphenols improve down-regulation of mitochondrial functional gene expression induced by bisphenol a in Zebrafish ovaries. *Frontiers in Marine Science, 9*, 915415.

Dametto, F. S., Fior, D., Idalencio, R., Rosa, J. G. S., Fagundes, M., Marqueze, A., . . . Barcellos, L. J. (2018). Feeding regimen modulates Zebrafish behavior. *PeerJ, 6*, e5343.

Darrow, K. O., & Harris, W. A. (2004). Characterization and development of courtship in Zebrafish. *Danio rerio. Zebrafish, 1*(1), 40–45.

Dijoux, E., Legendre, L., & Sohm, F. (20179). U.S. Patent Application No. 16/074,486.

Falahatkar, B., Poursaeid, S., Langroudi, H. E., Efatpanah, I., Meknatkhah, B., & Rahmati, M. (2013). Spawning induction in Kutum, Rutilus frisii kutum (Kamensky), with different hormones: Analysis of hormone profiles and induced spawning success. *Fisheries & Aquatic Life, 21*(4), 271–281.

Faria, M., Garcia-Reyero, N., Padrós, F., Babin, P. J., Sebastián, D., Cachot, J., . . . Raldúa, D. (2015). Zebrafish models for human acute organophosphorus poisoning. *Scientific Reports, 5*(1), 15591.

Franěk, R., Tichopád, T., Fučíková, M., Steinbach, C., & Pšenička, M. (2019). Production and use of triploid Zebrafish for surrogate reproduction. *Theriogenology, 140*, 33–43.

Fu, W., Zhang, J., Xu, W., He, S., Long, M., Liao, Q., . . . Xiao, Y. (2021). Characteristics of hatching enzymes and egg envelope in cross progenies from crucian carp (*Carassius auratus* var.) and Zebrafish (*Barchydanio rerio* var.). *Reproduction and Breeding, 1*(2), 81–88.

Fuentes, R., Letelier, J., Tajer, B., Valdivia, L. E., & Mullins, M. C. (2018). Fishing forward and reverse: Advances in Zebrafish phenomics. *Mechanisms of Development, 154*, 296–308.

Gao, D., Lin, J., Ou, K., Chen, Y., Li, H., Dai, Q., . . . Wang, C. (2018). Embryonic exposure to benzo (a) pyrene inhibits reproductive capability in adult female Zebrafish and correlation with DNA methylation. *Environmental Pollution, 240*, 403–411.

Ge, G., Long, Y., Song, G., Li, Q., Cui, Z., & Yan, H. (2022). Transcriptomic profiling revealed signaling pathways associated with the spawning of female Zebrafish under cold stress. *International Journal of Molecular Sciences, 23*(14), 7494.

Geisler, R., Borel, N., Ferg, M., Maier, J. V., & Strähle, U. (2016). Maintenance of Zebrafish lines at the European Zebrafish Resource Center. *Zebrafish, 13*(S1), S-19.

Gharaei, A., Rahdari, A., & Ghaffari, M. (2011). Induced spawning of Schizothorax zarudnyi (Cyprinidae) by using synthetic hormones (Ovaprim and HCG). *World Journal of Fish and Marine Sciences, 3*(6), 518–522.

Glickman, N. S., Kimmel, C. B., Jones, M. A., & Adams, R. J. (2003). Shaping the Zebrafish notochord. *Development, 130*, 873–887.

Goessling, W., & North, T. E. (2014). Repairing quite swimmingly: Advances in regenerative medicine using Zebrafish. *Disease Models & Mechanisms, 7*(7), 769–776.

Golan, M., Zelinger, E., Zohar, Y., Levavi-Sivan, B. (2015). Architecture of GnRH-gonadotrope-vasculature reveals a dual mode of gonadotropin regulation in fish. *Endocrinology, 156*, 4163–4173.

Hagedorn, M., & Carter, V. L. (2011). Zebrafish reproduction: Revisiting in vitro fertilization to increase sperm cryopreservation success. *PLoS One, 6*(6), e21059.

Hammer, H. S. (2020). Water quality for Zebrafish culture. In *The Zebrafish in biomedical research* (pp. 321–335). Academic Press.

Harper, C., & Lawrence, C. (2016). *The laboratory Zebrafish*. CRC Press. https://doi.org/10.1201/b13588.

Høgset, H., Horgan, C. C., Armstrong, J. P., Bergholt, M. S., Torraca, V., Chen, Q., . . . Stevens, M. M. (2020). In vivo biomolecular imaging of Zebrafish embryos using confocal Raman spectroscopy. *Nature Communications, 11*(1), 6172.

Horn, M. E., Ferrari, M. C., & Chivers, D. P. (2019). Retention of learned predator recognition in embryonic and juvenile rainbow trout. *Behavioral Ecology*, *30*(6), 1575–1582.

Howe, D. G., Bradford, Y. M., Conlin, T., Eagle, A. E., Fashena, D., Frazer, K., . . . Westerfield, M. (2012). ZFIN, the Zebrafish Model Organism Database: Increased support for mutants and transgenics. *Nucleic Acids Research*, *41*(D1), D854–D860.

Howe, D. G., Bradford, Y. M., Eagle, A., Fashena, D., Frazer, K., Kalita, P., . . . Westerfield, M. (2017). The Zebrafish Model Organism Database: New support for human disease models, mutation details, gene expression phenotypes and searching. *Nucleic Acids Research*, *45*(D1), D758–D768.

Huan, J., Wan, K., Liu, Y., Dong, W., & Wang, G. (2013). Removing PCR for the elimination of undesired DNA fragments cycle by cycle. *Scientific Reports*, *3*(1), 2303.

Institute of Laboratory Animal Resources (US), Committee on Care & Use of Laboratory Animals. (1986). *Guide for the care and use of laboratory animals* (No. 86). US Department of Health and Human Services, Public Health Service, National Institutes of Health.

Ishida, J. (1985). Hatching enzyme: Past, present and future. *Zoological Science*, *2*(1), 1–10.

Jiang, Y. X., Shi, W. J., Ma, D. D., Zhang, J. N., Ying, G. G., Zhang, H., & Ong, C. N. (2019). Dydrogesterone exposure induces Zebrafish ovulation but leads to oocytes over-ripening: An integrated histological and metabolomics study. *Environment International*, *128*, 390–398.

Jobling, M. (2012). National Research Council (NRC): *Nutrient requirements of fish and shrimp*. The National Academies Press. ISBN: 978-0-309-16338-5.

Jones, R. A., Renshaw, M. J., & Barry, D. J. (2023). Automated staging of Zebrafish embryos with deep learning. *Life Science Alliance*, *7*(1).

Jørgensen, L. V. G. (2020). Zebrafish as a model for fish diseases in aquaculture. *Pathogens*, *9*(8), 609.

Kanungo, J., Cuevas, E., Ali, S. F., & Paule, M. G. (2014). Zebrafish model in drug safety assessment. *Current Pharmaceutical Design*, *20*(34), 5416–5429.

Kaushik, S., Georga, I., & Koumoundouros, G. (2011). Growth and body composition of Zebrafish (*Danio rerio*) larvae fed a compound feed from first feeding onward: Toward implications on nutrient requirements. *Zebrafish*, *8*(2), 87–95.

Kent, M. L., Sanders, J. L., Spagnoli, S., Al-Samarrie, C. E., & Murray, K. N. (2020). Review of diseases and health management in Zebrafish *Danio rerio* (Hamilton 1822) in research facilities. *Journal of Fish Diseases*, *43*(6), 637–650.

Kimmel, C. B., Ballard, W. W., Kimmel, S. R., Ullmann, B., & Schilling, T. F. (1995). Stages of embryonic development of the Zebrafish. *Developmental Dynamics*, *203*(3), 253–310.

Kimmel, C. B., & Law, R. D. (1985). Cell lineage of Zebrafish blastomeres: II. Formation of the yolk syncytial layer. *Developmental Biology*, *108*(1), 86–93.

Konantz, J., & Antos, C. L. (2014). Reverse genetic morpholino approach using cardiac ventricular injection to transfect multiple difficult-to-target tissues in the Zebrafish larva. *JoVE (Journal of Visualized Experiments)*, (88), e51595.

Kusch, R. C., & Chivers, D. P. (2004). The effects of crayfish predation on phenotypic and life-history variation in fathead minnows. *Canadian Journal of Zoology*, *82*(6), 917–921.

Kütter, M. T., Barcellos, L. J. G., Boyle, R. T., Marins, L. F., & Silveira, T. (2023). Good practices in the rearing and maintenance of Zebrafish (*Danio rerio*) in Brazilian laboratories. *Ciência Animal Brasileira*, *24*.

Laurila, A., Pakkasmaa, S., Crochet, P. A., & Merilä, J. (2002). Predator-induced plasticity in early life history and morphology in two anuran amphibians. *Oecologia*, *132*, 524–530.

Lawrence, C. (2007). The husbandry of Zebrafish (*Danio rerio*): A review. *Aquaculture*, *269*, 1–20.

Lawrence, C. (2016). New Frontiers for Zebrafish management. In H. W. Detrich, M. Westerfield, & L. I. Zon (Eds.), *Methods in cell biology* (Vol. 135, pp. 483–508). Academic Press.

Li, J., Chu, L., Sun, X., Liu, Y., & Cheng, C. H. (2015). IGFs mediate the action of LH on oocyte maturation in Zebrafish. *Molecular Endocrinology*, *29*(3), 373–383.

Liew, W. C., & Orban, L. (2014). Zebrafish sex: A complicated affair. *Briefings in Functional Genomics*, *13*(2), 172e187. https://doi.org/ 10.1093/bfgp/elt041.

Liu, Y., Blackburn, H., Taylor, S. S., & Tiersch, T. R. (2019). Development of germplasm repositories to assist conservation of endangered fishes: Examples from small-bodied livebearing fishes. *Theriogenology*, *135*, 138–151.

Longkumer, S., Jamir, A., & Pankaj, P. P. (2020). Development and appraisal studies of chemically induced Zebrafish hyperglycemia model. *Journal of Experimental Zoology India*, *23*(2).

López-Olmeda, J. F., & Sanchez-Vazquez, F. J. (2011). Thermal biology of Zebrafish (*Danio rerio*). *Journal of Thermal Biology*, *26*, 91–104.

Mikut, R., Dickmeis, T., Driever, W., Geurts, P., Hamprecht, F. A., Kausler, B. X., . . . Peyriéras, N. (2013). Automated processing of Zebrafish imaging data: A survey. *Zebrafish*, *10*(3), 401–421.

Milbury, C. A., Li, J., & Makrigiorgos, G. M. (2009). PCR-based methods for the enrichment of minority alleles and mutations. *Clinical Chemistry*, *55*(4), 632–640.

Mizrahi, N., & Levavi-Sivan, B. (2023). A novel agent for induced spawning using a combination of GnRH analog and an FDA-approved dopamine receptor antagonist. *Aquaculture*, *565*, 739095.

Montero, J. A., & Heisenberg, C. P. (2004). Gastrulation dynamics: Cells move into focus. *Trends in Cell Biology*, *14*(11), 620–627.

Murray, K. N., Lains, D., & Spagnoli, S. T. (2020). Water quality and idiopathic diseases of laboratory Zebrafish. In S. C. Cartner, J. S. Eisen, S. C. Farmer, K. J. Guillemin, M. L. Kent, & G. E. Sanders (Eds.), *The Zebrafish in biomedical research* (pp. 463–477). Academic Press.

Myers, D. C., Sepich, D. S., & Solnica-Krezel, L. (2002). Convergence and extension in vertebrate gastrulae: Cell movements according to or in search of identity? *Trends in Genetics*, *18*(9), 447–455.

Mylonas, C. C., & Zohar, Y. (2001). Use of GnRHa-delivery systems for the control of reproduction in fish. *Reviews in Fish Biology and Fisheries*, *10*, 463–491.

Nasiadka, A., & Clark, M. D. (2012). Zebrafish breeding in the laboratory environment. *ILAR Journal*, *53*(2), 161–168.

Newport, J., & Kirschner, M. (1982). A major developmental transition in early Xenopus embryos: I. Characterization and timing of cellular changes at the midblastula stage. *Cell*, *30*(3), 675–686.

Otani, S., Iwai, T., Nakahata, S., Sakai, C., & Yamashita, M. (2009). Artificial fertilization by intracytoplasmic sperm injection in a teleost fish, the medaka (Oryzias latipes). *Biology of Reproduction*, *80*(1), 175–183.

Parichy, D. M., Elizondo, M. R., Mills, M. G., Gordon, T. N., & Engeszer, R. E. (2009). Normal table of postembryonic Zebrafish development: Staging by externally visible anatomy of the living fish. *Developmental Dynamics*, *238*(12), 2975–3015.

Paullada-Salmerón, J. A., Cowan, M., Aliaga-Guerrero, M., Gómez, A., Zanuy, S., Mañanos, E., & Muñoz-Cueto, J. A. (2016). LPXRFa peptide system in the European sea bass: A molecular and immunohistochemical approach. *The Journal of Comparative Neurology*, *524*, 176–198.

Pereira, A., Carvalho, A. P., Cruz, C., & Saraiva, A. (2017). Histopathological changes and zootechnical performance in juvenile Zebrafish (*Danio rerio*) under chronic exposure to nitrate. *Aquaculture*, *473*, 197–205.

Phillips, J. B., & Westerfield, M. (2014). Zebrafish models in translational research: Tipping the scales toward advancements in human health. *Disease Models & Mechanisms*, *7*(7), 739–743.

Rieb, J. P. (1974). Une installation de microcinématographie pour oeufs de Téléostéens, avec prises de vues à intervalles de temps déterminés [An installation for time lapse microcinematography of teleost eggs with variable frame cycle (author's transl)]. *Acta Microscopica*, *75*(4), 338–345. (in French)

Ribas, L., & Piferrer, F. (2014). The Zebrafish (*Danio rerio*) as a model organism, with emphasis on applications for finfish aquaculture research. *Reviews in Aquaculture*, *6*(4), 209–240.

Ribas, L., Valdivieso, A., Diaz, N., & Piferrer, F. (2017). Appropriate rearing density in domesticated Zebrafish to avoid masculinization: Links with the stress response. *Journal of Experimental Biology*, *220*(Pt 6), 1056e1064.

Riesco, M. F., Valcarce, D. G., Martinez-Vazquez, J. M., & Robles, V. (2019). Effect of low sperm quality on progeny: A study on Zebrafish as model species. *Scientific Reports*, *9*, 1–10.

Rohde, L. A., & Heisenberg, C. P. (2007). Zebrafish gastrulation: Cell movements, signals, and mechanisms. *International Review of Cytology*, *261*, 159–192.

Schulz, R. W., Nóbrega, R. H., Morais, R. D. V. D. S., De Waal, P. P., França, L. R. D., & Bogerd, J. (2018). Endocrine and paracrine regulation of Zebrafish spermatogenesis: The sertoli cell perspective. *Animal Reproduction (AR)*, *12*(1), 81–87.

Shang, W., Wang, F., Fan, G., & Wang, H. (2017). Key elements for designing and performing a CRISPR/Cas9-based genetic screen. *Journal of Genetics and Genomics*, *44*(9), 439–449.

Sneddon, L. U., Halsey, L. G., & Bury, N. R. (2017). Considering aspects of the 3Rs principles within experimental animal biology. *Journal of Experimental Biology*, *220*(17), 3007–3016.

Spence, R., Gerlach, G., Lawrence, C., & Smith, C. (2008). The behaviour and ecology of the Zebrafish, *Danio rerio*. *Biological Reviews*, *83*(1), 13–34.

Spence, R., & Smith, C. (2006). Mating preference of female Zebrafish, *Danio rerio*, in relation to male dominance. *Behavioral Ecology*, *17*(5), 779–783.

Tanaka, S., Zmora, N., Levavi-Sivan, B., & Zohar, Y. (2022a). Chemogenetic depletion of hypophysiotropic GnRH neurons does not affect fertility in mature female Zebrafish. *International Journal of Molecular Sciences*, *23*(10), 5596.

Tanaka, S., Zmora, N., Levavi-Sivan, B., & Zohar, Y. (2022b). Vasoactive intestinal peptide indirectly elicits pituitary LH secretion independent of GnRH in female Zebrafish. *Endoc rinology*, *163*(2), bqab264.

Tavares, B., & Lopes, S. S. (2013). The importance of Zebrafish in biomedical research. *Acta medica portuguesa*, *26*(5), 583–592.

Thisse, C., Thisse, B., Schilling, T. F., & Postlethwait, J. H. (1993). Structure of the Zebrafish snail1 gene and its expression in wild-type, spadetail and no tail mutant embryos. *Development*, *119*(4), 1203–1215.

Trinkaus, J. P. (1984). Mechanism of fundulus epiboly—A current view. *American Zoologist*, *24*(3), 673–688.

Tsang, B., Zahid, H., Ansari, R., Lee, R. C. Y., Partap, A., & Gerlai, R. (2017). Breeding Zebrafish: A review of different methods and a discussion on standardization. *Zebrafish*, *14*(6), 561–573.

Van den Hurk, R., & Resink, J. W. (1992). Male reproductive system as sex pheromone producer in teleost fish. *Journal of Experimental Zoology*, *261*(2), 204–213.

Varga, Z. M. (2016). Aquaculture, husbandry, and shipping at the Zebrafish International Resource Center. In *Methods in cell biology* (Vol. 135, pp. 509–534). Academic Press.

Warga, R. M., & Kimmel, C. B. (1990). Cell movements during epiboly and gastrulation in Zebrafish. *Development*, *108*(4), 569–580.

Webb, S. E., & Miller, A. L. (2006). Ca2+ signaling and early embryonic patterning during the blastula and gastrula periods of Zebrafish and Xenopus development. *Biochimica et Biophysica Acta (BBA)-Molecular Cell Research*, *1763*(11), 1192–1208.

Wisenden, B. D., Paulson, D. C., & Orr, M. (2022). Zebrafish embryos hatch early in response to chemical and mechanical indicators of predation risk, resulting in underdeveloped swimming ability of hatchling larvae. *Biology Open*, *11*(12), bio059229.

Zahro, H., Anshori, K., Fransisco, S., Rosa, A. A., & Retnoaji, B. (2022, May). Reproductive aspect and embryonic development of wader pari fish (Rasbora lateristriata Bleeker 1854) from Malang East java. In *7th International Conference on Biological Science (ICBS 2021)* (pp. 540–544). Atlantis Press.

Zempo, B., Tanaka, N., Daikoku, E., & Ono, F. (2021). High-speed camera recordings uncover previously unidentified elements of Zebrafish mating behaviors integral to successful fertilization. *Scientific Reports*, *11*(1), 1–9.

Zhao, S., Huang, J., & Ye, J. (2015). A fresh look at Zebrafish from the perspective of cancer research. *Journal of Experimental & Clinical Cancer Research*, *34*, 1–9.

3 Neurobehavioral Assessment of Zebrafish

Shaimaa Nasr Amin and Dalia Azmy Elberry

3.1 INTRODUCTION

In scientific terminology, South Asian tropical zebrafish are known as *Danio rerio* Before George Streisinger introduced this 2.5–4 cm teleost as a model organism for developmental biology study in vertebrates at Oregon University in the 1970s, it was mostly used as an ornamental fish (Tan et al., 2022).

Neurobehavioral diseases' exact causes have always been difficult to detect; typically, these conditions are amenable to creating or discovering new pharmacological treatments. To explore neurobehaviors, a faultless experimental model is necessary (Mrinalini et al., 2023).

Zebrafish are gaining much attention as a model organism for the study of complex neurobehavioral processes. The conceptual complexity and the overall feasibility are two interrelated factors that significantly impact the efficacy of a model organism in the context of neuroscience research (de Abreu et al., 2021). Zebrafish models meet both criteria, as their complex larval and adult behaviors have been thoroughly studied and documented (Kalueff et al., 2013). A variety of behavioral tasks have been devised to assess these activities within the key brain regions associated with activity, perception, cognition, social interaction, and emotion (Kalueff et al., 2014).

Ex vivo fertilization, tiny size (adults are 2–5 cm long), ease of conservation, prolific nature (each fish produces more than 100 eggs), quick growth (larvae begin swimming five days after fertilization), and genetic alteration viability are a few of this species' good traits (Tan et al., 2022). In addition to the optical transparency of embryos and even some adult fish strains, non-invasive imaging techniques can be utilized to examine the development of particular genes (Tang et al., 2020; Araya et al., 2016). Due to their unchanging compatibility with the human genome and illnesses, cheap maintenance requirements, and other characteristics, zebrafish are a vital alternative animal model for scientific research (Mrinalini et al., 2023).

Zebrafish have a brain architecture akin to that of other vertebrates, including humans, and share many genetic similarities with human genes that play a role in Alzheimer's disease and other neurodegenerative disorders (Kumar et al., 2021; Razali et al., 2021).

Remarkably similar, robust, and seemingly evolutionarily conserved are the behavioral responses of zebrafish and mammals (Jesuthasan, 2012). In addition, zebrafish neurodegenerative models exhibit central nervous system (CNS) abnormalities with an early start, permanent neuronal loss, and early mortality (similar to humans). Nonetheless, zebrafish models of Huntington's disease exhibit less mortality and neuromorphological and neurophysiological abnormalities than do their mammalian counterparts (Shaw et al., 2018). Therefore, deadly diseases that are difficult to study in rodents can be thoroughly examined in vivo using mutant zebrafish (Bashirzade et al., 2022). Moreover, these organisms pose a substantial threat to high-throughput screening that utilizes advanced video-tracking apparatus designed for both larval and adult zebrafish (Cachat et al., 2011).

The immaturity of key regions of the zebrafish brain, such as the cortex, is one of its limitations as a neurobehavioral animal model. One significant limitation of this paradigm is the underdevelopment of the prefrontal cortex and telencephalon, the regions most frequently damaged in the bulk of mental diseases. It was once assumed that this structural abnormality rendered zebrafish incapable of performing a variety of cognitive processes. Prior research has demonstrated, however, that small-brained vertebrates may conduct a variety of cognitive processes and even make complex judgments without a fully developed telencephalon (Parker et al., 2013).

3.2 MAIN BEHAVIORAL DOMAINS IN ZEBRAFISH

Despite having a more diverse behavioral repertoire than mammals due to their role as aquatic vertebrates, zebrafish can easily summarize all the important neurobehavioral symptoms and endpoints studied in rodents (Kalueff et al., 2013). Several articles state that the most researched zebrafish behaviors are locomotor activity, "anxiety-like" emotion, social interaction (preference, aggressiveness), and cognition (learning, object identification, and spatial memory) (Bartholomew et al., 2015; Peng et al., 2016).

3.2.1 Locomotive Operation

Cognitive processes such as learning, memory, social interaction, self-defense, and anxiety are shared by zebrafish and humans. As the investigation's endpoint, the range of each of these behaviors is evaluated using specific metrics. In nearly all of its investigations, the locomotive behavior

DOI: 10.1201/9781003402893-4

of zebrafish, defined by swimming, is the most basic factor. This is quite obvious, given that this organism is naturally a fish that requires swimming for survival. The emergence of autonomous tracking techniques has augmented the benefits of using swimming as a primary indicator (Mrinalini et al., 2023).

Swimming can provide a range of additional factors for analyzing various behaviors in any study (de Abreu et al., 2019). A statistic called thigmotaxis is used to assess the anxiety and behavior spectrum. The innate tendency to maintain physical contact with vertical surfaces is known as "wall-hugging" (Lamprea et al., 2008). This conduct is regarded as an anxiety index because it occurs normally in stressful or worrisome situations (Colwill & Creton, 2011).

3.2.2 Anxiety-Like Behavior

Traditional animal models of anxiety have concentrated on how animals react to novel situations (Kurt et al., 2000). When exposed to unknown (possibly hazardous) surroundings, many animals with "anti-predatory" functions exhibit avoidance-related behaviors (Sousa et al., 2006). It is thought that novelty-seeking enhances behavioral order in an unfamiliar environment and reflects the emotional state of the animals (Stewart et al., 2012b).

Growing data suggests that zebrafish neurobehavioral phenotyping can help us comprehend the neurobiology of stress and anxiety. For example, novelty, exposure to predators, alarm pheromones, anxiogenic medications, and drug withdrawal are all environmental stresses to which zebrafish are especially susceptible. Strong and sensitive neuroendocrine responses related to anxiety are also evident in zebrafish, and these responses correlate strongly (and reciprocally) with behavioral endpoints. This explains why larval and adult zebrafish models are helpful and trustworthy for assessing acute and chronic anxiety-like states (Kalue, 2017).

As the animals become accustomed to their new environments, their initial active exploratory behaviors tend to subside throughout the trial. This attenuation and loss of habituation often indicate increased worry (Wong et al., 2010). Similar to rats, zebrafish novelty-based paradigms and the attendant phenotypes are significantly influenced by short-term and long-term stimuli. Consequently, various novelty-based tests previously established and employed for rats, such as the open-field and light-dark tests, have been used for zebrafish neurophenotyping (Mrinalini et al., 2023). Other tasks, including the new tank test, have been designed expressly for zebrafish (Chen et al., 2020).

3.2.3 Cognitive Phenotypes in Zebrafish

Survival requires learning and remembering, which is shared by many species, especially vertebrates. In order to convert results from cognitive studies, such as learning and memory, to humans, it is required to employ suitable model organisms (Tan et al., 2022). Rats and non-human primates are conventional animal models for studying behavior, particularly cognition (Stephan et al., 2019). Their neurobehavioral profiles are the most similar to those of humans, but they are the most expensive to study and are subject to the strictest ethical constraints of any model creature. The high cost of using rodents as genetic and pharmacological models in brain research has traditionally limited this type of study. The zebrafish is the most often utilized non-mammalian model for cognitive research because it is a vertebrate with neurobehavioral traits similar to humans. Confirmation in mammalian models is necessary before implementing zebrafish research into clinical trials (Tan et al., 2022).

The startle response, novel tank test, learnt spatial alternation test, three-chamber tank, and T-maze are some of the most popular zebrafish behavioral tests for investigating different aspects of learning and memory. Pharmacological, genetic, or experimental methods can shift any paradigm. Each test has its own set of procedures and tools for evaluating the efficacy of this aquatic model (Kalue, 2017).

3.2.4 Social Interaction

Various factors, including individual phenotypic differences, life-history phases, and social situations, are recognized to influence social conduct evaluation (Taborsky et al., 2015). Zebrafish can distinguish homo- and heterospecifics with the same body size and shape (Engeszer et al., 2007). Living close to conspecifics is considered an altruistic behavior known as "social preference" (Liu et al., 2016). The social phenotype of zebrafish is assessed by monitoring how the animal reacts to or interacts with a social trigger (Ogi et al., 2021).

Many endogenous and exogenous factors can influence social choice in zebrafish. There are two types of external elements: (1) age, sex, and personality of the fish being studied; and (2) character traits of the group members providing social stimulation (their number, size, phenotype, sex ratio, and kinship between individuals). However, the results of social preference tests may be affected by exogenous factors, such as the size and temperature of the tank, the brightness of the test room, the number and size of preference areas, the presence of environmental enrichment, and the duration of the habituation and interaction periods (Ogi et al., 2021).

Various neuropsychiatric and neurodevelopmental problems can be better understood with the aid of zebrafish social preferences. The "social preference test" could be used to investigate elements that may influence the social behavior of zebrafish in the laboratory, as well as the effects of hormones and pharmaceuticals from a translational standpoint. It may also help to construct a zebrafish model for human neuropsychiatric and neurodevelopmental diseases marked by social deficits (Ogi et al., 2021).

A further significant social activity is aggression. Aggression is a feature that has evolved in numerous species, including humans (Rhee & Waldman, 2002). It plays an important role in the social life cycle of zebrafish by facilitating shoal cohesion, reproduction, and safety from predators. Zebrafish are a promising model organism for studying human aggression's neurological and genetic basis since they exhibit aggressive behaviors and are easy to cultivate in the laboratory (Jones & Norton, 2015). To acquire social dominance (Dahlbom et al., 2011), ensure access to a mate (Howard et al., 2015; Paull et al., 2010), or control a disputed resource, such as territory (Spence & Smith, 2005) or food, shoal mates may engage in aggressive behavior (Basquill & Grant, 1998).

Aggressive behaviors include biting, lateral displays, charges or attacks, darts, or retreats. Despite substantial research in artificial stimulus-based laboratory examinations, more antagonistic behaviors have been observed during direct fish-to-fish contacts (Oliveira et al., 2011). This additional behavior includes circling, freezing, pursuing, and running (Kalue, 2017).

Two significant challenges for zebrafish social behavior that must be overcome are the automated quantification of social and collective action and the precise manipulation of the social stimuli utilized in the testing. Both restrictions are being addressed, and substantial growth is anticipated (Kalue, 2017).

3.2.5 Influence of Colors on Zebrafish Behaviors

According to Cuthill et al. (2017), colors are a significant environmental component that influences the behavior and physiology of animals and humans. They are usually linked to conditions that indirectly influence behavioral reactions (Wiercioch-Kuzianik & Bąbel, 2019). In addition, since colors can influence how individuals think and act, there is a two-way interaction between color and behavior. For example, "pleasant" or "calming" hues, such as blue, can evoke good memories, but "unpleasant" hues, such as brown, may induce negative thoughts (Palmer & Schloss, 2010). Cool hues such as blue or green are typically relaxing and lessen clinical anxiety. Warm hues, on the other hand, assist depressed individuals (Azeemi & Raza, 2005).

In addition, colors play a significant role in animal development by providing visual cues for numerous essential behaviors, such as sexual, social, and aggressive ones, in numerous species, such as mammals, birds, and fish (Bergman et al., 2009; Dijkstra et al., 2009). There is growing evidence that zebrafish could be utilized to research the physiological and behavioral effects of color (Bault et al., 2015). In contrast to rat models, zebrafish are active during the day, have polychromatic vision, and rely on visual and color cues to guide their behavior (Howe et al., 2018). Also, unlike rats, zebrafish can quickly change their body color in response to stress, medication, general health, social status, dominance/aggression, and other things (Cachat et al., 2013). Because zebrafish appear paler during sleep, stress, and illness (Singh & Nüsslein-Volhard, 2015), it can be inferred that they undergo a similar process. During aggressive or dominant behavior, animals tend to darken and display more vibrant colors; such alterations may alter the behavior of conspecifics (Piato et al., 2011).

Aquatic neurobehavioral research for zebrafish should also consider color an essential environmental component. For example, older fish placed in white tanks exhibit more anxiety-like behavior (Genario et al., 2019). However, anxiety and cortisol levels in zebrafish kept in black or blue aquariums are lower (de Abreu et al., 2021). These considerations should be considered when doing behavioral research (de Abreu et al., 2020). Zebrafish prefer yellow, dark-striped conspecifics when interacting with con- and heterospecifics, indicating that body color patterns may be an important cue (Abaid et al., 2012). Color substantially impacts cognitive functions in zebrafish (de Abreu et al., 2021). Color cues are often used in cognitive tests to measure zebrafish learning and memory and other conditioned site preference paradigms in drug misuse research (Cognato et al., 2012; Zhang et al., 2018).

Color preference in zebrafish is also affected by where it comes from and how it reproduces. The effect of these factors on learning shows that these behaviors are linked to development and evolution (Roy et al., 2019). Zebrafish inhabit large riverbanks with quickly moving clear water, dense woods with murky water, and still ponds with thick flora in the wild (Sundin et al., 2019). Blue may thus be favored by zebrafish due to its overall relaxing behavioral and cortisol-lowering effects (de Abreu et al., 2020). Perhaps because these fish live in microhabitats with clear water and little to no vegetation in the environment, allowing mostly short-wavelength light (such as blue) to flow through (Kageyama, 1999). Red, which is connected with food and may be preferred by laboratory zebrafish, is the predominant color of the synthetic pellet meal provided to lab fish (Roy et al., 2019).

Color preferences have also been linked to "personality." A fondness for the color red, for example, is associated with an outwardly integrated personality in Western countries. Conversely, green is often associated with clever and friendly people (Birren, 1973). Given how important personalized medicine is and how zebrafish research is finding new ways to use it, this makes these animals gain more attention regarding this field (Neff, 2020). Also, because there are clear differences in how each zebrafish acts, more research is needed to find out how personality affects the way zebrafish act when it comes to color (Volgin et al., 2019; de Abreu et al., 2021).

According to these results, color is important for how social, emotional, and smart different zebrafish models are. Zebrafish can be used as a model to study color preference in different neurobehavioral paradigms (de Abreu et al., 2021). However, there remain considerable gaps in zebrafish color neurobiology research. Before behavioral testing, wild-type zebrafish are periodically evaluated for their visual abilities. Naturally occurring color atypicalities in such "control" groups of zebrafish may arise (de Abreu et al., 2021). Water quality (such as salinity and turbidity) may influence how animals see color in the same way that zebrafish behavior data may be altered by water quality (Kirk, 1979). Furthermore, humans and animals have different UV spectrum eyesight, which affects the broad application of color studies. Cones in rats and zebrafish can detect UV sunlight and colors with short wavelengths of 350–362 nm, but humans cannot (Gouras & Ekesten, 2004).

3.3 ZEBRAFISH NEUROBEHAVIORAL REPERTOIRE

The zebrafish neurobehavioral repertoire is comparable to that of humans (Tan et al., 2022). By examining zebrafish's behavior, researchers can enhance their comprehension of the neural networks, physiological biomarkers, and genetic underpinnings associated with typical and atypical brain functioning. Utilizing diverse behavioral terminology in a laboratory-specific context presents challenges, as various research groups may employ distinct terms to denote identical actions exhibited by zebrafish. In contrast, identical terms may be employed to characterize disparate behaviors in varying biological circumstances. Furthermore, many zebrafish activities may have comparable or overlapping behavioral manifestations but with slight yet significant and observable differences. For example, one can easily distinguish between sleep-like/resting behavior, characterized by the absence of bodily movement accompanied by decreased opercular activity, and stress-induced freezing behavior, characterized by the cessation of all bodily motions except eye movements and limited fin undulation (Kalueff et al., 2013). For this distinction to be straightforward, however, the correct name, characterization, and measurement of both behavioral features are required (Kalue, 2017). Moreover, the number of new zebrafish models is on the rise. Thus, providing the necessary terminology and contextualizing zebrafish actions is essential for a better understanding of zebrafish phenotypes (Kalueff et al., 2013).

As mentioned earlier, the action served as the primary measure in developing a more extensive ethogram of zebrafish behavior. This advancement would enhance the understanding and analysis of research findings, promote utilizing behavioral models across different species, and stimulate novel research groups' adoption of zebrafish neurobehavioral paradigms (Kalueff et al., 2013).

3.4 ZEBRAFISH NEUROBEHAVIORAL TESTS

Several water anxiety tests show that zebrafish have distinct behavioral characteristics (Kysil et al., 2017), aggression (Zabegalov et al., 2019), and sociality (Pham et al., 2012), which are particularly vulnerable to changes in the environment and pharmaceuticals (de Abreu et al., 2019). Using zebrafish as an experimental model, the following fundamental neurobehavioral evaluations are conducted.

3.4.1 Locomotor Activity Test

The locomotor activity test (LAT) evaluates the spontaneous swimming activities of zebrafish (Chen et al., 2020). The LAT assesses baseline locomotor activity as well as how medicines and environmental pressures change locomotor activity, which can affect cognitive test outcomes (Maeda et al., 2021). In the LAT, a single zebrafish is briefly submerged in a trapezoidal or rectangular water tank. Zebrafish movement is frequently captured on camera from the front or top of the tank. LAT measures overall distance traveled, swimming speed (average or maximum), freezing time, and "thigmotaxis" (time spent and distance covered near the tank wall). To follow the geographic movement of zebrafish, a 3D LAT was constructed (Audira et al., 2018).

In a battery of behavioral tests, the LAT is usually administered before other, more complex paradigms. This prevents the animals from becoming tired and losing interest in the subsequent stages. There is no agreement on a typical habitat or the environmental conditions required for LAT. Most zebrafish behavioral tests provide general instruction rather than a rigorous approach. Zebrafish are extensively utilized as study models in laboratories all around the world. Therefore, standardization may be necessary to compare the findings of various investigations (Tan et al., 2022).

Due to its challenges, precisely adhering to a prescribed technique may require numerous changes. The pharmacokinetics of the medications being assessed (and, subsequently, the timing to research drug efficacy, taking the paradigm into account), as well as the test equipment and setting, may impact adaptability (e.g., water exchange frequency and temperature). To increase the accuracy of the test findings, it would be desirable to include as much information as possible about these challenges in the publication (Tan et al., 2022).

3.4.2 Novel Tank Test

This ubiquitous novelty-based test is unique to zebrafish and other aquatic animals; it is often used to profile their behavior and evaluate medication effects (Kalue, 2017). It is comparable to the LAT apparatus, which assesses anxiety-like behavior in zebrafish (Chen et al., 2020). During data processing, the tank is split into two horizontal parts. During the "geotaxis" test, zebrafish usually stay in the lower half of the tank before slowly moving up to check out the upper half. As evidenced by decreased time and distance spent in the tank's upper portion, zebrafish exhibit anxiety-like behavior (Tan et al., 2022).

Prior to starting the test, keep in mind that the zebrafish's living environment may influence their behavior. Zebrafish kept in a tank that is as small as the novel tank equipment, for instance, might not show a diving reflex or changes in swimming speed, although fish kept in a bigger tank did, a result mostly attributable to habituation or adaptation to novelty (Bencan et al., 2009).

The new tank test is a suitable assay for determining if a drug is anxiolytic or anxiogenic; for instance, chronic fluoxetine reduces geotaxis while acute coffee increases it. Such modifications have a profound and reciprocal effect on anxiety-like behavior in zebrafish (Cachat et al., 2010b). In an experiment, the cutting-edge tank test can also examine the anxiety-inducing effects of drug withdrawal. In a novel zebrafish tank experiment, chronic morphine withdrawal develops a powerful anxiogenic phenotype (Cachat et al., 2010a). Finally, novelty-based paradigms reveal that larval zebrafish react similarly to anxiolytic and anxiogenic stimuli (Kalue, 2017).

3.4.3 Open-Field Test

The open-field test is a well-known animal behavioral paradigm that has been used frequently since its creation in 1932 to examine mice's responses to unexpected stimuli that induce anxiety-like sensations (Katz et al., 1981; Prut & Belzung, 2003; Krook et al., 2019).

The time spent in the center, the time spent in the periphery (thigmotaxis), and the distance traveled are measured. This paradigm's conventional exploratory-based behavioral endpoint is average speed, which has been recently adapted for neurobehavioral phenotyping of larval (Ahmad & Richardson, 2013) and adult zebrafish (Stewart et al., 2012b). Different laboratories may characterize the open-field apparatus's center and periphery with varying terms. The large field might be divided into 16 equal squares by one group, with the center four squares representing the field's center and the remaining squares representing its edges (Champagne et al., 2010). There are no specific rules for defining the center in an open-field test because, according to other research, the "center" is defined as the region within 5 cm of the apparatus walls (Grossman et al., 2010). Zones must be uniform and standardized across the laboratory for this test to produce meaningful behavioral phenotypic data. Rodents and zebrafish initially exhibit thigmotaxic anxiety-like reactions during open-field testing, but these behaviors eventually go away, indicating intra-session habituation to novelty (Lynn & Brown, 2009). Experimental stressors can change open-field phenotypes. A 15-minute net constraint is an example of an acute stressor that increases zebrafish thigmotaxis and average velocity (Champagne et al., 2010).

The length of the test, the housing conditions before the test, and the gadget's size, form, and material vary among laboratories, whereas open-field studies examine exploratory behavior in a novel and open setting in exactly the same ways. Unexpectedly, zebrafish exhibit higher locomotor activity in larger open-field arenas than smaller ones despite their similar temporal activity patterns for exploratory behaviors (Stewart et al., 2012b).

During mouse open-field testing (Horev et al., 2007), an important differentiating trait was the creation of a "homebase" or "preferred reference point position." Zebrafish have also lately been found to possess this characteristic. By dividing the open-field arena into quadrants and calculating the average time spent, number of trips made, and distance traveled in each quadrant, it is possible to study the zebrafish's homebase behavior (Stewart et al., 2010).

3.4.4 The Light-Dark Test

A different paradigm has been explored in rats and is now commonly utilized for phenotyping zebrafish. The light-dark test tank consists of half light and half dark (Champagne et al., 2010; Grossman et al., 2010; Stewart et al., 2011). The test can be performed in a variety of ways, including the light-dark plus labyrinth, which includes two light arms, two dark arms, and a gray starting region in the center (Sackerman et al., 2010).

Scototaxis, or zebrafish's peculiar preference for darkness, can be seen. It is a type of "thigmotaxis" that zebrafish can use to gauge anxiety and fear (Maximino et al., 2010). Piato et al. (2011) discovered that stressed animals prefer to spend more time in a dark area than do stressed control animals. This research implies increased time spent in poorly illuminated areas may indicate worry. According to preliminary studies, zebrafish prefer the white area of the tank (Champagne et al., 2010; Gerlai et al., 2000). These alleged arguments are most likely the result of various dwelling conditions, lighting, sex, age, social position, and fish strains. Adult zebrafish strongly prefer darkness, whereas fish larvae show an "inverted" predilection for light (Kalueff et al., 2013). For example, fish raised in an enriched habitat for two months spend more time in the light than do fish raised in a bad environment (Maximino et al., 2010). Zebrafish behavior in the light-dark test may also be influenced by lighting intensity; at 500 lux versus 250 lux, they increased phototaxis and spent more time freezing (Stewart et al., 2011; Kalue, 2017).

3.4.5 Y-Maze Test

Zebrafish were examined to evaluate how well they remember their spatial location based on their natural urge to discover new areas (Capatina et al., 2021). Individual zebrafish were studied in a Y-shaped aquarium. Indicators of spatial memory were spontaneous alternation percentage (the tendency to switch arms every three entries), time spent, and distance traveled in the novel arm (Tan et al., 2022).

3.4.6 Novel Object Recognition Test

The novel object recognition test assesses recognition memory based on a fish's reaction to novelty (as in the Y-maze). In the experiment conducted by Capatina et al. (2021), a zebrafish was given 10 minutes to inspect a tank that included two identifiable objects (red cubes). After an hour, one of the red cubes was replaced with a green cube (a new object), and the fish's exploration was recorded on video. The examination required a distance of less than 2.5 millimeters. Exploration time and liking for the unknown object are used to evaluate recognition memory (Capatina et al., 2021).

3.4.7 T-Maze Test

Multiple aspects of associative learning, such as acquisition, extinction, and reversal, can be investigated using this test. It has a wide range of applications. The T-maze for zebrafish is commonly constructed of clear Plexiglas and consists of a long stem with two shorter arms emerging perpendicularly from one end, which may be covered in patterned or colorful sleeves. Additionally, holes are evenly located throughout the maze to form gates or walls to restrict zebrafish migration, particularly when building acclimation start boxes or obstructing specific places (Gould, 2011).

The T-maze was first utilized in zebrafish to account for other study variables, such as determining if changes in conditioned site preference were caused by medicine or genetic changes rather than memory problems in the patients. A deep section with substrate and plants can be added at the end of one arm of the T-maze" to evaluate how well experimentally changed animals can remember and recall the location of their preferred enriched environment (Darland & Dowling, 2001; Swain et al., 2004). Data must be taken with caution because pharmaceutical treatments may lower zebrafish motivation to approach the enriched arm of the T-maze rather than improving cognitive ability (Swain et al., 2004).

3.4.8 Plus Maze Test

Despite having four equal-length arms that extend from a center area that can be set on a rotating platform, the plus maze differs in shape from the T-maze although they are both roughly comparable in design. In order to assess how well rats learn spatial and non-spatial associations, the plus maze can be compared to the radial arm maze. In a cued memory test, zebrafish were employed to measure non-spatial learning (Sison & Gerlai, 2010). After learning to identify an internal maze cue (a red card) with a food reward, independent of the location, zebrafish preferred entering the red-card-marked arm (Kalue, 2017).

Similar to the Morris Water Maze, mice were also subjected to spatial memory tests using the plus maze (Morris, 1984). Zebrafish also excelled at this task (Sison & Gerlai, 2010). The chosen arm and the time spent outside the selection chamber are recorded in the task utilizing the same techniques to investigate how acute stress affects learning (Gaikwad et al., 2011). Both pressures had a negative effect on the spatial and non-spatial performance of Zebrafish.

3.4.9 Inhibitory Avoidance Test

The inhibitory avoidance test (IAT) is one of the most extensively used cognitive tests for testing the learning and memory of zebrafish. The test can be constructed using a two-chamber tank and a tank with a variety of mazes. In a two-chamber tank, one side is selected to combine the cue with unpleasant stimuli like an electric shock (ES) (Sarasamma et al., 2018), lowering weight (Kim et al., 2010), stirring, or color sleeves on the other side of the tank/maze (Rajesh et al., 2018). In the IAT, the CS-US pairing is normally selected from the darker side (CS: conditioned stimulus, US: unconditioned stimulus). A training phase is conducted to strengthen and increase memory (Braida et al., 2020). After the last training session, a probing step is conducted to gauge memory retention. The time between the training and testing stages can be changed to assess both short- and long-term memory. The unconditioned stimulus is not supplied during the probing phase. Based on the separation of the treated and untreated groups, the conditioned response is determined as a gauge of memory retention (Devidas et al., 2021).

While being trained, the zebrafish is placed in the light compartment and permitted to swim to the dark compartment. When the fish reaches the dark section, it receives an electric shock. The training phase can be repeated numerous times until a conditioned response manifests. Following the last training session, the fish are given 24 hours (for short-term memory) to explore the compartment without being shocked during the probing phase. The time it takes to enter the dark chamber is kept track of during training and probing sessions as a measure of memory recall. Other elements, such as the number of entrances and the length of time spent in the dark space, can be taken into account (Tan et al., 2022).

3.4.10 Appetitive Conditioning Test

Similar to the IAT, pairing the CS and US in a maze or rectangular tank requires repeated exposure. To act as visual signals, Chen et al. added red and green sleeves to the left and right arms of the T-maze (Chen et al., 2020). An enlarged compartment containing a brine shrimp on one arm served as an enticing stimulus. The task includes sessions for acclimatization, instruction, and investigation. The effectiveness of memory was assessed based on entry latency and time spent in the targeted chamber. They can also attract conspecifics in a pleasant environment with artificial grass and stones (Tan et al., 2022). It is important to remember that food incentives may not meet the criteria for an acceptable motive in an associative learning test for a variety of reasons; therefore, modifying the reward type and using conspecifics may be an appropriate substitute (Al-Imari & Gerlai, 2008). For instance, managing the amount of food zebrafish consume can be difficult because the fish's size and sex may affect how full they feel after previous feedings (Kalue, 2017).

3.4.11 Startle Response Test

Since habituation research depends on evaluating changes in reflexive responses, tests concentrating on the zebrafish startle reflex have been devised (Kalue, 2017). It is crucial for monitoring the integration of motor and sensory information (Mena et al., 2016). This response has survived evolution because it protects against predatory dangers, and because it occurs automatically, it is useful for studying habituation (Stewart et al., 2010). Zebrafish develop their startle response early, allowing researchers to study both larval and adult startle reflexes (Chanin et al., 2012). It is utilized because it is a universal sensory-motor response coupling system that enables animals to respond quickly and defensively to a variety of stimuli, such as touch (which develops as early as two days after fertilization), visible or white light (which develops in three days), and sound above 200 Hz frequency (develops by five dpf). In response to any of these triggers, a rapid increase in swimming speed is observed (Ganzen et al., 2017; Basnet et al., 2019).

For instance, the acoustic startle reaction (ASR) of 96 larval zebrafish can be measured simultaneously in a well plate when the fish are exposed to a sound stimulus. Changes in walking distance after a sound presentation can be used to measure ASR; an increase in walking distance denotes an increase in ASR. Larval zebrafish displayed a larger ASR after the stimulus was repeated, indicating habituation, and a higher ASR after the initial stimulus was delivered, indicating a stronger response. A single additional terrifying stimulus (such as a light pulse) can be used to trigger the first ASR response after habituation, signaling "dishabituation" (Best et al., 2008).

3.4.12 Learned Spatial Alternation Test

Zebrafish might pick things up via association. Zebrafish, for instance, can learn to link environmental signals and places with enjoyable stimuli such as food, medications, and other conspecifics (Mathur et al., 2011; Al-Imari & Gerlai, 2008). In associative learning, known as spatial alternation learning, zebrafish are assessed on their capacity to recall the spot in space where they were previously rewarded. The experimental design and behavioral paradigm used to evaluate zebrafish's acquired spatial alternation are explained in the next section (Williams et al., 2002).

The zebrafish were put in a 2-gallon aquarium with a red card on one side to aid in spatial orientation and an opaque barrier in the center that they could swim around to access the edges of the tank. Twenty-eight attempts to place food on the opposing sides of the tank were made when the partition was tapped as a trigger for 5 seconds. Following the cue but prior to food administration was the choice phase. The precise location of the fish at the moment the meal was administered was documented to ascertain the accuracy of the selection. When the fish showed greater than 75% accuracy over the course of the first 14 trials, it was determined that they had learned the task satisfactorily. Zebrafish can also recall how to complete this exercise after a 10-day delay (Kalue, 2017).

In these paradigms, additional rewards for the zebrafish may be applied, such as shoaling species in which conspecifics could be a motivating stimulus (Al-Imari & Gerlai, 2008; Saverino & Gerlai, 2008). It is possible to increase the difficulty of spatial alternation paradigms, for instance, by employing an elongated tank that makes it harder for zebrafish to swim to the proper side (Kalue, 2017).

3.4.13 Three-Chamber Tank Test

Typically, a tank separated into three equal portions by two transparent Plexiglas sides is used for the three-chamber task. The starting area is in the center while sliding doors on the partition walls provide access to the side choice compartments. This gadget is beneficial for researching numerous aspects of learning and memory because it may be used in a variety of settings (Kalue, 2017).

Using this apparatus, an early study examined the acquisition of spatial and non-spatial visual discrimination (Arthur & Levin, 2001). The choice accuracy and latency of zebrafish trained to avoid potentially dangerous stimuli (such as a moving fishnet) by traveling to the left or right side were measured. A dark-colored panel was employed to represent the back wall of the equipment as a reference point for directions. The net was removed if the fish was placed in the proper chamber but was placed in the choice chamber for 60 seconds if the incorrect chamber was selected. The zebrafish completed this learning task and improved with each test, as evidenced by the increasing number of accurate responses. When the regarded correct side was reversed, zebrafish engaged in reverse learning. This indicates the cognitive flexibility of zebrafish and promotes the assumption that they can learn spatial relationships (Kalue, 2017).

In the second task of this study, a non-spatial learning test requiring visual color discrimination was evaluated using the three-chamber tank. Instead of a specific side, fish were trained to enter the chamber with a particular color (such as red or blue). The zebrafish once more showed learning abilities and the capacity to relearn the task under newly altered color conditions. The use of medications or gene modifications to discern between the most likely reasons for success (or failure) in this experiment, according to the authors, could enhance the research. For instance, if a fish is able to enter a chamber and flee from net stimulation but is unable to complete the reversal task, it is presumed that cognitive decline is to blame rather than, say, a lack of drive or subpar motor skills (Kalue, 2017).

3.4.14 Prey Capture or Food Hunting Test

Since this complicated motor action involves recognition, approaching, decision-making, and capture, this test is also used to evaluate cognitive ability (Muto & Kawakami, 2013).

3.4.15 Social Preference Tests

When placed in a tank with another group of conspecifics or heterospecifics, zebrafish prefer to stick closer to one another than to heterospecifics (Saverino & Gerlai, 2008). The social preference test or shoal preference test has been used to determine how sociable zebrafish are on a large scale. In the two-choice preference test paradigm, documenting the time spent near the stimulus fish is considered a sign of preference. As social stimuli, a recorded video of live fish displayed on a computer screen, computer-animated graphics of zebrafish, or virtual reality technologies may be used (Gerlai, 2017; Stowers et al., 2017).

There are two test phases in social preference tests. During habituation, the tester zebrafish is left alone in a tank compartment to investigate its new surroundings. The interaction phase begins with the introduction of the social stimulus, which is usually two or three small groups of live conspecifics (Ogi et al., 2021).

The test tank comprises a central chamber (2) and two side chambers (1 and 4) separated by transparent glass. One side chamber (1) contains conspecific shoals, whereas the other contains an alternate stimulus or an empty side chamber (4). The random assignment of stimuli to compartments will aid in preventing spatial bias. Sealing the glass barriers to the tank's walls reduces the influence of confounding olfactory cues. It is possible to use a T-maze constructed expressly for this purpose; in this case, the horizontal component of the maze is divided by guillotine doors, which are then divided into three sections (Deacon & Rawlins, 2006).

During the test, the behaviors of zebrafish swimming freely in several chambers are captured on camera (Norton et al., 2019). This time is then recorded as a percentage of the overall amount of time the fish spent in the area of the social stimulus. As illustrated by Aslanzadeh et al. (2019) and Landin et al. (2020), the preference area can also be separated into zones of "strong" and "weak" social preference based on the distance from the chamber containing the social stimulus.

3.4.16 Tests to Assess Aggressive Behaviors

The mirror image stimulus (MIS), which was first employed in zebrafish to explore aggressive behavior using video-tracking techniques, is one of the most regularly used approaches for eliciting aggressive behavior in fish. Gerlai et al. used slanted mirrors to restrict the ability of the focus fish to perceive its own reflection to certain portions of the experimental arena (thereby confining the area of the fish tank where aggressive behaviors can be manifested). When positioned flat against the wall of an experimental arena, the mirror can also be utilized for MIS; however, the tested fish will always be able to see the stimulus in this configuration. The presence of a refuge (a place where the focused fish can stop responding to mirror stimuli) may influence the behavioral manifestation (Gerlai et al., 2000; Weber & Ghorai, 2013).

Because of its low cost, simplicity, low danger of harm to the focal fish (enabling the same fish to be tested in several assays), and high interactivity between the focal fish and the stimuli, the MIS design is frequently utilized. Antagonism in zebrafish can also be researched using live fish as a trigger. The use of live fish as a stimulus is significantly more natural than methods such as MIS. However, because the stimulus cannot be precisely regulated, the stimulus fish's behavior may differ from that of the focus fish over time (Kalue, 2017).

3.4.17 Tests to Assess Color-Related Behaviors

Several tests can be used to identify which colors zebrafish prefer (or avoid) in order to assess the significance of color in zebrafish behavior (Park et al., 2016). When comparing two or more colors, zebrafish color preference tests frequently elicit a choice response (e.g., plus-maze tests). The location preference test, which tracks the number of entries and the amount of time spent in each compartment, is conducted on zebrafish in a two-chamber box with colored walls (Avdesh et al., 2012; Pieróg et al., 2018). Zebrafish are similarly placed in a neutral chamber in a three-chamber experiment before being permitted to pick between the other two colors (Barba-Escobedo & Gould, 2012). In the T-maze colored-arm test, fish must select between two colors, and their decision is influenced by the quantity of entries and the amount of time spent in each arm. The plus-maze test measures the number of entry and time spent in each arm while giving fish four color options (Park et al., 2016; Zhang et al., 2018).

The new item identification test, the T-maze, and the multi-chamber tank test, which demonstrates zebrafish's preference for shorter wavelengths like blue and green over longer wavelengths like red and yellow, demonstrate how important colors are for zebrafish cognitive processes (Braida et al., 2014; Bault et al., 2015; Oliveira et al., 2015). According to other research, zebrafish avoided blue in the T-maze test compared to red, green, and yellow (Avdesh et al., 2012). The diverse approaches and materials employed in these studies may be to blame for these discrepancies in behavioral investigations of color preference and other color-related behaviors in zebrafish (de Abreu et al., 2021).

3.5 ANALYSIS OF ZEBRAFISH BEHAVIORAL TESTS

3.5.1 Automated Analysis

Approximately 190 monitored behaviors have been identified in larval and adult zebrafish models. More than 20 of them, such as "biting," "charging," "chasing," "fighting" (aggressively), "swimming in circles," "courting," "shoaling," and social interaction, are complex behaviors that are primarily evaluated manually (Kalueff et al., 2013).

Various tools for the automated study of animal behavior are accessible today, some of which are commercial, but the majority are open-source and free. These advancements have provided researchers with numerous benefits, including time and cost savings, improved precision, the ability to evaluate a larger number of animals, and the potential to produce highly quantitative and accurate findings, such as total length of displacement, speed, acceleration, turn angles, track trajectories, and the number of entries in a region of interest (ROI). These programs share real-time behavior analysis of previously recorded automated or semi-automatic footage. This software performs a semi-automatic behavioral analysis because users must alter the settings (distance measurements and ROI). The software must then carry out the user's instructions to follow a specific item. One of the key limitations of the present tools is their inability to track many animals simultaneously. This challenge is mostly caused by animal spatial overlap, which results in identity loss and muddled data (Dankert et al., 2009; Luyten et al., 2014).

The number of turns, distance, speed, freezing time, immobilization time, and turn angle (customizable) must all be examined at a minimum for automated behavior analysis (Franco-Restrepo et al., 2019). As stated previously, they utilized aquatic environment-adapted mazes and video recording systems in conjunction with zebrafish behavior analysis software (such as EthoVision or DanioScope from Noldus or the more current BonZeb). Recently, behavioral analysis of adolescents and adults has gotten simpler (Guilbeault et al., 2021). Due to their unique characteristics, evaluating the behavior of fish embryos and larvae is significantly more challenging (extremely small size, unique behavioral patterns and parameters). This requires extremely delicate logistics, such as ViewPoint Behavior Technology's ZebraBox in France. ZebraBox gives very precise tools for monitoring the behavior of zebrafish larvae, and it was recently used to evaluate sleep patterns, locomotion, social interaction, and the effects of maternal exposure to chemicals on behavior (Cao et al., 2019; Cadena et al., 2020).

3.5.2 Advanced Statistical Analysis Methods

Due to statistical challenges caused by shared correlation, unique distributions, and the sheer volume of data, modern, high-throughput research generates complex time-series data. In developmental neurotoxicological investigations, individual zebrafish may exhibit extraordinarily variable behavior, resulting in unanticipated behavioral patterns that do not fit the most often used statistical distributions. For a range of research objectives, unusual data analysis approaches are required to manage vast amounts of data, identify outliers, and model behavioral data (Kalue, 2017).

Complex methods for examining the impact of various conditions on zebrafish larvae activity include Hotelling's T-squared test and multivariate analysis of variance (MANOVA) (Liu et al., 2015). In addition to their interactions, these variables also include biological variations (zebrafish subspecies and developmental stages), treatment effects (repeated measurements and various light/dark settings), and experimental changes (physical location and independent replication) (Kalue, 2017).

3.5.3 Combining Behavioral and Morphological Data

When combined with morphological data in a testing environment with a variety of objectives, behavioral data from zebrafish offer a tremendous amount of potential utility. This is particularly true during the early stages of development, when environmental constraints may impair an intricate web of mechanisms that support healthy development without pressuring them to produce a certain end result (or to be assessed at a specific time in an experiment) (Kalue, 2017).

3.6 ZEBRAFISH NEUROBEHAVIOR CATALOG (ZBC)

The International Zebrafish Neuroscience Research Consortium (ZNRC) created the Zebrafish Neurobehavior Catalog (ZBC), a complete alphabetized collection of zebrafish phenotypes based on the core behavioral features of adult and larval zebrafish (Kalue, 2017):

- **"Abnormal body position" or "contortion of the body"** may be caused by diseases, genetic deficiencies, exposure to toxins, or aging. Neurological traits and neurodevelopmental disorders account for the majority of zebrafish larvae (Kalue, 2017; Oliveira et al., 2011).
- **"Aggression":** Complicated behaviors (approach, raising fin, undulating body movement, mouth opening, changing body color, biting, charging, chasing, and circling) directed to conspecifics (or other objects) in adult zebrafish may occur in the context of defending the territory (territorial behavior), preserving resources (e.g., females), and gaining dominance (Oliveira et al., 2011).
- **"Akinesia":** When zebrafish are older or after being exposed to certain medications, such as pharmaceuticals that lower dopamine levels, swimming slows down or typical motor function is impaired. This condition, which is frequently accompanied by a sagging tail, can be measured by a decrease in both swimming speed and distance covered. It is comparable to a "hypolocomotion" state (Kalue, 2017).
- **"Alarm reaction":** An anti-predator response that is adaptive and manifested in response to stimuli that cause fear. A reaction set, which is frequently referred to as erratic movement, is typically characterized by faster movement and more frequent directional shifts. Other alarm responses include diving, changing in shoaling (such as abrupt dips in shoal cohesiveness followed by longer-lasting increases of several minutes), and freezing with fast opercular movements (Kalue, 2017).
- **"Anxiety (anxiety-like) behavior" ["geotaxis," "thigmotaxis," "scototaxis," "freezing," "opercular movements," "changing the color**

of the body," and "zigzagging"] is a complicated response prompted by a risky or possibly dangerous environment/stimulus, such as limited exploration. Anxiolytic medications often diminish anxiety-like behaviors, whereas anxiogenic medications increase them. Anxiety-like behavior mainly corresponds with fear-related behavior in zebrafish (Kurt et al., 2000).

- **"Appetitive olfactory behavior":** To sample appetitive odor plumes, increased swimming speed and distance traversed with frequent orientation shifts (>90°). Once eye contact is established with food, "approach" and "nibbling" behaviors are demonstrated (Kalue, 2017).
- **"Approach," "manifest presence and movement toward the target":** Sexual: abrupt swimming movement shown, regardless of the actions taken by the females to court the male. Exploratory: the process of approaching something new rather than avoiding it (Kalue, 2017).
- **"Ataxia":** a widespread lack of normal body posture and movement coordination (Kalueff et al., 2013).
- **Attack (attacking):** Short bursts of fast swimming directed at an opponent, accompanied by jaw opening and biting; part of aggressive behavior (Kalue, 2017).
- **"Attraction":** An increase in proximity to or movement toward an item or chemical stimulation; the opposite of "avoidance" (Kalueff et al., 2013).
- **"Avoidance":** The inverse of "attraction" is the growth of distance and time spent away from a stimulus (such as a predator or bright light) (Kalueff et al., 2013).
- **"Beat-and-glide"** is a type of intermittent swimming that is distinguished by tail pounding followed by glides (Buss and Drapeau, 2001).
- **"Bend" (bending):** The discovered neurological phenotype is unusual, exhibiting a peculiar swimming habit in which the body is laterally twisted. This particular phenotype is acknowledged as a sign of seizure activity. This behavior is characterized by intermittent episodes of sudden and involuntary movements resembling twitching or spasms (Kalueff et al., 2013).
- **"Bite" (biting):** Rapid movement toward a target with the jaws expanding, shutting, and making physical contact. Zebrafish regularly bite/nip (nipping) one another in the region of the gills or fins during "fights" (Oliveira et al., 2011).
- **"Boldness":** Characterized by a strong personality trait, frequently manifested by a heightened attitude to new objects. Bold animals typically exhibit less anxious behavior, a shift in body color, and more inquisitive activity (Maximino et al., 2010).
- **"Buoyancy dysregulation":** A defect in regulating buoyancy. The inability to maintain a continuous elevation while swimming without exerting physical effort, which is most typically caused by anomalies in the swim bladder or other peripheral systems, displays as "surfacing," "vertical drifting," "Cartesian diver behavior," inclined swimming, and tilting (Kalue, 2017).
- **"Burst-and-coast":** Only fish larvae that are unable to swim continuously exhibit darting. Fish advance in a single action known as a "burst," coast to a stop, and then advance once more (Kalueff et al., 2013).
- **"Burst swim" (swimming):** Swimming in rapid succession with considerable bend angles, most pronounced in the mid-body of larval zebrafish; first detected in larvae two days after fertilization. It is typically connected with escape activities and includes larger amplitude bending, faster speeds, and greater yaw than slow swimming. The pectoral fins are retracted and dormant (Thorsen & Hale, 2007).
- **"Camouflage response" (background adaption):** When exposed to a darker or lighter background, melanophore dispersion (skin darkening) or aggregation causes a shift in body pigmentation (resulting in a darker or lighter appearance; paling). In the literature on fish phenotyping, a component of the body color change response is referred to as the "expanded melanophore phenotype" (Kalue, 2017).
- **"Cartesian diver behavior":** a fish with an abnormal phenotype that alternates between passive vertical drift and sinking when exposed to certain neuroactive substances (Abramson & Evans, 1954).
- **"Charge or charging":** Approaching a second fish while increasing speed, as the first fish avoids the second. It denotes the establishment of social power as well as the end of a zebrafish "war" (Oliveira et al., 2011).
- **"Chemotaxis" (chemoattraction):** Zebrafish movement or predilection toward particular chemical stimuli that operate as chemoattractants. "Appetitive olfactory behavior," "chemically mediated kin recognition," and "sexual aggregating activity" are examples of chemically mediated attraction behaviors (elicited by sex pheromones). When fish follow increasing chemoattractant concentrations, their velocity tends to increase, whereas it decreases when they find the source of the signal (Kalue, 2017).
- **"Circling":** On a regular basis, swim in a circular motion (often observed after seizures, neurological impairments, and in response to the action of the chosen medications). Circling is associated with typical behaviors such as exhibition (seen in lateral display activity) and courting (visible in sexual behavior), which may be recorded manually or

automatically using video-tracking equipment. Some therapies induce "tight" circling, as assessed by its diameter in body length. The number of full circles (360°) per trial, the percentage of animals circling, and the direction of circling (left or right rotations) are all circling endpoints; automated systems can additionally quantify turn angle and angular velocity (Spence et al., 2008; Riehl et al., 2011).

- **C-launch, "C-bend/turn, Mauthner reflex":** The fish's body curls into a C-shape as a quick escape or startle response. The fish then quickly swims away from where it is at an angle. Mauthner cells control how both adults and young fish look. Head stimulation makes zebrafish larvae C-start, but tail stimulation makes them C- and S-start (Kalue, 2017).
- **"Coast or coasting":** Similar to "drifting," passive sliding occurs when the body does not move (Kalueff et al., 2013). It could occur naturally or as a result of interaction. Single or alternating left-right bending of the complete body begins around 18 hours after conception in the embryo and thereafter decreases in frequency (Ahmad et al., 2012).
- **Color of the body variation, "coloration response":** An anxiety symptom, a natural response to lighting/environmental conditions, a "camouflage response," a social behavior, or a drug-induced dispersion (skin darkening) or aggregation (paling) of melanophores. Certain drugs produce significant skin darkening, while others cause skin paling in adult zebrafish. The color response can be evaluated manually (through visual inspection) or automatically (using luminescence-based instruments) (Cachat et al., 2013).
- **"Color preference":** A natural preference/bias toward specific colors; for example, zebrafish will avoid colors that induce an innate aversion (e.g., preferring black (scototaxis) or yellow, green, or red vs. blue), as these colors are most likely associated with natural risks, such as predators (this, however, may depend on the context). The color of fish objects is also important in adult zebrafish social interactions (de Abreu et al., 2020).
- **"Corkscrew swimming":** Increasingly fast spiraling swimming in an uncoordinated direction is a typical seizure phenotype (Stewart et al., 2012b).
- **"Courtship"** refers to the extremely organized actions that take place before reproduction. After making eye contact with the female (since he must first chemically detect her readiness and exhibit sexual aggregating behavior), the male will pursue or chase her and attempt to tail-nose touch her with his dorsal fin up. If the male does not immediately spawn, he may stand slightly above the substrate with his body slightly tipped downward and circle, occasionally circling the female. During this time, the male will keep looking for a female to mate with. The female is capable of leading, introducing, accompanying, and approaching. The female might attempt to push the male away if she finds his attempts offensive. The parameters that define zebrafish courtship behavior include the average distance between male and female, the number of contacts between male and female, the time spent by male and female in the spawning area, the number of entrances by male and female into the spawning area, the swimming distance and velocity inside and outside of the spawning area, the total swimming activity, and the turning rate by male and female (Kalueff et al., 2013).
- **"Creeping":** Prolonged swimming in which only the fish's pectoral fins move it forward (Sison & Gerlai, 2011).
- **"Dart or darting":** The caudal fin is used for a single, quick acceleration in a single direction. It could be a speedy or erratic/zigzagging movement (combined with a number of darts, symbolizing a series of bursts of acceleration where the darts' trajectories alter seemingly randomly from one to the next). Darting was referred to as leaping or jumping in other articles (Oliveira et al., 2011).
- **"Dashing"** refers to purposeful (propulsion-driven) rapid, jerky motions typical of an "escape reflex."
- **"Depth preference":** A natural aversion to swimming in shallow water (Cachat et al., 2010b).
- **"Dispersion":** Several fish make quick, evasive movements away from one another before returning to the group. This behavior is usually brought on by coming into contact with a large, potentially dangerous moving object, like a predator (Kalueff et al., 2013).
- **Dive (diving, geotaxis):** In response to danger, movement toward or preference for the tank's bottom. In general, latency to the bottom, length spent at the bottom, the number of times it is visited, the distance traveled there, and relative top-to-bottom ratios can be used to assess a sensitive anxiety/avoidance behavior index. It is frequently lessened by habituation or anxiolytic treatments, made worse by sedative/anxiogenic drugs, and unusually reversed by some hallucinogens (e.g., ibogaine). The zebrafish action known as "diving" differs from sinking or "Cartesian diver behavior" (which often involves a horizontal body position) and resting behavior in that it is active, quick, and deliberate (Cachat et al., 2010a).
- **"Dorsal light reflex (DLR)":** The natural tendency is to turn with the light. When exposed to horizontal lighting, the vestibular righting reflex

acts against the fish to correct the body's tilt along the dorsoventral axis (Kalueff et al., 2013).

- **"Droopy (drooping) tail":** Aging, motor deficits, genetic modification, and pharmaceutical treatments can all lead to motor phenotypes that are associated with neurological problems, "akinesia," and widespread hypolocomotion; severe phenotypes can lead to inclined swimming. In contrast to tail dip, the droopy tail is a persistent trait that can also signify a particular napping or resting activity (Yokogawa et al., 2007).
- **"Eight-figure"** is a unique swimming style seen in zebrafish after various medication treatments, which could be a component of natural mating behavior (Spence et al., 2008).
- **"Erratic movement":** A complex behavior marked by sudden changes in speed or direction and quick darting. Adult zebrafish commonly exhibit irregular motility for a variety of reasons (fast acceleration bouts in rapid succession in which the direction of movement also changes in an apparently stochastic manner between rapid darts; see also "zigzagging"). Commonly observed right before and after freezing episodes; usually caused by acute stressors (introduction to a predator, emission of an alarm cue) or reflecting a general baseline anxiety/fear state as part of the alarm response. Furthermore, zebrafish larvae can exhibit unpredictable behavior in reaction to rapid changes in light (Ahmad et al., 2012).
- **Escape:** The angular acceleration and displacement of a large animal in response to a shocking stimulus. The first stage is a body "C-bend" (C-start), followed by a contralateral bend and tail beat(s). The initial acceleration is usually followed by quick zigzagging near the tank's bottom; on rare occasions, escape can result in jumping behavior. It consists of quick turning followed by burst swimming in larval fish (Thorsen & Hale, 2007; Buss & Drapeau, 2001).
- **"Escort or escorting":** A female's courtship behaviors include swimming next to a male or staying still while being courted (Darrow & Harris, 2004).
- **"Exploratory activity"** is a sophisticated group of actions intended to explore new environments. Not reliant on parameters associated with anxiety and movement; for instance, the percentage of zebrafish activity in different horizontal and vertical zones of the tank can be used to calculate exploration profiles (Champagne et al., 2010).
- **"Fast turn" (turning):** In 12 milliseconds, a larval zebrafish twists its head 180 degrees, followed by a C-shaped counterbend and a frenetic swimming episode as it swims away at a 90–180 degree angle. Rapid, large-angle twists requiring the bending of the entire body at high angular velocity distinguish these escape-like movements. —Larvae reactions to evasion (Budick & O'Malley, 2000).
- **"Fear-like behavior":** Fear is a direct response to the sight or perception of potentially aversive stimuli that are not currently present, whereas anxiety is a feeling or reaction generated by potentially aversive stimuli that are currently present. As a result, anxiety prevails, whereas terror is more cue-driven. It is unclear how the two scenarios emerge in zebrafish behavior; however, certain instances (e.g., alarm reaction) appear to represent the pathological anxiety-like state, while others (withdrawal) appear to indicate fear (Kurt et al., 2000).
- **Feeding:** Specific food types may be associated with food consumption behaviors.
- **Fight (fighting):** The fish first assess each other by engaging in "display," "biting/nipping," "flicking," and "circling" activities, which last until the first chase/flee. All agonistic activities are initiated by the "winner" (chaser), whereas the "loser" exhibits "fleeing," "submission behavior," or "freezing" (Oliveira et al., 2011).
- **"Fin raise" (fin extension/erection):** Zebrafish typically extend their dorsal and caudal fins during conflict and courting (Gerlai, 2003).
- **"Flick or flicking":** A specific agonistic behavior is seen when two zebrafish swim toward each other, briefly brush their mouths, and then flick away in opposite directions. It happens during agonistic interactions (fights) (Kalueff et al., 2013).
- **"Flight (fleeing)" "flight behavior"** refers to a quick movement away from another fish or a stimulus (Oliveira et al., 2011).
- **"Floating":** Passive swimming differs from surfacing, which is usually more active movement at the water's surface; drifting, which is usually in the middle of the water layer; and sleeping or resting, which is usually near or at the bottom. It can be caused by neurological problems or problems controlling buoyancy (Kalueff et al., 2013).
- **Follow (the) subsequent:** This behavior, which is common during courtship and social contact, resembles prey pursuit but is normally non-aggressive **(Darrow & Harris, 2004)**.
- **"Food seeking"** is a typical foraging activity in zebrafish that is driven by hunger and affected by pathogenic diseases or specific psychiatric medications that act as appetite suppressants (Kalueff et al., 2013).
- **"Foraging":** Motions that involve searching and probing in response to sensory inputs. The animal's foraging behavior is distinguished by frequent directional changes as it samples chemical plumes (Kalueff et al., 2013).
- **"Freeze or freezing":** The complete cessation of movement among the fish at the tank's

bottom (other than gills and eyes). In general, submissive behavior due to excessive stress or anxiety or as part of submissive behavior can be quantified using freezing latency, frequency, length, and posture. Opercular motions frequently happen when freezing is brought on by tension. Immobility is frequently brought on by toxic/ sedative medications, which can produce animal hypolocomotion and akinesia. It can also result in sinking. Immobility is frequently not accompanied by increased opercular motions (Kalueff et al., 2013).

- **"Habituation"** includes habituation between trials (inter-session) and within trials (intra-session). The term "habituation" refers to the tendency to exhibit a significantly reduced response after repeated exposure to a novel stimulus/environment. It usually consists of more top exploration, less diving, and consistent irregular motions over time. Zebrafish habituation can be assessed by calculating the ratios of behavioral activity during the beginning vs. most recent trials or by analyzing the fish's behavioral profile throughout the trial (Bencan et al., 2009).
- **"Head-butting":** Pushing the head against a vertical surface once or more, as when responding to mirror stimulation, thrashing, or as a result of the effects of different medications (Kalueff et al., 2013).
- **"Head shake" movements:** Seizures/tremors are common in larval zebrafish and are frequently accompanied by convulsions (Saint-Amant & Drapeau, 1998).
- **"Hide (hiding)":** A fish's attempt to stay out of sight (Marks et al., 2012).
- **"Homebase formation/behavior":** After investigating a new area, the fish has a tendency to establish a critical "secure" place where it spends more time and returns frequently. A kind of innate place preference behavior that may be quantified by the amount of time, trips, and distance spent at one's homebase against other locations. potential sensitivity to medication interventions (Stewart et al., 2010).
- **"Hyperlocomotion" (hyperactivity):** Swimming that is unusually fast for an extended time, usually due to psychostimulant/convulsant effect or anxiety-like behavior (Maximino et al., 2010).
- **"Hypolocomotion" (hypoactivity):** Swimming that is abnormally sluggish for an extended time, frequently accompanied by drowsiness, neuromotor deficits, and akinesia (Kalueff et al., 2013).
- **Immobility:** A total cessation of activity at the tank's base; unlike freezing and resting, it is not necessarily followed by changes in opercular activity (either increased or decreased). Sedatives are a common cause of this illness; if left untreated, it can lead to paralysis, akinesia, or even a lack of movement in the animal (Kalueff et al., 2013).
- **"Inclined swimming":** Neuroactive or neurotoxic chemicals can cause a phenotypic aberration, such as swimming at an oblique angle to the water's surface. Possible causes include motor impairment or aging, both of which have been associated with abnormal tail buoyancy (Uusi-Heikkilä et al., 2012).
- **"Jaw movements":** After the administration of some medicines (e.g., convulsant substances like domoic acid), some animals displayed stereotypical, non-foraging mouth-opening behavior (Kalueff et al., 2013).
- **"J-bend" "J-turn":** A delicate reorientation turning manifested by a bend in the larval tail and a tiny curve in the body (Bianco et al., 2011).
- **"Jittery" swimming:** A distinct swimming pattern characterized by repeated brief "jerky" motions with decreased smoothness of swimming trajectories, as seen in some epileptic behaviors, can be induced by certain chemicals (Williams et al., 2012).
- **"Jump (jumping)":** A zebrafish activity that involves jumping out of the water/tank as part of an escape or panic reaction, frequently prompted by anxiogenic stimuli (Gerlai et al., 2006).
- **"Kin preference":** In response to chemical and visual stimuli, juvenile zebrafish show a markedly increased level of kin preference compared to larvae (Spence et al., 2008).
- **"Kin recognition":** The ability to recognize and seek out relatives (among otherwise unrelated zebrafish) using chemical and optical cues, including approach and attraction, leading to increased proximity to relatives (Spence et al., 2008).
- **"Lead or leading":** Females during courtship return to the same area of the tank at least three times (Darrow & Harris, 2004).
- **"Lethargy":** In adult zebrafish, a behavioral state characterized by decreased locomotor activity, diminished escape reaction, unusual body color, and sticking near the bottom, with fins frequently kept close to the body (Westerfield, 2007). Because of the chronic and independent nature of the social setting, it differs from social submissive behavior (Kalueff et al., 2013).
- **"Loop (looping)":** Swimming in a circular pattern in larvae around a virtual point outside the larva's body; begins as early as five days after fertilization and is common in mutants with visual feedback issues (Huang et al., 2010).
- **Loss of the body's natural posture:** This is common as a result of sedation or motor incoordination caused by neurotoxicity (Gerlai et al., 2006).
- **"Magnetic behavior":** Zebrafish behavioral sensitivity to magnetic fields (Takebe et al., 2012).
- **"Meandering":** Movement with no discernible direction or path can be heightened during times

of extreme anxiety, particularly when moving erratically (Cachat et al., 2011).

- **"Mirror stimulation response":** Exposure to a mirror causes complex actions in fish, which may be associated with aggression; frequently includes approaching, head-butting, biting (the mirror), and chasing (Gerlai et al., 2000).
- **"Motor incoordination":** A widespread loss of coordination of regular physical movements (such swimming backwards or in a corkscrew pattern), typically brought on by neurological diseases like neurotoxicity (Kalueff et al., 2013).
- **"Mouth opening":** Frequent mouth opening could be a sign of aggression/attack, snapping, or a specific stereotypic behavior observed after receiving certain medications. The rate of mouth opening decreases dramatically during rest/sleep (Zhdanova, 2006).
- **"Neophobia":** Avoidance of new experiences or foods (Yu et al., 2006); measurable by evaluating the latency as well as the amount of time spent with the novel object and the frequency with which it came into touch.
- **"Nibbling":** The non-aggressive biting of an object is known as nibbling (usually in food-seeking or as part of object exploration). Measurable by looking at delay and time spent/frequency of touch with the novel object.
- **"Nocifensive (pain-related) conduct":** Pain response is frequently characterized by enhanced swimming, escape, and tail-beating responses produced by chemical, thermal, or electrical stimulation (Sanchez-Simon & Rodriguez, 2009; Gómez-Laplaza & Gerlai, 2010). Initially, pain responses in fish were viewed with skepticism, but a growing body of evidence suggests that pain and pain-related behavior occur in zebrafish.
- **"O-bend":** Twisting of the body of a larval zebrafish in order to change its swimming orientation (about 180 degrees). Unlike the C-bend response, this reaction is slower, independent of Mauthner cells, and is frequently triggered by brief periods of darkness (Burgess & Granato, 2007).
- **Odorant-induced behavioral activity:** Alterations in swimming velocity and distance traveled, avoidance, and attraction are anticipated to occur (Spence et al., 2008).
- **"Optokinetic response/reflex" (OKR):** Oculomotor responses to the motion of objects. It is anticipated that alterations will occur in swimming velocity and distance, avoidance, and attraction. Encompassing the visual field, it consists of two parts: a saccadic movement that resets the eyes when the object exits the visual area, as well as a pursuit movement that follows the object (Easter & Nicola, 1997).
- **"Optomotor response/reflex" (OMR):** When confronted with a stimulus that moves in a pattern, zebrafish will swim in lockstep with it (Maaswinkel & Li, 2003).
- **"Oscillations of locomotor activity":** When exposed to unfamiliar surroundings, zebrafish locomotion/swimming takes on a sinusoidal pattern, with alternating phases of high and low velocity (Stewart et al., 2012a).
- **Oviposition** is the female's egg release during reproduction (Spence et al., 2008).
- **"Parallel or paralleling":** The male swims next to the female during spawning, in close proximity but somewhat behind it, with the head nearly level with the female's operculum (Spence & Smith, 2005).
- **"Paralysis":** Complete immobility, including the inability to move the eyes, gills, or fins (like ataxia but with more severe/global motor dysfunction and sometimes an odd posture, like lying on one side, floating upside down, or standing straight up) (Gerlai et al., 2006). It is usually caused by specific neuro-paralyzing drugs (Buss & Drapeau, 2001) or inherited neurological defects (Ono et al., 2001).
- **"Photokinesis":** Both positive (light-seeking/dark aversion, scotophobia) and negative (dark avoidance, light-seeking) movement in reaction to light (light avoidance, dark preference, scototaxis). Zebrafish are sensitive to visible light even though they are not susceptible to ultraviolet (UV) radiation (Kalueff et al., 2013).
- **"Photomotor response (PMR)":** In embryonic zebrafish, a typical sequence of motor responses to light stimulation comprises motor excitation and vigorous shaking, followed by a refractory period where animals are inhibited from baseline movement and do not respond to additional light pulses (Kokel & Peterson, 2011).
- **"Piping":** In addition to being an indication of discomfort (such as hypoxia or toxicity), gulping air at the surface of the water can also occur while the swim bladder volume is being adjusted dependent on depth (while diving) (Robertson et al., 2008).
- **"Place preference"** is a fish's proclivity to develop a preferred area in which it spends the majority of its time (Kalue, 2017).
- **"Polarization"** is an adult zebrafish behavioral feature that measures the degree to which group members travel in the same direction; it is high in zebrafish schools and low in shoals (Kalue, 2017).
- **"Predator inspection":** A risky exploratory/bravery behavior in fish characterized by a stronger or weaker inclination to approach a predator, potentially to learn more about the predator's identification, particular position, and current motivational state. It is commonly

found in shoaling fish. They swim away from the school, toward the predator, and then back (Kalue, 2017).

- **"Predatory attack":** A fish's increased or decreased willingness to approach a predator during a risky exploratory or daring behavior, perhaps in an effort to learn more about the identity, location, and current motivational state of the predator. In shoaling fish, a greater or lesser drive to approach a predator is regularly observed. They move away from the shoal in the direction of the predator before returning to the school (Muto & Kawakami, 2013).
- **"Present or presenting":** Female courtship practices; examples include stopping and showing the side to a male or swimming up and down in front of a male (Darrow & Harris, 2004).
- **"Prey capture" (capturing):** A complex pattern of activity displayed by larval zebrafish that involves finding prey either visually or chemically, tracking it with a succession of regular turns, forward slow swims, and J-turns, and then catching and digesting it. The early bends are lower in amplitude and more noticeable in the caudal regions; later bends start more rostrally, have a higher amplitude, and are followed by an increase in tail-beat frequency (Muto & Kawakami, 2013).
- **"Quiver or quivering":** During spawning, when a male is orientated against the side of a female, he produces a tail oscillation with a high frequency but small amplitude (Darrow & Harris, 2004; Spence et al., 2008).
- **"Reflection chase" (chasing):** Some psychotropic medicines might also trigger chasing one's own mirror behavior (Gerlai et al., 2000).
- **Reflexive opening and closure of the mouth:** High concentrations of appetitive stimuli signify the start of the ingestive phase during feeding (Baier & Korsching, 1994).
- **"Respiration":** A gill movement that can be seen via slow-motion video capture, controlled in both directions by various psychiatric medicines, and significantly increased during distress and diminished during rest/sleep (Zhdanova, 2006).
- **Resting habit of adult and juvenile zebrafish:** Larval zebrafish float with their heads lowered or remain horizontal near the tank's bottom (Zhdanova, 2011).
- **"Retreat or retreating":** A type of fleeing behavior in zebrafish where a subordinate fish flees from an attacker by swimming quickly away from them (Oliveira et al., 2011).
- **"Rheotaxis"** is the turning toward a flowing and tendency to swim upstream behavior displayed by both larval and adult zebrafish. In experiments, zebrafish steer clear of the sucking source to show this (Olszewski et al., 2012).
- **"Risk-taking behavior":** The proclivity of zebrafish to cooperate in potentially dangerous situations, such as when a prey fish approaches/inspects a predator or when a shoal member leaves the group (Kalueff et al., 2013).
- **"R-turning"** is the turning toward a flowing and tendency to swim upstream behavior displayed by both larval and adult zebrafish. In experiments, zebrafish steer clear of the sucking source to show this (Budick & O'Malley, 2000).
- **"Rub or rubbing":** A typical aberrant zebrafish activity that involves rubbing its body sides against the tank walls and is frequently brought on by pathogenic conditions (Westerfield, 2007).
- **"School or schooling":** The coordinated/synchronous swimming of numerous fish is required to create a moderately polarized school, a feature of aggregation behavior that enhances feeding efficiency and predator detection and avoidance abilities. Shoals that form within schools tend to be less polarized (Miller & Gerlai, 2012).
- **"Scotophobia":** Widely documented in fish larvae as a natural predilection for bright (or avoidance of dark) lighting or habitat; in adult fish, it is often supplanted by "scototaxis" (Maximino et al., 2010).
- **"Scototaxis"** is the adult zebrafish's innate preference for darkness. It typically measures anxiousness. Note that zebrafish larvae exhibit opposite behavior (scotophobia). This is quantifiable in light-dark box tests (Maximino et al., 2010).
- **"Seizure" (like epilepsy) behavior:** Ataxia, screw-like (spiral) swimming, hyperactivity, circling, spasms, "wavering," head shake movements, tremors, and jittery locomotion are examples of involuntary, rapid body movements observed in both zebrafish larvae and adults. In extreme situations, death is possible. It can be measured manually or automatically using software for video tracking (Hortopan et al., 2010; Hortopan & Baraban, 2011).
- **"Sexual aggregating behavior":** As an instinctual response to hormonal cues given by females during ovulation, males exhibit interest, approach, and courtship behaviors after spotting the female (Darrow & Harris, 2004).
- **"Shoal or shoaling":** A component of aggregation behavior, a shoal is a relatively unpolarized group (of adult zebrafish) that is kept together by social factors rather than by individual attraction to an external stimulus. The shoal of fish "tightens" up in response to fear and anxiety, possibly forming a school. Hunger and habit cause the shoal to grow more and more disorganized and chaotic. Manual or automatic video-tracking systems can measure the average inter-fish distance, shoal area size, proximity, nearest and farthest neighbor distances,

time spent in the shoal, time spent away from the shoal, and the number of animals leaving the shoal (Miller & Gerlai, 2011, 2012).

- **"Shyness":** Less exploratory activity, decreased overall activities in a new setting and responsiveness to stimuli, or diminished risk-taking behavior (Oswald et al., 2012).
- **"Sickness behavior":** A significant collection of symptoms that point to disease, such as lethargy, hypoactivity, inhibited exploration, feeding, or looking for food (regularly holding its fins against its body) (Ewald, 2009).
- **"Sink or sinking":** The fish exhibits freezing behavior when it remains stationary yet shifts position in the water column without moving any of its fins (Sison & Gerlai, 2011).
- **"Sleep":** Reversible hypolocomotion and immobility, an elevated arousal threshold, decreased respiratory rate, and a rebound after sleep deprivation are all traits of zebrafish rest behavior. Adult zebrafish have brief periods of inactivity, usually accompanied by a drooping caudal fin, which are quickly reversed by unanticipated stimuli like tapping, hearing, or a modest electric field (Thorsen & Hale, 2007; Wang et al., 2014).
- **"Social interaction":** The normal social behavior of zebrafish is defined as the reciprocal change in behavior induced by the presence or activities of other conspecifics. Fighting or aggression, shoaling or schooling, courtship, and reproduction are a few examples. They can also appear as boldness or approach, social recognition, or social preference (Ogi et al., 2021; Liu et al., 2016).
- **"Social recognition":** Zebrafish's capacity to distinguish between known and unknown zebrafish (Saverino & Gerlai, 2008).
- **"Spasm (twitch, twitching)":** Sudden, quick body movement sudden (typically caused by neurological/neurotoxic damage, such as a seizure) (Williams et al., 2012).
- **"Spatiotemporal stability":** When exploring a new area, the capacity to withstand changes in environmental factors, especially by adjusting locomotor activity to the size of the environment while maintaining temporal budgeting, as determined by the temporal distribution of locomotion and location in a test tank (Stewart et al., 2012a).
- **"Spawning":** During mating, the male zebrafish approaches the female and leans forward to place his genital pore next to hers (also see "parallel"). After that, the male trembles to induce ovulation in the female and releases sperm at the same moment to fertilize the eggs. Exposure to shallow water can improve zebrafish's spawning behavior (Budick & O'Malley, 2000).
- **"S-start/bend"** is a quick escape/startle response in which the fish bends its body in an S-shape while contracting its rostral and caudal fins. Head stimulation causes C-starts in larval zebrafish, while tail stimulation causes C- and S-starts (Liu et al., 2012).
- **"Startle reaction":** A "C-start" activity, during which coiling and rushing may be seen, is an evolutionarily conserved, adaptive reaction to rapid, often unpleasant stimuli, such as vibration, light, sound, or contact.
- **"Stereotypic behaviors":** A series of inflexible, continuous motions other than swimming (Kalue, 2017).
- **"Stereotypic locomotion":** A pattern of rigid, repetitive behaviors evoked in zebrafish under certain circumstances (Kyzar et al., 2012).
- **"Strike or striking":** When a Zebrafish swims quickly toward its opponent without making physical contact, it exhibits aggressive behavior. It is frequently much faster and more violent than an approach (Oliveira et al., 2011).
- **"Struggle or struggling":** A characteristic of zebrafish larvae distinguished by slower, longer motor bursts than swimming. Instead of swimming, during a conflict, motor bursts go along the body in the opposite direction (Liao & Fetcho, 2008).
- **"Submissive conduct":** A social trait shown by zebrafish following confrontational encounters. Submissive fish remain motionless (fins retracted), often near the bottom or near the surface of the aquarium, with the caudal body directed downward (Oliveira et al., 2011).
- **"Surfacing (surface touching)":** Neuroactive drugs, particularly serotonergic agonists and glutamatergic antagonists, frequently cause fish surface residence, which may potentially be connected to buoyancy dysregulation (Kalue, 2017).
- **"Swimming":** The duration of simple zebrafish motility can be classified as "prolonged," which can continue for hours or last for a few minutes. It comprises explosive swimming and slow swimming in larvae. Swimming on the side, swimming vertically, swimming upside-down, and Cartesian diver behavior are some examples of aberrant swimming (Mrinalini et al., 2023). Simple shifts in swimming direction are seen in adult and juvenile zebrafish. Larval zebrafish exhibit a variety of diverse turning behaviors, such as quick turns and sluggish routine turns (Budick & O'Malley, 2000).
- **"Swimming on the side":** Ataxia causes loss of normal body position and is frequently brought on by sedation and motor incoordination brought on by neurotoxicity. Neuroactive/neurotoxic medicines

are frequently to blame for the condition known as swimming backward (Gerlai et al., 2006).

- **"Tail beating":** Characterized by waves of peripheral nerve discharges that proceed rostro-caudally and then reverse direction. When viewed from above, tail thumping looks to be a sweeping motion from side to side—using the head or nose to touch another fish's side or tail. They are rituals that males engage in during social interactions, most notably courtship (Darrow & Harris, 2004).
- **"Tail dipping":** An agonistic activity shown by a fish during a battle to signal surrender and cease the encounter (Kalue, 2017).
- **"Terrestrial jump(ing)":** A synchronized jump in response to being placed on a wet surface, whose height and rate of repetition can be measured (Gibb et al., 2011).
- **"Territoriality":** Having complete control over and fiercely protecting a small space within a habitat or tank. The dominating fish may chase away or bite at any interlopers that dare to "battle" against it (Spence et al., 2006).
- **"Thigmotaxis"** is a propensity for remaining along the edge/side; it is typically used to quantify zebrafish nervousness (Lamprea et al., 2008).
- **"Thrashing":** Swimming violently using the caudal fin while in physical contact with the side or bottom wall of the tank; swimming against the glass wall of the tank as if trying to swim through the glass barrier. By thrashing either towards the direction of an enticing cue or away from an unpleasant stimulus, this action can resemble head-butting (Sison & Gerlai, 2011).
- **"Tilt or tilting":** A common occurrence in inclined swimming is a deviation from the horizontal posture (Kalue, 2017).
- **"Top dwelling":** Swimming unnaturally close to the water's surface is one form of dwelling on the tank's upper half (surfacing).
- **"Touch response":** An embryo with a startle behavior-related phenotype that responds to contact by quickly coiling its trunk and bending its head (Kalue, 2017).
- **"Trance-like swim(ming)":** A fish's slow swimming motion brought on by a certain psychoactive substance (Braida et al., 2007).
- **"Tremor behavior":** When neurotoxic/convulsant drugs like domoic acid are administered to adult or larval zebrafish, specific shivering-like activity is frequently observed in the tail area (Kalue, 2017).
- **"Undulating body movement":** Particularly when two equal opponents are involved, a fight typically starts with a wave-like or snake-like action. A characteristic of aggressive behaviors is this motion. This tendency is common among fish species and is probably caused by the fact that fish use lateral lines to determine the size of an opponent based on the waves that it makes when it moves. Zebrafish's less pronounced undulating body movement than other fish can be seen in high-resolution footage (Gerlai, 2003).
- **"Vertical drifting":** The reverse of sinking, this aberrant phenotype is characterized by passive vertical floating and is frequently brought on by neuroactive/neurotoxic chemicals or agents connected to buoyancy dysregulation (Grossman et al., 2010).
- **"Vertical swimming":** A fish phenotype with vertical swimming that is aberrant, usually brought on by neuroactive or neurotoxic drugs or chemicals that cause buoyancy dysregulation (Kalue, 2017).
- **"Vestibulo-ocular reflex (VOR)":** In reaction to linear/angular acceleration and changes in head position with respect to gravity, zebrafish make eye movements that stabilize the retinal image. The semicircular canals in zebrafish's vestibular system are what cause these motions (Mo et al., 2010).
- **"Waver or wavering":** Characteristic that resembles abnormal head shaking and tremors, frequently brought on by some neurotoxic/convulsant medications.
- **"Withdrawal-related behavior":** A behavioral syndrome that zebrafish exhibit after stopping their drug use, which is frequently marked by increased anxiety-like behavior (Cachat et al., 2010a; Maximino et al., 2010).

3.7 CONCLUSION AND DIRECTIONS FOR THE FUTURE

In the study of behavioral neuroscience, the zebrafish has become a widely used model. As a neurobehavioral model, zebrafish has progressed without setbacks since its inception. It has opened up new avenues for progress in neuroresearch that could lead to useful discoveries.

Novel insights into zebrafish biology and behavior can be gained by correlating physiological and behavioral data with biomolecular indicators such as gene and protein expression. The use of zebrafish in cross-domain modeling is also growing quickly, which should help bring this model closer to perfection. In conclusion, despite its many benefits, zebrafish has yet to have its unique abilities completely investigated, calling for extensive further research into novel biomarkers, processes, and therapies for neuropsychiatric illnesses using this fascinating animal model.

REFERENCES

Abaid, N., Bartolini, T., Macrì, S., & Porfiri, M. (2012). Zebrafish responds differentially to a robotic fish of varying aspect ratio, tail beat frequency, noise, and color. *Behavioural Brain Research*, *233*(2), 545–553.

Abramson, H. A., & Evans, L. T. (1954). Lysergic acid diethylamide (LSD 25): II. Psychobiological effects on the Siamese fighting fish. *Science*, *120*(3128), 990–991.

Ahmad, F., Noldus, L. P., Tegelenbosch, R. A., & Richardson, M. K. (2012). Zebrafish embryos and larvae in behavioural assays. *Behaviour*, *149*(10–12), 1241–1281.

Ahmad, F., & Richardson, M. K. (2013). Exploratory behaviour in the open field test adapted for larval Zebrafish: Impact of environmental complexity. *Behavioural Processes*, *92*, 88–98.

Al-Imari, L., & Gerlai, R. (2008). Sight of conspecifics as reward in associative learning in Zebrafish (*Danio rerio*). *Behavioural Brain Research*, *189*(1), 216–219.

Araya, C., Ward, L. C., Girdler, G. C., & Miranda, M. (2016). Coordinating cell and tissue behavior during Zebrafish neural tube morphogenesis. *Developmental Dynamics*, *245*(3), 197–208.

Arthur, D., & Levin, E. D. (2001). Spatial and non-spatial visual discrimination learning in Zebrafish (*Danio rerio*). *Animal Cognition*, *4*, 125–131.

Aslanzadeh, M., Ariyasiri, K., Kim, O. H., Choi, T. I., Lim, J. H., Kim, H. G., . . . Kim, C. H. (2019). The body size of stimulus conspecifics affects social preference in a binary choice task in wild-type, but not in dyrk1aa mutant, Zebrafish. *Zebrafish*, *16*(3), 262–267.

Audira, G., Sampurna, B. P., Juniardi, S., Liang, S. T., Lai, Y. H., & Hsiao, C. D. (2018). A simple setup to perform 3D locomotion tracking in Zebrafish by using a single camera. *Inventions*, *3*(1), 11.

Avdesh, A., Martin-Iverson, M. T., Mondal, A., Chen, M., Askraba, S., Morgan, N., . . . Martins, R. N. (2012). Evaluation of color preference in Zebrafish for learning and memory. *Journal of Alzheimer's Disease*, *28*(2), 459–469.

Azeemi, S. T. Y., & Raza, M. (2005). A critical analysis of chromotherapy and its scientific evolution. *Evidence-Based Complementary and Alternative Medicine*, *2*, 481–488.

Baier, H., & Korsching, S. (1994). Olfactory glomeruli in the Zebrafish form an invariant pattern and are identifiable across animals. *Journal of Neuroscience*, *14*(1), 219–230.

Barba-Escobedo, P. A., & Gould, G. G. (2012). Visual social preferences of lone Zebrafish in a novel environment: Strain and anxiolytic effects. *Genes, Brain and Behavior*, *11*(3), 366–373.

Bartholomew, N. R., Burdett, J. M., VandenBrooks, J. M., Quinlan, M. C., & Call, G. B. (2015). Impaired climbing and flight behaviour in Drosophila melanogaster following carbon dioxide anaesthesia. *Scientific Reports*, *5*(1), 15298.

Bashirzade, A. A., Zabegalov, K. N., Volgin, A. D., Belova, A. S., Demin, K. A., de Abreu, M. S., . . . Kalueff, A. V. (2022). Modeling neurodegenerative disorders in Zebrafish. *Neuroscience & Biobehavioral Reviews*, *138*, 104679.

Basnet, R. M., Zizioli, D., Taweedet, S., Finazzi, D., & Memo, M. (2019). Zebrafish larvae as a behavioral model in neuropharmacology. *Biomedicines*, *7*(1), 23.

Basquill, S. P., & Grant, J. W. (1998). An increase in habitat complexity reduces aggression and monopolization of food by zebra fish (*Danio rerio*). *Canadian Journal of Zoology*, *76*(4), 770–772.

Bault, Z. A., Peterson, S. M., & Freeman, J. L. (2015). Directional and color preference in adult Zebrafish: Implications in behavioral and learning assays in neurotoxicology studies. *Journal of Applied Toxicology*, *35*(12), 1502–1510.

Bencan, Z., Sledge, D., & Levin, E. D. (2009). Buspirone, chlordiazepoxide and diazepam effects in a Zebrafish model of anxiety. *Pharmacology Biochemistry and Behavior*, *94*(1), 75–80.

Bergman, T. J., Ho, L., & Beehner, J. C. (2009). Chest color and social status in male geladas (Theropithecus gelada). *International Journal of Primatology*, *30*, 791–806.

Best, J. D., Berghmans, S., Hunt, J. J., Clarke, S. C., Fleming, A., Goldsmith, P., & Roach, A. G. (2008). Non-associative learning in larval Zebrafish. *Neuropsychopharmacology*, *33*(5), 1206–1215.

Bianco, I. H., Kampff, A. R., & Engert, F. (2011). Prey capture behavior evoked by simple visual stimuli in larval Zebrafish. *Frontiers in Systems Neuroscience*, *5*, 101.

Birren, F. (1973). Color preference as a clue to personality. *Art Psychotherapy*, *1*(1), 13–16.

Braida, D., Limonta, V., Pegorini, S., Zani, A., Guerini-Rocco, C., Gori, E., & Sala, M. (2007). Hallucinatory and rewarding effect of salvinorin A in Zebrafish: κ-opioid and CB 1-cannabinoid receptor involvement. *Psychopharmacology*, *190*, 441–448.

Braida, D., Ponzoni, L., Martucci, R., & Sala, M. (2014). A new model to study visual attention in Zebrafish. *Progress in Neuro-Psychopharmacology and Biological Psychiatry*, *55*, 80–86.

Braida, D., Ponzoni, L., Moretti, M., Viani, P., Pallavicini, M., Bolchi, C., . . . Sala, M. (2020). Behavioural and pharmacological profiles of Zebrafish administrated pyrrolidinyl benzodioxanes and prolinol aryl ethers with high affinity for heteromeric nicotinic acetylcholine receptors. *Psychopharmacology*, *237*, 2317–2326.

Budick, S. A., & O'Malley, D. M. (2000). Locomotor repertoire of the larval Zebrafish: Swimming, turning and prey capture. *Journal of Experimental Biology*, *203*(17), 2565–2579.

Burgess, H. A., & Granato, M. (2007). Modulation of locomotor activity in larval Zebrafish during light adaptation. *Journal of Experimental Biology*, *210*(14), 2526–2539.

Buss, R. R., & Drapeau, P. (2001). Synaptic drive to motoneurons during fictive swimming in the developing Zebrafish. *Journal of Neurophysiology*, *86*(1), 197–210.

Cachat, J., Canavello, P., Elegante, M., Bartels, B., Hart, P., Bergner, C., . . . Kalueff, A. V. (2010a). Modeling withdrawal syndrome in Zebrafish. *Behavioural Brain Research*, *208*(2), 371–376.

Cachat, J., Kyzar, E. J., Collins, C., Gaikwad, S., Green, J., Roth, A., . . . Kalueff, A. V. (2013). Unique and potent effects of acute ibogaine on Zebrafish: The developing utility of novel aquatic models for hallucinogenic drug research. *Behavioural Brain Research*, *236*, 258–269.

Cachat, J., Stewart, A., Grossman, L., Gaikwad, S., Kadri, F., Chung, K. M., . . . Kalueff, A. V. (2010b). Measuring behavioral and endocrine responses to novelty stress in adult Zebrafish. *Nature Protocols*, *5*(11), 1786–1799.

Cachat, J., Stewart, A., Utterback, E., Hart, P., Gaikwad, S., Wong, K., . . . Kalueff, A. V. (2011). Three-dimensional neurophenotyping of adult Zebrafish behavior. *PLoS One*, *6*(3), e17597.

Cadena, P. G., Cadena, M. R. S., Sarmah, S., & Marrs, J. A. (2020). Folic acid reduces the ethanol-induced morphological and behavioral defects in embryonic and larval Zebrafish (*Danio rerio*) as a model for fetal alcohol spectrum disorder (FASD). *Reproductive Toxicology*, *96*, 249–257.

Cao, Y., Yan, H., Yu, G., & Su, R. (2019). Flumazenil-insensitive benzodiazepine binding sites in GABAA receptors contribute to benzodiazepine-induced immobility in Zebrafish larvae. *Life Sciences*, *239*, 117033.

Capatina, L., Napoli, E. M., Ruberto, G., & Hritcu, L. (2021). Origanum vulgare ssp. hirtum (Lamiaceae) essential oil prevents behavioral and oxidative stress changes in the scopolamine Zebrafish model. *Molecules*, *26*(23), 7085.

Champagne, D. L., Hoefnagels, C. C., De Kloet, R. E., & Richardson, M. K. (2010). Translating rodent behavioral repertoire to Zebrafish (*Danio rerio*): Relevance for stress research. *Behavioural Brain Research*, *214*(2), 332–342.

Chanin, S., Fryar, C., Varga, D., Raymond, J., Kyzar, E., Enriquez, J., . . . Kalueff, A. V. (2012). Assessing startle responses and their habituation in adult Zebrafish. *Zebrafish Protocols for Neurobehavioral Research*, 287–300.

Chen, J., Fan, R., Wang, Y., Huang, T., Shang, N., He, K., . . . Zhang, Q. (2020). Progressive impairment of learning and memory in adult Zebrafish treated by Al_2O_3 nanoparticles when in embryos. *Chemosphere*, *254*, 126608.

Cognato, G. D. P., Bortolotto, J. W., Blazina, A. R., Christoff, R. R., Lara, D. R., Vianna, M. R., & Bonan, C. D. (2012). Y-Maze memory task in Zebrafish (*Danio rerio*): The role of glutamatergic and cholinergic systems on the acquisition and consolidation periods. *Neurobiology of Learning and Memory*, *98*(4), 321–328.

Colwill, R. M., & Creton, R. (2011). Imaging escape and avoidance behavior in zebrafish larvae. *Reviews in the Neurosciences*, *22*(1), 63–73.

Cuthill, I. C., Allen, W. L., Arbuckle, K., Caspers, B., Chaplin, G., Hauber, M. E., . . . Caro, T. (2017). The biology of color. *Science*, *357*(6350), eaan0221.

Dahlbom, S. J., Lagman, D., Lundstedt-Enkel, K., Sundström, L. F., & Winberg, S. (2011). Boldness predicts social status in Zebrafish (*Danio rerio*). *PLoS One*, *6*(8), e23565.

Dankert, H., Wang, L., Hoopfer, E. D., Anderson, D. J., & Perona, P. (2009). Automated monitoring and analysis of social behavior in Drosophila. *Nature Methods*, *6*(4), 297–303.

Darland, T., & Dowling, J. E. (2001). Behavioral screening for cocaine sensitivity in mutagenized Zebrafish. *Proceedings of the National Academy of Sciences*, *98*(20), 11691–11696.

Darrow, K. O., & Harris, W. A. (2004). Characterization and development of courtship in Zebrafish, *Danio rerio*. *Zebrafish*, *1*(1), 40–45.

de Abreu, M. S., Giacomini, A. C., Genario, R., Dos Santos, B. E., da Rosa, L. G., Demin, K. A., . . . Kalueff, A. V. (2019). Neuropharmacology, pharmacogenetics and pharmacogenomics of aggression: The Zebrafish model. *Pharmacological Research*, *141*, 602–608.

de Abreu, M. S., Giacomini, A. C., Genario, R., Dos Santos, B. E., Marcon, L., Demin, K. A., & Kalueff, A. V. (2020). The impact of housing environment color on Zebrafish anxiety-like behavioral and physiological (cortisol) responses. *General and Comparative Endocrinology*, *294*, 113499.

de Abreu, M. S., Giacomini, A. C., Genario, R., Dos Santos, B. E., Marcon, L., Demin, K. A., . . . Kalueff, A. V. (2021). Color as an important biological variable in Zebrafish models: Implications for translational neurobehavioral research. *Neuroscience & Biobehavioral Reviews*, *124*, 1–15.

Deacon, R. M., & Rawlins, J. N. P. (2006). T-maze alternation in the rodent. *Nature Protocols*, *1*(1), 7–12.

Devidas, S. B., Rahmatkar, S. N., Singh, R., Sendri, N., Purohit, R., Singh, D., & Bhandari, P. (2021). Amelioration of cognitive deficit in Zebrafish by an undescribed anthraquinone from Juglans regia L.: An in-silico, in-vitro and in-vivo approach. *European Journal of Pharmacology*, *906*, 174234.

Dijkstra, P. D., Hemelrijk, C., Seehausen, O., & Groothuis, T. G. (2009). Color polymorphism and intrasexual competition in assemblages of cichlid fish. *Behavioral Ecology*, *20*(1), 138–144.

Easter Jr, S. S., & Nicola, G. N. (1997). The development of eye movements in the Zebrafish (*Danio rerio*). *Developmental Psychobiology: The Journal of the International Society for Developmental Psychobiology*, *31*(4), 267–276.

Engeszer, R. E., Da Barbiano, L. A., Ryan, M. J., & Parichy, D. M. (2007). Timing and plasticity of shoaling behaviour in the Zebrafish, *Danio rerio*. *Animal Behaviour*, *74*(5), 1269–1275.

Ewald, H. S. (2009). *A Zebrafish model of schizophrenia and sickness behavior: MK-801 and endogenous NMDAR antagonism*. University of Louisville.

Franco-Restrepo, J. E., Forero, D. A., & Vargas, R. A. (2019). A review of freely available, open-source software for the automated analysis of the behavior of adult Zebrafish. *Zebrafish*, *16*(3), 223–232.

Gaikwad, S., Stewart, A., Hart, P., Wong, K., Piet, V., Cachat, J., & Kalueff, A. V. (2011). Acute stress disrupts performance of Zebrafish in the cued and spatial memory tests: The utility of fish models to study stress–memory interplay. *Behavioural Processes*, *87*(2), 224–230.

Ganzen, L., Venkatraman, P., Pang, C. P., Leung, Y. F., & Zhang, M. (2017). Utilizing Zebrafish visual behaviors in drug screening for retinal degeneration. *International Journal of Molecular Sciences*, *18*(6), 1185.

Genario, R., Giacomini, A. C., Demin, K. A., Dos Santos, B. E., Marchiori, N. I., Volgin, A. D., . . . Kalueff, A. V. (2019). The evolutionarily conserved role of melatonin in CNS disorders and behavioral regulation: Translational lessons from Zebrafish. *Neuroscience & Biobehavioral Reviews*, *99*, 117–127.

Gerlai, R. (2003). Zebra fish: An uncharted behavior genetic model. *Behavior Genetics*, *33*, 461–468.

Gerlai, R. (2017). Animated images in the analysis of Zebrafish behavior. *Current Zoology*, *63*(1), 35–44.

Gerlai, R., Lahav, M., Guo, S., & Rosenthal, A. (2000). Drinks like a fish: Zebra fish (*Danio rerio*) as a behavior genetic model to study alcohol effects. *Pharmacology Biochemistry and Behavior*, *67*(4), 773–782.

Gerlai, R., Lee, V., & Blaser, R. (2006). Effects of acute and chronic ethanol exposure on the behavior of adult Zebrafish (*Danio rerio*). *Pharmacology Biochemistry and Behavior*, *85*(4), 752–761.

Gibb, A. C., Ashley-Ross, M. A., Pace, C. M., & Long Jr, J. H. (2011). Fish out of water: Terrestrial jumping by fully aquatic fishes. *Journal of Experimental Zoology Part A: Ecological Genetics and Physiology*, *315*(10), 649–653.

Gómez-Laplaza, L. M., & Gerlai, R. (2010). Latent learning in Zebrafish (*Danio rerio*). *Behavioural Brain Research*, *208*(2), 509–515.

Gould, G. G. (2011). Modified associative learning T-maze test for Zebrafish (*Danio rerio*) and other small teleost fish. *Zebrafish Neurobehavioral Protocols*, 61–73.

Gouras, P., & Ekesten, B. (2004). Why do mice have ultra-violet vision? *Experimental Eye Research*, *79*(6), 887–892.

Grossman, L., Utterback, E., Stewart, A., Gaikwad, S., Chung, K. M., Suciu, C., . . . Kalueff, A. V. (2010). Characterization of behavioral and endocrine effects of LSD on Zebrafish. *Behavioural Brain Research*, *214*(2), 277–284.

Guilbeault, N. C., Guerguiev, J., Martin, M., Tate, I., & Thiele, T. R. (2021). BonZeb: Open-source, modular software tools for high-resolution Zebrafish tracking and analysis. *Scientific Reports*, *11*(1), 8148.

Horev, G., Benjamini, Y., Sakov, A., & Golani, I. (2007). Estimating wall guidance and attraction in mouse free locomotor behavior. *Genes, Brain and Behavior*, *6*(1), 30–41.

Hortopan, G. A., & Baraban, S. C. (2011). Aberrant expression of genes necessary for neuronal development and notch signaling in an epileptic mind bomb Zebrafish. *Developmental Dynamics*, *240*(8), 1964–1976.

Hortopan, G. A., Dinday, M. T., & Baraban, S. C. (2010). Spontaneous seizures and altered gene expression in GABA signaling pathways in a mind bomb mutant Zebrafish. *Journal of Neuroscience*, *30*(41), 13718–13728.

Howard, R. D., Rohrer, K., Liu, Y., & Muir, W. M. (2015). Mate competition and evolutionary outcomes in genetically modified Zebrafish (*Danio rerio*). *Evolution*, *69*(5), 1143–1157.

Howe, H. B., McIntyre, P. B., & Wolman, M. A. (2018). Adult Zebrafish primarily use vision to guide Piscivorous foraging behavior. *Behavioural Processes*, *157*, 230–237.

Huang, Y. Y., Tschopp, M., Straumann, D., & Neuhauss, S. C. (2010). Vestibular deficits do not underlie looping behavior in achiasmatic fish. *Communicative & Integrative Biology*, *3*(4), 379–381.

Jesuthasan, S. (2012). Fear, anxiety, and control in the Zebrafish. *Developmental Neurobiology*, *72*(3), 395–403.

Jones, L. J., & Norton, W. H. (2015). Using Zebrafish to uncover the genetic and neural basis of aggression, a frequent comorbid symptom of psychiatric disorders. *Behavioural Brain Research*, *276*, 171–180.

Kageyama, C. J. (1999). *What fish see: Understanding optics and color shifts for designing lures and flies*. F. Amato. https://books.google.com.eg/books?id=O3MBAAAACAAJ

Kalue, A. V. (2017). *The rights and wrongs of Zebrafish: Behavioral phenotyping of Zebrafish*. Springer. https://doi.org/101007978-3

Kalueff, A. V., Gebhardt, M., Stewart, A. M., Cachat, J. M., Brimmer, M., Chawla, J. S., . . . The Zebrafish Neuroscience Research Consortium, H. (2013). Towards a comprehensive catalog of Zebrafish behavior 1.0 and beyond. *Zebrafish*, *10*(1), 70–86.

Kalueff, A. V., Stewart, A. M., & Gerlai, R. (2014). Zebrafish as an emerging model for studying complex brain disorders. *Trends in Pharmacological Sciences*, *35*(2), 63–75.

Katz, R. J., Roth, K. A., & Carroll, B. J. (1981). Acute and chronic stress effects on open field activity in the rat: Implications for a model of depression. *Neuroscience & Biobehavioral Reviews*, *5*(2), 247–251.

Kim, Y. H., Lee, Y., Kim, D., Jung, M. W., & Lee, C. J. (2010). Scopolamine-induced learning impairment reversed by physostigmine in Zebrafish. *Neuroscience Research*, *67*(2), 156–161.

Kirk, J. T. O. (1979). Spectral distribution of photosynthetically active radiation in some south-eastern Australian waters. *Marine and Freshwater Research*, *30*(1), 81–91.

Kokel, D., & Peterson, R. T. (2011). Using the Zebrafish photomotor response for psychotropic drug screening. In *Methods in cell biology* (Vol. 105, pp. 517–524). Academic Press.

Krook, J. T., Duperreault, E., Newton, D., Ross, M. S., & Hamilton, T. J. (2019). Repeated ethanol exposure increases anxiety-like behaviour in Zebrafish during withdrawal. *PeerJ*, *7*, e6551.

Kumar, V., Singh, C., & Singh, A. (2021). Zebrafish an experimental model of Huntington's disease: Molecular aspects, therapeutic targets and current challenges. *Molecular Biology Reports*, 1–14.

Kurt, M., Arik, A. C., & Celik, S. (2000). The effects of sertraline and fluoxetine on anxiety in the elevated plus-maze test. *Journal of Basic and Clinical Physiology and Pharmacology*, *11*(2), 173–180.

Kysil, E. V., Meshalkina, D. A., Frick, E. E., Echevarria, D. J., Rosemberg, D. B., Maximino, C., . . . Kalueff, A. V. (2017). Comparative analyses of Zebrafish anxiety-like behavior using conflict-based novelty tests. *Zebrafish*, *14*(3), 197–208.

Kyzar, E. J., Collins, C., Gaikwad, S., Green, J., Roth, A., Monnig, L., . . . Kalueff, A. V. (2012). Effects of hallucinogenic agents mescaline and phencyclidine on Zebrafish behavior and physiology. *Progress in Neuro-Psychopharmacology and Biological Psychiatry*, *37*(1), 194–202.

Lamprea, M. R., Cardenas, F. P., Setem, J., & Morato, S. (2008). Thigmotactic responses in an open-field. *Brazilian Journal of Medical and Biological Research*, *41*(2), 135–140.

Landin, J., Hovey, D., Xu, B., Lagman, D., Zettergren, A., Larhammar, D., . . . Westberg, L. (2020). Oxytocin receptors regulate social preference in Zebrafish. *Scientific Reports*, *10*(1), 5435.

Liao, J. C., & Fetcho, J. R. (2008). Shared versus specialized glycinergic spinal interneurons in axial motor circuits of larval Zebrafish. *Journal of Neuroscience*, *28*(48), 12982–12992.

Liu, X., Zhang, Y., Lin, J., Xia, Q., Guo, N., & Li, Q. (2016). Social preference deficits in juvenile Zebrafish induced by early chronic exposure to sodium valproate. *Frontiers in Behavioral Neuroscience*, *10*, 201.

Liu, Y., Carmer, R., Zhang, G., Venkatraman, P., Brown, S. A., Pang, C. P., . . . Leung, Y. F. (2015). Statistical analysis of Zebrafish locomotor response. *PLoS One*, *10*(10), e0139521.

Liu, Y. C., Bailey, I., & Hale, M. E. (2012). Alternative startle motor patterns and behaviors in the larval Zebrafish (*Danio rerio*). *Journal of Comparative Physiology A*, *198*, 11–24.

Luyten, L., Schroyens, N., Hermans, D., & Beckers, T. (2014). Parameter optimization for automated behavior assessment: Plug-and-play or trial-and-error? *Frontiers in Behavioral Neuroscience*, *8*, 28.

Lynn, D. A., & Brown, G. R. (2009). The ontogeny of exploratory behavior in male and female adolescent rats (Rattus norvegicus). *Developmental Psychobiology*, *51*(6), 513–520.

Maaswinkel, H., & Li, L. (2003). Spatio-temporal frequency characteristics of the optomotor response in Zebrafish. *Vision Research*, *43*(1), 21–30.

Maeda, H., Fukushima, N., & Hasumi, A. (2021). Standardized method for the assessment of behavioral responses of Zebrafish larvae. *Biomedicines*, *9*(8), 884.

Marks, C., Kaut, K. P., Moore, F. G., & Bagatto, B. (2012). Ontogenetic oxygen changes alter zebra fish size, behavior, and blood glucose. *Physiological and Biochemical Zoology*, *85*(6), 635–644.

Mathur, P., Lau, B., & Guo, S. (2011). Conditioned place preference behavior in Zebrafish. *Nature Protocols*, *6*(3), 338–345.

Maximino, C., Marques de Brito, T., Dias, C. A. G. D. M., Gouveia Jr, A., & Morato, S. (2010). Scototaxis as anxiety-like behavior in fish. *Nature Protocols*, *5*(2), 209–216.

Mena, A., Ruiz-Salas, J. C., Puentes, A., Dorado, I., Ruiz-Veguilla, M., & De la Casa, L. G. (2016). Reduced prepulse inhibition as a biomarker of schizophrenia. *Frontiers in Behavioral Neuroscience*, *10*, 202.

Miller, N. Y., & Gerlai, R. (2011). Shoaling in Zebrafish: What we don't know. *Reviews in the Neurosciences*, *22*(1), 17–25.

Miller, N. Y., & Gerlai, R. (2012). From Schooling to shoaling: Patterns of collective motion in Zebrafish (*Danio rerio*). *PLoS One*, *7*(11), e48865.

Mo, W., Chen, F., Nechiporuk, A., & Nicolson, T. (2010). Quantification of vestibular-induced eye movements in Zebrafish larvae. *BMC Neuroscience*, *11*(1), 1–11.

Morris, R. (1984). Developments of a water-maze procedure for studying spatial learning in the rat. *Journal of Neuroscience Methods*, *11*(1), 47–60.

Mrinalini, R., Tamilanban, T., Kumar, V. N., & Manasa, K. (2023). Zebrafish–The neurobehavioural model in trend. *Neuroscience*, *520*, 95–118.

Muto, A., & Kawakami, K. (2013). Prey capture in Zebrafish larvae serves as a model to study cognitive functions. *Frontiers in Neural Circuits*, *7*, 110.

Neff, E. P. (2020). Building avatar armies with fish and flies. *Lab Animal*, *49*(4), 101–104.

Norton, W. H., Manceau, L., & Reichmann, F. (2019). The visually mediated social preference test: A novel technique to measure social behavior and behavioral disturbances in Zebrafish. *Psychiatric Disorders: Methods and Protocols*, 121–132.

Ogi, A., Licitra, R., Naef, V., Marchese, M., Fronte, B., Gazzano, A., & Santorelli, F. M. (2021). Social preference tests in Zebrafish: A systematic review. *Frontiers in Veterinary Science*, *7*, 590057.

Oliveira, J., Silveira, M., Chacon, D., & Luchiari, A. (2015). The Zebrafish world of colors and shapes: Preference and discrimination. *Zebrafish*, *12*(2), 166–173.

Oliveira, R. F., Silva, J. F., & Simoes, J. M. (2011). Fighting Zebrafish: Characterization of aggressive behavior and winner–loser effects. *Zebrafish*, *8*(2), 73–81.

Olszewski, J., Haehnel, M., Taguchi, M., & Liao, J. C. (2012). Zebrafish larvae exhibit rheotaxis and can escape a continuous suction source using their lateral line. *PLoS One*, *7*(5), e36661.

Ono, F., Higashijima, S. I., Shcherbatko, A., Fetcho, J. R., & Brehm, P. (2001). Paralytic Zebrafish lacking acetylcholine receptors fail to localize rapsyn clusters to the synapse. *Journal of Neuroscience*, *21*(15), 5439–5448.

Oswald, M. E., Drew, R. E., Racine, M., Murdoch, G. K., & Robison, B. D. (2012). Is behavioral variation along the bold-shy continuum associated with variation in the stress axis in Zebrafish? *Physiological and Biochemical Zoology*, *85*(6), 718–728.

Palmer, S. E., & Schloss, K. B. (2010). An ecological valence theory of human color preference. *Proceedings of the National Academy of Sciences*, *107*(19), 8877–8882.

Park, J. S., Ryu, J. H., Choi, T. I., Bae, Y. K., Lee, S., Kang, H. J., & Kim, C. H. (2016). Innate color preference of Zebrafish and its use in behavioral analyses. *Molecules and Cells*, *39*(10), 750.

Parker, M. O., Brock, A. J., Walton, R. T., & Brennan, C. H. (2013). The role of Zebrafish (*Danio rerio*) in dissecting the genetics and neural circuits of executive function. *Frontiers in Neural Circuits*, *7*, 63.

Paull, G. C., Filby, A. L., Giddins, H. G., Coe, T. S., Hamilton, P. B., & Tyler, C. R. (2010). Dominance hierarchies in Zebrafish (*Danio rerio*) and their relationship with reproductive success. *Zebrafish*, *7*(1), 109–117.

Peng, M., Zhang, C., Dong, Y., Zhang, Y., Nakazawa, H., Kaneki, M., . . . Xie, Z. (2016). Battery of behavioral tests in mice to study postoperative delirium. *Scientific Reports*, *6*(1), 29874.

Pham, M., Raymond, J., Hester, J., Kyzar, E., Gaikwad, S., Bruce, I., . . . Kalueff, A. V. (2012). Assessing social behavior phenotypes in adult Zebrafish: Shoaling, social preference, and mirror biting tests. *Zebrafish Protocols for Neurobehavioral Research*, 231–246.

Piato, Â. L., Capiotti, K. M., Tamborski, A. R., Oses, J. P., Barcellos, L. J., Bogo, M. R., . . . Bonan, C. D. (2011). Unpredictable chronic stress model in Zebrafish (*Danio rerio*): Behavioral and physiological responses. *Progress in Neuro-Psychopharmacology and Biological Psychiatry*, *35*(2), 561–567.

Pieróg, M., Guz, L., Doboszewska, U., Poleszak, E., & Wlaź, P. (2018). Effects of alprazolam treatment on anxiety-like behavior induced by color stimulation in adult Zebrafish. *Progress in Neuro-Psychopharmacology and Biological Psychiatry*, *82*, 297–306.

Prut, L., & Belzung, C. (2003). The open field as a paradigm to measure the effects of drugs on anxiety-like behaviors: A review. *European Journal of Pharmacology*, *463*(1–3), 3–33.

Rajesh, V., Mridhulmohan, M., Jayaseelan, S., Sivakumar, P., & Ganesan, V. (2018). Mefenamic acid attenuates chronic alcohol induced cognitive impairment in Zebrafish: Possible role of cholinergic pathway. *Neurochemical Research*, *43*(7), 1392–1404.

Razali, K., Othman, N., Mohd Nasir, M. H., Doolaanea, A. A., Kumar, J., Ibrahim, W. N., . . . Mohamed, W. M. (2021). The promise of the Zebrafish model for Parkinson's disease: Today's science and tomorrow's treatment. *Frontiers in Genetics*, *12*, 655550.

Rhee, S. H., & Waldman, I. D. (2002). Genetic and environmental influences on antisocial behavior: A meta-analysis of twin and adoption studies. *Psychological Bulletin*, *128*(3), 490.

Riehl, R., Kyzar, E., Allain, A., Green, J., Hook, M., Monnig, L., . . . Kalueff, A. V. (2011). Behavioral and physiological effects of acute ketamine exposure in adult Zebrafish. *Neurotoxicology and Teratology*, *33*(6), 658–667.

Robertson, G. N., Lindsey, B. W., Dumbarton, T. C., Croll, R. P., & Smith, F. M. (2008). The contribution of the swimbladder to buoyancy in the adult Zebrafish (*Danio rerio*): A morphometric analysis. *Journal of Morphology*, *269*(6), 666–673.

Roy, T., Suriyampola, P. S., Flores, J., López, M., Hickey, C., Bhat, A., & Martins, E. P. (2019). Color preferences affect learning in Zebrafish, *Danio rerio*. *Scientific Reports*, *9*(1), 14531.

Sackerman, J., Donegan, J. J., Cunningham, C. S., Nguyen, N. N., Lawless, K., Long, A., . . . Gould, G. G. (2010). Zebrafish behavior in novel environments: Effects of acute exposure to anxiolytic compounds and choice of *Danio rerio* line. *International Journal of Comparative Psychology/ISCP; Sponsored by the International Society for Comparative Psychology and the University of Calabria*, *23*(1), 43.

Saint-Amant, L., & Drapeau, P. (1998). Time course of the development of motor behaviors in the Zebrafish embryo. *Journal of Neurobiology*, *37*(4), 622–632.

Sanchez-Simon, F. M., & Rodriguez, R. E. (2009). Expression of the nociceptin receptor during Zebrafish development: Influence of morphine and nociceptin. *International Journal of Developmental Neuroscience*, *27*(4), 315–320.

Sarasamma, S., Audira, G., Juniardi, S., Sampurna, B. P., Liang, S. T., Hao, E., . . . Hsiao, C. D. (2018). Zinc chloride exposure inhibits brain acetylcholine levels, produces neurotoxic signatures, and diminishes memory and motor activities in adult Zebrafish. *International Journal of Molecular Sciences*, *19*(10), 3195.

Saverino, C., & Gerlai, R. (2008). The social Zebrafish: Behavioral responses to conspecific, heterospecific, and computer animated fish. *Behavioural Brain Research*, *191*(1), 77–87.

Shaw, M. P., Higginbottom, A., McGown, A., Castelli, L. M., James, E., Hautbergue, G. M., . . . Ramesh, T. M. (2018). Stable transgenic C9orf72 Zebrafish model key aspects of the ALS/FTD phenotype and reveal novel pathological features. *Acta Neuropathologica Communications*, *6*(1), 1–16.

Singh, A. P., & Nüsslein-Volhard, C. (2015). Zebrafish stripes as a model for vertebrate colour pattern formation. *Current Biology*, *25*(2), R81–R92.

Sison, M., & Gerlai, R. (2010). Associative learning in Zebrafish (*Danio rerio*) in the plus maze. *Behavioural Brain Research*, *207*(1), 99–104.

Sison, M., & Gerlai, R. (2011). Behavioral performance altering effects of MK-801 in Zebrafish (*Danio rerio*). *Behavioural Brain Research*, *220*(2), 331–337.

Sousa, N., Almeida, O. F. X., & Wotjak, C. T. (2006). A hitchhiker's guide to behavioral analysis in laboratory rodents. *Genes, Brain and Behavior*, *5*, 5–24.

Spence, R., Gerlach, G., Lawrence, C., & Smith, C. (2008). The behaviour and ecology of the Zebrafish, *Danio rerio*. *Biological Reviews*, *83*(1), 13–34.

Spence, R., Jordan, W. C., & Smith, C. (2006). Genetic analysis of male reproductive success in relation to density in the Zebrafish, *Danio rerio*. *Frontiers in Zoology*, *3*, 1–6.

Spence, R., & Smith, C. (2005). Male territoriality mediates density and sex ratio effects on oviposition in the Zebrafish, *Danio rerio*. *Animal Behaviour*, *69*(6), 1317–1323.

Stephan, M., Volkmann, P., & Rossner, M. J. (2019). Assessing behavior and cognition in rodents, nonhuman primates, and humans: Where are the limits of translation? *Dialogues in Clinical Neuroscience*, *21*(3), 249–259.

Stewart, A. M., Desmond, D., Kyzar, E., Gaikwad, S., Roth, A., Riehl, R., . . . Kalueff, A. V. (2012a). Perspectives of Zebrafish models of epilepsy: What, how and where next? *Brain Research Bulletin*, *87*(2–3), 135–143.

Stewart, A. M., Gaikwad, S., Kyzar, E., & Kalueff, A. V. (2012b). Understanding spatio-temporal strategies of adult Zebrafish exploration in the open field test. *Brain Research*, *1451*, 44–52.

Stewart, A., Cachat, J., Wong, K., Gaikwad, S., Gilder, T., DiLeo, J., . . . Kalueff, A. V. (2010). Homebase behavior of Zebrafish in novelty-based paradigms. *Behavioural Processes*, *85*(2), 198–203.

Stewart, A., Maximino, C., Marques de Brito, T., Herculano, A. M., Gouveia, A., Morato, S., . . . Kalueff, A. V. (2011). Neurophenotyping of adult Zebrafish using the light/dark box paradigm. *Zebrafish Neurobehavioral Protocols*, 157–167.

Stowers, J. R., Hofbauer, M., Bastien, R., Griessner, J., Higgins, P., Farooqui, S., . . . Straw, A. D. (2017). Virtual reality for freely moving animals. *Nature Methods*, *14*(10), 995–1002.

Sundin, J., Morgan, R., Finnøen, M. H., Dey, A., Sarkar, K., & Jutfelt, F. (2019). On the observation of wild Zebrafish (*Danio rerio*) in India. *Zebrafish*, *16*(6), 546–553.

Swain, H. A., Sigstad, C., & Scalzo, F. M. (2004). Effects of dizocilpine (MK-801) on circling behavior, swimming activity, and place preference in Zebrafish (*Danio rerio*). *Neurotoxicology and Teratology*, *26*(6), 725–729.

Taborsky, M., Hofmann, H. A., Beery, A. K., Blumstein, D. T., Hayes, L. D., Lacey, E. A., . . . Rubenstein, D. R. (2015). Taxon matters: Promoting integrative studies of social behavior: NESCent working group on integrative models of vertebrate sociality: Evolution, mechanisms, and emergent properties. *Trends in Neurosciences*, *38*(4), 189–191.

Takebe, A., Furutani, T., Wada, T., Koinuma, M., Kubo, Y., Okano, K., & Okano, T. (2012). Zebrafish respond to the geomagnetic field by bimodal and group-dependent orientation. *Scientific Reports*, *2*(1), 727.

Tan, J. K., Nazar, F. H., Makpol, S., & Teoh, S. L. (2022). Zebrafish: A pharmacological model for learning and memory research. *Molecules*, *27*(21), 7374.

Tang, W., Davidson, J. D., Zhang, G., Conen, K. E., Fang, J., Serluca, F., . . . Fishman, M. C. (2020). Genetic control of collective behavior in Zebrafish. *IScience*, *23*(3).

Thorsen, D. H., & Hale, M. E. (2007). Neural development of the Zebrafish (*Danio rerio*) pectoral fin. *Journal of Comparative Neurology*, *504*(2), 168–184.

Uusi-Heikkilä, S., Böckenhoff, L., Wolter, C., & Arlinghaus, R. (2012). Differential allocation by female Zebrafish (Danio rerio) to different-sized males–an example in a fish species lacking parental care. *PLoS One*, *7*(10), e48317.

Volgin, A. D., Yakovlev, O. A., Demin, K. A., de Abreu, M. S., Alekseeva, P. A., Friend, A. J., . . . Kalueff, A. V. (2019). Zebrafish models for personalized psychiatry: Insights from individual, strain and sex differences, and modeling gene x environment interactions. *Journal of Neuroscience Research*, *97*(4), 402–413.

Wang, Y. N., Hou, Y. Y., Sun, M. Z., Zhang, C. Y., Bai, G., Zhao, X., & Feng, X. Z. (2014). Behavioural screening of Zebrafish using neuroactive traditional Chinese medicine prescriptions and biological targets. *Scientific Reports*, *4*(1), 5311.

Weber, D. N., & Ghorai, J. K. (2013). Experimental design affects social behavior outcomes in adult Zebrafish developmentally exposed to lead. *Zebrafish*, *10*(3), 294–302.

Westerfield, M. (2007). *The Zebrafish book: A guide for the laboratory use of Zebrafish (Danio rerio)* (5th ed). University of Oregon.

Wiercioch-Kuzianik, K., & Bąbel, P. (2019). Color hurts. The effect of color on pain perception. *Pain Medicine*, *20*(10), 1955–1962.

Williams, F. E., White, D., & Messer Jr, W. S. (2002). A simple spatial alternation task for assessing memory function in Zebrafish. *Behavioural Processes*, *58*(3), 125–132.

Williams, L. R., Wong, K., Stewart, A., Suciu, C., Gaikwad, S., Wu, N., . . . Kalueff, A. V. (2012). Behavioral and physiological effects of RDX on adult Zebrafish. *Comparative Biochemistry and Physiology Part C: Toxicology & Pharmacology*, *155*(1), 33–38.

Wong, K., Elegante, M., Bartels, B., Elkhayat, S., Tien, D., Roy, S., . . . Kalueff, A. V. (2010). Analyzing habituation responses to novelty in Zebrafish (*Danio rerio*). *Behavioural Brain Research*, *208*(2), 450–457.

Yokogawa, T., Marin, W., Faraco, J., Pézeron, G., Appelbaum, L., Zhang, J., . . . Mignot, E. (2007). Characterization of sleep in Zebrafish and insomnia in hypocretin receptor mutants. *PLoS Biology*, *5*(10), e277.

Yu, L., Tucci, V., Kishi, S., & Zhdanova, I. V. (2006). Cognitive aging in Zebrafish. *PLoS One*, *1*(1), e14.

Zabegalov, K. N., Kolesnikova, T. O., Khatsko, S. L., Volgin, A. D., Yakovlev, O. A., Amstislavskaya, T. G., . . . Kalueff, A. V. (2019). Understanding Zebrafish aggressive behavior. *Behavioural Processes*, *158*, 200–210.

Zhang, S., Liu, X., Sun, M., Zhang, Q., Li, T., Li, X., . . . Feng, X. (2018). Reversal of reserpine-induced depression and cognitive disorder in Zebrafish by sertraline and Traditional Chinese Medicine (TCM). *Behavioral and Brain Functions*, *14*, 1–14.

Zhdanova, I. V. (2006). Sleep in Zebrafish. *Zebrafish*, *3*(2), 215–226.

Zhdanova, I. V. (2011). Sleep and its regulation in Zebrafish. *Reviews in the Neurosciences*, *22*(1), 27–36.

4 Cellular and Genetic Methods in Zebrafish

Dalia Hamza, Elshaimaa Ismael, and Hala Zaher

4.1 INTRODUCTION

The zebrafish (*Danio rerio*) has emerged as a useful animal model for biomedical research, drug testing, drug toxicity, and safety evaluation, with its use growing dramatically due to many appealing advantages. Importantly, the remarkable genetic conservation between zebrafish and humans, with approximately 70% of their protein-coding genes having human orthologs, enables researchers to study human genes and genetic mutations in a simplified organism, and over 80% of human disease-related genes can be detected in the zebrafish genome [1]. Furthermore, the transparency of zebrafish embryos allows for real-time observation of internal processes, rendering it a powerful tool for developmental biology and regenerative medicine [2].

The reproductive rate of zebrafish is prolific, with females capable of producing hundreds of eggs weekly, streamlines large-scale experiments, genetic screens, and the generation of numerous experimental replicates [1]. Furthermore, their optical transparency, which persists throughout life stages, including adulthood, enables non-invasive imaging techniques such as confocal and two-photon microscopy, offering insights into dynamic cellular processes and disease models in vivo [2]. Zebrafish's amenability to various genetic manipulation techniques, including transgenesis and CRISPR-Cas9 genome editing, facilitates the creation of specific gene mutations, transgenic lines, and knockdown models for targeted studies [3].

Disease modeling has flourished in zebrafish, with models established for a wide spectrum of human diseases, encompassing cancer, cardiovascular disorders, neurodegenerative diseases, and infectious diseases [4]. Cancer became one of the most thoroughly investigated illnesses using zebrafish as a model organism [5]. With advances in molecular biology, particularly at the biomolecular and gene levels, the process of creating zebrafish models has steadily shifted from chemical carcinogen-induction to genetic engineering. This type of model is often generated by combining several forward and reverse genetic procedures in the entire zebrafish organism or particular organs, tissues, and glands of zebrafish with novel DNA recombination, mutation, and genetic operation technology [6].

The most frequent genetically modified zebrafish model is the transgenic zebrafish. This kind of zebrafish embryo is frequently utilized to assess the anti-angiogenic effect of natural compounds. Angiogenesis is a feature of over 50 distinct disease states, and its malfunction is linked to a variety of illnesses, including cancer [7]. Angiogenesis is an essential step in tumor development, invasion, and metastasis. Only in the capillary terminal can oxygen diffuse freely, which is required for tumor development. As a result, tumors require the formation of new blood vessels in order to recruit oxygen and nutrients and spread to distant areas, thus preventing aberrant angiogenesis that can impede tumor growth [8]. Therefore, one emphasis of cancer therapy is the use of natural compounds to block angiogenesis.

These models not only elucidate disease mechanisms but also serve as platforms for drug screening and therapeutic development. Zebrafish's increasing role in drug discovery is attributed to their adaptability to high-throughput screening, with large-scale chemical screens in zebrafish embryos leading to the identification of potential drug candidates and elucidation of drug mechanisms of action [9].

Ethically, zebrafish offer advantages as a model organism compared to mammals, owing to their lower position on the phylogenetic scale. Ethical frameworks and guidelines ensure their responsible and humane use in research [10]. In summary, zebrafish's genetic conservation, optical transparency, ease of genetic manipulation, disease modeling capabilities, and contributions to drug discovery position them as an accommodating and valuable tool for advancing our understanding of human health and disease.

The application of cellular and genetic methods to zebrafish research has significantly enriched our comprehension of various biological phenomena, ranging from developmental processes and genetics to disease mechanisms, all of which hold direct relevance to human health. Among the powerful techniques in zebrafish research are transgenesis, gene editing, cell transplantation, and cell labeling. These approaches are essential for understanding complicated biological processes and their consequences for human health and disease.

Transgenesis, gene editing, cell transplantation, and cell labeling are all potent zebrafish research approaches that help us understand numerous biological processes and their significance to human health and illness. Transgenesis involves inserting foreign DNA into the zebrafish genome, allowing reporter genes to be expressed for observing gene expression patterns, cell types, and biological processes. This approach has been employed in the development of

DOI: 10.1201/9781003402893-5

zebrafish models of human illnesses such as cancer, diabetes, and neurodegeneration [11].

CRISPR-Cas9 gene editing tools allow for precise modifications to the zebrafish genome. Creating gene knockouts or knock-ins to examine the activities and interactions of certain genes is one example. Continuous developments in zebrafish CRISPR editing have permitted successful targeting of many genes at the same time, resulting in the rapid generation of mosaic (G0) or stable mutant lines and subsequent phenotypic characterization [12]. Following that, similar mutants were used to investigate putative genes associated with human disorders and developmental traits [13].

Cell transplantation is the process of transplanting donor cells into zebrafish embryos at various developmental stages, allowing researchers to explore cell fate specification, differentiation, migration, and integration. In zebrafish, this approach has revealed knowledge regarding tissue regeneration, stem cell activity, and even interspecies communication [14].

Cell labeling is another important method that involves labeling cells with different markers such as fluorescent proteins or dyes. This allows for the study of cell lineages, clonal origins, and behavior over time and space. Cell-labeled zebrafish have been used to track neural activity, blood flow, metabolic changes, and other factors [15].

As a result, these methodologies have helped researchers to gain insights into a variety of biological phenomena and processes related to human health and illness. Developmental biology, genetics, neurology, pharmacology, toxicity, and regenerative medicine are all applications of zebrafish research in fields. The following cellular and genetic approaches are commonly employed in zebrafish research:

Transgenesis: Transgenesis in zebrafish is an effective method for introducing foreign DNA into the genome, allowing researchers to explore gene expression and control. The Tol2 transposon technology, initially identified by [16], has greatly improved transgene integration efficiency. To facilitate visualization of transgenic cells or tissues, fluorescent proteins such as GFP could be incorporated in the transgene construct.

Morpholino Knockdown: Morpholinos are commonly used in zebrafish studies to temporarily inhibit the expression of certain genes. Morpholinos are synthetic compounds that can interfere with the translation of a specific mRNA sequence, essentially 'knocking down' the gene's activity. They are injected into single-cell zebrafish embryos. This technique provides valuable insights into gene function during development [17, 18].

CRISPR-Cas9 Mediated Genome Editing: CRISPR-Cas9 has revolutionized genetic manipulation in zebrafish. Researchers can create precise gene knockouts, introduce specific mutations, or tag genes of interest for visualization. This method has expanded possibilities for genetic studies in zebrafish [3]. In zebrafish, new and innovative CRISPR-based techniques address biological issues regarding developmental processes (e.g., cell-lineage tracking) as well as gene functions (e.g., epigenome editing and targeted mutagenesis) [11]. In the latter application, essential factors in creating CRISPR gene knockouts comprise predicting/maximizing 'on target' Cas9 cleavage activity, predicting/minimizing unintended 'off-target' editing events, and rapidly detecting small insertions or deletions (indels). The presence of indels at candidate loci can be determined in a cost-effective manner using a variety of methods, ranging from simple identification of heteroduplexes—arising from multiple alleles coexisting in the sampled DNA visualized using polyacrylamide gel electrophoresis (PAGE)—to more advanced sequencing approaches that exactly determine and measure mutant alleles.

Cell Transplantation: Cell transplantation involves injecting donor cells into host embryos. This can be used to trace cell lineages or to study cell-cell interactions during development. Zebrafish embryos' optical transparency allows direct visualization of transplanted cell behavior, making it a valuable tool for developmental research [12].

Whole-Mount In-Situ Hybridization (WISH): WISH is employed in zebrafish to visualize the spatial distribution of gene expression during development. This technique provides insights into where specific mRNA molecules are expressed within the organism [13].

Genetic Screens: Zebrafish are an excellent for genetic screens, which are systematic searches for individuals with phenotypic mutations. Forward and reverse genetic screening are both possible. A forward genetic screen involves randomly inducing mutations and screening individuals with interesting phenotypes. Researchers use a reverse genetic screen to check for phenotypic alterations when a specific gene of interest is disrupted [19].

These cellular and genetic methodologies add to our understanding of developmental biology, genetics, and disease mechanisms, making zebrafish a valuable model organism across several scientific fields.

Recent advances in single-cell RNA sequencing (scRNA-seq) have the ability to characterize the zebrafish cell landscape at high resolution, revealing the cellular heterogeneity and molecular dynamics of various tissue types during embryonic, adult, and tissue regeneration stages [20]. One of the difficulties of scRNA-seq research is determining cell types and gene expression patterns from big and complicated datasets.

4.2 INSIGHTS AND APPLICATIONS OF CELLULAR AND GENETIC METHODS IN ZEBRAFISH RESEARCH

The zebrafish (*Danio rerio*) has evolved as an extraordinary model organism in biological study, providing a plethora of cellular and genetic tools that enable scientists to dive deeply into the complexities of development, genetics, and disease. These adaptable tools have improved our understanding of zebrafish biology, and hence its significance to human health and biology. Zebrafish research has carved a niche at the forefront of recent biological investigation, from live imaging, fluorescence-based reporter assays, and cell ablation techniques that enable dynamic observation of cellular processes to forward and reverse genetics, transgenesis, and conditional gene expression strategies that provide molecular insights.

In this overview, we will investigate these cellular and genetic methods, emphasizing their applications in a variety of scientific domains such as developmental biology, disease modelling, and behavioral studies, highlighting the critical role that zebrafish play in advancing our understanding of fundamental biological phenomena. These are only a handful of several cellular and genetic techniques used by researchers to examine zebrafish. Because of the development of these approaches, the zebrafish has become a valuable model organism in many fields of biological research.

4.2.1 Cellular Methods in Zebrafish Research

1. Live Imaging Techniques: Live imaging techniques are essential for examining dynamic cellular processes in zebrafish embryos. Researchers can investigate cell behaviors such as migration, division, and differentiation in real-time using confocal microscopy, multiphoton microscopy, and light sheet microscopy [21].
2. Fluorescence-Based Reporter Assays: Fluorescent proteins such as Green Fluorescent Protein (GFP) are employed as reporters to mark certain cell types or proteins. Fluorescent tags are attached to promoters of genes of interest, allowing visualization of gene expression patterns and cellular processes [22].
3. Cell Ablation Techniques: Laser ablation and genetic approaches, such as cell-specific production of diphtheria toxin, allow for the selective destruction of certain cell populations. This enables researchers to better understand the effects of specific cell types on development and physiology [23].

4.2.2 Genetic Methods in Zebrafish Research:

1. Forward and Reverse Genetics: Forward genetics entails randomly introducing mutations and then finding phenotypic changes and relating them to specific genes. Reverse genetics uses tools such as CRISPR-Cas9 to manipulate particular genes and examine the phenotypic changes that occur [24].
2. Transgenesis: Transgenic zebrafish are created by inserting foreign DNA, which often includes fluorescent markers or disease-related genes. This method assists in better understanding of gene function, promoter activity, and disease modeling [16].
3. Knockout and Knock-In Strategies: Knockout involves altering the function of a specific gene, clarifying its role in development or illness. Knock-in methods insert specific DNA sequences to examine gene mutations and protein location [25].
4. Conditional Gene Expression: Techniques like the Cre-Lox system enable controlled gene expression in specific tissues or developmental stages. This allows researchers to understand gene function in a temporally and spatially restricted manner [14].

4.2.3 Applications

1. Developmental Biology: Cellular and genetic methods facilitate the investigation of organogenesis, tissue regeneration, and embryonic patterning. Insights gained from zebrafish research often extend to understanding human developmental processes [15].
2. Disease Modeling: Zebrafish serve as disease models for conditions like cancer, cardiovascular disorders, and neurodegenerative diseases. Cellular and genetic methods aid in elucidating disease mechanisms and screening potential therapeutics [16].
3. Behavioral Studies: Genetic manipulation combined with behavioral assays provides insights into the genetic basis of behavior. Studying zebrafish behavior in response to various stimuli does indeed provide valuable insights into neurological processes, making this a common approach in neuroscience and behavioral research [26].
4. *In Vivo* Toxicity Testing: When studying the *in vivo* toxicity of natural compounds, one of the most often used tools is the zebrafish embryo. The following are the benefits of zebrafish embryos over traditional rats and mice as suitable tools for *in vivo* toxicity testing: (1) ease of rearing, small space needed, and low maintenance cost; (2) short growth cycle and rapid evaluation; (3) transparent and easy to observe; (4) drugs could be added directly into the water (or culture medium) for absorption; and (5) simple to operate and less demand for the types of drugs [27]. Due to these characteristics, zebrafish embryos are an effective tool for toxicity assessment. Previous studies have employed zebrafish to investigate neurotoxicity,

cardiotoxicity, hepatotoxicity, liver toxicity, renal toxicity, developmental toxicity, and reproductive toxicity, etc. [28, 29].

4.3 ZEBRAFIŞH: A DYNAMIC ISSUE IN BIOLOGY AND MEDICINE

Zebrafish (*Danio rerio*) has emerged as a cornerstone of scientific research across a spectrum of disciplines, offering a versatile and indispensable model for investigating fundamental biological processes and their relevance to medicine. This diminutive aquatic organism's unique attributes, including genetic similarity to humans, optical transparency during development, ease of genetic manipulation, and applicability to high-throughput assays, have positioned it at the forefront of biological and medical studies. From developmental biology and genetics to disease modeling, drug discovery, regenerative medicine, infectious diseases, behavioral research, and environmental studies, zebrafish research has yielded a wealth of insights and applications, shaping our understanding of critical aspects of biology and opening new avenues for advancements in medicine and environmental science.

4.3.1 Developmental Biology

Zebrafish have left an indelible mark on the landscape of biological and medical research. Their unique attributes have enabled groundbreaking contributions to a multitude of scientific domains, firmly establishing their relevance to zebrafish-based investigations. Within the field of developmental biology, zebrafish embryos, transparent and externally developing, have unlocked windows into fundamental processes such as gastrulation, organogenesis, and tissue regeneration, providing unprecedented insights into these intricate mechanisms. Moreover, zebrafish have been instrumental in unraveling the mysteries of gene function during development. The techniques like transgenesis and gene knockout allow scientists to decipher the roles of specific genes [1].

Zebrafish as a model of multi-ciliated cells (MCC) ontogeny: MCCs are excellent cells responsible for variety of vital physiological processes [30]. The zebrafish model is an amazing tool for studying MCC formation. Because teleost fish and humans have a high degree of genetic conservation, it is probable that they share core processes of MCC ontogeny. Zebrafish achieve sexual maturity in around three months and breed prolifically to produce single clutches of 100–250 embryos or more for research purposes [19]. Furthermore, zebrafish embryological characteristics, such as optical transparency and big size, enable them to be used for physical manipulation, such as microinjection and robust real-time visualization of morphogenic processes [10].

4.3.2 Genetics and Genomics

In the realm of genetics and genomics, zebrafish have been used in gene discovery, aiding in the identification of novel genes and regulatory elements intertwined with development and disease. Comparative genomics studies between zebrafish and humans have further illuminated conserved genes and pathways, shedding light on the intricacies of human genetics and development [20].

4.3.3 Disease Modeling

The application of zebrafish extends boldly into disease modeling, as these models have proven substantial in deciphering the enigma of cancer progression, metastasis, and drug screening [4]. Cardiovascular research benefits from zebrafish studies, delving into heart development, regeneration, and the intricate web of cardiovascular diseases [21]. In neurobiology, zebrafish research has launched our understanding of neurodevelopment, neurodegenerative disorders, and behavioral intricacies, advancing the frontiers of neuroscience [22]

4.3.4 Drug Discovery and Toxicology

In the spheres of drug discovery and toxicology, zebrafish have been fundamental in high-throughput screens that identify potential drug candidates and elucidate mechanisms of action, propelling the field of pharmacology [23]. Additionally, their unique suitability for assessing the toxicity of chemicals, pollutants, and environmental toxins has great implications for environmental biology and toxicology [10].

4.3.5 Regenerative Medicine

Moreover, zebrafish serve as an extraordinary model for understanding tissue regeneration and potential applications in regenerative medicine [24].

4.3.6 Infectious Diseases

In the battle against infectious diseases, zebrafish have donned the mantle of infection models, enabling the study of host-pathogen interactions and immune responses, offering critical insights into infectious disease research [25].

4.3.7 Behavioral and Psychiatric Research

Zebrafish, with their behavioral intricacies, also contribute significantly to the fields of behavioral and psychiatric research, providing a unique lens through which to study the genetic solid structure of behavior [22].

4.3.8 Environmental Biology

In the field of environmental biology, these remarkable organisms aid in the assessment of the impact of environmental changes on aquatic ecosystems and species, further underlining their versatility [31].

4.4 EXPLORING NEW CHALLENGES IN CELLULAR AND GENETIC METHODS IN ZEBRAFISH RESEARCH

Zebrafish (*Danio rerio*) continues to be an excellent model organism for studying various biological processes due to its unique characteristics. The integration of cellular and genetic methods has significantly contributed to a better understanding of development, disease, and behavior. As researchers delve deeper into these methodologies, they encounter new challenges that necessitate innovative approaches. This section discusses emerging challenges in cellular and genetic research using zebrafish, highlighting the need for novel solutions.

4.4.1 Challenges in Cellular Methods

1. Single-Cell Analysis in Complex Tissues: The study of single-cell behavior within complex tissues presents challenges due to the intricacies of cell interactions. Techniques like single-cell RNA sequencing (scRNA-seq) are becoming popular, but applying them to zebrafish tissues with diverse cell types remains a challenge [32].
2. Quantitative Live Imaging at High Resolution: Capturing dynamic cellular processes at high resolution over extended periods requires sophisticated imaging setups. Techniques like light sheet microscopy face challenges in maintaining both temporal and spatial resolution [33].
3. Labeling Specific Cell Populations: Precise and selective labeling of specific cell populations is essential for tracking cell behaviors. Developing new genetic tools or enhancing existing ones to achieve accurate labeling remains an ongoing challenge [34].

4.4.2 Challenges in Genetic Methods

1. Off-Target Effects of Genome Editing: Although CRISPR-Cas9 revolutionized genetic manipulation, off-target effects still pose a challenge. Ensuring high specificity while targeting genes demands advanced computational algorithms and experimental validation [35].
2. Functional Analysis of Non-Coding Elements: A significant portion of the genome comprises non-coding elements with regulatory roles. Unraveling their functions using genetic methods requires innovative approaches beyond traditional knockout strategies [36].
3. Modeling Complex Multigenic Traits: Many diseases result from interactions among multiple genes. Creating zebrafish models that accurately recapitulate multigenic diseases is challenging due to the complexity of genetic interactions [37].

4.4.3 Innovative Approaches and Future Directions

1. Advanced Imaging Technologies: Emerging imaging technologies, such as lattice light-sheet microscopy and adaptive optics, hold promise for overcoming resolution and tissue penetration limitations in live imaging [38].
2. Enhanced Genetic Editing Tools: Developing improved CRISPR-Cas systems with enhanced specificity, as well as refining base editing techniques, could mitigate off-target effects and enable precise genetic modifications [39].
3. Integration of Multi-Omics Data: Combining genetic and cellular data with multi-omics approaches allows for a more comprehensive understanding of biological processes. Integrative analyses can provide insights into gene regulation, protein interactions, and metabolomics [40].

The intersection of cellular and genetic methods in zebrafish research has yielded remarkable insights into biology. However, new challenges continuously emerge as researchers aim to unravel the intricacies of cellular behaviors and genetic interactions. Innovative technologies and multidisciplinary approaches will be instrumental in overcoming these challenges, propeling zebrafish research into a new era of discovery.

4.5 ZEBRAFISH AS A MODEL ORGANISM FOR ASSESSING CLIMATE CHANGE

Climate change is often regarded as a major environmental threat, and when combined with environmental degradation caused by chemical pollution and habitat loss, it has the potential to have a significant influence on wildlife populations [41], especially freshwater ecosystems. Climate change is expected to induce a rise in worldwide water temperatures as well as an increase in the frequency and intensity of extreme heat waves [42]. This will put stress on aquatic ectotherms in some regions as temperatures approach their thermal limits. Thermal stress causes the organism to undergo a variety of physiological acclimatization responses to enhance overall performance at sub-optimal temperatures [43]. Environmental stressor combinations have been demonstrated to have interaction impacts on individual health in numerous species, including fish [44], amphibians [45], and reptiles [46]. Migration to habitats with more optimum temperatures is a common response of fish to temperature variation [47]. However, if a fish is unable to reach a habitat with optimal thermal conditions owing to natural or anthropogenic factors or due to habitat loss as a result of climate change, it will perceive this temperature as stressful, activating compensatory mechanisms to restore homeostasis. Acute exposure to sub-optimal temperature (hours or days) causes a stress reaction, whereas longer-term exposure (weeks or months) causes an acclimatization response [48]. The long-term survival of aquatic products, such as seafood, is partly linked to the human response to climate change. For example, the temperature of the sea surface modifies the distribution of fish stock [49], thus influencing where fish can sustainably be hunted [50].

Zebrafish can be used in experimental research as a model species to study how other marine or freshwater species will be influenced by climate change, particularly global warming [51]. While the number of reports on these species has grown rapidly, more specific research on climate change is still lacking. Zebrafish can live in both freshwater and saltwater habitats [7], making them ideal organisms for evaluating organism function changes within climate change–related investigations in either environment. Zebrafish are very useful for exploring the impacts of thermal fluctuation given its wide thermal tolerance (6.7–41.7°C) [52]. Wild zebrafish live in slow-moving or standing water basins characterized by fast daily (0.1–5.6°C) and seasonal (6°C in winter to above 38°C in summer) temperature variations [53]. This allows experimental animals to be acclimatized to high or low temperatures in aquaria while still remaining in the range tolerated by the animal. Importantly, the sex-determining process in zebrafish is known to be influenced by environmental conditions, and independent exposures to increased water temperatures have been found to be beneficial.

Zebrafish are one of the most commonly used laboratory species. A wide range of domestication effects has been detected in zebrafish, both between laboratory-strain and wild-caught fish [54], as well as in wild-caught zebrafish after being kept in the laboratory for only nine generations. These domestication effects induce behavioral, physiological, and genetic changes. Laboratory zebrafish can acclimatize to a broad range of temperatures, and under normal holding temperatures (28°C), they have an upper acute thermal tolerance that exceeds 40°C [55]. This makes it a useful species for investigating thermal biology questions. Since the 1970s, the zebrafish AB-WT laboratory population has been kept in a controlled laboratory setting with ideal temperatures (26–28°C), low thermal variation, and few thermal extremes [45]. As a result of decades of adaptation to laboratory circumstances with no adaptive value for high thermal tolerance, laboratory populations may have some of their high temperature tolerance. Wild populations, on the other hand, face daily and seasonal temperature fluctuations, as well as potentially harmful extreme temperatures [53], implying that high thermal tolerance is a significant adaptation trait for wild zebrafish.

4.6 ZEBRAFISH AND ENVIRONMENT

The exponential growth of the world's population, technological advancement, and industry have increased the chemical pollution of water, soil, and air with toxic substances. Pollutants such as heavy metals, particulate matter, endocrine-disrupting chemicals (EDCs), and industrial waste can pollute the water environment and deteriorate the quality of aquatic life. Toxic pollutants affect aquatic ecosystems and the behavior of aquatic organisms by direct and indirect mechanisms. Water toxic pollutants significantly impact fish reproduction, lifespan, embryonic and larval development, and sexual differentiation. Accordingly, these toxins exert harmful influences on both wildlife and human health. Environmental monitoring and toxin surveillance tools are required to determine toxin exposure risk, teratogenicity of water pollutants, and the biological, chemical, and genetic mechanisms underlying those effects [56, 57]. Zebrafish (*Danio rerio*) is a teleost fish that has recently emerged as a uniquely powerful experimental model. For nearly two decades, zebrafish have been used for biomonitoring. Over the last few decades, the implication of zebrafish in the field of environmental health has significantly increased. The culture tests were replaced by a whole vertebrate model using zebrafish. Zebrafish are being employed to investigate the presence of certain toxins and to determine how environmental exposures influence both animal and human health. It offers to utilize the combination of embryological and genomic approaches. Zebrafish are regarded as a real-time in vivo animal model for investigating the short- and long-term risks of environmental toxins to public health. The capability to visualize the effects of toxins *in vivo* permits the direct assessment of toxicity, as well as measurements of absorption, distribution, metabolism, and elimination. Furthermore, they can be used to evaluate therapies that potentially reduce adverse effects.

Moreover, climate change is a serious environmental threat that has a great impact on freshwater ecosystems. Fish can escape areas of high temperatures by shifting into deeper and cooler aquatic strata or by migration, which isn't possible for freshwater fish, as they are restricted to their local freshwater bodies. Moreover, aquatic species with temperature-dependent sex determination are negatively influenced, as the numbers of females, rather than males, largely determine the reproductive output of a population. In some reptiles, warmer temperatures during incubation induce female-biased populations, while in fish and some amphibians, elevated temperatures induce male-biased populations, which can dramatically alter population balance. Testing the physiology parameters of acclimation of aquatic animals to environmental changes and thermal tolerance is required to evaluate how and to which degree could these creatures resist climatic changes and avoid extinction [58–60].

Zebrafish are commonly used in scientific research related to aquatic environments. These freshwater and tropical fish are often referred to as the laboratory rats of this field. In fact, scientists have recently been using zebrafish to study the evolution of thermal tolerance in tropical species that live near their thermal limits. One study by Morgan et al. (2020) involved selecting a group of wild zebrafish that were able to withstand higher temperatures and breeding them in a laboratory setting for over six generations during a three-year experimental trial. The researchers revealed that with each new generation, the maximum limit of temperature tolerance jumped by 0.04°C. However, they also reported that the rate of change was slower than expected, indicating that evolution towards higher environmental temperatures may take longer than previously thought.

Zebrafish have an array of features that make them an ideal model organism for research studies. First, their fertilization is external, and the embryos develop rapidly. Second, a single female can produce between 100 and 500 embryos per week, allowing for longitudinal studies to be conducted on the developmental effects of environmental exposures at a lower cost than with rodents. Their big eggs (0.7mm in diameter) are highly manipulable, making them suitable for cell and tissue transplantation. In addition, zebrafish embryos and larvae are transparent, rendering them appropriate for creating transgenic animals and performing high-resolution analysis of living tissues with differential interference contrast (DIC) microscopy. The optically clear embryos also enable monitoring of individual cell behavior in vivo from the earliest stages using time-lapse microscopy. Zebrafish can also be used to determine the regenerative capacity of potential stem cells [57, 61]. The sex-determining mechanism in zebrafish can be modified by environmental factors, as evidenced by skewed sex ratios toward males in some experiments [41]. Additionally, zebrafish are freshwater and tropical aquatic species found in India, Nepal, and Bangladesh, inhabiting shallow streams and pools that can rapidly warm during heat waves. They are assumed to be tolerant to a wide temperature range, with laboratory zebrafish having an upper acute thermal tolerance exceeding 40°C [55, 60, 62].

4.7 EXPLORING ZEBRAFISH: UNDERSTANDING THEIR GENETIC DIVERSITY AND ADAPTATIONS IN LABS AND THE WILD

The zebrafish is a widely used model organism in scientific research. It is commonly used in developmental genetics, neurophysiology, and biomedical research. The zebrafish is easily captured and can be bred in captivity or studied under field conditions in South Asia. Researchers in more than 1,000 laboratories worldwide use zebrafish from established lab strains for their work. To further study the ecology and evolution of zebrafish, it is important to understand the existing genetic diversity and evolutionary relationships among natural populations and lab strains. Zebrafish have a wide geographic range, which means there may be major differences among them. However, past breeding practices and limited collection of natural populations may have caused significant differences between lab strains and wild populations. The substantial genetic variation that exists in the larger zebrafish population is missing from the lab strains. Inbreeding and the small effective population size of zebrafish lab strains lead to reduced phenotypic and genetic variations within and among strains. As a strain becomes more homogenous during breeding in the lab, variants become fixed or lost [54, 63, 64].

The wild populations are more diverse than the laboratory strains, but the knowledge about the genetic diversity in wild zebrafish is still insufficient. Most of the variants identified in the laboratory strains are also present in the wild. However, nearly 60% of all variants are not present in any of the laboratory strains; many of these are associated with protein-coding genes [64]. These genetic variations between wild and lab zebrafish make generalizations from specific populations to the species as a whole difficult [65]. In the wild, zebrafish were found living at elevations ranging from 8–1576 m and in an array of water temperatures of 12–39°C, water pH of 5.9–9.8, and water salinities of ~0.01–0.8 [65]. However, surveys suggest that wild zebrafish populations favor warm and slow-flowing waters with temperature ranges between 24–35°C, slightly alkaline (pH 7–8), highly transparent, and with plant coverage. Wild zebrafish live in groups with variable sizes according to the rate of water flow. In slow-flowing water, zebrafish exist in small groups of 7 to 10 fish, while large groups are found in fast-flowing waters with numbers up to 300 individuals [66].

4.8 ZEBRAFISH: ACHIEVING THE 3RS OF EXPERIMENTAL ANIMALS

Zebrafish have gained prominence in scientific research for their ability to align with the principles of the 3Rs: Replacement, Reduction, and Refinement, which promote ethical treatment of experimental animals. Let's explore how zebrafish meet these criteria with reference to their use in scientific experiments.

Replacement: Replacement refers to finding alternative methods to replace the use of sentient animals whenever possible. Zebrafish serve as a valuable replacement model in various experiments that previously involved more complex and sentient animals. For example, they have been employed to study human diseases, toxicology, and drug testing, reducing the reliance on mammals like mice or rats [67].

Reduction: Reduction entails minimizing the number of animals used in experiments. Zebrafish are advantageous in this regard because they are prolific breeders, and their embryos are transparent, allowing for non-invasive observation and manipulation. This means that fewer zebrafish are needed for experiments compared to other animal models. The reduction principle is emphasized in numerous studies, including [68].

Refinement: Refinement seeks to enhance the welfare and minimize suffering of animals used in experiments. Zebrafish are well-suited for refinement due to their small size, ease of handling, and lower sensitivity to pain. They are also amenable to the development of humane endpoints and monitoring techniques [69].

In the context of the 3Rs, zebrafish provide a valuable resource for researchers to ethically conduct experiments. Their ability to replace higher vertebrates, reduce the number of animals required, and be subjected to refinement measures makes them a model organism aligned with the principles of animal welfare and ethical research.

Zebrafish research is mainly conducted on 5-day-old embryos and larvae. So, zebrafish is a linkage between in vivo and in vitro experimental methods. Additionally, it complies with the European Commission standards of experimental animal use (EU Directive 2010/63/EU). Zebrafish as a "lower vertebrate" are considered a great alternative for reducing the use of higher-order animal models for research, and consequently achieving the replacement concept of the 3Rs of animal welfare in research ethics to reduce animal suffering (Replacement, Reduction, and Refinement). The welfare of zebrafish is assessed through their reproductive performance, health status, and absence of indicators of stress [66].

4.9 NEW PERSPECTIVE ON CELLULAR AND GENETIC METHODS IN ZEBRAFISH

Cellular and genetic methods in zebrafish research have advanced dramatically over the years, owing to the incorporation of cutting-edge technologies and novel experimental procedures. These approaches have used the unique advantages of the zebrafish model organism, *Danio rerio*, to unravel fundamental biological processes and create insights into different aspects of developmental biology, disease modelling, and drug discovery.

1. Transgenesis and Gene Editing Techniques: Zebrafish have shown to be an excellent tool for investigating gene function and regulation. Transgenic zebrafish lines, created through methods such as Tol2 transposon-mediated transgenesis [70], enable the visualization of specific cell types and tissues by driving fluorescent protein expression under the control of tissue-specific promoters. Furthermore, the utilization of CRISPR-Cas9 technology [3] for gene knockout, knock-in, and precise editing has revolutionized zebrafish genetics. These techniques have sped up the investigation of gene function and the modeling of human genetic disorders.
2. Live Imaging and Microscopy: Zebrafish embryos are visually translucent, allowing non-invasive observation of cellular and developmental processes in real time. High-resolution insights into embryogenesis, tissue regeneration, and disease development are provided by advanced microscopy methods such as confocal and two-photon microscopy [47]. Researchers can track cell motions, monitor cell signaling pathways, and investigate tissue morphogenesis *in vivo*.
3. Functional Genomics: For zebrafish, genome-wide techniques such as RNA sequencing (RNA-seq) and chromatin immunoprecipitation sequencing (ChIP-seq) have been applied. These methods permit the comprehensive study of gene expression patterns, epigenetic changes, and regulatory networks during development and illness [6]. Comparative genomics with humans has detected conserved pathways and genes related to human health.
4. Disease Modeling: Zebrafish have become known as an effective model for the study of a wide range of diseases, including cancer, cardiovascular problems, neurological diseases, as well as infectious diseases [1]. Genetic modification enables the creation of zebrafish models that mimic human disease phenotypes, hence offering significant platforms for drug screening and therapeutic discovery.
5. High-Throughput Screening: Robotics and automated systems have been incorporated into zebrafish research, allowing for high-throughput drug screening [42]. Large-scale chemical screens in zebrafish embryos resulted in the discovery of prospective therapeutic candidates as well as a better knowledge of drug mechanisms of action.
6. Gene-Environment Interactions: Because of zebrafish's aquatic habitat, they provide a unique opportunity to study gene-environment interactions. Researchers can investigate how genetic factors interact with environmental signals, such as toxins, pollutants, and drugs, which have developmental and physiological consequences.
7. Ethical Considerations: While zebrafish research has yielded significant insights, ethical concerns about the use of animals in research remain critical. Ethical frameworks and regulations, such as the 3Rs (Replacement, Reduction, Refinement), are fundamental for ensuring ethical and humane use of zebrafish in experimental studies [18].

REFERENCES

1. Haffter P, Granato M, Brand M, Mullins MC, Hammerschmidt M, Kane DA, Odenthal J, van Eeden FJM, Jiang Y-J, Heisenberg C-P: **The identification of genes with unique and essential functions in the development of the zebrafish, *Danio rerio*.** *Development* 1996, **123**(1):1–36.
2. Muto A, Kawakami K: **Prey capture in zebrafish larvae serves as a model to study cognitive functions**. *Frontiers in Neural Circuits* 2013, 7:110.
3. Hwang WY, Fu Y, Reyon D, Maeder ML, Tsai SQ, Sander JD, Peterson RT, Yeh JJ, Joung JK: **Efficient genome editing in zebrafish using a CRISPR-Cas system**. *Nature Biotechnology* 2013, **31**(3):227–229.
4. White R, Rose K, Zon L: **Zebrafish cancer: The state of the art and the path forward**. *Nature Reviews Cancer* 2013, **13**(9):624–636.
5. Al-Thani HF, Shurbaji S, Yalcin HC: **Zebrafish as a model for anticancer nanomedicine studies**. *Pharmaceuticals* 2021, **14**(7):625.

6. Shen Y, Sheng R, Guo R: **Application of zebrafish as a model for anti-cancer activity evaluation and toxicity testing of natural products**. *Pharmaceuticals* 2023, **16**(6):827.
7. Hsu S-Y, Wen Z-H, Shih P-C, Kuo H-M, Lin S-C, Liu H-T, Lee Y-H, Wang Y-J, Chen W-F, Chen N-F: **Sinularin induces oxidative stress-mediated apoptosis and mitochondrial dysfunction, and inhibits angiogenesis in glioblastoma cells**. *Antioxidants* 2022, **11**(8):1433.
8. Jiang X, Zhou J, Lin Q, Gong G, Sun H, Liu W, Guo Q, Feng F, Qu W: **Anti-angiogenic and anticancer effects of baicalein derivatives based on transgenic zebrafish model**. *Bioorganic & Medicinal Chemistry* 2018, **26**(15):4481–4492.
9. Dash SN, Patnaik L: **Flight for fish in drug discovery: A review of zebrafish-based screening of molecules**. *Biology Letters* 2023, **19**(8):20220541.
10. Lieschke GJ, Currie PD: **Animal models of human disease: zebrafish swim into view**. *Nature Reviews Genetics* 2007, **8**(5):353–367.
11. Uribe-Salazar JM, Kaya G, Sekar A, Weyenberg K, Ingamells C, Dennis MY: **Evaluation of CRISPR gene-editing tools in zebrafish**. *BMC Genomics* 2022, **23**(1):1–16.
12. Haldi M, Ton C, Seng WL, McGrath P: **Human melanoma cells transplanted into zebrafish proliferate, migrate, produce melanin, form masses and stimulate angiogenesis in zebrafish**. *Angiogenesis* 2006, **9**:139–151.
13. Thisse C, Thisse B: **High-resolution in situ hybridization to whole-mount zebrafish embryos**. *Nature Protocols* 2008, **3**(1):59–69.
14. Mosimann C, Kaufman CK, Li P, Pugach EK, Tamplin OJ, Zon LI: **Ubiquitous transgene expression and Cre-based recombination driven by the ubiquitin promoter in zebrafish**. *Development* 2011, **138**(1):169–177.
15. Stemple DL: **Structure and function of the notochord: An essential organ for chordate development**. *Development* 2005, **132**:2503–2512.
16. Adhish M, Manjubala I: **Effectiveness of zebrafish models in understanding human diseases—A review of models**. *Heliyon* 2023, **9**:e14557.
17. Nasevicius A, Ekker SC: **Effective targeted gene 'knockdown' in zebrafish**. *Nature Genetics* 2000, **26**(2):216–220.
18. Zakaria ZZ, Eisa-Beygi S, Benslimane FM, Ramchandran R, Yalcin HC: **Design and microinjection of morpholino antisense oligonucleotides and mRNA into zebrafish embryos to elucidate specific gene function in heart development**. *JoVE (Journal of Visualized Experiments)* 2022 (186):e63324.
19. Wesselman HM, Nguyen TK, Chambers JM, Drummond BE, Wingert RA: **Advances in understanding the genetic mechanisms of zebrafish renal multiciliated cell development**. *Journal of Developmental Biology* 2022, **11**(1):1.
20. Howe K, Clark MD, Torroja CF, Torrance J, Berthelot C, Muffato M, Collins JE, Humphray S, McLaren K, Matthews L: **The zebrafish reference genome sequence and its relationship to the human genome**. *Nature* 2013, **496**(7446):498–503.
21. Yanik MF, Rohde CB, Pardo-Martin C: **Technologies for micromanipulating, imaging, and phenotyping small invertebrates and vertebrates**. *Annual Review of Biomedical Engineering* 2011, **13**:185–217.
22. Kalueff AV, Stewart AM, Gerlai R: **Zebrafish as an emerging model for studying complex brain disorders**. *Trends in Pharmacological Sciences* 2014, **35**(2):63–75.
23. Zon LI, Peterson RT: **In vivo drug discovery in the zebrafish**. *Nature Reviews Drug Discovery* 2005, **4**(1):35–44.
24. Goessling W, North TE: **Repairing quite swimmingly: Advances in regenerative medicine using zebrafish**. *Disease Models & Mechanisms* 2014, **7**(7):769–776.
25. Saralahti AK, Harjula S-KE, Rantapero T, Uusi-Mäkelä MI, Kaasinen M, Junno M, Piippo H, Nykter M, Lohi O, Rounioja S: **Characterization of the innate immune response to Streptococcus pneumoniae infection in zebrafish**. *PLoS Genetics* 2023, **19**(1):e1010586.
26. Guo S: **Linking genes to brain, behavior and neurological diseases: What can we learn from zebrafish?** *Genes, Brain and Behavior* 2004, **3**(2):63–74.
27. Miyawaki I: **Application of zebrafish to safety evaluation in drug discovery**. *Journal of Toxicologic Pathology* 2020, **33**(4):197–210.
28. Lai KP, Gong Z, Tse WKF: **Zebrafish as the toxicant screening model: Transgenic and omics approaches**. *Aquatic Toxicology* 2021, **234**:105813.
29. Zhang Y, Xia Q, Wang J, Zhuang K, Jin H, Liu K: **Progress in using zebrafish as a toxicological model for traditional Chinese medicine**. *Journal of Ethnopharmacology* 2022, **282**:114638.
30. Spassky N, Meunier A: **The development and functions of multiciliated epithelia**. *Nature Reviews Molecular Cell Biology* 2017, **18**(7):423–436.
31. Bopp S, Lettieri T: **Gene expression profile assessment in Zebrafish (*Danio rerio*)**. *JRC Scientific and Technical Reports Ispra: JRC,[EUR 23094 EN]* 2007, **24**.
32. Packer JS, Zhu Q, Huynh C, Sivaramakrishnan P, Preston E, Dueck H, Stefanik D, Tan K, Trapnell C, Kim J: **A lineage-resolved molecular atlas of C. elegans embryogenesis at single-cell resolution**. *Science* 2019, **365**(6459):eaax1971.
33. Truong TV, Supatto W, Koos DS, Choi JM, Fraser SE: **Deep and fast live imaging with two-photon scanned light-sheet microscopy**. *Nature Methods* 2011, **8**(9):757–760.
34. Auer TO, Duroure K, De Cian A, Concordet J-P, Del Bene F: **Highly efficient CRISPR/Cas9-mediated knock-in in zebrafish by homology-independent DNA repair**. *Genome Research* 2014, **24**(1):142–153.
35. Frock RL, Hu J, Meyers RM, Ho Y-J, Kii E, Alt FW: **Genome-wide detection of DNA double-stranded breaks induced by engineered nucleases**. *Nature Biotechnology* 2015, **33**(2):179–186.
36. Catarino RR, Stark A: **Assessing sufficiency and necessity of enhancer activities for gene expression and the mechanisms of transcription activation**. *Genes & Development* 2018, **32**(3–4):202–223.
37. Gómez-García F, Martínez-Pulleiro R, Carrera N, Allegue C, Garcia-Gonzalez MA: **Genetic Kidney Diseases (GKDs) modeling using genome editing technologies**. *Cells* 2022, **11**(9):1571.
38. Chen B-C, Legant WR, Wang K, Shao L, Milkie DE, Davidson MW, Janetopoulos C, Wu XS, Hammer III JA, Liu Z: **Lattice light-sheet microscopy: Imaging molecules to embryos at high spatiotemporal resolution**. *Science* 2014, **346**(6208):1257998.
39. Anzalone AV, Randolph PB, Davis JR, Sousa AA, Koblan LW, Levy JM, Chen PJ, Wilson C, Newby GA, Raguram A: **Search-and-replace genome editing without double-strand breaks or donor DNA**. *Nature* 2019, **576**(7785):149–157.
40. Saha A, Kim Y, Gewirtz AD, Jo B, Gao C, McDowell IC, Engelhardt BE, Battle A, Aguet F, Ardlie KG: **Co-expression networks reveal the tissue-specific**

regulation of transcription and splicing. *Genome Research* 2017, **27**(11):1843–1858.
41. Brown AR, Owen SF, Peters J, Zhang Y, Soffker M, Paull GC, Hosken DJ, Wahab MA, Tyler CR: **Climate change and pollution speed declines in zebrafish populations**. *Proceedings of the National Academy of Sciences* 2015, **112**(11):E1237–E1246.
42. Seneviratne SI, Donat MG, Mueller B, Alexander LV: **No pause in the increase of hot temperature extremes**. *Nature Climate Change* 2014, **4**(3):161–163.
43. Angilletta MJ: **Thermal acclimation**. *Thermal Adaptation: A Theoretical and Empirical Synthesis* 2009:126–155.
44. Quinn AL: *The impacts of agricultural chemicals and temperature on the physiological stress response in fish*. Lethbridge, AB: University of Lethbirdge, Faculty of Arts and Science, 2007.
45. Bancroft BA, Baker NJ, Blaustein AR: **A meta-analysis of the effects of ultraviolet B radiation and its synergistic interactions with pH, contaminants, and disease on amphibian survival**. *Conservation Biology* 2008, **22**(4):987–996.
46. Willingham EJ: **The effects of atrazine and temperature on turtle hatchling size and sex ratios**. *Frontiers in Ecology and the Environment* 2005, **3**(6):309–313.
47. Chen I-C, Hill JK, Ohlemüller R, Roy DB, Thomas CD: **Rapid range shifts of species associated with high levels of climate warming**. *Science* 2011, **333**(6045):1024–1026.
48. Angiulli E, Pagliara V, Cioni C, Frabetti F, Pizzetti F, Alleva E, Toni M: **Increase in environmental temperature affects exploratory behaviour, anxiety and social preference in *Danio rerio***. *Scientific Reports* 2020, **10**(1):5385.
49. Sunday JM, Bates AE, Dulvy NK: **Thermal tolerance and the global redistribution of animals**. *Nature Climate Change* 2012, **2**(9):686–690.
50. Free CM, Cabral RB, Froehlich HE, Battista W, Ojea E, O'Reilly E, Palardy JE, García Molinos J, Siegel KJ, Arnason R: **Expanding ocean food production under climate change**. *Nature* 2022, **605**(7910):490–496.
51. Nor Azra M, Mohd Noor MI, Pau Tan M, Dawood M, Amin M, Zekker I, Fuad Abdullah M, Abd Latif Z, Pardi F: **Zebrafish and Medaka as model organisms for climate change research: Global literature scientometric analysis**. *Frontiers in Earth Science* 2022, **10**:988710.
52. Schaefer J, Ryan A: **Developmental plasticity in the thermal tolerance of zebrafish *Danio rerio***. *Journal of Fish Biology* 2006, **69**(3):722–734.
53. Spence R, Gerlach G, Lawrence C, Smith C: **The behaviour and ecology of the zebrafish, *Danio rerio***. *Biological Reviews* 2008, **83**(1):13–34.
54. Whiteley AR, Bhat A, Martins EP, Mayden RL, Arunachalam M, Uusi-Heikkila S, Ahmed A, Shrestha J, Clark M, Stemple D: **Population genomics of wild and laboratory zebrafish (*Danio rerio*)**. *Molecular Ecology* 2011, **20**(20):4259–4276.
55. Morgan R, Sundin J, Finnøen MH, Dresler G, Vendrell MM, Dey A, Sarkar K, Jutfelt F: **Are model organisms representative for climate change research? Testing thermal tolerance in wild and laboratory zebrafish populations**. *Conservation Physiology* 2019, **7**(1):coz036.
56. Jan A, Shah TH, Nissa NU: **Impact of aquatic pollution on fish fauna**. In: *Bacterial Fish Diseases*. Cambridge: Academic Press, 2022: 103–112.
57. Bambino K, Chu J: **Zebrafish in toxicology and environmental health**. *Current Topics in Developmental Biology* 2017, **124**:331–367.
58. Rankin DJ, Kokko H: **Do males matter? The role of males in population dynamics**. *Oikos* 2007, **116**(2):335–348.
59. Bates BC, Kundzewicz ZW, Wu S, Arnell N, Burgett V, Räisänen J (eds): *Climate change and water: Technical paper of the intergovernmental panel on climate change*. Geneva: IPCC Secretariat, 2008. http://www.ipcc.ch/pdf/technical-papers/climate-change-water-en.pdf.
60. Morgan R, Finnøen MH, Jensen H, Pélabon C, Jutfelt F: **Low potential for evolutionary rescue from climate change in a tropical fish**. *Proceedings of the National Academy of Sciences* 2020, **117**(52):33365–33372.
61. Sun Y, Wloga D, Dougan ST: **Embryological manipulations in zebrafish**. *Vertebrate Embryogenesis: Embryological, Cellular, and Genetic Methods* 2011:139–184.
62. Sundin J, Morgan R, Finnøen MH, Dey A, Sarkar K, Jutfelt F: **On the observation of wild zebrafish (*Danio rerio*) in India**. *Zebrafish* 2019, **16**(6):546–553.
63. Aleström P, D'Angelo L, Midtlyng PJ, Schorderet DF, Schulte-Merker S, Sohm F, Warner S: **Zebrafish: Housing and husbandry recommendations**. *Laboratory Animals* 2020, **54**(3):213–224.
64. Suurväli J, Whiteley AR, Zheng Y, Gharbi K, Leptin M, Wiehe T: **The laboratory domestication of zebrafish: From diverse populations to inbred substrains**. *Molecular Biology and Evolution* 2020, **37**(4):1056–1069.
65. Parichy DM: **Advancing biology through a deeper understanding of zebrafish ecology and evolution**. *Elife* 2015, **4**:e05635.
66. Ogi A, Licitra R, Naef V, Marchese M, Fronte B, Gazzano A, Santorelli FM: **Social preference tests in zebrafish: A systematic review**. *Frontiers in Veterinary Science* 2021, **7**:590057.
67. Takesono A, Kudoh T, Tyler CR: **Application of transgenic zebrafish models for studying the effects of estrogenic endocrine disrupting chemicals on embryonic brain development**. *Frontiers in Pharmacology* 2022, **13**:718072.
68. Phillips JB, Westerfield M: **Zebrafish models in translational research: Tipping the scales toward advancements in human health**. *Disease Models & Mechanisms* 2014, **7**(7):739–743.
69. Muniandy Y: **The role of zebrafish larvae for studying anxiety-like behaviour**. PhD dissertation, Leiden, 2019.
70. Kawakami K, Takeda H, Kawakami N, Kobayashi M, Matsuda N, Mishina M: **A transposon-mediated gene trap approach identifies developmentally regulated genes in zebrafish**. *Developmental Cell* 2004, **7**(1):133–144.

5 Repeated Blood Sampling from an Adult *Danio Rerio*

Manar A. Eissa, Raihan Abdullah, Faez Sharif, and Mohd Hamzah Mohd Nasir

5.1 INTRODUCTION

Zebrafish (*Danio rerio*) has gained popularity as a powerful tool for *in vivo* model in biomedical research, including developmental biology, toxicity, genetics, and neurobehavioral science (Binesh, 2021; Kobar et al., 2021; Mrinalini et al., 2023; Silva Brito et al., 2022; Trigueiro et al., 2020). The small size fish is relatively easy to handle, cheap to maintain, and requires less space compared to other vertebrates' models. Not limited to the logistics advantage, the complete genome of zebrafish has been successfully sequenced (Broughton et al., 2001; Chowdhury et al., 2022; Knapik et al., 1998). Upon annotations of the genome, other studies found that a minimum 70% of human genes possess at least one ortholog in the zebrafish genome (Howe et al., 2013). The versatility of zebrafish as a research tool is evident in the diverse range of studies that have been conducted. Since then, the use of zebrafish in biomedical research has become mainstream and indirectly increased the number of validated models for various study areas, including preclinical testing, developmental biology, immunological, toxicological, and environmental monitoring (Al-Dahmani et al., 2022; Binesh, 2021; Capiotti et al., 2014; Chowdhury et al., 2022; Eissa et al., 2021; Evans et al., 2004; Guo et al., 2020; Kobar et al., 2021; Mrinalini et al., 2023; Ota & Kawahara, 2014; Silva Brito et al., 2022; Trigueiro et al., 2020).

In the preclinical studies, adult zebrafish are normally used as a model to investigate the effect of therapeutic drugs or natural compounds against multiple pathological body conditions. Furthermore, zebrafish possess a highly conserved hematopoietic system, hemostasis as well as main metabolite homeostasis, which are comparable to humans (A. T. Chen & Zon, 2009; S. Y. Chen et al., 2013; Eames et al., 2010; Evans et al., 2004; Kobayashi et al., 2020). Although the volume of blood from this tiny fish is far smaller than that of other vertebrate models, the advancement in analysis of a small volume of body fluids has expanded the possibilities for zebrafish analysis; for instance, LC-MS/MS in the untargeted metabolomics assay (Langebeck-Jensen et al., 2019; Lößner et al., 2012; Lucitt et al., 2008). In addition to blood, zebrafish also have plasma, interstitial fluid, and cerebrospinal fluid, each with their own important functions. Zebrafish blood, like human blood, is composed of plasma, red blood cells, white blood cells, and platelets (Iyer & Jagadeeswaran, 2021). Likewise, there is a high degree of similarity between zebrafish and human plasma proteins (C. Li et al., 2016). This makes zebrafish a valuable model organism for studying hematologic and metabolic disorders. In other words, understanding the similarities and differences between zebrafish and human body fluids can provide valuable comparable animal models in drug design research.

Realizing the significance of blood fluid analysis in zebrafish, access to the limited volume of blood is precious. To proceed with a high-resolution analysis of associated functional components in the zebrafish, it is necessary to note that appropriate collection of blood samples is important. In the next section, various methods for blood sampling in zebrafish will be introduced, each with its own strengths and limitations. In general, blood sampling, when done correctly, can be a safe and effective procedure in zebrafish research. The optimum approach for zebrafish blood sample is, however, determined by the research purposes considering the required volume of blood, and the need for single or repeated measurement.

5.2 BLOOD COLLECTION METHODS IN ZEBRAFISH

Despite the advantages of using zebrafish in blood research, obtaining blood samples from these small aquatic organisms can be challenging. An adult zebrafish with an average length of 3.5–4.0 cm and weight between 0.4–0.5 g can only produce 0.05–0.07 ml of whole blood (Hohn & Petrie-Hanson, 2013; C. Li et al., 2016). Several methods were developed to withdraw blood from zebrafish, such as tail clipping (Avagyan & Zon, 2016; Hanumanthaiah et al., 2002; Hartig et al., 2016; Jagadeeswaran & Sheehan, 1999; Murtha et al., 2003; Velasco-Santamaría et al., 2011; Wu et al., 2018), cardiac puncture (Da'as et al., 2015; Moss et al., 2009; Oliveira et al., 2009; Rainieri et al., 2016), decapitation (Blazhenko et al., 2020; Eames et al., 2010; Foessl et al., 2021; Kitambi et al., 2012; Pedroso et al., 2012), lateral incision in the dorsal aorta (Jagadeeswaran & Sheehan, 1999; Zhang & Yeh, 2012), collection of blood through gills (Fakhlaei et al., 2021), and caudal vasculature (Abdullah et al., 2022; Félix et al., 2013; Zang et al., 2013, 2015). Some of these techniques are severely invasive or even lethal, while others necessitate the use of specialized equipment and proper training. Velasco-Santamaría et al. (2011) assessed the plasma cholesterol and steroid levels in zebrafish by collecting the blood from the fish tail by inserting a heparinized

DOI: 10.1201/9781003402893-6

microcapillary collection tube. Meanwhile, a modified tail ablation method was employed by Li et al. (2016), claiming its quick processing time and consistently high yield. On the other hand, Moss et al. (2009) intended to assess the level of blood glucose but were convinced that blood withdrawal from the tail vein caused more long-term damage than heart puncture did. Therefore, they collected blood using a cardiac puncture technique by inserting a pulled capillary pipet into the atria (Moss et al., 2009). Although cardiac puncture is minimally invasive, the amount of blood collected by this method is limited, and it requires high skills to insert the needle accurately and avoid damaging the heart.

The decapitation with scissors followed by inserting a heparinized microcapillary collecting tube into the pectoral articulation was another procedure applied by Eames et al. (2010) to assess the whole blood glucose level. This method is considered to be the most invasive and does not allow repeated measurements. Yet, it is a fast and reliable method, and an adequate amount of blood can be sampled. Other researchers preferred to collect zebrafish blood from the dorsal aorta, which is a large blood vessel that extends along the midline of the zebrafish body, from the head to the tail. This method was first implemented by Jagadeeswaran and Sheehan (1999) to examine the thrombocytes of the zebrafish but was found to be highly invasive. Therefore, Zang et al. (2015) developed a modified and reliable method for repeated blood collection from the dorsal aorta, which was less invasive. Blood samples were also withdrawn from the heart by inserting a needle through the gills (Fakhlaei et al., 2021). The use of a needle imitates the 'venipuncture' in a larger animal. Blood samples were also collected through insertion of a needle in the caudal vein (Félix et al., 2013). Both methods require high skills to insert the needle accurately. Recently, a method utilizing a fine glass capillary needle was used to allow repeated blood sampling from an adult zebrafish, enabling the multiple points measurement of blood glucose and triglycerides without sacrificing the fish (Abdullah et al., 2022; Foessl et al., 2021; Park et al., 2019).

The aforementioned methods were shown to yield maximal volume of blood from pooled or single fish for some analysis; however, they require sacrifice of the fish. Immunological studies such as ELISA and flow cytometry require a bigger volume of blood in which not to the advantage of zebrafish has forced the use of a pooled blood analysis strategy and limits only for single point measurement studies. Pooled blood analysis is superior in volume; however, the minimal change may be hindered by the interindividual variations.

TABLE 5.1
Summary of Blood Collection Method from Adult Zebrafish Highlighting its Advantages and Disadvantages

Method	Volume	Pooled/ Individual	Type of Analysis	Ref.	Advantages	Disadvantages
Tail clipping	NS	Pooled	Plasma lipid analysis	Velasco-Santamaría et al., 2011	Convenient	Lethal
Modified tail clipping	5-20 µL	Both	Plasma protein analysis	C. Li et al., 2016	Short processing time High yield	Lethal
Cardiac puncture	50 nL	Pooled	Blood glucose measurement	Moss et al., 2009	Least invasive, can be used for repeated sampling. Only venous blood can be collected.	Requires accurate insertion of the needle into heart chamber and needs profound skills and anatomical knowledge. Low yield
Decapitation	5-10 µL	Individual	Blood glucose measurement	Eames et al., 2010	Easy, quick, and reliable	Most invasive. Does not allow repeated sampling.
Lateral incision in dorsal aorta	1-5 µL	Pooled	Identification and characterization of zebrafish thrombocytes	Jagadeeswaran & Sheehan, 1999	High yield	Can cause severe bleeding. Invasive
Modified lateral incision in dorsal aorta	1-25 µL	Individual	Blood glucose and plasma triacylglycerol analysis	Zang et al., 2013	Repeated sampling Low mortality rate Minimally invasive High yield Less invasive	NS
Through gills	10 µL	Individual	Toxicity assay	Fakhlaei et al., 2021	Can collect a large amount of blood.	Requires high skills to insert the needle into the heart through the gills.
Through Caudal vein	1-5 µL	Individual	Steroid hormones analysis	Félix et al., 2013	Minimally invasive	Difficult to locate the caudal vein. Stressful for the fish and may cause bleeding. Low yield.

Note: NS = not specified

The methods of blood sampling from adult zebrafish are outlined in Table 5.1. As previously mentioned, no single method fits for all types of experiment. Hence, the best method is depending on the type of analysis, volume of blood required, single or repeated sampling, individual or pooled specimen. Because certain analyses call for a repeated blood sample, attempts have been made towards the development of repeated blood collection techniques, which entail cutting a small hole in the side of the zebrafish, where a catheter is introduced into the dorsal aorta. Over time, the catheter is utilized to collect blood.

The maximum volume of blood taken from an adult zebrafish is 0.07 mL, and it is classified as a terminal point where repeated sampling from the same zebrafish is impossible. Although tail vein ablation with centrifugation can yield a relatively higher volume of blood, the method restricts longitudinal studies and requires a larger sample size. Repeated blood sampling enables a better understanding of the hemodynamic parameters of an individual fish, but it does not provide a luxury of blood volume for more accurate analysis.

By using the microinjection on caudal vein method, the changes in the blood glucose levels of an individual zebrafish can be monitored over time, with reduced time intervals between sampling. In accordance with the 3Rs (Replacement, Reduction, and Refinement) of animal research, repeated blood collection from adult zebrafish is important to reduce the number of fish employed in a research study. In longitudinal studies, it is not necessary to use a large number of fish because researchers can collect repeated blood samples from the same fish throughout time. Moreover, by reducing interindividual disparities, this strategy makes it possible to analyze data in a more precise and reliable manner. Besides, repeated blood sampling permits the evaluation of temporal changes in blood parameters, providing important knowledge on the development of the disease and the effectiveness of treatment. The volume of blood that can be collected from an adult zebrafish is limited. It is recommended to collect ≤0.4% of the body weight per week for repeated measurements (Foessl et al., 2021). The amount of blood collected from each adult fish can vary between 1–10 µL (F. Li et al., 2016). The volume of plasma that can be collected is typically 40–50% of the whole blood volume (C. Li et al., 2016).

5.3 REPEATED BLOOD SAMPLING

Multi-point blood collection from an adult zebrafish is important in most blood and metabolic studies. The method allows researchers to profile the metabolic dynamic from an individual fish, which contributes to a high resolution of dataset. Currently, several devices have been developed for the purpose of blood withdrawal from adult zebrafish with good survival rate and sufficient yield. However, limited accessibility to the expensive instrument as well as the requirement of a higher skillset is hindering the log phase development of research utilizing zebrafish blood. This chapter demonstrates that multiple blood collection from an adult zebrafish can be achieved to the highest survival rate (94.75%) using the in-house fabricated blood withdrawal device called a Z-fish needle. The fabricated device is cheaper and can be accessible by a new or developing zebrafish laboratory. By utilizing the device, Abdullah et al. (2022) show that the dynamic of blood glucose can be monitored over 2–3 hours from the same fish without killing them. The device contains two main components: (1) the sampling syringe and (2) the microcapillary needle. Both components play an important role in contributing to the successful blood collection procedure.

5.4 Z-FISH NEEDLE

The Z-fish needle is a "do-it-yourself" microinjection device, assembled as shown in Figure 5.1a. In general, two major steps are required to fabricate a Z-fish needle, which are (1) the assembly of the sampling syringe and (2) the preparation of the microcapillary needle. The sampling syringe is a setup consisting of a dispensing syringe and a glass capillary holder. This part is responsible for controlling the volume of blood drawn from an adult zebrafish, as well as releasing the blood out for glucometer measurement. On the other hand, the microcapillary needle forms a sharp interface that is capable of penetrating zebrafish skin and flesh, making a puncture on the caudal vein without causing major damage to the neighboring tissues.

5.4.1 Step 1: Fabrication of the Z-Fish Needle Sampling Syringe

Similar to the microinjection apparatus, a Z-fish needle sampling syringe must be able to aspirate and dispense blood in a controlled mechanism, as well as holding the pulled microcapillary needle. Hence, two units of 1 mL syringe are required and fabricated as in Figure 5.1a. The sampling syringe consists of three components, namely, (1) a suction and dispense syringe, (2) a connecting tube, and (3) a needle holder (Figure 5.2b). The materials needed for the fabrication of the sampling needle are two units of 1 mL syringe (BD 1 ml Syringe Slip Tip with BD PrecisionGlide™), 20–30 cm long of rigid wall PVC tube with the outer diameter of 7 mm, and a pair of scissors, as demonstrated in Figure 5.2b.

Figure 5.2a-f shows the main steps in Z-fish needle sampling syringe fabrication. First, one unit of 1 mL syringe is disassembled, including the removal of plunger and rubber stopper (Figure 5.2b). Then, on the same syringe, the barrel flange must be removed to allow a perfect attachment of PVC tube (Figure 5.2c). To form an airtight seal of the microcapillary needle and the syringe, the rubber stop was reused as the gasket on the luer lock (Figure 5.2d). The needle holder allows for an ergonomic holding during the blood sampling procedure and functions as an interface between the syringe and the microcapillary needle. Due to the size difference between the body of the needle holder and the connecting tube, a blow torch was used to heat and soften the tube to allow an airtight attachment between the

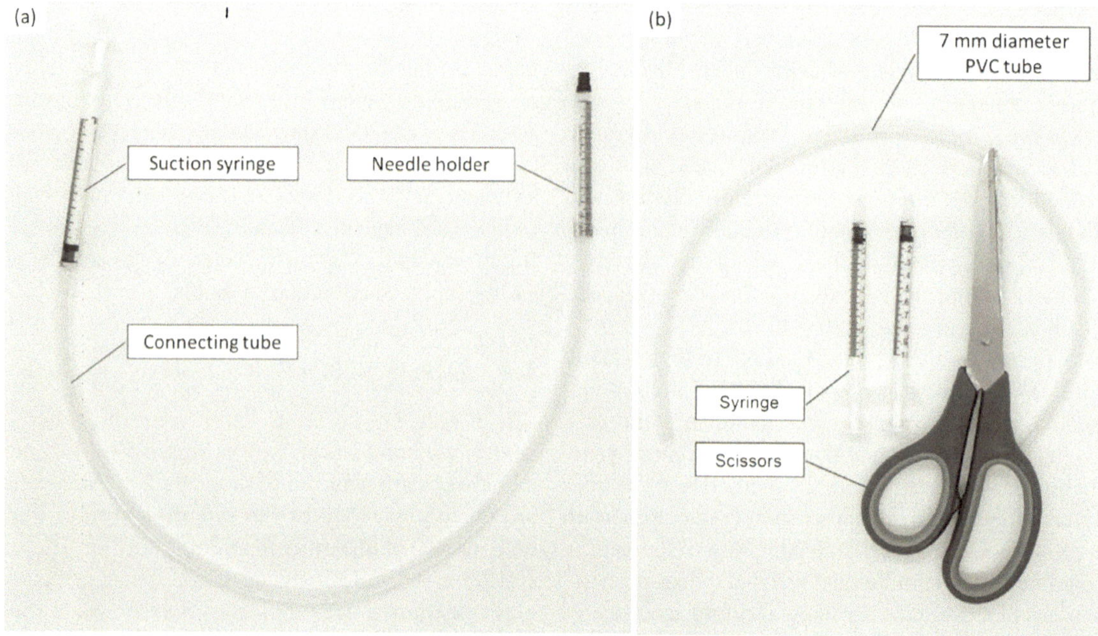

FIGURE 5.1 (a) The components of the sampling syringe, (b) the materials needed for fabrication of the sampling syringe.

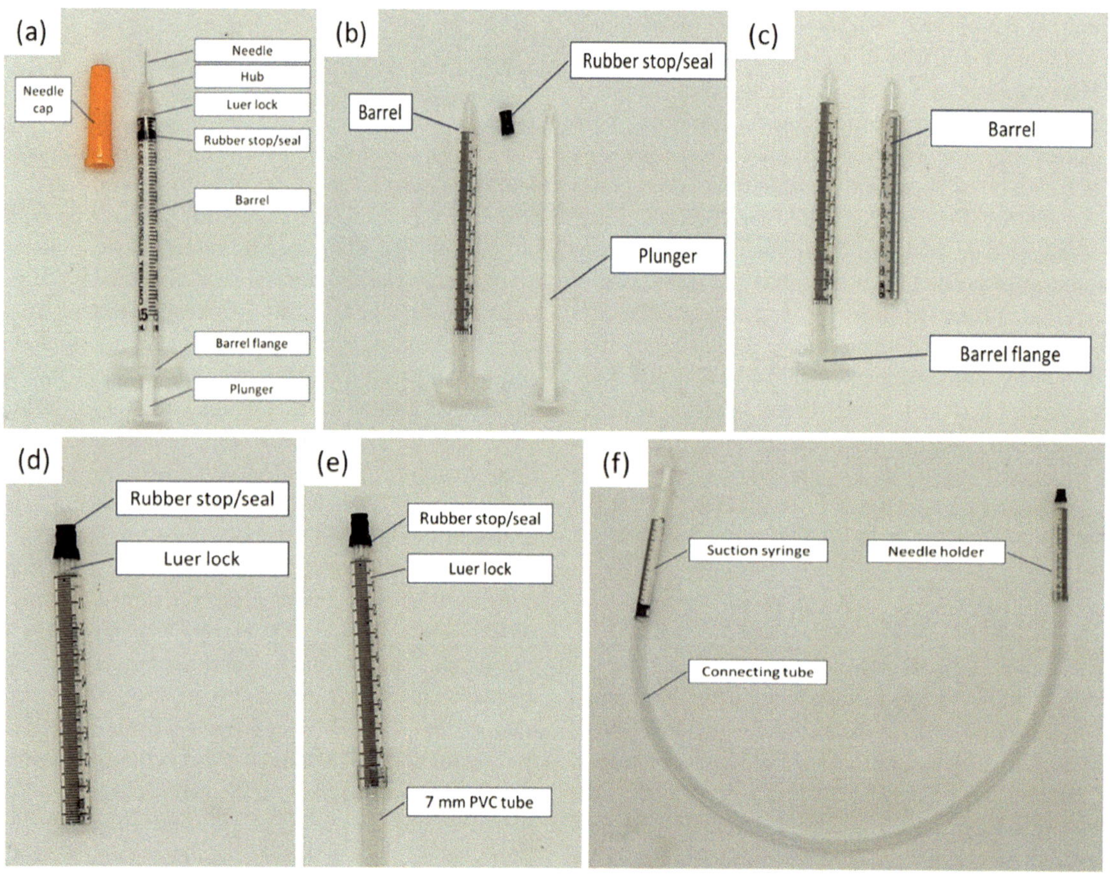

FIGURE 5.2 (a) The typical parts and components of a syringe, (b) removal of plunger and rubber stop from syringe barrel, (c) removal of barrel flange, (d) reinsert the rubber seal to the luer lock as an interface for microcapillary needle, (e) connecting the PVC tube to the trimmed syringe barrel, and (f) the other end of PVC tube connected to the luer lock of suction syringe.

connecting tube and needle holder (Figure 5.2e). The connecting tube links the syringe and the microneedle, so that the blood sampling procedure can be handled conveniently. Thus, the length of the connecting tube is determined according to the user preferences. The tube in this study was a clear high-pressurized polyurethane tubing (5 mm internal diameter, 7 mm outside diameter). This unit will be used as the needle holder of the Z-fish needle device. Finally, the prepared needle holder needs to be connected to another syringe to serve as the suction syringe, as shown in Figure 5.2f. This setup is made for a single operator procedure.

5.4.2 Step 2: Preparation of the Microcapillary Needle

The microcapillary needle is the component that penetrates skin and punctures the target blood vessel of a zebrafish, to access the capillary blood. To serve this purpose, commercially available needles cannot be used due to the cost, size, and materials used. Hence, a custom microneedle is required. The outer diameter of the tip of the microcapillary needle must be smaller than the target blood vessel. This is to allow a complete puncture but with less damage to the blood vessels. The inner diameter, however, must be bigger than the blood cells (oval shape with average of 9 µm long and 7 µm wide, see Figure 5.11) to reduce the risk for hemolysis. To fabricate the microneedle without using a commercial device, a set of apparatus as shown in Figure 5.3 is required.

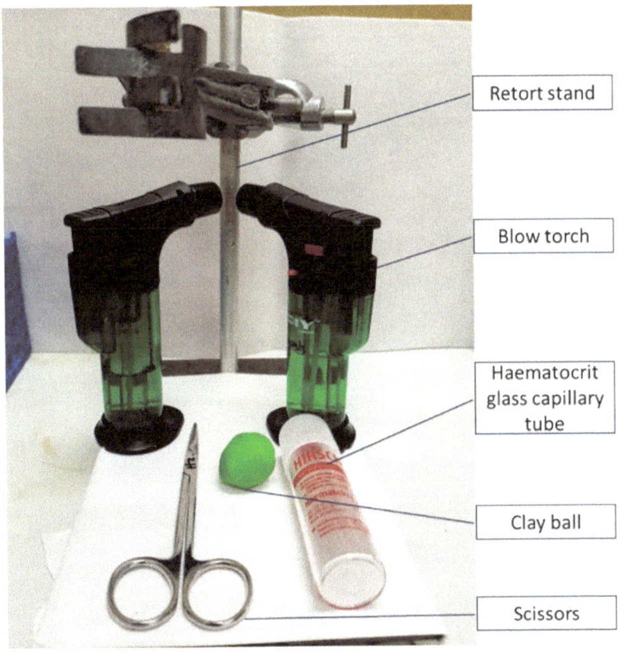

FIGURE 5.3 Haematocrit capillary tube Hirschmann (heparinized, 75 ul, 75 × 1.6 mm, 10''s/vial), two unit of torch blower, retort stand, modeling clay, storage box, and a pair of scissors.

First, a retort stand needs to be adjusted to where the position of the clamp is at the center of the retort base, with the height of 35 cm. This is to allow sufficient distance for the pulled needle to travel during the process. The microcapillary tube must be perpendicular to the gravity, as the alignment is crucial to ensure the capillary tube is pulled in a straight line. To reduce the sudden drop, a cotton wool cushion as shown in Figure 5.4 can be used.

Retort clamp has limited grip to a small material. To assist the grip, a small clay padding as shown in Figure 5.5a can be used. A consistent, controlled pulled gravitational force can be achieved by adding 2 g of clay at the bottom end of the microcapillary tube (Figure 5.5b). The needle can only form a fine thin tube when heated. To achieve homogeneous heating and shrinkage of the capillary tube, the flames on both the left and right sides must be blown at the same time as in Figure 5.5c.

FIGURE 5.4 The setup for the preparation of a microcapillary needle. The microcapillary tube is aligned straight to the gravity, and cotton wool serves as a receiving cushion of the pulled tube.

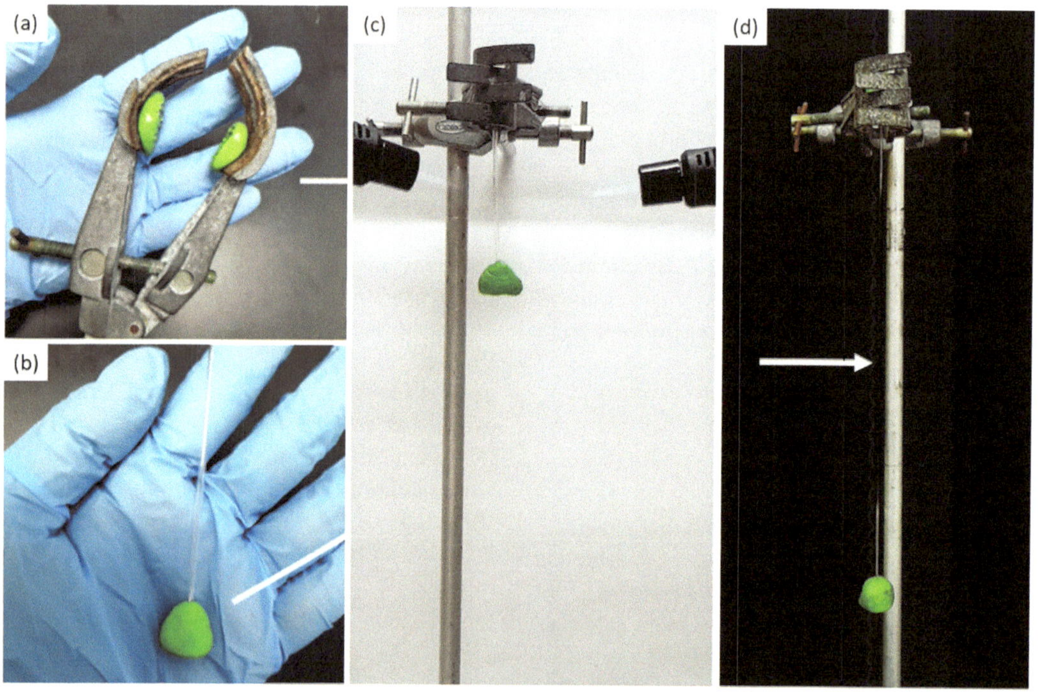

FIGURE 5.5 The preparation of retort stand: (a) the retort clamp was layered with a soft clay (green in color); (b) at one end of the microcapillary tube, clay was used to serve as weight; (c) two units of blowtorch were used to evenly heat the microcapillary tube; (d) the arrow shows an even shrinkage of the microcapillary tube after the heating.

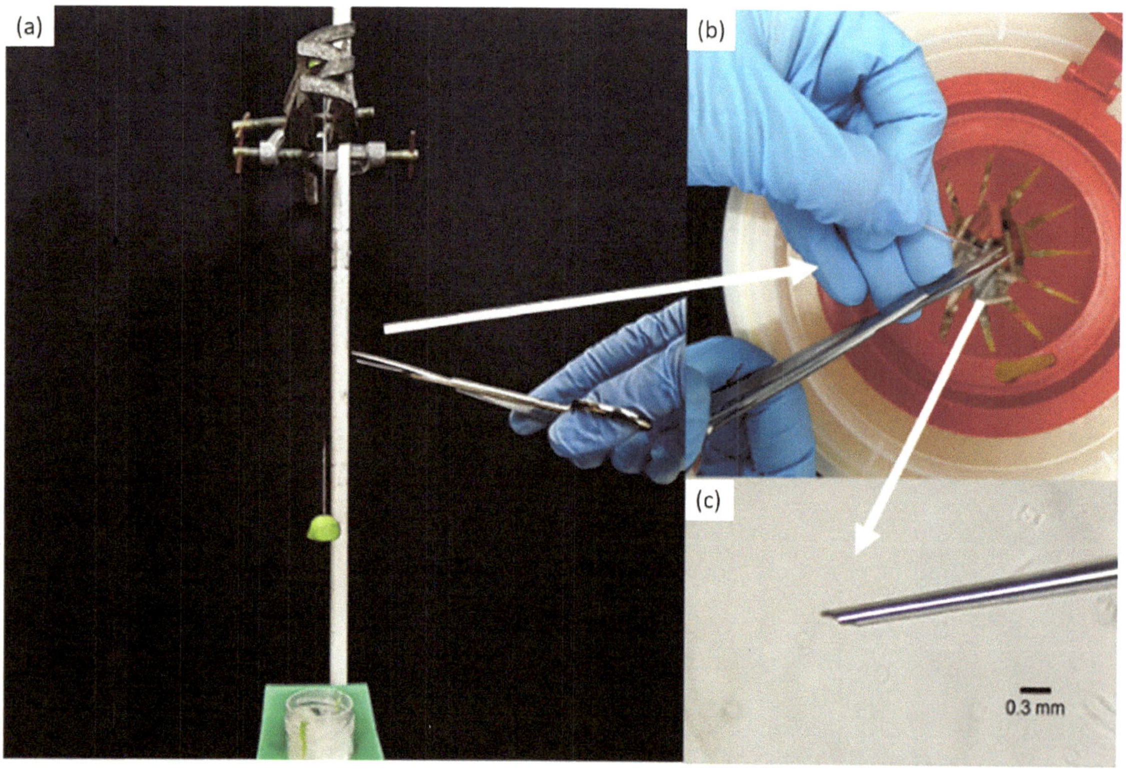

FIGURE 5.6 The finishing process of a microcapillary needle: (a) cutting at the middle of even shrinkage of the capillary tube using the scissors, (b) adjusting and refining the length and end point of the microcapillary tube, and (c) observing the prepared microcapillary needle under a microscope for bevel angle and diameter measurement.

The pulled needle must be cut and trimmed to form a functional needle with a sharp end. To cut, a pair of clean scissors can be used. The pointed bevel is achieved by cutting the needle at a 45° angle and smoothed by using a rubber lathe (Figure 5.6). A box with a clay block can be an alternative to secure a storage box of the fabricated microcapillary needle (Figure 5.7).

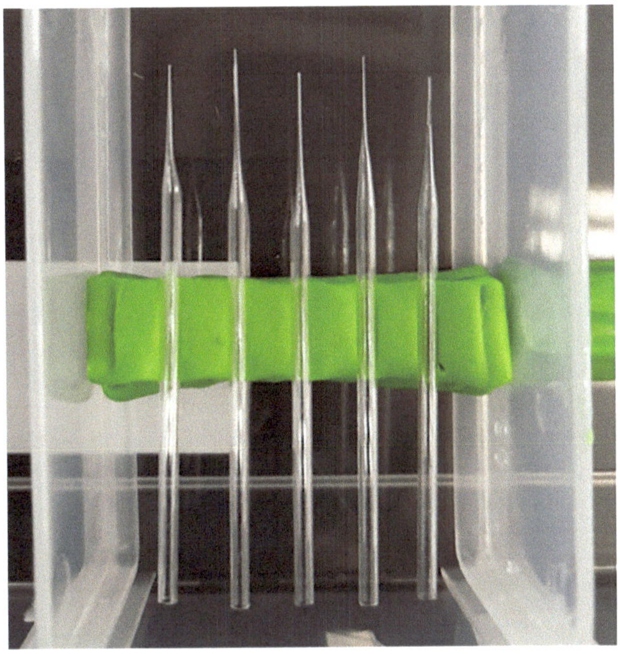

FIGURE 5.7 The prepared microcapillary needles were attached to the soft clay (green color) in a storage box for secure protection.

5.4.3 Volume Calibration of Microcapillary Needle

To ensure a uniform size and volume of the fabricated microcapillary needle, a size and volume calibration is required. The outer diameter and inner diameter of the needle's tip can be measured by image analysis software on a calibrated image (Figure 5.8). Open-source software such as ImageJ is recommended. The average outer diameter of 0.214 ± 0.07 mm is suitable for caudal vein puncture.

Volume calibration is important to reduce the event of over-withdrawing of blood, which can be fatal to the fish. Simple volume calibration can be achieved by placing 3 μL of sterile water on a clean parafilm, to form a water button. The water button is then aspirated using the microcapillary needle, and the level where the water stops on the needle is marked.

5.4.4 Heparin Coating and Sterilization

Clot is an immediate event when blood is taken out of a vessel. The clotting can be avoided by coating the microcapillary needle with 5 mg/mL of heparin. The treated needle needs to be air dried to allow a uniform coating of heparin. To minimize the risk of infection on the injected fish, the needle must be sterilized using UV light.

5.4.5 Repeated Blood Sampling from an Adult Zebrafish

As summarized in Table 5.1, it is important to consider the various methods of blood sampling to suit the requirement

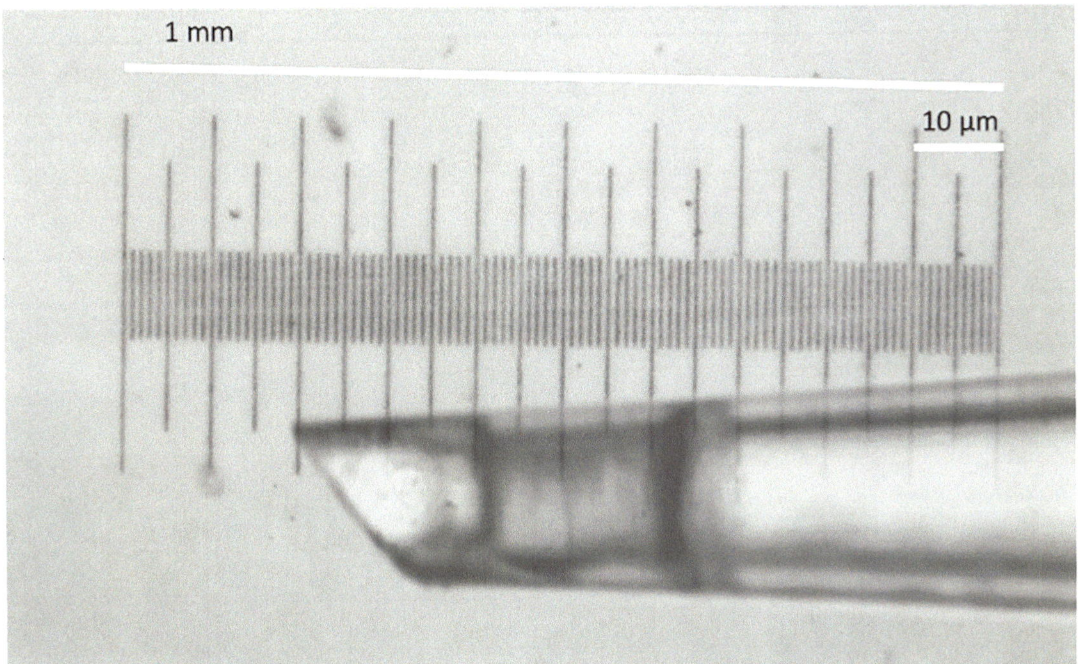

FIGURE 5.8 Microcapillary needle over a stage micrometer observed under 20x objective using a compound microscope.

of a study. By using a Z-fish needle, a protocol of repeated blood sampling has been established. In this example, a method by Abdullah et al. (2022) is referred. This was a modified method developed by Zang et al. (2013, 2015) but using the in-house fabricated Z-fish needle.

The method started with fish preparation. First, the zebrafish is transferred into an isolation tank containing system water to allow easy handling of the fish (Figure 5.9a). From the isolation tank, the fish is then anesthetized in a cold-water bath (10–14°C) until the fish loses its buoyancy balance and does not respond to the tail pinch (Figure 5.9b). The anesthetized fish is then transferred onto a cold sponge that serves as a fish holder and temperature stabilizer to keep the fish anesthetized throughout the venipuncture procedure (Figure 5.9c).

On the cold sponge, locate the anatomical site of the caudal vein as shown in Figure 5.10a. By using the fabricated device, insert the microcapillary needle at a 30–45° angle at the identified site (Figure 5.10b) to the depth of 3–5 mm, or until some resistance can be felt. The successful venipuncture of the caudal vein is demonstrated by the presence of blood in the microcapillary needle due to the capillary action as well as the blood pressure from the fish (Figure 5.10c). Immediately remove the needle once the blood reaches the calibration marks of the needle. It is important to note that a little compression must be applied for 5 to 10 seconds on the injection site to help stop the bleeding. To measure the blood glucose, for instance, dispense the collected blood onto a clean parafilm as shown in Figure 5.10d and read the glucose level using the appropriate electronic glucometer (Figure 5.10e).

Upon successful blood collection, the fish must be immediately returned to a recovery tank containing an appropriate concentration of antibacterial and antifungal agent (e.g., API MelaFix). The rate of successful recovery of the fish depends on many factors, especially (1) the temperature of water in the recovery tank, (2) the amount of blood withdrawn, and (3) the duration of the fish outside of water after anesthesia. Hence, it is crucial to control these factors in order to increase the success rate of repeated blood sampling from an adult zebrafish (Figure 5.11) less relevant as this is to show the size comparison and no hemolysis observed.

5.5 FISH INJECTION AND SURVIVAL RATE

Abdullah et al. (2022) showed that, of 75 fishes used in the study, only five fish died immediately due to the procedure. Interestingly, six times of injection were conducted over eight days, where three samplings were done in the first seven days and three samplings were done at 30- to 60-minute intervals on day 8 (Figure 5.12). By using this method, over 93% of survival was recorded, which is higher in comparison to other methods outlined in Table 5.1.

5.5.1 Ethical Statement

The in vivo study of adult zebrafish was approved by the Institutional Animal Care and Use Committee (IACUC)

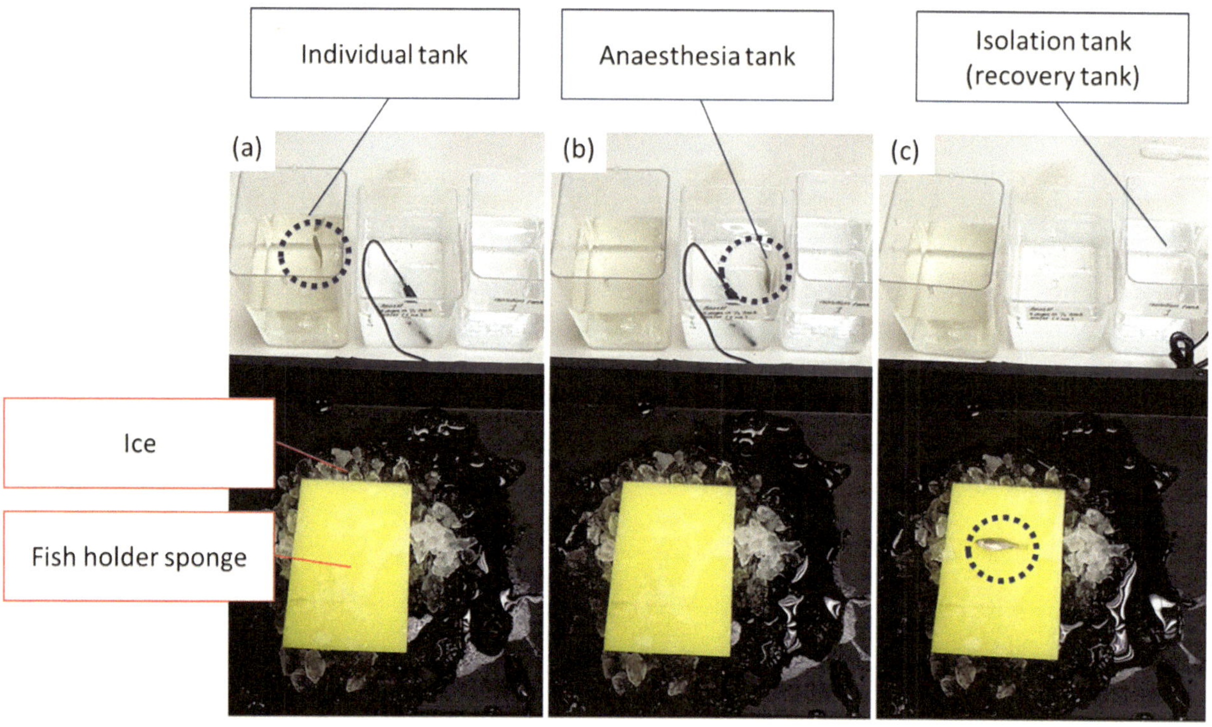

FIGURE 5.9 The preparation of fish for blood sampling: (a) individual fish is isolated from the community tank; (b) the fish is then anesthetized in a cold-water bath; and (c) the anesthetized fish is then transferred onto the cold sponge for the venipuncture procedure.

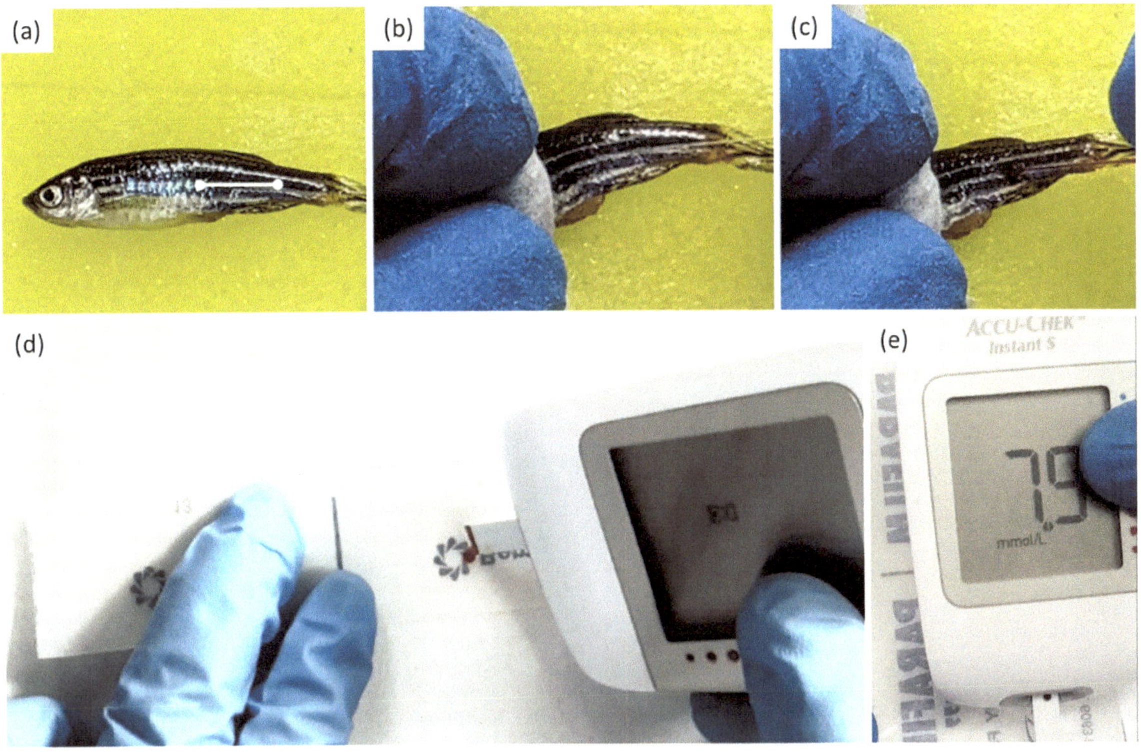

FIGURE 5.10 Blood sampling from the caudal vein of an adult zebrafish: (a) the blood collection site along the body axis and posterior to the anus, where caudal vein can be easily accessed; (b) insertion of the microcapillary needle (Z-fish needle) into the target injection site; (c) the blood will enter the microcapillary needle due to capillary action and blood pressure of the fish. The collected blood is dispensed on a clean parafilm (d), for the measurement of blood glucose as in (e).

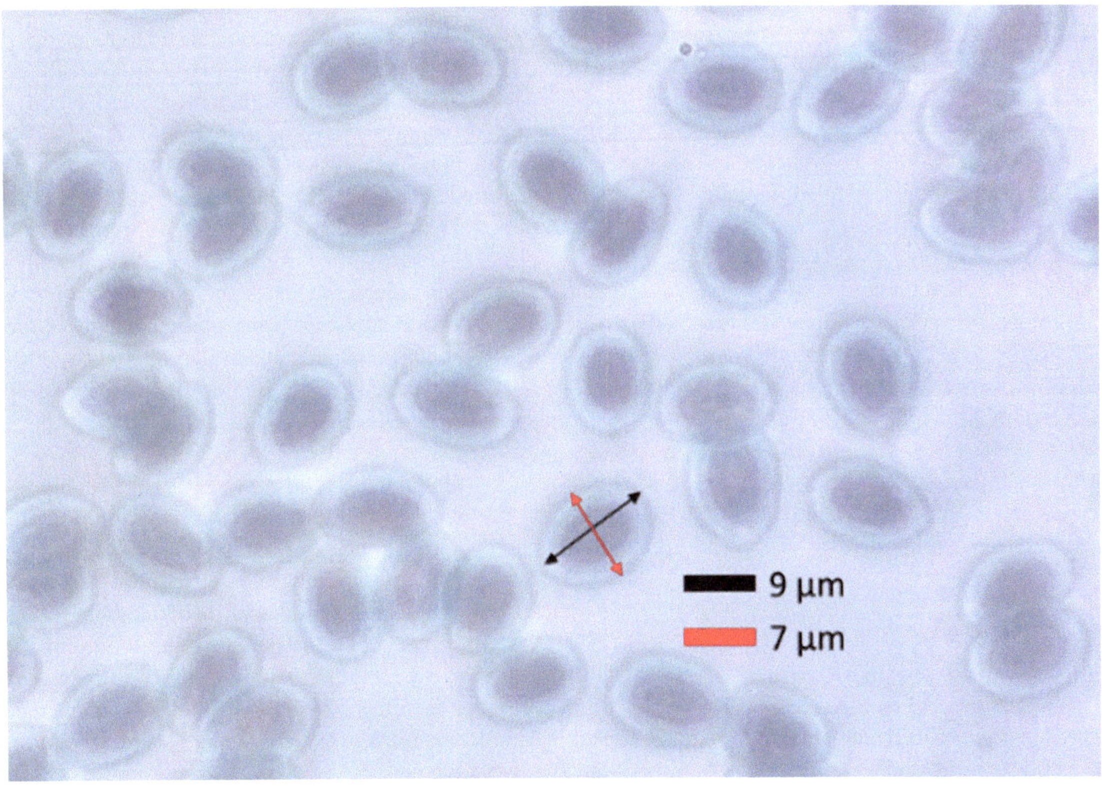

FIGURE 5.11 The red blood cells of zebrafish under 40x objective using compound microscope.

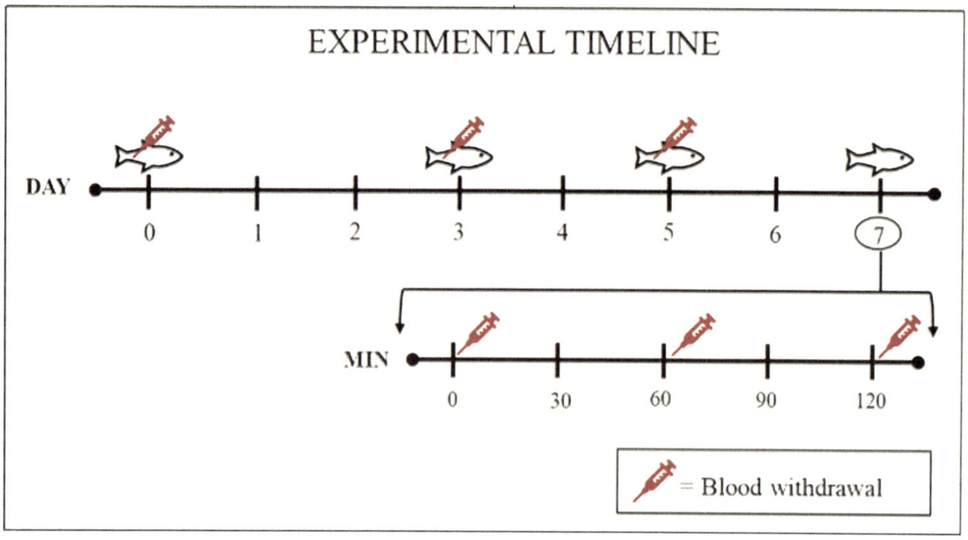

FIGURE 5.12 The experimental timeline showed blood specimens were successfully collected six times from each fish over eight days.

of the International Islamic University Malaysia (IIUM) (Reference: IIUM/504/14/2/IACUC), and the study was carried out in accordance with the Malaysian Animal Welfare Act 2015 for the care and use of animals for scientific purposes. The zebrafish were purchased from a local pet store in Kuantan and acclimated in standard laboratory conditions using a photoperiod of 14 hours light and 10 hours dark at a temperature of 26 + 2°C for 10 days. During the acclimation period, the fish were treated with an antifungal (API MelaFix) and an antibacterial (API MelaFix) to avoid any disease transfer into the facility. The experiment was carried out at Kulliyyah of Science, with each fish placed in a group of 15 in an 8L acrylic fish tank filled with filtered facility water and aerated with air bubbles and biofilter stones. The zebrafish were fed with commercial fish diet Otohime B2 (Marubeni Nisshin Feed, Japan).

5.5.2 Pain Management and Recovery

The fish were anesthetized using ice water (gradually from 10°C to 14°C) for 30 seconds. After injection, the fish will be immediately transferred back to the recovery tank (18°C to reduce the pain after injection) (Kinkel & Prince, 2009; Zang et al., 2017). To relieve wound healing and prevent any infection at the injection site, antibacterial (API Melafix) was added at the ratio of 1 mL Melafix to every 200 mL of water to the recovery tanks.

5.6 CONCLUSION

Blood sampling and analysis in zebrafish (*Danio rerio*) have emerged as pivotal tools in biomedical research due to their physiological and genetic resemblance to humans, coupled with their rapid reproduction and transparent embryos. The zebrafish is considered an excellent model organism for elucidating basic biological processes and investigating the mechanisms of several diseases (Lieschke & Currie, 2007). Due to their small size, ease of maintenance, and possibility of genetic manipulation, zebrafish became ideal for studying a wide range of diseases, including developmental disorders, cardiovascular diseases, and cancer (Lawson & Weinstein, 2002). The presence of several biomarkers, hormones, and cellular components in the blood enables researchers to identify molecular pathways, explore potential drugs, and better comprehend the disease progression through analyzing zebrafish blood.

It is important to remember that choosing the right zebrafish blood collection method is crucial for getting accurate and trustworthy data. Repeated blood sampling may result in physiological changes brought on by stress, which may then distort the experimental data. Single or terminal procedures, such as dorsal aorta and caudal vein puncture, have evident advantages over repeated blood sampling. Contrarily, the dorsal aorta and caudal vein puncture techniques reduce the stress and potential problems associated with repeated sampling and ensure the collection of adequate blood volumes for in-depth analysis. These methods provide the precise temporal control of sample sites, allowing longitudinal studies and obviating the necessity for experimental animals.

A wide range of tests is used in zebrafish blood analysis to provide useful information on several physiological and pathological processes (Goldsmith, 2004). Red and white blood cell counts, hemoglobin concentrations, and hematocrit measures are examples of hematological tests that provide information on the health of the blood, including, for example, the blood's capacity to carry oxygen. Blood tests that look at things like glucose, cholesterol, and electrolytes can also be used to evaluate renal and metabolic processes. Innovative methods such as flow cytometry can shed light on how infections and disorders affect the immune system's response in zebrafish blood. Analyzing blood components in a molecular fashion can also identify biomarkers and gene expression patterns associated with specific diseases, according to Veneman et al. (2014) and Zon and Peterson (2005).

Utilizing blood tests on zebrafish provides a strong foundation for uncovering medicines and studying harmful substances. These analyses serve as tools in understanding how drugs get absorbed, dispersed, metabolized, and finally leave the bloodstream of the zebrafish. By observing developing embryos of this fish species under several medicinal exposures, this gives us hints about prospective healing agents. Additionally, we also employ these assays to check the drug's poisonous nature or any damaging consequences they might bear with them. Such experiments aid researchers nail down compounds that cast positive impacts upon certain ailments while minimizing collateral damage to normal physiological functions.

Finally, various methods in the blood sampling from an adult zebrafish have become an important tool in biomedical investigation, connecting the theoretical findings and applied sciences. These indispensable tools serve as a crucial link, connecting the dots between fundamental science's theoretical principles and practicable clinical applications. To ensure data integrity and dependable results, selecting an appropriate method for collecting blood samples is critical. Fortunately, the numerous testing and analytical techniques applied to zebrafish blood can be used to get a thorough understanding of the physiological processes, disease mechanisms, and therapeutic activities of medications. Assays using the blood of zebrafish also hold great promise for improving toxicological research and drug development investigations. Blood analysis is still a cornerstone of research projects aimed at improving human health since zebrafish continue to serve as versatile and effective model organisms.

REFERENCES

Abdullah, N. R., Mohd Nasir, M. H., Azizan, N. H., Wan-Mohtar, W. A. A. Q. I., & Sharif, F. (2022). Bioreactor-grown exo- and endo-β-glucan from Malaysian Ganoderma lucidum: An in vitro and in vivo study for potential antidiabetic treatment. *Frontiers in Bioengineering and Biotechnology*, *10*. https://doi.org/10.3389/fbioe.2022.960320

Al-Dahmani, Z. M., Li, X., Wiggenhauser, L. M., Ott, H., Kruithof, P. D., Lunev, S., Batista, F. A., Luo, Y., Dolga, A. M., Morton, N. M., Groves, M. R., Kroll, J., & van Goor, H. (2022). Thiosulfate sulfurtransferase prevents hyperglycemic damage to the zebrafish pronephros in an experimental model for diabetes. *Scientific Reports*, *12*(1). https://doi.org/10.1038/s41598-022-16320-1

Avagyan, S., & Zon, L. I. (2016). Fish to Learn: Insights into blood development and blood disorders from Zebrafish hematopoiesis. *Human Gene Therapy*, *27*(4). https://doi.org/10.1089/hum.2016.024

Binesh, A. (2021). Decades-long involvement of signalling pathways in cardiovascular research using Zebrafish model and its global trends. *Reviews in Aquaculture*, *13*(1). https://doi.org/10.1111/raq.12486

Blazhenko, A. A., Khokhlov, P. P., Tissen, I. Y., Devyashin, A. S., Lebedev, A. A., Proshin, S. N., Bychkov, E. R., & Shabanov, P. D. (2020). Benzodiazepine tranquilizers abolish the stress-induced increase of the brain ghrelin level in *Danio rerio*. *Reviews on Clinical Pharmacology and Drug Therapy*, *18*(4). https://doi.org/10.17816/rcf184327-332

Broughton, R. E., Milam, J. E., & Roe, B. A. (2001). The complete sequence of the zebrafish (*Danio rerio*) mitochondrial genome and evolutionary patterns in vertebrate mitochondrial DNA. *Genome Research*, *11*(11). https://doi.org/10.1101/gr.156801

Capiotti, K. M., Antonioli, R., Kist, L. W., Bogo, M. R., Bonan, C. D., & Da Silva, R. S. (2014). Persistent impaired glucose metabolism in a zebrafish hyperglycemia model. *Comparative Biochemistry and Physiology Part B: Biochemistry and Molecular Biology*, *171*(1). https://doi.org/10.1016/j.cbpb.2014.03.005

Chen, A. T., & Zon, L. I. (2009). Zebrafish blood stem cells. *Journal of Cellular Biochemistry*, *108*(1). https://doi.org/10.1002/jcb.22251

Chen, S. Y., Shih, H. Y., Lin, S. J., Hsiao, C. Der, Li, Z. C., & Cheng, Y. C. (2013). Etv5a regulates the proliferation of ventral mesoderm cells and the formation of hemato-vascular derivatives. *Journal of Cell Science*, *126*(24). https://doi.org/10.1242/jcs.132613

Chowdhury, K., Lin, S., & Lai, S. L. (2022). Comparative study in Zebrafish and medaka unravels the mechanisms of tissue regeneration. *Frontiers in Ecology and Evolution*, *10*. https://doi.org/10.3389/fevo.2022.783818

Da'as, S. I., Balci, T. B., & Berman, J. N. (2015). Mast cell development and function in the Zebrafish. *Methods in Molecular Biology*, *1220*. https://doi.org/10.1007/978-1-4939-1568-2_3

Eames, S. C., Philipson, L. H., Prince, V. E., & Kinkel, M. D. (2010). Blood sugar measurement in zebrafish reveals dynamics of glucose homeostasis. *Zebrafish*, *7*(2). https://doi.org/10.1089/zeb.2009.0640

Eissa, M. A., Hashim, Y. Z. H. Y., Mohd Nasir, M. H., Nor, Y. A., Salleh, H. M., Isa, M. L. M., Abd-Azziz, S. S. S., Abd Warif, N. M., Ramadan, E., & Badawi, N. M. (2021). Fabrication and characterization of Agarwood extract-loaded nanocapsules and evaluation of their toxicity and anti-inflammatory activity on RAW 264.7 cells and in zebrafish embryos. *Drug Delivery*, *28*(1). https://doi.org/10.1080/10717544.2021.2012307

Evans, T., Gupta, S., & McReynolds, L. M. (2004). New animal models reveal stage-specific hematopoietic functions for the BMP signaling pathway. *Blood*, *104*(11). https://doi.org/10.1182/blood.v104.11.133.133

Fakhlaei, R., Selamat, J., Razis, A. F. A., Sukor, R., Ahmad, S., Babadi, A. A., & Khatib, A. (2021). In vivo toxicity evaluation of sugar adulterated heterotrigona itama honey using zebrafish model. *Molecules*, *26*(20), 1–20. https://doi.org/10.3390/molecules26206222

Félix, A. S., Faustino, A. I., Cabral, E. M., & Oliveira, R. F. (2013). Noninvasive measurement of steroid hormones in zebrafish holding-water. *Zebrafish*, *10*(1), 110–115. https://doi.org/10.1089/zeb.2012.0792

Foessl, I., Bassett, J. H. D., Bjørnerem, Å., Busse, B., Calado, Â., Chavassieux, P., Christou, M., Douni, E., Fiedler, I. A. K., Fonseca, J. E., Hassler, E., Högler, W., Kague, E., Karasik, D., Khashayar, P., Langdahl, B. L., Leitch, V. D., Lopes, P., Markozannes, G., . . . Obermayer-Pietsch, B. (2021). Bone phenotyping approaches in human, mice and Zebrafish—Expert overview of the EU cost action GEMSTONE ("GEnomics of MusculoSkeletal traits TranslatiOnal

NEtwork"). *Frontiers in Endocrinology*, *12*(December), 1–32. https://doi.org/10.3389/fendo.2021.720728

Goldsmith, P. (2004). Zebrafish as a pharmacological tool: The how, why and when. *Current Opinion in Pharmacology*, *4*(5). https://doi.org/10.1016/j.coph.2004.04.005

Guo, J. N., Liu, F., & Wang, L. (2020). Zebrafish blood disease models and applications. *Yi chuan = Hereditas*, *42*(8). https://doi.org/10.16288/j.yczz.20-148

Hanumanthaiah, R., Day, K., & Jagadeeswaran, P. (2002). Comprehensive analysis of blood coagulation pathways in teleostei: Evolution of coagulation factor genes and identification of Zebrafish factor VIIi. *Blood Cells, Molecules, and Diseases*, *29*(1). https://doi.org/10.1006/bcmd.2002.0534

Hartig, E. I., Zhu, S., King, B. L., & Coffman, J. A. (2016). Cortisol-treated zebrafish embryos develop into pro-inflammatory adults with aberrant immune gene regulation. *Biology Open*, *5*(8). https://doi.org/10.1242/bio.020065

Hohn, C., & Petrie-Hanson, L. (2013). Evaluation of visible implant elastomer tags in zebrafish (*Danio rerio*). *Biology Open*, *2*(12). https://doi.org/10.1242/bio.20136460

Howe, K., Clark, M. D., Torroja, C. F., Torrance, J., Berthelot, C., Muffato, M., Collins, J. E., Humphray, S., McLaren, K., Matthews, L., McLaren, S., Sealy, I., Caccamo, M., Churcher, C., Scott, C., Barrett, J. C., Koch, R., Rauch, G. J., White, S., . . . Stemple, D. L. (2013). The zebrafish reference genome sequence and its relationship to the human genome. *Nature*, *496*(7446). https://doi.org/10.1038/nature12111

Iyer, N., & Jagadeeswaran, P. (2021). Microkinetic coagulation assays for human and zebrafish plasma. *Blood Coagulation and Fibrinolysis*, *32*(1). https://doi.org/10.1097/MBC.0000000000000975

Jagadeeswaran, P., & Sheehan, J. P. (1999). Analysis of blood coagulation in the zebrafish. *Blood Cells, Molecules, and Diseases*, *25*(3–4). https://doi.org/10.1006/bcmd.1999.0249

Kinkel, M. D., & Prince, V. E. (2009). On the diabetic menu: Zebrafish as a model for pancreas development and function. *BioEssays*, *31*(2). https://doi.org/10.1002/bies.200800123

Kitambi, S. S., Nilsson, E. S., Sekyrova, P., Ibarra, C., Tekeoh, G. N., Andäng, M., Ernfors, P., & Uhlén, P. (2012). Small molecule screening platform for assessment of cardiovascular toxicity on adult zebrafish heart. *BMC Physiology*, *12*(1). https://doi.org/10.1186/1472-6793-12-3

Knapik, E. W., Goodman, A., Ekker, M., Chevrette, M., Delgado, J., Neuhauss, S., Shimoda, N., Driever, W., Fishman, M. C., & Jacob, H. J. (1998). A microsatellite genetic linkage map for zebrafish (*Danio rerio*). *Nature Genetics*, *18*(4). https://doi.org/10.1038/ng0498-338

Kobar, K., Collett, K., Prykhozhij, S. V., & Berman, J. N. (2021). Zebrafish cancer predisposition models. *Frontiers in Cell and Developmental Biology*, *9*. https://doi.org/10.3389/fcell.2021.660069

Kobayashi, I., Kobayashi-Sun, J., Hirakawa, Y., Ouchi, M., Yasuda, K., Kamei, H., Fukuhara, S., & Yamaguchi, M. (2020). Dual role of Jam3b in early hematopoietic and vascular development. *Development (Cambridge)*, *147*(1). https://doi.org/10.1242/dev.181040

Langebeck-Jensen, K., Shahar, O. D., Schuman, E. M., Langer, J. D., & Ryu, S. (2019). Larval Zebrafish proteome regulation in response to an environmental challenge. *Proteomics*, *19*(14). https://doi.org/10.1002/pmic.201900028

Lawson, N. D., & Weinstein, B. M. (2002). In vivo imaging of embryonic vascular development using transgenic zebrafish. *Developmental Biology*, *248*(2). https://doi.org/10.1006/dbio.2002.0711

Li, C., Tan, X. F., Lim, T. K., Lin, Q., & Gong, Z. (2016). Comprehensive and quantitative proteomic analyses of zebrafish plasma reveals conserved protein profiles between genders and between zebrafish and human. *Scientific Reports*, *6*. https://doi.org/10.1038/srep24329

Li, F., Wang, H., Liu, J., Lin, J., Zeng, A., Ai, W., Wang, X., Dahlgren, R. A., & Wang, H. (2016). Immunotoxicity of β-diketone antibiotic mixtures to zebrafish (*Danio rerio*) by transcriptome analysis. *PLoS One*, *11*(4). https://doi.org/10.1371/journal.pone.0152530

Lieschke, G. J., & Currie, P. D. (2007). Animal models of human disease: Zebrafish swim into view. *Nature Reviews Genetics*, *8*(5). https://doi.org/10.1038/nrg2091

Lößner, C., Wee, S., Ler, S. G., Li, R. H. X., Carney, T., Blackstock, W., & Gunaratne, J. (2012). Expanding the zebrafish embryo proteome using multiple fractionation approaches and tandem mass spectrometry. *Proteomics*, *12*(11). https://doi.org/10.1002/pmic.201100576

Lucitt, M. B., Price, T. S., Pizarro, A., Wu, W., Yocum, A. K., Seiler, C., Pack, M. A., Blair, I. A., FitzGerald, G. A., & Grosser, T. (2008). Analysis of the zebrafish proteome during embryonic development. *Molecular and Cellular Proteomics*, *7*(5). https://doi.org/10.1074/mcp.M700382-MCP200

Moss, J. B., Koustubhan, P., Greenman, M., Parsons, M. J., Walter, I., & Moss, L. G. (2009). Regeneration of the pancreas in adult zebrafish. *Diabetes*, *58*(8), 1844–1851. https://doi.org/10.2337/db08-0628

Mrinalini, R., Tamilanban, T., Naveen Kumar, V., & Manasa, K. (2023). Zebrafish—The neurobehavioural model in trend. *Neuroscience*, *520*. https://doi.org/10.1016/j.neuroscience.2022.12.016

Murtha, J. M., Qi, W., & Keller, E. T. (2003). Hematologic and serum biochemical values for zebrafish (*Danio rerio*). *Comparative Medicine*, *53*(1).

Oliveira, R., Domingues, I., Grisolia, C. K., & Soares, A. M. V. M. (2009). Effects of triclosan on zebrafish early-life stages and adults. *Environmental Science and Pollution Research*, *16*(6). https://doi.org/10.1007/s11356-009-0119-3

Ota, S., & Kawahara, A. (2014). Zebrafish: A model vertebrate suitable for the analysis of human genetic disorders. *Congenital Anomalies*, *54*(1). https://doi.org/10.1111/cga.12040

Park, K. H., Ye, Z. W., Zhang, J., & Kim, S. H. (2019). Palmitic acid-enriched diet induces hepatic steatosis and injury in adult Zebrafish. *Zebrafish*, *16*(6). https://doi.org/10.1089/zeb.2019.1758

Pedroso, G. L., Hammes, T. O., Escobar, T. D. C., Fracasso, L. B., Forgiarini, L. F., & da Silveira, T. R. (2012). Blood collection for biochemical analysis in adult zebrafish. *Journal of Visualized Experiments*, *63*. https://doi.org/10.3791/3865

Rainieri, S., Conlledo, N., Larsen, B. K., Granby, K., & Barranco, A. (2016). Effects of dietary microplastic exposure on the organ toxicity of a mixture of chemical contaminants in zebrafish (*Danio rerio*). *Toxicology Letters*, *258*. https://doi.org/10.1016/j.toxlet.2016.06.1995

Silva Brito, R., Canedo, A., Farias, D., & Rocha, T. L. (2022). Transgenic zebrafish (*Danio rerio*) as an emerging model system in ecotoxicology and toxicology: Historical review, recent advances, and trends. *Science of the Total Environment*, *848*. https://doi.org/10.1016/j.scitotenv.2022.157665

Trigueiro, N. S. D. S., Canedo, A., Braga, D. L. D. S., Luchiari, A. C., & Rocha, T. L. (2020). Zebrafish as an emerging model system in the global south: Two decades of research in Brazil. *Zebrafish*, *17*(6). https://doi.org/10.1089/zeb.2020.1930

Velasco-Santamaría, Y. M., Korsgaard, B., Madsen, S. S., & Bjerregaard, P. (2011). Bezafibrate, a lipid-lowering pharmaceutical, as a potential endocrine disruptor in male zebrafish (*Danio rerio*). *Aquatic Toxicology*, *105*(1–2), 107–118. https://doi.org/10.1016/j.aquatox.2011.05.018

Veneman, W. J., Marín-Juez, R., de Sonneville, J., Ordas, A., Jong-Raadsen, S., Meijer, A. H., & Spaink, H. P. (2014). Establishment and optimization of a high throughput setup to study staphylococcus epidermidis and mycobacterium marinum infection as a model for drug discovery. *Journal of Visualized Experiments*, *88*. https://doi.org/10.3791/51649

Wu, T. S., Lin, Y. T., Huang, Y. T., Cheng, Y. C., Yu, F. Y., & Liu, B. H. (2018). Disruption of liver development and coagulation pathway by ochratoxin A in embryonic zebrafish. *Toxicology and Applied Pharmacology*, *340*. https://doi.org/10.1016/j.taap.2017.12.012

Zang, L., Shimada, Y., & Nishimura, N. (2017). Development of a novel Zebrafish model for type 2 diabetes mellitus. *Scientific Reports*, *7*(1). https://doi.org/10.1038/s41598-017-01432-w

Zang, L., Shimada, Y., Nishimura, Y., Tanaka, T., & Nishimura, N. (2013). A novel, reliable method for repeated blood collection from aquarium fish. *Zebrafish*, *10*(3), 425–432. https://doi.org/10.1089/zeb.2012.0862

Zang, L., Shimada, Y., Nishimura, Y., Tanaka, T., & Nishimura, N. (2015). Repeated blood collection for blood tests in adult Zebrafish. *Journal of Visualized Experiments*, *102*. https://doi.org/10.3791/53272-v

Zhang, Y., & Yeh, J. R. J. (2012). In vivo chemical screening for modulators of hematopoiesis and hematological diseases. *Advances in Hematology*, *2012*. https://doi.org/10.1155/2012/851674

Zon, L. I., & Peterson, R. T. (2005). In vivo drug discovery in the zebrafish. *Nature Reviews Drug Discovery*, *4*(1). https://doi.org/10.1038/nrd1606

6 Zebrafish Culture

Zahra Morshed, Fatemeh Delkhosh-Kasmaie, Mahsa Valizadeh, Mohammad Rezaei, and Yaser Tahamtani

6.1 INTRODUCTION TO ZEBRAFISH AS A MODEL ORGANISM

The zebrafish (*Danio rerio*) has established itself as a model organism of exceptional value in biological research. Its genetic proximity to humans, rapid development, and transparent embryos make it a versatile tool for scientists studying a variety of biological questions. Because more than 70% of human genes have functional counterparts in zebrafish, researchers can explore fundamental biological processes and disease mechanisms with direct relevance to human health. External fertilization and transparent embryos enable real-time imaging of development and provide important insights into organ formation, cell differentiation, and morphogenesis. The zebrafish's compact size, large clutches, and easy maintenance make it inexpensive for experiments, while its regenerative abilities captivate researchers studying tissue repair. This model organism closes the gap between basic research and clinical applications and is therefore an invaluable asset for the understanding of biology and the further development of medical science (Belyaeva et al., 2009; Briggs, 2002; Dooley & Zon, 2000).

Zebrafish offer a constellation of unique advantages in the study of development, genetics, and disease that distinguish them as a valued model organism. Their genetic similarity to humans, with significant overlap in functional genes, provides a platform for studying critical genetic and molecular pathways involved in development and disease (Rubinstein, 2003). Zebrafish embryos develop externally and are optically transparent, allowing researchers to observe developmental processes in real time. This transparency is particularly useful for tracking organogenesis and understanding cellular events. Additionally, their rapid development, small size, and high fecundity make zebrafish ideal for large-scale genetic and drug testing. Zebrafish also have remarkable regenerative abilities and provide insight into tissue repair and regeneration, a property not commonly seen in other vertebrate models. This unique combination of genetic proximity, transparency, rapid development, and regenerative capabilities positions zebrafish at the forefront of developmental, genetic, and disease research, with profound implications for understanding human biology and developing new therapeutic strategies (Liu & Leach, 2011).

Cell lines are an indispensable part of zebrafish research and fulfill important requirements in both basic biological investigations and translational studies. While zebrafish provide a remarkable whole-organism model, the use of cell lines allows researchers to bridge the gap between in vivo complexity and controlled in vitro experiments. These cell lines provide a platform for analyzing the molecular and cellular mechanisms underlying developmental processes, disease states, and drug responses. In addition, they enable high-throughput screening, genetic manipulations, and toxicological assessments that are often difficult to perform in live animals. Zebrafish cell lines serve as important tools for researchers to unlock the intricacies of zebrafish biology, improving our understanding of human health and disease. Their role in zebrafish research spans diverse areas, including developmental biology, genetics, neuroscience, and drug development, further underscoring the critical need for these cell models to realize the full potential of zebrafish as a model organism (Choorapoikayil et al., 2013).

6.2 ORIGINS AND ESTABLISHMENT OF ZEBRAFISH CELL LINES

The historical development of zebrafish cell lines represents a dynamic journey marked by significant milestones in the field of zebrafish research. Originally, zebrafish cell lines were established to complement whole-organism studies and provide a controlled system for in vitro experiments. In the early 1990s, pioneering efforts were made to derive zebrafish embryonic cell lines, laying the foundation for the study of early developmental processes (Driever & Rangini, 1993). Subsequently, advances in cell culture techniques, including the development of immortalization strategies, led to the creation of stable and continuously growing zebrafish cell lines. These milestones opened new avenues for research, enabling studies of drug metabolism, toxicology, gene function, and disease modeling. Today, there is a diverse range of zebrafish cell lines with unique properties and applications, reflecting the rich history of innovation and continuous development of zebrafish cell line research (Collodi et al., 1992; Jin et al., 2017).

The isolation and establishment of zebrafish cell lines requires a series of precise and careful laboratory procedures. First, researchers dissect specific tissues or organs from zebrafish embryos or adults under sterile conditions. The removed tissues are then subjected to enzymatic digestion or mechanical dissociation to release individual cells from their natural matrix. After dissociation, the cells are cultured in special growth media that provide essential nutrients and growth factors to support their survival and proliferation.

DOI: 10.1201/9781003402893-7

The process may involve using selective culture conditions to enrich specific cell types. To ensure their longevity, some zebrafish cell lines are immortalized through genetic modifications or viral transformation, protecting them from senescence or cell death. Overall, establishing zebrafish cell lines requires precision, patience, and expertise, with each step carefully tailored to the specific tissue source and research goals (Eide et al., 2014; Vallone et al., 2007, 2008).

Immortalization techniques are central to zebrafish cell line research and provide a way to ensure the sustainable growth and long-term viability of cultured cells. These techniques are of utmost importance as they enable the creation of stable and reproducible cell lines, thus providing researchers with a continuous and controlled platform for a variety of scientific applications. Immortalized zebrafish cell lines serve as workhorses in high-throughput screening assays, accelerating drug discovery and toxicology studies. They facilitate genetic manipulation and allow scientists to explore the role of specific genes in cellular processes (Lakra et al., 2011). Furthermore, these immortalized lineages play an important role in disease modeling, enabling the study of disease mechanisms and the evaluation of potential therapeutics. Immortalization ensures the consistency of experiments and long-term maintenance of unique cell lines and highlights their indispensable role in advancing our understanding of zebrafish biology and its impact on human health (Takakura et al., 1999).

6.3 CHARACTERIZATION OF ZEBRAFISH CELL LINES

Zebrafish cell lines have several key characteristics that make them invaluable tools in scientific research. These cell lines typically come from specific zebrafish tissues, allowing researchers to focus on specific aspects of zebrafish biology. Importantly, zebrafish cell lines often share genetic and functional similarities with their in vivo counterparts, reflecting genetic conservation between zebrafish and humans. They are suitable for in vitro manipulation and allow researchers to conduct controlled experiments such as genetic modifications or drug screening. Zebrafish cell lines are well suited for high-throughput testing and facilitate large-scale studies. Furthermore, their maintenance in culture provides a stable and reproducible cell source for experiments over extended periods of time. These key features collectively underscore the importance of zebrafish cell lines in advancing our understanding of zebrafish biology and its relevance to human health and disease (Ho et al., 2014).

Authentication and strict quality control are of utmost importance when characterizing zebrafish cell lines. To ensure the reliability and reproducibility of research results, researchers use a variety of techniques to authenticate these cell lines. DNA profiling, particularly through short tandem repeat (STR) analysis, is a common method for verifying the identity of zebrafish cells and ensuring that they match the original source tissue. Quality control measures include regular assessments of cell morphology, growth rates, and behavior with the aim of confirming the stability and authenticity of the cell line over time. In addition, researchers routinely test for the absence of contaminants such as mycoplasma or cross-species cell line contamination that could affect experimental results. By adhering to strict authentication and quality control protocols, researchers can safely use zebrafish cell lines as reliable tools in various scientific investigations, thereby protecting the integrity of their research results (Fan et al., 2004; He et al., 2006).

The care and maintenance of zebrafish cell lines are crucial aspects of their usefulness in scientific research. Continuous care includes periodic subculture or passage to prevent excessive confluence and ensure a steady supply of cells for experiments. Culture conditions must be carefully controlled, including temperature, humidity, and the composition of growth media tailored to specific cell types. To prevent contamination and maintain cell line purity, sterile techniques are essential. Cryopreservation is an important long-term storage strategy that allows researchers to preserve zebrafish cell lines for extended periods of time. The cells are frozen in cryoprotective solutions and stored at extremely low temperatures to protect them from genetic drift and ensure a reliable cell source for future studies. To maintain the authenticity and functionality of zebrafish cell lines, proper maintenance and preservation protocols are critical to enable researchers to utilize their full potential for scientific investigation over time (Lakra et al., 2011; Şerifoğlu, 2019).

6.4 PROMINENT ZEBRAFISH CELL LINES AND THEIR APPLICATIONS

Several significant zebrafish cell lines have been established, each offering unique properties and applications. The ZF4 cell line derived from zebrafish embryos serves as a model for developmental biology, toxicology, and gene expression studies and preserves the characteristics of embryonic cells. PAC2 cells, also derived from embryos, play a central role in neurodevelopmental research and provide insights into the complex processes of neuronal development. The ZFL cell line isolated from zebrafish liver tissue is a cornerstone in the study of drug metabolism, toxicology, and the effects of environmental pollutants on the liver. AB9 cells derived from zebrafish brain tissue are invaluable for neurobiological studies, enabling studies of neuronal development and function. These diverse zebrafish cell lines provide researchers with specialized tools to explore a wide range of biological processes and contribute significantly to our understanding of zebrafish biology and its relevance to human health and disease. We introduce these cell lines in more detail in the following sections.

6.4.1 ZF4 (Zebrafish Fin Cell Line)

The ZF4 cell line, short for Zebrafish Fin Cell Line 4, is a remarkable and well-established zebrafish cell line derived from zebrafish fin tissue. These cells are characterized by their rapid growth, ease of maintenance in culture, and ability to maintain genetic and functional similarities to their in vivo counterparts. ZF4 cells have attracted considerable attention in the field of developmental biology and toxicology due to their versatility and utility as a model system. These cells retain several key features of embryonic cells, making them particularly valuable for studies related to early developmental processes. Researchers have used ZF4 cells to study tissue morphogenesis, gene expression patterns, and responses to environmental toxins, shedding light on critical aspects of zebrafish biology and their relevance to human health. One of the notable features of ZF4 cells is their suitability for in vitro experiments and high-throughput screening. This property has made them indispensable for toxicology studies, where they serve as a cost-effective and ethical alternative to whole-organism assessments. Researchers can expose ZF4 cells to various compounds and pollutants to quickly assess their effects on cellular processes, gene expression, and cell viability. This not only advances our understanding of toxicological mechanisms but also helps identify potential environmental hazards and develop safer chemicals. Essentially, ZF4 cells have become an important tool in zebrafish research, providing a unique insight into early development and environmental toxicology. In summary, the ZF4 cell line is a versatile and valuable tool in zebrafish research. Due to its properties, it is particularly suitable for developmental biology studies, toxicological studies, and high-throughput screening assays. By providing a controlled in vitro system, ZF4 cells contribute significantly to our understanding of zebrafish biology and its broader implications for human health and environmental science (Driever & Rangini, 1993).

6.4.2 PAC2 (Zebrafish Primary Alar Cholinergic Neurons)

The PAC2 cell line, known as the Zebrafish Primary Alar Cholinergic Neurons Cell Line, is a remarkable zebrafish cell line derived from zebrafish embryos. PAC2 cells have become a crucial tool in the field of neurodevelopmental research due to their unique properties and applications. These cells come from the wing plate of the zebrafish embryo, a region of the developing nervous system. Therefore, PAC2 cells are mainly used to study neuronal development and function and provide a valuable platform for understanding the intricacies of the zebrafish nervous system. One of the key advantages of PAC2 cells is their ability to mimic key aspects of neuronal development. Researchers are using PAC2 cells to study the differentiation of stem cells into specialized cholinergic neurons, thereby exploring the molecular and cellular mechanisms that control this essential process. This knowledge has far-reaching implications for understanding both normal neuronal development and how disruptions in this process can lead to neurodevelopmental disorders in humans. In addition, PAC2 cells have been instrumental in studying the effects of various factors on neuronal development. Researchers can expose these cells to different compounds, drugs, or genetic changes to assess how these factors affect the development and function of cholinergic neurons. Such studies are critical for identifying potential therapeutic targets for neurodevelopmental disorders and for improving our understanding of the complex processes that control neuronal development in both zebrafish and humans. In summary, the PAC2 zebrafish cell line is a valuable resource that contributes significantly to the field of neurodevelopmental research, shedding light on critical aspects of neuronal development and providing insights into potential therapies for neurological diseases (Amsterdam et al., 1999; Chen et al., 2002).

6.4.3 ZFL (Zebrafish Liver Cell Line)

The ZFL cell line (or Zebrafish Liver Cell Line) is a crucial resource in the field of zebrafish cell line research. These cells, derived from zebrafish liver tissue, have attracted considerable attention due to their central role in the study of drug metabolism, toxicology, and the effects of environmental pollutants on liver function. Like humans, zebrafish have a liver with essential functions in detoxification, metabolism, and lipid regulation. ZFL cells mimic these functions very well, making them a valuable model for studying how chemicals and compounds interact with the liver and influence its cellular processes. One of the main applications of ZFL cells is toxicology studies. Researchers expose these cells to various environmental pollutants, drugs, or compounds to assess their effects on liver function and health. This is critical for identifying potential hazards, assessing the safety of chemicals, and improving our understanding of how environmental factors can lead to liver-related health problems. The insights gained from ZFL cell studies contribute to risk assessment, environmental monitoring, and the development of safer chemicals and medicines. In addition, ZFL cells play a central role in drug metabolism research. The liver is a major site for drug metabolism in both zebrafish and humans. By studying ZFL cells, scientists can study how different drugs are processed and metabolized. This information is invaluable for optimizing drug dosage, identifying potential drug interactions, and ensuring the safety and effectiveness of pharmaceutical compounds. Overall, the ZFL zebrafish cell line is a critical tool for advancing our understanding of liver biology, toxicology, and drug metabolism with implications for human health and environmental sciences (Ghosh et al., 1994; Miranda et al., 1993).

6.4.4 AB9 (Zebrafish Brain Cell Line)

The AB9 cell line, derived from zebrafish brain tissue, is a significant gain in neurobiological research. These cells have gained importance in the study of neuronal development and function due to their unique properties and applications. Like humans, zebrafish have a complex nervous system, making AB9 cells an invaluable model for exploring the intricacies of neurobiology. One of the main applications of AB9 cells is to elucidate the development of neurons. Researchers use AB9 cells to study the differentiation of neuronal progenitor cells into mature neurons, thereby exploring the genetic and molecular mechanisms that control this complicated process. The insights gained from these studies contribute to our understanding of normal neuronal development and provide potential insights into the origins of neurodevelopmental disorders in humans. Furthermore, AB9 cells serve as an ideal model to study the functional aspects of neurons. Researchers can manipulate these cells to study processes such as synaptic transmission, neuronal signals, and how neurons respond to different connections or conditions. By exposing AB9 cells to various stimuli, scientists gain valuable insight into how neurons function, adapt, and respond to their environment. This knowledge has profound implications for neurobiology, neuropharmacology, and our understanding of neurological diseases and disorders. In summary, the AB9 zebrafish cell line is an important resource in neurobiological research, shedding light on fundamental aspects of neuronal development and function with relevance to human health and disease (Bhadra et al., 2015).

6.5 DEVELOPMENTAL BIOLOGY WITH ZEBRAFISH CELL LINES

Zebrafish cell lines play a crucial role in unraveling the mysteries of embryonic development. These cell lines provide a unique insight into the earliest stages of life, allowing researchers to analyze the intricate processes that control tissue and organ formation. Using zebrafish cell lines, scientists can study cell differentiation, tissue morphogenesis, and dynamic gene expression patterns during embryogenesis. This information is critical to understanding not only normal development but also how developmental processes can go awry and lead to congenital disorders and diseases in humans. Zebrafish cell lines, coupled with in vivo studies, contribute significantly to our knowledge of embryonic development and provide insights with far-reaching implications for regenerative medicine, birth defects, and developmental biology in general. Case studies using zebrafish cell lines have provided profound insights into cell differentiation, tissue morphogenesis, and gene expression dynamics during development. These studies have shed light on the molecular mechanisms that control the formation of complex organs and structures in zebrafish embryos. For example, researchers have used zebrafish cell lines to study the differentiation of stem cells into specialized cell types such as neurons or muscle cells, shedding light on the factors that control this process. Tissue morphogenesis studies have revealed how zebrafish cells organize to produce the various tissues and structures critical for embryonic development. Gene expression profiling using zebrafish cell lines has revealed the temporal and spatial regulation of genes and provided valuable data on the genetic control of development. These case studies illustrate the invaluable contribution of zebrafish cell lines in unraveling the intricacies of embryonic development and advance our understanding of fundamental biology and its relevance to human health and disease (Lungu-Mitea et al., 2020).

6.6 DISEASE MODELING AND GENE FUNCTION ANALYSIS

Zebrafish cell lines have become indispensable tools for modeling human diseases. Researchers can manipulate these cell lines to mimic specific genetic mutations or introduce disease-associated factors, effectively creating cellular models of human disorders. These models allow scientists to explore the underlying molecular and cellular mechanisms of disease in a controlled laboratory environment. For example, zebrafish cell lines have been used to study genetic mutations associated with neurological disorders, cardiovascular disease, and various cancers. By recapitulating disease-related cell phenotypes, zebrafish cell lines provide a platform for studying disease progression, drug response, and potential therapeutic interventions. These models bridge the gap between in vitro cell culture and in vivo animal models, providing a cost-effective and genetically controllable system for studying human diseases, ultimately improving our understanding of pathologies and facilitating the development of targeted treatments (Choorapoikayil et al., 2013; Tonon & Grassi, 2023).

6.7 GENE FUNCTION ANALYSIS USING ZEBRAFISH CELL LINES FOR UNDERSTANDING CELLULAR PROCESSES

Gene function analysis using zebrafish cell lines has proven crucial for elucidating the intricacies of cellular processes. Researchers can manipulate gene expression within these cell lines to assess the effects on cellular functions and thus decipher the role of specific genes in different biological contexts. Whether studying genes related to developmental pathways, signal transduction, or disease mechanisms, zebrafish cell lines provide a controlled environment for genetic perturbations. This approach allows researchers to pinpoint the genes responsible for critical cellular processes, helping to elucidate their functions and regulatory networks. Such insights are critical not only for understanding basic biology but also for uncovering the molecular basis of disease and advancing therapeutic development, making zebrafish cell lines invaluable tools for gene function analysis (Heilmann et al., 2015).

6.8 DRUG DISCOVERY AND TOXICOLOGY

Zebrafish cell lines play a central role in drug discovery, revolutionizing the way potential drugs are identified and developed. These cell lines are essential for high-throughput screening assays, allowing researchers to quickly and efficiently evaluate the effects of thousands of compounds.

By exposing zebrafish cells to different drugs, researchers can assess their effectiveness, toxicity, and mechanisms of action. Zebrafish cell lines are particularly valuable in identifying promising drug candidates for further investigation and significantly accelerate the drug development process. Additionally, these cell lines serve as a bridge between in vitro studies and in vivo animal models, allowing researchers to identify compounds that have the potential for success in whole-organism studies. The versatility and cost-effectiveness of zebrafish cell lines make them indispensable tools in modern drug discovery, contributing to the rapid development of novel therapies and the advancement of medicine (Zhang et al., 2003).

High-throughput screening (HTS) and drug testing using zebrafish cell lines represents an innovative approach to drug discovery and development. These cell lines, suitable for large-scale culture and automated testing, allow researchers to quickly determine the effects of numerous compounds on cellular processes to rate. This streamlined process significantly accelerates the identification of potential drug candidates, making it a cost-effective and time-efficient strategy. Zebrafish cell lines also serve as a bridge between in vitro experiments and whole-organism studies, providing a controlled environment for assessing drug toxicity and efficacy. By facilitating HTS and drug testing, zebrafish cell lines are driving innovation in the pharmaceutical industry and ultimately leading to the discovery of novel drugs and therapeutic interventions for various diseases and conditions (George et al., 2011).

Toxicological studies and compound toxicity assessment are important applications of zebrafish cell lines. These cell lines provide a controlled platform to evaluate the potential toxic effects of various compounds, including environmental pollutants and drugs. By exposing zebrafish cells to these substances, researchers can study their effects on cell viability, metabolism, and gene expression profiles. This information is critical to determining the safety of chemicals and compounds and helps regulatory agencies set guidelines for exposure levels. Zebrafish cell lines also enable high-throughput toxicological screening, which is critical for quickly and efficiently assessing the safety of numerous compounds. In addition, their genetic traceability allows researchers to investigate the molecular mechanisms underlying toxicity, contributing to our understanding of how certain chemicals affect cellular processes. In this way, zebrafish cell lines play a central role in advancing toxicology research and ensuring the safety of products and environments (Au-Rodrigues de Souza et al., 2023; Quevedo et al., 2021; Zhang et al., 2003).

6.9 CROSS-SPECIES COMPARISONS AND TRANSLATIONAL RESEARCH

The study of zebrafish cell lines provides invaluable insights into the conserved molecular and cellular processes shared by zebrafish and humans. Despite the evolutionary distance between the two species, zebrafish and humans share remarkable genetic similarities, with significant overlap in their functional genes (Howe et al., 2013). This genetic conservation underpins the conserved cellular processes observed in zebrafish cell lines and provides a powerful model system to study fundamental biological phenomena that apply to both zebrafish and humans. Researchers can explore conserved pathways related to development, cell signaling, and disease mechanisms and uncover the deep-rooted principles that govern these processes across species (Lam et al., 2011; Morrison et al., 2022). Such findings have far-reaching implications as they not only advance our understanding of zebrafish biology but also provide crucial parallels to human biology and help identify potential therapeutic targets, disease mechanisms, and drug candidates for various human diseases. Thus, zebrafish cell lines serve as a bridge that connects the intricacies of zebrafish biology to the broader field of human health and disease research.

Zebrafish cell lines hold enormous potential for translational research and drug development. Their genetic proximity to humans, coupled with their suitability for genetic manipulation and high-throughput screening, makes them a crucial tool for bridging the gap between basic research and clinical applications. Zebrafish cell lines enable rapid evaluation of drug candidates for efficacy and toxicity, thereby accelerating the drug development process. In addition, they serve as platforms for disease modeling and help researchers analyze the molecular basis of various human diseases. The insights gained from zebrafish cell line studies can guide the development of targeted therapies and serve as the basis for clinical trials. In addition, these cell lines offer a cost-effective and ethical alternative to traditional animal models in preclinical drug testing. In this way, zebrafish cell lines hold tremendous promise for advancing translational research and accelerating the development of innovative treatments for a broad range of human diseases.

6.10 CHALLENGES AND ADVANCES IN ZEBRAFISH CELL LINE RESEARCH

Maintaining cell lines, including zebrafish cell lines, poses several challenges that researchers must address diligently. Contamination is an ongoing problem, with mycoplasma and cross-species contamination posing a significant threat to cell line purity. Regular testing and strict adherence to sterile techniques are critical in combating these problems. Another challenge is maintaining the genetic stability of cell lines over time, as genetic drift can occur, potentially altering cell properties. Proper recording and regular authentication through techniques such as short tandem repeat (STR) analysis are critical to ensuring the identity and stability of cell lines. Additionally, avoiding excessive confluence and monitoring culture conditions are critical to prevent cell stress and maintain cell health. Researchers must also implement strict quality control measures to consistently monitor cell morphology, growth rates, and viability. Finally, cryopreservation protocols should be carefully optimized to ensure long-term viability of cell lines. Addressing these challenges is critical to preserving the authenticity and functionality of cell lines and ensuring their reliability in scientific research (Vallone et al., 2007).

Recent advances in zebrafish cell line research have significantly advanced the field in several ways. Improvements in scalability and stability have been achieved in cultivation techniques, as well as in the development of specific media for different cell types. These advances have improved the reliability and versatility of zebrafish cell lines for various applications. In addition, genetic manipulation methods have become more precise and efficient, allowing researchers to conduct more sophisticated experiments more easily. Techniques such as CRISPR-Cas9 have revolutionized the ability to generate targeted gene knockouts and modifications in zebrafish cell lines, further expanding their utility in gene function studies. These advances have opened new avenues in developmental biology, genetics, disease modeling, and drug development, making zebrafish cell lines a cornerstone of modern biological research (Goswami et al., 2022; Thangaraj et al., 2021).

6.11 FUTURE DIRECTIONS AND CONCLUSION

The future of zebrafish cell line research is extremely promising and offers exciting prospects. Continued advances in genetic manipulation techniques such as CRISPR-Cas9 will enable researchers to precisely engineer zebrafish cell lines for highly specific studies and expand their applications in gene function studies and disease modeling. The development of more sophisticated culture systems will provide improved platforms for mimicking complex tissue microenvironments and improve our ability to study organ development and tissue regeneration. Additionally, zebrafish cell lines are likely to play a growing role in personalized medicine and drug screening, tailoring treatments to individual genetic profiles. Additionally, as the field of regenerative medicine continues to advance, zebrafish cell lines will provide crucial insights into tissue repair and regeneration. The continued synergy between zebrafish cell line research and translational applications will undoubtedly lead to groundbreaking discoveries with profound implications for human health and biology. In any case, let's dive deeper into the future prospects of zebrafish cell line research:

Advances in gene manipulation techniques: The continued development of gene editing tools such as CRISPR-Cas9 will enable researchers to perform complex genetic modifications in zebrafish cell lines with unprecedented precision. This will facilitate the creation of cell lines that closely mimic specific disease states or cellular functions, providing an invaluable resource for studying gene functions and disease mechanisms at the molecular level (Li et al., 2021).

Enhanced culture systems: The future promises more advanced culture systems that better replicate the complex microenvironments of living organisms. These systems will allow researchers to recreate the conditions required for the development and function of various tissues and organs. This advance is crucial for studying organogenesis, cellular interactions, and tissue-specific responses in zebrafish cell lines and leads to deeper insights into developmental biology and disease modeling.

Personalized medicine and drug screening: Zebrafish cell lines are poised to play a critical role in personalized medicine by facilitating testing of potential treatments on cells with specific genetic profiles. This tailoring of therapies to individual patients can lead to more effective and targeted treatments, reducing side effects and improving patient outcomes. In addition, the high-throughput capabilities of zebrafish cell lines will further accelerate drug testing efforts, accelerate the discovery of novel therapeutic compounds, and shorten drug development timelines (Baxendale et al., 2017).

Regenerative medicine insights: As regenerative medicine advances, zebrafish cell lines will continue to be at the forefront of tissue repair and regeneration research. These cell lines can be used to elucidate the cellular and molecular mechanisms underlying zebrafish's extraordinary regenerative abilities and may provide solutions to promote tissue repair in humans and treat degenerative diseases (Arjmand et al., 2020).

In summary, the future of zebrafish cell line research is characterized by an exciting variety of possibilities driven by advances in genetic manipulation, culture systems, personalized medicine, and regenerative medicine. These prospects hold great promise for addressing complex biological questions, advancing disease therapies, and expanding the boundaries of our understanding of fundamental biological processes.

Zebrafish cell lines are indispensable tools in the field of scientific discovery and provide a wealth of insight into biology and medicine. Their genetic proximity to humans, coupled with their versatility in genetic manipulation, has advanced our understanding of conserved molecular and cellular processes. These cell lines have not only revolutionized our understanding of embryonic development, disease mechanisms, and gene function, but also hold enormous potential for drug development, toxicology, and personalized medicine. Looking forward, the prospects for zebrafish cell line research are incredibly promising and offer the potential to reshape how we approach critical questions in biology and medicine. Their continued development promises to open new horizons in regenerative medicine, therapeutic development and personalized healthcare, ultimately improving human health and our understanding of the natural world. With their unique properties, zebrafish cell lines will undoubtedly continue to be at the forefront of cutting-edge research, paving the way to groundbreaking scientific discoveries and innovative medical solutions.

REFERENCES

Amsterdam, A., Burgess, S., Golling, G., Chen, W., Sun, Z., Townsend, K., . . . Hopkins, N. (1999). A large-scale insertional mutagenesis screen in zebrafish. *Genes & Development*, *13*(20), 2713–2724. https://doi.org/10.1101/gad.13.20.2713

Arjmand, B., Tayanloo-Beik, A., Foroughi Heravani, N., Alaei, S., Payab, M., Alavi-Moghadam, S., . . . Larijani, B. (2020). Zebrafish for personalized regenerative medicine; a more predictive humanized model of endocrine disease. *Frontiers in Endocrinology, 11*, 396.

Au-Rodrigues de Souza, I., Au-Wilke Sivek, T., Au-Vaz de Oliveira, J. B., Au-Di Pietro Micali Canavez, A., Au-de Albuquerque Vita, N., Au-Cigaran Schuck, D., . . . Au-Leme, D. M. (2023). Cytotoxicity assays with Zebrafish cell lines. *Journal of Visualized Experiments,* (191), e64860. https://doi.org/10.3791/64860

Baxendale, S., van Eeden, F., & Wilkinson, R. (2017). The power of zebrafish in personalised medicine. In S. El-Khamisy (Ed.), *Personalised medicine: Lessons from neurodegeneration to cancer*, pp. 179–197. Cham: Springer International Publishing.

Belyaeva, N., Kashirtseva, V., Medvedeva, N., Khudoklinova, Y. Y., Ipatova, O., & Archakov, A. (2009). Zebrafish as a model system for biomedical studies. *Biochemistry (Moscow) Supplement Series β: Biomedical Chemistry*, *3*, 343–350.

Bhadra, J., Banerji, R., Singh, J., Sallada, N., Eble, D., & Iovine, M. (2015). The Zebrafish fibroblast cell line AB9 as a tool to complement gene regulation studies. *Musculoskeletal Regeneration*, *2*. https://doi.org/10.14800/mr.992

Briggs, J. P. (2002). The zebrafish: A new model organism for integrative physiology. *American Journal of Physiology-Regulatory, Integrative and Comparative Physiology*, *282*(1), R3–R9.

Chen, W., Burgess, S., Golling, G., Amsterdam, A., & Hopkins, N. (2002). High-throughput selection of retrovirus producer cell lines leads to markedly improved efficiency of germ line-transmissible insertions in Zebrafish. *Journal of Virology*, *76*(5), 2192–2198. https://doi.org/10.1128/jvi.76.5.2192-2198.2002

Choorapoikayil, S., Overvoorde, J., & den Hertog, J. (2013). Deriving cell lines from zebrafish embryos and tumors. *Zebrafish*, *10*(3), 316–325. https://doi.org/10.1089/zeb.2013.0866

Collodi, P., Kamei, Y., Ernst, T., Miranda, C., Buhler, D. R., & Barnes, D. W. (1992). Culture of cells from zebrafish (*Brachydanio rerio*) embryo and adult tissues. *Cell Biology and Toxicology*, *8*(1), 43–61. https://doi.org/10.1007/bf00119294

Dooley, K., & Zon, L. I. (2000). Zebrafish: A model system for the study of human disease. *Current Opinion in Genetics & Development*, *10*(3), 252–256.

Driever, W., & Rangini, Z. (1993). Characterization of a cell line derived from zebrafish (*Brachydanio rerio*) embryos. *In Vitro Cellular & Developmental Biology – Animal, 29a*(9), 749–754. https://doi.org/10.1007/bf02631432

Eide, M., Rusten, M., Male, R., Jensen, K. H. M., & Goksøyr, A. (2014). A characterization of the ZFL cell line and primary hepatocytes as in vitro liver cell models for the zebrafish (*Danio rerio*). *Aquatic Toxicology*, *147*, 7–17.

Fan, L., Alestrom, A., Alestrom, P., & Collodi, P. (2004). Development of cell cultures with competency for contributing to the zebrafish germ line. *Critical Reviews in Eukaryotic Gene Expression*, *14*(1–2), 43–51.

George, S., Xia, T., Rallo, R., Zhao, Y., Ji, Z., Lin, S., . . . Schoenfeld, D. (2011). Use of a high-throughput screening approach coupled with in vivo zebrafish embryo screening to develop hazard ranking for engineered nanomaterials. *ACS Nano*, *5*(3), 1805–1817.

Ghosh, C., Zhou, Y. L., & Collodi, P. (1994). Derivation and characterization of a zebrafish liver cell line. *Cell Biology and Toxicology*, *10*(3), 167–176. https://doi.org/10.1007/bf00757560

Goswami, M., Yashwanth, B. S., Trudeau, V., & Lakra, W. S. (2022). Role and relevance of fish cell lines in advanced in vitro research. *Molecular Biology Reports*, *49*(3), 2393–2411. https://doi.org/10.1007/s11033-021-06997-4

He, S., Salas-Vidal, E., Rueb, S., Krens, S. F., Meijer, A. H., Snaar-Jagalska, B. E., & Spaink, H. P. (2006). Genetic and transcriptome characterization of model zebrafish cell lines. *Zebrafish*, *3*(4), 441–453. https://doi.org/10.1089/zeb.2006.3.441

Heilmann, S., Ratnakumar, K., Langdon, E. M., Kansler, E. R., Kim, I. S., Campbell, N. R., . . . Van Rooijen, E. (2015). A quantitative system for studying metastasis using transparent zebrafish. *Cancer Research*, *75*(20), 4272–4282.

Ho, S. Y., Goh, C. W., Gan, J. Y., Lee, Y. S., Lam, M. K., Hong, N., . . . Shu-Chien, A. C. (2014). Derivation and long-term culture of an embryonic stem cell-like line from zebrafish blastomeres under feeder-free condition. *Zebrafish*, *11*(5), 407–420. https://doi.org/10.1089/zeb.2013.0879

Howe, K., Clark, M. D., Torroja, C. F., Torrance, J., Berthelot, C., Muffato, M., . . . Matthews, L. (2013). The zebrafish reference genome sequence and its relationship to the human genome. *Nature*, *496*(7446), 498–503.

Jin, Y., Chen, L., Le, Y., Li, Y., Hong, Y., Jia, K., & Yi, M. (2017). Establishment of a cell line with high transfection efficiency from zebrafish *Danio rerio* embryos and its susceptibility to fish viruses. *Journal of Fish Biology*, *91*(4), 1018–1031.

Lakra, W., Swaminathan, T. R., & Joy, K. (2011). Development, characterization, conservation and storage of fish cell lines: A review. *Fish Physiology and Biochemistry*, *37*, 1–20.

Lam, S. H., Lee, S. G., Lin, C. Y., Thomsen, J. S., Fu, P. Y., Murthy, K. R., . . . Bourque, G. (2011). Molecular conservation of estrogen-response associated with cell cycle regulation, hormonal carcinogenesis and cancer in zebrafish and human cancer cell lines. *BMC Medical Genomics*, *4*, 1–15.

Li, Y., Jia, Z., Zhang, S., & He, X. (2021). Progress in gene-editing technology of Zebrafish. *Biomolecules*, *11*(9). https://doi.org/10.3390/biom11091300

Liu, S., & Leach, S. D. (2011). Zebrafish models for cancer. *Annual Review of Pathology: Mechanisms of Disease*, *6*, 71–93.

Lungu-Mitea, S., Vogs, C., Carlsson, G., Montag, M., Frieberg, K., Oskarsson, A., & Lundqvist, J. (2020). Modeling bioavailable concentrations in zebrafish cell lines and embryos increases the correlation of toxicity potencies across test systems. *Environmental Science & Technology*, *55*(1), 447–457.

Miranda, C. L., Collodi, P., Zhao, X., Barnes, D. W., & Buhler, D. R. (1993). Regulation of cytochrome P450 expression in a novel liver cell line from zebrafish (*Brachydanio rerio*). *Archives of Biochemistry and Biophysics*, *305*(2), 320–327. https://doi.org/10.1006/abbi.1993.1429

Morrison, J. K., DeRossi, C., Alter, I. L., Nayar, S., Giri, M., Zhang, C., . . . Chu, J. (2022). Single-cell transcriptomics reveals conserved cell identities and fibrogenic phenotypes in zebrafish and human liver. *Hepatology Communications*, *6*(7), 1711–1724.

Quevedo, A. C., Ellis, L. A., Lynch, I., & Valsami-Jones, E. (2021). Mechanisms of silver nanoparticle uptake by embryonic Zebrafish cells. *Nanomaterials (Basel)*, *11*(10). https://doi.org/10.3390/nano11102699

Rubinstein, A. L. (2003). Zebrafish: From disease modeling to drug discovery. *Current Opinion in Drug Discovery and Development*, *6*(2), 218–223.

Şerifoğlu, N. (2019). *Investigating the mechanisms of telomere maintenance in zebrafish tissues and human brain cancer cell lines*. Bilkent Universitesi (Turkey).

Takakura, M., Kyo, S., Kanaya, T., Hirano, H., Takeda, J., Yutsudo, M., & Inoue, M. (1999). Cloning of human telomerase catalytic subunit (hTERT) gene promoter and identification of proximal core promoter sequences essential for transcriptional activation in immortalized and cancer cells. *Cancer Research*, *59*(3), 551–557.

Thangaraj, R. S., Narendrakumar, L., Prasannan Geetha, P., Shanmuganathan, A. R., Dharmaratnam, A., & Nithianantham, S. R. (2021). Comprehensive update on inventory of finfish cell lines developed during the last decade (2010–2020). *Reviews in Aquaculture*, *13*(4), 2248–2288.

Tonon, F., & Grassi, G. (2023). *Zebrafish as an experimental model for human disease* (Vol. 24, p. 8771). MDPI.

Vallone, D., Santoriello, C., Gondi, S. B., & Foulkes, N. S. (2007). Basic protocols for zebrafish cell lines: Maintenance and transfection. *Circadian Rhythms: Methods and Protocols*, 429–441.

Vallone, D., Santoriello, C., Gondi, S. B., & Foulkes, N. S. (2008). Basic protocols for zebrafish cell lines. *Circadian Rhythms: Methods and Protocols*, *362*, 429.

Zhang, C., Willett, C., & Fremgen, T. (2003). Zebrafish: An animal model for toxicological studies. *Current Protocols in Toxicology*. https://doi.org/10.1002/0471140856.tx0107s17

7 Calcium Imaging of Neuronal Activity in Free-Swimming Larval Zebrafish

Rui Peng, Xiuyun Liu, and Meijun Pang

7.1 INTRODUCTION

Parkinson's disease is a neurodegenerative disorder. Patients experience motor and non-motor impairments (Aarsland et al., 2021). In the realm of animal modeling, the ability to execute motor actions becomes a pivotal indicator of model success, and deciphering the neural underpinnings of motor behavior through neuronal activity visualization is paramount (Carvalho et al., 2013). Notably, even seemingly simple reflexes are orchestrated by neural circuits spanning vast brain regions, underscoring the need for comprehensive brain imaging to illuminate the intricate nexus between movement and neural dynamics (Inagaki et al., 2022).

When studying neural activity, understanding how neurons interconnect to generate complex behaviors is of paramount importance. Calcium imaging, a crucial tool, is widely employed for monitoring neuronal activity. This method capitalizes on the variations in intracellular calcium ion concentrations during neural activity, which are closely tied to neuronal excitability (Ghosh & Greenberg, 1995). Utilizing fluorescent probes, particularly calcium indicators, calcium imaging translates neuronal activity into visible fluorescence signals. As neurons become activated, there is a rapid increase in calcium ion concentration, leading to a corresponding rise in fluorescence signals, enabling real-time observation of neuronal dynamics. The originally silent movement of neurons is transformed into a colorful picture by calcium imaging. This technique not only offers a gateway to exploring the functionality and abnormalities of neural circuits but also provides profound insights into the mechanisms underlying neurodegenerative diseases.

Zebrafish, a widely employed model for studying neurological disorders, possesses a brain that exhibits high homology to that of mammals, sharing similar structural characteristics and undergoing rapid development. Notably, zebrafish display complex natural behaviors at a remarkably young age, making them an invaluable resource for neurological research (Panula et al., 2010). The inherent transparency of zebrafish embryos and larvae confers a substantial imaging advantage compared to mammals. This transparency facilitates the simultaneous imaging of neural activity throughout the entire brain, a pivotal capability in unraveling conditions spanning from autism spectrum disorders to Parkinson's and Alzheimer's diseases (Nishimura et al., 2015). Moreover, the rapid developmental trajectory of zebrafish offers distinct advantages. For instance, behaviors such as hunting can be observed within just five days post-fertilization, while socialization tendencies emerge after a mere fifteen days. This accelerated timeline grants zebrafish researchers the efficiency of studying these behaviors with a promptness that surpasses traditional rodent models, which are often constrained by cost and extended experimental timelines (Gahtan et al., 2005; Larsch & Baier, 2018).

In traditional neurobiological imaging experiments, zebrafish are often immobilized using agar-agar or FEP tubes for microscopic imaging, raising concerns about potential stress-induced artifacts that might influence neurobehavioral assessments (Kaufmann et al., 2012). Addressing these concerns, the concept of studying freely behaving zebrafish emerges as a pivotal avenue within neurobiology research, providing a more holistic comprehension of neurobehavioral dynamics.

In this chapter, we provide a deep insight into the nuances of calcium imaging in free-swimming zebrafish and its transformative impact on neurodegenerative disease research, and the myriad challenges and prospects it brings to the forefront. Through this exploration, we aim to provide a comprehensive guide to employing calcium imaging techniques in the pursuit of deciphering neural activity in free-swimming zebrafish, ultimately shedding light on the intricate pathways underlying neurological disorders like Parkinson's disease.

7.2 CALCIUM INDICATORS

In recent years, significant progress has been made in the development of calcium indicator technology. The choice of calcium indicator depends on its intended application and can be broadly categorized into four groups: dextran-coupled indicators, AM calcium fluorescent probes, Aequorin, and genetically encoded calcium sensors (Figure 7.1).

7.2.1 Dextran-Coupled Indicators

The dynamic nature of calcium signaling during neural activity necessitates precise measurements via indicators with low affinity and rapid kinetics. Dextrans, high molecular weight polysaccharides, are characterized by biological inertness. When labeled with fluorescein, dextrans can remain within cells, thus avoiding cellular excretion (De Nadai et al., 2021). The fluorescein moiety is attached to

DOI: 10.1201/9781003402893-8

the dextran through a stable thiocarbamoyl bond, a process that does not result in hydrolysis of the dextran, which is selected with different molecular weight sizes depending on the need. The combination of calcium ions and dextran-coupled indicator results in a change in the emitted fluorescence (Fig. 7.1A). The first to apply this technique to zebrafish was the team of Joseph R, who labeled the neural tissue of the spinal cord by injecting a dextran-coupled indicator into the muscles of the zebrafish (Fetcho & O'Malley, 1995). Since labeling of nerves with dextran-coupled indicators is accomplished by injection, making this method of labeling neurons confined to a portion of smaller areas and tissues, only a limited number of cells will be labeled.

7.2.2 AM Calcium Fluorescent Probe

The AM ester form of fluorescent dye probes is also a common method for labeling neurons. AM ester solution, a non-polar ester, easily traverses living cell membranes and is initially non-luminous. It fluoresces upon rapid hydrolysis by endogenous esterases and chelation with cytoplasmic calcium, obviating the need for additional aids (Fig. 7.2B). AM ester is widely used to non-invasively load a variety of polar fluorescent probes into living cells (Barreto-Chang & Dolmetsch, 2009). In 2003 scientists labeled most of the neurons in the zebrafish spinal cord for the first time with an AM calcium probe (Brustein et al., 2003).

7.2.3 Aequorin Indicators

Aequorin is a biological calcium indicator from jellyfish, where the binding of calcium ions to photoproteins leads to an enzymatic reaction that results in the emission of blue light (Shimomura et al., 1962). Aequorin has three calcium binding sites, and the rate of the reaction depends on the concentration of calcium ions (Fig. 7.1C). Since Aequorin is self-luminous, it does not require excitation by external light, thus avoiding the problems of phototoxicity, photobleaching, autofluorescence, and undesirable stimulation of photobiological processes (Xu et al., 2007). However, since extracted Aequorin cannot pass through cell membranes, it can only be loaded into cells by microinjection (Chiesa et al., 2001).

7.2.4 Genetically Encoded Calcium Sensors

Both dextran-coupled indicators and AM probes entail intricate manipulation of zebrafish, where indicators are manually injected into muscles or spinal cords. However, this method risks neural circuit damage and necessitates skilled operators. Genetic engineering has enabled the creation of calcium indicators that can be expressed in neurons through gene insertion. The advent of genetically encoded calcium sensors revolutionized calcium signal labeling, enhancing precision and efficiency. This innovation laid the groundwork for large-scale imaging.

The discovery that Green Fluorescent Protein (GFP) could fluoresce after circular rearrangement and insertion by Roger Y. Tsien's team led to circularly permuted GFP (cpGFP). Moreover, the insertion of proteins into cpGFP revealed changes in fluorescence intensity, offering a novel strategy for genetically encoded physiological indicators. Notably, the GCaMP system, based on cpGFP, was a breakthrough (Baird et al., 1999). In 2001, Junichi Nakai's team introduced G-CaMP, a high-affinity Ca2+ probe based on single GFP molecules. By fusing M13 and calmodulin (CaM) to cpEGFP, the resultant Ca2+-CaM-M13 interaction caused conformational changes and fluorescence intensity alterations upon calcium binding, allowing dynamic calcium ion changes in myotubes to be detected (Nakai et al., 2001) (Fig. 7.1D). The advent of GCaMP6, jointly developed by the Howard Hughes Institute team of Karel Svoboda and Douglas S. Kim, addressed sensitivity and kinetics limitations. GCaMP6s, GCaMP6m, and GCaMP6f variants demonstrated improved signal-to-noise ratios and diverse kinetics, subsequently applied to zebrafish, *Drosophila*, and mice (T.-W. Chen et al., 2013).

In zebrafish, pan-neuronal expression can be achieved using promoters like elavl3 or HuC, offering insights into neuronal development and function (C. H. Kim et al., 1996; Turrini et al., 2017). Further optimization led to GCaMP nuclear targeting via fusion with human histone H2B protein. In zebrafish neurons, the nucleus occupies almost the entire soma, which makes it a very effective technique as it helps in image segmentation compared to targeting GCaMP. In sum, cpEGFP's sensitivity to conformational changes drives imaging system development. The mature optimization process has yielded imaging systems with high signal-to-noise ratios and rapid sensitivity, prominently employed in neurobiology.

7.3 CALCIUM IMAGING SYSTEM

In the dynamic landscape of calcium imaging, not only have calcium indicators evolved, but imaging equipment has also undergone continuous refinement (Fig. 7.2). This evolution has paved the way for more precise, comprehensive, and real-time observation of neural activity, particularly in free-swimming larval zebrafish. In this section, we delve into the introduction and evolution of calcium imaging systems, shedding light on their transformative impact on neuroscience research (Figure 7.2).

7.3.1 Confocal Microscopy: High Special Resolution and Selective Imaging

Confocal microscopy operates on the principle of single-photon excitation. The excitation light is focused on a specific focal plane, while a pinhole or slit in the image conjugate stage eliminates out-of-focus fluorescence, enhancing image clarity. This technique offers high spatial resolution and is well-suited for detailed imaging of neural structures.

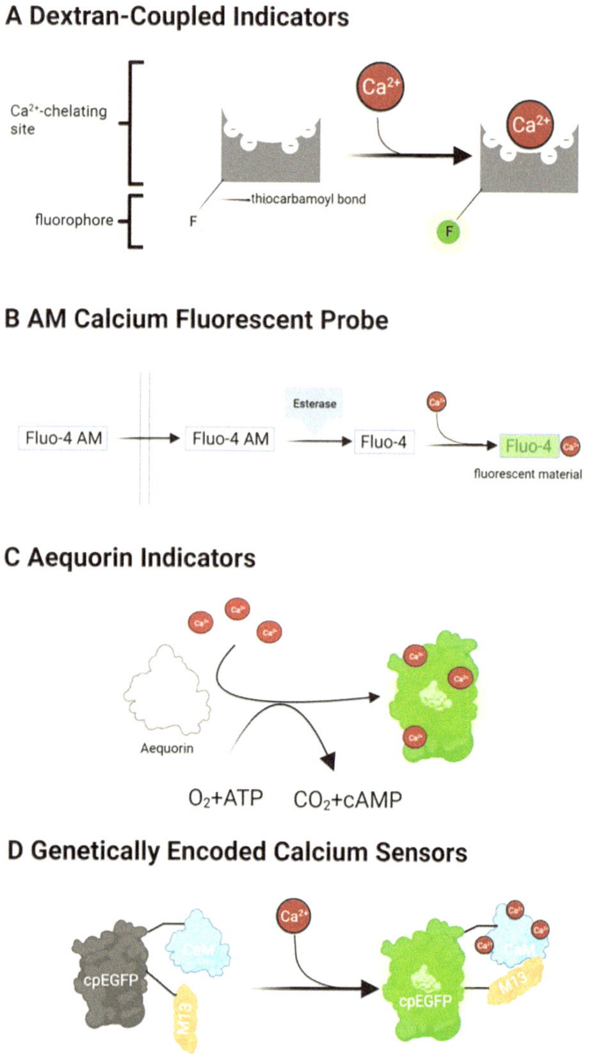

FIGURE 7.1 (A) Dextran-coupled indicators. Fluorescein molecules are attached to dextran by stable thiocarbamoyl bonds. The combination of calcium ions and dextran-coupled indicator results in a change in the emitted fluorescence. (B) AM calcium fluorescent probe. AM ester solution is a nonpolar ester that fluoresces upon rapid hydrolysis by endogenous esterases and chelation with cytoplasmic calcium. (C) Aequorin indicators. Aequorin is a biological calcium indicator from jellyfish, where the binding of calcium ions to photoproteins leads to an enzymatic reaction that produces blue light emission. (D) Genetically encoded calcium sensors. Upon binding of calcium ions to GCaMP, the intramolecular conformation changes and the emitted fluorescence increases.

Confocal microscopy is invaluable for capturing localized neural activity and intricate neural connections. It has been employed to study calcium dynamics in specific brain regions of zebrafish larvae, providing insights into neural circuitry and responses to stimuli. Zhao et al. used confocal microscopy for calcium imaging of zebrafish larvae and discovered the dynamic calcium activity of β-cells in zebrafish embryos, which revealed the calcium activity pattern related to pancreatic islets, and is applicable to calcium imaging of other tissues or organs (Zhao et al., 2021).

7.3.2 Two-Photon Microscopy: Deep Penetration and Low Photobleaching

Two-photon microscopy capitalizes on two-photon excitation, enabling deep tissue penetration while minimizing photodamage. The nonlinear relationship between excitation light and matter ensures that only a single point in space is excited, reducing photobleaching and out-of-focus backgrounds.

Two-photon microscopy is the workhorse of neuronal imaging and has been used in many neuroscience model organisms. Two-photon microscopy is particularly advantageous for imaging neural activity in deeper brain layers of free-swimming zebrafish (Portugues et al., 2014). It has been pivotal in capturing the calcium dynamics of large populations of neurons simultaneously, providing insights into complex brain-wide neural processes. The greatest limitation of the two-photon microscope is the imaging speed, as it can only scan one pixel point at a time. In the case of fixed zebrafish heads being imaged, imaging the calcium dynamics of 500 to 1,500 neurons in the apical hemisphere to monitor neural activity allowed for volumetric data consisting of five focal planes to be collected at a volumetric rate of 9.7 Hz (Sainsbury et al., 2023). This problem can be solved by the addition of an optical module, which can be integrated into a two-photon microscope, designed by Lu et al. The module allows for an axially elongated Bessel focus that converts the frame rate to a volume rate when scanning in 2D. The retrofitted two-photon microscope can image reticulospinal neurons in a volume of 270 μm × 270 μm × 100 μm at 30 Hz, which greatly improves conventional imaging performance (Lu et al., 2017).

7.3.3 Light Sheet Imaging: Rapid Volumetric Visualization

Light sheet microscopy, also referred to as selective plane illumination microscopy (SPIM), has emerged as a formidable technique for visualizing biological specimens, including the neural activity in free-swimming larval zebrafish. This approach operates on the principle of illuminating a thin, optical section of the specimen perpendicular to the detection objective. By doing so, it mitigates photodamage and reduces out-of-focus fluorescence, enabling rapid volumetric imaging while retaining cellular resolution. The orthogonal arrangement of the excitation light sheet and detection objective enhances spatial resolution while minimizing the specimen's exposure to excitation light.

The hallmark features of light sheet microscopy render it highly suitable for investigating neural dynamics in free-swimming larval zebrafish (Panier et al., 2013). These features encompass high-speed, volumetric imaging capabilities, diminished phototoxicity, superior spatial resolution, and the capacity to penetrate deep within tissues. The method's adeptness in capturing swift and dynamic neural processes at the cellular level is achieved by illuminating a thin optical section and minimizing the interference of out-of-focus fluorescence. Consequently, this approach preserves the integrity of surrounding tissues and facilitates prolonged imaging of delicate neural circuits, ensuring their viability throughout the observation period. Giuseppe et al. detected calcium imaging in a zebrafish epilepsy model by modified light-sheet microscopy and described for the first time the appearance of caudo-rostral ictal waves (de Vito et al., 2022). Moreover, the spatial resolution afforded by light sheet microscopy facilitates the visualization of intricate cellular structures and interactions within neural circuits. Misha et al. recorded calcium cell imaging of the entire brain of zebrafish using light-sheet microscopy at 0.8 HZ, achieving single-cell resolution imaging of 80% of the neurons and revealing the mode of operation of the brain circuits (Ahrens et al., 2013). In zebrafish larvae, this technique enables the imaging of approximately three brain volumes per second, capturing the dynamics of around 100,000 neurons per second, with the potential for further enhancements in the future. The advantage of rapidly imaging volumes lies in the simultaneous visualization of distant neurons, permitting the exploration of collective neuronal activity in response to stimuli or tasks. This significantly enhances data acquisition efficiency while maintaining satisfactory imaging resolution.

Although whole-brain light-sheet imaging with single-cell resolution has been realized, the fact that excitation light is directed from the side of the sample during light-sheet microscopy imaging makes it difficult to achieve tracking of sample motion with this imaging facility. Light-field microscopy compensates for this shortcoming.

7.3.4 Light Field Imaging (LFM): Efficient 3D Imaging and Low Photobleaching

Light Field Imaging (LFM), often referred to as plenoptic imaging, is a sophisticated imaging technique that capitalizes on microlens array (MLA) optics and intricate computational algorithms to capture and reconstruct three-dimensional images. At its core, LFM operates on the concept of capturing both the spatial and directional information of light rays emanating from a scene. This is achieved through an array of tiny lenses, known as a microlens array, positioned in front of an image sensor. Each microlens captures the light rays from a slightly different perspective, resulting in multiple angular views of the same scene within a single snapshot.

LFM provides several key advantages, including efficient single-shot 3D imaging, reduced phototoxicity, rapid imaging capabilities, simplified single-camera setup, and improved spatial resolution achieved through angle-based reconstruction. These attributes collectively establish LFM as a potent instrument for capturing intricate neural dynamics within free-swimming larval zebrafish. Using confocal light field microscopy, Zhang et al. achieved rapid volumetric imaging of the brain at a depth of hundreds of micrometers, providing optical slicing capabilities, improved spatial resolution and sensitivity for imaging in thick tissue, and captured calcium transients in the brain of a zebrafish during free-swimming, and the activity of a single neuron during predation (Z. Zhang et al., 2021). Nonetheless, LFM does come with certain limitations. The process of reconstructing the imaged volume entails a substantial computational burden, potentially posing challenges for real-time analysis. Additionally, the presence of background noise interference

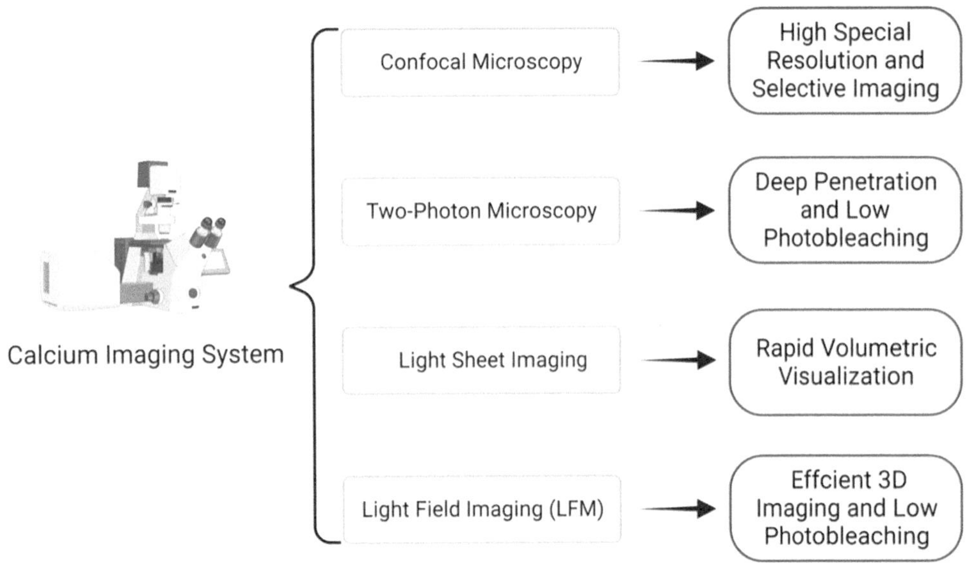

FIGURE 7.2 Calcium imaging system. Confocal microscopy: high special resolution and selective imaging. Two-photon microscopy: deep penetration and low photobleaching. Light sheet imaging: rapid volumetric visualization. Light field imaging (LFM): efficient 3D imaging and low photobleaching.

between out-of-focus and in-focus light can impact signal-to-noise ratios, leading to the creation of reconstruction artifacts. While LFM offers respectable spatial resolution, it falls slightly short of the capabilities of techniques like two-photon or light-sheet imaging, with current transverse spatial resolution hovering around 3 μm compared to 0.5 μm in the latter methods. This resolution is influenced by the numerical aperture (NA), and while improving spatial resolution by increasing NA is feasible, it may reduce axial coverage due to the inverse relationship between NA and depth of focus. Future advancements in light collection capabilities hold promise for overcoming this limitation. Furthermore, LFM's temporal resolution is primarily determined by camera performance and photon counting, making it especially suitable for 3D voltage imaging due to its high temporal resolution (Prevedel et al., 2014).

7.4 DESIGN AND PREPARATION OF EXPERIMENTS

7.4.1 Zebrafish Breeding

In accordance with the husbandry guidelines, the zebrafish's housing conditions were carefully set to maintain a constant temperature of 28.5°C with a cycle of 14 hours of light and 10 hours of darkness to mimic the natural circadian rhythm. Meanwhile, the zebrafish had free access to food in this environment, which provided them with the necessary nutrients for growth and development. In calcium imaging experiments, zebrafish larvae are usually used as research subjects, and their small and transparent body size makes them ideal for microscopic observation and imaging processing. To ensure the accuracy and reliability of the imaging results, it is necessary to perform special treatments on the samples before imaging (Westerfield, 2000).

For non-transgenic zebrafish pre-imaging preparations, it is first necessary to place the collected eggs in E3 culture solution containing PTU (1-phenyl-2-thiourea) for incubation. The purpose of this step is to inhibit the synthesis of melanin by the addition of PTU and to avoid the influence of melanin on the imaging results. Thereafter, in order to achieve the goal of calcium imaging, the dye needs to be introduced into the region of interest by some means, such as electroporation or microinjection. Such treatment ensures that the dye accurately enters the target area. The use of progeny from stabilized zebrafish for subsequent experiments similarly requires that the collected eggs be incubated in E3 culture medium containing PTU to avoid the effects of melanin on imaging results.

As for the preparation process of transgenic zebrafish, it also requires a series of steps. First, zebrafish embryos at the one-cell stage need to be collected in order to introduce the genetic modification during the early embryonic developmental stages. In this process, plasmids carrying genes for calcium indicators and fluorescent proteins are injected into fertilized eggs. As development proceeds, over a period of approximately 3 months, researchers will screen the offspring zebrafish to select genetically stable individuals that will carry the targeted genetic modifications, providing suitable research subjects for subsequent calcium imaging experiments.

7.4.2 Dye-Loading Approaches

Selection of the appropriate dye entry method should be based on the experimental needs, cell type, characteristics of the dye, and the requirements of the experimental equipment. The dye entry method should be as non-invasive to the cells as possible and should ensure the stability and reliability of the dye.

The most traditional method for labeling individual neurons is by means of a sharp microelectrode. The tip of a fine glass microelectrode is placed close to the cell membrane, creating a temporary "cut hole" through which the dye enters the cell. The chemical dye is loaded onto the target nerve through the sharp microelectrode. In recent years, the delivery of dye via whole-cell membrane-clamp micropipettes has become a standard procedure for single-cell dye loading in many applications. There are also methods of delivering dyes into neurons by electroporation, where a brief electrical pulse can create temporary holes in the cell membrane, allowing the dye to enter the cell.

For the whole nervous system, or for bulk loading of localized nerves, these methods are obviously not efficient enough. Some cell-permeable dyes, such as membrane-permeable calcium acetoxymethyl (AM) ester dyes, are commonly used. These include impulses of air pressure to the tissue. In addition to AM calcium dyes, dextran-coupled chemical calcium indicators can also be used for network loading, mainly by pressurized injection into the axonal pathway, where the dye molecules are taken up and transported forward and backward to the axon terminals and the cell body, respectively. In addition to the aforementioned methods, AM calcium dyes and dextran-coupled chemical calcium indicators can likewise be used to enter neurons by electroporation.

And as the most widely used genetically encoded calcium indicators, transduction using plasmids is the most common method. A selected calcium receptor gene is fused to a reporter gene (e.g., fluorescent protein GFP) gene. A promoter specific to the target neuron is selected to drive the expression of the fusion gene. The constructed fusion gene is then inserted into a gene vector, commonly a plasmid. Preferably, the constructed plasmid is injected into fertilized eggs of zebrafish, and the zebrafish are screened as adults to identify individuals carrying the fusion gene.

7.4.3 Sample Preparation and Image Acquisition

The first step in image acquisition is to prepare the samples according to the needs of the experiment. In the imaging process of fixed zebrafish, zebrafish first need to be anesthetized prior to the experiment to fix the zebrafish on the platform using appropriate fixation materials, such as agarose or FEP tubes. These materials should not only ensure fixation but also avoid adverse effects on the fish. Then, according to the experimental needs, the zebrafish is fixed at a specific imaging position to ensure that the region of interest is aligned with the imaging system so that the desired image information can be obtained. As for the imaging process of free-moving zebrafish, it is only necessary to place the zebrafish into the prepared chamber so that it can move freely.

Subsequently, the required conditions for the experiment were set up to ensure that clear images were obtained by adjusting the parameters of the microscope, including magnification, exposure time, and fluorescence channel settings. Subsequently, time-series acquisition was initiated to record the fluorescence signal emitted by the zebrafish neurons over time. This can be a brief pulsed signal or a continuous time series, depending on the experimental design and the purpose of the study. If desired, you can introduce stimuli or manipulations during image acquisition to elicit specific neural activity and probe the neuron's response patterns in depth. The data obtained will be stored to provide a basis for subsequent image processing and analysis.

7.4.4 Image Processing and Analysis

In the process of handling and analyzing zebrafish calcium imaging images, the initial step involves capturing images using fluorescent microscopy or similar tools. Background correction and image registration are conducted to eliminate noise and address motion-induced shifts. Subsequently, regions of interest (ROIs) are defined, and calcium signals are extracted. Background subtraction and signal filtering are applied to enhance data quality. Using thresholding techniques or alternative approaches, calcium events are detected and recorded, while features such as peak intensity and duration are extracted. Furthermore, spatial correlations between different ROIs and temporal associations among calcium events are analyzed to study neuronal activity patterns and networks. Statistical methods like analysis of variance (ANOVA) or t-tests are employed to assess differences in calcium signals under varying conditions. Ultimately, through data visualization techniques such as charts, images, and heatmaps, analysis results are presented in an intuitive manner, providing a scientific foundation for interpreting research questions and preparing reports.

7.5 APPLICATION OF CALCIUM IMAGING IN ZEBRAFISH

7.5.1 Application of Calcium Imaging in Neuronal Activity and Behavior

Calcium imaging in free-swimming zebrafish can help researchers reveal the relationship between neural activity and behavior. By recording neuronal activity, researchers can determine changes in neuronal activity during specific behavioral patterns, thus exploring the link between neurons and behavior. For example, researchers can observe zebrafish activity in different environments to understand how they make adaptive decisions.

In the zebrafish's virtual reality swimming attempts, visual feedback can be preserved so that swimming fails to trigger the expected visual flow and the zebrafish gives up on continuing to swim after repeated attempts. The combination of retrying and preserving stamina is strategic. The study utilized whole-brain calcium imaging to reveal the unique response of noradrenergic neurons to failed swimming attempts. In addition, neuronal calcium levels increased as the number of failures accumulated. Astrocyte calcium was found to be elevated before and during the passive state (abandonment). This study extends the understanding of astrocyte function and reveals the mechanisms behind futile swimming (Mu et al., 2019).

Escape behavior and rhythmic swimming in zebrafish are distinct behaviors mediated by two different neurons, with primary neurons having an inhibitory role and secondary neurons a facilitatory role. Primary neurons provide strong, single C-bends to initiate escape, while secondary neurons confer weaker, rhythmic contractions. Using zebrafish calcium imaging, the researchers found that despite the functional and structural differences between primary and secondary neurons, synapses on each individual muscle cell were shared by a primary neuron and at least one secondary neuron. This sharing provides a way to achieve both behaviors at the same synapse (Wen et al., 2020).

Affiliation with conspecifics is a central component of social behavior, providing benefits to the collective, such as detecting food or avoiding natural predators. Animals need to effectively recognize signals from their own kind in a cluttered environment. Completion of these actions depends on a balance of like-for-like distances. Researchers used zebrafish calcium imaging to reveal the neural circuitry behind social behavior. Electron microscopic reconstruction of dorsal thalamic neurons revealed synaptic inputs from the optic ganglion and projections from hypothalamic regions that have conserved social functions. Resection of the optic ganglia or dorsal thalamus selectively disrupts social attraction without affecting short-range rejection. Thus, this tectothalamic pathway facilitates visual recognition of conspecifics and separates neuronal control of attraction from repulsion in social affiliation, revealing the circuitry underlying collective behavior (Kappel et al., 2022).

Odors are thought to be represented by distributed codes on glomerular modules in the olfactory bulb. Researchers visualized synaptic responses induced by a class of natural odorants at the zebrafish olfactory bulb site by calcium imaging, directly demonstrating that the olfactory bulb can encode recognition of the type and concentration of odorants (Friedrich & Korsching, 1997).

7.5.2 Application of Calcium Imaging in Neuronal Development and Maturation Studies

As zebrafish grow, their nervous system changes. Calcium imaging of free-swimming zebrafish can be used to study the process of neuronal development and maturation. By comparing patterns of neural activity at different stages, researchers can explore changes and remodeling of neuronal connections, which in turn can reveal mechanisms of neural development.

Using calcium imaging in zebrafish, the researchers confirmed that the zebrafish central nervous system (CNS) does indeed host astrocyte populations that are very similar to those of mammals and *Drosophila*, further supporting the idea that astrocytes are an ancient, well-preserved CNS cell type. This work establishes zebrafish as a new system for exploring astrocyte biology and lays the groundwork for future research (J. Chen et al., 2020).

Using calcium imaging of zebrafish in combination with other means, the researchers showed that the zebrafish spinal cord consists of oligodendrocyte precursor cell subtypes with distinct functions related to the regulation of oligodendrocyte precursor cell numbers and differentiation into myelinating oligodendrocytes. This specialization may be intrinsically determined or caused by extrinsic factors (Marisca et al., 2020).

7.5.3 Application of Calcium Imaging in Mechanistic Research and Modeling of Diseases

Calcium imaging is also widely used in neurological disease research and can help researchers reveal the relationship between abnormal neuronal activity and neural network connectivity and delve into the pathogenesis of these diseases. These studies are expected to provide important clues for the development of new treatments and drugs to improve the quality of life of patients with neurological disorders.

By means of calcium imaging, one has also examined the entire neural network of zebrafish during epileptogenesis, revealing the propagation of neural signals from the forebrain to the hindbrain and finding that microcollections of neuronal subpopulations are active during the interictal period in a manner similar to that seen in human electrophysiology datasets, and that macroscopic and microscopic network calcium motifs may contribute to a better understanding of epilepsy (Liu & Baraban, 2019). The same goes for epilepsy, which was previously thought to be caused by neural over-excitation, but a recent discovery disproves this idea. Using calcium imaging in a zebrafish model of epilepsy, researchers looked at excitatory and inhibitory neurons and found that seizures always start in the midbrain's circumventricular zone before spreading to other brain regions. Excitatory neurons were more prevalent and more likely to be recruited than were inhibitory neurons in the initiation zone compared to the spreading zone. These findings support the mechanism that seizures begin in areas of hyperexcitability and then propagate more widely once the surrounding inhibitory suppression has been overcome (Niemeyer et al., 2022).

Calcium ions are found not only in neurons in the brain, but also have an irreplaceable role in muscle cells. In the heart, binding of calcium ions to myofilaments leads to systolic contraction and blood ejection. Visualizing this through calcium imaging, the researchers verified the relationship between calcium ions and cardiac contraction, generating transgenic zebrafish of the corresponding strain, which became a powerful tool for screening for heart disease (Salgado-Almario et al., 2022).

7.6 THE DEVELOPMENT OF CALCIUM IMAGING IN FREE-SWIMMING ZEBRAFISH

In previous imaging of neuronal calcium ions in zebrafish, zebrafish were generally encased in agarose. Since zebrafish skin can also breathe, zebrafish can survive for hours in an agarose-encased environment. As the study progressed, neurocalcium imaging in restrained zebrafish no longer met the needs of the experiment. In restraint, the fish does not receive the visual, proprioceptive, vestibular, or gustatory feedback it would experience during unconstrained movement and prey hunting. To more realistically simulate the reality of a zebrafish under free-swimming conditions, the researchers released the tail of the zebrafish while keeping the head immobilized. In further studies, recordings of zebrafish neural activity began to be integrated with the virtual reality paradigm, where the closed-loop virtual behavior of the immobilized zebrafish could be monitored and controlled through visual feedback (Ahrens et al., 2012). Yet similar to the pinnacle moment in a hunting situation, the moment of catching a grasshopper is difficult to easily replicate (Portugues & Engert, 2011). We need to observe more neural calcium imaging in zebrafish under free swimming. Examples include predatory and social behaviors in common zebrafish.

An early method of imaging free-swimming zebrafish was to place them in a small dish and observe them under low magnification, which allowed observation during swimming intervals because zebrafish swim in an intermittent manner (Figure 7.3). This method was applied to record neural activity in the optic parietal during zebrafish pursuit of mealybugs (Muto et al., 2013). However, the quality of the images obtained did not meet the requirements of single-cell resolution. Subsequently, the team improved the equipment to a 9 mm diameter × 0.8 mm deep chamber, and observed the larvae through a drop-shot microscope, manually moving the carrier stage each time the larvae swam out of the field of view to keep it in the center of the field of view. With this imaging device, Muto et al. revealed the important role played by the anterior hypothalamic pathway in transmitting visual information to the feeding center (Muto et al., 2017).

In recent years, scientists have made important breakthroughs in calcium imaging in free-swimming zebrafish, leading to significant advances in neuroscience research (Fig. 7.3). This technique allows us to image free-swimming zebrafish volumetrically at cellular or even near-cellular resolution, opening up new possibilities in the field. However, obtaining clear neuronal imaging in the context of high-speed zebrafish motion is not easy, requiring the imaging system to first address high-speed tracking and motion cancellation. Second, fluorescence image capture must be robust enough to adapt to changes in the animal's posture as well as motion blur caused by high-speed motion.

Kim et al. developed a new imaging platform that allows the tracking of freely moving zebrafish. Zebrafish larvae were allowed to move freely in a chamber with a diameter of 50 mm and a height of 750 μm. For optical illumination and image acquisition, the top and bottom of the chamber were made of transparent glass. The microscope used infrared imaging to track the moving zebrafish and a mechanical carrier stage to offset the three-dimensional motion of the zebrafish. And this tracking system was combined with a variant of the HiLo microscope, the differential illumination focus filtering (DIFF) microscope. Using this imaging platform, Kim et al. were able to image freely moving zebrafish brains at single-cell resolution for an hour without interruption. This work enriches the library of natural behaviors that can be studied with cell-resolution calcium imaging (D. H. Kim et al., 2017). Marques and colleagues utilized this platform to explore the neural underpinnings of two distinct larval behaviors during free swimming: exploration behavior characterized by efficient long-distance swimming, and exploitation behavior involving localized predation. Exploration and exploitation exert contrasting effects on different animals, where the former presents the potential for resource discovery at a higher metabolic cost, while the latter serves as an effective strategy for predatory feeding, contingent upon the presence of prey. These behavioral modes are influenced by external inputs and hunger levels, but they also exhibit spontaneous alternations, underscoring the potential role of internal brain states in determining which behavior is chosen. The researchers captured videos of freely swimming larvae, meticulously categorizing each movement to generate comprehensive behavior descriptions during exploration and exploitation phases (Marques et al., 2020). LFM is a promising imaging device with broad prospects. Cong et al. have enhanced LFM by introducing a new version called extended field-of-view LMF (XLMF). This advancement integrates high-speed cameras and computer software to accommodate the rapid movement of zebrafish. The zebrafish are allowed to freely swim within a water-filled chamber with a diameter of 42 millimeters and a depth of 800 micrometers. To track the fish's movement, they employ a high-speed three-axis stage, an autofocus camera, and lateral tracking optics. The autofocus camera captures light-field images and monitors the axial movement of the fish's head. Customized machine vision software rapidly estimates the fish's movements in the *xy* and *z* directions. By utilizing feedback control signals, a high-speed electric *xy* stage (300 Hz) and a piezoelectric *z* stage maintain the entire fish's head within the field of view of a high numerical aperture objective (25×, NA = 1.05). This setup can also adapt to sudden fish movements that often result in image blurring. This system ensures that the zebrafish's brain remains in focus. Cong et al. utilized this apparatus to successfully capture, for the first time, the activity of all neurons in the zebrafish larvae's brain during the process of chasing and capturing prey (Cong et al., 2017).

In the past two years, Zhang et al. achieved another iteration of the LFM imaging system, known as confocal LFM. Confocal LFM is equipped with a novel generalized confocal detection scheme, ensuring the selective and efficient collection of signals from the focal volume. This design significantly differs from traditional confocal detection setups that extract signals from only one imaging plane at a time. The background elimination feature, also referred to as optical sectioning, preserves a high signal-to-noise ratio and reduces reconstruction artifacts when imaging thick tissues. This technology allows for clear and artifact-free whole-brain imaging of freely moving zebrafish larvae within a range of 2 × 2 × 2.5 μm (Z. Zhang et al., 2021).

Both of these teams of scientists offset zebrafish movement by moving the behavioral arena, which undoubtedly raises a problem. Moving the water-filled behavioral arena at very high speeds can cause some interference with the zebrafish's own locomotor behavior. Symvoulidis et al. developed the NeuBtracker, a system with two imaging channels, a static infrared-sensitive camera used to observe and locate the fish, which receives a 1× image as an input to localize the fish's head, and a second channel used to track the free-swimming zebrafish larvae to provide a magnified fluorescent image. Its field of view (FOV) is 15 × 15 mm and the working distance to the sample is roughly 10 cm, so any clear plastic arena with a diameter less than 1.5 cm and a height of 2 mm–1 cm can be used for imaging. The tracking error distance peaked at around 190 μm when the fish speed was around 30 mm/s. For FOVs ranging from the entire body of the larva to zooming in on the larval brain only, a resolution of 150line pairs per mm (lp/mm) could be achieved (Symvoulidis et al., 2017).

In addition to the devices commonly used for calcium imaging mentioned previously, photoacoustic imaging has made important breakthroughs in calcium imaging. So far, photoacoustic-enabled brain imaging has mainly been used to detect hemodynamic and blood oxygenation changes, which can only indirectly reflect the slow and delayed process of neural activation. Ben et al. found that GCAMP5Gs can provide suitable calcium-dependent photoacoustic contrast, a property that is facilitated by the fact that they have strong calcium-dependent extinction changes (Akerboom et al., 2012). On this basis, a sender multispectral tomography platform was developed by Deán-Ben et al. The neural activity of the entire brain of a freely moving zebrafish can be imaged. The researchers placed the free-moving zebrafish in a 6×6×3-mm chamber filled with water in the focal region of a photoacoustic matrix detector. The platform recorded 100 volumetric images per second in a scalable field of view ranging from 50–1,000 cubic millimeters, with a spatial resolution of 35–200 micrometers, respectively. The first demonstration of the potential of photoacoustic imaging for functional whole-brain neural detection of calcium dynamics will open new horizons for studying neural activity and neurovascular coupling in health and disease (Deán-Ben et al., 2016). Recording the behavior of free-swimming zebrafish or the neural activity of immobilized zebrafish are traditional methods of recording neurodegenerative disease. Breakthroughs in imaging free-swimming zebrafish have allowed the recording of neural activity in unrestrained zebrafish during pathogenesis. Cagla Özsoy et al. employed optoacoustic tomography (OAT). OAT imaging utilizes a customized spherical matrix array probe consisting of 512 individual chips. The probe is capable of detecting signals with a bandwidth of up to 35 MHz and achieves nearly isotropic 3D resolution of less than 30 μm. Particularly noteworthy is its ability to efficiently cover an observation range of 5 × 5 × 5 cm³. Their experimental results emphasize that neural activity and associated behavioral changes in zebrafish larvae experiencing epileptic seizures can be effectively captured by volumetric optoacoustic imaging. The high temporal resolution of this technique enables real-time visualization of rapid calcium kinetic processes in the brain. The ability of this imaging method to track neuronal activation within unrestrained animals offers the possibility to gain insight into the intricate relationship between neural activity and behavioral changes during seizures. This innovative approach promises to shed light on the underlying mechanisms of epilepsy and may pave the way for the development of more effective antiepileptic drugs (Özsoy et al., 2022).

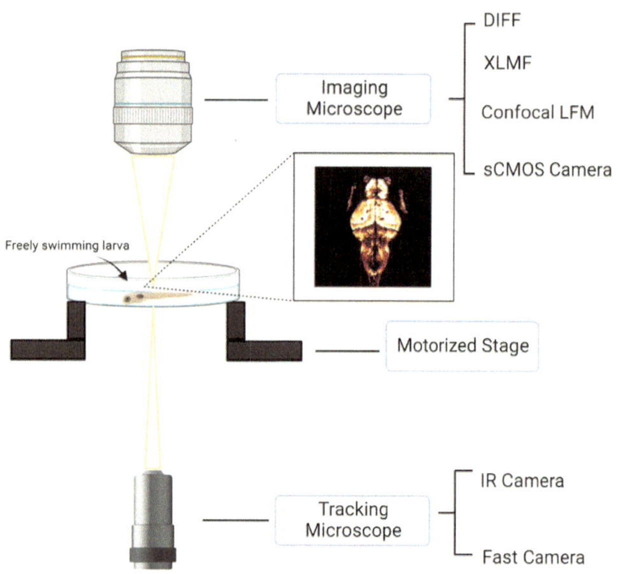

FIGURE 7.3 The calcium imaging equipment for free-moving zebrafish consists of three main parts: an imaging microscope, a motorized carrier stage, and a tracking microscope. The imaging microscope: differential illumination focus filtering (DIFF) microscope, extended field-of-view LMF (XLMF), confocal LFM, and sCMOS camera. Tracking microscope: infrared (IR) camera and fast camera.

7.7 DISCUSSION AND PROSPECTS

In conclusion, the utilization of calcium imaging techniques has illuminated a multitude of insights into the intricate neural dynamics within unrestrained larval zebrafish. As explored in this chapter, the deployment of various calcium indicators has enabled researchers to gain access to a comprehensive view of neuronal activity.

By capitalizing on the principles underlying these indicators, scientists have ventured into a realm of understanding that transcends traditional boundaries. The arsenal of calcium imaging systems, ranging from conventional wide-field fluorescence microscopes to Light Field Imaging, has provided a platform for observing neural activity at diverse spatial and temporal scales. These technologies have revolutionized our capacity to visualize and analyze the dynamic interplay of neurons within intricate neural circuits. This chapter has endeavored to shed light on the mechanisms that underpin these imaging systems, elucidating their strengths and limitations in the context of studying larval zebrafish. A cornerstone of successful calcium imaging lies in meticulous experimental design. The meticulous planning of preparatory phases, coupled with thoughtful considerations of image acquisition parameters and robust data analysis techniques, ensures that the acquired neural activity data faithfully reflect the physiological processes under scrutiny. The comprehensive insights derived from such an approach deepen our understanding of neural circuitry and enhance the credibility of research findings.

While remarkable progress has been achieved in calcium imaging of freely moving zebrafish, several challenges remain to be addressed. In terms of imaging indicators, prevailing imaging techniques primarily center around calcium indicators, the existing indicators often fall short in capturing the rapid dynamics of neural activity, with decay extending over several seconds (Y. Zhang et al., 2023). This temporal limitation becomes particularly pronounced when attempting to correlate swift behaviors like hunting with neural activity, highlighting the challenge of synchronizing sensorimotor integration timeframes. Additionally, calcium indicators represent just one facet of neuronal activity. The promise of more targeted genetically encoded sensors shines brightly for future studies. The groundbreaking work by Yulong Li's team at Peking University exemplifies this trend, showcased by the development of a dopamine imaging system (Sun et al., 2018). Their meticulous approach involved the identification of effective fusion sites within the dopamine receptor, optimization of N- and C-terminal linkers, and point mutations to enhance affinity between the receptor and cpEGFP. Notably, this system demonstrated exceptional signal-to-noise ratios and sensitivity across diverse contexts including in vitro settings, *Drosophila*, zebrafish, and mice. Moreover, their contributions extended to creating endogenous cannabinoid, ATP, and acetylcholine imaging systems, each tailored to specific applications.

In terms of imaging technology, current methodologies facilitate the observation of free-swimming zebrafish larvae within confined behavioral spaces. Nevertheless, these settings may not fully encapsulate the intricacies of their behaviors in more naturalistic environments. The expansion of behavioral arenas, encompassing size and depth, emerges as an imperative in genuinely capturing the responses of neural circuits orchestrating zebrafish behavior during authentic movements. Furthermore, ongoing research has predominantly focused on imaging juvenile zebrafish larvae, whose behavioral patterns evolve as they mature, accompanied by the progressively intricate regulation of underlying neural circuits. Consequently, the exploration of calcium imaging in freely swimming adult zebrafish stands as a promising direction for unveiling the complexities of their neural dynamics.

In summary, calcium imaging has propelled us forward in decoding neural dynamics within zebrafish. The fusion of neuroscience and imaging technology has carved new pathways for exploration, and the evolving landscape of these techniques holds promise in revealing deeper insights into the intricate interplay between neurons, circuits, and behaviors. Yet, a tapestry of challenges endures, urging us to continually innovate technologically and refine methodologies to attain a more comprehensive and precise understanding of the intricate interplay between neural circuits and behavior. As researchers continue to push the boundaries of our knowledge, the zebrafish model system remains an indispensable ally in unraveling the enigmas of the nervous system. This not only contributes to the fundamental understanding of neuroscience but also paves the way for advancing therapeutic strategies targeting neurological disorders such as Parkinson's disease.

REFERENCES

Aarsland, D., Batzu, L., Halliday, G. M., Geurtsen, G. J., Ballard, C., Ray Chaudhuri, K., & Weintraub, D. (2021). Parkinson disease-associated cognitive impairment. *Nature Reviews. Disease Primers*, *7*(1), 47. doi:10.1038/s41572-021-00280-3

Ahrens, M. B., Li, J. M., Orger, M. B., Robson, D. N., Schier, A. F., Engert, F., & Portugues, R. (2012). Brain-wide neuronal dynamics during motor adaptation in zebrafish. *Nature*, *485*(7399), 471–477. doi:10.1038/nature11057

Ahrens, M. B., Orger, M. B., Robson, D. N., Li, J. M., & Keller, P. J. (2013). Whole-brain functional imaging at cellular resolution using light-sheet microscopy. *Nature Methods*, *10*(5), 413–420. doi:10.1038/nmeth.2434

Akerboom, J., Chen, T.-W., Wardill, T. J., Tian, L., Marvin, J. S., Mutlu, S., . . . Looger, L. L. (2012). Optimization of a GCaMP calcium indicator for neural activity imaging. *The Journal of Neuroscience: The Official Journal of the Society for Neuroscience*, *32*(40), 13819–13840. doi:10.1523/JNEUROSCI.2601-12.2012

Baird, G. S., Zacharias, D. A., & Tsien, R. Y. (1999). Circular permutation and receptor insertion within green fluorescent proteins. *Proceedings of the National Academy of Sciences of the United States of America*, *96*(20), 11241–11246. Retrieved from https://pubmed.ncbi.nlm.nih.gov/10500161

Barreto-Chang, O. L., & Dolmetsch, R. E. (2009). Calcium imaging of cortical neurons using Fura-2 AM. *Journal of Visualized Experiments: JoVE*, (23). doi:10.3791/1067

Brustein, E., Marandi, N., Kovalchuk, Y., Drapeau, P., & Konnerth, A. (2003). "In vivo" monitoring of neuronal network activity in zebrafish by two-photon Ca(2+) imaging. *Pflugers Archiv: European Journal of Physiology*, *446*(6), 766–773. Retrieved from https://pubmed.ncbi.nlm.nih.gov/12883893

Carvalho, M. M., Campos, F. L., Coimbra, B., Pêgo, J. M., Rodrigues, C., Lima, R., . . . Salgado, A. J. (2013). Behavioral characterization of the 6-hydroxidopamine model of Parkinson's disease and pharmacological rescuing of non-motor deficits. *Molecular Neurodegeneration*, *8*, 14. doi:10.1186/1750-1326-8-14

Chen, J., Poskanzer, K. E., Freeman, M. R., & Monk, K. R. (2020). Live-imaging of astrocyte morphogenesis and function in zebrafish neural circuits. *Nature Neuroscience*, *23*(10), 1297–1306. doi:10.1038/s41593-020-0703-x

Chen, T.-W., Wardill, T. J., Sun, Y., Pulver, S. R., Renninger, S. L., Baohan, A., . . . Kim, D. S. (2013). Ultrasensitive fluorescent proteins for imaging neuronal activity. *Nature*, *499*(7458), 295–300. doi:10.1038/nature12354

Chiesa, A., Rapizzi, E., Tosello, V., Pinton, P., de Virgilio, M., Fogarty, K. E., & Rizzuto, R. (2001). Recombinant aequorin and green fluorescent protein as valuable tools in the study of cell signalling. *The Biochemical Journal*, *355*(Pt 1). Retrieved from https://pubmed.ncbi.nlm.nih.gov/11256942

Cong, L., Wang, Z., Chai, Y., Hang, W., Shang, C., Yang, W., . . . Wen, Q. (2017). Rapid whole brain imaging of neural activity in freely behaving larval zebrafish (*Danio rerio*). *ELife*, *6*. doi:10.7554/eLife.28158

Deán-Ben, X. L., Sela, G., Lauri, A., Kneipp, M., Ntziachristos, V., Westmeyer, G. G., . . . Razansky, D. (2016). Functional optoacoustic neuro-tomography for scalable whole-brain monitoring of calcium indicators. *Light, Science & Applications*, *5*(12), e16201. doi:10.1038/lsa.2016.201

De Nadai, A., Vajente, N., Pendin, D., & Mattarei, A. (2021). Mt-fura-2, a ratiometric mitochondria-targeted Ca2+ sensor. Determination of spectroscopic properties and Ca2+ imaging assays. *Methods in Molecular Biology (Clifton, N.J.)*, *2275*, 187–215. doi:10.1007/978-1-0716-1262-0_12

de Vito, G., Turrini, L., Müllenbroich, C., Ricci, P., Sancataldo, G., Mazzamuto, G., . . . Pavone, F. S. (2022). Fast whole-brain imaging of seizures in zebrafish larvae by two-photon light-sheet microscopy. *Biomedical Optics Express*, *13*(3), 1516–1536. doi:10.1364/BOE.434146

Fetcho, J. R., & O'Malley, D. M. (1995). Visualization of active neural circuitry in the spinal cord of intact zebrafish. *Journal of Neurophysiology*, *73*(1), 399–406. Retrieved from https://pubmed.ncbi.nlm.nih.gov/7714582

Friedrich, R. W., & Korsching, S. I. (1997). Combinatorial and chemotopic odorant coding in the zebrafish olfactory bulb visualized by optical imaging. *Neuron*, *18*(5), 737–752. Retrieved from https://pubmed.ncbi.nlm.nih.gov/9182799

Gahtan, E., Tanger, P., & Baier, H. (2005). Visual prey capture in larval zebrafish is controlled by identified reticulospinal neurons downstream of the tectum. *The Journal of Neuroscience: The Official Journal of the Society for Neuroscience*, *25*(40), 9294–9303. Retrieved from https://pubmed.ncbi.nlm.nih.gov/16207889

Ghosh, A., & Greenberg, M. E. (1995). Calcium signaling in neurons: Molecular mechanisms and cellular consequences. *Science (New York, N.Y.)*, *268*(5208), 239–247. Retrieved from https://pubmed.ncbi.nlm.nih.gov/7716515

Inagaki, H. K., Chen, S., Daie, K., Finkelstein, A., Fontolan, L., Romani, S., & Svoboda, K. (2022). Neural algorithms and circuits for motor planning. *Annual Review of Neuroscience*, *45*, 249–271. doi:10.1146/annurev-neuro-092021-121730

Kappel, J. M., Förster, D., Slangewal, K., Shainer, I., Svara, F., Donovan, J. C., . . . Larsch, J. (2022). Visual recognition of social signals by a tectothalamic neural circuit. *Nature*, *608*(7921), 146–152. doi:10.1038/s41586-022-04925-5

Kaufmann, A., Mickoleit, M., Weber, M., & Huisken, J. (2012). Multilayer mounting enables long-term imaging of zebrafish development in a light sheet microscope. *Development (Cambridge, England)*, *139*(17), 3242–3247. doi:10.1242/dev.082586

Kim, C. H., Ueshima, E., Muraoka, O., Tanaka, H., Yeo, S. Y., Huh, T. L., & Miki, N. (1996). Zebrafish elav/HuC homologue as a very early neuronal marker. *Neuroscience Letters*, *216*(2), 109–112. Retrieved from https://pubmed.ncbi.nlm.nih.gov/8904795

Kim, D. H., Kim, J., Marques, J. C., Grama, A., Hildebrand, D. G. C., Gu, W., . . . Robson, D. N. (2017). Pan-neuronal calcium imaging with cellular resolution in freely swimming zebrafish. *Nature Methods*, *14*(11), 1107–1114. doi:10.1038/nmeth.4429

Larsch, J., & Baier, H. (2018). Biological motion as an innate perceptual mechanism driving social affiliation. *Current Biology*, *28*(22). doi:10.1016/j.cub.2018.09.014

Liu, J., & Baraban, S. C. (2019). Network properties revealed during multi-scale calcium imaging of seizure activity in Zebrafish. *ENeuro*, *6*(1). doi:10.1523/ENEURO.0041-19.2019

Lu, R., Sun, W., Liang, Y., Kerlin, A., Bierfeld, J., Seelig, J. D., . . . Ji, N. (2017). Video-rate volumetric functional imaging of the brain at synaptic resolution. *Nature Neuroscience*, *20*(4), 620–628. doi:10.1038/nn.4516

Marisca, R., Hoche, T., Agirre, E., Hoodless, L. J., Barkey, W., Auer, F., . . . Czopka, T. (2020). Functionally distinct subgroups of oligodendrocyte precursor cells integrate neural activity and execute myelin formation. *Nature Neuroscience*, *23*(3), 363–374. doi:10.1038/s41593-019-0581-2

Marques, J. C., Li, M., Schaak, D., Robson, D. N., & Li, J. M. (2020). Internal state dynamics shape brainwide activity and foraging behaviour. *Nature*, *577*(7789), 239–243. doi:10.1038/s41586-019-1858-z

Mu, Y., Bennett, D. V., Rubinov, M., Narayan, S., Yang, C.-T., Tanimoto, M., . . . Ahrens, M. B. (2019). Glia accumulate evidence that actions are futile and suppress unsuccessful behavior. *Cell*, *178*(1). doi:10.1016/j.cell.2019.05.050

Muto, A., Ohkura, M., Abe, G., Nakai, J., & Kawakami, K. (2013). Real-time visualization of neuronal activity during perception. *Current Biology*, *23*(4), 307–311. doi:10.1016/j.cub.2012.12.040

Muto, A., Lal, P., Ailani, D., Abe, G., Itoh, M., & Kawakami, K. (2017). Activation of the hypothalamic feeding centre upon visual prey detection. *Nature Communications*, *8*, 15029. doi:10.1038/ncomms15029

Nakai, J., Ohkura, M., & Imoto, K. (2001). A high signal-to-noise Ca(2+) probe composed of a single green fluorescent protein. *Nature Biotechnology*, *19*(2), 137–141. Retrieved from https://pubmed.ncbi.nlm.nih.gov/11175727

Niemeyer, J. E., Gadamsetty, P., Chun, C., Sylvester, S., Lucas, J. P., Ma, H., . . . Aksay, E. R. F. (2022). Seizures initiate in zones of relative hyperexcitation in a zebrafish epilepsy model. *Brain: A Journal of Neurology*, *145*(7), 2347–2360. doi:10.1093/brain/awac073

Nishimura, Y., Murakami, S., Ashikawa, Y., Sasagawa, S., Umemoto, N., Shimada, Y., & Tanaka, T. (2015). Zebrafish as a systems toxicology model for developmental neurotoxicity testing. *Congenital Anomalies*, *55*(1). doi:10.1111/cga.12079

Özsoy, Ç., Hotz, A. L., Rieser, N. N., Chen, Z., Deán-Ben, X. L., Neuhauss, S. C. F., & Razansky, D. (2022). Volumetric optoacoustic neurobehavioral tracking of epileptic seizures in freely-swimming zebrafish larvae. *Frontiers in Molecular Neuroscience*, *15*, 1004518. doi:10.3389/fnmol.2022.1004518

Panier, T., Romano, S. A., Olive, R., Pietri, T., Sumbre, G., Candelier, R., & Debrégeas, G. (2013). Fast functional imaging of multiple brain regions in intact zebrafish larvae using selective plane illumination microscopy. *Frontiers in Neural Circuits*, *7*, 65. doi:10.3389/fncir.2013.00065

Panula, P., Chen, Y. C., Priyadarshini, M., Kudo, H., Semenova, S., Sundvik, M., & Sallinen, V. (2010). The comparative neuroanatomy and neurochemistry of zebrafish CNS systems of relevance to human neuropsychiatric diseases. *Neurobiology of Disease*, *40*(1), 46–57. doi:10.1016/j.nbd.2010.05.010

Portugues, R., & Engert, F. (2011). Adaptive locomotor behavior in larval zebrafish. *Frontiers in Systems Neuroscience*, *5*, 72. doi:10.3389/fnsys.2011.00072

Portugues, R., Feierstein, C. E., Engert, F., & Orger, M. B. (2014). Whole-brain activity maps reveal stereotyped, distributed networks for visuomotor behavior. *Neuron*, *81*(6), 1328–1343. doi:10.1016/j.neuron.2014.01.019

Prevedel, R., Yoon, Y.-G., Hoffmann, M., Pak, N., Wetzstein, G., Kato, S., . . . Vaziri, A. (2014). Simultaneous whole-animal 3D imaging of neuronal activity using light-field microscopy. *Nature Methods*, *11*(7), 727–730. doi:10.1038/nmeth.2964

Sainsbury, T. T. J., Diana, G., & Meyer, M. P. (2023). Topographically localized modulation of tectal cell spatial tuning by complex natural scenes. *ENeuro*, *10*(1). doi:10.1523/ENEURO.0223-22.2022

Salgado-Almario, J., Vicente, M., Molina, Y., Martinez-Sielva, A., Vincent, P., Domingo, B., & Llopis, J. (2022). Simultaneous imaging of calcium and contraction in the beating heart of zebrafish larvae. *Theranostics*, *12*(3), 1012–1029. doi:10.7150/thno.64734

Shimomura, O., Johnson, F. H., & Saiga, Y. (1962). Extraction, purification and properties of aequorin, a bioluminescent protein from the luminous hydromedusan, Aequorea. *Journal of Cellular and Comparative Physiology*, *59*, 223–239. Retrieved from https://pubmed.ncbi.nlm.nih.gov/13911999

Sun, F., Zeng, J., Jing, M., Zhou, J., Feng, J., Owen, S. F., . . . Li, Y. (2018). A genetically encoded fluorescent sensor enables rapid and specific detection of dopamine in flies, fish, and mice. *Cell*, *174*(2). doi:10.1016/j.cell.2018.06.042

Symvoulidis, P., Lauri, A., Stefanoiu, A., Cappetta, M., Schneider, S., Jia, H., . . . Westmeyer, G. G. (2017). NeuBtracker-imaging neurobehavioral dynamics in freely behaving fish. *Nature Methods*, *14*(11), 1079–1082. doi:10.1038/nmeth.4459

Turrini, L., Fornetto, C., Marchetto, G., Müllenbroich, M. C., Tiso, N., Vettori, A., . . . Vanzi, F. (2017). Optical mapping of neuronal activity during seizures in zebrafish. *Scientific Reports*, *7*(1), 3025. doi:10.1038/s41598-017-03087-z

Wen, H., Eckenstein, K., Weihrauch, V., Stigloher, C., & Brehm, P. (2020). Primary and secondary motoneurons use different calcium channel types to control escape and swimming behaviors in zebrafish. *Proceedings of the National Academy of Sciences of the United States of America*, *117*(42), 26429–26437. doi:10.1073/pnas.2015866117

Westerfield, M. (2000). *The Zebrafish book. A guide for the laboratory use of Zebrafish (Danio rerio)*, 4th ed.; University of Oregon Press: Eugene, OR, 2000. Retrieved from https://zfin.org/zf_info/zfbook/zfbk.html

Xu, X., Soutto, M., Xie, Q., Servick, S., Subramanian, C., von Arnim, A. G., & Johnson, C. H. (2007). Imaging protein interactions with bioluminescence resonance energy transfer (BRET) in plant and mammalian cells and tissues. *Proceedings of the National Academy of Sciences of the United States of America*, *104*(24), 10264–10269. Retrieved from https://pubmed.ncbi.nlm.nih.gov/17551013

Zhao, J., Liang, S., Zhao, Y., Zong, W., Tran, E., Chen, L., & Liu, Y. (2021). In vivo imaging of calcium activities from pancreatic β-cells in Zebrafish embryos using spinning-disc confocal and two-photon light-sheet microscopy. *Bio-Protocol*, *11*(23), e4245. doi:10.21769/BioProtoc.4245

Zhang, Y., Rózsa, M., Liang, Y., Bushey, D., Wei, Z., Zheng, J., . . . Looger, L. L. (2023). Fast and sensitive GCaMP calcium indicators for imaging neural populations. *Nature*, *615*(7954), 884–891. doi:10.1038/s41586-023-05828-9

Zhang, Z., Bai, L., Cong, L., Yu, P., Zhang, T., Shi, W., . . . Wang, K. (2021). Imaging volumetric dynamics at high speed in mouse and zebrafish brain with confocal light field microscopy. *Nature Biotechnology*, *39*(1), 74–83. doi:10.1038/s41587-020-0628-7

8 Anesthesia, Analgesia, and Euthanasia Techniques for Laboratory Zebrafish

Azhwar Raghunath

8.1 INTRODUCTION

The zebrafish (*Danio rerio*) emerged as a highly valuable model organism in the 1980s and has facilitated studies in diverse fields such as development, host-microbe interactions, neurobehavior, and human disease (Ablain and Zon, 2013). Notably, the zebrafish genome sequencing project revealed that around 70% of human protein-coding genes possess orthologs in fish (Howe et al., 2013). With the advent of modern genetic manipulation technologies, the past decade has witnessed a rapid acceleration in the development of zebrafish disease models (Hwang et al., 2013; Adamson et al., 2018). Moreover, zebrafish play a pivotal role in drug testing and toxicant screening, contributing to both an understanding of disease mechanisms and the development of novel therapies, complementing the use of murine models (Horzmann and Freeman, 2018). Isogenicity is unattainable within the zebrafish population due to its inherent genetic diversity (Orger and de Polavieja, 2017). This genetic diversity, stemming from their extensive clutch size, serves as a valuable attribute for researchers to reproduce experimental data. Basic and biomedical research involving zebrafish has witnessed an exponential increase in recent times, largely attributable to their ease of housing and management within laboratory settings. Within the realm of bony (teleost) fish, they manifest behavioral and physiological responses when subjected to noxious stimuli (Sneddon et al., 2003). Consequently, it becomes imperative to prioritize the welfare of zebrafish employed in research, with the overarching aim of mitigating pain and stress during experimental procedures (Iwama, 2007). Ethical guidelines, complementing scientific requirements, assert that zebrafish must be maintained in a state of optimal well-being throughout the course of experiments. There is a prevalent misconception surrounding fish that pertains to their purported insensitivity to pain. However, recent investigations have demonstrated that zebrafish are indeed capable of experiencing nociception, pain, and stress (Sneddon, 2015). While no universally standard protocol exists for identifying zebrafish experiencing pain or distress, common indicators include a reduction in swimming or overall activity, coupled with a propensity to linger at the bottom of the tank (Collymore et al., 2016). When exposed to noxious stimuli, zebrafish display a range of conspicuous responses, such as heightened ventilation characterized by rapid opercular movement, decreased swimming activity, a preference for staying close to the bottom, an increased inclination to seek cover when it is accessible, and elevated cortisol levels (Maximino, 2011). Hence, there exists a compelling need for anesthesia, analgesia, and euthanasia in the context of zebrafish welfare, with paramount consideration for procedures that are painless and reproducible without compromising the experimental outcomes. This chapter summarizes frequently employed agents for anesthesia, analgesia, and euthanasia in zebrafish research.

8.2 ANESTHESIA

With the burgeoning and exponential utilization of zebrafish to solve scientific problems, the importance of appropriate anesthesia cannot be overstated. The comprehensive analysis of zebrafish disease models encompasses a spectrum of techniques, including genetic screening, small-molecule screening, and staining techniques (Gibert et al., 2013; Rennekamp and Peterson, 2015), all of which necessitate the use of effective anesthesia protocols. Establishing a suitable anesthetic regimen is paramount in zebrafish research, as it serves to minimize pain and distress during experimental procedures. Neglecting the significance of a proper anesthesia protocol can profoundly impact zebrafish and introduce unwarranted variability into experimental data.

Anesthetic agents ensure that animals do not endure pain or distress during invasive procedures while allowing them to regain consciousness afterwards without causing any short- or long-term side effects (Figure 8.1). The fundamental attributes of an optimal anesthetic agent for fish include seamless administration, efficacy at low doses, initiation of anesthesia with minimal stress, rapid recovery of consciousness, and negligible perturbations to physiology and behavior (Martins et al., 2016). Additionally, it should be cost-effective, widely available, user-friendly, and ensure safety for all personnel involved.

8.2.1 Tricaine Methanesulfonate (MS222) (Ethyl 3-Aminobenzoate; Methanesulfonic Acid)

MS222, a derivative of benzocaine, is a widely used immersion anesthetic agent in various aquatic species (Topic Popovic et al., 2012). This ester-type agent exerts systemic effects and stands as the sole fish anesthesia approved by

DOI: 10.1201/9781003402893-9

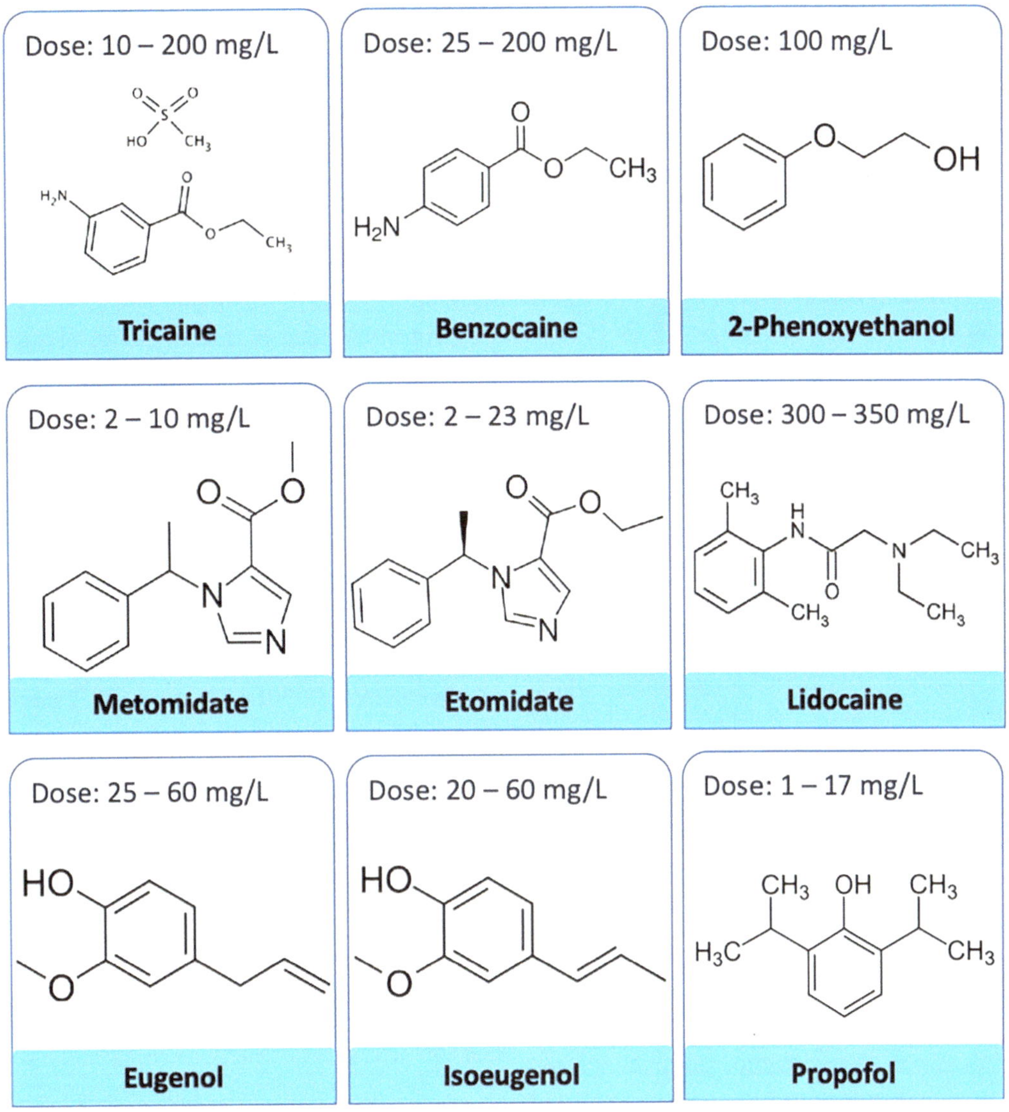

FIGURE 8.1 Anesthetic agents used in zebrafish show the range of doses.

the US Food and Drug Administration. MS222 is absorbed through both gills and skin, achieving distribution across the bloodstream to act on nervous system and muscles. Its subsequent elimination primarily occurs through the gills, complemented by excretion through urine and bile (Stenger and Maren, 1974; Carter et al., 2011). The effective dose of MS222 hinges upon the age, size, and metabolic state of the zebrafish (Rombough, 2007). Facilitated by its lipid-soluble nature, MS222 traverses membranes with remarkable ease (Spears et al., 2014). The mechanism of MS222 involves blocking sodium channels, thereby impeding the transmission of action potentials, and leading to muscle relaxation (Frazier and Narahashi, 1975). Additionally, it obstructs potassium currents within nerve membranes.

The features that make MS222 an ideal fish anesthetic include its versatility for administration through immersion or injection, potent effectiveness, high solubility in various matrices, and complete recovery potential (Thienpont and Niemegeers, 1965). However, the usage of MS222 at elevated doses and extended exposure durations gives rise to aversive reactions, epidermal and corneal lesions, bradycardia, hypoxemia, stress, and eventual mortality (Huang et al., 2010; Carter et al., 2011; Wong et al., 2014). On the contrary, when administered at optimal dose, MS222 is safe for zebrafish. The concomitant use of MS222 with isoflurane diminishes bradycardia, prolongs anesthesia, and accelerates recovery (Huang et al., 2010; Lockwood et al., 2017). Within the zebrafish research community, MS222 reigns as the most widely employed anesthetic agent and as the agent of choice for invasive procedures (Collymore et al., 2014; Priborsky and Velisek, 2018).

When exposing fish to MS222, a well-aerated tank serves as the recommended setting. Initial moments of exposure may elicit aversive behaviors like excitatory responses, heightened opercular movement, and hypertaxia characterized by erratic swimming, thrashing, twitching, and piping

(Matthews and Varga, 2012). These behaviors tend to subside once equilibrium is lost, typically within seconds. Due to its acidic nature, MS222 can induce a pH drop of 5 in weakly buffered or unbuffered solutions, which necessitates prior buffering. This acidity can be countered using sodium hydroxide, sodium bicarbonate (Harper and Lawrence, 2011), or Tris buffer (Westerfield, 2007), thereby enhancing MS222's effectiveness and safety. Other buffering agents such as calcium carbonate, imidazole, and sodium hydrogen phosphate also demonstrate efficacy. pH modulation influences MS222's efficiency (Wedemeyer, 1970). In cases where unbuffered MS222 is employed, prolonged stress and inter-renal acid depletion can occur (Carter et al., 2011).

For practicality, it is advisable to prepare concentrated stock solutions of MS222, reducing time, labor, and potential exposure risks. Monitoring potency is crucial, because the appearance of an oily residue or discoloration signifies reduced effectiveness, likely due to desulfonation product formation, necessitating replacement (Neiffer and Stamper, 2009). Prolonged preservation of the concentrated MS222 stock solution (100 mg/mL) can be achieved through refrigeration at 4°C or freezing at -20°C within amber jars or aluminum foil-wrapped bottles, ensuring stability for up to 6 months (Katz et al., 2020). Given its irritant properties, the handling of tricaine powder mandates preparation in a fume hood while wearing protective gloves and goggles.

8.2.2 Benzocaine (Ethyl 4-Aminobenzoate)

Benzocaine, the parent compound of MS222, is an odorless and tasteless anesthetic that is less acidic than MS222. Despite its structural similarity to MS222, it lacks a sulfate moiety and shares a common mechanism of action by inhibiting action potentials by blocking the sodium channels. Due to its pronounced lipophilic nature, it exhibits reduced solubility in water, necessitating the use of ethanol or acetone as solvents when preparing stock solutions for immersion anesthesia. It induces anesthesia rapidly, with recovery typically occurring in under 10 minutes. Benzocaine is highly aversive, and its efficacy is influenced by the age of the fish and the water temperature (Ross and Ross, 2008; Readman et al., 2013). Additionally, acetone and propylene glycol are viable options for preparing stock solutions of benzocaihe. These solutions can be safely stored in amber bottles at room temperature for up to a year. Notably, benzocaine is environmentally friendly due to its reduced toxicity for personnel handling it, and it undergoes disintegration in water within a mere 4 hours. Benzocaine hydrochloride, a crystalline salt, is water-soluble but exhibits higher acidity similar to MS222. To bring the solution's pH to 7.0, sodium bicarbonate buffering is required when working with benzocaine hydrochloride.

8.2.3 2-Phenoxyethanol (2-PE)

2-PE serves as a valuable agent in various applications, particularly in the transportation of fish and as an immersion anesthesia (Hseu et al., 1997). This versatile compound exists as a slightly water-soluble, colorless, and cost-effective oily liquid. When utilized with ethanol as a solvent, 2-PE's solubility increases significantly, making it an excellent choice for immersion anesthesia. It does not change pH when added to seawater and produces hypoventilation. Prolonged and frequent exposure to 2-PE can lead to a neuropsychological syndrome in individuals handling it (Hseu et al., 1998). While 2-PE induces deep anesthesia, it is not without side effects. These side effects include bradycardia, alterations in blood parameters, cortisol stress responses, and immunomodulatory effects (Cuesta et al., 2004; Witeska et al., 2015; Akbary et al., 2016). Nevertheless, the precise mechanism underlying its action is not completely understood.

8.2.4 Metomidate (Methyl 3-(1-Phenylethyl) Imidazole-4-Carboxylate)

Metomidate, an imidazole-based nonbarbiturate water-soluble compound, serves as a valuable hypnotic agent (Collymore et al., 2014). Metomidate is used during handling and transportation of fish. It exerts its hypnotic effects by functioning as a gamma aminobutyric acid type A (GABAA) receptor agonist and disrupts cortisol production by suppressing 11-betahydroxylase (Dorr et al., 1984). It effectively dampens the cortisol response by obstructing steroidogenesis through the inhibition of adrenocorticotropic hormone (Iversen et al., 2003; Small, 2003). Methyl 3-[1-phenyl-ethyl] imidazole-4-carboxylate hydrochloride, commonly known as metomidate hydrochloride, anesthetizes zebrafish rapidly and is less aversive than MS222, but has a recovery time exceeding half an hour (Wong et al., 2014). Metomidate maintains opercular ventilation for extended duration when compared to other immersion anesthetics (Mattson and Riple, 1989). The side effects include muscle twitching and transient darkening of the fish color due to reduced cortisol synthesis. Additionally, its use may lead to diminished respiration and circulation, ultimately resulting in hypoxemia and an increase in blood acidity (Iwama et al., 1989). It is not an ideal agent for use with zebrafish larvae. It is advisable to store stock solutions in light-protected bottles to maintain its stability. The working solution of metomidate remains viable for only 3 days.

8.2.5 Etomidate (Ethyl 3-[(1R)-1-Phenylethyl] Imidazole-4-Carboxylate)

Etomidate, akin to metomidate, is a potent sedative agent, yet its efficacy remains extensively unstudied in zebrafish. The toxicity of etomidate diminishes at higher temperatures. It is less aversive to zebrafish and effective in alkaline waters (Readman et al., 2013). Etomidate is recommended for mitigating long-term stress and the subsequent delayed immunosuppression (Witeska et al., 2015). Neither metomidate nor etomidate provide analgesic effects.

8.2.6 Lidocaine (2-(Diethylamino)-N-(2,6-Dimethylphenyl)Acetamide)

Lidocaine, unlike tricaine, is not an ideal immersion anesthetic agent. Similar to benzocaine, lidocaine exists in two forms: a non-water-soluble basic form and a crystalline salt. In the context of local anesthetic, lidocaine functions by blocking sodium channel conduction. In recent years, researchers have turned to lidocaine hydrochloride as a water-soluble anesthetic agent for zebrafish. It elicits a moderately aversive response in zebrafish, offering rapid onset and recovery (Martins et al., 2016). Lidocaine induces surgical anesthesia and provides perioperative analgesia. Its efficacy is enhanced when used in combination with propofol and with the addition of sodium bicarbonate (Valentim et al., 2016). The combination with sodium bicarbonate not only boosts potency but also reduces toxicity levels (Carrasco et al., 1984).

8.2.7 Clove Oil Compounds

Clove oil is an essential oil primarily derived from the cloves of the *Syzygium aromaticum* (clove tree) (Haro-González et al., 2021). It consists predominantly of eugenol (2-methoxy-4-prop-2-enylphenol), comprising 70–90% of its weight, along with isoeugenol (2-methoxy-4-[(E)-prop-1-enyl]phenol) and methyleugenol (1,2-dimethoxy-4-prop-2-enylbenzene). This aromatic oil finds application as an inhalant anesthetic. Owing to its highly lipophilic nature, it allows rapid absorption through both gills and skin. Compared to MS222, it offers quicker induction and extended recovery time (Detar and Mattingly, 2004; Sladky et al., 2001). The use of clove oil for anesthesia also impacts various physiological aspects, including food intake, electrolyte balance, and stress responses (Pirhonen & Schreck, 2003; Cooke et al., 2004). The advantages of clove oil include its affordability, widespread availability, and effectiveness across range of temperatures (Neiffer and Stamper, 2009). However, repeated exposures may result in mild gill necrosis (Afifi et al., 2001). To use it as an immersion anesthetic, it needs to be mixed with ethanol. Eugenol, which is a key component of clove oil, is readily available, less expensive, and works at a range of temperatures. Nevertheless, high eugenol doses can lead to cardiac complications (Hill and Forster, 2004). Eugenol is known to modulate cardiac functions, causing side effects (Kiessling et al., 2009). Isoeugenol (2-methoxy-4-prop-1-enyl-phenol), another constituent of clove oil structurally similar to eugenol, influences calcium, potassium, and sodium channels, potentiates GABAA receptors, and suppresses N-methyl D-aspartate (NMDA) receptors (Aoshima and Hamamoto, 1999). Isoeugenol may also trigger a stress response, leading to elevated catecholamines and hematocrit levels (Hill and Forster, 2004). When handling eugenol, isoeugenol, and methyleugenol, it is essential to prioritize personal protective measures. Both eugenol and isoeugenol can sensitize skin, while methyleugenol is considered a potential carcinogen (Bertrand et al., 1997; Herrmann et al., 2013).

8.2.8 Propofol (2,6-Di(Propan-2-yl)Phenol)

Propofol, a highly lipophilic anesthetic drug, exhibits limited solubility in water (Steinbacher, 2001). This compound acts as a GABAA receptor agonist and is commonly employed for immersion anesthetics. Its lipophilic nature facilitates rapid and smooth induction of anesthesia, with a correspondingly swift and complete recovery (Gholipourkanani and Ahadizadeh, 2013). In the context of fish transportation, propofol emerges as an ideal agent (Felix et al., 2021). While it lacks reported analgesic properties, propofol has been proven to be a safe and potent anesthetic agent in various fish species (Fassoulaki, 2011). Due to its limited solubility, propofol should be used within 30 minutes of preparation.

8.2.9 Ketamine (2-(2-Chlorophenyl)-2-(Methylamino)Cyclohexan-1-One)

Ketamine, a cyclohexamine primarily employed to induce and maintain anesthesia, exhibits properties as a dissociative anesthetic. However, it has been observed to induce developmental abnormalities in zebrafish and to exert neurotoxic effects on zebrafish larvae (Kanungo et al., 2013; Félix et al., 2014). To ensure the safe and effective anesthesia of zebrafish, it is advisable to utilize ketamine in combination with MS222.

8.2.10 Combination of Drugs for Anesthesia

A comparative investigation of the safety and efficacy of MS222 (168 mg/L) and isoeugenol (20 mg/L) for zebrafish anesthesia during caudal fin clipping demonstrated similar effects (Musk et al., 2020). Despite the dissimilar pharmacokinetic properties of these two compounds, there was no discernible difference in their ability to anesthetize zebrafish. Isoeugenol exhibited a more rapid onset of anesthesia compared to MS222; however, the recovery from anesthesia was shorter with MS222. A similar pattern of effects was also observed when utilizing a combination of propofol and lidocaine. The propofol/lidocaine combination resulted in a rapid loss of equilibrium, but prolonged recovery compared to the quick recovery observed with MS222 (Valentim et al., 2016).

8.3 ANALGESIA

Analgesia is the loss of pain sensation, which occurs when there is a disruption in communication between the brain and the sense organs due to interruptions in the nervous system. While our understanding of analgesia in clinical practice is well-established, the comprehension of analgesia in fish is scarce and often misleading, as the experience of pain in animals, particularly in fish, differs from that in humans (Criado, 2010). Reports indicate that fish are sentient beings capable of experiencing pain. The following evidence supports this: Fish possess nociceptors through which they respond to noxious stimuli. They prioritize behavioral changes under toxic stimuli, and they exhibit physiological responses such as increased respiration through their gills, reduced food intake, and decreased swimming (Ashley et al., 2009; Sneddon, 2009). All of these parameters can be mitigated through the use of analgesics (Sneddon, 2003; Harms, 2005). Furthermore, the presence of mu (μ) and kappa (κ) opioid receptors in the brains of teleost fish supports the existence of an endogenous opioid system and a pain response (Alvarez et al., 2006). Teleost fish display abnormal behavior indicative of distress in response to pain (Reilly et al., 2008). However, the use of analgesics to alleviate abnormal behavior is limited in fish. Measuring pain sensation is challenging due to the variability in pain responses among different species, and there is no globally accepted or validated set of parameters (Short, 1998). Therefore, no single dosage regimen is suitable for all species. In cases involving large fish species, intramuscular, intracelomic, and subcutaneous administration of perioperative analgesia is possible, but it is unfeasible for small fish species like zebrafish. In such cases, water-soluble analgesics ensure pain relief as an immersion agent (Schroeder and Sneddon, 2017). Anesthesia drugs induce a loss of consciousness by depressing the central nervous system, whereas analgesic agents modulate the signal transmission (Table 8.1). Similar to anesthetic agents, combining different classes of analgesics offers lower dosages and fewer side effects, a practice known as balanced analgesia. Further research is warranted for the appropriate use of analgesics in zebrafish, as limited information is currently available. The most commonly categorized analgesic drugs are local anesthetics, non-steroidal anti-inflammatory drugs (NSAIDs), and opioids (Figure 8.2) (Sneddon, 2012).

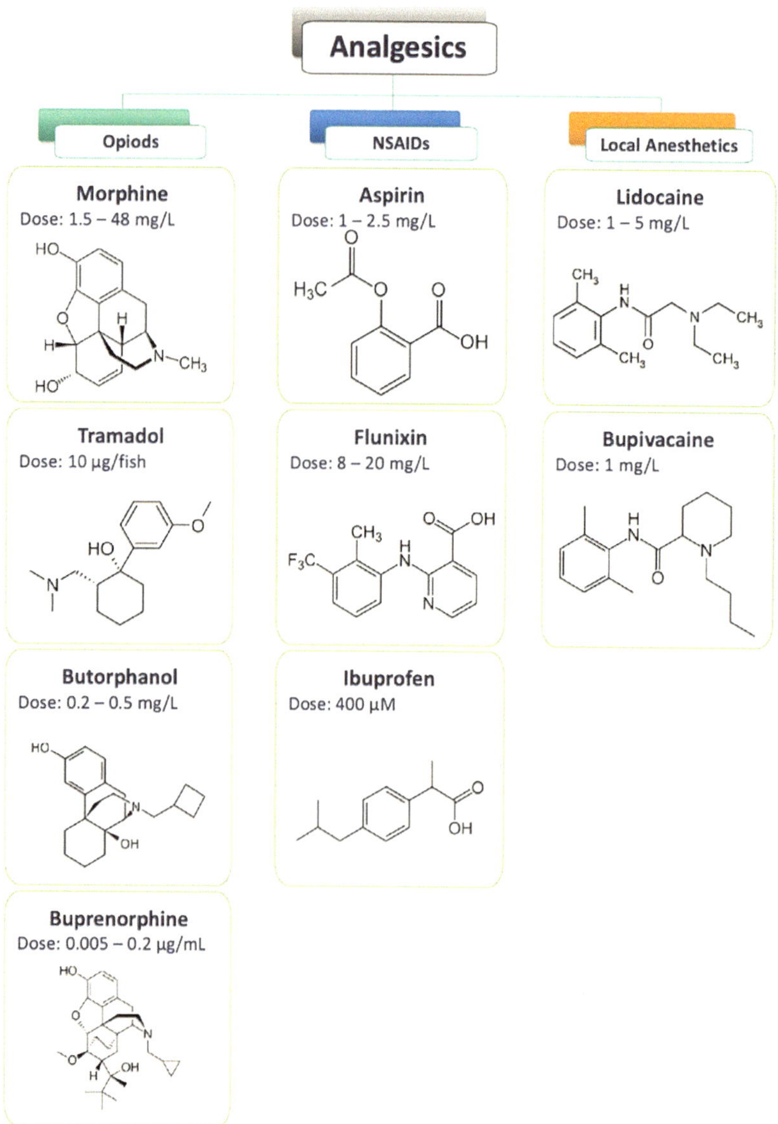

FIGURE 8.2 Analgesic agents used in zebrafish show the range of doses.

TABLE 8.1
Difference between Anesthesia and Analgesia

Anesthesia	Analgesia
Discovered by William T. G. Morton (1853)	Discovered by John W. Hughes and Hans W. Kosterlitz (Kosterlitz and Hughes, 1977)
The absence or loss of physical sensation, whether accompanied by a loss of consciousness or not	The alleviation of pain devoid of the cessation of consciousness or sensory perception
Inhibits the sodium channels located on the nerves	Hinders the synthesis of prostaglandins, pivotal mediators of the inflammatory cascade
Used during invasive procedures	Used for pain management in minimally invasive treatments

8.3.1 Opioids

Opioids constitute one class of analgesic agents that inhibit and alter pain signals (Raffa, 2014). However, their utilization in the context of zebrafish is still limited. Opioids inhibit and modulate pain signals by activating delta (δ), mu (μ), and kappa (κ) opioid receptors (Raffa, 2014; Pergolizzi et al., 2017). Opioids exert their presynaptic effects by suppressing the release of neurotransmitters. This action not only dampens nociceptor activity but also effectively hinders the transmission of signals within the central nervous system. Morphine, a μ-opioid receptor agonist, is a strictly regulated substance used as an analgesic agent in fish (Sneddon, 2003). Its use and access are limited, primarily because it is addictive in nature to humans. Morphine is used as an analgesic agent alone or in combination with other analgesic agents in fish. In species such as goldfish (*Carassius auratus*) and koi (*Cyprinus carpio*), morphine induces analgesia (Ward et al., 2012; Baker et al., 2013). Morphine reduces pain in goldfish after temperature threshold testing by ameliorating drastic behavioral changes, which suggests the presence of thermonociception in fish (Nordgreen et al., 2009). In rainbow trout (*Oncorhynchus mykiss*), morphine attenuated pain-related anomalous behavior and reduced opercular beat rate following exposure to a noxious acetic acid treatment (Sneddon, 2003). Morphine also ameliorated the behavioral changes induced by noxious acetic acid stimuli in 5 days post-fertilization zebrafish larvae, a response similar to that observed in adult zebrafish (Lopez-Luna, 2017a). Morphine as analgesia implied that larval zebrafish could be used as a model similar to adult zebrafish in pain and nociception investigations. Though the pharmacokinetics of morphine in fish is similar to that in mammals, morphine tends to persist longer in fish before being completely eliminated. This characteristic opens up the possibility of using morphine for chronic pain conditions (Newby et al., 2006, 2009). However, it is important to recognize that the elimination of morphine varies among varied species of fish. Tramadol (1R,2R)-2-[(dimethylamino) methyl]-1-(3-methoxyphenyl)cyclohexan-1-ol) is a μ-opioid receptor agonist and a central analgesic (Duthie, 1998). Its analgesic effect is achieved through the synergistic action of opioid activity and enhanced monoaminergic transmission. In carp, the dose-dependent analgesic effect of tramadol in response to electric shock indicates the presence of antinociceptive system in fish (Chervova and Lapshin, 2000). A high dose of tramadol was lethal in zebrafish and even at half the lethal dose, it caused zebrafish to respire at the oxygen-rich surface and modulate their activity (Zhuo et al., 2016). In zebrafish embryo toxicity tests involving tramadol, there was a hatching delay without any discernible impact on development (Sehonova et al., 2016). Butorphanol is another opioid, and its analgesic effect is limited due to its dual action as a κ-opioid receptor agonist and μ-opioid receptor antagonist. In a study involving chain dogfish (*Scyliorhinus retifer*), butorphanol hardly displayed any analgesic effect in response to noxious stimuli (Davis et al., 2006). Buprenorphine exerts its analgesic effect by acting as a μ-opioid receptor agonist. It is widely used in veterinary medicine due to its lower addiction potential compared to other μ agonists. However, buprenorphine exhibited poor analgesic effects in response to the noxious treatments (Mettam et al., 2011) and post-surgical analgesia (Grans et al., 2014) in the rainbow trout.

8.3.2 NSAIDs

NSAIDs work by inhibiting cyclooxygenase 1 and 2 (COX-1 and COX-2), enzymes responsible for catalyzing the conversion of arachidonic acid into prostaglandins. This action provides NSAIDs with analgesic, anti-inflammatory, and antipyretic properties (Papich, 2000). Prostaglandins mediate pain signals in the spinal cord and peripheral tissues (Kawabata, 2011). Aspirin (2-acetyloxybenzoic acid), a commonly used NSAID, reduces pain by inhibiting COX-2 and modifying COX-1 activity. It has been shown to hinder behavioral changes induced by acetic acid in 5 dpf zebrafish larvae (Lopez-Luna, 2017a). However, aspirin did not exhibit significant advantages in response to high temperatures (Lopez-Luna, 2017b). Aspirin immersion offered favorable outcomes for adult zebrafish following fin clipping (Schroeder and Sneddon, 2017). Flunixin (2-[2-methyl-3-(trifluoromethyl)anilino]pyridine-3-carboxylic acid), by inhibiting COX, finds application in the management of post-operative pain in dogs and serves as an effective analgesic in variety of species, including cats, pigs, sheep, cattle, and horses (Reid and Nolan, 1991). In adult zebrafish, flunixin demonstrated a preventive effect by reducing activity and swimming distance following fin clipping (Deakin et al., 2019). Ibuprofen (2-[4-(2-methylpropyl)phenyl]propanoic acid) is a nonselective COX inhibitor. In fish, it serves as an anti-inflammatory drug. While its effect in adult zebrafish remains relatively well-explored, it has been observed that ibuprofen is ineffective in reversing sensitization to noxious temperature in larvae (Curtright et al., 2015). Ketoprofen (2-(3-benzoylphenyl)propanoic acid) is a medication that targets both COX-1 and COX-2, although it is primarily considered a highly selective COX-1 drug (Streppa et al., 2002). It possesses both central and peripheral activity and has the ability to cross the blood-brain barrier (Herrero et al., 1997). In a model assessing the minimum anesthetic concentration, ketoprofen may provide analgesia in goldfish (Ward et al., 2012). Ketoprofen also demonstrated positive results in a surgical model involving Koi fish, reducing muscle damage due to its anti-inflammatory properties (Harms, 2005). However, in a rainbow trout exploratory celiotomy model, ketoprofen did not appear to influence healing (Rizzo et al., 2017). Carprofen (2-(6-chloro-9H-carbazol-2-yl)propanoic acid) is a perioperative analgesia used for managing both acute and chronic pain. In contrast to ketoprofen, carprofen is a COX-2 selective analgesia with a relatively weak COX-1 activity (Lees et al., 2004). Carprofen exerts robust analgesia and anti-inflammatory effects (Fox and Johnston, 1997). Based on observations of feeding resumption following noxious stimuli in the trout, carprofen may exhibit analgesic properties but with a depressed activity (Mettam et al., 2011).

8.3.3 Local Anesthetics

Local anesthetic agents obstruct the propagation of action potentials through two mechanisms: they block sodium ions from entering through voltage-gated sodium channels and disrupt membrane function. Consequently, these local anesthetics suppress the transmission of nociceptive signals, leading to the alleviation of pain sensation. However, only a limited number of studies have explored the use of local anesthetics as analgesics in fish. The potential of lidocaine as an analgesic agent was demonstrated in rainbow trout, where it attenuated detrimental physiological and behavioral responses to pain (Mettam et al., 2011). The administration of lidocaine at a concentration of 5 mg/L for 30 minutes leads to increased activity and a quicker restoration of opercular beat rates to baseline in adult zebrafish following fin clipping, compared to fish that do not receive analgesic treatment. In zebrafish larvae, lidocaine promoted activity levels when exposed to heat and mitigated behavior altered by chemical treatment (Lopez-Luna, 2017a, 2017b). The analgesic properties of bupivacaine have indicated its potential in reducing pain when administered prior to fin clipping in adult zebrafish (Deakin et al., 2019).

8.4 EUTHANASIA

Euthanasia refers to the humane and painless killing of animals, and it is necessary in various situations, such as when fish experience immense pain, suffering, serious health issues, tissue collection at the end of the experiments, and when they no longer serve their intended purpose (Close et al., 1997). When conducting euthanasia, several important principles should be followed: it should (1) induce rapid unconsciousness, (2) prioritize a swift and humane process with minimal stress, (3) ensure safety for the individual performing the procedure, and (4) critically, the method should be irreversible. Euthanasia in fish typically occurs due to hypoxia, depression, and destruction of neurons. When designing euthanasia methods for different fish species, it is crucial to consider their unique anatomical, physiological, and metabolic characteristics, as well as factors like age, quantity, and overall health (Underwood et al., 2013). Euthanasia can be achieved through either the administration of drugs or the use of physical methods. Unlike in mammals, where the cessation of heartbeat is a reliable indicator of death, this is not as dependable in fish. Therefore, confirming complete cessation of life is essential in fish. A recommended guideline for confirming death is to observe a 10-to-20-minute period with the absence of any opercular movement (Matthews and Varga, 2012). While comprehensive guidelines for zebrafish euthanasia may still be under development, it is imperative for personnel to stay updated on the latest discoveries and research in this field. Following the implementation of any euthanasia procedure, it is essential to confirm the death by checking for the absence of reflexes and respiration. In the case of embryos and larvae, treatment with proteinase K, bleach, or maceration is recommended. Subsequently, the carcass should be appropriately refrigerated until it can be disposed of in accordance with established guidelines and regulations.

8.4.1 Drug Overdose

Overdosing with anesthetics is often a preferred method for euthanizing fish. This involves immersing the fish in water containing a higher dose of anesthetic agents sufficient to induce rapid anesthesia, followed by euthanasia. It is recommended to keep the fish in the treated water for approximately 10 minutes after all signs of movement have ceased. The mechanism of death in this process is primarily hypoxia, which can cause stress in fish and potentially impact the experimental results. Embryos and larvae are more tolerant to anesthetics due to their underdeveloped or absence of gills, and anesthetics are absorbed through diffusion and metabolic processes (Kohler et al., 2017). To use anesthetics for euthanasia in embryos and larvae, exceedingly high concentrations and prolonged exposure are typically required.

8.4.2 CO_2

The bubbling of CO_2 gas into water or the generation of CO_2 by chemical means in water euthanizes fish. The utilization of CO_2 for euthanasia in fish is recognized as a humane practice, similar to its use in mammals, albeit with certain caveats such as the relatively slow process and the potential for causing significant stress (Marking and Meyer, 1985). Therefore, further investigations are warranted to explore the effect of CO_2 on fish during euthanasia. Chemicals like carbonating acids, when introduced into water, generate and elevate CO_2 in water, leading to the death of fish from hypoxemia. Exposure to high concentrations of CO_2 in water results in a rapid reduction in blood pH levels, which impairs the efficient transport of oxygen to the brain and other tissues. This, in turn, triggers cerebral hypoxia, leading to the suppression of activity within the central nervous system (Gelwicks et al., 1998). One drawback of using CO_2 for euthanasia is that it can alter enzyme activity, making it unreliable for experiments involving enzyme assays. Researchers should be aware of this limitation when considering CO_2 euthanasia in experimental setups.

8.4.3 Rapid Chilling

Rapid chilling is accomplished by transferring the fish into ice-cold water, typically ranging from 0–4°C, while ensuring that there is no direct contact with any ice. This rapid change in temperature results in an immediate and complete cessation of all vital signs, providing a 100% certainty of death due to hypothermal shock. Over the last decade, research findings have contributed to the acceptance of rapid chilling as an effective method for euthanizing zebrafish and other similar tropical fish species (Wilson et al., 2009; Underwood et al., 2013). Embryos and larvae may sometimes exhibit greater resistance to rapid chilling. In such cases, it is recommended to use alternatives such as proteinase K, bleach, or liquid nitrogen.

8.5 CONCLUSION

As the role of zebrafish in both basic and biomedical research continues to expand, it becomes increasingly vital to establish appropriate protocols for anesthesia, analgesia, and euthanasia. In the recent decades, our understanding of the effects of anesthetics in zebrafish has grown significantly, encompassing parameters such as anesthetic duration, potential side effects, induction characteristics, and the quality and duration of post-anesthesia recovery. The application of analgesics in zebrafish has the potential to enhance welfare, yet realizing these benefits requires further research to assess their effectiveness and advantages, as well as deeper understanding of the pharmacodynamics of analgesic agents in zebrafish. Presently, the guidelines for euthanizing zebrafish remain somewhat indefinite, relying heavily on individual research facility practices. This lack of standardization can lead to disparities in experimental outcomes across different laboratories. Therefore, the implementation of a standardized euthanasia approach would prove highly advantageous. To establish robust and scientifically valid protocols for analgesia and euthanasia, additional research is imperative, as the current literature in these areas remains relatively limited. The selection of precise agents and methodologies for anesthesia, analgesia, and euthanasia in zebrafish research hinges on a range of factors, including the specific scientific questions being addressed, the developmental stage of the zebrafish, and the potential impact of the chosen agents or procedure on the particular organ systems under investigation. These considerations underscore the need for a tailored approach in determining the most appropriate techniques and substances to ensure the welfare and integrity of zebrafish research.

REFERENCES

Ablain, J., & Zon, L. I. (2013). Of fish and men: Using zebrafish to fight human diseases. *Trends in Cell Biology*, 23(12), 584–586.

Adamson, K. I., Sheridan, E., & Grierson, A. J. (2018). Use of zebrafish models to investigate rare human disease. *Journal of Medical Genetics*, 55(10), 641–649.

Afifi, S. H., Al-Thobaiti, S., & Rasem, B. M. (2001). Multiple exposure of Asian sea bass, (Lates calcarifer, Centropomidae) to clove oil: A histopathological study. *Journal of Aquaculture in the Tropics*, 16(2), 131–138.

Akbary, P., Pirbeigi, A., & Jahanbakhshi, A. (2016). Analysis of primary and secondary stress responses in bighead carp (Hypophthalmichthys nobilis) by anesthetization with 2-phenoxyethanol. *International Journal of Environmental Science and Technology*, 13(4), 1009–1016.

Alvarez, F. A., Rodriguez-Martin, I., Gonzalez-Nuñez, V., de Velasco, E. M. F., Gonzalez Sarmiento, R., & Rodríguez, R. E. (2006). New kappa opioid receptor from zebrafish *Danio rerio*. *Neuroscience Letters*, 405(1–2), 94–99.

Aoshima, H., & Hamamoto, K. (1999). Potentiation of GABAA receptors expressed in Xenopus oocytes by perfume and phytoncid. *Bioscience, Biotechnology, and Biochemistry*, 63(4), 743–748.

Ashley, P. J., Ringrose, S., Edwards, K. L., Wallington, E., McCrohan, C. R., & Sneddon, L. U. (2009). Effect of noxious stimulation upon antipredator responses and dominance status in rainbow trout. *Animal Behaviour*, 77(2), 403–410.

Baker, T. R., Baker, B. B., Johnson, S. M., & Sladky, K. K. (2013). Comparative analgesic efficacy of morphine sulfate and butorphanol tartrate in koi (Cyprinus carpio) undergoing unilateral gonadectomy. *Journal of the American Veterinary Medical Association*, 243(6), 882–890.

Bertrand, F., Basketter, D. A., Roberts, D. W., & Lepoittevin, J. P. (1997). Skin sensitization to eugenol and isoeugenol in mice: Possible metabolic pathways involving ortho-quinone and quinone methide intermediates. *Chemical Research in Toxicology*, 10(3), 335–343.

Carrasco, S., Sumano, H., & Navarro-Fierro, R. (1984). The use of lidocaine-sodium bicarbonate as anaesthetic in fish. *Aquaculture*, 41(4), 395–398.

Carter, K. M., Woodley, C. M., & Brown, R. S. (2011). A review of tricaine methanesulfonate for anesthesia of fish. *Reviews in Fish Biology and Fisheries*, 21(1), 51–59.

Chervova, L. S., & Lapshin, D. N. (2000). Opioid modulation of pain threshold in fish. In *Doklady Biological Sciences* (Vol. 375, No. 1–6, pp. 590–591). Kluwer Academic/Plenum Publishers.

Close, B., Banister, K., Baumans, V., Bernoth, E. M., Bromage, N., Bunyan, J., . . . Warwick, C. (1997). Recommendations for euthanasia of experimental animals: Part 2. *Laboratory Animals*, 31(1), 1–32.

Collymore, C., Crim, M. J., & Lieggi, C. (2016). Recommendations for health monitoring and reporting for Zebrafish research facilities. *Zebrafish*, 13(S1), S–138.

Collymore, C., Tolwani, A., Lieggi, C., & Rasmussen, S. (2014). Efficacy and safety of 5 anesthetics in adult zebrafish (*Danio rerio*). *Journal of the American Association for Laboratory Animal Science*, 53(2), 198–203.

Cooke, S. J., Suski, C. D., Ostrand, K. G., Tufts, B. L., & Wahl, D. H. (2004). Behavioral and physiological assessment of low concentrations of clove oil anaesthetic for handling and transporting largemouth bass (Micropterus salmoides). *Aquaculture*, 239(1–4), 509–529.

Criado, A. (2010). Recognition and alleviation of pain in laboratory animals. *Laboratory Animals*, 44(4), 380.

Cuesta, A., Meseguer, J., & Esteban, M. (2004). Total serum immunoglobulin M levels are affected by immunomodulators in seabream (Sparus aurata L.). *Veterinary Immunology and Immunopathology*, 101(3–4), 203–210.

Curtright, A., Rosser, M., Goh, S., Keown, B., Wagner, E., Sharifi, J., . . . Dhaka, A. (2015). Modeling nociception in zebrafish: A way forward for unbiased analgesic discovery. *PLoS One*, 10(1), e0116766.

Davis, M. R., Mylniczenko, N., Storms, T., Raymond, F., & Dunn, J. L. (2006). Evaluation of intramuscular ketoprofen and butorphanol as analgesics in chain dogfish (Scyliorhinus retifer). *Zoo Biology: Published in Affiliation with the American Zoo and Aquarium Association*, 25(6), 491–500.

Deakin, A. G., Buckley, J., AlZu'bi, H. S., Cossins, A. R., Spencer, J. W., Al'Nuaimy, W., . . . Sneddon, L. U. (2019). Automated monitoring of behaviour in zebrafish after invasive procedures. *Scientific Reports*, 9(1), 9042.

Detar, J. E., & Mattingly, H. T. (2004). Response of southern redbelly dace to clove oil and MS-222: Effects of anesthetic concentration and water temperature. *Proceedings of the Annual Conference/Southeastern Association of Fish and Wildlife Agencies*, 58, 219–227.

Dorr, H. G., Kuhnle, U., Holthausen, H., Bidlingmaier, F., & Knorr, D. (1984). Etomidate: A selective adrenocortical 11β-hydroxylase inhibitor. *Klinische Wochenschrift*, 62(21), 1011–1013.

Duthie, D. J. (1998). Remifentanil and tramadol. *British Journal of Anaesthesia*, 81(1), 51–57.

Fassoulaki, A. (2011). Is propofol an analgesic? *European Journal of Anaesthesiology*, 28(7), 481–482.

Felix, L., Correia, R., Sequeira, R., Ribeiro, C., Monteiro, S., Antunes, L., . . . Valentim, A. (2021). MS-222 and propofol sedation during and after the simulated transport of Nile tilapia (Oreochromis niloticus). *Biology*, 10(12), 1309.

Félix, L. M., Antunes, L. M., & Coimbra, A. M. (2014). Ketamine NMDA receptor-independent toxicity during zebrafish (*Danio rerio*) embryonic development. *Neurotoxicology and Teratology*, 41, 27–34.

Fox, S. M., & Johnston, S. A. (1997). Use of carprofen for the treatment of pain and inflammation in dogs. *Journal of the American Veterinary Medical Association*, 210(10), 1493–1498.

Frazier, D. T., & Narahashi, T. (1975). Tricaine (MS-222): Effects on ionic conductances of squid axon membranes. *European Journal of Pharmacology*, 33(2), 313–317.

Gelwicks, K. R., Zafft, D. J., & Bobbitt, J. P. (1998). Efficacy of carbonic acid as an anesthetic for rainbow trout. *North American Journal of Fisheries Management*, 18(2), 432–438.

GholipourKanani, H., & Ahadizadeh, S. (2013). Use of propofol as an anesthetic and its efficacy on some hematological values of ornamental fish *Carassius auratus*. *Springerplus*, 2, 1–5.

Gibert, Y., Trengove, M., & Ward, A. (2013). Zebrafish as a genetic model in pre-clinical drug testing and screening. *Current Medicinal Chemistry*, 20(19), 2458–2466.

Grans, A., Sandblom, E., Kiessling, A., & Axelsson, M. (2014). Post-surgical analgesia in rainbow trout: Is reduced cardioventilatory activity a sign of improved animal welfare or the adverse effects of an opioid drug? *PLoS One*, 9(4), e95283.

Harms, C. A. (2005). Surgery in fish research: Common procedures and postoperative care. *Lab Animal*, 34(1), 28–34.

Harms, C. A., Lewbart, G. A., Swanson, C. R., Kishimori, J. M., & Boylan, S. M. (2005). Behavioral and clinical pathology changes in koi carp (Cyprinus carpio) subjected to anesthesia and surgery with and without intra-operative analgesics. *Comparative Medicine*, 55(3), 221–226.

Haro-González, J. N., Castillo-Herrera, G. A., Martínez-Velázquez, M., & Espinosa-Andrews, H. (2021). Clove essential oil (Syzygium aromaticum L. Myrtaceae): Extraction, chemical composition, food applications, and essential bioactivity for human health. *Molecules*, 26(21), 6387.

Harper, C., & Lawrence, C. (2011). *The Laboratory Zebrafish*. CRC Press.

Herrero, J. F., Parrado, A., & Cervero, F. (1997). Central and peripheral actions of the NSAID ketoprofen on spinal cord nociceptive reflexes. *Neuropharmacology*, 36(10), 1425–1431.

Herrmann, K., Schumacher, F., Engst, W., Appel, K. E., Klein, K., Zanger, U. M., & Glatt, H. (2013). Abundance of DNA adducts of methyleugenol, a rodent hepatocarcinogen, in human liver samples. *Carcinogenesis*, 34(5), 1025–1030.

Hill, J. V., & Forster, M. E. (2004). Cardiovascular responses of Chinook salmon (Oncorhynchus tshawytscha) during rapid anaesthetic induction and recovery. *Comparative Biochemistry and Physiology Part C: Toxicology & Pharmacology*, 137(2), 167–177.

Horzmann, K. A., & Freeman, J. L. (2018). Making waves: New developments in toxicology with the Zebrafish. *Toxicological Sciences*, 163(1), 5–12.

Howe, K., Clark, M. D., Torroja, C. F., Torrance, J., Berthelot, C., Muffato, M., . . . Stemple, D. L. (2013). The zebrafish reference genome sequence and its relationship to the human genome. *Nature*, 496(7446), 498–503.

Hseu, J. R., Yeh, S. L., Chu, Y. T., & Ting, Y. Y. (1997). Different anesthetic effects of 2-phenoxyethanol on four species of teleost. *Journal of the Fisheries Society of Taiwan*, 24(3), 185–191.

Hseu, J. R., Yeh, S. L., Chu, Y. T., & Ting, Y. Y. (1998). Comparison of efficacy of five anesthetics in goldlined sea bream, Sparus sarba. *Acta Zoologica Taiwanica*, 9(1), 35–41.

Huang, W. C., Hsieh, Y. S., Chen, I. H., Wang, C. H., Chang, H. W., Yang, C. C., . . . Chuang, Y. J. (2010). Combined use of MS-222 (Tricaine) and isoflurane extends anesthesia Time and minimizes cardiac rhythm side effects in adult Zebrafish. *Zebrafish*, 7(3), 297–304.

Hwang, W. Y., Fu, Y., Reyon, D., Maeder, M. L., Tsai, S. Q., Sander, J. D., . . . Joung, J. K. (2013). Efficient genome editing in zebrafish using a CRISPR-Cas system. *Nature Biotechnology*, 31(3), 227–229.

Iversen, M., Finstad, B., McKinley, R. S., & Eliassen, R. A. (2003). The efficacy of metomidate, clove oil, Aqui-S™

and Benzoak® as anaesthetics in Atlantic salmon (Salmo salar L.) smolts, and their potential stress-reducing capacity. *Aquaculture*, 221(1–4), 549–566.

Iwama, G. K. (2007). The welfare of fish. *Diseases of Aquatic Organisms*, 75(2), 155–158.

Iwama, G. K., McGeer, J. C., & Pawluk, M. P. (1989). The effects of five fish anaesthetics on acid–base balance, hematocrit, blood gases, cortisol, and adrenaline in rainbow trout. *Canadian Journal of Zoology*, 67(8), 2065–2073.

Kanungo, J., Cuevas, E., Ali, S. F., & Paule, M. G. (2013). Ketamine induces motor neuron toxicity and alters neurogenic and proneural gene expression in zebrafish. *Journal of Applied Toxicology*, 33(6), 410–417.

Katz, E. M., Chu, D. K., Casey, K. M., Jampachaisri, K., Felt, S. A., & Pacharinsak, C. (2020). The stability and efficacy of tricaine methanesulfonate (MS222) solution after long-term storage. *Journal of the American Association for Laboratory Animal Science*, 59(4), 393–400.

Kawabata, A. (2011). Prostaglandin E2 and pain—an update. *Biological and Pharmaceutical Bulletin*, 34(8), 1170–1173.

Kiessling, A., Johansson, D., Zahl, I. H., & Samuelsen, O. B. (2009). Pharmacokinetics, plasma cortisol and effectiveness of benzocaine, MS-222 and isoeugenol measured in individual dorsal aorta-cannulated Atlantic salmon (Salmo salar) following bath administration. *Aquaculture*, 286(3–4), 301–308.

Kohler, A., Collymore, C., Finger-Baier, K., Geisler, R., Kaufmann, L., Pounder, K. C., . . . Strähle, U. (2017). Report of workshop on euthanasia for zebrafish—a matter of welfare and science. *Zebrafish*, 14(6), 547–551.

Kosterlitz, H. W., & Hughes, J. (1977). Opiate receptors and endogenous opioid peptides in tolerance and dependence. In: Gross, M. M. (eds) *Alcohol Intoxication and Withdrawal—IIIb. Advances in Experimental Medicine and Biology* (Vol. 85B, pp. 141–154). Springer.

Lees, P., Giraudel, J., Landoni, M. F., & Toutain, P. L. (2004). PK–PD integration and PK–PD modelling of nonsteroidal anti-inflammatory drugs: Principles and applications in veterinary pharmacology. *Journal of Veterinary Pharmacology and Therapeutics*, 27(6), 491–502.

Lockwood, N., Parker, J., Wilson, C., & Frankel, P. (2017). Optimal anesthetic regime for motionless three-dimensional image acquisition during longitudinal studies of adult nonpigmented Zebrafish. *Zebrafish*, 14(2), 133–139.

Lopez-Luna, J., Al-Jubouri, Q., Al-Nuaimy, W., & Sneddon, L. U. (2017a). Reduction in activity by noxious chemical stimulation is ameliorated by immersion in analgesic drugs in zebrafish. *Journal of Experimental Biology*, 220(8), 1451–1458.

Lopez-Luna, J., Al-Jubouri, Q., Al-Nuaimy, W., & Sneddon, L. U. (2017b). Impact of analgesic drugs on the behavioural responses of larval zebrafish to potentially noxious temperatures. *Applied Animal Behaviour Science*, 188, 97–105.

Marking, L. L., & Meyer, F. P. (1985). Are better anesthetics needed in fisheries? *Fisheries*, 10(6), 2–5.

Martins, T., Valentim, A. M., Pereira, N., & Antunes, L. M. (2016). Anaesthesia and analgesia in laboratory adult zebrafish: A question of refinement. *Laboratory Animals*, 50(6), 476–488.

Matthews, M., & Varga, Z. M. (2012). Anesthesia and euthanasia in Zebrafish. *ILAR Journal*, 53(2), 192–204.

Mattson, N. S., & Riple, T. H. (1989). Metomidate, a better anesthetic for cod (Gadus morhua) in comparison with benzocaine, MS-222, chlorobutanol, and phenoxyethanol. *Aquaculture*, 83(1–2), 89–94.

Maximino, C. (2011). Modulation of nociceptive-like behavior in zebrafish (*Danio rerio*) by environmental stressors. *Psychology & Neuroscience*, 4, 149–155.

Mettam, J. J., Oulton, L. J., McCrohan, C. R., & Sneddon, L. U. (2011). The efficacy of three types of analgesic drugs in reducing pain in the rainbow trout, Oncorhynchus mykiss. *Applied Animal Behaviour Science*, 133(3–4), 265–274.

Morton WTG. Statements, Supported by Evidence, of Wm. T. G. Morton, M.D. on his Claim to the Discovery of the Anaesthetic Properties of Ether, Submitted to the Honorable the Select Committee Appointed by the Senate of the United States. Presented at 32nd Congress, 2nd Session, Washington DC, January 21, 1853. [1853]:397–435.

Musk, G. C., Ezzy, B. J., Kenchington, L. M., Hopper, W. A., & Callahan, L. M. (2020). A Comparison of buffered tricaine methanesulfonate (MS-222) and isoeugenol anesthesia for caudal fin clipping in Zebrafish (*Danio rerio*). *Journal of the American Association for Laboratory Animal Science*, 59(6), 732–736.

Neiffer, D. L., & Stamper, M. A. (2009). Fish sedation, anesthesia, analgesia, and euthanasia: Considerations, methods, and types of drugs. *ILAR Journal*, 50(4), 343–360.

Newby, N. C., Mendonça, P. C., Gamperl, K., & Stevens, E. D. (2006). Pharmacokinetics of morphine in fish: Winter flounder (*Pseudopleuronectes americanus*) and seawater-acclimated rainbow trout (*Oncorhynchus mykiss*). *Comparative Biochemistry and Physiology Part C: Toxicology & Pharmacology*, 143(3), 275–283.

Newby, N. C., Wilkie, M. P., & Stevens, E. D. (2009). Morphine uptake, disposition, and analgesic efficacy in the common goldfish (*Carassius auratus*). *Canadian Journal of Zoology*, 87(5), 388–399.

Nordgreen, J., Garner, J. P., Janczak, A. M., Ranheim, B., Muir, W. M., & Horsberg, T. E. (2009). Thermonociception in fish: Effects of two different doses of morphine on thermal threshold and post-test behaviour in goldfish (*Carassius auratus*). *Applied Animal Behaviour Science*, 119(1–2), 101–107.

Orger, M. B., & de Polavieja, G. G. (2017). Zebrafish behavior: Opportunities and challenges. *Annual Review of Neuroscience*, 40, 125–147.

Papich, M. G. (2000). Pharmacologic considerations for opiate analgesic and nonsteroidal anti-inflammatory drugs. *Veterinary Clinics: Small Animal Practice*, 30(4), 815–837.

Pergolizzi, J. V., LeQuang, J. A., Berger, G. K., & Raffa, R. B. (2017). The basic pharmacology of opioids informs the opioid discourse about misuse and abuse: A review. *Pain and Therapy*, 6, 1–16.

Pirhonen, J., & Schreck, C. B. (2003). Effects of anaesthesia with MS-222, clove oil and CO_2 on feed intake and plasma cortisol in steelhead trout (Oncorhynchus mykiss). *Aquaculture*, 220(1–4), 507–514.

Priborsky, J., & Velisek, J. (2018). A review of three commonly used fish anesthetics. *Reviews in Fisheries Science & Aquaculture*, 26(4), 417–442.

Raffa, R. B. (2014). On subclasses of opioid analgesics. *Current Medical Research and Opinion*, 30(12), 2579–2584.

Readman, G. D., Owen, S. F., Murrell, J. C., & Knowles, T. G. (2013). Do fish perceive anaesthetics as aversive? *PLoS One*, 8(9), e73773.

Reid, J., & Nolan, A. M. (1991). A comparison of the postoperative analgesic and sedative effects of flimixin and pap aver etum in the dog. *Journal of Small Animal Practice*, 32(12), 603–608.

Reilly, S. C., Quinn, J. P., Cossins, A. R., & Sneddon, L. U. (2008). Behavioural analysis of a nociceptive event in fish: Comparisons between three species demonstrate specific responses. *Applied Animal Behaviour Science*, 114(1–2), 248–259.

Rennekamp, A. J., & Peterson, R. T. (2015). 15 years of zebrafish chemical screening. *Current Opinion in Chemical Biology*, 24, 58–70.

Rizzo, A. L., Wooster, G. A., Guanzini, L. E., Peterson, C. M., Fenderson, K., Erb, H. N., . . . Martin, M. E. (2017). Biochemical, histopathologic, physiologic, and behavioral effects of nonsteroidal antiinflammatory drugs in rainbow trout (Oncorhynchus mykiss). *Comparative Medicine*, 67(2), 106–111.

Rombough, P. J. (2007). Ontogenetic changes in the toxicity and efficacy of the anaesthetic MS222 (tricaine methanesulfonate) in zebrafish (*Danio rerio*) larvae. *Comparative Biochemistry and Physiology Part A: Molecular & Integrative Physiology*, 148(2), 463–469.

Ross, L. G., & Ross, B. (2008). *Anaesthetic and Sedative Techniques for Aquatic Animals*. John Wiley & Sons.

Schroeder, P. G., & Sneddon, L. U. (2017). Exploring the efficacy of immersion analgesics in zebrafish using an integrative approach. *Applied Animal Behaviour Science*, 187, 93–102.

Sehonova, P., Plhalova, L., Blahova, J., Berankova, P., Doubkova, V., Prokes, M., . . . Svobodova, Z. (2016). The effect of tramadol hydrochloride on early life stages of fish. *Environmental Toxicology and Pharmacology*, 44, 151–157.

Short, C. E. (1998). Fundamentals of pain perception in animals. *Applied Animal Behaviour Science*, 59(1–3), 125–133.

Sladky, K. K., Swanson, C. R., Stoskopf, M. K., Loomis, M. R., & Lewbart, G. A. (2001). Comparative efficacy of tricaine methanesulfonate and clove oil for use as anesthetics in red pacu (Piaractus brachypomus). *American Journal of Veterinary Research*, 62(3), 337–342.

Small, B. C. (2003). Anesthetic efficacy of metomidate and comparison of plasma cortisol responses to tricaine methanesulfonate, quinaldine and clove oil anesthetized channel catfish Ictalurus punctatus. *Aquaculture*, 218(1–4), 177–185.

Sneddon, L. U. (2003). The evidence for pain in fish: The use of morphine as an analgesic. *Applied Animal Behaviour Science*, 83(2), 153–162.

Sneddon, L. U. (2009). Pain perception in fish: Indicators and endpoints. *ILAR Journal*, 50(4), 338–342.

Sneddon, L. U. (2012). Clinical anesthesia and analgesia in fish. *Journal of Exotic Pet Medicine*, 21(1), 32–43.

Sneddon, L. U. (2015). Pain in aquatic animals. *The Journal of Experimental Biology*, 218(7), 967–976.

Sneddon, L. U., Braithwaite, V. A., & Gentle, M. J. (2003). Do fishes have nociceptors? Evidence for the evolution of a vertebrate sensory system. *Proceedings of the Royal Society of London. Series B: Biological Sciences*, 270(1520), 1115–1121.

Spears, J., Kamunde, C., & Stevens, E. D. (2014). Effect of TRIS and bicarbonate as buffers on anesthetic efficacy of tricaine methane sulfonate in Zebrafish (*Danio rerio*). *Zebrafish*, 11(6), 590–596.

Steinbacher D. M. (2001). Propofol: A sedative-hypnotic anesthetic agent for use in ambulatory procedures. *Anesthesia Progress*, 48(2), 66–71.

Stenger, V. G., & Maren, T. H. (1974). The pharmacology of MS 222 (ethyl-m-aminobenzoate) in squalus acanthias. *Comparative and General Pharmacology*, 5(1), 23–35.

Streppa, H. K., Jones, C. J., & Budsberg, S. C. (2002). Cyclooxygenase selectivity of nonsteroidal anti-inflammatory drugs in canine blood. *American Journal of Veterinary Research*, 63(1), 91–94.

Thienpont, D., & Niemegeers, C. J. E. (1965). Propoxate (R 7464): A new potent anaesthetic agent in cold-blooded vertebrates. *Nature*, 205(4975), 1018–1019.

Topic Popovic, N., Strunjak-Perovic, I., Coz-Rakovac, R., Barisic, J., Jadan, M., Persin Berakovic, A., & Sauerborn Klobucar, R. (2012). Tricaine methane-sulfonate (MS-222) application in fish anaesthesia. *Journal of Applied Ichthyology*, 28(4), 553–564.

Underwood, W., Anthony, R., Cartner, S., Corey, D., Grandin, T., Greenacre, C., . . . Yanong, R. (2013). *AVMA Guidelines for the Euthanasia of Animals: 2013 Edition*. American Veterinary Medical Association.

Valentim, A. M., Félix, L. M., Carvalho, L., Diniz, E., & Antunes, L. M. (2016). A new anaesthetic protocol for adult Zebrafish (*Danio rerio*): Propofol combined with lidocaine. *PLoS One*, 11(1), e0147747.

Ward, J. L., McCartney, S. P., Chinnadurai, S. K., & Posner, L. P. (2012). Development of a minimum-anesthetic-concentration depression model to study the effects of various analgesics in goldfish (*Carassius auratus*). *Journal of Zoo and Wildlife Medicine*, 43(2), 214–222.

Wedemeyer, G. (1970). Stress of anesthesia with M.S. 222 and benzocaine in rainbow trout (Salmo gairdneri). *Journal of the Fisheries Research Board of Canada*, 27(5), 909–914.

Westerfield, M. (2007). The Zebrafish book. In *A Guide for the Laboratory Use of* Zebrafish *(Danio rerio)*. University of Oregon Press.

Wilson, J. M., Bunte, R. M., & Carty, A. J. (2009). Evaluation of rapid cooling and tricaine methanesulfonate (MS222) as methods of euthanasia in zebrafish (*Danio rerio*). *Journal of the American Association for Laboratory Animal Science*, 48(6), 785–789.

Witeska, M., Dudyk, J., & Jarkiewicz, N. (2015). Haematological effects of 2-phenoxyethanol and etomidate in carp (*Cyprinus carpio* L.). *Veterinary Anaesthesia and Analgesia*, 42(5), 537–546.

Wong, D., von Keyserlingk, M. A. G., Richards, J. G., & Weary, D. M. (2014). Conditioned place avoidance of Zebrafish (*Danio rerio*) to three chemicals used for euthanasia and anaesthesia. *PLoS One*, 9(2), e88030.

Zhuo, H., Jin, H., Peng, H., & Huang, H. (2016). Distribution, pharmacokinetics and primary metabolism model of tramadol in zebrafish. *Molecular Medicine Reports*, 14(6), 5644–5652.

9 Zebrafish Guidelines, Policies, and Regulations in Research

Yashwanth Bomma, Rajesh Pamanji, and Joseph Selvin

9.1 INTRODUCTION

Toxicologists are actively exploring innovative approaches to address the challenges posed by the increasing demand for safety testing of chemicals. With mounting pressures on animal welfare and ethical concerns, finding alternatives to traditional testing methods has become a priority (Silva & Tamburic, 2022). The European Union's Registration, Evaluation, Authorization, and Restriction of Chemicals (REACH) policy exemplifies this urgency, mandating the testing of a substantial number of chemicals, with estimates indicating that more than 7.2 million laboratory animals would be required for the assessment of 30,000 chemicals within a 12- to 15-year timeframe (Rudén et al., 2010). A significant portion of these animals would be utilized for reproductive and developmental toxicity testing, underscoring the need for more efficient and ethical methods (Lee et al., 2012).

Furthermore, the pharmaceutical and chemical industries are grappling with the limitations of animal testing, which can be both resource-intensive and time-consuming. The cost of testing a single new drug entity through animal experimentation can exceed $1 billion. Given that most of these compounds do not progress beyond this stage, there is a pressing need for alternative models and in vitro tests that can streamline safety assessments.

The Organisation for Economic Co-operation and Development (OECD) has taken the lead in coordinating global efforts to enhance developmental neurotoxicity (DNT) testing, an area that has remained relatively underexplored in regulatory contexts. Collaborative workshops have led to a consensus that a comprehensive DNT testing battery should encompass in vitro endpoints and alternative species assays. This integrated approach aims to improve testing and assessment practices and is guided by initiatives such as the development of an OECD guidance document, the creation of reference sets of chemicals, and case studies illustrating the interpretation of DNT in vitro data.

The zebrafish has emerged as a valuable tool in various research applications, including large-scale genetic mutation screens and disease modeling for biomedical research (Giardoglou et al., 2019). Leveraging revolutionary technologies like CRISPR and next-generation sequencing, zebrafish offer distinct advantages over other vertebrate models for simulating human diseases. Particularly, the genomic similarity between zebrafish and humans has enabled researchers to gain insights into the molecular mechanisms underlying human genetic disorders (Li et al., 2021; Choi et al., 2021). With approximately 70% of human genes having evident zebrafish orthologs, the zebrafish model accelerates our understanding of genetic illnesses, offering potential breakthroughs in precision medicine, including novel treatments and diagnostic approaches (Postlethwait et al., 1998; Howe et al., 2013).

In the realm of reproductive toxicity testing, a significant focus lies in developing alternatives that can replace or reduce the use of animals. The mouse embryonic stem cell test (EST), under study since the late 1990s, has emerged as a promising alternative. However, the implementation of such alternatives has been hindered by uncertainties surrounding their applicability and effectiveness. Addressing these challenges, discussions have been held to review the progress of various in vitro models for predicting developmental toxicity. To pave the way forward, efforts are underway to characterize the responses and effects observed in rat and rabbit studies, optimize in vitro assays, and understand the ideal conditions and applicability of these alternative tests (van der Laan et al., 2012).

9.2 ZEBRAFISH STATISTICS

The analysis of zebrafish-related research publications extracted from PubMed paints a comprehensive picture of the growing interest and significance of this model organism across various scientific fields. The data spans from the late 1940s to 2024, revealing a clear trend of increasing research output in recent years.

In the year 2023, an impressive total of 2,895 research articles were dedicated to zebrafish-related studies. This marked the continuation of a pattern observed over the previous years, with 4,491 publications in 2022 and 4,717 in 2021, demonstrating a consistent upward trajectory in zebrafish research. The preceding years also displayed remarkable engagement in this area, with 4,421 publications in 2020, 4,043 in 2019, and 3,661 in 2018. This suggests a sustained and growing interest in understanding zebrafish's biology, genetics, behavior, and applications in various research domains

This increased attention aligns with the broader trend in the scientific community to leverage zebrafish as a model organism due to its unique characteristics, such as rapid development, transparency of embryos, and genetic

DOI: 10.1201/9781003402893-10

similarities to humans. Zebrafish have proven to be instrumental in advancing our understanding of developmental processes, disease mechanisms, drug screening, and toxicology assessment.

Remarkably, the surge in zebrafish research is not confined to recent years. The trend becomes even more evident when analyzing data from the early 2000s onwards. Starting from the year 2002, the number of publications on zebrafish experiences a noticeable rise, indicating a heightened interest in this model organism. While the volume of research was comparatively smaller in earlier decades, even during the 1950s and 1960s, the presence of zebrafish in scientific discourse was evident, albeit to a more limited extent.

This comprehensive collection of zebrafish-related publications underscores the enduring impact and relevance of zebrafish in scientific investigation. The model's versatility and utility continue to foster ground-breaking discoveries and innovations, extending across diverse fields including genetics, developmental biology, neuroscience, toxicology, and drug discovery. The growing body of research signifies the sustained effort to unlock the secrets held by zebrafish, making it an invaluable tool in advancing our knowledge of fundamental biological processes and their applications in human health and disease.

9.3 ZEBRAFISH OECD TEST GUIDELINES

ASTM International, formerly known as the American Society for Testing and Materials, stands as a distinguished global organization renowned for its pivotal role in the development and dissemination of voluntary consensus standards. These standards encapsulate a wide spectrum of materials, products, systems, and services, and are widely acknowledged as ASTM test methods. This eminent body of standards finds ubiquitous application across industries, governmental bodies, and regulatory entities, collectively ensuring the caliber, safety, and efficacy of diverse products and materials. These guidelines epitomize a pivotal contribution to the realm of standardized testing practices, offering meticulously detailed testing protocols, methodologies, and criteria. The intrinsic value lies in their ability to foster uniformity and reliability in testing endeavors conducted across varied laboratories and settings.

Concurrently, the OECD is a prominent international entity steadfastly dedicated to advancing policies that augment the global socio-economic well-being. A cornerstone of this endeavor is the formulation of comprehensive guidelines and recommendations across diverse domains, including those pertaining to environmental and health-related testing.

The OECD Test Guidelines (TGs) are acclaimed as the preeminent global reference in the arena of chemical and material testing. These guidelines afford intricate and comprehensive procedural directives for executing tests aimed at gauging the potential risks and hazards that chemicals pose to both human health and the environment at large.

In synthesis, the symbiotic roles of both ASTM and OECD test guidelines resonate profoundly across a gamut of industries. By furnishing standardized protocols for the assessment of materials, products, and chemicals, these guidelines bestow an invaluable gift of assured quality, reliability, and comparability to testing outcomes. This collective effort substantially bolsters the safety, excellence, and sustainability of products and materials that permeate global markets.

9.4 OECD TEST GUIDELINES RELATED TO ZEBRAFISH

The OECD test guidelines are foundational documents that provide standardized methodologies for conducting a

Guideline	Scope & Focus	Endpoints Assessed	Types of Fish	Study Duration
OECD TG 203	Fish, Acute Toxicity Test	Mortality	Various fish species	OECD TG 203 specifically deals with acute toxicity in fish for 96 hours (OECD (2019), *Test No. 203*).
OECD TG 210	Fish, Early-Life Stage Toxicity Test	Survival, growth, development	Early-life stage fish	OECD TG 210 focuses on assessing toxicity during the early life stages of fish for 30 days post-hatch (OECD (2013a), *Test No. 210*).
OECD TG 212	Fish, Short-Term Toxicity Test on Embryo and Sac-Fry Stages	Development, deformities	Embryo and sac-fry stages	OECD TG 212 specifically targets the embryo and sac-fry stages for short-term toxicity testing for 5 hpf (OECD (1998), *Test No. 212*).
OECD TG 215	Fish, Juvenile Growth Test	Growth, condition factor	Juvenile fish	OECD TG 215 specifically assesses the growth of juvenile fish (OECD (2000), *Test No. 215*).
OECD TG 236	Fish Embryo Acute Toxicity (FET) Test	Mortality, hatching success	Fish embryos	OECD TG 236 focuses on acute toxicity testing using fish embryos for 96 hours (OECD (2013b), *Test No. 236*).
OECD TG 234	Fish Sexual Development Test	Effects on early-life stages and sexual development	Embryo and sac-fry stages	OECD TG 234 specifically targets the embryo and sac-fry stages toxicity testing for 60 days (OECD (2011), *Test No. 234*).
OECD TG 305	Bioaccumulation of Aquatic Organism	Bioaccumulation	Various fish species	OECD TG 305 assesses bioaccumulation in aquatic organisms for minimum of 28 days uptake and minimum 14 days depuration (OECD (2012), *Test No. 305*).

diverse array of aquatic toxicity, ecotoxicity, and environmental fate studies (https://one.oecd.org/document/env/jm/mono(2020)8/en/pdf). These guidelines are pivotal in maintaining the consistency, reliability, and comparability of data generated through testing, thereby contributing significantly to effective risk assessment and informed regulatory decision-making. The following sections delve into several crucial dimensions that define the OECD test guidelines, emphasizing their profound significance in aquatic studies involving model organisms such as zebrafish.

9.4.1 Validity Criteria

Validity criteria occupy a paramount role within the OECD test guidelines, acting as critical benchmarks that ensure the integrity, accuracy, and credibility of experimental outcomes in toxicity and safety assessments. The incorporation of validity criteria is indispensable for achieving harmonization and standardization of testing procedures across different laboratories and jurisdictions. This standardization, in turn, facilitates meaningful data comparison and enables regulatory bodies to make well-informed decisions regarding substances' potential risks. The rationale behind the inclusion of validity criteria is multifaceted:

- **Reliable Data Generation**: Validity criteria establish a framework that experimental data must adhere to in order to be considered reliable and truly reflective of the impacts of substances on organisms.
- **Quality Assurance**: By setting forth specific criteria, these guidelines enhance the quality assurance of testing protocols, minimizing errors and variations that could compromise data accuracy.
- **Bias Minimization**: Validity criteria play a pivotal role in identifying and mitigating sources of bias or confounding factors that could potentially influence results, thus ensuring the fidelity of observed effects.
- **Reproducibility Enhancement**: The capacity for reproducibility is a cornerstone of scientific inquiry. Validity criteria offer clear guidelines for replicating experiments, enabling disparate laboratories to produce comparable results.
- **Facilitating Regulatory Decisions**: Regulatory authorities necessitate robust, accurate, and reliable data to make informed decisions. Validity criteria guarantee that data submitted for evaluation meet rigorous standards.
- **Consistency Promotion**: The presence of validity criteria fosters consistency in testing protocols across diverse laboratories and entities, facilitating the amalgamation of data from various sources to comprehensively assess substance risks.
- **Confidence Building**: Adherence to validity criteria bolsters confidence in the reliability of generated data, forming the bedrock of sound scientific decisions and public health protection.
- **Harmonization Facilitation**: The establishment of uniform validity criteria supports international harmonization efforts, ensuring that consistent standards are applied globally, thereby promoting cooperation and data exchange (Kaltenhäuser et al., 2017).

9.4.2 Chemical Properties

A comprehensive understanding of a substance's chemical properties is fundamental in aquatic studies. These properties encompass a wide range of attributes, including solubility, volatility, reactivity, bioaccumulation potential, and persistence, each of which shapes the substance's behavior in aquatic environments. The intricate interplay of these properties influences a substance's toxicity and mode of action.

- **Chemical Structure**: The precise arrangement of atoms within a substance's chemical structure exerts a profound influence on its behavior and interactions in aquatic environments. The chemical structure determines the potential for interactions with biological systems, ultimately dictating its toxicity and mode of action (Brini et al., 2017).
- **Solubility and Dissociation**: A substance's solubility in water is a pivotal determinant of its bioavailability for exposure to aquatic organisms. Substances with low solubility may exhibit limited availability, potentially resulting in less pronounced toxic effects. Additionally, understanding the degree of dissociation of ionizable substances in water is critical as it influences their bioaccumulation potential and toxicity (Gad and Gad, 2005).
- **Acidity and Basicity (pKa)**: The pKa value of a substance characterizes its propensity to donate or accept protons in an aqueous environment. This attribute significantly influences the substance's speciation, solubility, and interactions with biological systems (Reijenga et al., 2013).
- **Volatility**: The volatility of a substance dictates its potential to evaporate from water into the air, thereby impacting the dynamics of exposure and aquatic toxicity. Vapor pressure, a key parameter in this context, provides crucial insights into the potential pathways of substance transfer between water and air.
- **Photolysis and Degradation**: Some substances are susceptible to degradation under the influence of sunlight, a process known as photolysis. The sensitivity of a substance to photolysis is instrumental in understanding its environmental stability and potential persistence.
- **Reactivity**: The reactivity of a substance with other molecules, such as reactive oxygen species,

holds implications for its toxicity and its capacity to induce oxidative stress in aquatic organisms.

- **Chelating Properties**: Substances possessing chelating properties can form stable complexes with metal ions present in water. This phenomenon significantly affects the bioavailability of metals and subsequently influences pathways of toxicity in aquatic organisms.
- **Bioaccumulation Potential**: A substance's bioaccumulation potential is closely linked to its lipophilicity. Lipophilic substances have a heightened propensity to accumulate in lipid-rich tissues of organisms. The bioconcentration factor (BCF) aids in predicting the potential bioaccumulation within aquatic food chains (Patel et al., 2019).
- **Persistence and Degradability**: A thorough understanding of a substance's persistence or degradability in aquatic environments is imperative for predicting its long-term impact. Persistent substances can accumulate over time, potentially leading to chronic effects.
- **Metabolites and Transformation Products**: Certain substances can undergo metabolic transformations within aquatic organisms, leading to the creation of metabolites or transformation products. Accounting for these transformed compounds is critical in conducting a comprehensive toxicity assessment.

9.4.3 Toxicokinetics and Bioavailability

Toxicokinetics, encompassing processes such as absorption, distribution, metabolism, and excretion, deeply influence the behavior and toxicity of substances within aquatic organisms (Tan et al., 2021).

- **Absorption and Distribution**: The capacity of a substance to be absorbed and distributed within an organism's tissues significantly influences its bioavailability and subsequent toxic effects. The substance's ability to traverse biological barriers and access vital organs dictates its bioavailability.
- **Metabolism and Excretion**: Understanding the metabolic pathways and excretion mechanisms of a substance is pivotal in predicting its persistence within organisms and its potential to induce chronic effects. Metabolism can lead to the formation of transformation products, adding complexity to toxicity assessments.

This level of detail in considering toxicokinetics and bioavailability provides nuanced insights into how substances interact with aquatic organisms, enhancing the accuracy and relevance of toxicity assessments and risk evaluations. The incorporation of toxicokinetic considerations aids researchers in comprehending the mechanisms underlying observed toxic effects, enabling more informed regulatory decisions and effective ecological protection.

9.4.4 Exposure Routes

Understanding the various routes through which substances can enter organisms is vital in constructing realistic exposure scenarios that mirror real-world situations.

- **Waterborne Exposure**: Many aquatic organisms, including zebrafish, experience direct and continuous exposure to substances dissolved in water. Examining how substances enter organisms through gills or other surfaces and their subsequent distribution is paramount. Furthermore, considering different exposure durations can shed light on varying toxic effects. Short-term exposures can unveil immediate toxicity, whereas chronic exposures may reveal effects that manifest over extended timeframes.
- **Diet-Based Exposure**: Exposure via the food chain, whereby substances accumulate in prey organisms, introduces the possibility of indirect effects through the ingestion of contaminated organisms. Investigating dietary exposure provides insights into potential trophic transfer and magnification.
- **Dermal Absorption**: Aquatic organisms with permeable skin, such as fish, can absorb substances directly through their skin. Understanding dermal absorption sheds light on the potential impacts of substances present in water or sediment (OECD (2012), *Test No. 305*).

By scrutinizing these diverse exposure routes, researchers construct a comprehensive understanding of the potential pathways through which substances can come into contact with aquatic organisms. This understanding, in turn, guides the selection of relevant exposure scenarios in studies, enabling the accurate assessment of risks and an enhanced grasp of the ecological implications of substances on aquatic ecosystems and model organisms like zebrafish.

9.4.5 Test Concentrations

The selection of appropriate test concentrations is a cornerstone of sound aquatic toxicity testing, ensuring that studies capture relevant and meaningful data.

- **Nominal Concentrations**: These represent the intended exposure levels of the test substance that researchers aim to subject aquatic organisms to during the study. Nominal concentrations are typically prepared based on initial calculations and dilutions, serving as a reference point for exposure.
- **Measured Concentrations**: In contrast, measured concentrations denote the actual concentrations of

the test substance determined through analytical methods in the exposure media. Measured concentrations provide accurate data for correlating exposure levels with observed effects.

- **Concentration Range**: The selection of an appropriate concentration range is crucial, encompassing environmentally realistic levels as well as higher concentrations. This range ensures that the study captures effects across a spectrum of exposure scenarios, including both ecologically relevant and potentially extreme conditions (OECD (2019), *Test No. 203*).
- **Dilution and Dispersion**: Adequate mixing and dispersion of the test substance within the exposure media are essential. These factors prevent localized high concentrations and ensure uniform exposure of organisms.
- **Control Group Concentrations**: In addition to exposure groups, control groups exposed solely to the solvent or vehicle used to dissolve the test substance are included. These control group concentrations serve as a baseline to differentiate the effects of the substance itself from those of the exposure medium.
- **Acute and Chronic Exposure**: Different concentrations may be employed for acute and chronic exposure scenarios. Acute exposure concentrations are often higher and are designed to elicit more immediate effects, whereas chronic exposure concentrations provide insights into longer-term impacts.
- **Equilibrium Partitioning**: In scenarios involving water-sediment interactions, equilibrium partitioning adjusts concentrations based on the distribution of a substance between these phases. This adjustment ensures that the exposure scenario accurately reflects conditions in the water column.
- **Dynamic Exposure Scenarios**: Certain studies may involve dynamic exposure scenarios, where concentrations change over time to mimic fluctuating environmental conditions. Pulsed exposure represents episodic events, while continuous exposure simulates sustained exposure conditions.
- **Multiple Concentration Levels**: The testing of multiple concentration levels, encompassing both lower and higher levels, is instrumental in establishing dose-response relationships. Such relationships are vital for discerning threshold effects, determining no-observed-effect concentrations, and evaluating the potential for adverse effects (OECD guidance document on aqueous-phase aquatic toxicity testing of difficult test chemicals).

The meticulous selection and consideration of test concentrations ensure that aquatic studies encompass a breadth of exposure scenarios, enhancing their relevance, accuracy, and ecological validity. A thoughtful approach to concentration selection contributes to robust risk assessments and facilitates well-informed regulatory decision-making processes.

9.4.6 Duration of Exposure

The duration of exposure within aquatic studies is a crucial parameter that requires careful consideration. The appropriate exposure duration should be aligned with the likely timeframe of exposure within natural aquatic environments. This duration is subject to variation based on factors such as the substance's degradation rate and the life cycle of the studied organisms.

- **Short-Term Exposure**: Short-term exposure studies involve subjecting aquatic organisms to the test substance for relatively brief periods. These studies, often spanning from a few hours to a few days, are designed to assess immediate and potentially severe effects that may manifest rapidly.
- **Long-Term Exposure**: In contrast, long-term exposure studies entail subjecting organisms to the test substance over extended durations, typically ranging from weeks to months. These studies are well-suited for evaluating effects that develop gradually or become apparent after sustained exposure.
- **Early-Life Stage Exposure**: In certain cases, substances may exert more pronounced effects on the early-life stages of aquatic organisms. Early-life stage exposure studies focus on critical developmental windows, during which substances may induce distinct outcomes compared to exposures during other life stages (OECD (2019), *Test No. 203;* OECD (2013a), *Test No. 210;* OECD (1998), *Test No. 212;* OECD (2000), *Test No. 215;* OECD (2013b), *Test No. 236;* OECD (2012), *Test No. 305*).

By thoughtfully varying the duration of exposure in aquatic studies, researchers can achieve a comprehensive understanding of the temporal dynamics of substance impacts. This level of temporal detail is crucial for capturing both immediate and delayed effects, assessing potential sub-lethal impacts, and uncovering implications for population dynamics and ecosystem health. The careful consideration of exposure duration, as guided by OECD test guidelines, contributes to well-rounded risk assessments and informed decision-making processes concerning substances' effects on aquatic organisms, including model organisms like zebrafish.

9.4.7 Mode of Action

An in-depth exploration of a substance's mode of action constitutes a pivotal aspect of the OECD test guidelines.

Investigating how substances exert toxic effects at molecular and cellular levels is instrumental in comprehending their overall impacts on organisms. Such insights guide the selection of appropriate endpoints for assessment and provide a framework for predicting potential adverse outcomes.

- **Mechanism of Toxicity**: Understanding the specific mechanisms through which a substance induces toxic effects is central to elucidating its overall impact on organisms. This knowledge enables researchers to identify the pathways and processes that are disrupted (National Research Council (US) Committee on Developmental Toxicology, 2020).
- **Receptor Interaction**: Some substances interact with specific cellular receptors, either activating or blocking their functions. Understanding receptor interactions is essential for deciphering the mechanism through which the substance exerts its effects (Marino et al., 2023).
- **Enzyme Inhibition**: Substances can influence the activity of enzymes that play crucial roles in various biochemical processes. Identifying the enzymes affected by a substance helps reveal the biochemical pathways that are disrupted (Robinson, 2015).
- **Oxidative Stress**: Certain substances can induce the production of reactive oxygen species (ROS), leading to oxidative stress. This stress can damage cellular structures and result in adverse effects on aquatic organisms (Song et al., 2023).
- **DNA Damage**: Some substances may cause damage to DNA, potentially leading to mutations or other genetic abnormalities. Investigating genotoxicity aids in assessing the potential for long-term genetic impacts (Phillips and Arlt, 2009).
- **Endocrine Disruption**: Substances can interfere with the endocrine system by mimicking or blocking hormones, leading to disruptions in growth, development, and reproduction. Understanding endocrine effects is especially relevant in species sensitive to hormonal changes (Diamanti-Kandarakis et al., 2009).
- **Neurotoxicity**: Neurotoxic substances can impact the nervous system of aquatic organisms, resulting in behavioral changes, impaired locomotion, or other neurological abnormalities. Research into neurotoxicity provides insights into the effects on behavior and sensory functions (Legradi et al., 2018).
- **Cellular and Subcellular Effects**: Investigating effects at the cellular and subcellular levels, such as disruptions in cellular membranes, organelle damage, or intracellular signaling pathways, helps identify the primary sites of impact within organisms.
- **Whole Organism Effects**: Observing the overall physiological responses of aquatic organisms to substances provides insights into how multiple systems are affected and how these effects may propagate throughout the organism.
- **Behavioral Changes**: Substances can influence the behavior of aquatic organisms, affecting critical behaviors such as feeding, mating, predator avoidance, and more. Assessing behavioral changes provides insights into how substances impact an organism's fitness and survival.
- **Teratogenic Effects**: Certain substances may induce developmental abnormalities in aquatic organisms, leading to malformations or impaired growth. Research into teratogenic effects is crucial for understanding potential impacts on population dynamics (Yashwanth et al., 2016; Pamanji et al., 2015a, 2015b).

The thorough examination of a substance's mode of action provides researchers with nuanced insights into the precise mechanisms through which it exerts toxic effects on aquatic organisms. This knowledge facilitates the selection of appropriate endpoints, enhances the accuracy of risk assessments, and offers a deeper understanding of the potential ecological impacts of substances. Informed by such information, regulatory decisions can more effectively protect aquatic ecosystems and the organisms they harbor, including model organisms like zebrafish.

In conclusion, the OECD test guidelines serve as a foundational framework for conducting aquatic toxicity studies that underpin effective risk assessment and informed regulatory decision-making. By meticulously considering a spectrum of substance-specific parameters alongside general physical and chemical characteristics, these guidelines ensure that aquatic studies provide a comprehensive and accurate assessment of substances' potential impacts on aquatic organisms and ecosystems. The holistic approach fostered by the OECD guidelines, spanning validity criteria, chemical properties, toxicokinetics, exposure routes, test concentrations, duration of exposure, and mode of action, represents a rigorous and scientifically sound methodology that enhances our understanding of substances' effects and contributes to the protection of aquatic environments and their inhabitants, including model organisms such as zebrafish.

9.5 REGULATIONS

Regulations can be categorized into different types based on the industries or sectors they address. Here are some common categories of regulations:

1. Food Safety Regulations: These regulations ensure the safety of food products for consumption. They cover aspects such as hygiene, labeling, additives, contaminants, and allergens.

2. Drug Approval and Registration Regulations: These regulations outline the requirements for the approval, registration, and marketing authorization of pharmaceutical drugs.
3. Air Quality Regulations: These regulations set limits on air pollutants to protect human health and the environment.
4. Chemical Safety Regulations: These regulations govern the safe production, use, and disposal of chemicals to protect human health and the environment.
5. Workplace Safety Regulations: These regulations establish requirements to ensure the safety and well-being of workers in various industries.
6. Hazardous Materials Transportation Regulations: These regulations govern the safe transportation of hazardous materials, including chemicals and dangerous goods.
7. Product Safety Regulations: These regulations set safety standards for consumer products to prevent injuries and ensure product quality.

9.5.1 Food Regulations

Global food regulations encompass a comprehensive framework of standards and guidelines that are instrumental in safeguarding the safety, quality, labeling, and international trade of food products. In a concerted effort to ensure consumer health and facilitate equitable commerce, various prominent global food regulations and standards have been established. One of the most noteworthy among them is the Codex Alimentarius (www.fao.org/fao-who-codexalimentarius/en/), an initiative jointly created by the Food and Agriculture Organization (FAO) and the World Health Organization (WHO) (www.who.int/). The Codex is entrusted with the development of international food standards, guidelines, and codes of practice that span a multitude of critical domains, including food safety, additives, contaminants, labeling, and more.

Furthermore, the Food Safety Modernization Act (FSMA) (www.fda.gov/food/guidance-regulation-food-and-dietary-supplements/food-safety-modernization-act-fsma) stands as a pivotal regulation in the United States. Designed to counteract the prevalence of foodborne illnesses, the FSMA proactively implements preventive controls throughout the entirety of the food supply chain. Its comprehensive scope encompasses preventive controls for both human and animal food, produce safety, and the implementation of the Foreign Supplier Verification Program (FSVP).

On the European front, the European Food Safety Authority (EFSA) (www.efsa.europa.eu/en) takes on a crucial role. It serves as the authoritative source of scientific advice for the European Commission and member states, evaluating risks associated with diverse facets of the food supply chain. This includes in-depth assessments of food additives, pesticides, contaminants, and novel foods, enabling the formulation of well-informed regulatory decisions.

Within the European Union (EU), the **EU** Regulation 1169/2011 plays a pivotal role in shaping the landscape of food labeling and information provision to consumers. Commonly referred to as the EU Food Information Regulation, it establishes a comprehensive framework for food labeling, mandating essential elements such as nutrition labeling, allergen information, and other consumer-relevant details.

In the Southern Hemisphere, the Food Standards Australia New Zealand (FSANZ) (www.foodstandards.gov.au/Pages/default.aspx) takes the lead in developing harmonized food standards for both Australia and New Zealand. With an all-encompassing purview, its standards span food safety, labeling, additives, and contaminants, ensuring uniformity in regulatory practices across these nations.

The China Food Safety Law sets a benchmark for comprehensive food safety regulations within China. By addressing critical areas including food production, processing, distribution, and international trade, this law prioritizes risk prevention and management, thereby bolstering consumer confidence and public health.

In Japan, the Japanese Food Sanitation Law takes center stage in setting stringent standards for food additives, contaminants, labeling, and hygiene practices. Designed to uphold the safety and quality of food products, this law plays a pivotal role in maintaining the nation's robust food safety landscape.

Across the northern border, the Canadian Food Inspection Agency (CFIA) (https://inspection.canada.ca/eng/1297964599443/1297965645317) assumes the responsibility of overseeing food safety, labeling, and quality standards within Canada. Its comprehensive oversight spans all aspects of the food supply chain, ensuring strict adherence to regulatory requirements.

In the Gulf Cooperation Council (GCC) region, GCC Food Regulations (https://inspection.canada.ca/eng/1297964599443/1297965645317) facilitate seamless trade by harmonizing food regulations across member countries. By covering a diverse array of dimensions such as food safety, labeling, and standards, these regulations contribute to a unified approach to food regulation in the region.

On an international trade front, the World Trade Organization (WTO) Sanitary and Phytosanitary (SPS) Agreement (www.wto.org/english/tratop_e/sps_e/spsund_e.htm) sets forth a framework for the application of food safety, animal health, and plant health regulations. This agreement stands as a pivotal instrument in preventing undue barriers to trade arising from disparate regulatory practices.

The International Organization for Standardization (ISO) takes a holistic approach to food-related standards, offering a comprehensive suite of guidelines. This includes standards such as ISO 22000, which focuses on food safety management systems, as well as standards concerning traceability and quality management.

Addressing specific dietary requirements, Halal and Kosher regulations (Regenstein et al., 2003) play an integral role in guiding the preparation and labeling of food products for specific religious communities. These regulations cater to the dietary needs of Muslim and Jewish populations, respectively.

These highlighted global food regulations represent a mere fraction of the intricate network of standards and guidelines that collectively ensure the safety, quality, and credibility of food products on a global scale. Tailored to meet the unique requirements of each region and country, these regulations stand as bulwarks of public health, consumer confidence, and international trade integrity.

9.5.2 Pharmaceutical Regulations

Global pharmaceutical regulations lies a labyrinthine matrix of standards and guidelines meticulously sculpted to safeguard the realms of safety, quality, efficacy, and ethical utilization of pharmaceutical products. Emerging as a sentinel amidst this intricate regulatory panorama, zebrafish studies have assumed an indispensable role, bestowing their influence upon diverse facets of pharmaceutical research, development, and safety assessment. As the curtains rise, a detailed synopsis is revealed, capturing the interplay between zebrafish and global pharmaceutical regulations, each harmoniously complementing the other.

a. **International Council for Harmonisation of Technical Requirements for Pharmaceuticals for Human Use (ICH) Guidelines:** As the symphony of regulatory harmonization swells, the ICH guidelines (www.ich.org) take center stage, uniting regulatory requisites across borders. Here, zebrafish studies unfold their wings, contributing to the overture of preclinical drug development. The canvas of their contribution extends to efficacy testing, toxicity screening, and the manifold dimensions of safety assessments, sculpting the pharmacopeia of tomorrow.
b. **United States Food and Drug Administration (FDA) Regulations:** A symposium of drug development, approval, and vigilance finds its venue within the embrace of FDA regulations. In this theater, zebrafish models unveil their myriad talents, adroitly navigating the contours of drug discovery and target validation. They lend their insight to the investigation of drug-induced toxicities and the exploration of potential off-target effects, thus substantiating the stages preceding pharmaceutical fruition (United States Food and Drug Administration (FDA) Regulations).
c. **European Medicines Agency (EMA) Regulations:** The EMA regulations exert their vigilant guardianship over the evaluation and supervision of medicinal products within the European Union. Within this echelon of oversight, the role of zebrafish studies is writ large. From illuminating safety assessment to disclosing mechanistic insights and meticulously profiling the terrain of pharmaceutical toxicity, zebrafish studies stand as torchbearers, aligning their strides with the regulatory tides that shape the European pharmaceutical landscape (European Medicines Agency (EMA) Regulations).
d. **China National Medical Products Administration (NMPA) Regulations:** As the curtains rise upon the canvas of NMPA regulations, zebrafish models find their niche. Within their embrace, these models take their place in the cadence of early drug screening, safety assessment, and the unveiling of potential mechanisms of action. The zebrafish, a sentinel within the pages of regulatory compliance, contributes its chapter to the narrative of pharmaceutical oversight in China (China National Medical Products Administration (NMPA) Regulations).

Amidst the intricate fabric of pharmaceutical regulations, a complex framework of guidelines and standards is meticulously woven to ensure the sanctity of safety, quality, efficacy, and ethical employment of pharmaceutical products. Amidst this labyrinthine landscape, the emergence of zebrafish studies stands as a beacon of versatility and value, contributing across the spectrum of pharmaceutical research, development, and safety assessment. The zebrafish, endowed with genetic parallels to humans, embryos as translucent as truth, and a tempo of development as swift as innovation, offers an extraordinary gateway to fathom drug effects, mechanisms of action, toxicities, and potential risks. In this symphony of science and oversight, the role of zebrafish studies plays like a sonnet of significance, exemplifying a harmony of exploration and compliance.

9.5.3 Environmental Regulations

Global environmental concerns—a mosaic of standards, policies, and agreements—have been meticulously woven to tackle the myriad challenges of pollution, habitat degradation, biodiversity erosion, and the specter of climate change. A litany of significant global environmental regulations and agreements unfolds, each bearing testament to humanity's resolve to safeguard the Earth's fragile balance

Within the realm of international cooperation, the United Nations Framework Convention on Climate Change (UNFCCC) takes center stage as it orchestrates a collaborative effort to combat the looming threat of climate change. Its blueprint for cooperation charts a course toward emission reduction targets, guiding the global symphony of climate negotiations that culminate in the powerful crescendo of the Paris Agreement (https://unfccc.int/process-and-meetings/the-paris-agreement/the-paris-agreement). This accord stands as a testament to collective determination, uniting nations in a harmonious commitment to curbing global warming. The harmonies of the Paris Agreement resound with the pledge to restrict temperature rise to 2 degrees Celsius above pre-industrial levels, with an audacious goal of limiting it to a mere 1.5 degrees Celsius. As countries harmonize their efforts to reduce greenhouse gas emissions, they forge a resilient future in the face of climate challenges.

Within the Convention on Biological Diversity (CBD), the anthem of harmonious coexistence among species finds its voice. Its verses emphasize the conservation of biodiversity, the sustainable utilization of nature's gifts, and the equitable sharing of genetic treasures, resonating across ecosystems. The Nagoya Protocol adds a counterpoint to this ballad, ensuring harmony in the equitable sharing of genetic wealth. As the Earth's ozone layer seeks protection, the Montreal Protocol on Substances that Deplete the Ozone Layer takes up the anthem of safeguarding. Its chords echo with the solemn commitment to cease the use of ozone-depleting agents, crafting a melody that preserves the delicate balance of the skies.

The rhythm of responsible waste management reverberates through the Basel Convention (www.basel.int/) on the Control of Transboundary Movements of Hazardous Wastes and Their Disposal. This symphony of cooperation oversees the seamless movement of hazardous wastes, ensuring that the strains of pollution do not mar the harmony of distant lands. The Stockholm Convention on Persistent Organic Pollutants (POPs) (www.pops.int/) unveils an anthem within the realm of toxic chemicals that linger. Its verses caution against persistent pollutants while harmonizing with the need for vigilant monitoring and controlled release, creating an ode to environmental health.

Amidst the trade winds of commerce, the stanzas of conservation resonate in the Convention on the Trade in Endangered Species of Wild Fauna and Flora (CITES). This pact safeguards the harmonious coexistence of life on Earth, protecting endangered species in a cadence that transcends borders. The World Heritage Convention (https://whc.unesco.org), a paean to cultural and natural treasures, immortalizes a legacy of protection. Its harmonies resound through the preservation of exceptional sites, ensuring the sanctity of heritage for generations to come.

In the European Union's Water Framework Directive (https://ec.europa.eu/environment/water/water-framework/index_en.html), the melody of water management unfolds. Each note strives for the pristine cadence of good ecological and chemical water quality, orchestrating the sustenance of aquatic life. Across American skies, the notes of clean air regulations resound with the Clean Air Act in the United States (Clean Air Act). This melody of emission regulation harmonizes with the health of the environment and human life, crafting a sonnet of ecological well-being. Similarly, in the United States, the Safe Drinking Water Act lends voice to the resonance of pure water. This act composes a tune of drinking water quality, weaving an anthem of health and sustenance for the nation.

Finally, the Environmental Impact Assessment (EIA) regulations create a symphony of caution and care that echoes across various lands. Through their crescendo of assessments, they orchestrate a prelude to responsible decisions for projects and policies, with each note resounding in harmony with the well-being of the environment.

Environmental regulations contain a complex interplay of standards, policies, and global agreements meticulously designed to safeguard the delicate equilibrium of our natural world, stave off the encroachment of pollution, and uphold the sustainable utilization of Earth's precious resources. Amid this intricate framework, zebrafish studies have arisen as a potent and versatile instrument, poised to contribute significantly to the realm of environmental research, risk assessment, and the comprehensive evaluation of the impacts wrought by pollutants and contaminants.

9.5.4 Chemical Regulations

At the core of global chemical regulations lies a sophisticated framework of standards, policies, and international agreements meticulously designed to guide the responsible production, utilization, and disposal of chemicals. This intricate web of regulations serves as a shield against potential risks to both human health and the delicate balance of the environment. Amidst this labyrinth of chemical management, several prominent global chemical regulations emerge as pivotal players, contributing distinct threads to the tapestry of safeguarded chemical practices.

REACH (Registration, Evaluation, Authorization, and Restriction of Chemicals) takes center stage as a fundamental pillar of the European Union's commitment to chemical safety. It mandates that manufacturers and importers within the EU meticulously register chemicals, providing essential data while facilitating risk assessment and management. Meanwhile, the Globally Harmonized System of Classification and Labelling of Chemicals (GHS) orchestrates an international harmony by standardizing the classification, labeling, and communication of chemical hazards. This unified language of chemical danger transcends borders, fostering seamless comprehension across nations.

In the United States, the Toxic Substances Control Act (TSCA) steps forward as a sentinel of chemical governance. With a sweeping mandate, TSCA oversees the manufacturing, distribution, import, and use of chemicals. It takes on the responsibility of identifying and managing potential risks posed by both established and novel substances. These regulations are complemented by other noteworthy initiatives, such as KKDIK in Turkey (KKDIK), Australian Industrial Chemicals Introduction Scheme (AICIS) in Australia, MEE Order 7 in China (MEE Order 7), and Canada's Chemicals Management Plan (CMP), each adding a distinct note to the global symphony of chemical safety.

As these global chemical regulations converge, they collectively cultivate an ethos of responsible stewardship. Each regulation, with its unique scope and focus, contributes a distinct refrain to the symphony of chemical management.

This harmonious collaboration resonates as a testament to international cooperation and the shared commitment to protect both human well-being and the integrity of our shared environment.

9.5.5 Occupational Health and Safety Regulations

Global occupational health and safety regulations intricately weave a comprehensive tapestry of standards, laws, and guidelines meticulously designed to safeguard the physical well-being, safety, and holistic welfare of workers across diverse industries. This intricate fabric is imbued with the collective vision of fostering secure workplaces and nurturing the health of the labor force. Engraved within this fabric are some pivotal global occupational health and safety regulations, serving as beacons of protection.

At the core of this protective framework lies the International Labour Organization (ILO) Conventions, a constellation of international labor standards. Among them, Convention No. 155 casts its protective mantle over occupational safety and health, while Convention No. 187 crafts a framework for the promotion of occupational safety and health, exemplifying international collaboration. Meanwhile, within the United States, the Occupational Safety and Health Administration (OSHA) takes the reins, diligently orchestrating the realm of workplace safety and health. Crafting meticulous standards, OSHA extends its vigilant gaze over industries and orchestrates inspections to ensure unwavering compliance.

Across the Atlantic, the United Kingdom reveals its Health and Safety at Work Act (Health and Safety at Work Act UK), an embodiment of employer responsibilities to safeguard the well-being of their workforce. This act embraces the triple duty of ensuring health, safety, and welfare, extending its embrace to encompass those tangentially affected by labor activities. Similarly, the symphony of occupational health and safety resonates within the Canada Labour Code Part II harmonizing protections for federal employees and workplaces.

Within the European Union, the harmonizing chords of occupational safety and health reverberate under the aegis of EU-OSHA. Orchestrating a symphony of initiatives, EU-OSHA champions the cause of safety and health across the diverse array of its member states. From down under, Australia's clarion call for safety echoes through the Work Health and Safety Act (Work Health and Safety Act (Australia)) (www.legislation.gov.au/Details/C2011A00137). This legal edifice erects a fortified citadel of safety, its regulations shifting in harmony with the diverse rhythms of its states and territories.

The global symphony continues with South Africa's Occupational Health and Safety Act (www.labour.gov.za/DOL/legislation/acts/occupational-health-and-safety), Singapore's Workplace Safety and Health Act (Workplace Safety and Health Act (Singapore)), India's Factories Act and Mines Act (https://labour.gov.in/), Japan's Occupational Safety and Health Law, the UAE's Occupational Safety and Health Administration (OSHAD), and the United States' National Institute for Occupational Safety and Health (NIOSH). These regulations resonate as steadfast sentinels, woven into a global tapestry, each a testament to the commitment of nations and regions to nurturing secure workplaces, preventing workplace injuries and illnesses, and kindling the flame of well-being for workers across the globe. These regulations, laws, and entities intertwine as interconnected threads, collectively embracing the profound importance of occupational health and safety on a global scale.

9.5.6 Transportation Regulations

Global transportation regulations encompass a complex network of rules, laws, and agreements that govern various facets of transportation, including safety, efficiency, environmental impact, and international trade. This comprehensive framework plays a crucial role in ensuring the smooth and secure movement of goods and people across borders. A multitude of significant global transportation regulations contribute to this intricate system.

The International Maritime Organization (IMO) establishes global standards for safety, security, and environmental performance in international shipping through conventions like SOLAS and MARPOL. The International Civil Aviation Organization (ICAO) sets international aviation standards covering safety, security, efficiency, and environmental protection, while the International Air Transport Association (IATA) provides guidelines for safe and efficient air transportation.

On the roads, the United Nations Economic Commission for Europe (UN/ECE) (https://unece.org/trans/main/wp29/wp29regs.html) lays down regulations for road vehicles, components, and road safety through agreements like the 1958 and 1968 Agreements. International Rail Transport Regulations (OTIF) regulate international rail transportation, covering technical, safety, and legal aspects. Customs regulations and World Trade Organization (WTO) agreements establish rules for customs procedures and tariffs, facilitating international trade.

Within the European Union (EU) (https://ec.europa.eu/transport/index_en), extensive regulations encompass road, rail, maritime, and aviation transport, ensuring safety, environmental protection, and standards harmonization. In North America, agreements like North American Free Trade Agreement (NAFTA) (https://ustr.gov/usmca) and USMCA (United States-Mexico-Canada Agreement (USMCA)) regulate trade and transportation among the US, Canada, and Mexico, outlining customs rules and transportation safety.

For specific aspects, regulations such as the European Agreement concerning the International Carriage of Dangerous Goods by Road (ADR) ensure the safe international road transportation of dangerous goods. The International Railway Safety Guide (IRSG) offers railway safety guidelines developed by the International Union of Railways (UIC). International Regulations for the

Prevention of Collision at Sea (COLREGs) govern ship navigation rules, preventing collisions.

Even in the realm of trade, the International Agreement on the Harmonized Commodity Description and Coding System (HS) administered by the World Customs Organization (WCO) provides a standardized system for classifying goods, streamlining customs procedures and international trade.

Collectively, these regulations address various transportation modes and dimensions, aiming to ensure safety, efficiency, environmental responsibility, and the facilitation of international trade and travel. Each regulation contributes to the global framework that governs the intricate movement of people and goods across borders, fostering a safer, more efficient, and interconnected global transportation network.

9.5.7 Consumer Product Safety Regulations

Global consumer product safety regulations form a comprehensive framework comprising laws, standards, and guidelines meticulously crafted to uphold the well-being of consumers and the integrity of products available in the market. These regulations serve as protective shields, erected to mitigate potential hazards that consumer goods might pose in daily life. A constellation of significant global consumer product safety regulations shines light on this critical sphere.

The Consumer Product Safety Act (CPSA) in the United States establishes the Consumer Product Safety Commission (CPSC) to oversee and regulate the safety of consumer products (CPSA, US). The European Union's General Product Safety Directive outlines fundamental safety requirements for products within the EU market, especially those not governed by specific regulations. While the REACH Regulation (https://echa.europa.eu/) mainly addresses chemicals, it also extends its protective mantle to articles containing chemicals, including consumer goods.

Canada's Consumer Product Safety Act (CCPSA) diligently regulates the safety aspects of consumer products, from their manufacturing to distribution and recall processes. Similarly, the Consumer Goods (Product Safety) Act in Australia, ACCC (www.productsafety.gov.au/) enforces stringent safety standards for products circulating within its market. China Product Quality Law (http://english.samr.gov.cn/) safeguards consumers by setting quality and safety standards for products sold within its jurisdiction.

The Japan Consumer Product Safety Act ensures the safety of consumer products through the establishment of robust safety standards and requirements for manufacturers and importers. The United Kingdom's Consumer Protection Act (www.gov.uk/) fortifies safety regulations for consumer products and defines manufacturers' and suppliers' liabilities. Brazil's INMETRO regulations (INMETRO) mandate the adherence of consumer products to strict safety and quality criteria before they enter the Brazilian market.

Further fortifying this global tapestry, the International Organization for Standardization (ISO) develops comprehensive international standards that encompass diverse dimensions of consumer product safety, including testing methods, labelling, and risk assessment. The World Health Organization (WHO) lends its expertise by providing guidelines that address consumer product safety issues, particularly those with potential public health implications, such as pharmaceuticals, cosmetics, and medical devices.

Harmonizing efforts across Asia-Pacific economies, the APEC Product Safety Initiatives foster collaboration, knowledge exchange, and capacity building to enhance product safety standards. Ultimately, these regulations orchestrate an intricate ensemble designed to ensure the safety of consumer products, spanning from common household items to intricate electronics, toys, and cosmetics. Each country and region contributes its unique regulations, all united by the shared commitment to safeguarding consumer interests and nurturing a marketplace where safety is paramount.

These regulations are designed to ensure that consumer products are safe for use and to minimize the risks of injury or harm to consumers. They cover a wide range of products, from household items to children's toys, electronics, cosmetics, and more. Each country or region has its own set of regulations aimed at safeguarding consumer interests and promoting the safety of products in the marketplace.

9.6 ZEBRAFISH RESEARCH

Within the intricate web of global regulations lies a multifaceted network of standards, guidelines, and policies meticulously designed to uphold the safety, quality, and trustworthiness of products consumed on a global scale. Amidst this complex regulatory landscape, zebrafish studies have emerged as an indispensable tool, wielding their profound influence in the realms of assessment and management of food safety risks, additives, contaminants, and beyond. The versatility inherent in zebrafish studies, coupled with their genetic likeness to humans and transparent nature during the embryonic stages, casts them as pivotal model organisms that extend their contributions to the evaluation of potential repercussions stemming from food-related compounds. Thus, the symbiotic relationship between zebrafish studies and global food regulations plays out in a profoundly impactful manner, orchestrating a harmonious symphony of scientific exploration and regulatory vigilance.

1. **Toxicity Assessment:** Venturing beyond the realm of conventional laboratory settings, zebrafish studies are paramount in the meticulous assessment of toxicity associated with food-related compounds. Whether it be additives, contaminants, or residual substances, the practice of exposing zebrafish embryos or adults to varying concentrations allows an opportunity to meticulously observe the emergence of developmental abnormalities,

patterns of mortality, and shifts in behavioral paradigms. This intricate dance of experimentation facilitates the establishment of safe thresholds of exposure for these compounds within the realm of food products, thereby ensuring seamless conformance with the overarching realm of food safety regulations (Chahardehi et al., 2020).

2. **Additive Evaluation:** The nuanced evaluation of food additives—ranging from preservatives to colorants and flavor enhancers—transpires as an essential prerogative prior to their approval for incorporation into consumable items. Zebrafish studies deftly illuminate the shadows cast by potential adverse effects stemming from these additives. Their invaluable insights fortify the decision-making processes undertaken by regulatory bodies, thereby steering their judgments with a profound appreciation for the safety and permissible concentrations of these additives within the fabric of food products (Duy-Thanh et al., 2023).
3. **Contaminant Detection:** The sentinel nature of zebrafish studies is strategically harnessed for the purpose of detecting and deciphering the presence of contaminants in food, a cohort inclusive of heavy metals, pesticides, and mycotoxins. By virtue of these studies, the pursuit to establish the boundaries of permissible contamination levels within the tapestry of food is fostered, thus erecting steadfast bastions that safeguard consumers from the insidious specter of inadvertent exposure to deleterious substances (Tye et al., 2019).
4. **Allergen Assessment:** Within the labyrinthine realm of allergenicity assessment, zebrafish studies show their formidable prowess. Aided by these studies, researchers effectively decipher the enigmatic complexities inherent in the allergenic potential of food products. The application of zebrafish models brings to the fore a heightened ability to identify potential allergens, all the while meticulously gauging cross-reactivity. The confluence of these insights culminates in the articulation of precise allergen labeling, bolstering compliance with regulations that orbit the intricate cosmos of allergen-related protocols (Contreras et al., 2023).
5. **Sustainable Aquaculture:** In an era where the sustainable facets of aquaculture are venerated, zebrafish studies stand as torchbearers in the crusade for responsible fish farming. The examination of feed additives, growth stimulants, and the ecological ramifications of aquacultural endeavors transcends the laboratory realm to usher in a paradigm of enlightened food production. In tandem with sustainability regulations, these studies underscore the ethical dimensions that govern food production, fostering an ecosystem where responsible practices are harmoniously entwined with regulatory mandates (May, 2021; Ulloa et al., 2014).
6. **Novel Foods and GMOs:** The frontier of novel foods, punctuated by genetically modified organisms (GMOs), invites the comprehensive gaze of zebrafish studies. These studies demonstrate insights into the potential ramifications these products could hold for human consumption. The intricate exploration of impacts on the health, development, and behavioral propensities of zebrafish adds a profound layer of understanding that underpins the regulatory bedrock dictating the safety thresholds of these novel foods (Sissener et al., 2010).
7. **Ecotoxicity and Environmental Impact:** Zebrafish studies transcend their boundaries to encompass the evaluation of environmental implications stemming from food production paradigms. The painstaking analysis of how food-related compounds resonate within aquatic ecosystems is a testament to their unparalleled utility. Such evaluations resonate with the tenets of sustainable food production regulations, igniting a profound ripple effect where every step towards food safety reverberates into safeguarding the sanctity of the environment.
8. **Risk Assessment and Public Health:** The profound ramifications of zebrafish studies extend their embrace to the echelons of risk assessment within the realm of food safety. The scientific constellations that these studies unveil pave the way for the formulation of limits, guidelines, and recommendations within the crucible of food consumption. Regulatory agencies, including the custodians of Codex Alimentarius guidelines, glean from the wealth of zebrafish research to promulgate dictates that resonate with the preservation of public health (Ben Chabchoubi et al., 2023).
9. **Drug Discovery and Target Validation:** The prologue to drug discovery is revealed beneath the gaze of zebrafish models. With their dexterity, these models traverse the terrain of drug identification, heralding potential therapeutic compounds and validating the very targets that lay the foundation of pharmaceutical endeavors. A cost-effective mantle of efficacy screening accompanies this journey, a testament to the agility with which zebrafish models shape the annals of drug discovery (Omprakash et al., 2015).
10. **Disease Modeling and Mechanistic Studies:** Zebrafish models extend their embrace to mimicry—mimicry of human diseases, a dimension that burgeons under the mantle of mechanistic exploration. Within the silent depths of their aqueous abode, these models mirror human conditions, thus proffering a vantage point from which the mechanisms underlying diseases may be gleaned.

As such, zebrafish studies form a crucible wherein disease modeling and the potency of potential drug candidates converge (Adhish and Manjubala, 2023).

11. **Mechanism of Action Studies:** The dance of pharmaceutical compounds within zebrafish waters embarks on a journey of revelation—a journey to decipher the intricate footprints of mechanisms of action. Here, zebrafish studies illuminate pathways, unravel signaling cascades, and unlock cellular processes, mirroring the complexities of human biology. As such, these studies don the mantle of guides, leading drug development through the labyrinthine corridors of precision (Zakaria et al., 2018; Cassar et al., 2020).
12. **Efficacy Testing:** In the amphitheater of pharmaceutical efficacy, zebrafish models assume their rightful place. As scrutinizers of potential drug candidates, they cast their gaze upon disease-related phenotypes and physiological orchestras. Their observations manifest as the first notes of a symphony, as preliminary evidence of therapeutic promise serenading the future (Ordas et al., 2015).
13. **Drug Metabolism and Pharmacokinetics:** The lexicon of zebrafish studies encompasses the language of metabolism and pharmacokinetics. Their enzymatic pathways mirror the symphony of humans, enabling the unraveling of drug fate and journey. Within these waters, the predictive notes of drug behavior in the human body are whispered, offering insights that echo beyond their aqueous confines (Park et al., 2020).
14. **Preclinical Assessments**: Zebrafish studies stand as the conduit between the realms of in vitro and in vivo. Their swift metamorphosis and malleable nature usher in preclinical data with swiftness and finesse. As precursors to animal studies, they stand as sentinels of drug effects, the embodiment of early-stage understanding that harmonizes with the rhythm of development (Parng et al., 2002).
15. **Ecotoxicity and Pollution Assessment:** Zebrafish studies stand as a linchpin in the assessment of toxicity and the potential environmental ramifications of pollutants and contaminants. By exposing zebrafish to varying concentrations of these agents, researchers can discern behavioral anomalies, deviations in growth, disruptions in reproduction, and developmental deviations. These studies unveil critical insights into the potential threats posed to aquatic life and the intricate balance of ecosystems (Saiki et al., 2021).
16. **Bioindicators of Environmental Health:** Within the fluid mosaic of aquatic ecosystems, zebrafish assume the role of living gauges, finely attuned to the pulse of environmental vitality. Minute alterations in their behavior, growth trajectories, and physiological responses serve as indicators, heralding the presence of pollutants or stressors within the water. As these bioindicators, zebrafish offer a clarion signal, illuminating the narrative of environmental stress and steering the course of regulatory response (Bambino et al., 2017).
17. **Water Quality Assessment:** In the aqueous expanse, zebrafish metamorphose into living sentinels of water quality. Their dynamic responses, encompassing mortality patterns, shifts in behavior, and reproductive performances, weave a vivid tapestry that reflects the symphony of water quality metrics: pH, temperature, dissolved oxygen levels, and the nuanced spectrum of pollutants (Shao et al., 2019).
18. **Endocrine Disruption Studies:** Zebrafish studies emerge as a linchpin in the scrutiny of endocrine-disrupting compounds, those elusive agents that insinuate themselves into hormonal systems of aquatic organisms. In this intricate dance of interactions, these studies trace the intricate repercussions, casting shadows over fish populations and illuminating pathways that resonate through the web of the food chain, with potential implications for human health (Tao et al., 2022).
19. **Microplastics and Nanoparticles Research:** Within the fluid currents of aquatic ecosystems, zebrafish studies assume the mantle of interpreters, unraveling the impact of microplastics and nanoparticles. As these modern entrants make their mark, zebrafish function as artistic conduits, decoding the narrative of health, behavior, and physiology, illuminating the impact on aquatic environments and potentially beyond (Gao et al., 2021).
20. **Environmental Impact Assessment (EIA):** The zebrafish, akin to the environmental oracles, occupy a pivotal position in the realm of Environmental Impact Assessments. Here, they wield the power to foresee the repercussions of human actions upon aquatic ecosystems, enabling regulators to anticipate and navigate the potential impacts of various projects or activities (Brown et al., 2022).
21. **Chemical Safety Assessment:** At the forefront of chemical safety assessment stand zebrafish studies, instrumental in evaluating the safety of chemical substances. By exposing zebrafish embryos or adults to various concentrations of chemicals, researchers can unveil developmental anomalies, shifts in behavior, and physiological reactions. These empirical insights hold profound significance, underpinning the formulation of comprehensive safety evaluations (Ben Chabchoubi et al., 2023).
22. **Hazard Identification and Classification:** Within zebrafish studies lies the power to illuminate

chemical hazards and allocate them to appropriate classifications. Through meticulous examinations of chemical impacts on zebrafish embryonic development, researchers discern whether a substance triggers teratogenic, embryotoxic, or neurotoxic effects. This clarity is pivotal for assigning hazard classifications within universally recognized systems like the Globally Harmonized System (GHS).

23. **Alternative Testing Methods:** In the zebrafish's rapid progression and manipulability, the promise of an alternative to conventional animal testing for chemical safety assessments emerges. A pivotal model to identify potential hazards and decipher toxicological mechanisms, the zebrafish streamlines the search for empirical answers (Falcão et al., 2018).
24. **Regulatory Compliance and Risk Management:** Zebrafish studies cast a luminary light upon regulatory compliance by unveiling the potential risks associated with chemical substances. Regulatory bodies tap into these revelations to set admissible exposure thresholds, mold safe handling practices, and steer the course toward prudent risk management (Lawrence et al., 2009).
25. **Early-Life Stage Toxicity Assessment:** Zebrafish embryos are transparent, allowing researchers to observe the effects of substances on early developmental stages. This is particularly important for consumer products that may affect pregnant women, fetuses, or young children.
26. **Reproductive and Developmental Effects:** Zebrafish studies help evaluate the impact of consumer products on reproductive and developmental processes. By studying how products influence zebrafish reproduction, researchers gain insights into potential effects on human fertility and prenatal development (He et al., 2014).

In summary, zebrafish studies are an indispensable tool in advancing consumer product safety regulations by contributing essential data on chemical toxicity, developmental effects, and environmental impacts. Their role in hazard identification, risk assessment, and the search for safer alternatives ultimately supports the development of regulations that prioritize the health and safety of consumers while promoting sustainable and responsible product development and use.

9.7 ACKNOWLEDGEMENT

The author Rajesh Pamanji is thankful to the Vice Chancellor and Registrar, Pondicherry University, Pondicherry, India, for providing the facilities for ongoing research work and to UGC, Government of India, for providing financial assistance in the form of DSK postdoctoral fellowship BL/19-20/0302.

REFERENCES

Adhish M, Manjubala I. Effectiveness of zebrafish models in understanding human diseases—A review of models. *Heliyon* 2023;9(3):e14557. https://doi.org/10.1016/j.heliyon.2023.e14557

AICIS. (Australian Industrial Chemicals Introduction Scheme). www.industrialchemicals.gov.au/

APEC. Asia-Pacific Economic Cooperation. Product Safety Initiatives: The APEC's official website provides information on regional initiatives for product safety. www.apec.org/

Australia Consumer Goods (Product Safety). Act: The Australian Competition and Consumer Commission (ACCC) enforces this act. www.productsafety.gov.au/

Bambino K, Chu J. Zebrafish in Toxicology and Environmental Health. *Curr Top Dev Biol.* 2017;124:331–367. https://doi.org/10.1016/bs.ctdb.2016.10.007

Basel convention on the control of transboundary movements of hazardous wastes and their disposal. www.basel.int/

Ben Chabchoubi I, Lam SS, Pane SE, et al. Hazard and health risk assessment of exposure to pharmaceutical active compounds via toxicological evaluation by zebrafish. *Environ Pollut.* 2023;324:120698. https://doi:10.1016/j.envpol.2022.120698.

Bhusnure, OG, Mane JM, Gholve SB, et al. Drug target screening and its validation by zebrafish as a novel tool. *Pharm Anal Acta* 2015;6:10 https://doi:10.4172/2153-2435.1000426.

Brazil National Institute of Metrology, Quality, and Technology (INMETRO) Regulations. https://www.gov.br/inmetro/pt-br/assuntos/regulamentacao/modelo-regulatorio-do-inmetro/inmetro-regulatory-model-published-march-2022.pdf

Brini E, Fennell CJ, Fernandez-Serra M, et al. How water's properties are encoded in its molecular structure and energies. *Chem Rev.* 2017;117(19):12385–12414. https://doi:10.1021/acs.chemrev.7b00259.

Brown CJM, Curry RA, Gray MA, et al. Considering fish as recipients of ecosystem services provides a framework to formally link baseline, development, and post-operational monitoring programs and improve aquatic impact assessments for large scale developments. *Environ Manage.* 2022;70(2):350–367. https://doi:10.1007/s00267-022-01665-0.

Canada Labour Code Part II (Canada). https://laws-lois.justice.gc.ca/eng/acts/L-2/page-1.html

Cassar S, Adatto I, Freeman JL, et al. Use of zebrafish in drug discovery toxicology. *Chem Res Toxicol.* 2020;33(1):95–118. https://doi:10.1021/acs.chemrestox.9b00335.

Canadian Food Inspection Agency (CFIA). https://inspection.canada.ca/eng/1297964599443/1297965645317

CBD. Convention on Biological Diversity. www.cbd.int/

Chahardehi AM, Arsad H, Lim V. Zebrafish as a successful animal model for screening toxicity of medicinal plants. *Plants (Basel).* 2020;9(10):1345. https://doi:10.3390/plants9101345.

China Food Safety Law. https://resourcehub.bakermckenzie.com/en/resources/asia-pacific-food-law-guide/asia-pacific/china/topics/food-product-and-safety-regulation#:~:text=The%20PRC%20Food%20Safety%20Law%20sets%20out%20comprehensive%20statutory%20requirements,different%20stages%20of%20food%20safety

China National Medical Products Administration (NMPA) Regulations. https://english.nmpa.gov.cn/lawsandregulations.html

China Product Quality Law: The State Administration for Market Regulation (SAMR) oversees product quality regulations in China. https://english.samr.gov.cn/

Choi TY, Choi TI, Lee YR, et al. Zebrafish as an animal model for biomedical research. *Exp Mol Med*. 2021;53:310–317. https://doi:10.1038/s12276-021-00571-5.

CITES. Convention on the Trade in Endangered Species of Wild Fauna and Flora. www.cites.org/

Clean Air Act (United States). www.epa.gov/clean-air-act-overview

CMP. Canada's Chemicals Management Plan. www.canada.ca/en/health-canada/services/chemical-substances/chemicals-management-plan.html

Codex Alimentarius. www.fao.org/fao-who-codexalimentarius/en/

COLREGs. International Regulations for the Prevention of Collision at Sea. www.imo.org/en/About/Conventions/Pages/COLREG.aspx

Contreras M, Vaz-Rodrigues R, Mazuecos L, et al. Allergic reactions to tick saliva components in zebrafish model. *Parasites Vectors* 2023;16:242. https://doi:10.1186/s13071-023-05874-2.

CPSA, US. Consumer Product Safety Act (United States). The Consumer Product Safety Commission (CPSC) oversees the implementation of this act. www.cpsc.gov/

Customs Regulations and World Trade Organization (WTO) Agreements. www.wto.org/

Diamanti-Kandarakis E, Bourguignon JP, Giudice LC, et al. Endocrine-disrupting chemicals: An Endocrine Society scientific statement. *Endocr Rev*. 2009;30(4):293–342. https://doi:10.1210/er.2009-0002.

Duy-Thanh D, Bich-Ngoc N, Van den Bossche F, et al. Discovering novel bioactivities of controversial food additives by means of simple zebrafish embryotoxicity (ZET) assays. *Toxics* 2023;11:8. https://doi:10.3390/toxics11010008.

EIA. Environmental Impact Assessment Regulations. https://eur-lex.europa.eu/legal-content/EN/TXT/PDF/?uri=CELEX:31985L0337

EU-OSHA (European Union Agency for Safety and Health at Work). https://osha.europa.eu/en

European Agreement concerning the International Carriage of Dangerous Goods by Road (ADR). www.unece.org/trans/danger/publi/adr/adr_e.html

European Food Safety Authority. www.efsa.europa.eu/en

European Medicines Agency (EMA) Regulations. www.ema.europa.eu/en/human-medicines-regulatory-information

European Union General Product Safety Directive. The European Commission is responsible for overseeing consumer product safety regulations in the EU. https://ec.europa.eu/consumers/consumers_safety/safety_products/rapex/alerts/repository/content/pages/index_en.htm

European Union REACH Regulation: The European Chemicals Agency (ECHA) administers the REACH regulation. https://echa.europa.eu/

European Union Transport Regulations. https://ec.europa.eu/transport/index_en

Falcão MAP, de Souza LS, Dolabella SS, et al. Zebrafish as an alternative method for determining the embryo toxicity of plant products: A systematic review. *Environ Sci Pollut Res Int*. 2018;25(35):35015–35026. https://doi:10.1007/s11356-018-3399-7.

Food Safety Modernization Act (FSMA). www.fda.gov/food/guidance-regulation-food-and-dietary-supplements/food-safety-modernization-act-fsma

Food Standards Australia New Zealand (FSANZ). www.foodstandards.gov.au/Pages/default.aspx

Gad SE, Gad SC. Toxicity testing. *Dermal Encycloped Toxicol*. 2005;2005:252–262. https://doi.org/10.1016/B0-12-369400-0/00966-2

Gao N, Huang Z, Xing J, et al. Impact and molecular mechanism of microplastics on zebrafish in the presence and absence of copper nanoparticles. *Front Mar Sci*. 2021;8:762530. https://doi:10.3389/fmars.2021.76253

GCC Food Regulations. www.gcc-sg.org/en/us/CognitiveSources/DigitalLibrary/Lists/DigitalLibrary/Agriculture%20Livestock%20and%20Fisheries/1429075065.pdf

GHS. Globally Harmonized System of Classification and Labelling of Chemicals. www.unece.org/trans/danger/publi/ghs/ghs_welcome_e.html

Giardoglou P, Beis D. On zebrafish disease models and matters of the heart. *Biomedicines*. 2019;7(1):15. https://doi:10.3390/biomedicines7010015.

He JH, Gao JM, Huang CJ, Li CQ. Zebrafish models for assessing developmental and reproductive toxicity. *Neurotoxicol Teratol*. 2014;42:35–42. https://doi:10.1016/j.ntt.2014.01.006.

Health and Safety at Work Act (United Kingdom). www.hse.gov.uk/legislation/hswa.htm

Howe K, Clark M, Torroja C, et al. The zebrafish reference genome sequence and its relationship to the human genome. *Nature* 2013;496:498–503. https://doi:10.1038/nature12111.

IATA. International Air Transport Association regulations. www.iata.org/

ICAO. International Civil Aviation Organization regulations. www.icao.int/

ICH Guidelines. www.ich.org

ILO. International Labour Organization conventions. www.ilo.org/global/standards/introduction-to-international-labour-standards/conventions-and-recommendations/lang--en/index.htm

IMO. International Maritime Organization regulations. www.imo.org/

INMETRO. The INMETRO oversees the quality and safety of products in Brazil. www.inmetro.gov.br/

International Agreement on the Harmonized Commodity Description and Coding System (HS). www.wcoomd.org/en/topics/nomenclature/instrument-and-tools/hs-nomenclature-2022/hs-nomenclature-2022-edition.aspx

International Road Traffic and Vehicle Regulations (UN/ECE). https://unece.org/trans/main/wp29/wp29regs.html

IRSG. International Railway Safety Guide. www.uic.org/safety

ISO. International Organization for Standardization (ISO) Standards: The official website of ISO provides information on international standards, including those related to consumer product safety. www.iso.org/

Japan Consumer Product Safety Act: The Ministry of Economy, Trade and Industry (METI) of Japan is responsible for this act. www.meti.go.jp/english/

Japanese Food Sanitation Law. www.japaneselawtranslation.go.jp/en/laws/view/3687/en#:~:text=Food%20Sanitation%20Act%20%2D%20English%20%2D%20Japanese%20Law%20Translation&text=Article%201The%20purpose%20of,thereby%20to%20protect%20citizens%27%20health

Kaltenhäuser J, Kneuer C, Marx-Stoelting P, et al. Relevance and reliability of experimental data in human health risk assessment of pesticides panel. *Roland Solecki Regulat Toxicol Pharmacol*. 2017;88:227–237. https://doi.org/10.1016/j.yrtph.2017.06.010.

KKDIK (Turkey REACH). www.kimyaverensarji.gov.tr/en/

Lawrence C, Sanders GE, Varga ZM, et al. Regulatory compliance and the zebrafish. *Zebrafish*. 2009 Dec;6(4):453–456. https://doi:10.1089/zeb.2009.0595.

Lee HY, Inselman AL, Kanungo J, et al. Alternative models in developmental toxicology. *Syst Biol Reprod Med*.

2012;58(1):10–22. https://doi:10.3109/19396368.2011.648302.

Legradi JB, Di Paolo C, Kraak MHS, et al. An ecotoxicological view on neurotoxicity assessment. *Environ Sci Eur.* 2018;30(1):46. https://doi:10.1186/s12302-018-0173-x.

Li Y, Jia Z, Zhang S, et al. Progress in gene-editing technology of zebrafish. *Biomolecules.* 2021;11(9):1300. https://doi:10.3390/biom11091300.

Marino M, Jamal Z, Zito PM. Pharmacodynamics. [Updated 2023 Jan 29]. In: *StatPearls* [Internet]. StatPearls Publishing, Treasure Island, FL, 2023. www.ncbi.nlm.nih.gov/books/NBK507791/

May C. Using zebrafish (*Danio rerio*) to evaluate the use of food waste based feed in sustainable small-scale aquaculture (unpublished master's thesis). Georgia Southern University, 2021.

MEE Order 7 (China). Website: The official website might be in Chinese; you can search for relevant information through China's Ministry of Ecology and Environment website.

Montreal Protocol on Substances that Deplete the Ozone Layer. https://ozone.unep.org/treaties/montreal-protocol-substances-deplete-ozone-layer

NAFTA. North American Free Trade Agreement (NAFTA)/United States-Mexico-Canada Agreement (USMCA): NAFTA has been replaced by USMCA. https://ustr.gov/usmca

National Research Council (US) Committee on Developmental Toxicology. *Scientific Frontiers in Developmental Toxicology and Risk Assessment.* National Academies Press, Washington, DC, 2000. 4, Mechanisms of Developmental Toxicity. www.ncbi.nlm.nih.gov/books/NBK225674/

NIOSH. National Institute for Occupational Safety and Health, United States. www.cdc.gov/niosh/

Occupational Health and Safety Act (India). https://labour.gov.in/

Occupational Health and Safety Act (South Africa). www.labour.gov.za/DOL/legislation/acts/occupational-health-and-safety

Occupational Safety and Health Law (Japan). www.jniosh.go.jp/

OECD. https://one.oecd.org/document/env/jm/mono(2020)8/en/pdf

OECD, Test No. 203: Fish, acute toxicity test, in *OECD Guidelines for the Testing of Chemicals, Section 2*, OECD Publishing, Paris, 2019. http://doi:10.1787/9789264069961-en.

OECD, Test No. 210: Fish, early-life stage toxicity test, in *OECD Guidelines for the Testing of Chemicals, Section 2*, OECD Publishing, Paris, 2013a. http://doi:10.1787/9789264203785-en.

OECD, Test No. 212: Fish, short-term toxicity test on embryo and sac-fry stages, in *OECD Guidelines for the Testing of Chemicals, Section 2*, OECD Publishing, Paris, 1998. https://doi:10.1787/9789264070141-en.

OECD, Test No. 215: Fish, juvenile growth test, in *OECD Guidelines for the Testing of Chemicals, Section 2*, OECD Publishing, Paris, 2000. https://doi:10.1787/9789264070202-en.

OECD, Test No. 234: Fish sexual development test, in *OECD Guidelines for the Testing of Chemicals, Section 2*, OECD Publishing, Paris, 2011. https://doi:10.1787/9789264122369-en.

OECD, Test No. 236: Fish embryo acute toxicity (FET) test, in *OECD Guidelines for the Testing of Chemicals, Section 2*, OECD Publishing, Paris, 2013b. https://doi:10.1787/9789264203709-en.

OECD, Test No. 305: Bioaccumulation in fish: Aqueous and dietary exposure, in *OECD Guidelines for the Testing of Chemicals, Section 3*, OECD Publishing, Paris, 2012. http://doi:10.1787/9789264185296-en.

Ordas A, Raterink RJ, Cunningham F, et al. Testing tuberculosis drug efficacy in a zebrafish high-throughput translational medicine screen. *Antimicrob Agents Chemother.* 2015;59(2):753–762. https://doi:10.1128/AAC.03588-14.

OSHA. Occupational Safety and Health Administration (United States). www.osha.gov/

OSHAD. Abu Dhabi Occupational Safety and Health System, UAE. https://oshad.ae/

OTIF. International Rail Transport regulations. https://otif.org/

Pamanji R, Bethu MS, Yashwanth B, et al. Developmental toxic effects of monocrotophos, an organophosphorous pesticide, on zebrafish *(Danio rerio)* Embryos. *Environ Sci Pollut Res.* 2015a;22(10):7744–7753. https://doi:10.1007/s11356-015-4120-8

Pamanji R, Yashwanth B, Bethu MS, et al. Toxicity Effects of Profenofos on Embryonic and Larval Development of Zebrafish (*Danio rerio*). *Environ Toxicol Pharmacol.* 2015b;39(2):887–897. https://doi:10.1016/j.etap.2015.02.020.

Paris Agreement. https://unfccc.int/process-and-meetings/the-paris-agreement/the-paris-agreement

Park YM, Meyer MR, Müller R, et al. Drug administration routes impact the metabolism of a aynthetic cannabinoid in the zebrafish larvae model. *Molecules.* 2020;25(19):4474. https://doi:10.3390/molecules25194474.

Parng C, Seng WL, Semino C, et al. Zebrafish: A preclinical model for drug screening. *Assay Drug Dev Technol.* 2002;1(1 Pt 1):41–48. https://doi:10.1089/154065802761001293.

Patel M, Kumar R, Kishor K, et al. Pharmaceuticals of emerging concern an aquatic systems: Chemistry, occurrence, effects, and removal methods. *Chem Rev.* 2019;119(6):3510–3673. https://doi.org/10.1021/acs.chemrev.8b00299.

Phillips DH, Arlt VM. Genotoxicity: Damage to DNA and its consequences. *EXS.* 2009;99:87–110. https://doi:10.1007/978-3-7643-8336-7_4.

Postlethwait JH, Yan YL, Gates MA, et al. Vertebrate genome evolution and the zebrafish gene map. *Nat Genet.* 1998;18(4):345–349. https://doi:10.1038/ng0498-345.

REACH. Registration, evaluation, authorization, and restriction of chemicals. https://echa.europa.eu/

Regenstein JM, Chaudry MM, Regenstein CE. The kosher and halal food laws. *Compr Rev Food Sci Food Saf.* 2003;2(3):111–127. https://doi:10.1111/j.1541-4337.2003.tb00018.x.

Regulation (EU) No 1169/2011 of the European parliament and of the council of 25 October 2011.

Reijenga J, van Hoof A, van Loon A, et al. Development of methods for the determination of pKa values. *Anal Chem Insights.* 2013;8:53–71. https://doi:10.4137/ACI.S12304.

Robinson PK. Enzymes: Principles and biotechnological applications. *Essays Biochem.* 2015;59:1–41. https://doi:10.1042/bse0590001. Erratum in: Essays Biochem. 2015;59:75.

Rudén C, Hansson SO. Registration, evaluation, and authorization of chemicals (REACH) is but the first step-how far will it take us? Six further steps to improve the European Chemicals Legislation. *Environ Health Perspect.* 2010;118(1):6–10. https://doi:10.1289/ehp.0901157.

Saiki P, Mello-Andrade F, Gomes T, et al. Sediment toxicity assessment using zebrafish (*Danio rerio*) as a model system: Historical review, research gaps and trends. *Sci Total Environ.* 2021;793:148633. https://doi:10.1016/j.scitotenv.2021.148633.

Shao Y, Xiao H, DiPaolo C, et al. Integrated zebrafish-based tests as an investigation strategy for water quality assessment panel. *Seiler Water Res.* 2019;150:252–260. https://doi:10.1016/j.watres.2018.11.039

Silva RJ, Tamburic S. A state-of-the-art review on the alternatives to animal testing for the safety assessment of cosmetics. *Cosmetics* 2022;9:90. https://doi:10.3390/cosmetics9050090.

Sissener N, Johannessen L, Hevrøy E, et al. Zebrafish (*Danio rerio*) as a model for investigating the safety of GM feed ingredients (soya and maize); performance, stress response and uptake of dietary DNA sequences. *Br J Nutr.* 2010;103(1):3–15. https://doi:10.1017/S0007114509991401.

Song C, Sun C, Liu B, Xu P. Oxidative stress in aquatic organisms. *Antioxidants (Basel).* 2023 Jun 6;12(6):1223. https://doi:10.3390/antiox12061223.

Stockholm Convention on Persistent Organic Pollutants (POPs). www.pops.int/

Tan Y-M, Barton HA, Boobis A, et al. Opportunities and challenges related to saturation of toxicokinetic processes: Implications for risk assessment. *Michelle Embry Regul. Toxicol Pharmacol.* 2021;127:105070. https://doi.org/10.1016/j.yrtph.2021.105070.

Tao Y, Li Z, Yang Y, et al. Effects of common environmental endocrine-disrupting chemicals on zebrafish behavior. *Water Res.* 2022;208:117826. https://doi.org/10.1016/j.watres.2021.117826.

TSCA. Toxic Substances Control Act. www.epa.gov/tsca

Tye M, Masino MA. Dietary contaminants and their effects on zebrafish embryos. *Toxics.* 2019;7(3):46. https://doi:10.3390/toxics7030046.

Ulloa PE, Medrano JF and Feijoo CG. Zebrafish as animal model for aquaculture nutrition research. *Front Genet.* 2014;5:313. https://doi:10.3389/fgene.2014.00313.

UNFCCC. United Nations Framework Convention on Climate Change (UNFCCC). https://unfccc.int/

United States Food and Drug Administration (FDA) Regulations. www.fda.gov/regulatory-information/fda-rules-and-regulations

van der Laan JW, Chapin RE, Haenen B, et al. Testing strategies for embryo-fetal toxicity of human pharmaceuticals. Animal models vs. in vitro approaches: A workshop report. *Regul Toxicol Pharmacol.* 2012, 63(1):115–123. https://doi:10.1016/j.yrtph.2012.03.009.

Water Framework Directive (EU). https://ec.europa.eu/environment/water/water-framework/index_en.html

WHO. World Health Organization Guidelines. The WHO provides guidelines for various aspects of consumer product safety that may impact public health. www.who.int/

Work Health and Safety Act (Australia). www.legislation.gov.au/Details/C2011A00137

Workplace Safety and Health Act (Singapore). https://sso.agc.gov.sg/Act/WSHA2006

World Heritage Convention. https://whc.unesco.org/

World Trade Organization (WTO) Sanitary and Phytosanitary (SPS) agreement. www.wto.org/english/tratop_e/sps_e/spsund_e.htm

Yashwanth B, Pamanji R, Venkateswara Rao J. Toxicomorphomics and toxicokinetics of quinalphos on embryonic development of zebrafish (Danio rerio) and its binding affinity towards hatching enzyme, ZHE1. *Aqua. Toxicol.* 2016;180:155–163. https://doi.org/10.1016/j.aquatox.2016.09.018.

Zakaria ZZ, Benslimane FM, Gheyath K, et al. Using zebrafish for investigating the molecular mechanisms of drug-induced cardiotoxicity. *Biomed Res Int.* 2018;Article 1642684. https://doi.org/10.1155/2018/1642684.

Section II

Zebrafish as a Model of Parkinson's Disease

10 The Role of Neurotransmitters in the Pathophysiology of Parkinson's Disease

Shaimaa Nasr Amin and Dalia Azmy Elberry

10.1 INTRODUCTION

Parkinson's disease (PD) is a common neurodegenerative disorder characterized by progressive neurodegeneration, resulting in various motor and non-motor symptoms. The cardinal motor manifestations encompass a range of symptoms, namely resting tremors, bradykinesia, rigidity, and postural abnormalities. In contrast, the illness is associated with non-motor symptoms, including sleep difficulties, dementia, visual impairments, and digestive-related disorders (Lees et al., 2009).

The basal ganglia (BG) are the primary brain region exhibiting the most pathological manifestations. The BG is a subcortical center consisting of five nuclei, namely the putamen, caudate, globus pallidus (GP), subthalamic nucleus (STN), and substantia nigra (SN). The GP comprises two distinct components: the globus pallidus pars externa (GPe) and the globus pallidus pars interna (GPi). Similarly, the SN consists of two subdivisions, namely the substantia nigra pars reticulata (SNpr) and the substantia nigra pars compacta (SNpc). The BG exhibit complex connections with the brainstem and the cerebral cortex. A substantial quantity of excitatory and inhibitory neurotransmitters regulate the intracellular and extracellular communication in the BG. The neurotransmitters that have been identified as playing a role in the pathogenesis of PD and have been found to exhibit abnormalities include classical neurotransmitters, neuromodulators, neuropeptides, and neurosteroids (Vu et al., 2012).

The diagnostic observations in the brains of individuals with PD are characterized by the pathological reduction in dopaminergic neurons in the SN and the development of intraneuronal protein aggregates referred to as Lewy bodies (LB). One significant element of intraneuronal LB is the presence of α-synuclein (αSyn) aggregates, which induce endoplasmic reticulum (ER) stress in dopaminergic neurons and other neurons that release various neurotransmitters within the central nervous system and the enteric nervous system (Colla, 2019).

The neurotransmission of excitatory and inhibitory signals is compromised in PD, resulting in an imbalance between dopaminergic and cholinergic neurotransmission. Additional imbalance exists in the neurotransmission of GABAergic and glutaminergic systems. The manifestation of PD does not consistently exhibit a dominating state of excitatory or inhibitory neurotransmission. In other words, specific manifestations of PD display a prevailing state of excitation, such as tremors, while others demonstrate a predominance of inhibitory neurotransmission, such as bradykinesia (Lees et al., 2009).

10.2 THE NEURAL CIRCUITS IN THE BASAL GANGLIA AND THEIR ALTERATIONS IN PARKINSON'S DISEASE

The BG play a role in initiating and planning voluntary movements. They receive feedback during movement and modulate motor commands accordingly, either facilitating or suppressing them. It should be noted that the BG are not solely responsible for motor functions, as they also have cognitive and affective functions. In this sense, they can be seen as the conductors that coordinate and communicate the cortical intention of movement with the desired sequence of thoughts (Brown & Marsden, 1998).

The BG form intricate circuits encompassing cortical areas and the thalamus. The striatum, consisting of the caudate and putamen, serves as the first entrance point for the BG in these neural circuits. At the same time, the GP and the SNpr serve as the primary departure points for the BG. The striatum is the recipient of glutamatergic inputs originating from the cortex. Specifically, the axons from the cerebral cortex establish synaptic connections with the dendrites of a specific population of striatal neurons known as medium spiny neurons (MSNs). The axons of the MSNs exhibit convergence onto neurons of the GB and the SNpr, releasing inhibitory gama-amino-butyric acid (GABA) in a phasic transitory pattern (Purves et al., 2018).

The neural connections linking the BG and the cerebral cortex consist of direct and indirect pathways. The neurons of the striatum project directly to the GPi and the SNr, which are the output nuclei of the BG. The direct route neurons comprise D1 dopaminergic receptors that possess additional receptors for peptide neurotransmitters. The direct pathway exerts phasic inhibition on the neurons of the GPi and SNr through the neurotransmission of GABA. This inhibition leads to the release of the thalamus from the tonic inhibition imposed by the GP, a process known as disinhibition. Ultimately, this disinhibition facilitates movement (Purves et al., 2018).

In contrast, the indirect pathway encompasses the GPe and the STN within the neural circuit. D2 dopaminergic

DOI: 10.1201/9781003402893-12

receptors are present on neurons within the indirect pathway. The process of striatal stimulation in the indirect pathway involves the inhibition of the GPe and the subsequent disinhibition of the STN. This disinhibition leads to the excitation of the GPi/SNr, which continues to inhibit the thalamus. As a result, movement is inhibited while a braking action is exerted. The movement process is influenced by the contrasting net effects of the direct and indirect pathways, which operate through distinct dopaminergic receptors. Specifically, the direct pathway is mediated by the D1 receptor, while the D2 receptor regulates the indirect system (Gerfen et al., 1995).

According to the conventional paradigm that delineates the direct and indirect pathways within the BG-cortical circuits, the reduction of dopamine (DA) in PD results in a drop in the level of excitation within the direct pathway and a decrease in the level of inhibition within the indirect pathway. Nevertheless, the conventional model fails to comprehensively explain the underlying mechanisms of the motor and non-motor symptoms observed in PD. Consequently, this limitation has prompted researchers to develop and explore alternative BG circuit models. One of the models under consideration examines the internal connections within the BG by focusing on the role of DA neurotransmission as a stabilizing factor for the BG-cortical circuits. In PD, the depletion of DA destabilizes these circuits, disrupting movement's harmonious synchronization (Hammond et al., 2007).

In addition to the circuits connecting the BG and the cortex that regulates voluntary movement, separate circuits originate from the anterior and posterior regions of the striatum. These circuits control saccadic eye movement and movement's cognitive and behavioral aspects. Each circuit connects the BG to a specific cortical area, leading to the division of the neostriatum into two parts based on its function and cortical projection: (1) the posterior neostriatum, responsible for motor function, and (2) the anterior neostriatum, responsible for cognitive and behavioral functions. According to O'Doherty et al. (2003), the suggested subdivisions of the neostriatum can enhance comprehension of the symptoms associated with PD. Moreover, Koerts et al. (2007) have provided evidence indicating that cognitive failure in PD is linked to a defect in the dopamine transporter (DAT) inside the head of the caudate and putamen.

Graybiel (1991) proposed an alternative BG model that delineates the circuits into three pathways. This model also incorporates a division of the striatum into two components: striosomes and extrastriosomal matrix. The extrastriosomal matrix receives information from the motor cortex and DA-releasing neurons in the ventral tegmental area. The striosomes receive afferent connections from the prelimbic cortex and dopaminergic neurons originating from the SN. Graybiel (1991) examined the potential existence of a third channel originating from the striosomal region and connecting it to the SN. Graybiel (1991) stated that the striosomes are primarily involved in emotional, affective, and behavioral processes, whereas the matrix component is primarily associated with sensorimotor functions.

According to a study conducted by Lozza et al. (2004), positron emission tomography (PET) scans have revealed that individuals with PD exhibit dysfunction in the orbitofrontal region of the brain, as well as increased activity in the striatum and specific nuclei of the thalamus (namely, the medial and dorsal nuclei). Furthermore, this dysfunction in the orbitofrontal region is associated with reduced motor and cognitive functioning (Lozza et al., 2004).

Losing dopaminergic neurons in the nigrostriatal pathway reduces dopamine levels inside the basal ganglia in individuals with PD. The molecular mechanisms underlying the development of degenerative alterations in PD are complex and can involve neurotransmitters other than dopamine. These mechanisms disrupt BG circuits' synchronization and normal interaction (Calabresi et al., 2006).

10.3 MOTOR AND NON-MOTOR MANIFESTATIONS OF PARKINSON'S DISEASE

10.3.1 Motor Manifestations

The primary motor symptoms of PD are resting tremors, bradykinesia, and rigidity. The most notable manifestation of PD is tremor, although it is not consistently observed and can be mild or nonexistent in certain individuals. PD tremors result from oscillatory movements in the agonist and antagonist muscle groups, particularly in the distal regions of the limbs. In severe stages of PD, these tremors may also impact the lips or tongue (Bartels & Leenders, 2009).

In addition to static tremors, the motor manifestations observed in individuals with PD encompass akinesia, bradykinesia, and hypokinesia. Akinesia is the inability to initiate voluntary movement, while bradykinesia denotes a slower speed in making voluntary movements. Hypokinesia, conversely, pertains to a decrease in the amplitude of movement. In addition to bradykinesia and hypokinesia, individuals with PD may exhibit other manifestations, such as a masked facial expression, reduced vocal volume (hypophonia), and difficulties with communication (Bartels & Leenders, 2009).

Patients with PD may exhibit many characteristics, including a shuffling gait, walking with tiny steps, and reducing related movements such as arm swinging, an arched back, and gait freezing. The freezing of gait harms the overall well-being of those diagnosed with PD. To mitigate this impairment, patients have to develop strategies to alleviate freezing episodes by redirecting their attention toward external stimuli (Giladi & Hausdorff, 2006).

10.3.2 Non-Motor Manifestations

Non-motor symptoms may appear several years before motor symptoms (Chaudhuri et al., 2015). These symptoms include cognitive dysfunctions, rapid eye movement (REM) sleep behavior disorder, depression, pain, fatigue, olfactory abnormalities, visual abnormalities, and gastrointestinal symptoms (Pfeiffer, 2018).

Cognitive dysfunction has been observed in certain individuals with PD, manifesting as various forms of cognitive impairment. These impairments may include deficits in planning, cognitive flexibility, task-switching executive functions, and spatial memory. According to Williams-Gray et al. (2006), cognitive impairment in individuals with severe PD can progress to dementia or psychosis. This progression contributes to a further decline in the quality of life and places an additional mental burden on PD patients. The cognitive impairment observed can be attributed to the disrupted neural pathways associated with cognitive and emotional processes in the BG and the decline in cholinergic activity in the nucleus basalis of Meynert (Factor et al., 2017).

Parkinson's disease psychosis (PDP) exhibits heterogeneity in its type and underlying pathophysiology. PDP can be induced by drugs or associated with dysfunction in various neurotransmission systems, namely DA, serotonin, and acetylcholine. PDP is categorized into four distinct classes, although it is common to observe overlap between these classes and the underlying causes of the condition. Class I is attributed to non-dopaminergic pharmaceuticals or other medical causes, such as surgical procedures. Class II is called dopaminergic PDP and is attributed to using dopaminergic drugs. The manifestation of psychosis in this context can be alleviated by reducing the dosage or discontinuing the specific medication responsible for the symptoms. Patients with Class III or serotonergic type should be treated using antidepressants while making necessary adjustments to dopaminergic drugs. Cholinergic psychosis, or Class IV psychosis, is attributed to the loss of cholinergic neurons (Friedman & Akbar, 2016).

According to Zahodne and Fernandez (2008), the presence of other comorbidities in individuals with PD, such as visual impairments caused by retinal degeneration and sleep disturbances, can potentially exacerbate symptoms of psychosis.

*The **speech difficulties*** observed in individuals with PD, such as reduced volume (hypophonia) and a lack of variation in pitch and tone (monotonous speech), can be attributed to the impact of pathophysiological changes on motor circuits and motor control. As stated earlier, it is subordinate to bradykinesia and hypokinesia. In addition, cognitive impairment is a contributing factor to speaking difficulties. The study conducted by Castner et al. (2007) establishes a correlation between DA depletion and the worsening of semantic processing.

*The **gastrointestinal manifestations*** are believed to arise within the gastrointestinal tract due to degenerative changes in the neurons of the enteric nervous system. These manifestations propagate from the periphery to the central nervous system via the spinal cord, ultimately affecting the SN and the cortex (Bartels & Leenders, 2009).

According to Mertsalmi et al. (2017), many individuals diagnosed with PD exhibit the clinical characteristics associated with irritable bowel syndrome (IBS). There has been a documented association between the lateralization of DAT availability in the cerebral hemispheres and gastrointestinal symptoms. Specifically, individuals with PD and IBS exhibited increased DAT binding in the right hemisphere, indicating the potential involvement of the dopaminergic system in the gut–brain axis among PD patients (Murtomäki et al., 2023).

Pain is experienced by individuals with PD as a result of motor impairment. However, it is worth noting that many PD patients also encounter "primary pain" prior to the manifestation of noticeable motor symptoms (Lin et al., 2013). The categorization of pain experienced by individuals with PD includes musculoskeletal, neuropathic, dystonia-related, akathisia-related, and central pain. Furthermore, a classification system based on the King's Parkinson's Disease Pain Scale is extensively employed in clinical practice (Chaudhuri et al., 2015).

The topic of discussion pertains to the central and peripheral pain processes. Central reasons encompass several factors, such as a decreased pain threshold, modified pain processing, and motor and non-motor fluctuations. The involvement of the dopaminergic system in pain regulation has been suggested by Mylius et al. (2015), who proposed that this system may exert its effects through limbic and sensory cortex connections to the BG. PD was found to be linked to a reduction in GABAergic activity in the periaqueductal grey matter (PAG) on the opposite side of the brain, as well as a drop in enkephalin and -opioid receptor activities in the dorsal horns of the spinal cord on both sides. According to Domenici et al. (2019), the ability to regulate descending analgesia was compromised in individuals with PD.

The parabrachial nucleus (PBN) is a crucial recipient for transmitting ascending nociceptive signals from the spinal cord, which are then relayed to the BG. The projection of PBN neurons extends to the STN, and these neurons also get a direct projection from the SNr, which serves as the primary output component of the BG. According to Pautrat et al. (2023), an association exists between PD and central neuropathic pain.

Peripheral factors include changes in inflammatory signaling and levodopa-induced vitamin B12 insufficiency. The study conducted by Donadio et al. (2014) reveals that individuals in the early stages of PD exhibit degeneration of small-diameter sensory nerves and the accumulation of phosphorylated synuclein in cutaneous sensory and autonomic nerves.

Furthermore, it was observed that individuals with PD who experienced pain exhibited notably elevated levels of plasma interleukin-1 (IL-1). This suggests a potential correlation between peripheral inflammatory cytokines and central pain modulation in PD patients. According to Park et al. (2017).

Prolonged exposure to levodopa contributes to the development of polyneuropathy, a condition characterized by the depletion of vitamin B12. The precursor of vitamin B12 is known as methylmalonic acid (MMA). According to Andréasson et al. (2017), there is a recognized correlation between the breakdown of levodopa and the occurrence of vitamin B12 insufficiency, increased levels of blood homocysteine, and elevated levels of serum MMA. A study conducted by Park et al. (2017) showed a positive correlation between levels of MMA and the neuropathy pain scale in individuals with PD and peripheral neuropathy. However, no significant correlation was found between Vitamin B12 or homocysteine levels and the neuropathy pain scale.

Sleep disturbances in PD, such as insomnia, sleep fragmentation, restless legs syndrome, excessive daytime sleepiness, and REM sleep behavior disorder, can be attributed to the degeneration of serotonergic neurons. This degeneration leads to decreased serotonin levels, contributing to depression and reduced melatonin secretion (Pinna et al., 2021). In addition, there is evidence of disturbance in GABAergic neurotransmission, as reported by Wong et al. (2014).

Visual impairments and abnormalities in the retina can be detected by reduced visual acuity, atypical spatial contrast sensitivity, or challenges encountered when performing intricate visual activities. DA serves as a crucial neurotransmitter within the retina, and DA depletion in the retinal region has been associated with specific visual manifestations observed in individuals diagnosed with PD. The synthesis of DA occurs within the retinal amacrine cells. According to Bodis-Wollner (1990), DA may have a role in light adaptation inside the retina by influencing photoreceptor transduction and interacting with retinal circuitry to improve it.

PD has the potential to affect GABAergic circuits. The neurotransmitter GABA plays a crucial role in the organization of the receptive fields of the ganglion and bipolar cells, as well as in modulating the activity of photoreceptors. According to Witkovsky (2004), DA impacts other neurotransmitters in the retina, such as glutamatergic and glycinergic pathways. According to Alves et al. (2023), individuals diagnosed with PD have an extended duration for signal transmission between the retina and cortex, correlated to observable anatomical modifications in the neurotransmission within the retina.

10.4 THE NEUROTRANSMITTERS INVOLVED IN THE PATHOPHYSIOLOGY OF PARKINSON'S DISEASE

10.4.1 Dopamine

DA is a catecholamine synthesized from the amino acid tyrosine. DA is significant in various physiological processes, including motor function, cognition, and drug addiction. The ventral tegmental area (VTA) and SN are the primary origins of DA release within the central nervous system (Farahimanesh et al., 2018).

The primary motor symptoms of PD are associated with the gradual deterioration of dopaminergic neurons in the nigrostriatal pathway, resulting in decreased dopamine levels in the striatum (Girault & Greengard, 2004). The metabolism and auto-oxidation of DA have been identified as potential risk factors for dementia, as discussed by Jinsmaa et al. (2018). The interaction between DA, or its oxidized metabolites, and α-Syn has been shown to enhance the neurotoxicity of the protein, as demonstrated by Mor et al. (2019). Furthermore, it has been observed that both enzymatic processes and auto-oxidation of DA results in the generation of cytotoxic chemicals that have the potential to cause damage to proteins, lipids, and DNA (Raina et al., 2021).

DA affects a set of G-protein-coupled receptors (GPCRs), known as D1-D5. The interaction between DA and its receptor can elicit a stimulatory or inhibitory response, contingent upon the specific receptor type and subsequent biochemical processes within the cell triggered by the resulting second messenger. The receptors for DA can be classified into two primary groups, D1-like and D2-like. The D1 receptors, specifically D1 and D5, are associated with GS proteins and elicit cAMP release. The D2-like receptors, including D2, D3, and D4, belong to the class of Gi proteins and exert inhibitory effects on the production of cyclic adenosine monophosphate (cAMP) (Gurevich et al., 2016).

Activating adenyl cyclase (AC) leads to an elevation in the intracellular levels of cyclic cAMP, which serves as a second messenger. The activation of protein kinase A (PKA) is facilitated by elevated levels of cAMP resulting from the activation of the D1 receptor. One of the phosphorylation targets of PKA includes the subunits of the N-methyl-D-aspartate (NMDA) receptor, an ionotropic glutamate receptor. Upon phosphorylation of these subunits, there is an observed elevation in calcium ion (Ca2+) influx, leading to heightened neuronal excitability (Traynelis et al., 2010).

In contrast, reducing cAMP during D2 stimulation weakens the synaptic transmission of glutamatergic signals from the cortex to the striatum (Hallett et al., 2006). According to Rizzi and Tan (2017), the activation of DA receptors in the direct and indirect routes can modulate the transmission of signals from the cortex to the striatum, resulting in either signal amplification or attenuation.

While most drugs are designed to affect D1 and D2 receptors, it is important to note that D3 receptors (D3Rs) also have a role in the pathogenesis of PD (Björklund & Dunnett, 2007). The D3 receptors are both postsynaptically and presynaptically, serving as autoreceptors that regulate neurotransmitter release through a negative feedback loop in response to extracellular neurotransmitter levels (Beaulieu & Gainetdinov, 2011).

On certain occasions, the activation of D3Rs has been observed to impede the reuptake of DA, the breakdown of DA, and the aggregation of α-synuclein. The consequences of stimulating D3Rs elucidate the potential processes by which D3Rs agonists alleviate symptoms associated with PD. Additionally, the activation of D3Rs has been observed to enhance α-syn autophagy and promote the release of brain-derived neurotrophic factor (BDNF). The BDNF plays a crucial role in supporting the viability and development of dopaminergic neurons (Yang et al., 2020).

Sexual dysfunction (SD) is an overlooked non-motor sign of PD. DA regulates sexual behavior by hypothalamic DA, and any disruption in the levels of DA might lead to atypical sexual behavior. SD manifests prior to the occurrence of motor symptoms and degradation of DA neurons (Koza et al., 2023).

According to Kehagia et al. (2013), cognitive impairment (CI) in PD is primarily caused by deficiencies in dopaminergic function. Furthermore, as the disease progresses, more severe abnormalities in cognitive function can be attributed to dysfunction in the cholinergic system inside the cortex. Moreover, it is worth noting that individuals with PD who misuse dopaminergic medications may experience a neuropsychiatric condition known as dopamine dysregulation syndrome (DDS). This illness is associated with a decrease in DAT uptake in the striatum (Liang et al., 2023).

Moreover, individuals with PD who undergo treatment with dopamine medications may exhibit maladaptive behaviors, manifesting as a tendency to engage compulsively in activities that provide natural rewards. These behaviors, commonly known as impulsive and compulsive behaviors (ICB), often involve increased time dedicated to hobbies, compulsive eating, gambling, or hypersexuality (Weintraub et al., 2015).

10.4.2 Acetylcholine

During the initial phases of PD, PET scans have revealed a disparity in acetylcholine (ACh) and dopamine signaling within the striatum among individuals diagnosed with PD (Sanchez-Catasus et al., 2022).

ACh has a binding affinity towards ionotropic nicotinic (nACh) receptors, characterized by their ionotropic properties, and muscarinic receptors (M1-M5), which belong to the class of metabotropic GPCRs. The nACh receptors are primarily in presynaptic glutamatergic terminals, but upon activation, these receptors facilitate the influx of Ca2+ and sodium (Na+), which amplifies these terminals' depolarization. Consequently, this heightened depolarization promotes the release of neurotransmitters and neuromodulators. The muscarinic receptors, similar to the DA receptors, can be categorized into two primary groups: the M1-like receptors (M1, M3, and M5) that are connected to Gq and initiate molecular events resulting in elevated intracellular calcium levels and heightened neuronal excitability and the M2-like receptors (M2 and M4) that are coupled to Gi/o. The AC signaling cascade is suppressed, reducing neuronal excitability (Kreitzer, 2009).

DA and ACh are the primary neuromodulators that regulate synaptic transmission inside the BG. The specific localizations of DA and ACh receptors facilitate this regulation. Most MSNs receive dopaminergic and glutamatergic inputs, creating a triad structure. This triad allows DA to selectively regulate the convergence of glutamate-releasing neurons onto certain spines (Yao et al., 2008).

PD is characterized by an imbalance in DA and ACh neurotransmitters, including a deficit of DA acting on D2 receptors and an overactivity of cholinergic neurons acting on M4 receptors. The connection between dopaminergic and muscarinic cholinergic neurons is characterized by antagonism, mediated by presynaptic GABAergic neurons through GABAA receptors and glutaminergic neurons through NMDA receptors. Nevertheless, it has been observed that nACh neurons can stimulate dopaminergic neurons located in the putamen nucleus (Pérez et al., 2011).

According to Yarnall et al. (2011), cognitive impairment and hallucinations in PD are linked to deficiencies in cholinergic neurotransmission resulting from the degeneration of ascending projections originating from the nucleus basalis of Meyner. Additionally, there is a decrease in the activity of choline acetyltransferase (ChAT) in the hippocampus and cortical regions (van Laar et al., 2010).

10.4.3 Glutamate

An imbalance of GABAergic and glutaminergic neurotransmitters characterizes PD. This imbalance manifests as elevated levels of GABA in the pons, basal ganglia, and thalamus, reduced levels of cortical glutamate, and an increased GABA/glutamate ratio in the substantia nigra. Previous studies have established a correlation between aberrant levels of GABA and glutamate in many neurodegenerative conditions, which exhibit similar clinical manifestations to PD, such as cognitive decline and depressive symptoms (Riese et al., 2015).

Within the putamen, it has been observed that glutaminergic neurons could exert potent inhibitory effects on presynaptic D2 dopaminergic neurons through the activation of NMDA receptors; consequently, exacerbating deficiencies in DA levels. As a result of this mechanism, antagonists of the NMDA receptor exhibit therapeutic properties in mitigating dyskinetic symptoms (Yao et al., 2008).

Furthermore, Masilamoni et al. (2011) state that a population of metabotropic glutaminergic neurons exhibits heightened activation in response to adenosine neurons through A2A receptors within the putamen.

10.4.4 Gamma-Aminobutyric Acid (GABA)

Based on the model proposed by Hamani et al. (2006), DA depletion in PD leads to elevated levels of GABA in the thalamus and a diminished ability to facilitate the motor cortex. To be more precise, DA leads to the release of activity in the indirect pathway, which in turn causes an elevation in the inhibitory drive of GABA from the GPi onto the ventral lateral (VL) nucleus of the thalamus. The VL nucleus is a conduit for transmitting signals to the motor cortex. Simultaneously, individuals diagnosed with PD exhibited notably elevated levels of GABA in the BG. This finding demonstrated a positive correlation with the severity of gait impairment (O'Gorman Tuura et al., 2018).

PD is not linked to higher levels of thalamic GABA. However, there is a negative correlation between the severity of the disease and GABA levels in the motor cortex, regardless of the use of dopaminergic medication. This correlation suggests that GABA may have a regulatory role in the underlying mechanisms of PD, separate from the effects of dopaminergic denervation (van Nuland et al., 2020).

According to Firbank et al. (2018), there is an association between reduced levels of GABA and the occurrence of visual hallucinations in individuals with PD. Patients with PD undergoing treatment with dopaminergic medicines and experiencing ICB have a reduced thalamic GABA response. This observation implies a heightened stimulation of the direct pathway or a diminished activation of the indirect pathway (Trujillo et al., 2022).

Sleep disturbances are considered the most prevalent non-motor symptoms of PD, and their occurrence may be attributed to the disruption of GABAergic neurotransmission during the prodromal stages of PD, as shown by Wong et al. (2014). According to Iranzo et al. (2017), there is a notable and precise degeneration of GABA neurons in the ventral medulla, and these neurons are directly connected to spinal motor neurons, which produce atonia during REM sleep.

PD commonly impacts fundamental visual functions, including low-contrast visual acuity, contrast sensitivity, and color vision (Weil et al., 2016). The neurotransmission of GABA has a role in both the peripheral retina and the central visual cortex, contributing to normal vision. In the retina, amacrine cells are responsible for the co-release of DA and GABA. It has been proposed that the degeneration of these specialized cells may lead to primary visual impairment (Nguyen-Legros, 1988). Moreover, the signaling of GABA plays a crucial role in the core processes of visual perception. Any disruption within this circuit occurring at any stage of the visual pathway can impact the accurate processing of sensory information.

According to a study conducted by Firbank et al. (2018), individuals diagnosed with PD who experience visual hallucinations exhibit a decreased concentration of GABA in the occipital region; this finding adds a central contribution to GABA dysfunction to the peripheral retinal mechanism.

The cognitive impairment found in PD can be mostly attributed to impairments in DA and ACh within the frontostriatal pathways. However, it is worth noting that there are also noticeable alterations in global transcriptional activity related to GABAergic neurotransmission in the frontal cortex, as reported by Santpere et al. (2018).

The gastrointestinal abnormalities observed in people with PD have been linked to the presence of α-synuclein, which causes damage to the tiny nerve fibers in the enteric nervous system (ENS). Additionally, neurodegeneration of the enteric branches of the vagus nerve in the brainstem has also been implicated in these disturbances. The role of GABA as a significant regulator of gastrointestinal tract function has been established (Pfeiffer, 2018).

Conversely, intestinal inflammation manifests in PD and has been associated with the onset and advancement of the condition. The purinergic system exerts regulatory control over enteric inflammation, whereas GABA also plays a significant role in modulating immune cell function and inflammatory processes (Murueta-Goyena et al., 2019).

10.4.5 Adenosine

Adenosine is a nucleoside consisting of the purine base adenine and the sugar ribose. The chemical in question does not function primarily as a transmitter but as a typical metabolite that also plays a role in signaling. Therefore, it may be argued that adenosine occupies a pivotal position where multiple metabolic pathways intersect. In all physiological conditions, all cells possess intracellular adenosine. The maintenance of equilibrium between extra- and intracellular adenosine concentrations is facilitated by equilibriative transporters. The presence of adenosine within the cell ensures the existence of a measurable concentration of adenosine outside the cell (Fredholm, 2007).

Adenosine levels can be elevated through two primary mechanisms: enhanced intracellular synthesis or augmented release of adenosine nucleotides, leading to their swift breakdown. Notably, the release of ATP from glial cells significantly contributes to this process (Bekar et al., 2008).

Adenosine exerts its effects by activating four GPCRs, namely A1, A2A, A2B, and A3. The A1 and A3 receptors exhibit a primary association with the Gi family of G proteins, resulting in the reduction of cAMP levels, augmentation of K+ conductance, and attenuation of transient Ca^{2+} conductance, which plays a role in transmitter release. The predominant signaling pathway of A2A and A2B receptors involves the activation of G proteins belonging to the Gs family, leading to increased cAMP levels. Adenosine exhibits comparable potency on human A1, A2A, and A3 receptors, while demonstrating decreased potency on A2B receptors (Fredholm et al., 2001).

Besides, it has been observed that extracellular ATP can indirectly transmit signals by interacting with A2AR adenosine receptors. This signaling is initiated by converting extracellular ATP into Adenosine by the enzyme CD73 (Zimmermann et al., 2012). Furthermore, it has been observed that the ATP-CD73-A2AR pathway is excessively active in individuals with presymptomatic PD, leading to enhanced plasticity at glutamatergic synapses in the striatum. This phenomenon is believed to have a role in initiating motor symptoms in PD (Gonçalves et al., 2023).

The localization of the adenosine A2A receptor in the BG is limited to the GABA-producing neurons of the indirect pathway originating from the striatum and projecting to the GPe. Furthermore, the GPe selectively expresses the D2 dopamine receptor and the peptide enkephalin. In contrast, it has been observed that neurons belonging to the direct pathway, which project to the SNr or the GPi, exhibit specific expression of the D1 dopamine receptor and the peptide dynorphin. Notably, these neurons do not possess significant quantities of A2A receptors (Augood & Emson, 1994).

The spatial arrangement of A2A adenosine and D2 dopamine receptors in the human BG and STN was examined, and it was observed that these receptors coexist in certain regions, indicating the potential existence of receptor-receptor interactions in the form of A2AR-D2R heteromers. Adenosine receptors not only interact with dopaminergic receptors, according to Emmi et al. (2022), but A2AR exhibits direct interactions with NMDAR, mGlu5R, and cannabinoid CB1R, resulting in a synergistic effect (Köfalvi et al., 2020). Furthermore, it should be noted that A2A receptors located in the striatum are expressed by cholinergic nerve terminals, hence exerting control over the release of acetylcholine (Jenner et al., 2009).

A2AR are also found at corticostriatal terminals, where they exert influence over the release of glutamate, hence regulating the activity of A1R, cannabinoid type 1 receptors (CB1R), and metabotropic glutamate receptor 5 (mGlu5R). According to Ferré et al. (2023), A2AR receptors can integrate incoming glutamate impulses and modulate neuronal sensitivity to these signals through DA signaling.

The unique characteristics of striatopallidal A2AR, which allow for the precise modulation of dopamine and glutamate signaling, facilitate its role as a regulatory mechanism that limits diverse behavioral components (Chen et al., 2023).

Moreover, the simultaneous activation of adenosine A2AR and dopamine D2 receptors leads to a shift in the plasticity of MSNs in the striatopallidal pathway. This shift involves transitioning from long-term depression (LTD) to long-term potentiation (LTP). It influences a specific type of LTD in the striatum that relies on the production of endocannabinoids and the activation of D2R. As a result, this modulation of striatal plasticity affects instrumental learning (Ferré & Ciruela, 2019).

According to Fernández-Dueñas et al. (2019), there is a documented increase in the oligomerization status of D2R/A2AR. Additionally, the availability of A2A receptors has decreased the caudate among patients in the early stages of PD. At the same time, it increases the pallidum among patients in the moderate stages of PD (Waggan et al., 2023).

Furthermore, apart from the observed striatal alterations in PD, there was also an elevated presence of A2A receptors in the parietal white matter of the cerebral cortex (Laansma et al., 2021). Additionally, a noteworthy reduction in grey matter volumes was observed in the occipital, parietal, and temporal cortices before frontal atrophy.

Wang et al. (2023) report that decreased serum concentration of adenosine is associated with an increased risk of sleep disturbances in individuals with PD.

10.4.6 Serotonin

Serotonergic neurons start in the raphe nuclei and extend their projections to the BG, encompassing the SNpc and the striatum (Hornung, 2003). The serotonergic system regulates various cognitive processes, including cognition, mood, emotion, sleep, and eating behavior. Hence, it is plausible to suggest that PD's motor and non-motor symptoms may be associated with impaired serotonergic neurotransmission (Schapira et al., 2017).

LB have been observed within raphe neurons in PD progression, coinciding with the manifestation of anxiety and sadness. These symptoms are also associated with disrupted melatonin secretion and sleep disturbances, typically occurring prior to the start of motor symptoms. According to de Natale et al. (2021), in preclinical models of PD, there is a reduction in the levels of serotonin and its metabolite 5-hydroxy-indoleacetic acid (5-HIAA), as well as a decrease in the expression of the serotonin transporter (SERT) and its binding in several BG nuclei, thalamus, and hypothalamus.

Additionally, the impairment of serotonergic neurotransmission may play a role in resting tremor, which is not fully mitigated by dopaminergic treatment. Serotonergic therapies may assist patients with PD with tremor symptoms linked to mainly malfunctioning within the raphe region, as shown by Pasquini et al. (2018).

Serotonergic neurons are equipped with an enzyme mechanism that enables the conversion of levodopa into DA. These neurons can store DA alongside serotonin within neuronal vesicles. Furthermore, they can release a combination of DA and serotonin in response to action potentials. Therefore, the release of DA as a "false neurotransmitter" from the serotonergic terminals appears to play a role in the elevated synaptic levels of extracellular DA observed after levodopa administration. Furthermore, serotonergic terminal sprouting has been found to exacerbate dyskinesia, as Beaudoin-Gobert et al. (2018) demonstrated.

Additionally, it has been observed that patients with PD who have a longer disease duration tend to exhibit a higher ratio of serotonergic-to-dopaminergic terminals in the striatum. This may be attributed to a slower degeneration of serotonergic terminals than dopaminergic terminals or an abnormal growth of serotonergic innervation (Roussakis et al., 2016).

Serotonergic neurotransmission is a multifaceted process regulated by the specialized transporter SERT and seven unique classes of receptors (5-HT1 to 5-HT7), encompassing a minimum of 15 receptor subtypes (Hoyer, 2020).

The 5-HT1 autoreceptors are present in the serotonergic raphe nuclei, where these receptors can be found on both presynaptic and postsynaptic neurons, as well as on striatal MSNs and glutamatergic cortico-striatal terminals (Miguelez et al., 2014). According to Cenci et al. (2022), the 5-HT1 autoreceptors regulate neurotransmitters' release from serotonergic terminals. This regulation helps decrease the excessive release of DA after levodopa administration, ultimately reducing dyskinesia and relieving symptoms associated with PD.

The serotonergic system additionally influences the motivational impairments observed in PD; according to a study by Nobis et al. (2023), DA and serotonin exhibit interdependence in encoding signals related to reward and punishment.

10.4.7 Cocaine-Amphetamine Transcriptional Regulation (CART)

The cocaine-amphetamine regulated transcript (CART) is a neuropeptide that significantly protects against ischemic brain injury. Consequently, CART may be a valuable biomarker for specific human diseases, including dementia in PD. CART possesses multiple antioxidant properties. Developmental pathways involving the generation of dopaminergic neurons reveal a relationship between CART and dopaminergic neurons. Furthermore, CART transcripts have been detected in mesocorticolimbic dopaminergic regions of the brain (Bharne et al., 2013).

Synaptic connectivity and distribution of CART peptides were abundant in the SN, where they overlapped with the dendrites of dopaminergic neurons. In addition, CART influences locomotor activity by simultaneously activating D1 and D2 dopaminergic receptors. Neurons that produce CART may also aid in migrating dopaminergic SN neurons. CART functions may be a potential therapeutic target for protecting dopaminergic neurons in PD patients (Khairnar et al., 2022).

10.4.8 Neurosteroids

Neurosteroids (NS) are a type of steroid that acts on the nervous system; they are synthesized in the mitochondria of neurons or glial cells using cholesterol or other steroidal precursors, such as estrogen, progesterone, and vitamin D (VD) (Eser et al., 2006). NS modulate the DA system directly or indirectly via the GABAergic system. NS levels and their synthetic enzyme expression are altered in PD patients (Ratner et al., 2019).

10.4.9 Substance P

The primary mechanism of action of Substance P is mediated by its interaction with the neurokinin 1 receptor (NK-1R). There exists a positive correlation between the release of DA and the concentration of substance P in the SN. According to Werner and Coveñas (2013), a population of neurons in the caudate nucleus contains substance P. These neurons are subjected to inhibition by dynorphin neurons, as a result, they only weakly stimulate GABAergic neurons in the GPi.

10.4.10 Dynorphin

The dynorphin neurons situated in the caudate nucleus get a relatively modest activating input from the D1 dopaminergic neurons found in the SN. These dynorphinergic neurons, through the use of kappa receptors, exert presynaptic inhibition on substance P neurons. There is an observed negative relationship between DA insufficiency and the heightened functioning of dynorphin neurons in the SN. These neurons exhibit a strong affinity for mu and delta opioid receptors (Ljungdahl et al., 2011).

10.4.11 Cholecystokinin (CCK)

Elevated cholecystokinin (CCK) levels have been observed in the striatum of rat models of PD (Nilsson et al., 2009).

10.4.12 Neurotensin (NT)

Neurotensin (NT) is a neuron peptide and a neuromodulator for many neurotransmitters, including ACh, serotonin, GABA, and DA. NT is predominantly localized in dense nuclear vesicles within neurons. An elevated level of NT in the striatum has been observed in PD, leading to heightened glutamate excitotoxicity through NMDA receptors and a DA deficit (Ferraro et al., 2009).

10.5 NERVE-DRIVEN IMMUNITY AND NEUROTRANSMITTERS INTERACTION IN PARKINSON'S DISEASE

Inflammasomes are a class of multiprotein complexes that serve as sensors for detecting diseased or dangerous signals. The release of pro-inflammatory cytokines is regulated by inflammasome complexes found in microglia and other immune cell types (Haque et al., 2020).

Microglial cells serve as the resident macrophages within the central nervous system (CNS) and function as the initial protective barrier against external pathogens. In physiological circumstances, microglia engage in a diverse array of tissue-supportive functions. The presence of activated microglia in the SN region of postmortem brain tissue from patients with PD indicates the involvement of microglia in the pathogenesis of PD. Moreover, it has been observed that there is a decline in the regulatory and neurotrophic capabilities of microglia, along with an intensified pro-inflammatory condition throughout the progression of the disease. Microglia activation in the brain affected by PD is linked to various immune responses. The activation of microglia triggers the activation of the inflammasome, the release of pro-inflammatory cytokines that can impact neighboring cells, the production of reactive oxygen species (ROS), the activation of complement proteins, and the disruption of phagocytic activity in microglia (Weiss et al., 2022).

The induction of pro-inflammatory cytokine release by microglia also facilitates the transition of astrocytes toward a neurotoxic phenotype (Cai et al., 2022). Astrocytes, the predominant glial cells in the central nervous system, can be categorized into two primary subtypes: protoplasmic and fibrous. Their cellular morphology and anatomical positioning determine this classification. Protoplasmic astrocytes are distributed ubiquitously within the grey matter surrounding neuronal cell bodies and synapses. In contrast, fibrous astrocytes are primarily located in white matter and possess a shape characterized by many fiber-like processes that establish contact with the nodes of Ranvier (Halliday & Stevens, 2011).

Several studies have documented favorable results regarding reactive astrogliosis in the substantia nigra of postmortem brain tissue affected by PD. Reactive astrocytes can transition between two distinct phenotypes, namely the neurotoxic A1 phenotype and the neuroprotective A2 phenotype, in a manner that is particular to the stimuli they encounter (Sathe et al., 2012).

According to Liddelow et al. (2017), A1 astrocytes exhibit upregulation of several conventional complement cascade genes with pro-inflammatory properties. Conversely, A2 astrocytes demonstrate an upregulation of neurotrophic factors, indicating their role in neuroprotection. The involvement of activated microglia and invading immune cells in promoting astrocyte reactivity towards the neurotoxic phenotype, resulting in a decline in their capacity to support neuronal survival (Yun et al., 2018).

Furthermore, astrocytes have a role in the uptake and degradation of neuronal α-synuclein, hence assisting in the removal of accumulated α-synuclein. Nevertheless, in cases where the degradation process is unsuccessful, an excessive accumulation of α-syn is observed in astrocytes, leading to its transmission to adjacent astrocytes and contributing to the proliferation of disease. Astrocytes inside the brain affected by PD exhibit the expression of major histocompatibility complex class II (MHCII) receptors, which signifies their involvement in the process of antigen presentation and the activation of CD4 T cells. According to Weiss et al. (2022), the release of pro-inflammatory cytokines enhances and prolongs neuroinflammation, impairing their typical neuroprotective roles and subsequent degeneration of DA neurons in Parkinson's disease.

Previous studies have demonstrated that elevated levels of inflammatory cytokines, such as TNF and IL-1β, are released by activated inflammatory cells, including microglia, astroglia, and macrophages. This release increases synaptic glutamate levels and glutamate overflow into the extrasynaptic space. This overflow occurs due to a reduction in the capacity of glial transporters to regulate glutamate levels. Consequently, the excess glutamate activates intrasynaptic α-amino-3-hydroxy-5-methyl-4-isoxazolepropionic acid (AMPA) and NMDA receptors, as well as extrasynaptic NMDA receptors, leading to the degeneration and reduction in the size of dendritic spines and processes, and deterioration of the synaptic structure, ultimately resulting in the death of neurons. Additionally, it inhibits the protective effects of neurotrophic factors, such as BDNF, on neuronal health (Oshaghi et al., 2023).

The potential impact of glutamate on the progression of PD involves mitigating the degradation of DA neurons. The activation of glutamate induces a regulatory T-cell reaction that mitigates neuroinflammation. The upregulation of regulatory T-cells is facilitated by the expression of mGluR4 on dendritic cells, which play a vital role in promoting immunological responses. Hence, glutamate may serve a protective function in the context of inflammatory neurodegenerative disorders. According to Hansen and Caspi (2010), a mGluR4 enhancer mitigates glutamate release within the brain, presenting a potential therapeutic approach for PD.

T-lymphocytes possess functional receptors for various neurotransmitters, such as DA. The direct interaction between T-cells and these neurotransmitters can elicit positive or negative signals, modulating crucial T-cell-mediated processes, including adhesion, migration, and extravasation. This phenomenon holds significance in a wide range of physiological and pathological contexts. The CD8 T-lymphocytes increased within the SN of postmortem brain tissues obtained from patients with PD (Levite, 2000).

In contrast, B-lymphocytes are crucial in facilitating the humoral immune response. They are accountable for producing antibodies targeted towards a specific antigen following activation by antigen-presenting cells (APCs) and T-cells. The presence of complement components C1q, C3d, C4d, C7, and C9 in the SN brain samples of patients with PD provides evidence for activating the classical complement cascade initiated by antibodies. Moreover, a significant correlation exists between IgG immunoreactive DA neurons in the brains of patients with PD and activated microglia expressing FcγRI. This finding suggests that inflammation mediated by antibodies may contribute to the progression of inflammation and degeneration of neurons in PD (Depboylu et al., 2012).

It has been observed that specific immune cells can synthesize GABA and exhibit the presence of GABA receptors. The interaction between GABA and these receptors triggers biochemical signaling processes, facilitating immune regulatory responses that reduce inflammatory reactions and inhibit autoimmune disorders. The presence of a significant GABAergic influence in the gastrointestinal tract indicates that targeting the GABAergic system may offer a promising approach for the development of pharmaceutical interventions aimed at restoring normal gastrointestinal motility in various motor gastrointestinal disorders, such as gastroparesis, ileus, and acute colonic pseudo-obstruction. There has been a notable association between GABAergic signaling and its potential impact on the immune system. This association has prompted researchers to investigate its potential role in the well-established neuroimmune interaction in inflammatory gastrointestinal diseases. It is also believed that GABAergic signaling may synergize with other enteric immunomodulatory mediators (Auteri et al., 2015).

The modulation of immune responses by GABA signaling is a complex process, primarily including the negative regulation of T-cell proliferation and the production of pro-inflammatory cytokines. This regulation is achieved by downregulating specific signaling pathways associated with these immune responses. Therefore, it can be inferred that the GABAergic system possesses significant potential to suppress inflammatory reactions. GABA is synthesized by the bacterial strains *Lactobacillus* and several species belonging to the *Bifidobacterium* genus. The dysregulation of neurotransmitter levels within the CNS can give rise to disruptions in cognitive function, as a result of the impact of these neurotransmitters on the vagus and peripheral neurons (Wiatrak et al., 2022).

The anti-inflammatory effect of ACh is mediated by the activation of α7-nicotinic acetylcholine receptors (α7nAChRs), which are present in inflammatory macrophages. Moreover, ACh reduces inflammation by inhibiting the translocation of NF-κB to the nucleus and decreasing the synthesis of inflammatory cytokines (Butkovich et al., 2018).

10.6 NEUROTRANSMITTERS AS DIAGNOSTIC MARKERS IN PARKINSON'S DISEASE

The classification of biomarkers for PD encompasses six primary groups, including neurotransmitters, oxidative stress markers, inflammatory and immunological markers, growth factors, proteins implicated in PD pathogenesis, and miscellaneous biomarkers (Andersen et al., 2017). The consideration of biomarkers holds significant importance, particularly in patients exhibiting premotor symptoms at an early stage of the disease (Meles et al., 2021).

The neurotransmitter systems of ACh and DA have been identified as the primary factors contributing to cognitive dysfunction in PD. The most frequently utilized method for investigating changes in these neurotransmitter systems associated with cognitive dysfunction in PD is molecular imaging using positron emission tomography (PET) or single-photon emission computed tomography (SPECT) (Aarsland et al., 2017).

Losing dopaminergic neurons in the SN is a crucial pathological feature in PD. Additionally, the depletion of noradrenergic neurons in the locus ceruleus has been reported, along with metabolites such as dihydroxyphenyl acetic acid (DOPAC) and homovanillic acid (HVA). Nevertheless, the origins and interpretations of these compounds vary. The relationship between cerebrospinal fluid (CSF) homovanillic acid and neuronal dopamine reserves is somewhat distant, indicating the involvement of numerous intermediate steps. The absence of catechol-O-methyltransferase in dopaminergic neurons results in the reliance of CSF homovanillic acid on absorption and intracellular O-methylation in cells that are not dopaminergic. According to Goldstein et al. (2012), reducing striatal homovanillic acid levels in PD is not as pronounced as in DA levels.

The measurement of DA levels in CSF may not accurately represent central DA insufficiency. The primary source of DA in the extracellular fluid is the exocytotic release triggered by route traffic and the efflux of DA due to neuronal reuptake mediated by the cell membrane DAT. Losing dopaminergic neurons is associated with a compensatory increase in pathway traffic to the remaining terminals. This traffic increase enhances dopamine delivery from those terminals to the extracellular fluid. Consequently, CSF dopamine levels may not accurately reflect the extent of neuronal DA storage loss and can be influenced by levodopa intake (Goldstein et al., 2012). In addition, it is essential to note that despite the strong connection between CSF and the brain, the invasiveness of the sampling procedure represents a significant constraint (Cocco et al., 2023).

PET imaging is utilized to evaluate the dopaminergic system through the utilization of markers that bind to either aromatic amino acid decarboxylase (AADC) or vesicular monoamine transporter 2 (VMAT2) in presynaptic dopaminergic neuron terminals. This imaging technique has revealed slight variations in dopaminergic denervation within the anterior cingulate, ventral striatum, and caudate regions among patients with PD, with or without accompanying dementia. Using PET or SPECT to visualize DAT has revealed diminished striatum levels among individuals with PDD. Furthermore, a decline in DAT uptake within the caudate region has been linked to executive dysfunction, as Petrou et al. (2012) indicated.

According to Katunina et al. (2023), compared to the control group, no significant alterations were observed in the plasma levels of adrenaline, norepinephrine, DA, serotonin, and DOPAC in patients at risk of developing PD.

Examining the cholinergic system involves using PET scans, which employ ligands capable of measuring the activity of AChE, revealing that cholinergic denervation of the cerebral cortex manifests at an early stage in PD. A decrease in cortical AChE activity is linked to diminished cognitive performance in attention, memory, and executive function domains. In PDD, there is a notable increase in the severity and distribution of diminished AChE activity. This reduction affects various regions, including the temporal, frontal, and medial occipital cortices and the thalamus (Müller & Bohnen, 2013).

One of the issues associated with utilizing neurotransmitter anomalies as a biomarker for cognitive impairment in PD is the issue of specificity, as the cholinergic system is also implicated in other neurodegenerative dementias, including Alzheimer's disease (AD) and dementia with Lewy bodies (DLB) (Kalia, 2018).

10.7 CONCLUSIONS AND FUTURE DIRECTIONS

PD, the second common neurodegenerative disease, with diverse manifestations not limited to the brain, is historically linked to DA deficiency in the nigrostriatal fibers. However, the pathophysiology of PD is not that straightforward to explain. Numerous neurotransmitters are involved in the pathogenesis and interact with each other on multiple levels to develop the whole picture of motor and non-motor features of PD. Despite the complexity of neurotransmission in the BG and the diverse correlations between neurotransmitters, this may provide the opportunity to attenuate the manifestations by several management lines for each involved circuit or receptor. Recent techniques overcoming the issues of specificity or invasiveness, such as electrochemical detection, nano techniques, or biosensors, can provide us with more details and uncover more about the pathophysiology and neurotransmission dysfunction in PD, which may pave the way for new therapeutic options.

REFERENCES

Aarsland, D., Creese, B., Politis, M., Chaudhuri, K. R., Ffytche, D. H., Weintraub, D., & Ballard, C. (2017). Cognitive decline in Parkinson disease. Nature Reviews Neurology, 13(4), 217–231. https://doi.org/10.1038/nrneurol.2017.27

Alves, J. N., Westner, B. U., Højlund, A., Weil, R. S., & Dalal, S. S. (2023). Structural and functional changes in the retina in Parkinson's disease. Journal of Neurology, Neurosurgery & Psychiatry, 94, 448–456.

Andersen, A. D., Binzer, M., Stenager, E., & Gramsbergen, J. B. (2017). Cerebrospinal fluid biomarkers for Parkinson's disease - a systematic review. Acta Neurologica Scandinavica, 135, 34–56.

Andréasson, M., Brodin, L., Laffita-Mesa, J. M., & Svenningsson, P. (2017). Correlations between methionine cycle metabolism, COMT genotype, and polyneuropathy in L-dopa treated Parkinson's disease: a preliminary cross-sectional study. Journal of Parkinson's Disease, 7(4), 619–628. https://doi.org/10.3233/JPD-171127

Augood, S. J., & Emson, P. C. (1994). Adenosine A2a receptor mRNA is expressed by enkephalin cells but not by somatostatin cells in rat striatum: a co-expression study. Brain Research. Molecular Brain Research, 22(1–4), 204–210.

Auteri, M., Zizzo, M. G., & Serio, R. (2015). GABA and GABA receptors in the gastrointestinal tract: from motility to inflammation. Pharmacological Research, 93, 11–21. https://doi.org/10.1016/j.phrs.2014.12.001

Bartels, A. L., & Leenders, K. L. (2009). Parkinson's disease: the syndrome, the pathogenesis and pathophysiology. Cortex, 45(8), 915–921. https://doi.org/10.1016/j.cortex.2008.11.010

Beaudoin-Gobert, M., Météreau, E., Duperrier, S., Thobois, S., Tremblay, L., & Sgambato, V. (2018). Pathophysiology of levodopa-induced dyskinesia: insights from multimodal imaging and immunohistochemistry in non-human primates. Neuroimage, 183, 132–141.

Beaulieu, J. M., & Gainetdinov, R. R. (2011). The physiology, signaling, and pharmacology of dopamine receptors. Pharmacological Reviews, 63, 182–217.

Bekar, L., Libionka, W., Tian, G. F., Xu, Q., Torres, A., Wang, X., Lovatt, D., Williams, E., Takano, T., Schnermann, J., Bakos, R., & Nedergaard, M. (2008). Adenosine is crucial for deep brain stimulation-mediated attenuation of tremor. Nature Medicine, 14(1), 75–80.

Bharne, A. P., Upadhya, M. A., Shelkar, G. P., Singru, P. S., Subhedar, N. K., & Kokare, D. M. (2013). Neuroprotective effect of cocaine- and amphetamine-regulated transcript peptide in spinal cord injury in mice. Neuropharmacology, 67, 126–135. https://doi.org/10.1016/j.neuropharm.2012.10.028

Björklund, A., & Dunnett, S. B. (2007). Dopamine neuron systems in the brain: an update. Trends in Neurosciences, 30(5), 194–202. https://doi.org/10.1016/j.tins.2007.03.006

Bodis-Wollner, I. (1990). Visual deficits related to dopamine deficiency in experimental animals and Parkinson's disease patients. Trends in Neurosciences, 13(7), 296–302. https://doi.org/10.1016/0166-2236(90)90113-o

Brown, P., & Marsden, C. D. (1998). What do the basal ganglia do? Lancet, 351, 1801–1804.

Butkovich, L. M., Houser, M. C., & Tansey, M. G. (2018). α-synuclein and noradrenergic modulation of immune cells in Parkinson's disease pathogenesis. Frontiers in Neuroscience, 12, 626.

Cai, Y., Liu, J., Wang, B., Sun, M., & Yang, H. (2022). Microglia in the neuroinflammatory pathogenesis of Alzheimer's disease and related therapeutic targets. Frontiers in Immunology, 13, 856376. https://doi.org/10.3389/fimmu.2022.856376

Calabresi, P., Picconi, B., Parnetti, L., & Di, F. M. (2006). A convergent model for cognitive dysfunctions in Parkinson's disease: the critical dopamine-acetylcholine synaptic balance. Lancet Neurology, 5, 974–983.

Castner, J. E., Chenery, H. J., Copland, D. A., Coyne, T. J., Sinclair, F., & Silburn, P. A. (2007). Semantic and affective priming as a function of stimulation of the subthalamic nucleus in Parkinson's disease. Brain, 130, 1395–1407.

Cenci, M. A., Skovgård, K., & Odin, P. (2022). Non-dopaminergic approaches to the treatment of motor complications in Parkinson's disease. Neuropharmacology, 210, 109027.

Chaudhuri, K. R., Rizos, A., Trenkwalder, C., Rascol, O., Pal, S., Martino, D., Carroll, C., Paviour, D., Falup-Pecurariu, C., Kessel, B., Silverdale, M., Todorova, A., Sauerbier, A., Odin, P., Antonini, A., Martinez-Martin, P., & EUROPAR and the IPMDS Non Motor PD Study Group. (2015). King's Parkinson's disease pain scale, the first scale for pain in PD: an international validation. Movement Disorders, 30(12), 1623–1631. https://doi.org/10.1002/mds.26270

Chen, J. F., Choi, D. S., & Cunha, R. A. (2023). Striatopallidal adenosine A2A receptor modulation of goal-directed behavior: homeostatic control with cognitive flexibility. Neuropharmacology, 226, 109421. https://doi.org/10.1016/j.neuropharm.2023.109421

Cocco, C., Manai, A. L., Manca, E., & Noli, B. (2023). Brain-biomarker changes in body fluids of patients with Parkinson's disease. International Journal of Molecular Sciences, 24(13), 10932. https://doi.org/10.3390/ijms241310932

Colla, E. (2019). Linking the endoplasmic reticulum to Parkinson's disease and alpha-synucleinopathy. Frontiers in Neuroscience, 13, 560.

de Natale, E. R., Wilson, H., Politis, M. (2021). Serotonergic imaging in Parkinson's disease. Progress in Brain Research, 261, 303–338.

Depboylu, C., Stricker, S., Ghobril, J.-P., Oertel, W. H., Priller, J., & Höglinger, G. U. (2012). Brain-resident microglia predominate over infiltrating myeloid cells in activation, phagocytosis and interaction with T-lymphocytes in the MPTP mouse model of Parkinson disease. Experimental Neurology, 238, 183–191.

Domenici, R. A., Campos, A. C. P., Maciel, S. T., Berzuino, M. B., Hernandes, M. S., Fonoff, E. T., & Pagano, R. L. (2019). Parkinson's disease and pain: modulation of nociceptive circuitry in a rat model of nigrostriatal lesion. Experimental Neurology, 315, 72–81. https://doi.org/10.1016/j.expneurol.2019.02.007

Donadio, V., Incensi, A., Leta, V., Giannoccaro, M. P., Scaglione, C., Martinelli, P., Capellari, S., Avoni, P., Baruzzi, A., & Liguori, R. (2014). Skin nerve α-synuclein deposits: a biomarker for idiopathic Parkinson disease. Neurology, 82(15), 1362–1369. https://doi.org/10.1212/WNL.0000000000000316

Emmi, A., Antonini, A., Sandre, M., Baldo, A., Contran, M., Macchi, V., Guidolin, D., Porzionato, A., & De Caro, R. (2022). Topography and distribution of adenosine A2A and dopamine D2 receptors in the human subthalamic nucleus. Frontiers in Neuroscience, 16, 945574. https://doi.org/10.3389/fnins.2022.945574

Eser, D., Schüle, C., Baghai, T. C., Romeo, E., Uzunov, D. P., & Rupprecht, R. (2006). Neuroactive steroids and affective disorders. Pharmacology, Biochemistry, and Behavior, 84, 656–666. https://doi.org/10.1016/j.pbb.2006.05.020

Factor, S. A., McDonald, W. M., & Goldstein, F. C. (2017). The role of neurotransmitters in the development of Parkinson's disease-related psychosis. European Journal of Neurology, 24(10), 1244–1254. https://doi.org/10.1111/ene.13376

Farahimanesh, S., Moradi, M., Nazari-Serenjeh, F., Zarrabian, S., & Haghparast, A. (2018). Role of D1-like and D2-like dopamine receptors within the ventral tegmental area in stress-induced and drug priming-induced reinstatement of morphine seeking in rats. Behavioural Pharmacology, 29(5), 426–436.

Fernández-Dueñas, V., Gómez-Soler, M., Valle-León, M., Watanabe, M., Ferrer, I., & Ciruela, F. (2019). Revealing adenosine A2A-dopamine D2 receptor heteromers in Parkinson's disease postmortem brain through a new AlphaScreen-based assay. International Journal of Molecular Sciences, 20(14), 3600. https://doi.org/10.3390/ijms20143600

Ferraro, L., Tomasini, M. C., Beggiato, S., Guerrini, R., Salvadori, S., Fuxe, K., Calzà, L., Tanganelli, S., & Antonelli, T. (2009). Emerging evidence for neurotensin receptor 1 antagonists as novel pharmaceutics in neurodegenerative disorders. Mini-Reviews in Medicinal Chemistry, 9, 1429–1438.

Ferré, S., & Ciruela, F. (2019). Functional and neuroprotective role of striatal adenosine A2A receptor heterotetramers. Journal of Caffeine and Adenosine Research, 9(3), 89–97.

Ferré, S., Sarasola, L. I., Quiroz, C., & Ciruela, F. (2023). Presynaptic adenosine receptor heteromers as key modulators of glutamatergic and dopaminergic neurotransmission in the striatum. Neuropharmacology, 223, 109329.

Firbank, M. J., Parikh, J., Murphy, N., Killen, A., Allan, C. L., Collerton, D., Blamire, A. M., & Taylor, J. P. (2018). Reduced occipital GABA in Parkinson disease with visual hallucinations. Neurology, 91(7), e675–e685. https://doi.org/10.1212/WNL.0000000000006007

Fredholm, B. B. (2007). Adenosine, an endogenous distress signal, modulates tissue damage and repair. Cell Death & Differentiation, 14(7), 1315–1323.

Fredholm, B. B., Irenius, E., Kull, B., & Schulte, G. (2001). Comparison of the potency of adenosine as an agonist at human adenosine receptors expressed in Chinese hamster ovary cells. Biochemical Pharmacology, 61(4), 443–448.

Friedman, J. H., & Akbar, U. (2016). Psychosis in Parkinson's disease: unexplained observations in a seemingly simple model. Expert Review of Neurotherapeutics, 16, 595–596.

Gerfen, C. R., Keefe, K. A., & Gauda, E. B. (1995). D1 and D2 dopamine receptor function in the striatum: coactivation of D1- and D2-dopamine receptors on separate populations of neurons results in potentiated immediate early gene response in D1-containing neurons. Journal of Neuroscience, 15, 8167–8176.

Giladi, N., & Hausdorff, J. M. (2006). The role of mental function in the pathogenesis of freezing of gait in Parkinson's disease. Journal of Neurological Science, 248, 173–176.

Girault, J. A., & Greengard, P. (2004). The neurobiology of dopamine signaling. Archives of Neurology, 61(5), 641–644. doi: 10.1001/archneur.61.5.641. Erratum in: Archives of Neurology, 61(8), 1180. PMID: 15148138.

Goldstein, D. S., Holmes, C., & Sharabi, Y. (2012). Cerebrospinal fluid biomarkers of central catecholamine deficiency in Parkinson's disease and other synucleinopathies. Brain, 135(Pt 6), 1900–1913. https://doi.org/10.1093/brain/aws055

Gonçalves, F. Q., Matheus, F. C., Silva, H. B., Real, J. I., Rial, D., Rodrigues, R. J., Oses, J. P., Silva, A. C., Gonçalves, N., Prediger, R. D., Tomé, Â. R., & Cunha, R. A. (2023). Increased ATP release and higher impact of adenosine A_{2A} receptors on corticostriatal plasticity in a rat model of presymptomatic Parkinson's disease. Molecular Neurobiology, 60(3), 1659–1674. https://doi.org/10.1007/s12035-022-03162-1

Graybiel, A. M. (1991). Basal ganglia--input, neural activity, and relation to the cortex. Current Opinion in Neurobiology, 1(4), 644–651. https://doi.org/10.1016/s0959-4388(05)80043-1

Gurevich, E. V., Gainetdinov, R. R., & Gurevich, V. V. (2016). G protein-coupled receptor kinases as regulators of dopamine receptor functions. Pharmacological Research, 111, 1–16. https://doi.org/10.1016/j.phrs.2016.05.010

Hallett, P. J., Spoelgen, R., Hyman, B. T., Standaert, D. G., & Dunah, A. W. (2006). Dopamine D1 activation potentiates striatal NMDA receptors by tyrosine phosphorylation-dependent subunit trafficking. Journal of Neuroscience 26, 4690–4700. https://doi.org/10.1523/JNEUROSCI.0792-06.2006

Halliday, G. M., & Stevens, C. H. (2011). Glia: initiators and progressors of pathology in Parkinson's disease. Movement Disorders, 26(1), 6–17.

Hamani, C., Dostrovsky, J. O., & Lozano, A. M. (2006). The motor thalamus in neurosurgery. Neurosurgery, 58(1), 146–158. https://doi.org/10.1227/01.NEU.0000192166.62017.C1

Hammond, C., Bergman, H., & Brown, P. (2007). Pathological synchronization in Parkinson's disease: networks, models and treatments. Trends in Neurosciences, 30, 357–364.

Hansen, A. M., & Caspi, R. R. (2010). Glutamate joins the ranks of immunomodulators. Nature Medicine, 16(8), 856–858.

Haque, M. E., Akther, M., Jakaria, M., Kim, I., Azam, S., & Choi, D.-K. (2020). Targeting the microglial NLRP3 inflammasome and its role in Parkinson's disease. Movement Disorders, 35 (1), 20–33.

Hornung, J. P. (2003). The human raphe nuclei and the serotonergic system. Journal of Chemical Neuroanatomy, 26(4), 331–343.

Hoyer, D. (2020). Targeting the 5-HT system: potential side effects. Neuropharmacology, 179, 108233.

Iranzo, A., Stefani, A., Serradell, M., Marti, M. J., Lomena, F., Mahlknecht, P., Stockner, H., Gaig, C., Fernández-Arcos, A., Poewe, W., Tolosa, E., Högl, B., & Santamaria, J. (2017). Characterization of patients with longstanding idiopathic REM sleep behavior disorder. Neurology, 89(3), 242–248. https://doi.org/10.1212/wnl.0000000000004121

Jenner, P., Mori, A., Hauser, R., Morelli, M., Fredholm, B. B., & Chen, J. F. (2009). Adenosine, adenosine A 2A antagonists, and Parkinson's disease. Parkinsonism & Related Disorders, 15(6), 406–413. https://doi.org/10.1016/j.parkreldis.2008.12.006

Jinsmaa, Y., Sharabi, Y., Sullivan, P., Isonaka, R., & Goldstein, D. S. (2018). 3,4-Dihydroxyphenylacetaldehyde-induced protein modifications and their mitigation by N-acetylcysteine. Journal of Pharmacology and Experimental Therapeutics, 366, 113–124.

Kalia, L. V. (2018). Biomarkers for cognitive dysfunction in Parkinson's disease. Parkinsonism & Related Disorders, 46(Suppl 1), S19–S23. https://doi.org/10.1016/j.parkreldis.2017.07.023

Katunina, E. A., Blokhin, V., Nodel, M. R., Pavlova, E. N., Kalinkin, A. L., Kucheryanu, V. G., Alekperova, L., Selikhova, M. V., Martynov, M. Y., & Ugrumov, M. V. (2023). Searching for biomarkers in the blood of patients at risk of developing Parkinson's disease at the prodromal stage. International Journal of Molecular Sciences, 24(3), 1842. https://doi.org/10.3390/ijms24031842

Kehagia, A. A., Barker, R. A., & Robbins, T. W. (2013). Cognitive impairment in Parkinson's disease: the dual syndrome hypothesis. Neurodegenerative Disease, 11(2), 79–92.

Khairnar, R. C., Parihar, N., Prabhavalkar, K. S., & Bhatt, L. K. (2022). Emerging targets signaling for inflammation in Parkinson's disease drug discovery. Metabolic Brain Disease, 37, 2143–2161.

Koerts, J., Leenders, K. L., Koning, M., Portman, A. T., & Van Beilen, M. (2007). Striatal dopaminergic activity (FDOPA-PET) associated with cognitive items of a depression scale (MADRS) in Parkinson's disease. European Journal of Neuroscience, 25, 3132–3136.

Köfalvi, A., Moreno, E., Cordomí, A., Cai, N. S., Fernández-Dueñas, V., Ferreira, S. G., Guixà-González, R., Sánchez-Soto, M., Yano, H., Casadó-Anguera, V., Cunha, R. A., Sebastião, A. M., Ciruela, F., Pardo, L., Casadó, V., & Ferré, S. (2020). Control of glutamate release by complexes of adenosine and cannabinoid receptors. BMC Biology, 18(1), 9.

Koza, Z., Ayajuddin, M., Das, A., Chaurasia, R., Phom, L., & Yenisetti, S. C. (2023). Sexual dysfunction precedes motor defects, dopaminergic neuronal degeneration, and impaired dopamine metabolism: insights from Drosophila model of Parkinson's disease. Frontiers in Neuroscience, 17, 1143793. https://doi.org/10.3389/fnins.2023.1143793

Kreitzer, A. C. (2009). Physiology and pharmacology of striatal neurons. Annual Review of Neuroscience, 32, 127–147. https://doi.org/10.1146/annurev.neuro.051508.135422

Laansma, M. A., Bright, J. K., Al-Bachari, S., Anderson, T. J., Ard, T., Assogna, F., Baquero, K. A., Berendse, H. W., Blair, J., Cendes, F., Dalrymple-Alford, J. C., de Bie, R. M. A., Debove, I., Dirkx, M. F., Druzgal, J., Emsley, H. C. A., Garraux, G., Guimarães, R. P., Gutman, B. A., . . . ENIGMA-Parkinson's Study. (2021). International multicenter analysis of brain structure across clinical stages of Parkinson's Disease. Movement Disorders, 36(11), 2583–2594. https://doi.org/10.1002/mds.28706

Lees, A. J., Hardy, J., & Revesz, T. (2009). Parkinson's disease. Lancet, 373, 2055–2066.

Levite, M. (2000). Nerve-driven immunity. The direct effects of neurotransmitters on T-cell function. Annals of the New York Academy of Sciences, 917, 307–321. https://doi.org/10.1111/j.1749-6632.2000.tb05397.x

Liang, K., Li, X., Ma, J., Yang, H., Shi, X., Fan, Y., Yang, D., Guo, D., Liu, C., Dong, L., Chang, Q., Gu, Q., Chen, S., & Li, D. (2023). Predictors of dopamine dysregulation syndrome in patients with early Parkinson's disease. Neurological Sciences, 44(12), 4333–4342. https://doi.org/10.1007/s10072-023-06956-w

Liddelow, S. A., & Barres, B. A. (2017). Reactive astrocytes: production, function, and therapeutic potential. Immunity, 46(6), 957–967.

Lin, C. H., Wu, R. M., Chang, H. Y., Chiang, Y. T., & Lin, H. H. (2013). Preceding pain symptoms and Parkinson's disease: a nationwide population-based cohort study. European Journal of Neurology, 20(10), 1398–1404. https://doi.org/10.1111/ene.12197

Ljungdahl, A., Hanrieder, J., Fälth, M., Bergquist, J., & Andersson, M. (2011). Imaging mass spectrometry reveals elevated nigral levels of dynorphin neuropeptides in L-DOPA-induced dyskinesia in rat model of Parkinson's disease. PLoS One, 6, e25653.

Lozza, C., Baron, J. C., Eidelberg, D., Mentis, M. J., Carbon, M., & Marie, R. M. (2004). Executive processes in Parkinson's disease: FDG-PET and network analysis. Human Brain Mapping, 22, 236–245.

Masilamoni, G. J., Bogenpohl, J. W., Alagille, D., Delevich, K., Tamagnan, G., Votaw, J. R., Wichmann, T., & Smith, Y. (2011). Metabotropic glutamate receptor 5 antagonist protects dopaminergic and noradrenergic neurons from degeneration in MPTP-treated monkeys. Brain, 134, 2057–2073.

Meles, S. K., Oertel, W. H., & Leenders, K. L. (2021). Circuit imaging biomarkers in preclinical and prodromal Parkinson's Disease. Molecular Medicine, 27, 111.

Mertsalmi, T. H., Aho, V. T. E., Pereira, P. A. B., Paulin, L., Pekkonen, E., Auvinen, P., & Scheperjans, F. (2017). More than constipation—bowel symptoms in Parkinson's disease and their connection to gut microbiota. European Journal of Neurology, 24(11), 1375–1383. https://doi.org/10.1111/ene.13398

Miguelez, C., Morera-Herreras, T., Torrecilla, M., Ruiz-Ortega, J. A., & Ugedo, L. (2014). Interaction between the 5-HT system and the basal ganglia: functional implication and therapeutic perspective in Parkinson's disease. Frontiers in Neural Circuits, 8, 21.

Mor, D. E., Daniels, M. J., & Ischiropoulos, H. (2019). The usual suspects, dopamine and alpha-synuclein, conspire to cause neurodegeneration. Movement Disorders, 34, 167–179.

Müller, M. L., & Bohnen, N. I. (2013). Cholinergic dysfunction in Parkinson's disease. Current Neurology and Neuroscience Reports, 13(9), 377. https://doi.org/10.1007/s11910-013-0377-9

Murtomäki, K., Joutsa, J., Mertsalmi, T., Jaakkola, E., Mäkinen, E., Levo, R., Eklund, M., Nuuttila, S., Pekkonen, E., Noponen, T., Ihalainen, T., Kaasinen, V., & Scheperjans, F. (2023). Dopamine transporter binding in the brain is linked to irritable bowel syndrome in Parkinson's disease. Brain and Behavior, 13(7), e3097. https://doi.org/10.1002/brb3.3097

Murueta-Goyena, A., Andikoetxea, A., Gómez-Esteban, J. C., & Gabilondo, I. (2019). Contribution of the GABAergic system to non-motor manifestations in premotor and early stages of Parkinson's disease. Frontiers in Pharmacology, 10, 1294. https://doi.org/10.3389/fphar.2019.01294

Mylius, V., Ciampi de Andrade, D., Cury, R. G., Teepker, M., Ehrt, U., Eggert, K. M., Beer, S., Kesselring, J., Stamelou, M., Oertel, W. H., Möller, J. C., & Lefaucheur, J. P. (2015). Pain in Parkinson's disease: current concepts and a new diagnostic algorithm. Movement Disorders Clinical Practice, 2(4), 357–364. https://doi.org/10.1002/mdc3.12217

Nguyen-Legros, J. (1988). Functional neuroarchitecture of the retina: hypothesis on the dysfunction of retinal dopaminergic circuitry in Parkinson's disease. Surgical and Radiologic Anatomy, 10(2), 137–144. https://doi.org/10.1007/BF02307822

Nilsson, A., Fälth, M., Zhang, X., Kultima, K., Sköld, K., Svenningsson, P., & Andrén, P. E. (2009). Striatal alterations of secretogranin-1, somatostatin, prodynorphin, and cholecystokinin peptides in an experimental mouse model of Parkinson disease. Molecular & Cellular Proteomics, 8, 1094–1104.

Nobis, L., Maio, M. R., Saleh, Y., Manohar, S., Kienast, A., McGann, E., & Husain, M. (2023). Role of serotonin in modulation of decision-making in Parkinson's disease. Journal of Psychopharmacology, 37(4), 420–431.

O'Doherty, J., Critchley, H., Deichmann, R., & Dolan, R. J. (2003). Dissociating valence of outcome from behavioral control in human orbital and ventral prefrontal cortices. Journal of Neuroscience, 23, 7931–7939.

O'Gorman Tuura, R. L., Baumann, C. R., & Baumann-Vogel, H. (2018). Beyond dopamine: GABA, glutamate, and the axial symptoms of Parkinson disease. Frontiers in Neurology, 9, 806. https://doi.org/10.3389/fneur.2018.00806

Oshaghi, M., Kourosh-Arami, M., & Roozbehkia, M. (2023). Role of neurotransmitters in immune-mediated inflammatory disorders: a crosstalk between the nervous and immune systems. Neurological Sciences, 44, 99–113.

Park, J. S., Park, D., Ko, P. W., Kang, K., & Lee, H. W. (2017). Serum methylmalonic acid correlates with neuropathic pain in idiopathic Parkinson's disease. Neurological Sciences, 38(10), 1799–1804. https://doi.org/10.1007/s10072-017-3056-9

Pasquini, J., Ceravolo, R., Qamhawi, Z., Lee, J. Y., Deuschl, G., Brooks, D. J., Bonuccelli, U., & Pavese, N. (2018). Progression of tremor in early stages of Parkinson's disease: a clinical and neuroimaging study. Brain, 141(3), 811–821.

Pautrat, A., Al Tannir, R., Pernet-Gallay, K., Soutrenon, R., Vendramini, E., Sinniger, V., Overton, P. G., David, O., & Coizet, V. (2023). Altered parabrachial nucleus nociceptive processing may underlie central pain in Parkinson's disease. NPJ Parkinson's Disease, 9(1), 78. https://doi.org/10.1038/s41531-023-00516-x

Pérez, L. M., Farriols, C., Puente, V., Planas, J., & Ruiz, I. (2011). The use of subcutaneous scopolamine as a palliative treatment in Parkinson's disease. Palliative Medicine, 25, 92–93.

Petrou, M., Kotagal, V., & Bohnen, N. I. (2012). An update on brain imaging in Parkinsonian dementia. Imaging in Medicine, 4(2), 201–213. https://doi.org/10.2217/iim.12.10

Pfeiffer, R. F. (2018). Gastrointestinal dysfunction in Parkinson's Disease. Current Treatment Options in Neurology, 20(12), 54. https://doi.org/10.1007/s11940-018-0539-9

Pinna, A., Costa, G., Serra, M., Contu, L., & Morelli, M. (2021). Neuroinflammation and Ldopa-induced abnormal involuntary movements in 6-hydroxydopamine-lesioned rat model of Parkinson's disease are counteracted by combined administration of a 5-HT1A/1B receptor agonist and A2A receptor antagonist. Neuropharmacology, 196, 108693.

Purves, D., Augustine, G. J., Fitzpatrick, D., Hall, W. C., LaMantia, A.-S. Mooney, R. D., Platt, M. L., & White, L. E. (2018). Neuroscience. 6th Edition, Sinauer Associates, New York.

Raina, A., Leite, K., Guerin, S., Mahajani, S. U., Chakrabarti, K. S., Voll, D., Becker, S., Griesinger, C., Bähr, M., & Kügler, S. (2021). Dopamine promotes the neurodegenerative potential of β-synuclein. Journal of Neurochemistry, 156(5), 674–691. https://doi.org/10.1111/jnc.15134

Ratner, M. H., Kumaresan, V., & Farb, D. H. (2019). Neurosteroid actions in memory and neurologic/neuropsychiatric disorders. Frontiers in Endocrinology, 10, 169. https://doi.org/10.3389/fendo.2019.00169

Riese, F., Gietl, A., Zolch, N., Henning, A., O'Gorman, R., Kalin, A. M., Leh, S. E., Buck, A., Warnock, G., Edden, R. A., Luechinger, R., Hock, C., Kollias, S., & Michels, L. (2015). Posterior cingulate gamma-aminobutyric acid and glutamate/glutamine are reduced in amnestic mild cognitive impairment and are unrelated to amyloid deposition and apolipoprotein E genotype. Neurobiology of Aging, 36, 53–59. https://doi.org/10.1016/j.neurobiolaging.2014.07.030

Rizzi, G., & Tan, K. R. (2017). Dopamine and acetylcholine, a circuit point of view in Parkinson's disease. Frontiers in Neural Circuits, 11, 110. https://doi.org/10.3389/fncir.2017.00110

Roussakis, A. A., Politis, M., Towey, D., & Piccini, P. (2016). Serotonin-to-dopamine transporter ratios in Parkinson disease: relevance for dyskinesias. Neurology, 86(12), 1152–1158. https://doi.org/10.1212/WNL.0000000000002494

Sanchez-Catasus, C. A., Bohnen, N. I., D'Cruz, N., & Müller, M. L. T. M. (2022). Striatal acetylcholine-dopamine imbalance in Parkinson disease: in vivo neuroimaging study with dual-tracer PET and dopaminergic PET-informed correlational tractography. Journal of Nuclear Medicine, 63(3), 438–445. https://doi.org/10.2967/jnumed.121.261939

Santpere, G., Garcia-Esparcia, P., Andres-Benito, P., Lorente-Galdos, B., Navarro, A., & Ferrer, I. (2018). Transcriptional network analysis in frontal cortex in Lewy body diseases with focus on dementia with Lewy bodies. Brain Pathology, 28(3), 315–333. https://doi.org/10.1111/bpa.12511

Sathe, K., Maetzler, W., Lang, J. D., Mounsey, R. B., Fleckenstein, C., Martin, H. L., Schulte, C., Mustafa, S., Synofzik, M., Vukovic, Z., Itohara, S., Berg, D., & Teismann, P. (2012). S100B is increased in Parkinson's disease and ablation protects against MPTP-induced toxicity through the RAGE and TNF-α pathway. Brain, 135(Pt 11), 3336–3347.

Schapira, A. H. V., Chaudhuri, K. R., & Jenner, P. (2017). Non-motor features of Parkinson disease. Nature Reviews Neuroscience, 18(7), 435–450.

Traynelis, S. F., Wollmuth, L. P., McBain, C. J., Menniti, F. S., Vance, K. M., Ogden, K. K., Hansen, K. B., Yuan, H., Myers, S. J., & Dingledine, R. (2010). Glutamate receptor ion channels: structure, regulation, and function. Pharmacological Reviews, 62(3), 405–496. https://doi.org/10.1124/pr.109.002451

Trujillo, P., Song, A. K., Hay, K. R., Aumann, M., Yan, Y., Kang, H., Donahue, M. J., & Claassen, D. O. (2022). Dopamine-induced changes to thalamic GABA concentration in impulsive Parkinson disease patients. npj Parkinsons Disease, 8, 37.

van Laar, T., De Deyn, P. P., Aarsland, D., Barone, P., & Galvin, J. E. (2010). Effects of cholinesterase inhibitors in Parkinson's disease dementia: a review of clinical data. CNS Neuroscience & Therapeutics, 17, 428–441. https://doi.org/10.1111/j.1755-5949.2010.00166.x

van Nuland, A. J. M., den Ouden, H. E. M., Zach, H., Dirkx, M. F. M., van Asten, J. J. A., Scheenen, T. W. J., Toni, I., Cools, R., & Helmich, R. C. (2020). GABAergic changes in the thalamocortical circuit in Parkinson's disease. Human Brain Mapping, 41(4), 1017–1029. https://doi.org/10.1002/hbm.24857

Vu, T. C., Nutt, J. G., & Holford, N. H. (2012). Progression of motor and nonmotor features of Parkinson's disease and their response to treatment. British Journal of Clinical Pharmacology, 74, 267–283.

Waggan, I., Rissanen, E., Tuisku, J., Joutsa, J., Helin, S., Parkkola, R., Rinne, J. O., & Airas, L. (2023). Adenosine A2A receptor availability in patients with early- and moderate-stage Parkinson's disease. Journals of Neurology, 270(1), 300–310. https://doi.org/10.1007/s00415-022-11342-1

Wang, L., Gao, Z., Chen, G., Geng, D., & Gao, D. (2023). Low levels of adenosine and GDNF are potential risk factors for Parkinson's disease with sleep disorders. Brain Science, 13(2), 200. doi: 10.3390/brainsci13020200. PMID: 36831743; PMCID: PMC9953846.

Weil, R. S., Schrag, A. E., Warren, J. D., Crutch, S. J., Lees, A. J., & Morris, H. R. (2016). Visual dysfunction in Parkinson's disease. Brain, 139(11), 2827–2843. https://doi.org/10.1093/brain/aww175

Weintraub, D., David, A. S., Evans, A. H., Grant, J. E., & Stacy, M. (2015). Clinical spectrum of impulse control disorders in Parkinson's disease. Movement Disorders, 30, 121–127.

Weiss, F., Labrador-Garrido, A., Dzamko, N., & Halliday, G. (2022). Immune responses in the Parkinson's disease brain. Neurobiology of Disease, 168, 105700. https://doi.org/10.1016/j.nbd.2022.105700

Werner, F. M., & Coveñas, R. (2013). Classical neurotransmitters and neuropeptides involved in major depression in a multi-neurotransmitter system: A focus on antidepressant drugs. Current Medicinal Chemistry, 20(38), 4853–4858.

Wiatrak, B., Balon, K., Jawień, P., Bednarz, D., Jęśkowiak, I., & Szeląg, A. (2022). The role of the microbiota-gut–brain axis in the development of Alzheimer's disease. International Journal of Molecular Sciences, 23(9), 4862.

Williams-Gray, C. H., Foltynie, T., Lewis, S. J., & Barker, R. A. (2006). Cognitive deficits and psychosis in Parkinson's disease: a review of pathophysiology and therapeutic options. CNS Drugs, 20, 477–505.

Witkovsky, P. (2004). Dopamine and retinal function. Documenta Ophthalmologica, 108(1), 17–39.

Wong, J. C., Li, Y., Schwarzschild, M. A., Ascherio, A., & Gao, X. (2014). Restless legs syndrome: an early clinical feature of Parkinson disease in men. Sleep, 37(2), 369–372. https://doi.org/10.5665/sleep.3416

Yang, P., Perlmutter, J. S., Benzinger, T. L. S., Morris, J. C., & Xu, J. (2020). Dopamine D3 receptor: a neglected participant in Parkinson disease pathogenesis and treatment?. Ageing Research Reviews, 57, 100994. https://doi.org/10.1016/j.arr.2019.100994

Yao, W.-D., Spealman, R. D., & Zhang, J. (2008). Dopaminergic signaling in dendritic spines. Biochemical Pharmacology, 75, 2055–2069. https://doi.org/10.1016/j.bcp.2008.01.018

Yarnall, A., Rochester, L., & Burn, D. J. (2011). The interplay of cholinergic function, attention, and falls in Parkinson's disease. Movement Disorders, 26, 2496–2503.

Yun, S. P., Kam, T. I., Panicker, N., Kim, S., Oh, Y., Park, J. S., Kwon, S. H., Park, Y. J., Karuppagounder, S. S., Park, H., Kim, S., Oh, N., Kim, N. A., Lee, S., Brahmachari, S., Mao, X., Lee, J. H., Kumar, M., An, D., . . . Ko, H. S. (2018). Block of A1 astrocyte conversion by microglia is neuroprotective in models of Parkinson's disease. Nature Medicine, 24(7), 931–938.

Zahodne, L. B., & Fernandez, H. H. (2008). Pathophysiology and treatment of psychosis in Parkinson's disease. Drugs Aging, 25, 665–682.

Zimmermann, H., Zebisch, M., & Sträter, N. (2012). Cellular function and molecular structure of ecto-nucleotidases. Purinergic Signalling, 8, 437–502.

11 Diagnosis and Biomarkers of Parkinson's Disease

Abraham Olufemi Asuku, Maryam Tayo Ayinla, Tobiloba Samuel Olajide, Ebele Joan Ajagun, Kaosara Oyinola Ganiyu, Yusuf Olamilekan Hamza, Ayodeji Johnson Ajibare, and Taoheed Adedeji Abdulsalam

11.1 INTRODUCTION

Parkinson's disease (PD) comes after Alzheimer's disease (AD) in terms of prevalence among the disorders that cause decline and deterioration of the neurons (Poewe *et al.*, 2017), demonstrating an occurrence that grows with age and is more common in the male gender. PD is a neurodegenerative disorder that affects a little over 1% of all persons over the age of 55 (Noelker *et al.*, 2011). Motor deficits like bradykinesia, rigidity of the muscles, imbalance posture, and resting tremor are the primary signs of PD (Jankovic, 2008; Kouli *et al.*, 2018). Moreover, some symptoms that are non-motor are preclinically diagnosed before the commencement of the motor symptoms; these are sleep disorders, hallucinations, autonomic dysfunction, olfactory dysfunction, dementia, and depression (Greenland and Barker, 2018). The pathological characteristics of PD include the deterioration and decline of dopamine-producing neurons in a particular part of the basal ganglia of the brain called the pars compacta of the substantia nigra. It is believed that in mentally sound individuals, around 3–5% of dopaminergic (DA) neurons are deteriorated every decade, but in PD patients, the rate of deterioration is greater than that found in a mentally healthy individual. Research autopsy samples of PD patients' brains indicated that about 70–75% of DA neurons are damaged at the time the disease becomes manifest (Dauer and Przedborski, 2003). This reveals that DA neurons have a high degree of functional flexibility, i.e., plasticity. Aside from the decline and deterioration of these neurons, there is the development of neuronal proteinopathy inclusions dubbed Lewy bodies, made up of the aggregation of alpha-synuclein (Stefanis, 2012; Yasuda *et al.*, 2013). Hence, PD has been related with a sort of dementia dubbed Lewy body dementia. Over the years, PD has been seen as asymptomatic in its early phases of development.

In this review, for improved understanding of PD biomarkers and its diagnosis, the postulated aetiology of PD needs to be developed. In terms of aetiology, there are two forms of PD: Firstly, in around 70% of situations, the aggregations of abnormal alpha-synuclein proteins have been seen to be created in the brain. Secondly, for the other 30% of situations, this proteinopathy develops from the neurological system of the intestines—the enteric nerve system (Schaffrath *et al.*, 2023). In a review, intestinal dysbiosis in the gut may be the first stage of an event that is responsible for the development of Parkinson's disease because of changes in the gastrointestinal micro-organisms, as well as their products of metabolism that give rise to inflammation of the intestine. These changes may well affect how the brain barrier and the gut–brain communicate, which may give rise to inflammation of the brain and degeneration of neurons known as neurodegeneration (Xiromerisiou *et al.*, 2023). Following the Gut–Brain Axis hypothesis propounded by Braak and colleagues in 2003, further findings have been shown to corroborate the idea by the prion-like mechanism of alpha-synuclein in allowing them to increase and proliferate among brain cells in the nervous system and multiply through spreading great distances from the gastrointestinal tract as well as the brain through the vagus nerves (Hawkes *et al.*, 2009; Arotcarena *et al.*, 2020).

Nevertheless, a clinical trial investigation conducted lately shows that the aggregates of alpha-synuclein may be discharged gradually into stool using a fluorescence intensity distribution analysis assay that is surface-based in PD patients, which could support the detection of prodromal synucleinopathies in persons suffering from PD originating from the enteric nervous system (Schaffrath *et al.*, 2023). For many nations, PD is now a significant worry. Early diagnosis is essential for ensuring effective treatment and illness prevention. Although PD's management is symptomatic because clinical symptoms do not manifest until a significant number of dopamine-producing neurons have been lost, neuroprotective medications and putative disease-modifying agents would be of limited use at a late stage of the illness. Since it is difficult to find and develop neuroprotective and disease-modifying medications, Parkinsonism is both a socioeconomic issue and a major source of concern for the health conditions of many people. As a result, novel studies on specific and sensitive biomarkers for PD are required to facilitate early illness diagnosis, monitor disease development, as well as assess response to the existing and emerging medications (Schlossmacher and Mollenhauer, 2010).

Currently, the fundamental goal of most clinical research for PD is to increase early identification of disease and

DOI: 10.1201/9781003402893-13

pre-symptomatic detection of neuronal degeneration. Though we also need new tools to detect PD progression, over the years, various prospective disease-modifying drugs have been found and examined at its preclinical stage, which might lead to clinical trials in the near future for which biomarkers are critically necessary. Biomarkers are crucial signs of normal and abnormal biological processes. Specific abnormalities in pathology, biochemistry, and genetics in the human body may provide us with specific information about the nature of each unique sickness. Furthermore, biomarkers should not only be employed to assist in anticipating the beginning of the disorder but also to assist in monitoring the rate of their progress. Biomarkers are analysed by a range of various ways and are represented in varied categories, including genetic, biochemical, and imaging-based biomarkers. This study covers the many classes of biomarkers in people suffering from Parkinsonism with a focus on clinical non-movement-related symptoms, biofluid-based biomarkers, and neuroimaging biomarkers for PD, as this can further assist and reveal potential therapeutic methods at the initial phase of the disorder.

11.2 EPIDEMIOLOGY OF PARKINSON'S DISEASE

A thorough study of the Global Burden of Disease Study 2016 that was published in the *Lancet Neurology* in 2018 is one of the most significant and current studies on Parkinson's epidemiology. According to this study, there are between 41 and 100,000 patients with PD who are 50 years old and older, and there are up to 1,900 patients per 100,000 people who are 80 years of age or older. This shows that, like other neurodegenerative illnesses associated with aging, PD's incidence normally rises with age. Regardless of age, 4% or so of people have PD diagnoses before the age of 50. The worldwide occurrences of PD increased from over 2 million in the year 1990 to 5 million in the year 2016, with varying distributions in different nations. The increased prevalence of PD over time in the general population cannot solely be attributed to the rise in the number of elderly people in various nations during this time (GBD 2016 Neurology Collaborators, 2019).

Studies have indicated a varying increasing level of PD when taking into account gender and age categories. Various findings demonstrated that PD is less common in the female population in comparison to the male population; however, the reason for this disparity is unknown, given that the aetiology of PD is not well understood and that the disorder is a result of the interaction of environmental factors and some kind of genetic predisposition. Having known that environmental variables have a bigger influence, it is then reinforced by the fact that males are more typically employed in locations with toxins exposure (Kouli *et al.*, 2018). A range of recognised modified predisposing factors are available for PD, with varied rates of influence (Rajan and Kaas, 2022). Geographically, some of these studies have been disputed, which might be owing to incorrect sample choices, a myopic view of various data-gathering techniques as well as analysis, and varying life expectancies in numerous communities. Alternatively, these disparities may originate from the variability in environmental risk factors or genetic factors for PD occurrence in various geographical locations. Furthermore, considering those predisposing factors, the influence of disorders of the heart and metabolic activities on PD risk is yet to be thoroughly confirmed, while accumulating data in the Korean population has demonstrated that cardiometabolic syndrome is strongly connected to PD (Kang *et al.*, 2023).

11.3 MEANING OF BIOMARKERS

The National Institutes of Health (NIH) noted in 1998 that a "biomarker is a characteristic that is objectively measured and evaluated as an indicator of normal biological processes, pathogenic processes, or pharmacological responses to a therapeutic intervention." A biomarker is "Any substance, structure, or process that can be quantified in the body or its products and predict or influence the occurrence of disease or outcome," according to the International Programme on the Safety of Chemicals, which was championed by the World Health Organization in 2017 and in conjunction with the United Nations and the International Labour Organization. According to the functions they perform, biomarkers can be divided into three categories: (1) diagnostic biomarkers are utilised to confirm the presence of PD; (2) susceptibility risk biomarkers represent the potential for having PD (Parnetti *et al.*, 2019); and (3) prognostic biomarkers indicate the changes as a result of medication, progression of the diseases, or recurrence of the disorder (Perlis, 2011).

11.4 CLASSIFICATION OF BIOMARKERS

According to the Food and Drug Administration (FDA) of the United States, biomarkers have been classified based on two basic criteria, which are dependent on their "Application" or "Characteristics." Biomarkers may serve as several objectives in research and clinical practises, and they are generally categorised according to their intended use, i.e., application. These are the diagnostic, prognostic, and progressive biomarkers. Diagnostic biomarkers are the biomarkers that define patients as presented with or not presented with a given disease, or more broadly, measure how severe a disease is. Prognostic biomarkers involve time and forecast the possibility of the occurrence of the disease in the future. A progressive biomarker records the change of a disorder's severity or status through time. Even though the same biomarkers may be beneficial for many various purposes, assessing performance for each sort of use frequently necessitates utilising a new study strategy, data format, and technique. For instance, it is usually possible to evaluate the performance of diagnostic markers utilising one measurement at a time point, but on the other hand, prognostic marker assessment usually requires at least one measurement of response that are related to the disease sometimes after the marker has been measured, and progression markers usually involved many measurements of the marker and a response over time (Tevak *et al.*, 2010).

Based on their characteristics, biomarkers for neurodegenerative diseases could be divided into imaging biomarkers, which include magnetic resonance imaging and positron emission tomography, or molecular biomarkers: biopsy biomarkers, bodily fluid, and volatile breath (Subramaniam *et al.*, 2018). The term "molecular biomarkers" refers to biomarkers that are non-imaging that can be quantified in living samples (such as cerebrospinal fluid, plasma, biopsy, serum, as well as bronchoalveolar lavage) due to their biological and physical characteristics. These consist of biomarkers that are nucleic acid-based like polymorphisms, or mutations of genes, analysis of quantitative gene expression, lipid metabolites, proteins, peptides, and many other micromolecules.

11.5 BIOMARKERS OF PARKINSONISM

Parkinsonism is recognized as a complex and deteriorating disorder of the neurons, which also has to do with the gradual deterioration of voluntary motor function. Hence, the crucial and cardinal indicators of PD are the massive continuous loss of dopamine-producing neurons within the substantia nigra pars compacta (SNpc) of the basal ganglia, thus reflecting to a deficit of striatal dopaminergic, which underlies the motor symptoms that feature this disease (Kalia and Lang, 2015). The late detection of this progressive condition hindered the provision of an effective management strategy (Frisardi *et al.*, 2016; Mazzoni *et al.*, 2012).

Clinical practitioners rely on a mix of medical history, clinical examination, and certain biomarkers to reach an accurate diagnosis of PD, as there is no specific diagnostic test for it. PD currently lacks a single ambiguous diagnostic biomarker. However, scientists have been looking into a number biomarkers that might assist in diagnosing Parkinson's disease or differentiate it from other diseases. This review goes into further detail on PD diagnostic biomarkers. Among these biomarkers are the following.

11.5.1 Clinical Biomarkers (Non-Motor Symptoms)

Over the years, PD's diagnosis has primarily depended on clinical motor symptoms. Though these symptoms are not prodromal of PD, they are generally discovered in the terminal phase of the disorder (Mazzoni *et al.*, 2012). Recently, it has been commonly recognised that before classical movement-related symptoms develop, non-movement-related symptoms may already exist. Symptoms that are non-movement or non-motor such as constipation, rapid eye movement sleep behaviour disorder (RBD), olfactory dysfunction, and problems with mood have all been recognised as possible indicators in the prodromal PD's identification (Fullard *et al.*, 2017). Hence, aside from the well-known trio of movement-related symptoms in PD, significant improvements in research have addressed looking into the triad of

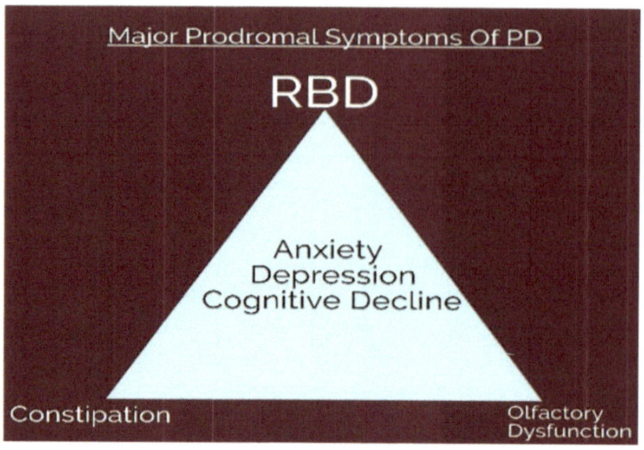

FIGURE 11.1 The figure shows the non-movement-related prodromal symptoms connected to PD, with the three most common non-movement-related clinical predictors of PD angularly represented.

Note: **(*RBD: Rapid eye movement sleep behaviour disorder, constipation and olfactory dysfunctions)**

non-motor clinical symptoms, notably constipation, problem with a sense of smell, and RBD as the greatest clinical indicators of PD. Figure 11.1 shows the non-movement-related prodromal symptoms connected to PD.

11.5.2 Abnormal Function of the Olfactory System

The olfactory nerve (CN1) comprises 6 to 10 million receptor cells whose cell bodies and ciliated dendrites are enclosed inside a specific neuroepithelium found within the posterior nasal cavity (Doty, 2017). Olfactory dysfunction, such as hyposmia, a decline in the sense of smell, has been recognised as one of the fundamental non-motor signs of PD (Zhou *et al.*, 2020). Hyposmia is among the fundamental non-movement clinical signs in Parkinsonism and is present in 9 out of 10 PD patients (Haehner *et al.*, 2009). In a finding, a considerable percentage of persons with PD have olfactory impairment even prior to the manifestation of movement-related symptoms like tremors and stiffness (Ross *et al.*, 2008; Doty, 2012).

The olfactory bulb's neurons deterioration is responsible for hyposmia in PD patients. According to the Braak staging theory, peripheral inputs at the epithelium of the sense of smell or the dorsal motor nucleus of the vagus nerve rising from the stomach cause Lewy bodies, which are endowed with inclusions of alpha-synuclein, to pile-up and spread (Braak *et al.*, 2003).

For the evaluation of loss of smell as a promising PD biomarker, Bohnen *et al.* (2010) examined the degree of precision of numerous clinical movement and non-movement tests for PD's diagnosis and concluded that odour recognition is the most effective PD predictor. Deeb *et al.* (2010) further assessed how accurate and precise is the identification of scent utilising the dopamine transporter scan (DaTSCAN) and University of Pennsylvania Smell Identification Test (UPSIT). They discovered that the DaTSCAN is equally sensitive as UPSIT, the two having a sensitivity of 90%. Another evaluation for hyposmia is the Sniffin's Sticks Test (Zhou *et al.*, 2020), though the Sniffin's Sticks and the UPSIT are the greatest extensively utilised tools to evaluating scent recognition. Specifically, Sniffin's Sticks may be used to assess smell threshold detection, identification of odour, and discrimination of odour by employing an instrument that looks like a pen (Casjens *et al.*, 2013). However, it is crucial to highlight that olfactory dysfunction alone is not sufficient for a conclusive diagnosis of PD since other disorders, such as sinusitis or other neurodegenerative diseases such as AD, may also induce olfactory impairment. Therefore, olfactory impairment should be addressed in combination with other clinical and diagnostic tests when examining patients for PD.

11.5.3 Constipation

The disruption of bowel motions in people suffering from PD may be influenced by several variables, such as intestinal bacteria, intestinal inflammation, neuro-humoral factors, medicines, as well as individual's lifestyle (Barichella *et al.*, 2019). Constipation in PD may arise owing to problems with the autonomic nervous system (ANS) of the gastrointestinal tract, i.e., the digestive tract may not work efficiently and produce constipation (Metta *et al.*, 2022). Among other autonomic problems in Parkinsonism, constipation has been the most prevalent sign of autonomic problems in patients diagnosed with PD (Frazzitta *et al.*, 2019). Constipation has a greater incidence in the Parkinson's prodromal phase and is connected with a greater susceptibility to PD compared to other risk factors, according to a systematic analysis correlation between constipation and other risk factors of PD through a quantitative statistical analysis of several separate but similar studies of 17 findings that involved over 3 million cases (Yao *et al.*, 2023).

The Gut–Brain Hypothesis, put forth by Braak *et al.* in 2003, contends that the enteric nervous system, notably cranial nerve X, is where PD's neuropathological problems first appear. This is the most compelling and commonly accepted evidence for the hypothesis. It has been established that the myenteric plexus of the colonic sympathetic is involved in the degradation and deterioration of alpha-synuclein of the autonomic nervous system (ANS) in PD patients. According to numerous studies and observations, constipation may start in patients with PD prior to the onset of movement-related symptoms (Cersosimo *et al.*, 2013; Bloem *et al.*, 2021). Precisely, constipation may come before PD's motor signs by at least a decade (Fasano *et al.*, 2015). The Rome criteria are the internationally established collection of objective observations that indicate functional constipation. This approach focuses on six symptoms: lumpy and/or hard stools, straining, two or fewer BMs in 7 days, incomplete evacuation, manual manipulation to promote removal of stools, and the sense of anorectal obstruction or blockage. With these criteria, constipation's diagnosis in PD patients can be established when one has two or more symptoms for at least one-quarter of the time throughout a duration of at least 3 months (Drossman, 2016). However, many treatment options for constipation prescribed for other diseases that involved constipation are equally helpful for the person suffering from PD (Travagli *et al.*, 2020).

11.5.4 Rapid Eye Movement Sleep Behaviour Disorder (RBD)

RBD was initially defined as parasomnia, which is a situation whereby one lacks normal muscular atonia, allowing the individual to behave abnormally during sleep (Cooray *et al.*, 2019). Although the cause is unclear, but fortunately, the Braak's hypothesis second mechanism (Braak *et al.*, 2003) in their suggestion stated that the problem of the preventive mechanisms mediated by the locus ceruleus, brainstem, and pedunculopontine nucleus gives rise to the muscle atonia loss in RBD, which leads to dreaming (Bollu and Sahota, 2017). It is generally linked with destructive and undesired actions such as falling or leaping out of bed, punching, kicking, and vocalisations and crying, raising the chance of physical harm and impacting the patients as well as their bed partners (Zhang *et al.*, 2019). RBD's connection with some neurodegenerative disorders is now obvious, notably Parkinsonism, multiple system atrophy, dementia with Lewy body, and so on (Calabresi *et al.*, 2023; Sixel-Döring *et al.*, 2016). Therefore, it is regarded as a first sign of many neurodegenerations that increase in severity. Clinical studies have revealed that Parkinson's patients with RBD exhibit more chronic motor symptoms, autonomic symptoms, and cognitive impairment than Parkinson's patients without RBD (Suzuki *et al.*, 2015; Baumann-Vogel *et al.*, 2020). Though, as the existence of RBD is highly connected to dementia (Boeve *et al.*, 1998), it could be a possible clinical biomarker for monitoring the progression of the symptoms of PD into dementia. In addition to RBD, blepharoclonus (eyelid fluttering) has been found in persons suffering from PD. In a clinical study conducted by Margolesky *et al.* (2020) in roughly 20 PD patients, 85% of the cohorts were shown to exhibit blepharoclonus, including individuals at early PD stages; however, it is not yet known whether it represents a possible prodromal symptom. Overall, RBD could be a helpful biomarker for the detection of prodromal PD, especially when employed in concert with other diverse biomarkers such as neuroimaging.

11.6 BIOFLUID BIOMARKERS (BIOCHEMICAL MARKERS)

The clinical problems of neurodegenerative diseases involve the difficulty to establish a definite early-stage diagnosis by the use of biofluid-based biomarkers (Hansson, 2021). In the treatment of PD, effective diagnostic and prognostic biofluid-based biomarkers are instantly required. Since the movement-related clinical symptoms of PD are detected late after the onset of the disease and have impeded the detection of the earliest phases of the disorder, i.e., the treatment period at which available disorder-modifying medications could have the greatest healing ability (Parnetti *et al.*, 2019). We are all aware that biomarkers, which may be found in almost all human biological cells and fluids, are crucial for the diagnosis of prodromal PD. Blood, cerebrospinal fluid (CSF), and saliva have all been found to include a variety of potential biofluid-based biomarkers in the past. Currently, combining a variety of biofluid-based indications is crucial to achieving high diagnostic sensitivity and specificity.

11.6.1 Alpha-Synuclein

Alpha-synuclein is largely present in the brain parts, notably in the hippocampus, thalamus, neocortex, substantia nigra (SN), and cerebellum, and is identified in Lewy bodies (McLean *et al.*, 2000). Alternatively, erythrocytes are also a primary source of peripheral alpha-synuclein, as they have currently been identified as a possible PD biomarker in blood (Tian *et al.*, 2019). Although the physiologic impact of alpha-synuclein is yet to be entirely recognized, this protein appears to be controlling synaptic plasticity as one of its major tasks. Alpha-synuclein could demonstrate prion-like behaviour—that is, the potential to propagate by itself through intercellular diffusion and templated misfolding (Peng *et al.*, 2020). The abnormal folding of the alpha-synuclein protein is crucial for the development of various neurological conditions, like Parkinsonism, dementia with Lewy bodies (DLB), and other neuropathologically linked diseases. Also, it is the main element of Lewy bodies, the characteristic protein inclusion identified in PD. Hence, most of the findings on diagnostic markers for these disorders have been focused on discovering pathology-specific versions of the protein, i.e., Lewy body disease (PD and DLB). In CSF It is reasonably straightforward to detect total alpha-synuclein level, but its concentration has no correlation with Lewy body disease. Rather, it is hypothesised that elevated amounts are a sign of neuronal deterioration (Parnetti *et al.*, 2019). Histological investigations examined the amounts of alpha-synuclein in CSF using ELISA to determine that the quantity of a PD patient's CSF alpha-synuclein is much smaller than that in controls (Mollenhauer *et al.*, 2011; Parnetti *et al.*, 2014). Although the overall alpha-synuclein content in CSF rises if there is blood contamination of the sample owing to the elevated amount of alpha-synuclein among erythrocytes, that further lowers its value as a promising biomarker (Barbour *et al.*, 2008), Also, in blood, due to the significant likelihood of serum and plasma sample contamination with alpha-synuclein from erythrocytes, it is expected that total alpha-synuclein should show dismal results as a diagnostic marker in plasma, whole blood, and serum (Parnetti *et al.*, 2019). The ratio of erythrocytic alpha-synuclein species to total protein may be a potential peripheral biomarker for the Parkinson's diagnosis, according to some investigations that have found it to be substantially more elevated in people with the disorder than in healthy individuals (Wang *et al.*, 2015; Tian *et al.*, 2019). Additionally, it has been discovered that as the severity of the condition worsens in PD, less non-phosphorylated and soluble alpha-synuclein is present in susceptible brain parts, and the protein gradually becomes more insoluble and phosphorylated (Foulds *et al.*, 2011). In order to diagnose PD patients, this phosphorylated protein may be utilised as a possible biological indicator. The phosphorylated form of alpha-synuclein has been the most prevalent post-translational protein alteration of alpha-synuclein. According to a study by Volpicelli-Daley *et al.*, phosphorylated alpha-synuclein was seen to be the major substance responsible for the deterioration of the dopaminergic system (Volpicelli-Daley *et al.*, 2011). Beyond that, a clinical investigation by Vivacqua *et al.* in 2019 revealed that salivary alpha-synuclein detection might be an auspicious and readily available PD biological indicator. They discovered decreased total alpha-synuclein and more oligomeric alpha-synuclein in the mouth fluid of PD participants compared to the non-PD participants.

11.6.2 Urate

Urate is a byproduct of purine metabolism, which has been reported to be a metabolomic biomarker related with PD (Trupp *et al.*, 2014). Metabolomics is the study of tiny molecules involved in metabolic processes. There is a connection between urate and a lower chance of developing PD. Small levels of blood urate have been studied as a possible biomarker. A research study done by Ascherio *et al.* (2009) explored the connection between urate levels of the serum and the degree of deterioration in PD. The researchers observed that greater levels of urate were related with slower disease development, indicating that urate may have a neuroprotective impact. Another research studied the connection between plasma urate levels and the likelihood of acquiring PD, and the probability of having PD is inversely proportional to the urate levels, indicating that greater urate levels may be related with a lower chance of getting the condition (Weisskopf *et al.*, 2007).

11.6.3 DJ-1 Protein

DJ-1 is a crucial protein that is engaged in numerous metabolic activities of the cells, which include protective functions for oxidative-induced degeneration of the neurons, mitochondrial control RNA binding, chaperones, and proteases (Saito, 2017). While most occurrences of PD are sporadic, there are certain hereditary forms of the disease caused by particular gene abnormalities. There is a linkage between DJ-1 and PD. However, genetic testing is often reserved for those with a great family history or early start of Parkinsonism (Cook *et al.*, 2021). As a biofluid-based biomarker, its levels in CSF and blood have been studied as a possible biomarker. A 2006 research study by Waragai *et al.*

evaluated the DJ-1 protein's concentration in the CSF of people with sporadic PD. They observed that the CSF levels of DJ-1 were considerably elevated in PD individuals compared to control persons. In addition, a urinalysis research done on a sample of patients from Korea showed that the amount of DJ-1 considerably rises in the Parkinsonism-affected men's urine in comparison to males without Parkinsonism, but findings for women and girls remain negligible (Ho *et al.*, 2014). However, several experiments have been observed to dispute the rise of DJ-1 protein in CSF (Hong *et al.*, 2010). The contradiction of these data has sparked the idea of its oxidised form in erythrocytes as a possible PD biological indicator. In an independent investigation, DJ-1 oxidised on cysteine-106 particularly was demonstrated to be elevated in the erythrocytes of persons with PD (Saito, 2017). A research examining oxidised DJ-1 protein amount in the brain and in the peripheral tissues of mice reveals that the former is richer in oxidised DJ-1 compared to the latter (Mita *et al.*, 2018), although the oxidised form of this protein is now being explored by researchers as a potential biomarker for PD.

11.7 INFLAMMATORY BIOMARKERS

Inflammatory biomarkers are also known to be classified as biofluid-based biomarkers in as much as they can be seen in the biological fluid. Neuroinflammation is the normal, natural, anticipated, and adaptive response of the brain to damage, with the objective of promoting both the elimination of the harmful stimulus and self-protection. Astrocytes and microglial cell activation as well as inflammation of the neurons are important hallmarks of most neurodegenerative disorders (Kwon and Koh, 2020). As a general rule, inflammation is only a problem when it fails to diminish within the necessary period (Nathan and Ding, 2010). For over ten years, questions have been continued to be asked about whether astrogliosis or neuroinflammation are key causes of degeneration of the neurons or a side consequence of the abnormal buildups of tau, alpha-synuclein, and amyloid proteins. Nevertheless, inflammation of the neurons plays a vital part in the pathophysiology of PD.

Increasing evidence has revealed that the inflammation of the central and peripheral nervous systems performs critical roles in the causes, characteristics, and symptoms of PD (Wang *et al.*, 2015). Varieties of inflammatory markers, including chemokines such as CX3 chemokine ligands (CX3CL) and (CCL), interleukins (ILs), tumour necrosis factor alpha (TNF-alpha), and C-reactive protein (CRP), have been noted as vital signaling chemicals of immune stimulation that exert their effects in the brain and spinal cord as well as in the periphery (Tabas and Glass, 2013). However, their performances as putative biomarkers vary substantially in various studies owing to variances in study locations and techniques. Numerous studies have shown that the exerted effect of inflammatory markers in the blood and CSF of persons with PD vary from those in healthy individuals (Kim *et al.*, 2018; Hall *et al.*, 2018; Zimmermann and Brockmann, 2022). Moreover, several of these indicators of inflammation could be biomarkers of clinical aspects, including non-motor as well as motor complaints. According to Qu *et al.*'s meta-analysis, people with PD had substantially more elevated levels of inflammatory cytokines like TNF, IL-6, and IL-1 in their peripheral blood and CSF than healthy controls do (Qu *et al.*, 2023). Earlier meta-analyses by Qin *et al.* (2016) and Zimmermann and Brockmann (2022) were in agreement with this finding. According to Qu *et al.* 2023's meta-analysis, PD patients had greater levels of the chemokine CCL2 in their blood and CSF than did controls. Although MCP-1 is linked to T-cell and monocyte migration to inflammatory sites in a number of neurodegenerative disorders (Bose and Cho, 2013), so MCP-1 might not be a specific and promising biomarker for PD. C-reactive protein (CRP) is an inflammatory biomarker, and its increased level in the blood has been seen to be related to predicted life prognosis of PD patients as well as the risk of death (Sawada *et al.*, 2015). Similarly, a high level of IL-6 also suggests an overall risk of death in PD-affected persons (Dufek *et al.*, 2015).

Based on the results of the research on neuroinflammation described in this review, neuroinflammation is connected with numerous pathogenic pathways of neurodegenerative diseases such as AD. Interestingly, comparable compounds as found in individuals affected with PD are similarly higher in CSF and/or blood in AD patients in comparison to healthy individuals (Compta and Revesz, 2021). Thus, neuroinflammatory processes appear to be relatively not specific with disease and induced by different protein types that are connected to diseases such as alpha-synuclein, amyloid beta, and tau. However, attempts are being undertaken to verify consistent neuroinflammatory markers for PD.

11.8 NEUROIMAGING BIOMARKERS

By examining the prospective structures, ultra-structure, and changes in perfusion pattern of PD patients, neuroimaging is now a dependable and advanced biomarker for the diagnosis of PD. Numerous neuroimaging techniques can be used to track the development of the disease and discover early alterations in people with PD. As an individual biomarker is yet to be established, neuroimaging can be combined with other biomarkers to determine the level of PD and how it progresses. However, neuroimaging biological indicators are continuously utilised in diagnosing PD to corroborate observations from the clinical stage due to the fact that evaluation of molecular, functional, and structural changes in the brains of persons with PD is non-invasive.

Table 11.1 summarises the various neuroimaging biomarkers of PD at structural and functional levels.

11.8.1 Magnetic Resonance Imaging (MRI)

MRI enhances accuracy in clinical diagnosis, hence minimising misdiagnosis. However, early diagnosis might be facilitated and can be utilised to track disease development

TABLE 11.1
Early-Stage Biomarkers for Diagnosis and Detection of PD

Structural Neuroimaging Biomarker	Functional Neuroimaging Biomarker
Magnetic Resonance Imaging (MRI)	Functional MRI (fMRI)
Dopamine Transporter Scan (DATSCAN)	Single-Photon Emission Computed Tomography (SPECT)
Trans-Cranial Sonography (TCS)	Positron Emission Tomography (PET)
Diffusion Tensor Imaging (DTI)	Glucose Metabolism Imaging (FDG-PET)

(Heim *et al.*, 2018). The improvement in MRI technology permits the research of brain morphology by quantifying segments of brain structure, volumes of areas, tissue density, and surface area (Despostovic *et al.*, 2015). There are different MRI sequences:

a) T1- and T2-weighted MRI sequences
b) Diffusion-weighted MRI
c) Fluid-Attenuated Inversion Recovery (FLAIR)

All of these images are utilised selectively or combined to acquire the volumes of brain anatomy, regional cortical thickness, i.e., white matter and grey matter, and to identify regional tissue anomalies. In MRI sequences linked with an automated surface reconstruction approach, it was observed that there was cortical thinning in ventrolateral, orbitofrontal, prefrontal, as well as the occipitoparietal cortical areas in PD participants, accompanied by subcortical striatal volumetric decreases (Tinaz *et al.*, 2011). A study has revealed that MRI-based grey matter density may identify an orderly pattern of atrophy across disease course (Blair *et al.*, 2019).

11.8.2 Dopamine Transporter Scan (DATSCAN)

This test works by injecting DATSCAN, a specific radioactive tracer, into the circulation. It then goes to the brain and connects to the molecules that are located on dopamine receptors. Afterward, sophisticated imaging equipment is utilised to scan the skull and assess how much DATSCAN is present. In the area of the brain where the neurons that produce dopamine are located, people with PD frequently have a diminished signal. Although a positive test could support the diagnosis of PD, a negative DATSCAN does not usually rule it out. DATSCAN has the benefit of being used in a variety of early PD diagnosis procedures, from non-PD individuals to PD patients with tremors and vascular Parkinsonism (Vlaar *et al.*, 2007). Lower binding ratios of the posterior putamen-to-caudate are observed in PD patients, as well as a decrease in DAT (or larger signal) in the caudate/putamen transitional area and the caudate head relative to the putamen (Im *et al.*, 2006). Depletion of DAT in the striatum has not, however, demonstrated dependability in differentiating PD in a particular instance (Saeed *et al.*, 2017).

11.8.3 Single-Photon Emission Computed Tomography (SPECT)

SPECT is an in vivo functional nuclear imaging method greatly employed to investigate the stability of the dopaminergic system of the nigostriatal and to spot changes in cerebral blood flow after the injection of a DATSCAN. SPECT uses molecules co-ordinated to dopamine D2 receptors, such as selective gamma-emitting cocaine analogs that stick to dopamine transporters (DAT) of the striatal and form a cohesive mass.

11.8.4 Trans-Cranial Sonography (TCS)

TCS is a B-mode neuroimaging procedure in which ultrasonography is utilised to examine the brain tissue's echogenicity through the whole skull. TCS is primarily administered utilising an array transducer that is 2.5×10^6 Hz phase, positioned at the pre-auricular pit over the trans-temporal bone window. A recent longitudinal research carried out by Chen *et al.* revealed that there was high, efficient, and effective prediction of cognitive impairment in Parkinsonism-affected individuals by TCS paired with features that are clinical, notably non-motor symptoms of PD (Chen *et al.*, 2023). Although, one disadvantage of using this procedure is that it might not be practicable in particular situations because of the bone passage or window being somewhat narrow in some individuals (Paes *et al.*, 2023).

11.8.5 Diffusion Tensor Imaging (DTI)

DTI assesses the intactness of the brain microscopic structure. This method examines the water molecules' displacement (Mean Diffusivity; MD) and the displacing rate (Fractional Anisotropy; FA). PD patients frequently experience decreased FA in the significant nigra (SN). However, its relationship with disease severity is still ambiguous (Cochrane and Ebmeier, 2013). More regions of the brain have been explored; for instance, a quantitative statistical analysis showed that DTI had the ability to deal with structure-related disparities between people suffering from PD and non-PD individuals, specifically in five regions of the cerebrum using FA as well as MD; these five regions are the cingulate cortice, substantia nigra, temporal cortice, corpus callosum, and corticospinal tract (Atkinson-Clement *et al.*, 2017). One of the longest longitudinal DTI studies on three time-point MD alterations for over 12 years suggests that a progressive degeneration of striatal neurons is connected with movement dysfunction in PD, which was in agreement with the idea of a network of interconnected neurons of sensory motor basal ganglia-thalamocortices (Shih *et al.*, 2023).

11.8.6 Functional Magnetic Resonance Imaging (fMRI)

fMRI is a non-invasive brain imaging method with a great temporal and spatial resolution that measures neuronal activities by examining the paramagnetic properties of the blood (blood oxygenation level-dependent-BOLD mechanism and blood flow fluctuations). fMRI technique is of two types; they are resting state and task or work-based fMRIs.

The resting state fMRI technique involves measuring neuronal activities at rest, requiring the patient to be awake and not doing any activity. In this state, there are small-frequency oscillations from 0.01 to 0.1 Hz of BOLD signals. While in task or work-based technique, the discharge of an assigned duty or work is required to aid the firing of neurons in specific areas such as sensory, motor, and visual areas.

Several investigations have examined the resting-state fMRI patterns that underline significant movement-related symptoms in PD patients, pointing to alteration in the cerebello-thalamo-cortical neurons network as a vital feature of the disease. This network's functional connectivity has been found to be enhanced or decreased in individuals suffering from PD in comparison to healthy persons (Akram *et al.*, 2017; Suo *et al.*, 2017; Tuovinen *et al.*, 2018). A longitudinal study over the course of a year has also found that the work-related fMRI signal reflects various forms of disorder progression in PD, progressive supranuclear palsy (PSP), and multiple system atrophy (MSA) using a motor control paradigm and found a reduction in fMRI signal in the cortex and putamen in persons suffering from PD in comparison to the controls. Notwithstanding, the findings indicate more extensive and distinct forms of functional abnormalities in MSA and PSP than in PD (Burciu *et al.*, 2016).

11.8.7 Positron Emission Tomography (PET)

PET uses a number of radionuclides to access the dopaminergic system, the cerebrum's metabolism, and the buildup of pathological proteins. With radiotracers like 18F-DOPA, the activity of the presynaptic and postsynaptic nigrostriatal projections can be examined. Given that it examines the nigro-striatal synthesis of dopamine, measuring the absorption of [18F]-DOPA by the striatum using PET is currently recognised as one of the most reliable and accurate methods for PD diagnosis in vivo (Snow *et al.*, 1993; Morrish *et al.*, 1996). So far, it has been used as a gauge for dopamine's terminal depletion (Depierreux *et al.*, 2021). Tau tracer in PD patients may assist in differentiating the disease from other neurodegenerative conditions, according to recent in vivo imaging investigations with PET (Zhang *et al.*, 2023).

11.8.8 Glucose Metabolism Imaging (FDG-PET)

The study of the glucose metabolism of the cerebrum using the tracer [18F] fluorodeoxyglucose positron emission tomography (FDG-PET) has made it possible to distinguish with high precision among PD, MSA, and PSP (Meyer *et al.*, 2017). This tracer frequently detects intact metabolic activities of glucose in PD patients' lenticular nucleus (LN) and thalamus (Zhao *et al.*, 2012). Also, it detects low metabolism in the premotor, bilateral parietal, and supplementary motor areas when compared to non-PD individuals (Poston and Eidelberg, 2010). Studies in persons suffering from PD showed that the substantia nigra pars compacta's metabolism is decreased, and this conforms to putaminal dopamine synthesis ability using PET (Ruppert *et al.*, 2020). The metabolic activities of glucose in the basal ganglia can separate persons suffering from PD from those with PSP and MSA (Schroter *et al.*, 2022) because of an observed corresponding metabolic decline. In addition, MSA patients show loss in the metabolic activities of glucose in the pons, putamen contralateral basal ganglia, as well as cerebellum in comparison to patients suffering from PD (Juh *et al.*, 2004).

Neuroimaging biomarkers give supporting data to aid in diagnosing and monitoring of therapy for PD. Functional and structural neuroimaging biomarker approaches, on the other hand, are critical in uncovering and assessing neuroanatomical and neurobiological alterations that occur at the beginning of PD. When PD is diagnosed early, it might lead to timely treatments and management of this degenerative condition if these alterations are appropriately identified. As a result, more validation and reliability studies, as well as continuing technological improvements, are required to fully exploit the promise of the aforementioned neuroimaging methods for diagnosing patients with PD and monitoring the success of their treatment intervention.

11.9 FUTURE PERSPECTIVES

Many studies have been done and more will investigate innovative techniques for early diagnosis and successful management of PD. The following possibilities are excellent but might potentially come with many obstacles:

1. *Identification of unique disease network and molecular pathway*: Given that PD has a complicated pathogenesis, this would assist in understanding the disease process and devising more effective biomarkers.
2. *Combination of biomarkers*: A combination of biomarkers is claimed to increase diagnosis accuracy since no single biomarker can supplant the clinical evaluation yet. However, integrating two or more biomarkers with clinical evaluation will serve as a superior diagnostic method.

3. *Salivary gland pharyngeal biopsy*: Saliva has some benefit compared to blood and CSF because it includes less protein content and is a fast and noninvasive collection technique, making it a suitable specimen for PD. Autopsy investigations on PD have showed that the salivary ducts contain relatively significant concentrations of aS aggregates. Based on this, Lewy bodies (LBs) were examined in samples collected from minor salivary glands and submandibular salivary glands. In samples obtained from the minor salivary gland, LBs were observed in all the Parkinson's group, but not in the non-Parkinson's group (Cersosimo *et al.*, 2013) samples retrieved from the submandibular salivary duct, in which LBs were identified in 75% of PD. Since dysphagia is one of the symptoms leading to significant difficulties in PD, pharyngeal autopsy investigations have been undertaken and phosphorylated alpha-synuclein (P-αS) aggregates have been observed in the sensory nerve axons of the pharynx.
4. *Advance imaging techniques*: Advancements in neuroimaging technologies like functional MRI (fMRI), PET scans, and diffusion tensor imaging (DTI) will enable researchers to view the brain's structure and function in greater detail. These approaches might assist in recognising minor alterations linked with Parkinson's disease.
5. *Artificial intelligence and machine learning*: These are being designed to assess intricate datasets, such as genetic information, brain imaging, and clinical data. These algorithms might potentially help in early detection and prediction of Parkinson's disease. Some machine learning technique enables for both directing therapeutic choices such as the commencement of neuroprotective measures and critically, the prospect of earlier diagnosis for clinical trials for disease-modifying medications.

11.10 CONCLUSION

PD is marked clinically by symptoms that are movement and non-movement-related. Over the years, the motor triad symptoms have become its main diagnostic sign. However, various non-motor symptoms (clinical biomarkers) have been established so far, as the most prevalent of all are olfactory dysfunction, constipation, and RBD. In addition, several biofluid-based biomarkers and structural and functional neuroimaging of its proteinopathies have also been found to monitor and trace the severity of PD. Among all, there is no unique singular biomarker that may substitute the clinical motor symptoms, although various neuroimaging techniques have created high differential diagnosis for PD among other neurodegenerative diseases. The cardinal indicator of PD is the clinical motor symptoms that arise at the latter end of the disease development when most of the neuronal cell has been destroyed. It is vital to give reliable biomarkers that will assist in early diagnosis of PD as this will help in giving therapeutic intervention that will halt or hinder the course of the disorder at its onset.

REFERENCES

Akram, H., Wu, C., Hyam, J., Foltynie, T., Limousin, P., De Vita, E., Yousry, T., Jahanshahi, M., Hariz, M., Behrens, T., Ashburner, J., & Zrinzo, L. (2017). L-Dopa responsiveness is associated with distinctive connectivity patterns in advanced Parkinson's disease. *Movement Disorders*, *32*(6), 874–883.

Arotcarena, M. L., Dovero, S., Prigent, A., Bourdenx, M., Camus, S., Porras, G., Thiolat, M. L., Tasselli, M., Aubert, P., Kruse, N., Mollenhauer, B., Trigo Damas, I., Estrada, C., Garcia-Carrillo, N., Vaikath, El-Agnaf, O. M. A., Herrero, Vila, Obeso, J. A., Derkinderen, P., & Bezard, E. (2020). Bidirectional gut-to-brain and brain-to-gut propagation of synucleinopathy in non-human primates. *Brain*, *143*(5), 1462–1475, https://doi.org/10.1093/brain/awaa096

Ascherio, A., LeWitt, P. A., Xu, K., Eberly, S., Watts, A., Matson, W. R., Marras, C., Kieburtz, K., Rudolph, A., Bogdanov, M. B., Schwid, S. R., Tennis, M., Tanner, C. M., Beal, M. F., Lang, A. E., Oakes, D., Fahn, Shoulson, I., & Schwarzschild, M. A. (2009). Urate as a predictor of the rate of clinical decline in Parkinson's disease. *Archives of Neurology*, *66*(12), 1460–1468, https://doi.org/10.1001/archneurol.2009.247

Atkinson-Clement, C., Pinto, S., Eusebio, A., & Coulon, O. (2017). Diffusion tensor imaging in Parkinson's disease: Review and meta-analysis. *NeuroImage: Clinical*, *16*, 98–110, https://doi.org/10.1016/j.nicl.2017.07.011

Barbour, R., Kling, K., Anderson, J. P., Banducci, K., Cole, T., Diep, L., Fox, M., Goldstein, J. M., Soriano, F., Seubert, P., & Chilcote, T. J. (2008). Red blood cells are the major source of alpha-synuclein in the blood. *Neuro-degenerative Diseases*, *5*(2), 55–59, https://doi.org/10.1159/000112832

Barichella, M., Severgnini, M., Cilia, R., Cassani, E., Bolliri, C., Caronni, S., Ferri, V., Cancello, R., Ceccarani, C., Faierman, S., Pinelli, G., De Bellis, G., Zecca, L., Cereda, E., Consolandi, C., & Pezzoli, G. (2019). Unravelling the gut microbiota in Parkinson's disease and atypical Parkinsonism. *Movement Disorders*, *34*(3), 396–405, https://doi.org/10.1002/mds.27581

Baumann-Vogel, H., Hor, H., Poryazova, R., Valko, P., Werth, E., & Baumann, C. R. (2020). REM sleep behaviour in Parkinson disease: Frequent, particularly with higher age. *PLoS One*, *15*(12), e0243454, https://doi.org/10.1371/journal.pone.0243454

Blair, J. C., Barrett, M. J., Patrie, J., Flanigan, J. L., Sperling, S. A., Elias, W. J., & Druzgal, T. J. (2019). Brain MRI reveals ascending atrophy in Parkinson's disease across severity *Frontiers in Neurology*, *10*, 1329, https://doi.org/10.3389/fneur.2019.01329

Bloem, B. R., Okun, M. S., & Klein, C. (2021). Parkinson's disease. *Lancet*, *397*(10291), 2284–2303, https://doi.org/10.1016/S0140-6736(21)00218-X

Boeve, B. F., Silber, M. H., Ferman, T. J., Kokmen, E., Smith, G. E., Ivnik, R. J., Parisi, J. E., Olson, E. J., & Petersen, R. C. (1998). REM sleep behaviour disorder and degenerative dementia: An association likely reflecting Lewy body disease. *Neurology*, *51*(2), 363–370, https://doi.org/10.1212/wnl.51.2.363

Bohnen, N. I., Müller, M. L., Kotagal, V., Koeppe, R. A., Kilbourn, M. A., Albin, R. L., & Frey, K. A. (2010). Olfactory dysfunction, central cholinergic integrity, and cognitive impairment in Parkinson's disease. *Brain*, *133*(Pt. 6), 1747–1754, https://doi.org/10.1093/brain/awq079

Bollu, P. C., & Sahota, P. (2017). Sleep and Parkinson disease. *Missouri Medicine*, *114*(5), 381–386.

Bose, S., & Cho, J. (2013). Role of the chemokine CCL2 and its receptor CCR2 in neurodegenerative diseases. *Archives of Pharmacal Research*, *36*(9), 1039–1050, https://doi.org/10.1007/s12272-013-0161-z

Braak, H., Del Tredici, K., Rüb, U., de Vos, R. A., Jansen Steur, E. N., & Braak, E. (2003). Staging of brain pathology related to sporadic Parkinson's disease. *Neurobiology of Ageing*, *24*(2), 197–211, https://doi.org/10.1016/s0197-4580(02)00065-9

Burciu, R. G., Chung, J. W., Shukla, P., Ofori, E., Li, H., McFarland, N. R., Okun, M. S., & Vaillancourt, D. E. (2016). Functional MRI of disease progression in Parkinson's disease and atypical Parkinsonian syndromes. *Neurology*, *87*(7), 709–717, https://doi.org/10.1212/WNL.0000000000002985

Calabresi, P., Mechelli, A., Natale, G., Volpicelli-Daley, L., Di Lazzaro, G., & Ghiglieri, V. (2023). Alpha-synuclein in Parkinson's disease and other synucleinopathies: From overt neurodegeneration back to early synaptic dysfunction. *Cell Death and Disease*, *14*(3), 176, https://doi.org/10.1038/s41419-023-05672-9

Casjens, S., Eckert, A., Woitalla, D., Ellrichmann, G., Turewicz, M., Stephan, C., Eisenacher, M., May, C., Meyer, H. E., Brüning, T., & Pesch, B. (2013). Diagnostic value of the impairment of olfaction in Parkinson's disease. *PLoS One*, *8*(5), e64735, https://doi.org/10.1371/journal.pone.0064735

Cersosimo, M. G., Raina, G. B., Pecci, C., Pellene, A., Calandra, C. R., Gutiérrez, C., Micheli, F. E., & Benarroch, E. E. (2013). Gastrointestinal manifestations in Parkinson's disease: Prevalence and occurrence before motor symptoms. *Journal of Neurology*, *260*(5), 1332–1338, https://doi.org/10.1007/s00415-012-6801-2

Chen, Z., Zhang, W., He, W., Guang, Y., Yu, T., Du, Y., & Li, R. (2023). Transcranial sonography with clinical and demographic characteristics to predict cognitive impairment in PD: A longitudinal study. *BMC Neurology*, *23*, 15, https://doi.org/10.1186/s12883-023-03057-1

Cochrane, C. J., & Ebmeier, K. P. (2013). Diffusion tensor imaging in Parkinsonian syndromes: A systematic review and meta-analysis. *Neurology*, *80*, 857–864.

Compta, Y., & Revesz, T. (2021). Neuropathological and biomarker findings in Parkinson's disease and Alzheimer's disease: From protein aggregates to synaptic dysfunction. *Journal of Parkinson's Disease*, *11*(1), 107–121, https://doi.org/10.3233/JPD-202323

Cook, L., Schulze, J., Naito, A., & Alcalay, R. N. (2021). The role of genetic testing for Parkinson's disease. *Current Neurology and Neuroscience Reports*, *21*(4), 17, https://doi.org/10.1007/s11910-021-01100-7

Cooray, N., Andreotti, F., Lo, C., Symmonds, M., Hu, M. T. M., & De Vos, M. (2019). Detection of REM sleep behaviour disorder by automated polysomnography analysis. *Clinical Neurophysiology*, *130*(4), 505–514, https://doi.org/10.1016/j.clinph.2019.01.011

Dauer, W., & Przedborski, S. (2003). Parkinson's disease: Mechanisms and models. *Neuron*, *39*(6), 889–909, https://doi.org/10.1016/s0896-6273(03)00568-3

Deeb, J., Shah, M., Muhammed, N., Gunasekera, R., Gannon, K., Findley, L. J., & Hawkes, C. H. (2010). A basic smell test is as sensitive as a dopamine transporter scan: Comparison of olfaction, taste, and DaTSCAN in the diagnosis of Parkinson's disease. *QJM*, *103*(12), 941–952, https://doi.org/10.1093/qjmed/hcq142

Depierreux, F., Parmentier, E., Mackels, L., Baquero, K., Degueldre, C., Balteau, E., Salmon, E., Phillips, C., Bahri, M. A., Maquet, P., & Garraux, G. (2021). Parkinson's disease multimodal imaging: F-DOPA PET, neuromelanin-sensitive, and quantitative iron-sensitive MRI. *NPJ Parkinson's Disease*, *7*(1), 57, https://doi.org/10.1038/s41531-021-00199-2

Despotovic, I., Goossens, B., & Philips, W. (2015). MRI segmentation of the human brain: Challenges, methods, and applications. *Computational and Mathematical Methods in Medicine*, *2015*, 450341, https://doi.org/10.1155/2015/450341

Doty, R. L. (2012). Olfaction in Parkinson's disease and related disorders. *Neurobiology of Disease*, *46*(3), 527–552, https://doi.org/10.1016/j.nbd.2011.10.026

Doty, R. L. (2017). Olfactory dysfunction in neurodegenerative diseases: Is there a common pathological substrate? *The Lancet: Neurology*, *16*(6), 478–488, https://doi.org/10.1016/S1474-4422(17)30123-0

Drossman, D. A. (2006). Functional gastrointestinal disorders and the Rome III process *Gastroenterology*, *130*(5), 1377–1390, https://doi.org/10.1053/j.gastro.2006.03.008

Dufek, M., Rektorova, I., Thon, V., Lokaj, J., & Rektor, I. (2015). Interleukin-6 may contribute to mortality in Parkinson's disease patients: A 4-year prospective study *Parkinson's Disease*, *2015*, 898192, https://doi.org/10.1155/2015/898192

Fasano, A., Visanji, N. P., Liu, L. W., Lang, A. E., & Pfeiffer, R. F. (2015). Gastrointestinal dysfunction in Parkinson's disease. *The Lancet: Neurology*, *14*(6), 625–639, https://doi.org/10.1016/S1474-4422(15)00007-1

Foulds, P. G., Mitchell, J. D., Parker, A., Turner, R., Green, G., Diggle, P., Hasegawa, M., Taylor, M., Mann, D., & Allsop, D. (2011). Phosphorylated synuclein can be detected in blood plasma and is potentially a useful biomarker for Parkinson's disease. *FASEB Journal*, *25*(12), 4127–4137, https://doi.org/10.1096/fj.10-179192

Frazzitta, G., Ferrazzoli, D., Folini, A., Palamara, G., & Maestri, R. (2019). Severe constipation in Parkinson's disease and in Parkinsonisms: Prevalence and affecting factors *Frontiers in Neurology*, *10*, 621, https://doi.org/10.3389/fneur.2019.00621

Frisardi, V., Santamato, A., & Cheeran, B. (2016). Parkinson's disease: New insights into pathophysiology and rehabilitative approaches. *Parkinson's Disease*, *2016*, 3121727, https://doi.org/10.1155/2016/3121727

Fullard, M. E., Morley, J. F., & Duda, J. E. (2017). Olfactory dysfunction as an early biomarker in Parkinson's disease. *Neuroscience Bulletin*, *33*(5), 515–525, https://doi.org/10.1007/s12264-017-0170-x

GBD 2016 Neurology Collaborators. (2019). Global, regional, and national burden of neurological disorders, 1990–2016: A systematic analysis for the global burden of disease study 2016. *The Lancet: Neurology*, *18*(5), 459–480, https://doi.org/10.1016/S1474-4422(18)30499-X

Greenland, J. C., & Barker, R. A. (2018). The differential diagnosis of Parkinson's disease. In T. B. Stoker, et al. (Eds.), *Parkinson's Disease: Pathogenesis and Clinical Aspects*. Codon Publications.

Haehner, A., Boesveldt, S., Berendse, H. W., Mackay-Sim, A., Fleischmann, J., Silburn, P. A., Johnston, A. N., Mellick, G. D., Herting, B., Reichmann, H., & Hummel, T. (2009). Prevalence of smell loss in Parkinson's disease: A multicenter study. *Parkinsonism & Related Disorders*, *15*(7), 490–494, https://doi.org/10.1016/j.parkreldis.2008.12.005

Hall, S., Janelidze, S., Surova, Y., Widner, H., Zetterberg, H., & Hansson, O. (2018). Cerebrospinal fluid concentrations of inflammatory markers in Parkinson's disease and atypical Parkinsonian disorders. *Scientific Reports, 8*, 13276, https://doi.org/10.1038/s41598-018-31517-z

Hansson, O. (2021). Biomarkers for neurodegenerative diseases. *Nature Medicine, 27*(6), 954–963, https://doi.org/10.1038/s41591-021-01382-x

Hawkes, C. H., Del Tredici, K., & Braak, H. (2009). Parkinson's disease: The dual hit theory revisited. *Annals of the New York Academy of Sciences, 1170*, 615–622, https://doi.org/10.1111/j.1749-6632.2009.04365.x

Heim, B., Krismer, F., & Seppi, K. (2018). Structural imaging in atypical Parkinsonism. *International Review of Neurobiology, 142*, 67–148.

Ho, D. H., Yi, S., Seo, H., Son, I., & Seol, W. (2014). Increased DJ-1 in the urine exosome of Korean males with Parkinson's disease. *BioMed Research International, 2014*, 704678, https://doi.org/10.1155/2014/704678

Hong, Z., Shi, M., Chung, K. A., Quinn, J. F., Peskind, E. R., Galasko, D., Jankovic, J., Zabetian, C. P., Leverenz, J. B., Baird, G., Montine, T. J., Hancock, A. M., Hwang, H., Pan, C., Bradner, J., Kang, U. J., Jensen, P. H., & Zhang, J. (2010). DJ-1 and alpha-synuclei in human cerebrospinal fluid as biomarkers of Parkinson's diseas. *Brain, 133*, 713–726.

Im, J.-H., Chung, S. J., Kim, J.-S., & Lee, M. C. (2006). Differential patterns of dopamine transporter loss in the basal ganglia of progressive supranuclear palsy and Parkinson's disease: Analysis with [(123) I] IPT single photon emission computed tomography. *Journal of the Neurological Sciences, 244*, 103–109.

Jankovic, J. (2008). Parkinson's disease: Clinical features and diagnosis. *Journal of Neurology, Neurosurgery, and Psychiatry, 79*(4), 368–376, https://doi.org/10.1136/jnnp.2007.131045

Juh, R., Kim, J., Moon, D., Choe, B., & Suh, T. (2004). Different metabolic patterns analysis of Parkinsonism on the 18 F-FDG PET. *European Journal of Radiology, 51*, 223–33.

Kalia, L. V., & Lang, A. E. (2015). Parkinson's disease. *Lancet, 386*(9996), 896–912, https://doi.org/10.1016/S0140-6736(14)61393-3

Kang, S. H., Moon, S.-J., Kang, M., Chung, S. J., Cho, G. J., & Koh, S.-B. (2023). Incidence of Parkinson's disease and modifiable risk factors in the Korean population: A longitudinal follow-up study of a nationwide cohort. *Frontiers in Aging Neuroscience, 15*, 1094778, https://doi.org/10.3389/fnagi.2023.1094778

Kim, R., Kim, H. J., Kim, A., Jang, M., Kim, A., Kim, Y., Yoo, D., Im, J. H., Choi, J. H., & Jeon, B. (2018). Peripheral blood inflammatory markers in early Parkinson's disease. *Journal of Clinical Neuroscience, 58*, 30–33, https://doi.org/10.1016/j.jocn.2018.10.079

Kouli, A., Torsney, K. M., & Kuan, W. L. (2018). Parkinson's disease: Aetiology, neuropathology, and pathogenesis In T. B. Stoker, et al. (Eds.), *Parkinson's Disease: Pathogenesis and Clinical Aspects*. Codon Publications.

Kwon, H. S., & Koh, S. H. (2020). Neuroinflammation in neurodegenerative disorders: The roles of microglia and astrocytes. *Translational Neurodegeneration, 9*(1), 42, https://doi.org/10.1186/s40035-020-00221-2

Margolesky, J., Fleming, N., Shpiner, D. S., Luca, C., Moore, H. P., Singer, C., Wuu, J., Haq, I. U., & Benatar, M. (2020). Blepharoclonus in Parkinson's disease: A prevalent and meaningful finding*? Neurology, 94*(15 Suppl), 2174.

Mazzoni, P., Shabbott, B., & Cortés, J. C. (2012). Motor control abnormalities in Parkinson's disease. *Cold Spring Harbour Perspectives in Medicine, 2*(6), a009282, https://doi.org/10.1101/cshperspect.a009282

McLean, P. J., Kawamata, H., Ribich, S., & Hyman, B. T. (2000). Membrane association and protein conformation of alpha-synuclein in intact neurons: Effects of Parkinson's disease-linked mutations. *The Journal of Biological Chemistry, 275*(12), 8812–8816, https://doi.org/10.1074/jbc.275.12.8812

Metta, V., Leta, V., Mrudula, K. R., Prashanth, L. K., Goyal, V., Borgohain, R., Chung-Faye, G., & Chaudhuri, K. R. (2022). Gastrointestinal dysfunction in Parkinson's disease: Molecular pathology and implications of the gut microbiome, probiotics, and faecal microbiota transplantation. *Journal of Neurology, 269*(3), 1154–1163, https://doi.org/10.1007/s00415-021-10567-w

Meyer, P. T., Frings, L., Rücker, G., & Hellwig, S. (2017). ^{18}F-FDG PET in Parkinsonism: Differential diagnosis and evaluation of cognitive impairment. *Journal of Nuclear Medicine, 58*(12), 1888–1898, https://doi.org/10.2967/jnumed.116.186403

Mita, Y., Kataoka, Y., Saito, Y., Kashi, T., Hayashi, K., Iwasaki, A., Imanishi, T., Miyasaka, T., & Noguchi, N. (2018). Distribution of oxidised DJ-1 in Parkinson's disease-related sites in the brain and in the peripheral tissues: Effects of ageing and a neurotoxin. *Scientific Reports, 8*, 12056, https://doi.org/10.1038/s41598-018-30561-z

Mollenhauer, B., Locascio, J. J., Schulz-Schaeffer, W., Sixel-Döring, F., Trenkwalder, C., & Schlossmacher, M. G. (2011). Synuclein and tau concentrations in the cerebrospinal fluid of patients presenting with Parkinsonism: A cohort study. *The Lancet: Neurology, 10*(3), 230–240, https://doi.org/10.1016/S1474-4422(11)70014-X

Morrish, P. K., Sawle, G. V., & Brooks, D. J. (1996). An [18F] dopa-PET and clinical study of the rate of progression in Parkinson's disease. *Brain, 119*(Pt 2), 585–591, https://doi.org/10.1093/brain/119.2.585

Nathan, C., & Ding, A. (2010). Nonresolving inflammation. *Cell, 140*(6), 871–882, https://doi.org/10.1016/j.cell.2010.02.029

Noelker, C., Hampel, H., & Dodel, R. (2011). Blood-based protein biomarkers for diagnosis and classification of neurodegenerative diseases: Current progress and clinical potential. *Molecular Diagnosis and Therapy, 15*(2), 83–102, https://doi.org/10.1007/BF03256398

Paes, M. A. S., Nicaretta, D. H., Alvarenga, R. M. P., Rosso, A. L. Z., Brisson, R. T., & Fernandes, R. C. L. (2023). Comparison between a handheld ultrasound device and a traditional ultrasound for performing transcranial sonography in patients with Parkinson's disease. *Brain and Behaviour, 13*(2), e2891, https://doi.org/10.1002/brb3.2891

Parnetti, L., Chiasserini, D., Persichetti, E., Eusebi, P., Varghese, S., Qureshi, M. M., Dardis, A., Deganuto, M., De Carlo, C., Castrioto, A., Balducci, C., Paciotti, S., Tambasco, N., Bembi, B., Bonanni, L., Onofrj, M., Rossi, A., Beccari, T., El-Agnaf, O., & Calabresi, P. (2014). Cerebrospinal fluid lysosomal enzymes and alpha-synuclein in Parkinson's disease. *Movement Disorders, 29*(8), 1019–1027, https://doi.org/10.1002/mds.25772

Parnetti, L., Gaetani, L., Eusebi, P., Paciotti, S., Hansson, O., El-Agnaf, O., Mollenhauer, B., Blennow, K., & Calabresi, P. (2019). CSF and blood biomarkers for Parkinson's disease. *The Lancet: Neurology, 18*(6), 573–586, https://doi.org/10.1016/S1474-4422(19)30024-9

Peng, C., Trojanowski, J. Q., & Lee, V. M. (2020). Protein transmission in neurodegenerative disease. *Nature Reviews:*

Neurology, *16*(4), 199–212, https://doi.org/10.1038/s41582-020-0333-7

Perlis, R. H. (2011). Translating biomarkers into clinical practise. *Molecular Psychiatry*, *16*(11), 1076–1087, https://doi.org/10.1038/mp.2011.63

Poewe, W., Seppi, K., Tanner, C. M., Halliday, G. M., Brundin, P., Volkmann, J., Schrag, A. E., & Lang, A. E. (2017). Parkinson disease. *Nature Reviews. Disease Primers*, *3*, 17013, https://doi.org/10.1038/nrdp.2017.13

Poston, K. L., & Eidelberg, D. (2010). FDG PET in the evaluation of Parkinson's disease. *PET Clinics*, *5*, 55–64.

Qin, X. Y., Zhang, S. P., Cao, C., Loh, Y. P., & Cheng, Y. (2016). Aberrations in peripheral inflammatory cytokine levels in Parkinson disease: A systematic review and meta-analysis. *JAMA Neurology*, *73*(11), 13161324, https://doi.org/10.1001/jamaneurol.2016.2742

Qu, Y., Li, J., Qin, Q., Wang, D., Zhao, J., An, K., Mao, Z., Min, Z., Xiong, Y., Li, J., & Xue, Z. (2023). A systematic review and meta-analysis of inflammatory biomarkers in Parkinson's disease. *NPJ Parkinson's Disease*, *9*, 18, https://doi.org/10.1038/s41531-023-00449-5

Rajan, S., & Kaas, B. (2022). Parkinson's disease: Risk factor modification and prevention. *Seminars in Neurology*, *42*(5), 626–638, https://doi.org/10.1055/s-0042-1758780

Ross, G. W., Petrovitch, H., Abbott, R. D., Tanner, C. M., Popper, J., Masaki, K., Launer, L., & White, L. R. (2008). Association of olfactory dysfunction with risk for future Parkinson's disease. *Annals of Neurology*, *63*(2), 167–173, https://doi.org/10.1002/ana.21291

Ruppert, M. C., Greuel, A., Tahmasian, M., Schwartz, F., Stürmer, S., Maier, F., Hammes, J., Tittgemeyer, M., Timmermann, L., van Eimeren, T., Drzezga, A., & Eggers, C. (2020). Network degeneration in Parkinson's disease: Multimodal imaging of nigro-striato-cortical dysfunction. *Brain*, *143*(3), 944–959, https://doi.org/10.1093/brain/awaa019

Saeed, U., Compagnone, J., Aviv, R. I., Strafella, A. P., Black, S. E., Lang, A. E., & Masellis, M. (2017). Imaging biomarkers in Parkinson's disease and Parkinsonian syndromes: Current and emerging concepts. *Translational Neurodegeneration*, *6*, 8, https://doi.org/10.1186/s40035-017-0076-6

Saito, Y. (2017). DJ-1 as a biomarker of Parkinson's disease. *Advances in Experimental Medicine and Biology*, *1037*, 149–171, https://doi.org/10.1007/978-981-10-6583-5_10

Sawada, H., Oeda, T., Umemura, A., Tomita, S., Kohsaka, M., Park, K., Yamamoto, K., & Sugiyama, H. (2015). Baseline C-reactive protein levels and life prognosis in Parkinson's disease. *PLoS One*, *10*(7), e0134118, https://doi.org/10.1371/journal.pone.0134118

Schaffrath, A., Schleyken, S., Seger, A., Jergas, H., Özdüzenciler, P., Pils, M., Blömeke, L., Cousin, A., Willbold, J., Bujnicki, T., Bannach, O., Fink, G. R., Willbold, D., Sommerauer, M., Barbe, M. T., & Tamgüney, G. (2023). Patients with isolated REM-sleep behaviour disorder have elevated levels of alpha-synuclein aggregates in stool. *NPJ Parkinson's Disease*, *9*, 14, https://doi.org/10.1038/s41531-023-00458-4

Schlossmacher, M. G., & Mollenhauer, B. (2010). Biomarker research in Parkinson's disease: Objective measures needed for patient stratification in future cause-directed trials *Biomarkers in Medicine*, *4*(5), 647–650, https://doi.org/10.2217/bmm.10.93

Schroter, N., Blazhenets, G., Frings, L., Jost, W. H., Weller, C., Rijntjes, M., Meyer, P. T., & Brumberg, J. (2022). Nigral glucose metabolism as a diagnostic marker of neurodegenerative parkinsonian syndromes. *NPJ Parkinsons Disease*, *8*(1), 123.

Shih, Y.-C., Ooi, L. Q. R., Li. H.-H., Allen, J. C., Hartono, S., Welton, T., Tan, E. K., & Chan, L. L. (2023). Serial deep grey nuclear DTI changes in Parkinson's disease over twelve years. *Frontiers in Aging Neuroscience*, *15*, 1169254, https://doi.org/10.3389/fnagi.2023.1169254

Sixel-Döring, F., Zimmermann, J., Wegener, A., Mollenhauer, B., & Trenkwalder, C. (2016). The evolution of REM sleep behaviour disorder in early Parkinson's disease. *Sleep*, *39*(9), 1737–1742, https://doi.org/10.5665/sleep.6102

Snow, B. J., Tooyama, I., McGeer, E. G., Yamada, T., Calne, D. B., Takahashi, H., & Kimura, H. (1993). Human positron emission tomographic [18F] fluorodopa studies correlate with dopamine cell counts and levels. *Annals of Neurology*, *34*(3), 324–330, https://doi.org/10.1002/ana.410340304

Stefanis, L. (2012). Synuclein in Parkinson's disease. *Cold Spring Harbour Perspectives in Medicine*, *2*(2), a009399, https://doi.org/10.1101/cshperspect.a009399

Subramaniam, N. S., Bawden, C. S., Waldvogel, H., Faull, R. M. L., Howarth, G. S., & Snell, R. G. (2018). Emergence of breath testing as a new non-invasive diagnostic modality for neurodegenerative diseases. *Brain Research*, *1691*, 75–86, https://doi.org/10.1016/j.brainres.2018.04.017

Suo, X., Lei, D., Li, N., Cheng, L., Chen, F., Wang, M., Kemp, G. J., Peng, R., & Gong, Q. (2017). Functional brain connectome and its relation to Hoehn and Yahr stages in Parkinson's disease. *Radiology, 285*(3), 904–913.

Suzuki, K., Miyamoto, M., Miyamoto, T., & Hirata, K. (2015). Restless legs syndrome and leg motor restlessness in Parkinson's disease. *Parkinson's Disease*, *2015*, 490938, https://doi.org/10.1155/2015/490938

Tabas, I., & Glass, C. K. (2013). Anti-inflammatory therapy in chronic disease: Challenges and opportunities. *Science*, *339*(6116), 166–172, https://doi.org/10.1126/science.1230720

Tevak, Z., Kondratovich, M. V., & Mansfield, E. (2010). The US FDA and personalised medicine: An in vitro diagnostic regulatory perspective *Personalised Medicine*, *7*(5), 517–530, https://doi.org/10.2217/pme.10.53

Tian, C., Liu, G., Gao, L., Soltys, D., Pan, C., Stewart, T., Shi, M., Xie, Z., Liu, N., Feng, T., & Zhang, J. (2019). Erythrocytic synuclein as a potential biomarker for Parkinson's disease. *Translational Neurodegeneration*, *8*, 15, https://doi.org/10.1186/s40035-019-0155-y

Tinaz, S., Courtney, M. G., & Stern, C. E. (2011). Focal cortical and subcortical atrophy in early Parkinson's disease. *Movement Disorders*, *26*, 436–441.

Travagli, R. A., Browning, K. N., & Camilleri, M. (2020). Parkinson's disease and the gut: New insights into pathogenesis and clinical relevance. *Nature Reviews: Gastroenterology & Hepatology*, *17*(11), 673–685, https://doi.org/10.1038/s41575-020-0339-z

Trupp, M., Jonsson, P., Ohrfelt, A., Zetterberg, H., Obudulu, O., Malm, L., Wuolikainen, A., Linder, J., Moritz, T., Blennow, K., Antti, H., & Forsgren, L. (2014). Metabolite and peptide levels in plasma and CSF differentiate healthy controls from patients with newly diagnosed Parkinson's disease. *Journal of Parkinson's Disease*, *4*(3), 549–560, https://doi.org/10.3233/JPD-140389

Tuovinen, N., Seppi, K., de Pasquale, F., Muller, C., Nocker, M., Schocke, M., Gizewski, E. R., Kremser, C., Wenning, G. K., Poewe, W., Djamshidian, A., Scherfler, C., & Seki, M. (2018). The reorganisation of functional architecture in the early stages of Parkinson's disease. *Parkinsonism and Related Disorders, 50*, 61–68.

Vivacqua, G., Suppa, A., Mancinelli, R., Belvisi, D., Fabbrini, A., Costanzo, M., Formica, A., Onori, P., Fabbrini, G., &

Berardelli, A. (2019). Salivary alpha-synuclein in the diagnosis of Parkinson's disease and progressive supranuclear palsy. *Parkinsonism and Related Disorders*, *63*, 143–148, https://doi.org/10.1016/j.parkreldis.2019.02.014

Vlaar, A. M. M., van Kroonenburgh, M. J. P. G., Kessels, A. G. H., & Weber, W. E. J. (2007). Meta-analysis of the literature on the diagnostic accuracy of SPECT in Parkinsonian syndromes. *BMC Neurology*, *7*, 27.

Volpicelli-Daley, L. A., Luk, K. C., Patel, T. P., Tanik, S. A., Riddle, D. M., Stieber, A., Meaney, D. F., Trojanowski, J. Q., & Lee, V. M. (2011). Exogenous synuclein fibrils induce Lewy body pathology, leading to synaptic dysfunction and neuron death. *Neuron*, *72*(1), 57–71, https://doi.org/10.1016/j.neuron.2011.08.033

Wang, Q., Liu, Y., & Zhou, J. (2015). Neuroinflammation in Parkinson's disease and its potential as a therapeutic target. *Translational Neurodegeneration*, *4*, 19, https://doi.org/10.1186/s40035-015-0042-0

Weisskopf, M. G., O'Reilly, E., Chen, H., Schwarzschild, M. A., & Ascherio, A. (2007). Plasma urate and the risk of Parkinson's disease. *American Journal of Epidemiology*, *166*(5), 561–567, https://doi.org/10.1093/aje/kwm127

Xiromerisiou, G., Marogianni, C., Androutsopoulou, A., Ntavaroukas, P., Mysiris, D., & Papoutsopoulou, S. (2023). Parkinson's disease, it takes guts: The correlation between intestinal microbiome and cytokine network with neurodegeneration. *Biology*, *12*(1), 93, https://doi.org/10.3390/biology12010093

Yao, L., Liang, W., Chen, J., Wang, Q., & Huang, X. (2023). Constipation in Parkinson's disease: A systematic review and meta-analysis. *European Neurology*, *86*(1), 34–44, https://doi.org/10.1159/000527513

Yasuda, T., Nakata, Y., & Mochizuki, H. (2013). Synuclein and neuronal cell death. *Molecular Neurobiology*, *47*(2), 466–483, https://doi.org/10.1007/s12035-012-8327-0

Zhang, J., Jin, J., Su, D., Feng, T., & Zhao, H. (2023). Tau-PET imaging in Parkinson's disease: A systematic review and meta-analysis. *Frontiers in Neurology*, *14*, 1145939, https://doi.org/10.3389/fneur.2023.1145939

Zhang, Z., Zhang, S., Fu, P., Zhang, Z., Lin, K., Ko, J. K., & Yung, K. K. (2019). Roles of glutamate receptors in Parkinson's disease. *International Journal of Molecular Sciences*, *20*(18), 4391, https://doi.org/10.3390/ijms20184391

Zhao, P., Zhang, B., & Gao, S. (2012). 18F-FDG PET study on idiopathic Parkinson's disease in several Parkinsonian-plus syndromes. *Parkinsonism & Related Disorders*, *18*(1), S60–S62.

Zhou, F., Yu, T., Du, R., Fan, G., Liu, Y., Liu, Z., Xiang, J., Wang, Y., Song, B., Gu, X., Guan, L., Wei, Y., Li, H., Wu, X., Xu, J., Tu, S., Zhang, Y., Chen, H., & Cao, B. (2020). Clinical course and risk factors for mortality in adult inpatients with COVID-19 in Wuhan, China: A retrospective cohort study. *Lancet*, *395*(10229), 1054–1062, https://doi.org/10.1016/S0140-6736(20)30566-3

Zimmermann, M., & Brockmann, K. (2022). Blood and cerebrospinal fluid biomarkers of inflammation in Parkinson's disease. *Journal of Parkinson's Disease*, *12*(s1), S183–S200, https://doi.org/10.3233/JPD-223277

12 Parkinson's Disease Models Pros and Cons

Sobanaa Murugesan, Rajesh Pamanji, and Joseph Selvin

12.1 INTRODUCTION

Parkinson's disease affects countless individuals worldwide, presenting as a progressive and chronic neurological disorder. Although great strides have been made in comprehending its roots and potential remedies, the development of successful therapeutic approaches remains a daunting task (Konnova and Swanberg, 2018). In order to expedite research and facilitate the evaluation of potential treatment options, the generation of Parkinson's disease models has become imperative (Chia et al., 2020). These models, whether generated in animals or through advanced cell culture systems, play a crucial role in mimicking the pathophysiological processes of the disease, allowing scientists to study its mechanisms, test novel therapeutic approaches, and, ultimately, improve the lives of those living with Parkinson's disease (Kin et al., 2019).

At present, there are various neurotoxic models of Parkinson's disease (PD) available, which cause significant degeneration in the substantial nigrostriatal region and effectively replicate the motor symptoms of PD. Each of these models comes with its own pros and cons. The choice of model depends on the specific research objectives (Rai and Singh, 2020). Generally, MPTP is considered the standard model for toxin-induced PD in animals because it successfully induces the formation of alpha-synuclein-positive inclusions in both rodent and non-human primate models (Breger and Fuzzati Armentero, 2019). However, creating an ideal experimental model that accurately reproduces all of PD's phenotypic and pathological features remains challenging. Such a model is crucial for obtaining a comprehensive understanding of the underlying mechanisms of PD. Thus, there is no perfect model of PD as no model can fully replicate the human condition; these models are only representations, each potentially having some significance level.

12.2 MPTP—PD MOUSE MODELS

The MPTP (1-methyl-4-phenyl-1,2,3,6-tetrahydropyridine) mouse model is commonly used in PD research because it closely mimics the symptoms seen in human PD and allows for further understanding of the molecular and cellular mechanisms involved. MPTP causes various insults to the brain, leading to damage in dopaminergic neurons and Parkinson-like symptoms. The MPTP-induced mouse model has been widely used in PD research due to its simplicity, practicality, affordability, and clinical correlation (Mustapha and Mat Taib, 2021). On the other hand, mice were given a substance called MPTP to examine the behavioural pattern after chronic treatment. The findings showed that three days of treatment with MPTP in mice led to alterations in protein expression and neuronal death comparable to those in the prodromal stage of PD. Unlike mice treated for an additional 21 days, which displayed noticeably more severe symptoms consistent with the clinical course of PD, mice treated for another 14 days revealed behavioural and histological abnormalities consistent with the early clinical stage (Chia et al., 2020; Tian et al., 2023). In addition, MPTP models in mice replicate the degenerative processes that PD patients experience in other brain areas and the periphery in addition to the nigrostriatal system, as done by Kozina et al. (2014). It also helps to ascertain how DHA affected the distribution of apelin and the apelin receptor (APJ) in the central nervous system in a PD model generated by 1-methyl-4-phenyl-1, 2, 3, and 6 tetrahydropyridine (MPTP) (Acar et al., 2019). Moreover, Acar et al. (2019) conducted a study to ascertain how DHA affected the distribution of apelin and apelin receptor (APJ) in the central nervous system in a model of PD brought on by 1-methyl-4-phenyl-1, 2, 3, and MPTP. DHA boosted the animals' ambulatory activity, distance travelled, and overall locomotor activity in the PD model.

Additionally, MPTP-induced behavioural abnormalities in rat models are catalogued using the rotarod, pole-climbing, and grasping tests to identify motor function deficits. MPTP-mediated behavioural differences are primarily caused by the degenerative death of dopaminergic neurons in the midbrain and striatal nerve degeneration. They also established that in rats treated with MPTP but not control (PD) rats, AuNPs produced from *Cinnamomum verum* reduced changes in the pathway of TLR/NF-B related molecules (Linga et al., 2019). Further, animals treated with MPTP significantly help to understand the drug treatment basis (Wang et al., 2022), when it reduced their capacity for exploration, while animals given PD mice's faecal microbiota exhibited comparable behaviours. Ceftriaxone intraperitoneally injected reversed the behavioural alterations brought on by MPTP and PD faecal microbiota (Zhou et al., 2021; Collyer et al., 2020). Ginseng, a natural treatment, has been found to possess positive effects on nerve growth and protection. These properties could potentially help in

DOI: 10.1201/9781003402893-14

preventing various types of neuronal cell decline, including the degeneration of the nigrostriatal pathway observed in PD and other similar conditions. In these studies, two rodent models of PD have been used to see if the ginseng extract G115 may have neuroprotective effects. Before and/or after exposure to the neurotoxin that causes Parkinsonism in mice, MPTP, or its toxic metabolite, 1-methyl-4-phenylpyridinium (MPP), in rats, G115 was given orally to the animals. The appearance of locomotor dysfunction was diminished and tyrosine hydroxylase-positive cell loss in the substantia nigra was significantly and dramatically blocked by such treatment. In rodent models of PD, ginseng appears to protect against neurotoxicity when taken orally. The prolonged oral treatment of the ginseng extract G115 dramatically reduces measures of functional asymmetry and dopaminergic cell death after exposure to a neurotoxic that causes Parkinsonism.

These results support a growing body of research showing that ginseng and each of its constituent parts prevent neuronal cell death (Kampen et al., 2003). Moreover, rats have been shown to have relatively high amounts of monoamine oxidase B in the blood-brain barrier (BBB), which makes them capable of converting MPTP to MPP+ at this location. This is significant since it indicates that rats are highly resistant to MPTP. MPP+ cannot cross the BBB and enter the brain because it is not lipophilic (Riachi et al., 1989). Mice are more susceptible to the systemic effects of MPTP than are rats because they are less effective at converting MPTP to MPP+ at the BBB. Thusly, IP injection of MPTP (30 mg/kg) to mice causes DA neuronal loss in the SN and a gradual decrease in the levels of DA and its metabolites, 3,4-dihydroxyphenylacetic acid (DOPAC), and HVA, in the CS (Heikkila et al., 1984). However, it's crucial to note that only specific strains of mice are vulnerable to MPTP injection.

12.2.1 Pros

It is a cost-effective and easily manageable model with fewer ethical concerns than other toxin-induced animal models. The MPTP-mouse model has helped identify non-motor symptoms of PD and has played a role in electrophysiological studies for deep brain stimulation-related therapy (Bobela et al., 2014). Another benefit of MPTP-mouse models is the ability to contrast their health with that of transgenic mice that replicate certain aspects of the PD pathophysiology, such as α-synucleinopathy. Induced MPTP models help in better tracking the physiological checkpoints and also in understanding the reciprocation response of the models during the treatment (Liu et al., 2020; Han et al., 2022b; Sun et al., 2022). It has also been instrumental in developing potential therapeutic strategies for protective and restorative treatments. Additionally, the model has furthered our understanding of mitochondrial dysfunction and autophagy's role in PD pathogenesis. Overall, the MPTP-mouse model has significantly contributed to advancements in clinical research for PD.

12.2.2 Cons

This model has limitations in inducing Lewy body formation and demonstrating behavioural features similar to human Parkinson's disease. Moreover, like other PD models, MPTP models in mice have several drawbacks, such as the infrequent occurrence of α-synucleinopathy. Moreover, unlike monkeys, there have been no successful attempts to reproduce the gradual decline of nigrostriatal dopaminergic neurons and the resulting motor impairment in mice once they reach a certain level of degradation. Despite extensive research into MPTP-induced NHP models, their high rates of symptom variation make it difficult to develop treatments. Moreover, administering this toxin does not result in a stable PD model in rats and does not help simulate PD-MCI (Marshall et al., 2019).

12.3 MPTP—CYNOMOLGUS MONKEYS (*MACACA FASCICULARIS*)

Although MPTP exhibits a strong preference for affecting dopaminergic neurons, there have been inconsistent reports regarding its impact on adrenergic, serotoninergic, and other systems. Lewy bodies, the pathologic sign of Parkinson's disease, have not been definitively confirmed after MPTP toxicity in humans or monkeys. Nevertheless, administering MPTP systematically to macaques, as commonly done, replicates significant cell loss in the substantia nigra pars compacta. This results in the typical motor symptoms of Parkinson's disease, such as abnormal movements and postures, affecting both the body axis and extremities. Additionally, it elicits the complete range of motor complications associated with long-term dopaminergic treatment (Potts et al., 2014). However, the varied sensitivity to MPTP, which entails varying reactions between primate species and within individual monkeys, was one of the study's significant results (Langston, 2017). Despite obtaining a lower MDS than CM, RM needed more MPTP, highlighting the significance of species choice. The varying sensitivity to the toxin across rodents, non-human primates, and human primates aligns with the differences in reaction amongst macaque species. This might be explained by changes in metabolism, such as how MPP+, the poisonous metabolite of MPTP, is converted or cleared. The same species of monkeys were given various MPTP doses, but equivalent (moderate) Parkinsonism was the intended outcome (Seo et al., 2019).

Furthermore, melatonin or CDK5 knockdown were able to completely correct the PD phenotype that MPTP induced in mice. According to a study by Su et al. (2015), CDK5-mediated autophagy is implicated in the onset of PD, and regulating autophagic pathways may be a useful marker for finding potential targets for new treatments. This study evaluated the potential therapeutic effects of CBP on both motor and non-motor symptoms using a rat model of PD. The researchers wanted to investigate how CBP affected the gut microbiota, which is linked to inflammation, after giving the rats MPTP through the gut to induce PD. For comparison, a control group was also included. The study's findings were significant. The PD rats that received treatment with CBP demonstrated considerably higher performance in tests like the rotarod test, colonic motility, and gastrointestinal transit (Cammarota et al., 2016). Moreover, CBP administration led to a notable decrease in dopaminergic cell death in the SNpc region of the brain and significantly enhanced colonic motility. The researchers suggest that combining CBP with cord blood stem cells may be a future avenue to explore for developing an innovative therapy for PD, as CBP currently shows promise in treating the inflammatory gut associated with PD (Lee et al., 2019). Li et al. (2021) found a cynomolgus monkey with spontaneous PD, according to four lines of evidence: typical PD clinical signs, therapeutic responses, pathology hallmarks, and genetic alterations. It shows that PD is not a condition that affects only humans, making monkeys prime candidates to produce accurate animal versions of PD, in which the etiology and pathology are conserved across evolutionary history.

12.3.1 Pros

Studies in monkeys help to understand the exhibiting akinesia and bradykinesia changes in GA levels, revealing the degree of movement abnormalities brought on by MPTP. The 18F-FP-CIT BP in all seven sub-regions of the striatum was favourably connected with GA assessments and negatively correlated with ratings of Parkinsonian behaviour in all animals. It enables comparison of the parallels and variations in PD development across species and aids in understanding PD etiology from an evolutionary standpoint.

12.3.2 Cons

Despite having drawbacks like a longer lifespan, expensive upkeep, and complex ethical issues, non-human primates like marmosets (*Callithrix jacchus*), cynomolgus macaques, and baboons help analyze the motor and non-motor symptoms of PD and preclinical studies of novel drug therapies. The models that would have been suitable would probably not have been able to match the data if the timeframes for developing the model were also taken into consideration. Because of this, even though it hasn't been demonstrated, the repeatability of primate results in clinical research and the generalizability of data from various model applications may be constantly questioned (Thirugnanam & Santhakumar, 2022). The MPTP-treated monkey has been acknowledged as the premier PD model for the past three decades since it exhibits many of the major pathophysiological and behavioural abnormalities found in PD patients. The absence of neuronal loss outside of the nigrostriatal dopaminergic system, as with other neurotoxin-based models of PD, is a recurrent complaint levelled at this model by several writers (Masilamoni and Smith, 2018).

12.4 MPTP—MICE MODEL

Numerous investigations have revealed a link between the stomach and the development of PD. Patients with PD have gastroparesis and intestinal abnormalities such as intestinal inflammation. It's interesting to note that, for many years, these atypical gastrointestinal symptoms precede damage to the CNS and motor abnormalities (Huo et al., 2021; Liang et al., 2021). According to mounting data, PD occurrence, development, and progression may be linked to gut microbiota dysbiosis (Hirayama et al., 2021; Tan et al., 2022). For diagnosing and treating PD, changing the makeup of the gut microbiota is a feasible target (Fan et al., 2022a). Intestinal inflammation, neuroinflammation, gut microbiota, and motor function (Qiu et al., 2020) were only a few of the conditions that the NaB therapy had no discernible impact on in normal animals. Nevertheless, NaB is crucial in the management of illness. Notably, NaB has anti-apoptotic, anti-inflammatory, and neuroprotective effects in PD animals (Kakoty et al., 2021; Zhang et al., 2022), although the precise mechanism behind these actions is yet unknown. The study's findings by Guo et al. (2023) suggested that NaB has neuroprotective benefits through enhancing gut microbial dysbiosis, reducing intestinal and neuroinflammation, and activating the TLR4/MyD88/NF-B pathway (Dutta et al., 2019). A work by Huo et al. (2021) assessed the impact of three dominant SCFAs on motor dysfunction, dopaminergic neuronal degeneration, and underlying neuroinflammatory-associated pathways in MPTP-induced PD mice, where the three dominant SCFAs are acetate, propionate, and butyrate.

12.4.1 PROS AND CONS

Different short-chain fatty acids' (SCFAs') effects on PD and its underlying processes are yet unknown.

12.5 6-OHDA—RAT MODELS

6-OHDA-induced rat models help better understand the complex system pathway's effect in mitochondria (Prajapati et al., 2017), oxidant stress (Smith et al., 2008), motor, emotional, and cognitive behaviours (Idrissi et al., 2023). It has been reported that the neuronal loss is mediated, at least in part, by oxidative stress brought on by increased iron release from other iron-storing cell compartments and from Fe-S cluster proteins of the mitochondrial respiratory chain, which in turn worsens reactive oxygen species (ROS) production via the Fenton's reaction (Onukwufor et al., 2022). Iron is a transition metal that is essential for cellular health and the metabolism of neurotransmitters, including dopamine (DA). Still, when it rises over homeostatic levels, it becomes toxic and causes oxidative damage, ultimately resulting in cell death (Nnah & Wessling-Resnick, 2018). When 6-hydroxydopamine (6-OHDA) is injected into a rat's medial forebrain bundle (MFB), it replicates the severe loss of dopaminergic neurons and also simulates the symptoms of PD (Guimarães et al., 2021). 6-OHDA-induced hemi PD paradigm with different striatal DA levels in rat models was taken to study the motor deficits of the model. The forelimb movement in the hemi-PD paradigm was thus the most adept at spotting the decline in striatal DA level.

Since compensating mechanisms exist, it has been observed that apparent motor abnormalities do not manifest until the striatal DA level has decreased to roughly 20% of normal. In contrast to the 6-OHDA-induced model, the LPS model showed a less pronounced drop in striatal DA levels. Moreover, the 6-hydroxydopamine (OHDA) and lipopolysaccharide (LPS)-induced hemi-PD model rats were treated with various striatal DA levels to conduct multiple behavioural tests (open-field, forepaw adjustment steps (FAS), cylinder, apomorphine-induced rotationrota-rod, and beam walking) to determine the most accurate method for predicting the DA level from earlier disease stages. In contrast to the 6-OHDA-induced model, the LPS model showed a less pronounced drop in striatal DA levels. The results of the behavioural tests that measured the forelimb movements had the strongest correlations with DA level drop. They also correlate well with reduced body weight increase (Miyanishi et al., 2019). A previous study by Haghparast et al. (2018, 2019) demonstrated that apelin-13 therapy significantly restored SN dopaminergic neurons, whereas 6-OHDA injection into SNc significantly reduced dopaminergic neurons in the SNc of lesion rats. However, in a more recent work by Esmaeili-Mahani et al. (2021), apelin was applied topically to the SNc after 6-OHDA was used to kill the dopaminergic neurons in the SNc by injecting it into the rats. Based on the findings from H&E staining, it was theorized that apelin injection can directly enhance dopaminergic neurons in the SNc and indirectly affect the hippocampus. The staining revealed that the 6-OHDA injection caused damage to SNc dopaminergic neurons, but apelin administration preserved these neurons. A study by Wang et al. (2018) described the gender variations in synaptic activity in a brain region essential for learning spatial information. According to findings, female rats have greater synaptic levels of oestrogen receptor (ER) (Hojo et al., 2008) and, in contrast to males, need membrane ER to activate signalling kinases that enable long-term potentiation (LTP), a kind of synaptic plasticity considered to be the basis of learning (Ogiue-Ikeda et al., 2008). The threshold for LTP and hippocampal field CA1-dependent spatial learning was raised due to the extra need for oestrogen signalling in females.

Some animal models of Parkinsonism and other neurodegenerative illnesses responded favourably to cerium oxide nanoparticles (CeO2NPs). The role of CeO2NPs in Parkinsonism caused by 6-OHDA has not yet been fully explored. Additionally, the 0.5 mg/kg dose of CeO2NPs almost restored the neurobiochemical derangements, but the 0.1 mg/kg dose was inadequate to change the biochemical measures in the striatum. Without substantially changing oxidative stress, 1 mg/kg of CeO2NPs was administered to somewhat improved striatal dopamine and lowered apoptosis (Hegazy et al., 2017). In 2016, Noor et al. found significant results of their study showing that CBL has a promising therapeutic impact in the 6-OHDA rat model of PD by reducing oxidative stress, replenishing dopamine levels, and improving behavioral outcomes.

Additionally, Bar-am and his team (2015) used the 6-OHDA-rat model to study the neuroprotective impact of VAR. They showed that VAR therapy reduced rotating behaviour in rats, prevented decreased striatal DA content, and increased DA turnover. In mice, VAR injection enhanced DA levels and protein expression. It is advantageous to research glucose metabolism, dopamine transporter (Dat) binding, behaviour (limb use asymmetry test), and dopaminergic innervation (validated by in vitro tyrosine hydroxylase staining) using the 6-OHDA lesion rat model, which is a well-known acute model for PD (Casteels et al., 2008). Furthermore, according to Lazarova et al.'s (2022) study employing the 6-OHDA model of PD, the recently synthesized amantadine derivative Tyr-Am significantly inhibited PD progression, where this model is used for the better understanding of the Tyr-Am's neuroprotective effects that were equivalent to those of amantadine, enhancing learning and memory capacity as well as neuromuscular coordination.

12.5.1 Pros

With some success, these models can mimic the main symptoms of PD, such as Lewy bodies, gradual loss of dopaminergic neurons in the substantia nigra pars compacta, and motor abnormalities. When examining how new medications affect motor abilities, researchers can use the 6-OHDA animal model. Overall, pharmacological animal models are repeatable and have significantly advanced our understanding of PD (Salari and Bagheri, 2019). Preclinical models can potentially improve our experience of the cellular and molecular mechanisms underlying pain in PD and serve as testing grounds for novel analgesics' effectiveness in treating this crippling non-motor symptom (Buhidma et al., 2020). However, the unilateral striatal 6-OHDA model has the benefit of accurately simulating nigrostriatal tract degeneration, which is also well tolerated by the animals (Penttinen et al., 2016, 2018a, 2018b).

12.5.2 Cons

However, 6-OHDA does not penetrate the BBB but destroys dopamine and norepinephrine-containing neurons by creating hydrogen peroxide. A study has found that MFB injuries in rats result in modest cognitive impairments associated with PD. Because PD-MCI is not associated with a whole range of abnormalities, 6-OHDA injuries in other brain regions, such as the PFC, may result in cognitive impairment that more closely matches the disease. Therefore, combined rat lesions in which the 6-OHDA toxin is bilaterally injected into the MFB and PFC should be employed to more precisely simulate PD-MCI and eventually develop new therapies to treat PD-MCI (Marshall et al., 2019). Moreover, the unilateral striatal 6-OHDA model's drawback is that it is not very good at simulating human PD because the loss of dopaminergic neurons advances more quickly in rats than in patients (days to weeks), and there are no telltale signs like Lewy bodies or other filamentous intraneuronal inclusions (Muma et al., 2001; Zeng et al., 2002; Airavaara et al., 2020).

12.6 PARAQUET—RAT MODELS

Following the discovery that MPTP/MPP+ may cause Parkinsonism, the herbicide paraquat, which has a structure similar to MPP+, has been utilized to imitate PD. The BBB is also crossable by paraquat (Shimizu et al., 2001). The pretreatment of L-valine, a high-affinity substrate for the neutral amino acid transporter, significantly decreased the penetration of paraquat through the BBB, indicating that the neutral amino acid transporter is involved in the penetration process. Due to paraquat's poor affinity for mitochondrial complex I, it is not inhibited by paraquat. As a result, complex I inhibition does not significantly contribute to paraquat's (PQ) neurotoxicity. The pentose phosphate pathway is manipulated by PQ to produce more NADPH-reducing equivalents and promote PQ redox cycling (Powers et al., 2017). It has been suggested that the commonly used herbicide PQ, which causes oxidative stress, may increase the likelihood of developing PD (Cristóvão et al., 2020).

Ahmad et al. (2021) conducted a study to explore the potential neuroprotective effects of NAR in a rat model with PQ-induced Parkinsonism in SH-SY5Y cells. This research aimed to address the gaps in knowledge regarding the neuroprotective role of NAR and the mechanisms behind it. Significant neuroprotection against PQ-induced behavioural impairments, oxidative stress, mitochondrial dysfunction, and astrocytosis was seen in rats after NAR therapy. A mouse model of progressive lung fibrosis has been developed using PQ intratracheally. This study helped to understand the top five pathways—chemical carcinogenesis, cytochrome P450-mediated drug metabolism, cytochrome P450-mediated xenobiotic metabolism, complement and coagulation cascades, and protein digestion and absorption—were active on days 7 and 28. On day 28, compared to day 7, several pathways were more enriched, including tight junction, focal adhesion, platelet activation, and ECM-receptor interaction. Spearman $R = 0.3950$ and 0.2477 on days 7 and 28, respectively; from an integrated analysis of the proteome and transcriptome, it showed a moderate correlation between quantitative protein abundance ratios and RNA abundance ratios, demonstrating the significance of post-transcriptional regulation in lung injury and repair was reported by Fan et al. (2022b). However, PQ-mediated ROS production in the SNpc, which is first generated by the fast influx of extracellular Zn^{2+} into dopaminergic neurons via AMPA receptor activation, causes nigrostriatal dopaminergic degeneration, which causes PQ-induced PD in rats. The SNpc pathogenicity caused by PQ is defended by blocking intracellular Zn^{2+} toxicity, which is caused by intracellular Zn^{2+} dysregulation in dopaminergic neurons (Tamano et al., 2019).

Additionally, injections of haloperidol and paraquat in rats have been shown to produce dopaminergic neuron degeneration in the substantia nigra (Kumar et al., 2012; Grandi et al., 2018). In both haloperidol and PQ-induced PD models, dicyclomine therapy seemed to raise the levels of brain neurotransmitters (Sanawar et al., 2020). Male Sprague-Dawley rats exhibited altered stomach tone and motility in response to brainstem dopamine microinjection after receiving three weekly IP injections of PQ (10 mg/kg) (Bove et al., 2019). These effects might be brought on by a nigrovagal pathway that has recently been revealed to regulate the motility of stomach tone and has been proven to be sensitive to PQ exposure (Anselmi et al., 2017). This adds more proof to the mounting body of data that PQ exposure may play a critical role in mediating gastrointestinal alterations in PD (El-Gamal et al., 2021). A study by Zuo et al. (2021) used a PQ-induced model to study the PQ related to increased PD risk and neuropathology resembling PD. PQ caused mice's tyrosine hydroxylase levels to drop, which led to movement damage. Through the ferritinophagy pathway produced by NCOA4, PQ significantly contributed to the iron buildup in the cytoplasm and mitochondria in vitro. By generating lipid ROS, iron excess induced lipid peroxidation via 12Lox, which subsequently caused ferroptosis. PQ dramatically elevated Cox2 expression while drastically downregulating SLC7A11 and GPX4 expression, vital indicators of ferroptosis. Fer-1, a ferroptosis inhibitor, might considerably reduce the ferroptosis brought on by PQ.

12.6.1 Pros

Despite many available animal models, using rat models helps to replicate several neuropathological features of PD (Cristóvão et al., 2020). According to a recent study conducted by Zeng et al. (2018), the primary cause of the neurotoxins' molecular effects is the disruption of mitochondrial function. This is achieved through the inhibition of mitochondrial complex I, increased oxidative stress, inflammation in the nervous system, interference with protein degradation, and programmed cell death.

12.6.2 Cons

Finding an animal model that consistently mimics human PD conditions regarding both motor behaviours and biochemical pathological outcomes is difficult. According to 2014 research by Minnema et al. (2014), PQ (10–50 ppm) similarly failed to activate microglia and astrocytes, destroy DA neurons in the SN or CS, or lose DA or its metabolites. The conclusions drawn from the PQ rodent model are highly varied. The contradictory information from clinical and epidemiological research is consistent with this (El-Gamal et al., 2021).

12.7 RETENONE—RAT MODEL

Although PD is the second most prevalent neurodegenerative disorder, none of the available therapies for PD can stop the illness's progression since the condition's pathophysiology is still poorly understood.

The term "microbiota-gut–brain axis" refers to the connection between the brain and the digestive system regulated by gut microbiota in developing PD (Mulak & Bonaz, 2015; Ullah et al., 2023). The specific processes of microbial dysbiosis in the development of PD have not yet been fully understood. Rotenone triggers in vivo loss of substantia nigra dopaminergic cells, a characteristic feature of PD neuropathology, according to several laboratory investigations. Additionally, rotenone causes α-synuclein to aggregate in vivo, another characteristic of the neuropathology associated with Parkinson's disease. α-synuclein is a crucial component of Lewy bodies and Lewy neurites, which were seen in the brains of PD patients. Some in vivo rotenone models also replicate peripheral symptoms of PD, including as decreased intestinal motility and peripheral α-synuclein aggregation, which are both believed to occur before to the characteristic symptoms of PD in people, such as cogwheel stiffness, bradykinesia, and resting tremor. The same rotenone exposure regimen for animal cohorts as well as across cohorts exposed to identical rotenone dosages have both shown variability (Innos and Hickey, 2021). Zhao et al. (2021a) demonstrated for the first time how FLZ therapy can significantly lessen the GI dysfunctions and motor symptoms that rotenone-induced mice develop. They reported that by preventing the activation of the TLR4/MyD88/NF-kB signalling pathway in both the gut and the brain, FLZ therapy influences the microbiota-gut–brain axis (Zhao et al., 2021b). According to recent studies, FLZ therapy reduces microbiota dysbiosis to safeguard the PD mouse model by blocking the TLR4 pathway, contributing to one of the fundamental processes underpinning its protective benefits. They also demonstrated the significance of the microbiota-gut–brain axis in the aetiology of PD, raising the possibility that it might serve as a brand-new therapeutic target for treating the disease.

In order to study the effect of dapagliflozin, Arab et al. (2021) looked at dapagliflozin's possible neuroprotective benefits against PD-related neurodegenerative abnormalities and motor dysfunction in a rat model of PD brought on by rotenone. In the open-field and rotarod tests, dapagliflozin (1 (mg/kg)/day, via gavage for 3 weeks) reduced PD motor dysfunction and enhanced motor coordination without inducing hypoglycemia. Additionally, it reduced the histopathological changes and α-synuclein expression and increased levels of tyrosine hydroxylase and dopamine (Lyu et al., 2021). By reducing lipid peroxides, dapagliflozin significantly reduced neuronal oxidative stress, which led to the repair of the disrupted DJ-1/Nrf2 pathway (Zhang et al., 2017a; Soni & Kumar, 2022; Han et al., 2022a; Li et al., 2023). Additionally, dapagliflozin increased GDNF and the PI3K/AKT/GSK-3 (Ser9) pathway downstream of GDNF, preventing ROS-dependent neuronal death (Fão et al., 2019; Niu et al., 2021). Meanwhile, it reduced neuroinflammation by reducing TNF- and NF-B pathway activity. Moreover, a study by Azar et al. (2022) indicated agmatine's unique function in PD and dyskinesia amelioration through modulating NMDA receptors, Nrf2, and HMGB1/RAGE/TLR4/NF-B signalling pathways, whereas cilostazol exhibits the action by inhibiting mTOR, also activates Nrf2/HO-1, suppresses the HMGB1/TLR4 axis, and upregulates PI3K/Akt in order to protect neurons against the neurodegenerative effects of rotenone (El-Sayed et al., 2023).

12.7.1 Pros

In rats given rotenone treatment in 2000, Betarbet et al. successfully mimicked most of the behavioural, biochemical, and clinical characteristics of PD (Radad et al., 2019). Rat models help to understand the neuropathic effect of any medicinal extract and give better knowledge on studying physiological pathways in PD models.

12.8 ROTENONE—MICE MODEL

According to research by Dodiya et al. (2020), the low-dose oral rotenone PD mice paradigm led us to theorize that intestinal barrier failure brought on by stress and microbial dysbiosis also aggravates the PD phenotype. Chronic stress has been observed to increase neuronal decline and motor deficits in animal models of PD. It has also been connected to intestinal dysbiosis and hyperpermeability. Additionally, in PD, altered gut microbiota composition and motor impairments frequently coexist with GI symptoms, including impaired intestinal barrier function (Zhang et al., 2017b). Using a mouse model to study the PD induced by rotenone, Bhattarai et al. (2021) explored the impact of gut microbiota and epithelial barrier disruption on developing motor symptoms in this work. The study showed that where tyrosine hydroxylase (TH) neurons were lost after six weeks of chronic oral gavage treatment of 10 mg/kg of rotenone in both germ-free (GF) and conventionally bred (CR) mice, only CR animals showed a decline in motor strength and coordination. While chronic rotenone therapy did not affect intestinal permeability in GF mice, it significantly altered the makeup of the gut microbiota and increased intestinal permeability in CR animals. These findings provide information on the gut microbiota's possible contribution to controlling PD's motor impairments and barrier dysfunction. The study by Lee et al. (2023) showed that administration of *Lactobacillus plantarum* PS128 (PS128) significantly reduced dopaminergic neuron loss, decreased microglial activation, decreased levels of inflammatory factors, and increased expression of neurotrophic factor in PD-like mice. Their research showed that PS128 regulates the gut microbiome and the miR-155–5p/SOCS1 pathway in rotenone-induced PD-like mice to alleviate motor impairments and display neuroprotective benefits. A recent investigation by Zhao et al. (2021b) shows that fecal microbiota transplantation therapy can improve the rotenone-induced PD mouse model and correct gut microbiota dysbiosis, with possible significant contributions from the inhibition of inflammation caused by the LPS-TLR4 signaling pathway in both the gut and the brain. We also demonstrate that the microbiota-gut–brain axis contributes to developing PD via rotenone-induced microbiota dysbiosis (Gupta and Khanna, 2017; Sun and Shen, 2018).

The natural polyphenolic substance curcumin, which is derived from the rhizomes of the ginger family member *Curcuma longa* (turmeric), has been utilized widely in southeast Asia and around the world (Wu et al., 2013). In a 2016 study, Khatri and Juvekar looked into the potential neuroprotective benefits of curcumin on rotenone-induced cognitive decline, oxidative stress, and mitochondrial dysfunction in mice. Three weeks of chronic rotenone therapy were given to albino mice by the researchers. As a result, according to many tests like the actophotometer, rotarod, and open-field tests, the mice's cognitive ability dramatically decreased. The rotenone-treated animals also displayed decreased activity of antioxidant enzymes such as superoxide dismutase, catalase, and glutathione as well as elevated levels of lipid peroxidation and nitrite, indicating increased oxidative stress. The rotenone treatment also resulted in a decrease in the activity of the enzymes II and III of the mitochondrial complexes. According to the study's findings, curcumin has the potential to be used as a therapeutic therapy for PD since it has neuroprotective effects against the PD that rotenone causes in mice (Cui et al., 2016).

12.8.1 Pros

Rotenone is a pesticide of wide usage that was found to be mimicking the pathophysiological condition of PD (Zhao et al., 2021b). Rotenone is a powerful tool for examining most genetic and cellular pathology in PD patients. Rotenone may thus effectively help evaluate new therapy approaches that might be advantageous for PD patients. Recent research has shown that PD patients' lungs alter with time, leading to airflow restriction and weak ventilatory muscles, among other symptoms.

12.8.2 Cons

Rotenone injection into the peritoneum was said to only cause behavioural and neurochemical abnormalities with a very high death rate (Cannon et al., 2009). Rotenone is a useful tool for studying Parkinsonism, but because it is not selective for the dopaminergic system, it may negatively affect other neuronal systems (Pingale and Gupta, 2020). Lee et al. (2023) reported that motor impairments are one of PD's main symptoms, and it was challenging to determine how PS128 affected control mice whose motions were typical.

12.9 RESERPINE-TREATED RATS AND MICE MODELS

Reserpine is a medication employed in PD modelling research, and it is administered to mouse models. At a usual dosage, reserpine depletes monoamines in the brain and peripheral locations, including dopamine. The aforementioned illness results in symptoms including hindlimb rigidity and loss of movement, which are similar to Parkinson's disease. The model's objective is to assess the effectiveness of various therapies in mitigating these symptoms, particularly concerning their capacity to reverse them. However, it is not acceptable to investigate the effects on individuals of heavy drug use (Duty & Jenner, 2011). The use of reserpine in Parkinson's disease research may be controversial due to its potential side effects, including drowsiness and blood pressure changes. The treatment's impact on behaviour and neurochemical changes may also be independent.

12.9.1 Pros

Treatment with reserpine causes decreased locomotor activity and muscular stiffness in rats, which are symptoms similar to Parkinson's disease. Researchers can examine possible therapeutic approaches and neurochemical pathways using this model. Reserpine depletes dopamine reserves, mimicking Parkinson's disease sufferers' dopaminergic deficiencies. Researchers may assess how well medications such as lithium and cerebrolysin work to reduce Parkinson's disease-like symptoms, which could lead to the creation of novel therapies (Leao et al., 2015). According to Tharwat et al. (2023), the PD model created by reserpine satisfies both face and pharmacological validity requirements, exhibiting neurochemical and molecular characteristics akin to the pathophysiology of the illness.

12.9.2 Cons

The reserpine model was created to assess the effectiveness of new Parkinson's disease medications as symptomatic treatments. Rats treated with rotenone develop Lewy's body-like inclusions and neurodegeneration, but the degree of neurodegeneration varies significantly compared to 6MPTP administration. Reserpine therapy mimics certain features of Parkinson's disease but does not fully represent the illness's range of symptoms, including the development of Lewy bodies and the involvement of non-dopaminergic systems (Tieu, 2011).

12.10 HALOPERIDOL—RAT MODELS

Haloperidol is a drug that blocks particular D1 and D2 dopamine receptors in the brain. Haloperidol acts by blocking the D2 and, to a lesser extent, the D1 receptors on medium-spiny neurons, which make up the direct and indirect routes of the motor circuit. Within 60 minutes of haloperidol (0.5–5 mg·kg-1, i.p.) injection, the inhibition of striatal dopamine transmission causes aberrant downstream firing within the basal ganglia circuits, which manifests as symptoms of muscular stiffness and catalepsy (Sanberg, 1980). Duty and Jenner's (2011) study assessed the effectiveness of new medications for Parkinson's disease (PD). The study used the haloperidol model, which measures the time it takes an animal to release its front paws from a bar, to determine the reversal of stiffness or catalepsy. However, comparing data from various research was difficult due to inconsistent test settings. Their concept was further verified by the effects of haloperidol on raising glutamate levels and pro-inflammatory cytokines in the brain. While sub-chronic haloperidol inhibits exploratory activity, which is similar to bradykinesia, acute haloperidol administration causes psychomotor deficits in rats known as catalepsy, which mimics akinesia in people and described as an animal model of acute Parkinsonism (Melo-Thomas et al., 2020).

12.10.1 Pros

Similar to the reserpine model, the haloperidol model exhibits none of the typical pathology linked to Parkinson's disease, which further restricts its application. However, it continues to be a widely used model to evaluate the possible symptomatic effectiveness of new non-dopaminergic medications, such as mGlu7 agonists, Adenosine A2A/A1 antagonists, and mGlu-positive allosteric modulators in Parkinson's disease (Waku et al., 2021).

12.10.2 Cons

Pharmacological models have limitations in finding PD symptomatic medications due to their temporary nature and inability to evaluate long-term symptom alleviation in animal models receptive to chronic dosing regimens. They are also ineffective for researching innovative neuroprotection or neuro repair approaches as they show no signs of disease. However, other PD animal models partially address these limitations.

12.11 HALOPERIDOL—MICE MODEL

Haloperidol, a typical neuroleptic medication, functions by inhibiting the postsynaptic dopamine D2 receptors in the mesolimbic system. This leads to an increase in dopamine turnover as a result of the blockade of D2 receptors (Saeed et al., 2017). Haloperidol likely achieves its antipsychotic outcome by strongly obstructing dopamine receptors in the brain, resulting in significant rigidity when administered. In animal research, the common approach to creating a Parkinson's disease experimental model involves using neurotoxins like MPTP and 6-OHDA, as well as haloperidol.

12.12 HALOPERIDOL—*C. ELEGANS* MODEL

Caenorhabditis elegans (*C. elegans*) is a small worm commonly used in research for studying drug mechanisms and animal behavior. It has 302 neurons, 8 of which are dopaminergic. Dopamine in *C. elegans* is transduced by G protein-coupled receptors, similar to vertebrates. Haloperidol, a drug used to treat certain psychiatric conditions, was found to antagonize the D2-like receptor dop-3 in *C. elegans*. This resulted in increased lifespan, decreased basal slowing response and locomotion speed, and lower dopamine levels. However, it did not affect gene expression or the structure of dopaminergic neurons. These findings provide insights into how haloperidol modulates dopamine signaling in *C. elegans* (Krum et al., 2020).

12.12.1 Pros

The life span of the *C. elegans* increased with the treated haloperidol groups to about 16–17 days of life span compared with the non-treated worms. The life span was extending towards the differentiation of the haloperidol dopamine concentration.

12.13 DROSOPHILA PD—MODELS

The fruit fly Drosophila is a helpful model organism for Parkinson's disease studies. Due to its short lifespan and ease of maintenance, it is ideal for research (Duty & Jenner, 2011). Many human genes, including those linked to Parkinson's disease (PD) like parkin (Wasner et al., 2020), UCH-L1, PINK1 (Park et al., 2005), DJ-1 (Lavara-Culebras & Paricio, 2007), and LRRK2 (Staveley, 2015), have highly conserved homologues in Drosophila (Bier et al., 2005). A significant number of human genes are shared by both Drosophila and humans. When chemicals like paraquat or rotenone are applied to fruit flies, it can lead to Parkinson's disease-like symptoms such oxidative stress, degeneration of dopaminergic neurons, and impaired movement. Genetic and environmental factors may work together to exacerbate Parkinson's disease progression. When exposed to rotenone, fruit flies with mutant synuclein have more severe PD symptoms. Additionally. these models are employed to test possible medicinal agents and investigate the consequences of mitochondrial malfunction and oxidative stress. Research has demonstrated that Drosophila DJ-1 mutants are more vulnerable to oxidative stress, which increases the mortality of dopaminergic neurons. This is comparable to the neurodegeneration seen in Parkinson's disease in humans, where oxidative stress is a major factor (Yang et al., 2005). DJ-1 is crucial for maintaining mitochondrial homeostasis, and its loss causes dysfunction in Drosophila models. Defective fly models exhibit abnormalities similar to those in Parkinson's disease patients (Clark et al., 2006). Additionally, fruit flies with the DJ-1 mutant exhibit more severe defects in their locomotion (Park et al., 2005).

12.13.1 Pros

Drosophila melanogaster is a valuable model organism for studying environmental and genetic factors, providing insights into Parkinson's disease pathogenesis and enabling the development of treatment options. Drosophila homologues are present in over 75% of all human disease genes (Pandey & Nichols, 2011), making them an effective tool for shedding light on pathophysiological pathways and advancing neurodegenerative illnesses. Drosophila models can benefit drug development in cellular and behavioural investigations (Aryal & Lee, 2019). Studies on DJ-1 in Drosophila models have improved our understanding of Parkinson's disease processes, particularly oxidative stress and mitochondrial dysfunction. These models are still helpful in determining potential therapy targets and evaluating novel Parkinson's disease therapies (Bellen et al., 2010).

12.13.2 Cons

Drosophila's simpler nervous systems and lack of intricacy in human brain anatomy make it difficult to fully model the spectrum of Parkinson's disease pathology (Shulman et al., 2011). Although dopaminergic neurons are present in Drosophila, they are less numerous and sophisticated than those found in humans, which may affect research on complex facets of dopaminergic transmission and neurodegeneration (Whitworth et al., 2005). Drosophila also has a smaller behavioural repertoire, which limits research on non-motor symptoms of PD, such as cognitive decline and psychiatric symptoms. Specific behavioural abnormalities associated with PD in Drosophila are challenging to detect, which could impact treatment interventions and disease causes reliably (Tain et al., 2009; Adel & Griffith, 2021). Variations in genetic background, environment, and experimental techniques can lead to significant variability in experimental results, necessitating strict standardization and replication of investigations to maintain trustworthiness (Hewitt & Whitworth, 2017).

12.14 SNAIL PD MODELS

Due to its bigger neurons than fruit flies, the pond snail, *Lymnaea stagnalis*, is frequently used in invertebrate neurobiology research. Dopaminergic neurons have been found in the buccal ganglia of snails. Rotenone, a chemical, has been used in studies to cause PD in snails. These studies have shown that rotenone reduces the impacts on the snails' central nervous system, such as dopamine levels, feeding, and movement (Hamre et al., 1999). Therefore, snails are a good model for researching PD and examining rotenone's behavioural and cellular effects on the dopaminergic system (Dehay & Bezard, 2011; Dehay & Fernagut, 2016).

12.14.1 Cons

There may be limitations to the direct application of findings to Parkinson's disease in humans due to the notable physiological differences between snails and people. Snails' cellular microenvironment and neurological systems differ from those of humans, which influences the illness processes. Absence of Dopaminergic System Complexity: Snails have a far simpler dopaminergic system than humans, which may limit research on dopaminergic neurodegeneration associated with Parkinson's disease. While snails have several advantages due to their simplicity, the intricate relationships and multifactorial nature of Parkinson's disease (PD) observed in people may not be entirely replicated in snails. It's possible that snail models don't show all aspects of Parkinson's disease (PD), particularly the motor and non-motor symptoms that are essential for thorough PD research. Even though they may be seen, snails' actions are far less sophisticated than those of vertebrates.

12.15 *C. ELEGANS* PD MODELS

C. elegans may be a model organism to investigate PD. *C. elegans* is a helpful model for understanding the human brain because it has a straightforward nervous system with 302 neurons, including a circuitry of eight dopaminergic neurons. Additionally, the worm has genes comparable to genetic risk factors for familial PD, including VPS35, LRRK2, PINK1, Parkin, and DJ-1 (Martinez et al., 2017). The protein α-synuclein, another one linked to PD in humans, is not expressed in *C. elegans*. In *C. elegans*, the formation of α-synuclein inclusions (similar to Lewy bodies identified in PD patients) and a decrease in baseline slowing response when wild-type or mutant forms of α-synuclein are overexpressed. Additionally, 6-OHDA, paraquat, rotenone, and MPP+, as well as other neurotoxins, cause damage to mitochondrial DNA and specifically degenerate dopaminergic neurons in *C. elegans* (Harrington et al., 2010). The loss of dopamine neurons, the disruption of dopamine-dependent behaviours, increased susceptibility to stress, age-dependent aggregation, and mobility deficiencies in these worms have all been observed (Cooper and Van Raamsdonk, 2018).

12.15.1 Pros

As an animal model for the study of PD, *C. elegans* offers several benefits. The availability of a wide range of genetic tools and the ability to quickly and affordably screen for disease modifiers are the two most significant benefits.

12.15.2 Cons

However, the main drawbacks of using *C. elegans* for the study of PD are the absence of an α-synuclein homolog, the difficulty of performing molecular biology on dopamine neurons specifically (8 out of 959 cells are dopamine neurons), and the fact that neuronal connectivity is different from that of humans. Despite these drawbacks, the model system has allowed for several significant advancements in our understanding of PD. For instance, even though *C. elegans* lacks an ortholog to α-synuclein, using this model to identify numerous.

12.16 NEURODEGENERATIVE MODELS

12.16.1 Pros

Neurotoxin-based models have been commonly used to replicate the selective neuronal death seen in PD, but there is now a shift towards using disease gene-based models. Utilising multiple models with different mechanisms of insult is important because dopamine-producing neurons can die through various pathways. The article emphasizes the need for clinically relevant designs in the preclinical development of new therapies before progressing to clinical trials (Bezard et al., 2013). The relationship between genes and the environment may contribute to the neurodegeneration seen in PD because the combined influence of hereditary and environmental variables enhances the death of dopaminergic neurons.

12.16.2 Cons

Unfortunately, none of the current vertebrate models are conducive for examining the processes behind selective neuronal vulnerability. However, as was already said, conditional methods have the potential to be especially instructive. The degeneration of DA neurons in *Drosophila* and *C. elegans* has not been studied using these models to understand selective neuronal susceptibility. They have also failed to offer comprehensive phenotypic details on the impacted dopaminergic neurons, and the impacts on nondopaminergic neurons have not been well described. Furthermore, the difference between cell-autonomous and non-cell-autonomous dementia has not been investigated using these models.

The above-discussed models were focused on neurotoxins such as 6-hydroxydopamine, 1-methyl-4-phenyl-1,2,3, 6-tetrahydropyridine, paraquat, and rotenone to recreate the characteristic features of the disease in animals, thereby providing an overview of the animal models and their pathogenic mechanisms, aiding researchers in choosing the most appropriate model for their studies and advancing the understanding of Parkinson's disease (Zeng et al., 2018).

12.17 ALPHA-SYNUCLEIN

The long-term impact of ischemic stroke in mice, which results in reduced blood supply to the brain, were studied for the onset of PD. The scientists gave mice that expressed a human gene linked to PD a stroke and then monitored their behaviour and brain pathology for up to a year. The researchers discovered that motor deficiencies started appearing 6 months after the stroke and got worse over time. The animals also exhibited inflammation, a loss of neurons, and a rise in the protein α-synuclein, which is linked to Parkinson's disease, aggregating.

12.17.1 Pros

It helped to study the late loss of dopaminergic neurons, which are critical in PD and occurred along with these alterations. This study by Lohmann et al. (2022) raised the possibility that stroke and PD are related, and it offered a framework for further investigation and treatment.

12.18 MUTANT PINK1, DJ1, AND LRRK2 GENE-INDUCED MOUSE MODELS

One can try to simulate PD in animals or cellular systems by genetically modifying them. This involves introducing mutated genes for autosomal-dominant genes like α-synuclein and leucine-rich repeat kinase 2 (LRRK2), or creating models where genes for autosomal-recessive genes like Parkin, DJ-1, and phosphatase and tensin homolog (PTEN)-induced novel kinase 1 (PINK1) are deactivated or suppressed. Other gene abnormalities, such as those in the glucocerebrosidase (GBA) and PD syndromes like ATP13A2, are also responsible for PD, but no established animal models have been developed for these mutations. It is possible to manipulate the genetic makeup of various species and cellular models. A lot of these models have been particularly effective for dissecting the pathogenic pathways behind PD. The LRRK2 gene is often mutated in PD, a neurological condition. Mature LRRK2 has been expressed in mouse models to understand the PD processes better. The animals showed axonopathy, hyperphosphorylated tau, dopamine transmission deficiency, and motor impairments, where the gene understands the disease at cellular levels (Dawson et al., 2010; Xu et al., 2012).

12.18.1 Cons

Sometimes information about the human situation that was not obvious from standard neuropathological evaluations is revealed by using animal models. Powerful models for quickly screening pharmacological and genetic therapies that might slow neurodegeneration in these species include *C. elegans* and *Drosophila*. They have some flaws in their PD modelling, such as the absence of α-synuclein expression and a small selection of cell death effectors. These models have the benefit of revealing evolutionarily conserved pathways, but employing them presents the issue of confirming that possible modifiers exist in human PD (Blesa et al., 2012).

12.19 MITOCHONDRIAL DYSFUNCTION IN PD

Even though extensive research has been conducted for many years, our understanding of PD remains limited. By examining genetic factors in rare familial cases, researchers have gained valuable insights and have found evidence suggesting a connection between mitochondrial dysfunction and the development of PD. The use of mouse models with altered mitochondrial genes has proven beneficial in studying PD and serves as a valuable tool for testing potential treatments (Terzioglu and Galter, 2008; Pickrell et al., 2013; Klivenyi and Vecsei, 2011; Anvret et al., 2012).

12.19.1 Pros

Familial variants of Parkinson's disease (PD) are linked to mutations in genes related to mitochondrial function, such as PINK1, PARKIN, and LRRK2. These mutations can affect mitochondrial quality control, resulting in malfunction and accelerating the ageing process. The main source of reactive oxygen species (ROS) is mitochondria. An overabundance of ROS can harm lipids, proteins, and DNA in cells, which exacerbates the pathogenesis of Parkinson's disease (PD) (Narendra et al., 2008; Dias et al., 2013). Research has indicated that PD patients' substantia nigras have decreased complex I of the mitochondrial electron transport chain function, which impairs ATP synthesis and increases oxidative stress (Schapira, 2008). Alpha-synuclein aggregates observed in Parkinson's disease patients are hypothesized to impede mitochondrial activity. According to Exner et al. (2012), alpha-synuclein can bind to mitochondrial membranes, impairing their integrity and functionality. The loss of dopaminergic neurons in Parkinson's disease (PD) can be attributed to the activation of cell death mechanisms, including apoptosis and mitophagy, by mitochondrial malfunction.

12.19.2 CONS

A combination of lifestyle, environmental, and genetic variables all have a role in Parkinson's disease (PD). They ignore other important factors, including neuroinflammation, proteasomal failure, and lysosomal storage diseases, and concentrate only on mitochondrial dysfunction (Beal, 2003). Personal Differences: Since not all PD patients show apparent mitochondrial dysfunction, this characteristic may not be present in all cases of the illness (Surmeier et al., 2017). In PD clinical trials, therapies that target mitochondrial dysfunction—such as antioxidants and mitochondrial protective agents—have shown varying degrees of effectiveness. Pharmacologically enhancing mitochondrial activity may have unanticipated adverse effects that might impair other cellular functions and result in toxicity (Wang & Zhang, 2019).

12.20 EFFECT OF BRAIN-DERIVED NEUROTROPHIC FACTOR LEVEL IN PD

The production of brain-derived neurotrophic factor (BDNF), a small dimeric protein structurally related to NGF, is widespread and abundant in the adult mammalian brain. It has been established that BDNF is advantageous for all of the major neuronal types affected by AD and PD, including nigral dopaminergic neurons, hippocampus, and neocortical neurons. When BDNF levels fall in the reserpine group, we may observe a progressive increase in neuronal death in all the locations that have been analyzed; these impairments may be irreversible. In a study by Seyedebrahimi et al. (2020), GNPs and BDNF were loaded onto chitosan nanoparticles, indicating a significant interaction between the two with an increase of up to 90% in loading efficiency. In vitro studies conducted as part of the study showed that neurotrophins like BDNF multiplied in large amounts in a variety of cell types when GNPs were present.

12.20.1 Pros

Neurogenesis, synaptic connection, memory improvement, and learning process enhancement are some of BDNF's crucial roles. Numerous studies have been conducted on how BDNF affects cell migration, differentiation, and proliferation, which might be useful in treating a variety of illnesses (Córneo et al., 2020).

12.21 GUT TO BRAIN STUDY IN MICE AND MOUSE MODELS

The Kim et al. (2019) study investigated the theory that pathogenic α-syn may go through the vagus nerve and stereotypically enter the brain from the gastrointestinal tract. This study supports the Braak theory. This approach also more precisely models the clinical state and manifestations of idiopathic Parkinson's disease (PD), including both motor and non-motor symptoms. These findings further support the role of the gut in Parkinson's disease and give us a framework to study the onset of the condition. Blocking the transfer of alpha-synuclein from the stomach to the brain could serve as a potential target for early intervention in the disease (Wood, 2019).

12.22.1 Pros

Focusing on the stomach, the findings may assist in uncovering early indicators for Parkinson's disease (PD), which might result in an earlier diagnosis and course of treatment. New treatment targets, such as gut-based therapies to stop or delay the course of Parkinson's disease (PD), become possible with a better understanding of the gut-brain propagation network.

12.22.2 Cons

PD is a complicated, multifaceted disease, and focusing only on the gut-brain route may oversimplify it. Additional channels and processes, including genetic alterations and environmental influences, also contribute significantly. Although PD is known for its α-synuclein pathology, its precise involvement in the disease's course is still under discussion. Early intervention is essential if the spread of α-synuclein begins in the gut. It's still difficult to detect at-risk people before severe brain damage manifests. Treatments aimed at the gut-brain axis may have unexpected effects on gut microbiota and health; therefore, it's essential to weigh safety and efficacy carefully.

12.22 CELLULAR LEVEL STUDY IN PD

In order to answer the question of whether the presence of aggregated species caused by nanoparticles in the media can trigger certain harmful effects on healthy neurons, Lyu et al. (2021) worked on cellular levels. Based on the findings of their investigation, it was shown that the medium's presence of CuO nanoparticles may considerably enhance the development of α-synuclein amyloid fibrils and the underlying toxicity to cells. The function of several internal organs, including plasma membrane and mitochondria, is disrupted when the structure of α-synuclein alters in a way that makes it vulnerable to the process known as fibrillation and the creation of α-sheets structures. The idea that the formation of protein amyloids caused by nanoparticles is crucial for neuronal toxicity and eventually cell death is supported by data from animal models as well as genetic, biochemical, and biophysical investigations. Amyloid aggregates can cause cytotoxicity in two ways: directly and indirectly. The impacts of intracellular pathways and metabolites, which ultimately result in cell death, are the cause of the indirect effects of amyloid accumulations on the cell. In fact, the cytotoxicity caused by protein amyloid fibrils led to an increase in reactive oxygen and nitrogen species in the cell, a decrease in protein function, and a rise in the phosphorylation of mediator proteins. As a result, the cell sustains significant damage, and the culmination of numerous intracellular occurrences will cause cell death. Extracellular vesicles contain α-synuclein, and therefore α-synuclein from the donor cell is discharged into the extracellular space, where it interacts with the receptor cell to create new hazardous clumps.

12.22.1 Pros

According to several research studies, cells have the ability to discharge α-synuclein into the environment. For instance, α-synuclein monomers and oligomers have been found in the plasma and CSF fluid of PD patients and healthy people. These types of α-synuclein are, however, highly concentrated in PD sufferers. This attests to the existence of the cell-secreted α-synuclein pathway.

12.23 TRANSGENIC MICE

The original identification of these 140 amino acid proteins as a significant component of LBs/LNs in a spectrum of neurodegenerative diseases now termed as synucleinopathies resulted from the finding that point mutations in α-Syn produce autosomal-dominant forms of PD. A sizeable amount of the α-Syn recovered from PD brains is insoluble and exhibits different post-translational changes, including proteolytic cleavage, hyperphosphorylation, ubiquitination, nitrosylation, and oxidation. This is in contrast to its highly soluble condition in healthy brains, where it presumably governs synaptic vesicle release. The finding that recipient M83 mice's illness did not progress faster in homogenates from asymptomatic transgenic mice without detectable α-Syn inclusions suggested that misfolded α-Syn within LBs/LNs seen in sick donors could be the pathogen in motion. The recombinant human α-Syn has been administered into the in vitro assembled into amyloid fibrils to healthy M83 mice to test this theory. This proves that misfolded α-Syn alone is sufficient to start pathology, as shown by the virtually same pattern of LB/LN-like pathology that these synthetic fibrils provoked. It's interesting to note that inclusions were also seen in astrocytes, which in this strain of transgenic mice produce the mutant α-Syn, suggesting that non-neuronal cells can promote α-Syn disease when exposed to it. When injected in vivo, misfolded α-Syn may also cause Lewy-like disease. The brainstem and spinal cord, where mutant α-Syn production is highest in this strain of transgenic mice and where the A53T familial PD mutation is present, are the two areas of the CNS where these mice spontaneously acquire a Sypathology. This condition often manifests between the ages of 8 and 16 months, associated with a range of motor and behavioural abnormalities and mortality not long after (Luk and Lee, 2014).

12.24 INSOMNIA STUDY IN HUMANS

This study examines the relationship between insomnia and PD in a group of 421 patients, both in terms of cross-sectional and longitudinal connections. The patients were followed for an average of 4.56 years. A significant study conducted by Zhu et al. (2016) found that 51% of participants experienced insomnia at some point, and 37% of those who initially did not have insomnia developed it during the follow-up period. However, these rates are lower than those reported in a previous population-based study, likely due to differences in study settings (hospital-based versus population-based) and the characteristics of the populations (the Norwegian study had a higher proportion of female patients). Furthermore, the questionnaire used in this study had different response options, and while both studies assessed similar aspects of insomnia, the criteria for diagnosis were broader in the previous study. In this study, patients were classified as having insomnia if they reported difficulty sleeping at night or used sleeping pills for at least one month, whereas the previous study had additional criteria. The number of patients with insomnia slightly increased over time and became more persistent, emphasising the importance of monitoring this symptom, especially in high-risk patients.

12.24.1 Pros

The subjective sleep disruption and worse HRQoL are related. Contrarily, HRQoL was only weakly correlated with the objective PSG indications of sleep-disordered breathing, periodic limb movement, REM-related behaviour, and sleep architectural characteristics. This could have occurred as a result of the weak correlation observed in PD patients between perceived sleep disruptions and objective sleep metrics. It has not been determined the reasons for this discrepancy between subjective and objective sleep disruptions in PD patients or what it means (Tsuru et al., 2022).

12.25 DROSOPHILA MODEL

The *Drosophila* model organism, which shares similarities with human genes associated with PD, has been extensively studied. Various *Drosophila* models with gene mutations related to PD have been created, but their reliability is questionable. Many of these models exhibit motor impairments, such as a premature decline in climbing ability. A specific *Drosophila* model expressing α-synuclein transgene was reported to display a loss of climbing ability between 20 and 40 days of age. Some studies found a roughly 50% decrease in dopaminergic neurons in 20-day-old flies, while others did not observe any loss (Kin et al., 2019). However, all of these models showed restoration of climbing ability after treatment with L-DOPA, dopamine agonists, or muscarinic antagonists. Consensus exists among most studies that loss-of-function mutant or Parkin-null *Drosophila* display a selective loss of dopaminergic neurons at 1 and 20 days of age, accompanied by a 60% reduction in brain dopamine levels and the characteristic decline in climbing ability. The behavioural phenotype in these flies can be partially restored with L-DOPA. Similarly, inactivating PINK1 in *Drosophila* leads to a reduction in the number of dopaminergic neurons and impaired climbing ability.

12.26 TREATMENT STUDY MODEL

12.26.1 Prp-αSynA53T Transgenic Mice

Lycorine has shown potential in treating PD by increasing the destruction of pathogenic α-syn species through PKA-mediated UPS activation. The α-syn gene, which has missense variants, is linked to familial PD. Lowering α-syn levels reverses neurodegeneration in PD models, demonstrating that α-syn expression is crucial. The double or triple SNCA is necessary for PD onset, and lycorine, an active α-syn degradation molecule, was discovered by screening chemicals that degrade it (Zhu et al., 2021).

12.26.2 Using Baicalein in Treating PD Sprague-Dawley (SD) Rats

The anti-inflammatory effects of baicalein have been carefully studied in a variety of illness models, such as ischemia/reperfusion, diabetes, acute lung injury, and rheumatoid arthritis. In the current study, Zhang et al. (2017b) connected baicalein's neuroprotective qualities to its anti-inflammatory powers in a rat model of PD caused by rotenone. Whereas in rats exposed to rotenone repeatedly, we found that baicalein prevented brain damage and reversed motor deficits. Baicalein also has a special role in preventing neuroinflammation, according to Zhang et al.'s (2017b) study, which included lowering the production of proinflammatory cytokines, inhibiting astrocyte and microglial activation, and blocking the NF-B and MAPK signaling pathways.

Baicalein also prevented IB phosphorylation and p65 translocation in a model of LPS-activated microglia, which reduced inflammatory mediators. Baicalein also reduced inflammatory mediators by downregulating the Toll-like receptor 4 (TLR4), which functions upstream of the NF-B signal, I-B phosphorylation, and p65 translocation in an LPS-activated microglia model. Therefore, our research suggests that baicalein's anti-inflammatory capabilities aid in reducing the risk of rotenone-induced brain injury. Inflammation appears to be a major player in the etiopathogenesis of PD, according to mounting data. There is growing evidence from human postmortems and PD animal models that a neuroinflammatory cascade is present in PD.

12.27 ACKNOWLEDGMENT

The authors are thankful to the Vice Chancellor and Registrar, Pondicherry University, Pondicherry, India, for providing the facilities for ongoing research work. Dr. Rajesh Pamanji is thankful to UGC, Government of India, for providing financial assistance in the form of DSK postdoctoral fellowship BL/19-20/0302 and Sobanaa M, is thankful to Savitribai Jyotirao Phule Fellowship for Single Girl Child, UGC (Ref. No.: 202223-UGCES-22-OB-PUD-F-SJSGC-4671).

REFERENCES

Acar, N., Parlak, H., Ozkan, A., Soylu, H., Avcı, S., Ustunel, I., Izgut-Uysal, V. N., & Agar, A. (2019). The effect of docosahexaenoic acid on apelin distribution of nervous system in the experimental mouse model of Parkinson's disease. *Tissue & Cell*, *56*, 41–51. https://doi.org/10.1016/j.tice.2018.12.002

Adel, M., & Griffith, L. C. (2021). The role of dopamine in associative learning in drosophila: An updated unified model. *Neuroscience Bulletin*, *37*, 831–852. https://doi.org/10.1007/s12264-021-00665-0

Ahmad, M. H., Fatima, M., Ali, M., Rizvi, M. A., & Mondal, A. C. (2021). Naringenin alleviates paraquat-induced dopaminergic neuronal loss in SH-SY5Y cells and a rat model of Parkinson's disease. *Neuropharmacology*, *201*, 108831. https://doi.org/10.1016/j.neuropharm.2021.108831

Airavaara, M., Parkkinen, I., Konovalova, J., Albert, K., Chmielarz, P., & Domanskyi, A. (2020). Back and to the future: From neurotoxin-induced to human Parkinson's disease models. *Current Protocols in Neuroscience*, *91*(1), e88. https://doi.org/10.1002/cpns.88

Anselmi, L., Toti, L., Bove, C., Hampton, J., & Travagli, R. A. (2017). A nigro–vagal pathway controls gastric motility and is affected in a rat model of parkinsonism. *Gastroenterology*, *153*(6), 1581–1593. https://doi.org/10.1053/j.gastro.2017.08.069

Anvret, A., Ran, C., Westerlund, M., Gellhaar, S., Lindqvist, E., Pernold, K., Lundströmer, K., Duester, G., Felder, M. R., Galter, D., & Belin, A. C. (2012). Adh1 and Adh1/4 knockout mice as possible rodent models for presymptomatic Parkinson's disease. *Behavioural Brain Research*, *227*(1), 252–257. https://doi.org/10.1016/j.bbr.2011.10.040

Arab, H. H., Safar, M. M., & Shahin, N. N. (2021). Targeting ROS-dependent AKT/GSK-3β/NF-κB and DJ-1/Nrf2 pathways by dapagliflozin attenuates neuronal injury and motor dysfunction in rotenone-induced Parkinson's disease rat model. *ACS Chemical Neuroscience*, *12*(4), 689–703. https://doi.org/10.1021/acschemneuro.0c00722

Aryal, B., & Lee, Y. (2019). Disease model organism for Parkinson disease: *Drosophila melanogaster*. *BMB Reports*, *52*(4), 250–258. https://doi.org/10.5483/bmbrep.2019.52.4.204

Azar, Y. O., Badawi, G. A., Zaki, H. F., & Ibrahim, S. M. (2022). Agmatine-mediated inhibition of NMDA receptor expression and amelioration of dyskinesia via activation of Nrf2 and suppression of HMGB1/RAGE/TLR4/MYD88/NF-κB signaling cascade in rotenone lesioned rats. *Life Sciences*, *311*(Pt A), 121049. https://doi.org/10.1016/j.lfs.2022.121049

Bar-Am, O., Amit, T., Kupershmidt, L., Aluf, Y., Mechlovich, D., Kabha, H., Danovitch, L., Zurawski, V. R., Youdim, M. B., & Weinreb, O. (2015). Neuroprotective and neurorestorative activities of a novel iron chelator-brain selective monoamine oxidase-A/monoamine oxidase-B inhibitor in animal models of Parkinson's disease and aging. *Neurobiology of Aging*, *36*(3), 1529–1542. https://doi.org/10.1016/j.neurobiolaging.2014.10.026

Beal, M. F. (2003). Bioenergetic approaches for neuroprotection in Parkinson's disease. *Annals of Neurology*, *53*(S3), S39–S47.

Bellen, H., Tong, C., & Tsuda, H. (2010). 100 years of Drosophila research and its impact on vertebrate neuroscience: A history lesson for the future. *Nature Reviews Neuroscience*, *11*, 514–522. https://doi.org/10.1038/nrn2839

Bier, E. (2005). Drosophila, the golden bug, emerges as a tool for human genetics. *Nature Reviews Genetics*, *6*, 9–23. https://doi.org/10.1038/nrg1503

Bezard, E., Yue, Z., Kirik, D., & Spillantini, M. G. (2013). Animal models of Parkinson's disease: Limits and relevance to neuroprotection studies. *Movement Disorders*, *28*(1), 61–70. https://doi.org/10.1002/mds.25108

Bhattarai, Y., Si, J., Pu, M., Ross, O. A., McLean, P. J., Till, L., Moor, W., Grover, M., Kandimalla, K. K., Margolis, K. G., Farrugia, G., & Kashyap, P. C. (2021). Role of gut microbiota in regulating gastrointestinal dysfunction and motor symptoms in a mouse model of Parkinson's disease. *Gut Microbes*, *13*(1), 1866974. https://doi.org/10.1080/19490976.2020.1866974

Blesa, J., Phani, S., Jackson-Lewis, V., & Przedborski, S. (2012). Classic and new animal models of Parkinson's disease. *Journal of Biomedicine & Biotechnology*, *2012*, 845618. https://doi.org/10.1155/2012/845618

Bobela, W., Zheng, L., & Schneider, B. L. (2014). Overview of mouse models of Parkinson's disease. *Current Protocols in Mouse Biology*, *4*(3), 121–139. https://doi.org/10.1002/9780470942390.mo140092

Bove, C., Anselmi, L., & Travagli, R. A. (2019). Altered gastric tone and motility response to brain-stem dopamine in a rat model of parkinsonism. *American Journal of Physiology-Gastrointestinal and Liver Physiology*, *317*(1), G1–G7. https://doi.org/10.1152/ajpgi. 00076.2019

Breger, L. S., & Fuzzati Armentero, M. T. (2019). Genetically engineered animal models of Parkinson's disease: From worm to rodent. *The European Journal of Neuroscience*, *49*(4), 533–560. https://doi.org/10.1111/ejn.14300

Buhidma, Y., Rukavina, K., Chaudhuri, K. R., & Duty, S. (2020). Potential of animal models for advancing the understanding and treatment of pain in Parkinson's disease. *npj Parkinson's Disease*, *6*(1), 1. https://doi.org/10.1038/s41531-019-0104-6

Cammarota, G., Pecere, S., Ianiro, G., Masucci, L., & Currò, D. (2016). Principles of DNA-based gut microbiota assessment and therapeutic efficacy of fecal microbiota transplantation in gastrointestinal diseases. *Digestive Diseases*, *34*(3), 279–285. https://doi.org/10.1159/000443362

Cannon, J. R., Tapias, V., Na, H. M., Honick, A. S., Drolet, R. E., & Greenamyre, J. T. (2009). A highly reproducible rotenone model of Parkinson's disease. *Neurobiology of Disease*, *34*(2), 279–290. https://doi.org/10.1016/j.nbd.2009.01.016

Casteels, C., Lauwers, E., Bormans, G., Baekelandt, V., & Van Laere, K. (2008). Metabolic-dopaminergic mapping of the 6-hydroxydopamine rat model for Parkinson's disease. *European Journal of Nuclear Medicine and Molecular Imaging*, *35*(1), 124–134. https://doi.org/10.1007/s00259-007-0558-3

Chia, S. J., Tan, E. K., & Chao, Y. X. (2020). Historical perspective: Models of Parkinson's disease. *International Journal of Molecular Sciences*, *21*(7), 2464. https://doi.org/10.3390/ijms21072464

Clark, I., Dodson, M., Jiang, C., Cao, J., Huh, R., Seol, J., Yoo, S., Hay, B., & Guo, M. (2006). Drosophila pink1 is required for mitochondrial function and interacts genetically with Parkin. *Nature*, *441*, 1162–1166. https://doi.org/10.1038/nature04779

Collyer, R., Clancy, A., & Borody, T. (2020). Faecal microbiota transplantation alleviates symptoms of depression in individuals with irritable bowel syndrome: A case series. *Medicine in Microecology*, *6*, 100029. https://doi.org/10.1016/j.medmic.2020.100029

Cooper, J. F., & Van Raamsdonk, J. M. (2018). Modeling Parkinson's disease in *C. elegans*. *Journal of Parkinson's Disease*, *8*(1), 17–32. https://doi.org/10.3233/JPD-171258

Córneo, E., De Bem Silveira, G., Scussel, R., Corrêa, M. E. A. B., Da Silva Abel, J., Luiz, G. P., Feuser, P. E., Silveira, P. C. L., & Machado-De-Ávila, R. A. (2020). Effects of gold nanoparticles administration through behavioral and oxidative parameters in animal model of Parkinson's disease. *Colloids and Surfaces B: Biointerfaces*, *196*, 111302. https://doi.org/10.1016/j.colsurfb.2020.111302

Cristóvão, A. C., Campos, F. L., Je, G., Esteves, M., Guhathakurta, S., Yang, L., Beal, M. F., Fonseca, B. M., Salgado, A. J., Queiroz, J., Sousa, N., Bernardino, L., Alves, G., Yoon, K. S., & Kim, Y. S. (2020). Characterization of a Parkinson's disease rat model using an upgraded paraquat exposure paradigm. *The European Journal of Neuroscience*, *52*(4), 3242–3255. https://doi.org/10.1111/ejn.14683

Cui, Q., Li, X., & Zhu, H. (2016). Curcumin ameliorates dopaminergic neuronal oxidative damage via activation of the Akt/Nrf2 pathway. *Molecular Medicine Reports*, *13*(2), 1381–1388. https://doi.org/10.3892/mmr.2015.4657

Dawson, T. M., Ko, H. S., & Dawson, V. L. (2010). Genetic animal models of Parkinson's disease. *Neuron*, *66*(5), 646–661. https://doi.org/10.1016/j.neuron.2010.04.034

Dehay, B., & Bezard, E. (2011). New animal models of Parkinson's disease. *Movement Disorders*, *26*(7), 1198–1205. doi: 10.1002/mds.23546.

Dehay, B., & Fernagut, P. O. (2016). Alpha-synuclein-based models of Parkinson's disease. *Revue Neurologique (Paris)*, *172*(6–7), 371–378. doi: 10.1016/j.neurol.2016.04.003.

Dias, V., Junn, E., & Mouradian, M. M. (2013). The role of oxidative stress in Parkinson's disease. *Journal of Parkinson's Disease*, *3*(4), 461–491.

Dodiya, H. B., Forsyth, C. B., Voigt, R. M., Engen, P. A., Patel, J., Shaikh, M., Green, S. J., Naqib, A., Roy, A., Kordower, J. H., Pahan, K., Shannon, K. M., & Keshavarzian, A. (2020). Chronic stress-induced gut dysfunction exacerbates Parkinson's disease phenotype and pathology in a rotenone-induced mouse model of Parkinson's disease. *Neurobiology of Disease*, *135*, 104352. https://doi.org/10.1016/j.nbd.2018.12.012

Dutta, S. K., Verma, S., Jain, V., Surapaneni, B. K., Vinayek, R., Phillips, L., & Nair, P. P. (2019). Parkinson's disease: The emerging role of gut dysbiosis, antibiotics, probiotics, and fecal microbiota transplantation. *Journal of Neurogastroenterology and Motility*, *25*(3), 363–376. https://doi.org/10.5056/jnm19044

Duty, S., & Jenner, P. (2011). Animal models of Parkinson's disease: A source of novel treatments and clues to the cause of the disease. *British Journal of Pharmacology*, *164*, 1357–1391. https://doi.org/10.1111/j.1476–5381.2011.01426.x

El-Gamal, M., Salama, M., Collins-Praino, L. E., Baetu, I., Fathalla, A. M., Soliman, A. M., Mohamed, W., & Moustafa, A. A. (2021). Neurotoxin-induced rodent models of parkinson's disease: Benefits and drawbacks. *Neurotoxicity Research*, *39*(3), 897–923. https://doi.org/10.1007/s12640-021-00356-8

El-Sayed, R. M., Abdelaziz, A. M., Zaki, H. F., & Abdel Rasheed, N. O. (2023). Cilostazol novel neuroprotective mechanism against rotenone-induced Parkinson's disease in rats: Correlation between Nrf2 and HMGB1/TLR4/PI3K/Akt/mTOR signaling. *International Immunopharmacology*, *117*, 109986. https://doi.org/10.1016/j.intimp.2023.109986

Esmaeili-Mahani, S., Haghparast, E., Nezhadi, A., Abbasnejad, M., & Sheibani, V. (2021). Apelin-13 prevents hippocampal synaptic plasticity impairment in Parkinsonism rats. *Journal of Chemical Neuroanatomy*, *111*, 101884. https://doi.org/10.1016/j.jchemneu.2020.101884

Exner, N., Lutz, A. K., Haass, C., & Winklhofer, K. F. (2012). Mitochondrial dysfunction in Parkinson's disease: Molecular mechanisms and pathophysiological consequences. *EMBO Journal*, *31*(14), 3038–3062.

Fan, H.-X., Sheng, S., & Zhang, F. (2022a). New hope for Parkinson's disease treatment: Targeting gut microbiota. *CNS Neuroscience & Therapeutics*, *28*, 1675–1688.

Fan, L., Li, Y., Zhang, X., Wu, Y., Song, Y., Zhang, F., Zhang, J., & Sun, H. (2022b). Time-resolved proteome and transcriptome of paraquat-induced pulmonary fibrosis. *Pulmonary Pharmacology & Therapeutics*, *75*, 102145. https://doi.org/10.1016/j.pupt.2022.102145

Fão, L., Mota, S. I., & Rego, A. C. (2019). Shaping the Nrf2-ARE-related pathways in Alzheimer's and Parkinson's diseases. *Ageing Research Reviews*, *54*, 100942. https://doi.org/10.1016/j.arr.2019.100942

Grandi, L. C., Di Giovanni, G., & Galati, S. (2018). Animal models of early-stage Parkinson's disease and acute dopamine deficiency to study compensatory neurodegenerative mechanisms. *Journal of Neuroscience Methods*, *308*, 205–218. https://doi.org/10.1016/j.jneumeth.2018.08.012

Guimarães, R. P., Ribeiro, D. L., Dos Santos, K. B., Godoy, L. D., Corrêa, M. R., & Padovan-Neto, F. E. (2021). The 6-hydroxydopamine rat model of Parkinson's disease. *Journal of Visualized Experiments*, (176), 10.3791/62923. https://doi.org/10.3791/62923

Guo, T. T., Zhang, Z., Sun, Y., Zhu, R. Y., Wang, F. X., Ma, L. J., Jiang, L., & Liu, H. D. (2023). Neuroprotective effects of sodium butyrate by restoring gut microbiota and inhibiting TLR4 signaling in mice with MPTP-induced Parkinson's disease. *Nutrients*, *15*, 930. https://doi.org/10.3390/nu15040930

Gupta, A., & Khanna, S. (2017). Fecal microbiota transplantation. *JAMA*, *318*(1), 102. https://doi.org/10.1001/jama.2017.6466

Haghparast, E., Esmaeili-Mahani, S., Abbasnejad, M., & Sheibani, V. (2018). Apelin-13 ameliorates cognitive impairments in 6-hydroxydopamine-induced substantia nigra lesion in rats. *Neuropeptides*, *68*, 28–35. https://doi.org/10.1016/j.npep.2018.01.001

Haghparast, E., Sheibani, V., Abbasnejad, M., & Esmaeili-Mahani, S. (2019). Apelin-13 attenuates motor impairments and prevents the changes in synaptic plasticity-related molecules in the striatum of Parkinsonism rats. *Peptides*, *117*, 170091. https://doi.org/10.1016/j.peptides.2019.05.003

Hamre, K., Tharp, R., Poon, K., Xiong, X., & Smeyne, R. J. (1999). Differential strain susceptibility following 1-methyl-4-phenyl-1,2,3,6-tetrahydropyridine (MPTP) administration acts in an autosomal dominant fashion: Quantitative analysis in seven strains of Mus musculus. *Brain Research Bulletin*, *828*(1–2), 91–103. https://doi.org/10.1016/s0006-8993(99)01273-1

Han, R., Yu, Y., Zhao, K., Wei, J., Hui, Y., & Gao, J. M. (2022a). Lignans from Eucommia ulmoides Oliver leaves exhibit neuroprotective effects via activation of the PI3K/Akt/GSK-3β/Nrf2 signaling pathways in H_2O_2-treated PC-12 cells. *Phytomedicine*, *101*, 154124. https://doi.org/10.1016/j.phymed.2022.154124

Han, X., Liu, Y., Dai, Y., Xu, T., Hu, Q., Yi, X., Rui, L., Hu, G., & Hu, J. (2022b). Neuronal SH2B1 attenuates apoptosis in an MPTP mouse model of Parkinson's disease via promoting PLIN4 degradation. *Redox Biology*, *52*, 102308. https://doi.org/10.1016/j.redox.2022.102308

Harrington, A. J., Hamamichi, S., Caldwell, G. A., & Caldwell, K. A. (2010). *C. elegans* as a model organism to investigate molecular pathways involved with Parkinson's disease. *Developmental Dynamics*, *239*(5), 1282–1295. https://doi.org/10.1002/dvdy.22231

Hegazy, M., Maklad, H. M., Samy, D. M., Abdelmonsif, D. A., Sabaa, B. M. E., & Elnozahy, F. Y. (2017). Cerium oxide nanoparticles could ameliorate behavioral and neurochemical impairments in 6-hydroxydopamine induced Parkinson's disease in rats. *Neurochemistry International*, *108*, 361–371. https://doi.org/10.1016/j.neuint.2017.05.011

Heikkila, R. E., Hess, A., & Duvoisin, R. C. (1984). Dopaminergic neurotoxicity of 1-methyl-4-phenyl-1,2,5,6-tetrahydropyridine in mice. *Science*, *224*(4656), 1451–1453. https://doi.org/10.1126/science.6610213

Hewitt, V. L., & Whitworth, A. J. (2017). Mechanisms of Parkinson's disease: Lessons from Drosophila. *Current Topics in Developmental Biology*, *121*, 173–200.

Hirayama, M., & Ohno, K. Parkinson's disease and gut microbiota. (2021). *Annals of Nutrition and Metabolism*, *77*(Suppl. 2), 28–35.

Hojo, Y., Murakami, G., Mukai, H., Higo, S., Hatanaka, Y., Ogiue-Ikeda, M., Ishii, H., Kimoto, T., & Kawato, S. (2008). Estrogen synthesis in the brain—Role in synaptic plasticity and memory. *Molecular and Cellular Endocrinology*, *290*(1–2), 31–43. https://doi.org/10.1016/j.mce.2008.04.017

Hou, Y., Li, X., Liu, C., Zhang, M., Zhang, X., Ge, S., & Zhao, L. (2021). Neuroprotective effects of short-chain fatty acids in MPTP induced mice model of Parkinson's disease. *Experimental Gerontology*, *150*, 111376. https://doi.org/10.1016/j.exger.2021.111376

Idrissi, S. E., Fath, N., Ibork, H., Taghzouti, K., Alamy, M., & Abboussi, O. (2023). Restraint stress exacerbates apoptosis in a 6-OHDA animal model of Parkinson disease. *Neurotoxicity Research*, *41*(2), 166–176. https://doi.org/10.1007/s12640-022-00630-3

Innos, J., & Hickey, M. A. (2021). Using rotenone to model Parkinson's disease in mice: A review of the role of pharmacokinetics. *Chemical Research in Toxicology*, *34*(5), 1223–1239. https://doi.org/10.1021/acs.chemrestox.0c00522

Kakoty, V., Sarathlal, K. C., Dubey, S. K., Yang, C.-H., & Taliyan, R. (2021). Neuroprotective effects of trehalose and sodium butyrate on preformed fibrillar form of α-synuclein-induced rat model of Parkinson's disease. *ACS Chemical Neuroscience*, *12*, 2643–2660.

Khatri, D. K., & Juvekar, A. R. (2016). Neuroprotective effect of curcumin as evinced by abrogation of rotenone-induced motor deficits, oxidative and mitochondrial dysfunctions in mouse model of Parkinson's disease. *Pharmacology, Biochemistry, and Behavior, 150–151*, 39–47. https://doi.org/10.1016/j.pbb.2016.09.002

Kim, S., Kwon, S., Kam, T. I., Panicker, N., Karuppagounder, S. S., Lee, S., Lee, J. H., Kim, W. R., Kook, M., Foss, C. A., Shen, C., Lee, H., Kulkarni, S., Pasricha, P. J., Lee, G., Pomper, M. G., Dawson, V. L., Dawson, T. M., & Ko, H. S. (2019). Transneuronal propagation of pathologic α-synuclein from the gut to the brain models Parkinson's disease. *Neuron, 103*(4), 627–641.e7. https://doi.org/10.1016/j.neuron.2019.05.035

Kin, K., Yasuhara, T., Kameda, M., & Date, I. (2019). Animal models for Parkinson's disease research: Trends in the 2000s. *International Journal of Molecular Sciences, 20*(21), 5402. https://doi.org/10.3390/ijms20215402

Klivenyi, P., & Vecsei, L. (2011). Pharmacological models of Parkinson's disease in rodents. *Methods in Molecular Biology, 793*, 211–227. https://doi.org/10.1007/978-1-61779-328-8_14

Konnova, E. A., & Swanberg, M. (2018). In animal models of Parkinson's disease. https://doi.org/10.15586/codonpublications.parkinsonsdisease.2018.ch5

Kozina, E. A., Khakimova, G. R., Khaindrava, V. G., Kucheryanu, V. G., Vorobyeva, N. E., Krasnov, A. N., Georgieva, S. G., Kerkerian-Le Goff, L., & Ugrumov, M. V. (2014). Tyrosine hydroxylase expression and activity in nigrostriatal dopaminergic neurons of MPTP-treated mice at the presymptomatic and symptomatic stages of parkinsonism. *Journal of the Neurological Sciences, 340*(1–2), 198–207. https://doi.org/10.1016/j.jns.2014.03.028

Krum, B. N., Martins, A. C., Queirós, L., Ferrer, B., Milne, G. L., Soares, F. A. A., Fachinetto, R., & Aschner, M. (2020). Haloperidol interactions with the dop-3 receptor in *Caenorhabditis elegans*. *Molecular Neurobiology, 58*(1), 304–316. https://doi.org/10.1007/s12035-020-02124-9

Kumar, A., Singh, B. K., Ahmad, I., Shukla, S., Patel, D. K., Srivastava, G., Kumar, V., Pandey, H. P., & Singh, C. (2012). Involvement of NADPH oxidase and glutathione in zinc-induced dopaminergic neurodegeneration in rats: Similarity with paraquat neurotoxicity. *Brain Research, 1438*, 48–64. https://doi.org/10.1016/j.brainres.2011.12.028

Langston, J. W. (2017). The MPTP story. *Journal of Parkinson's Disease, 7*(s1), S11–S19. https://doi.org/10.3233/JPD-179006

Lavara-Culebras, E., & Paricio, N. (2007). Drosophila DJ-1 mutants are sensitive to oxidative stress and show reduced lifespan and motor deficits. *Gene, 400*(1–2), 158–165.

Lazarova, M., Tancheva, L., Chayrov, R., Tzvetanova, E., Alexandrova, A., Popatanasov, A., Uzunova, D., Stefanova, M., Stankova, I., & Kalfin, R. (2022). Tyrosinyl-amantadine: A new amantadine derivative with an ameliorative effect in a 6-OHDA experimental model of Parkinson's disease in rats. *Journal of Molecular Neuroscience, 72*(4), 900–909. doi: 10.1007/s12031-021-01964-x.

Leao, A. H., Sarmento-Silva, A. J., Santos, J. R., Ribeiro, A. M., & Silva, R. H. (2015). Molecular, neurochemical, and behavioral hallmarks of reserpine as a model for Parkinson's disease: New perspectives to a long-standing model. *Brain Pathology, 25*(4), 377–390. https://doi.org/10.1111/bpa.12253

Lee, J. Y., Tuazon, J. P., Ehrhart, J., Sanberg, P. R., & Borlongan, C. V. (2019). Gutting the brain of inflammation: A key role of gut microbiome in human umbilical cord blood plasma therapy in Parkinson's disease model. *Journal of Cellular and Molecular Medicine, 23*, 5466–5474. https://doi.org/10.1111/jcmm.14429

Lee, Y. Z., Cheng, S. H., Chang, M. Y., Lin, Y. F., Wu, C. C., & Tsai, Y. C. (2023). Neuroprotective effects of *Lactobacillus plantarum* PS128 in a mouse model of Parkinson's disease: The role of gut microbiota and microRNAs. *International Journal of Molecular Sciences, 24*(7), 6794. https://doi.org/10.3390/ijms24076794

Li, H., Su, L. Y., Yang, L., Li, M., Liu, Q., Li, Z., Hu, Y., Li, H., Wu, S., Wang, W., Hu, Y., Wang, Z., Rizak, J. D., Huang, B., Xu, M., Wu, J., Lv, L. B., Turck, C. W., Yin, Y., Yao, Y. G., . . . Hu, X. (2021). A cynomolgus monkey with naturally occurring Parkinson's disease. *National Science Review, 8*(3), nwaa292. https://doi.org/10.1093/nsr/nwaa292

Li, J., Yu, J., Guo, J., Liu, J., Wan, G., Wei, X., Yang, X., & Shi, J. (2023). *Nardostachys jatamansi* and levodopa combination alleviates Parkinson's disease symptoms in rats through activation of Nrf2 and inhibition of NLRP3 signaling pathways. *Pharmaceutical Biology, 61*(1), 1175–1185. https://doi.org/10.1080/13880209.2023.2244176

Liang, Y., Cui, L., Gao, J., Zhu, M., Zhang, Y., & Zhang, H. L. (2021). Gut microbial metabolites in Parkinson's disease: Implications of mitochondrial dysfunction in the pathogenesis and treatment. *Molecular Neurobiology, 58*(8), 3745–3758. https://doi.org/10.1007/s12035-021-02375-0

Linga, L., Jiang, Y., Liu, Y., Li, H., Bari, A., Ullah, R., & Xue, J. (2019). Role of gold nanoparticle from Cinnamomum verum against 1-methyl-4-phenyl-1, 2, 3, 6-tetrahydropyridine (MPTP) induced mice model. *Journal of Photochemistry and Photobiology. B, Biology, 201*, 111657. https://doi.org/10.1016/j.jphotobiol.2019.111657

Liu, W. W., Wei, S. Z., Huang, G. D., Liu, L. B., Gu, C., Shen, Y., Wang, X. H., Xia, S. T., Xie, A. M., Hu, L. F., Wang, F., & Liu, C. F. (2020). BMAL1 regulation of microglia-mediated neuroinflammation in MPTP-induced Parkinson's disease mouse model. *FASEB Journal, 34*(5), 6570–6581. https://doi.org/10.1096/fj.201901565RR

Lohmann, S., Grigoletto, J., Bernis, M. E., Pesch, V., Ma, L., Reithofer, S., & Tamgüney, G. (2022). Ischemic stroke causes Parkinson's disease-like pathology and symptoms in transgenic mice overexpressing alpha-synuclein. *Acta Neuropathologica Communications, 10*(1), 26. https://doi.org/10.1186/s40478-022-01327-6

Luk, K. C., & Lee, V. M. (2014). Modeling Lewy pathology propagation in Parkinson's disease. *Parkinsonism & Related Disorders, 20*(Suppl 1), S85–S87. https://doi.org/10.1016/S1353-8020(13)70022-1

Lyu, J., Long, X., Xie, T., Jiang, G., Jiang, J., Ye, L., & Li, O. (2021). Copper oxide nanoparticles promote α-synuclein oligomerization and underlying neurotoxicity as a model of Parkinson's disease. *Journal of Molecular Liquids, 323*, 115051. https://doi.org/10.1016/j.molliq.2020.115051

Marshall, C. A., King, K. M., & Kortagere, S. (2019). Limitations of the rat medial forebrain lesion model to study prefrontal cortex mediated cognitive tasks in Parkinson's disease. *Brain Research, 1702*, 105–113. https://doi.org/10.1016/j.brainres.2018.03.035

Martinez, B. A., Caldwell, K. A., & Caldwell, G. A. (2017). C. elegans as a model system to accelerate discovery for Parkinson disease. *Current Opinion in Genetics & Development, 44*, 102–109. https://doi.org/10.1016/j.gde.2017.02.011

Masilamoni, G. J., & Smith, Y. (2018). Chronic MPTP administration regimen in monkeys: A model of dopaminergic and non-dopaminergic cell loss in Parkinson's disease. *Journal of Neural Transmission, 125*(3), 337–363. https://doi.org/10.1007/s00702-017-1774-z

Melo-Thomas, L., Tonelli, L. C., Müller, C. P., Wöhr, M., & Schwarting, R. K. W. (2020). Playback of 50-kHz ultrasonic vocalizations overcomes psychomotor deficits induced by sub-chronic haloperidol treatment in rats. *Psychopharmacology*, *237*(7), 2043–2053. https://doi.org/10.1007/s00213-020-05517-9

Minnema, D. J., Travis, K. Z., Breckenridge, C. B., Sturgess, N. C., Butt, M., Wolf, J. C., Zadory, D., Beck, M. J., Mathews, J. M., Tisdel, M. O., Cook, A. R., Botham, P. A. & Smith, L. L. (2014). Dietary administration of diquat for 13 weeks does not result in a loss of dopaminergic neurons in the substantia nigra of C57BL/6J mice. *Regulatory Toxicology and Pharmacology*, *68*(2), 250–258. https://doi.org/10.1016/j.yrtph.2013.12.010.

Miyanishi, K., Choudhury, M. E., Watanabe, M., Kubo, M., Nomoto, M., Yano, H., & Tanaka, J. (2019). Behavioral tests predicting striatal dopamine level in a rat hemi-Parkinson's disease model. *Neurochemistry International*, *122*, 38–46. https://doi.org/10.1016/j.neuint.2018.11.005

Mulak, A., & Bonaz, B. (2015). Brain-gut-microbiota axis in Parkinson's disease. *World Journal of Gastroenterology*, *21*(37), 10609–10620. https://doi.org/10.3748/wjg.v21.i37.10609

Muma, N. A., Lee, J. M., Gorman, L., Heidenreich, B. A., Mitrovic, I., & Napier, T. C. (2001). 6-hydroxydopamine-induced lesions of dopaminergic neurons alter the function of postsynaptic cholinergic neurons without changing cytoskeletal proteins. *Experimental Neurology*, *168*(1), 135–143. https://doi.org/10.1006/exnr.2000.75 82.

Mustapha, M., & Mat Taib, C. N. (2021). MPTP-induced mouse model of Parkinson's disease: A promising direction of therapeutic strategies. *Bosnian Journal of Basic Medical Sciences*, *21*(4), 422–433. http://doi.org/10.17305/bjbms.2020.5181.

Narendra, D., Tanaka, A., Suen, D.-F., & Youle, R. J. (2008). Parkin is recruited selectively to impaired mitochondria and promotes their autophagy. *Journal of Cell Biology*, *183*(5), 795–803.

Niu, Y., Zhang, J., & Dong, M. (2021). Nrf2 as a potential target for Parkinson's disease therapy. *Journal of Molecular Medicine*, *99*(7), 917–931. https://doi.org/10.1007/s00109-021-02071-5

Nnah, I. C., & Wessling-Resnick, M. (2018). Brain iron homeostasis: A focus on microglial iron. *Pharmaceuticals*, *11*(4), 129. https://doi.org/10.3390/ph11040129

Noor, N. A., Mohammed, H. S., Mourad, I. M., Khadrawy, Y. A., & Aboul Ezz, H. S. (2016). A promising therapeutic potential of cerebrolysin in 6-OHDA rat model of Parkinson's disease. *Life Sciences, 155*, 174–179. https://doi.org/10.1016/j.lfs.2016.05.022

Ogiue-Ikeda, M., Tanabe, N., Mukai, H., Hojo, Y., Murakami, G., Tsurugizawa, T., Takata, N., Kimoto, T., & Kawato, S. (2008). Rapid modulation of synaptic plasticity by estrogens as well as endocrine disrupters in hippocampal neurons. *Brain Research Reviews*, *57*(2), 363–375. https://doi.org/10.1016/j.brainresrev.2007.06.010

Onukwufor, J. O., Dirksen, R. T., & Wojtovich, A. P. (2022). Iron dysregulation in mitochondrial dysfunction and Alzheimer's disease. *Antioxidants*, *11*(4), 692. https://doi.org/10.3390/antiox11040692

Pandey, U. B., & Nichols, C. D. (2011). Human disease models in Drosophila melanogaster and the role of the fly in therapeutic drug discovery. *Pharmacological Reviews*, *63*(2), 411–436. https://doi.org/10.1124/pr.110.003293

Park, J., Kim, S. Y., Cha, G.-H., Lee, S. B., Kim, S., & Chung, J. (2005). Drosophila DJ-1 mutants show oxidative stress-sensitive locomotive dysfunction. *Gene*, *361*, 133–139. https://doi.org/10.1016/j.gene.2005.06.040

Park, J., Lee, S., Lee, S., Kim, Y., Song, S., Kim, S., Bae, E., Kim, J., Shong, M., Kim, J.-M., & Chung, J. (2006). Mitochondrial dysfunction in Drosophila PINK1 mutants is complemented by Parkin. *Nature*, *441*, 1157–1161. https://doi.org/10.1038/nature04788

Penttinen, A. M., Parkkinen, I., Blom, S., Kopra, J., Andressoo, J. O., Pitkanen, K., Voutilainen, M. H., Saarma, M., & Airavaara, M. (2018a). Implementation of deep neural networks to count dopamine neurons in substantia nigra. *European Journal of Neuroscience*, *48*(6), 2354–2361. https://doi.org/10.1111/ejn.14129.

Penttinen, A. M., Parkkinen, I., Voutilainen, M. H., Koskela, M., Back, S., Their, A., Richie, C. T., Domanskyi, A., Harvey, B. K., Tuominen, R. K., Nevalaita, L., Saarma, M., & Airavaara, M. (2018b). Pre-alpha-pro-GDNF and pre-betapro-GDNF isoforms are neuroprotective in the 6-hydroxydopamine rat model of Parkinson's disease. *Frontiers in Neurology*, *9*, 457. https://doi.org/10.3389/fneur.2018.00457

Penttinen, A. M., Suleymanova, I., Albert, K., Anttila, J., Voutilainen, M. H., & Airavaara, M. (2016). Characterization of a new low-dose 6- hydroxydopamine model of Parkinson's disease in rat. *Journal of Neuroscience Research*, *94*(4), 318–328. https://doi.org/10.1002/jnr.23708

Pickrell, A. M., Pinto, M., & Moraes, C. T. (2013). Mouse models of Parkinson's disease associated with mitochondrial dysfunction. *Molecular and Cellular Neuroscience*, *55*, 87–94. https://doi.org/10.1016/j.mcn.2012.08.002

Pingale, T., & Gupta, G. L. (2020). Classic and evolving animal models in Parkinson's disease. *Pharmacology, Biochemistry and Behavior*, *199*, 173060. https://doi.org/10.1016/j.pbb.2020.173060.

Potts, L. F., Wu, H., Singh, A., Marcilla, I., Luquin, M. R., & Papa, S. M. (2014). Modeling Parkinson's disease in monkeys for translational studies, a critical analysis. *Experimental Neurology*, *256*, 133–143. https://doi.org/10.1016/j.expneurol.2013.09.014

Powers R., Lei S., Anandhan A., Marshall D. D., Worley B., Cerny R. L., Dodds E. D., Huang Y., Panayiotidis M. I., Pappa A., & Franco R. (2017). Metabolic investigations of the molecular mechanisms associated with Parkinson's disease. *Metabolites*, *7*, E22.

Prajapati, S. K., Garabadu, D., & Krishnamurthy, S. (2017). Coenzyme Q10 prevents mitochondrial dysfunction and facilitates pharmacological activity of atorvastatin in 6-OHDA induced dopaminergic toxicity in rats. *Neurotoxicity Research*, *31*(4), 478–492. https://doi.org/10.1007/s12640-016-9693-6

Qiu, J., Liu, R., Ma, Y., Li, Y., Chen, Z., He, H., Chen, J., Tong, L., Huang, C., & You, Q. (2020). Lipopolysaccharide-induced depression-like behaviors is ameliorated by sodium butyrate via inhibiting neuroinflammation and oxido-nitrosative stress. *Pharmacology*, *105*, 550–560.

Radad, K., Al-Shraim, M., Al-Emam, A., Wang, F., Kranner, B., Rausch, W. D., & Moldzio, R. (2019). Rotenone: From modelling to implication in Parkinson's disease. *Folia Neuropathologica*, *57*(4), 317–326. https://doi.org/10.5114/fn.2019.89857

Rai, S. N., & Singh, P. (2020). Advancement in the modelling and therapeutics of Parkinson's disease. *Journal of Chemical Neuroanatomy*, *104*, 101752. https://doi.org/10.1016/j.jchemneu.2020.101752

Riachi, N. J., LaManna, J. C., & Harik, S. I. (1989). Entry of 1-methyl-4-phenyl1,2,3,6-tetrahydropyridine into the rat brain. *Journal of Pharmacology and Experimental Therapeutics*, *249*(3).

Saeed, A., Shakir, L., Khan, M. A., Ali, A., & Zaidi, A. A. (2017). Haloperidol induced Parkinson's disease mice model and motor-functiӧn modulation with Pyridine-3-carboxylic acid. *Biomedical Research and Therapy*, *4*(05), 1305. https://doi.org/10.15419/bmrat.v4i05.169

Salari, S., & Bagheri, M. (2019). In vivo, in vitro and pharmacologic models of Parkinson's disease. *Physiological Research*, *68*(1), 17–24. https://doi.org/10.33549/physiolres.933895

Sanawar, M., Saleem, U., Anwar, F., Nazir, S., Akhtar, M. F., Ahmad, B., & Ismail, T. (2022). Investigation of anti-Parkinson activity of dicyclomine. *The International Journal of Neuroscience*, *132*(4), 338–351. https://doi.org/10.1080/00207454.2020.1815732

Sanberg, P. R. (1980). Haloperidol-induced catalepsy is mediated by postsynaptic dopamine receptors. *Nature*, *284*(5755), 472–473. https://doi.org/10.1038/284472a0

Schapira, A. H. V. (2008). Mitochondria in the aetiology and pathogenesis of Parkinson's disease. *The Lancet Neurology*, *7*(1), 97–109.

Seo, J., Lee, Y., Kim, B. S., Park, J., Yang, S., Yoon, H. J., Yoo, J., Park, H. S., Hong, J. J., Koo, B. S., Baek, S. H., Jeon, C. Y., Huh, J. W., Kim, Y. H., Park, S. J., Won, J., Ahn, Y. J., Kim, K., Jeong, K. J., Kang, P., . . . Lee, S. R. (2019). A non-human primate model for stable chronic Parkinson's disease induced by MPTP administration based on individual behavioral quantification. *Journal of Neuroscience Methods*, *311*, 277–287. https://doi.org/10.1016/j.jneumeth.2018.10.037

Seyedebrahimi, R., Razavi, S., & Varshosaz, J. (2020). Controlled delivery of brain derived neurotrophic factor and gold-nanoparticles from chitosan/TPP nanoparticles for tissue engineering applications. *Journal of Cluster Science*, *31*, 99–108. doi: 10.1007/S10876-019-01621-9.

Shimizu, K., Ohtaki, K., Matsubara, K., Aoyama, K., Uezono, T., Saito, O., Suno, M., Ogawa, K., Hayase, N., Kimura, K., & Shiono, H. (2001). Carrier-mediated processes in blood–brain barrier penetration and neural uptake of paraquat. *Brain Research*, *906*, 135–142.

Shulman, J. M., De Jager, P. L., & Feany, M. B. (2011). Parkinson's disease: Genetics and pathogenesis. *Annual Review of Pathology: Mechanisms of Disease*, *6*, 193–222. https://doi.org/10.1146/annurev-pathol-011110-130242

Smith, L. K., Jadavji, N. M., Colwell, K. L., Perehudoff, K., & Metz, G. a. S. (2008). Stress accelerates neural degeneration and exaggerates motor symptoms in a rat model of Parkinson's disease. *European Journal of Neuroscience*, *27*(8), 2133–2146. https://doi.org/10.1111/j.1460-9568.2008.06177.x

Soni, D., & Kumar, P. (2022). GSK-3β-mediated regulation of Nrf2/HO-1 signaling as a new therapeutic approach in the treatment of movement disorders. *Pharmacological Reports*, *74*(4), 557–569. https://doi.org/10.1007/s43440-022-00390-z

Staveley, B. E. (2015). Drosophila models of Parkinson disease. In *Movement disorders* (2nd ed., pp. 345–354). Academic Press. https://doi.org/10.1016/B978-0-12-405195-9.00020-2

Su, L.-Y., Li, H., Lv, L., Feng, Y.-M., Li, G.-D., Luo, R., Zhou, H.-J., Lei, X.-G., Ma, L., Li, J.-L., Xu, L., Hu, X.-T., & Yao, Y.-G. (2015). Melatonin attenuates MPTP-induced neurotoxicity via preventing CDK5-mediated autophagy and SNCA/α-synuclein aggregation. *Autophagy*, *11*(10), 1745–1759. https://doi.org/10.1080/15548627.2015.1082020

Sun, C. P., Zhou, J. J., Yu, Z. L., Huo, X. K., Zhang, J., Morisseau, C., Hammock, B. D., & Ma, X. C. (2022). Kurarinone alleviated Parkinson's disease via stabilization of epoxyeicosatrienoic acids in animal model. *Proceedings of the National Academy of Sciences of the United States of America*, *119*(9), e2118818119. https://doi.org/10.1073/pnas.2118818119

Sun, M. F., & Shen, Y. Q. (2018). Dysbiosis of gut microbiota and microbial metabolites in Parkinson's disease. *Ageing Research Reviews*, *45*, 53–61. https://doi.org/10.1016/j.arr.2018.04.004

Surmeier, D. J., Obeso, J. A., & Halliday, G. M. (2017). Selective neuronal vulnerability in Parkinson disease. *Nature Reviews Neuroscience*, *18*(2), 101–113.

Tamano, H., Morioka, H., Nishio, R., Takeuchi, A., & Takeda, A. (2019). Blockade of rapid influx of extracellular Zn^{2+} into nigral dopaminergic neurons overcomes paraquat-induced Parkinson's disease in rats. *Molecular Neurobiology*, *56*(6), 4539–4548. https://doi.org/10.1007/s12035-018-1398-9

Tan, A. H., Lim, S. Y., & Lang, A. E. (2022). The microbiome-gut–brain axis in Parkinson disease—from basic research to the clinic. *Nature Reviews. Neurology*, *18*(8), 476–495. https://doi.org/10.1038/s41582-022-00681-2

Tain, L. S., Chowdhury, R. B., Tao, R. N., Plun-Favreau, H., & Martins, L. M. (2009). Drosophila HtrA2 is dispensable for apoptosis but acts downstream of PINK1 independently from Parkin. *PLoS Genetics*, *5*(7), e1000514. https://doi.org/10.1371/journal.pgen.1000514

Terzioglu, M., & Galter, D. (2008). Parkinson's disease: Genetic versus toxin-induced rodent models. *The FEBS Journal*, *275*(7), 1384–1391. https://doi.org/10.1111/j.1742-4658.2008.06302.x

Tharwat, E. K., Abdelaty, A. O., Abdelrahman, A. I., Elsaeed, H., Elgohary, A., El-Feky, A. S., Ebrahim, Y. M., Sakraan, A., Ismail, H. A., Khadrawy, Y. A., & Aboul Ezz H. S. (2023). Evaluation of the therapeutic potential of cerebrolysin and/or lithium in the male Wistar rat model of Parkinson's disease induced by reserpine. *Metabolic Brain Disease*, *38*(5), 1513–1529.

Thirugnanam, T., & Santhakumar, K. (2022). Chemically induced models of Parkinson's disease. *Comparative Biochemistry and Physiology Part C: Toxicology & Pharmacology*, *252*, 109213. https://doi.org/10.1016/j.cbpc.2021.109213

Tian, Y., Xian, J., Tian, Y., & Gao, X. (2023). Establishment of a staging model for Parkinson's disease in mice. *Cellular and Molecular Biology*, *69*(4), 141–146. https://doi.org/10.14715/cmb/2023.69.4.22

Tieu, K. (2011). A guide to neurotoxic animal models of Parkinson's disease. *Cold Spring Harbor Perspectives in Medicine*, *1*(1), a009316. https://doi.org/10.1101/cshperspect.a009316

Tsuru, A., Matsui, K., Kimura, A., Yoshiike, T., Otsuki, R., Nagao, K., Hazumi, M., Utsumi, T., Fukumizu, M., Mukai, Y., Takahashi, Y., Sakamoto, T., & Kuriyama, K. (2022). Sleep disturbance and health-related quality of life in Parkinson's disease: A clear correlation between health-related quality of life and subjective sleep quality. *Parkinsonism & Related Disorders*, *98*, 86–91. https://doi.org/10.1016/j.parkreldis.2022.04.014

Ullah, H., Arbab, S., Tian, Y., Liu, C. Q., Chen, Y., Qijie, L., Khan, M. I. U., Hassan, I. U., & Li, K. (2023). The gut microbiota-brain axis in neurological disorder. *Frontiers in Neuroscience*, *17*, 1225875. https://doi.org/10.3389/fnins.2023.1225875

Van Kampen, J., Robertson, H., Hagg, T., & Drobitch, R. (2003). Neuroprotective actions of the ginseng extract G115 in two rodent models of Parkinson's disease. *Experimental Neurology*, *184*(1), 521–529. https://doi.org/10.1016/j.expneurol.2003.08.002

Waku, I., Magalhães, M. S., Alves, C. O., & de Oliveira, A. R. (2021). Haloperidol-induced catalepsy as an animal model for parkinsonism: A systematic review of experimental studies. *EJN*, *53*(11), 3743–3767. https://doi.org/10.1111/ejn.15222

Wang, L., Zhao, Z., Zhao, L., Zhao, Y., Yang, G., Wang, C., Gao, L., Niu, C., & Li, S. (2022). *Lactobacillus plantarum* DP189 Reduces α-SYN aggravation in MPTP-induced Parkinson's disease mice via regulating oxidative damage, inflammation, and gut microbiota disorder. *Journal of Agricultural and Food Chemistry*, *70*(4), 1163–1173. https://doi.org/10.1021/acs.jafc.1c07711

Wang, W., Le, A. A., Hou, B., Lauterborn, J. C., Cox, C. D., Levin, E. R., Lynch, G., & Gall, C. M. (2018). Memory-related synaptic plasticity is sexually dimorphic in rodent hippocampus. *The Journal of Neuroscience*, *38*(37), 7935–7951. https://doi.org/10.1523/JNEUROSCI.0801-18.2018

Wang, X., & Zhang, J. (2019). Mitophagy: Mitophagy: The origin of mitochondrial autophagy. *Cell Research*, *29*(3), 192–198.

Wasner, K., Grünewald, A., & Klein, C. (2020). Parkin-linked Parkinson's disease: From clinical insights to pathogenic mechanisms and novel therapeutic approaches. *Neuroscience Research*, *159*, 34–39. https://doi.org/10.1016/j.neures.2020.09.001

Whitworth, A. J., Theodore, D. A., Greene, J. C., Benes, H., Wes, P. D., & Pallanck, L. J. (2005). Increased glutathione S-transferase activity rescues dopaminergic neuron loss in a Drosophila model of Parkinson's disease. *Proceedings of the National Academy of Sciences*, *102*(22), 8024–8029. https://doi.org/10.1073/pnas.0501078102

Wood, H. (2019). New models show gut–brain transmission of Parkinson disease pathology. *Nature Reviews Neurology*, *15*, 491. https://doi.org/10.1038/s41582-019-0241-x

Wu, J., Li, Q., Wang, X., Yu, S., Li, L., Wu, X., Chen, Y., Zhao, J., & Zhao, Y. (2013). Neuroprotection by curcumin in ischemic brain injury involves the Akt/Nrf2 pathway. *PLoS One*, *8*(3), e59843. https://doi.org/10.1371/journal.pone.0059843

Xu, Q., Shenoy, S., & Li, C. (2012). Mouse models for LRRK2 Parkinson's disease. *Parkinsonism & Related Disorders*, *18*(Suppl 1), S186–S189. https://doi.org/10.1016/S1353-8020(11)70058-X

Yang, Y., Gehrke, S., Haque, M. E., Imai, Y., Kosek, J., Yang, L., Beal, M. F., Nishimura, I., Wakamatsu, K., Ito, S., Takahashi, R., & Lu, B. (2005). Inactivation of Drosophila DJ-1 leads to impairments of oxidative stress response and phosphatidylinositol 3-kinase/Akt signaling. *Proceedings of the National Academy of Sciences of the United States of America*, *102*(38), 13670–13675. https://doi.org/10.1073/pnas.0504610102

Zeng, B. Y., Dass, B., Owen, A., Rose, S., Cannizzaro, C., Tel, B. C., & Jenner, P. (2002). 6-Hydroxydopamine lesioning differentially affects alpha-synuclein mRNA expression in the nucleus accumbens, striatum and substantia nigra of adult rats. *Neuroscience Letters*, *322*(1), 33–36. https://doi.org/10.1016/s0304-3940(02)00083-6

Zeng, X. S., Geng, W. S., & Jia, J. J. (2018). Neurotoxin-induced animal models of Parkinson disease: Pathogenic mechanism and assessment. *ASN Neuro*, *10*, 1759091418777438. https://doi.org/10.1177/1759091418777438

Zhang, X., Yang, Y., Du, L., Zhang, W., & Du, G. (2017b). Baicalein exerts anti-neuroinflammatory effects to protect against rotenone-induced brain injury in rats. *International Immunopharmacology*, *50*, 38–47. https://doi.org/10.1016/j.intimp.2017.06.007

Zhang, X. L., Yuan, Y. H., Shao, Q. H., Wang, Z. Z., Zhu, C. G., Shi, J. G., Ma, K. L., Yan, X., & Chen, N. H. (2017a). DJ-1 regulating PI3K-Nrf2 signaling plays a significant role in bibenzyl compound 20C-mediated neuroprotection against rotenone-induced oxidative insult. *Toxicology Letters*, *271*, 74–83. https://doi.org/10.1016/j.toxlet.2017.02.022

Zhang, Y., Xu, S., Qian, Y., He, X., Mo, C., Yang, X., & Xiao, Q. (2022). Sodium butyrate attenuates rotenone-induced toxicity by activation of autophagy through epigenetically regulating PGC-1α expression in PC12 cells. *Brain Research*, *1776*, 147749.

Zhao, Z., Li, F., Ning, J., Peng, R., Shang, J., Liu, H., Shang, M., Bao, X. Q., & Zhang, D. (2021a). Novel compound FLZ alleviates rotenone-induced PD mouse model by suppressing TLR4/MyD88/NF-κB pathway through microbiota-gut–brain axis. *Acta Pharmaceutica Sinica. B*, *11*(9), 2859–2879. https://doi.org/10.1016/j.apsb.2021.03.020

Zhao, Z., Ning, J., Bao, X. Q., Shang, M., Ma, J., Li, G., & Zhang, D. (2021b). Fecal microbiota transplantation protects rotenone-induced Parkinson's disease mice via suppressing inflammation mediated by the lipopolysaccharide-TLR4 signaling pathway through the microbiota-gut-brain axis. *Microbiome*, *9*(1), 226. https://doi.org/10.1186/s40168-021-01107-9

Zhou, X., Lu, J., Wei, K., Wei, J., Tian, P., Yue, M., Wang, Y., Hong, D., Li, F., Wang, B., Chen, T., & Fang, X. (2021). Neuroprotective effect of ceftriaxone on MPTP-induced Parkinson's disease mouse model by regulating inflammation and intestinal microbiota. *Oxidative Medicine and Cellular Longevity*, 9424582. https://doi.org/10.1155/2021/9424582

Zhu, K., van Hilten, J. J., & Marinus, J. (2016). The course of insomnia in Parkinson's disease. *Parkinsonism & Related Disorders*, *33*, 51–57. https://doi.org/10.1016/j.parkreldis.2016.09.010

Zhu, Q., Zhuang, X. X., Chen, J. Y., Yuan, N. N., Chen, Y., Cai, C. Z., Tan, J. Q., Su, H. X., & Lu, J. H. (2021). Lycorine, a natural alkaloid, promotes the degradation of alpha-synuclein via PKA-mediated UPS activation in transgenic Parkinson's disease models. *Phytomedicine*, *87*, 153578. https://doi.org/10.1016/j.phymed.2021.153578

Zuo, Y., Xie, J., Li, X., Li, Y., Thirupathi, A., Zhang, J., Yu, P., Gao, G., Chang, Y., & Shi, Z. (2021). Ferritinophagy-mediated ferroptosis involved in paraquat-induced neurotoxicity of dopaminergic neurons: Implication for neurotoxicity in PD. *Oxidative Medicine and Cellular Longevity*, *2021*, 9961628. https://doi.org/10.1155/2021/9961628

13 Pharmacotherapy of Parkinson's Disease and Their Limitations

Ebele Joan Ajagun, Abraham Olufemi Asuku, Maryam Tayo Ayinla, Taoheed Adedeji Abdulsalam, Ayodeji Johnson Ajibare, Tobiloba Samuel Olajide, Yusuf Olamilekan Hamza, and Kaosara Oyinola Ganiyu

13.1 BACKGROUND

Evidence of the possible existence of Parkinsonism can be found in earlier medical documents; however, the earliest documented medical description of Parkinson's disease (PD) reported to have occurred in the 1800s was by an English physician, James Parkinson, where he described in details the symptoms of a neurological disorder that he referred to as the "shaking palsy" (Duvoisin, 1992; Parkinson, 2002; Dorseya *et al.*, 2018). Several years later, Jean-Martin Charcot clearly defined the disease condition and its symptoms while internationally disseminating information on the disease, for which he coined the term *Parkinson's disease* (Dorseya et *al.*, 2018).

Parkinsonism can simply be referred to as a severe and progressive neurological dysfunction of the nervous system due to the loss, malfunction, or death of brain cells responsible for the production of dopamine and is characterised by the cardinal features of rigidity (arms, legs or neck), bradykinesia, tremor, and postural instability (Rizek *et al.*, 2016; Keener & Bordelon, 2016; Sveinbjornsdottir, 2016; Vázquez-Vélez & Zoghbi, 2021; Zanardi *et al.*, 2021).

Autonomic dysfunction, including difficulty swallowing and choking; excessive sweating, stooling/diarrhea, or fluctuations in blood pressure control; anxiety and/or depression (affective disorders); confusion, personality changes, and cognitive deficit; apathy; problems with sleeping; fatigue; and sensory disturbances or pain are examples of non-movement-related symptoms that are experienced at the terminal phases of the disorder (Rizek *et al.*, 2016; Vázquez-Vélez & Zoghbi, 2021; Mysiris *et al.*, 2022).

The aetiology of PD remains unknown; however, common causes include heredity, age, gender, head injuries, and/or exposure to environmental neurotoxins amongst others, resulting in basically three different types of Parkinsonism: idiopathic, early commencement, and PD tending to occur in more members of a family than expected by chance alone (Rizek *et al.*, 2016; Mysiris *et al.*, 2022). In general, the increasing and continuous dopamine neurotransmitters' loss, which causes the neurons to fire out of control, results in the key symptoms/features of PD. As the disease progresses, dopamine production in the brain reduces, which makes the maintenance of normal mobility impossible. Hence, PD is managed via the use of drugs (pharmacotherapy) and or surgical procedures, where the sole goal of therapy remains the maintenance of functioning remaining brain cells, exogenous supply of dopamine, and to nullify the occurrence of complications brought on by drugs (Rizek *et al.*, 2016; Poewe *et al.*, 2017; Armstrong & Okun, 2020; Bloem *et al.*, 2021).

13.2 PHARMACOTHERAPY OF PARKINSONISM

PD is generally known to follow a progressive course; hence, it is incurable. However, it is known that the initiation of timely drug therapy not only relieves the disorder's motor symptoms but also enhances the quality of life (QoL) of affected persons for many years (Rizek *et al.*, 2016; Bloem *et al.*, 2021). Pharmacotherapy is derived from two words "pharma" and "therapy" and refers to the use of pharmaceutical products such as medications (drugs) and drug products for the treatment and management of various disease conditions. Pharmacotherapy of Parkinsonism can thus be defined as the utilisation of pharmaceutical products/agents/drugs in the management of PD designed to replace, mimic, or potentiate the action of depleted endogenous dopamine in the striatum. This can be achieved by the administration of drugs/medications that are either converted to dopamine through metabolic process, stimulate the dopaminergic receptor, avoid the decomposition of endogenous dopamine, or brings the excitatory and inhibitory neurotransmitters at the striatum to equilibrium amongst others (Connolly & Lang, 2014; Rizek *et al.*, 2016).

Pharmacotherapy of PD is mainly symptomatic, where medications are used to either prevent or manage the movement-related symptoms of the disorder as they present. Hence, the goal of pharmacotherapy in Parkinsonism is basically to relieve the symptoms of the disorder, minimize the side effects of drugs, as well as improve the patient's QoL (Rizek *et al.*, 2016; Bloem *et al.*, 2021; Lee & Yankee, 2021).

Pharmacotherapy of PD is an individualised patient-centred care/therapy, where patient-specific factors such as level of disability, tolerability of drugs, functional and chronological age, temporary advantages, etc. as well as the

DOI: 10.1201/9781003402893-15

permanent repercussions of therapy are taken into consideration when choosing agents to initiate pharmacotherapy (Chen & Swope, 2012; Rizek *et al.*, 2016; Beckers *et al.*, 2022). Pharmacotherapy in drug-naïve patients begins with the initiation of a single drug agent—a dopamine agonist or a non-dopaminergic agent (mono-therapy). Since PD is chronic and continuously increasing in severity, the degree of loss, damage, and/or dysfunction of the brain cells may increase over time, making periodic adjustments to pharmacotherapy a necessity (Rizek *et al.*, 2016; Bloem *et al.*, 2021).

13.3 CLASSIFICATION OF ANTI-PARKINSONISM DRUGS

According to the American Academy of Neurology (AAN) indication, the relatively early treatment of Parkinsonism with a dopamine precursor remains the most effective in alleviating movement-related symptoms of the disorder in comparison to treatment with other dopaminergic agonists, with levodopa-carbidopa remaining the benchmark for the symptomatic management of Parkinsonism (Rizek *et al.*, 2016; AAN, 2021).

Dopamine can be described as a neurotransmitter, and its production takes place at the substantia nigra as well as ventral tegmental regions of the brain, associated with movement, attention, learning, and the brain's pleasure and rewards system essential for our daily activities. Therefore, a deficiency in dopamine levels associated with cellular death in the substantia nigra has been implicated as the major reason for Parkinsonism (Chen & Swope, 2012; Rizek *et al.*, 2016; AAN, 2021; Lee & Yankee, 2021).

An underlying reason for the utilisation of drugs in the management of Parkinsonism is to replace depleted levels of dopamine, stimulate nerve receptors to enable neurotransmission, increase the effect of dopamine on nerve receptors, and/or restore the imbalance between the excitatory (acetylcholine (Ach)) and inhibitory (dopamine) neurotransmitters within the brain (Patel & Chang, 2014; Rizek *et al.*, 2016; Bloem *et al.*, 2021; Beckers *et al.*, 2022). Dopamine in itself has no therapeutic effect when used in the management of Parkinsonism as it fails to pass through the blood-brain barrier (BBB) following its administration (Rizek *et al.*, 2016). Several medications are currently in the market for the treatment of Parkinsonism, and they can be broadly divided into two major groups: dopaminergic system-acting drugs and cholinergic system-acting drugs. The choice to initiate treatment is dependent upon the severity of the disease, how the symptoms affect the patient's QoL, and also the patient's perception of medication use for life (Rizek *et al.*, 2016; Bloem *et al.*, 2021).

13.4 DRUGS ACTING ON THE DOPAMINERGIC SYSTEM

13.4.1 Dopamine Precursor–Levodopa

Dopamine's metabolic precursor known as levodopa is the mainstay of the management/treatment of Parkinsonism, with the best results usually obtained within the first few years of treatment (Hornykiewicz, 2010; LeWitt, 2014; Chagraoui *et al.*, 2020). Following administration, levodopa crosses the BBB through the L-amino acid transporter (LAT), where it is decarboxylated to dopamine (Rizek *et al.*, 2016). This exogenous supply of dopamine increases the amount of dopamine available for uptake by dopaminergic neurons. Although treatment with levodopa does not halt the increasing and continuous severity of Parkinsonism, its early onset has been reported to delay disease progression and reduce mortality (Chagraoui *et al.*, 2020; Lee & Yankee, 2021).

Levodopa has been the most efficient and effective medication for the management of Parkinsonism symptoms to date, especially in patients presenting with bradykinesia and disabilities resulting from it; however, levodopa has been reported to also improve quite a number of the clinical motor characteristics of the disorder (Hornykiewicz, 2010; LeWitt, 2014; Rizek *et al.*, 2016; Tambasco *et al.*, 2018; Tran *et al.*, 2018). Therapy is usually initiated using low doses of rapid-acting formulations two to three times daily post-meals (to decrease the incidences of nausea). The dose is then stepwisely increased over many days to weeks depending on the patient's tolerance of the drug till the lowest dose that can control symptoms is reached (Rizek *et al.*, 2016; Tambasco *et al.*, 2018; Tran *et al.*, 2018). Table 13.2 lists the various examples of Levodopa combinations with their dosage regimens.

Over time, patients become less responsive to therapy, which has been attributed to the increasing and continuous loss of nerve endings of dopaminergic nigrostriatal or the indirect depletion of the dopamine receptors in the striatal as a result of disease conditions; hence, the impacts of levodopa therapy have been reported to diminish over time, irrespective of initial response (Connolly & Lang, 2014; Rizek *et al.*, 2016). Table 13.4 shows MAOB inhibitors with their dosage regimens for treating PD.

13.4.1.1 Pharmacokinetics

Approximately 1–3% of an administered dose of levodopa gets to the brain unchanged; the rest undergoes metabolic activities extra-cerebrally to dopamine by the decarboxylase enzyme. Hence, when used as mono-therapy, levodopa is administered in high doses. Its absorption from the small intestine is rapid; however, the extent through which it is absorbed is dependent upon the extent of gastric emptying and the gastric content's PH. Levodopa is a levorotatory stereoisomer of dopa, which is dopamine and norepinephrine's amino acid precursor. Therefore, absorption of the drug from the gut and transport to the brain can be compromised by the presence of food and certain amino acids in the stomach. A peak plasma concentration of levodopa happens 60–120 minutes after oral administration, with a half-life of 60–180 minutes in the plasma, which varies considerably from person to person. The main route of excretion is via the kidney, where two-thirds of an administered dose appears in urine before 8 hours after oral administration as dihydroxyphenylacetic acid (DOPAC) and homovanillic acid (HVA) (LeWitt, 2014; Rizek *et al.*, 2016; Tambasco *et al.*, 2018).

13.4.2 Peripheral Decarboxylase Inhibitors (PDIs)—Carbidopa and Benserazide

PDIs have been reported to inhibit levodopa from being converted into dopamine within the blood by inhibiting aromatic L-amino acid decarboxylase (AADC) enzyme (Whitfield *et al.*, 2014; AAN, 2021; van Rumund *et al.*, 2021).

Carbidopa and Benserazide are extra-cerebral dopa decarboxylase inhibitors; they do not inhibit the conversion of levodopa within the brain due to their inability to move through the BBB (Whitfield *et al.*, 2014; van Rumund *et al.*, 2021). PDIs are usually administered in combination with levodopa, except in patients who develop marked involuntary movements on administration of this two-drug combination (van Rumund *et al.*, 2021). Several therapeutic benefits are gained from the administration of the PDI+levodopa combination: the inhibition of the PDI enzyme prolongs levodopa's half-life in the plasma, allowing for the drastic reduction of measured quantity required to achieve therapeutic effect (60–80% reduction), and thus therapeutic doses are quickly attained (Whitfield *et al.*, 2014; Rizek *et al.*, 2016; van Rumund *et al.*, 2021). Other benefits include minimised cardiac complications, on-off effects, and the pyridoxine reversal of levodopa effects are prevented (Rizek *et al.*, 2016; van Rumund *et al.*, 2021). Benspar/Madopar (levodopa and benserazide) and Sinemet/Duopa/Co-careldopa (carbidopa and levodopa) are examples of available levodopa+PDIs combinations (Rizek *et al.*, 2016).

13.4.2.1 Adverse Effects

Common adverse effects of levodopa therapy include gastrointestinal, cardiovascular, and behavioural effects amongst others; the gastrointestinal adverse effects (nausea and vomiting) of levodopa can be minimised when the drug is administered in smaller doses, with meals or immediately after meals, or the slow gradual increase in total daily doses (Hauser, 2009; Calabresi *et al.*, 2010; Rizek *et al.*, 2016; van Rumund *et al.*, 2021). Anorexia, stomach distress, and an urge to discharge the contents of the stomach through the mouth commonly happen in over 80% of individuals on levodopa monotherapy; this effect can be reduced by administering levodopa in combination with a peripheral decarboxylase inhibitor-Carbidopa, allowing patients to tolerate higher doses of the drug. The use of antacids and antiemetics before the administration of levodopa has also been reported to be beneficial in limiting gastrointestinal adverse effects. However, the phenothiazines (an antiemetic) are contraindicated as it has been proven to reduce the healing impacts of levodopa while exacerbating the disease condition (Hauser, 2009; Rizek *et al.*, 2016).

Patients on levodopa have also reported a variety of cardiac arrhythmia—tachycardia, ventricular extrasystoles, and atrial fibrillation in rare cases—especially in cases of established cardiac diseases as a result of increased peripheral catecholamine formation. Asymptomatic postural hypotension is common, and it has been investigated and proven to reduce with continuation of treatment. In persons on large doses of levodopa, sympathomimetic or nonselective monoamine oxidase inhibitors, hypertension can also occur (Kalinderi *et al.*, 2011; Rizek *et al.*, 2016).

A variety of psychiatric/behavioural side effects, including depression, anxiety, agitation, insomnia, somnolence, sleep attacks, confusion, delusions, hallucinations, nightmares, euphoria, and personality changes, have been reported more in patients on levodopa-carbidopa therapy than in those on levodopa monotherapy (Kalinderi *et al.*,

DRUG HOLIDAYS AND ON-OFF PHENOMENON

A drug holiday refers to the discontinuation of the drug for a period of between 3–21 days. Drug holidays have been reported to temporarily improve response to levodopa and relieve some associated side effects with no effect on the on-off phenomenon.

Drug holidays are usually discouraged as they are associated with high risks of aspiration pneumonia, venous thrombosis, pulmonary embolism, and depression resulting from the immobility accompanying severe Parkinsonism.

The on-off/wearing-off phenomenon refers to a situation where the therapeutic effect of the drug wears off before the next scheduled dose.

The term "**punding**" refers to the performance of purposeless, complex, and stereotyped motor activities. Examples of such activities include repetitive self-grooming, colour/shape sorting, or lining up various objects and are managed either by dose reduction of dopaminergic agents or the use of atypical antipsychotic agents.

Dopamine dysregulation syndrome (DDS), an iatrogenic disturbance, is more common in patients on dopamine agonist therapy than levodopa and is characterised by compulsive overuse of dopaminergic medication and other impulsive behaviours.

Management involves the close regulation of dopaminergic intake.

2011; Rizek *et al.*, 2016). This has been attributed to the increased concentration of levodopa reaching the brain when there is co-administration of levodopa and carbidopa, a decarboxylase inhibitor, making the reduction in dose or complete withdrawal of the drug necessary. However, when the complete withdrawal of the drug is not possible, concomitant use with several commercially available atypical antipsychotics, such as clozapine, olanzapine, quetiapine, and risperidone (which have low affinity for dopamine D2 receptors), may be clinically beneficial to reduce/counteract adverse behavioural effects (Rizek *et al.*, 2016).

Dyskinesias refers to the abnormal involuntary movements of the limbs, causing rapid or slow jerking of the limbs and extended periods of muscle spasms. It is said to occur in approximately 80% of persons who have received long-term treatment with levodopa. The development of dyskinesia is usually dose-dependent and the pathogenesis is unclear, with individualized and varying clinical presentation in each patient. Although the cause is unclear, loss of nerve supply to the dopaminergic system and dopamine receptors severe activation following levodopa administration are responsible for dyskinesia development. The most common presentation is choreoathetosis of the face and distal extremities (Calabresi *et al.*, 2010; Huot *et al.*, 2013). The incidence of dyskinesia can be reduced by continuous intraduodenal or enteral/intra-jejunal administration of levodopa, allowing continuous administration of the dopaminergic drug. A reduction in the doses of levodopa administered relieves dyskinesia but aggravates the motor symptoms of Parkinsonism. Mild dyskinesia does not require treatment. Both Amantadine and Clozapine have been proven to attenuate the incidence of troublesome dyskinesia (Rizek *et al.*, 2016). Other adverse effects include mydriasis, which can trigger glaucoma attack in most patients. Uncommon side effects of levodopa therapy include worsening or precipitation of gout; various blood dyscrasias; an abnormal sense of smell or taste; hot flushes; brown discolouration of the saliva, urine, and or vaginal discharge; priapism; and transient mild increases in alkaline phosphatase, serum transaminases, bilirubin, and blood urea nitrogen (BUN) (Arora & Fletcher, 2013).

13.4.2.2 Drug Interactions

Pyridoxine (vitamin B6) has been reported to enhance the extra-cerebral breakdown of levodopa, therefore limiting its pharmacological effect when used in the absence of carbidopa. Furthermore, levodopa must not be used by patients taking monoamine oxidase A inhibitors (MOAI) until 2 weeks after discontinuation of therapy, as a combination might result in hypertensive crises, especially in hypertensive patients (Rizek *et al.*, 2016).

13.4.2.3 Contraindications

Levodopa is not advised for treatment in psychotic patients as its use could exacerbate mental disturbance. Levodopa may be used in persons with severe open-angle glaucoma where intra-ocular pressure is well controlled and properly monitored; however, its use in patients with angle-closure glaucoma is contraindicated. There is a need for proper monitoring in active peptic ulcer patients as gastrointestinal bleeding has been reported to occasionally occur on administration of levodopa.

Persons with a history of melanoma tending to occur in more members of a family than expected by chance alone and those suspected of having undiagnosed skin lesions should be careful, as levodopa can form cutaneous melanin and could stimulate melanoma that possess an ability to infiltrate, metastasize, and terminate fatally. Therefore, these patients ought to be regularly monitored by a dermatologist (Chen & Swope, 2014; Rizek *et al.*, 2016).

13.4.3 Dopamine Receptor Agonists (DRAs)—Bromocriptine, Pergolide, Ropinirole, Pramipexole, and Apomorphine

DRAs act directly on postsynaptic dopamine receptors; hence, they are of benefit in individuals who have no ability to produce, store, and release dopamine from levodopa, thereby making a beneficial addition to levodopa therapy (Rizek *et al.*, 2016; Lee & Yankee, 2021). DRAs need no conversion by enzymes to active metabolic products, therefore having no possible harmful metabolic products. They readily cross the BBB, and thereby have no competition with other compounds for active transport (Pagliaro & Pagliaro, 2020).

The ergot (ergoline) by-products pergolide and bromocriptine are not commonly utilised in the management of Parkinsonism and are associated with more severe adverse effects. The newer dopamine receptor agonists, which include pramipexole and ropinirole, now replace older drugs in this class (Kalinderi *et al.*, 2011; Rizek *et al.*, 2016; Cacabelos, 2017; Güneş & Karavana, 2021). Although the superiority of any of the dopamine receptor agonists is yet to be evidently proven, patients have been known to respond differently to agents within this group, with variations in duration of action prolonged on the administration of sustained/extended-release preparations.

Dopaminergic agonists play a vital role as first-line medication for Parkinsonism as their intake is connected with a reduced occurrence of response instability as well as dyskinesias, which usually happen with prolonged levodopa medication. Hence, they may also be administered in combination with levodopa, especially in patients experiencing the change between their relatively normal motor performances when levodopa is effective and when it has transiently lost its impact on motor responses and Parkinsonism re-emerges (Kalinderi *et al.*, 2011; Rizek *et al.*, 2016; Güneş & Karavana, 2021). When using dopamine agonists in combination therapy, it is important to reduce the levodopa's administered doses to avoid the development of unbearable adverse aftermath (Brooks, 2000). In levodopa-naive patients, the responses to dopamine agonists are usually discouraging, as this produces few symptomatic benefits and has greater chances to cause somnolence, psychiatric side effects, and oedema. In the treatment of Parkinsonism, the measured concentration of administered dopamine agonist is increased based on the individual tolerability and feedback till the maximum tolerated dose is reached (Brooks, 2000; Choi & Horner, 2023).

13.4.3.1 Bromocriptine

Bromocriptine, a derivative of ergot (ergoline), and a D2 agonist with partial D1 agonist/antagonist was used extensively to manage Parkinsonism in the past; its use has been supplanted by the development of newer dopamine agonists (Brooks, 2000; Cánovas *et al.*, 2014). Bromocriptine's daily dose for the management of Parkinsonism ranges from 7.5–30 mg, with dose titration over 60 to 90 days depending on the response to therapy and/or the occurrence of adverse reactions (Choi & Horner, 2023).

Bromocriptine mono-therapy requires very high doses and is connected with intolerable side effects including vomiting, hallucination, low blood pressure, and nasal stuffiness amongst others, with marked fall in blood pressure after first dose intake, especially in persons on antihypertensive medications (Via *et al.*, 2010; Choi & Horner, 2023).

Bromocriptine is now used as a supplementary therapy to levodopa in late cases of Parkinsonism, with doses as low as 1.25 mg nocte, gradually increased as required to a maximum dose of 5–10 mg three times daily. Bromocriptine has been reported to enhance motor control and the "end of dose" and "on-off" fluctuation effects of levodopa therapy (Brooks, 2000; Choi & Horner, 2023).

13.4.3.2 Pergolide

Pergolide acts by direct stimulation of both D1 and D2 receptors (Markham & Diamond, 1989). Its use has been discontinued in the United States following reports of the development of valvular heart disease connected to its use. However, it is still used to manage Parkinsonism in some nations to date (Lieberman *et al.*, 1981; Weiener & Buhimschi, 2009; van Camp *et al.*, 2004).

13.4.3.3 Pramipexole

Pramipexole is a selective non-ergoline D2/D3 receptor agonist with insignificant attraction for D1 and non-dopaminergic receptors, with preferential attraction for the D3 receptors (Cánovas *et al.*, 2014; Li *et al.*, 2022). Pramipexole is effective as mono-therapy (mild Parkinsonism) and as combination therapy towards the late phases of the disorder (Li *et al.*, 2022). When used in combination therapy, the dose of levodopa administered is reduced, thus reducing fluctuations in therapeutic response. Pramipexole has been reported to improve mood/affective symptoms, with possible neuroprotective effects due to its hydrogen peroxide removal and neurotrophic enhancing properties in the dopaminergic neurons of the mid-brain. Treatment with this drug starts at 0.125 mg thrice daily; based on patients' response and drug tolerance, further increments of 0.75 mg daily dose at weekly intervals are then added till the maximum dose of 0.5–1.5 mg three times daily is achieved (Li *et al.*, 2022). After oral administration, it is quickly absorbed, getting to maximum concentrations in the plasma within 120 minutes of administration and is removed through the urine majorly unchanged. Therefore, caution should be used in renal insufficiency; dose adjustments are often required in some patients. The currently available sustained-release formulations of Pramipexole, which are convenient for patients, prevent fluctuations in the plasma concentration of the drug by allowing once-daily dosing at doses equal to the patient's total daily dose (Li *et al.*, 2022; Choi & Horner, 2023).

13.4.3.4 Ropinirole

Ropinirole is a non-ergoline derivative dopaminergic agonist with absolute D2 receptor agonistic activity compared to others and is used as a single therapy in individuals with moderate cases of Parkinsonism as well as co-therapy with levodopa to reduce adverse effects and change in response, especially in individuals with more severe cases of the disorder (Jost & Angersbach, 2005; Choi & Horner, 2023).

Ropinirole therapy is initiated at an oral dose of 0.25 mg to be taken three times per day, with a 0.75 mg increase in total daily dose weekly till the fourth week and by 1.5 mg weekly afterwards titrated to a maximum measured concentrations of 4–8 mg three times daily. Early cases of Parkinsonism generally require 1–2 mg three times a day (Choi & Horner, 2023). The availability of preparations

Drugs for Management of Parkinsonism Mnemonic

C-SALAD
COMT inhibitors
(Entacapone)
Selegiline
Anticholinergics
Levodopa+Carbidopa
Amantadine
Dopamine receptor agonist
(Bromocriptine)

allow for once-daily dosing (Shill & Stacy, 2009). Ropinirole is rapidly absorbed following oral administration; 40% of the administered medication binds to plasma protein and is extensively subjected to metabolic process to convert it to inactive metabolites by liver CYP1A2. Hence, clearance is affected by the administration of drugs metabolized by this enzyme and removed with a half-life of 360 minutes (Shill & Stacy, 2009; Choi & Horner, 2023). Ropinirole has been recently approved for the management of a syndrome called restless leg (Jost & Angersbach, 2005; Choi & Horner, 2023).

13.4.3.5 Rotigotine

Rotigotine is a non-ergoline dopamine agonist approved for the management of early cases of Parkinsonism as well as restless leg syndrome; it has affinity for and can bind with all five subtypes of dopamine receptors, with D3 subtype most preferred in terms of affinity. (Wood *et al.*, 2015; Frampton, 2019; Choi & Horner, 2023). Rotigotine also acts as an antagonist at α-2-adrenergic receptors and an agonist at 5HT1A receptors; it also hinders the secretion of prolactin and uptake of dopamine (Scheller *et al.*, 2009).

Rotigotine is approved for mono-therapy of early cases and co-administration with levodopa through the progression of the disorder and effectively improves both motor and overall functions in Parkinsonism patients (Frampton, 2019; Choi & Horner, 2023).

Rotigotine is taken one time per day by supplying it in a form for absorption through the skin into the bloodstream, supposedly allowing for persistent dopaminergic stimulation in early Parkinsonism, with efficacy in more advanced cases unclear (Frampton, 2019; Choi & Horner, 2023).

13.4.3.6 Apomorphine

Apomorphine, used for the short-term management of "off-periods" of akinesia occurring in persons on effective and functional dopaminergic treatment, is a potent non-ergoline dopamine agonist, hence used as an adjunct to dopamine therapy (Katzenschlager *et al.*, 2018; Choi & Horner, 2023). Apomorphine acts by interacting with the putamen and caudate nucleus' D2 receptors in the postsynaptic region and is administered by subcutaneous injection (Gancher *et al.*, 1991; Carbone *et al.*, 2019; Choi & Horner, 2023). Treatment with Apomorphine begins with the intake of a test measured concentration of 2 mg; the dose is then titrated by 1 mg every few days till the greatest approved dose of 6 mg is achieved, based on individualized patient response and tolerance. Blood pressure (both standing and supine) should be monitored every 20 minutes from the initiation of therapy till an hour after the dose has been administered (Carbone *et al.*, 2019; Choi & Horner, 2023). Treatment of Parkinsonism with Apomorphine is marked by troublesome episodes of nausea; therefore, antiemetic therapy three days before the initiation of therapy and two months during therapy is necessary to prevent this adverse effect being the norm (Carbone *et al.*, 2019; Choi & Horner, 2023). Bruising at the site of injection, mental disturbances, orthostatic hypotension, dyskinesias, drowsiness, insomnia, chest pain, sweating, syncope, and panniculitis are other adverse effects that have been reported by patients on Apomorphine therapy (Skorvanek & Bhatia, 2017; Carbone *et al.*, 2019; Choi & Horner, 2023). Table 13.3 shows the common examples of dopamine agonists used in managing Parkinsonism.

13.4.3.7 Dopamine Agonists Adverse Effects

The dopamine agonists' general gastrointestinal adverse effects include anorexia, nausea, and vomiting; the likelihood of occurrence of these effects has been reported to be reduced when medications are taken with meals (Perez-Lloret & Rascol, 2010; Kalinderi *et al.*, 2011; Borovac, 2016; Ke *et al.*, 2022). Furthermore, bleeding has been reported in active peptic ulcer patients. Less common GI effects include constipation, dyspepsia, and symptoms of reflux esophagitis (Perez-Lloret & Rascol, 2010; Kalinderi *et al.*, 2011; Borovac, 2016; Choi & Horner, 2023).

Cardiovascular adverse effects include postural hypotension, digital vasospasm, cardiac arrhythmias, and peripheral oedema amongst others (Ke *et al.*, 2022; Choi & Horner, 2023). Postural hypotension may occur on initiation of therapy; patients should therefore be advised to look out for obvious signs and make gradual changes to their posture to reduce occurrences. Patients on prolonged therapy with ergot derivatives such as pergolide or bromocriptine have reported dose-related episodes of painless digital vasospasm. If, in rare cases, a cardiac arrhythmia occurs, this is a clear indication that treatment should be discontinued. Patients on Pergolide have also reported cases of cardiac valvulopathy (Tran *et al.*, 2015; Ke *et al.*, 2022; Choi & Horner, 2023).

Dyskinesia refers to having dysfunction in the limbs movement on initiation of therapy and is usually reversed on reduction of the total measured concentrations of dopaminergic agonist administered. Several abnormalities in mental functions have been reported and include confusion, hallucinations, delusions, as well as other psychiatric reactions that are more severe and common in persons on dopamine agonists than in persons on levodopa therapy (Perez-Lloret & Rascol, 2010; Kalinderi *et al.*, 2011; Choi & Horner, 2023).

These psychological reactions can occur as part of the disease or as a complication of therapy and are severe and more common with dopamine agonists than with levodopa therapy. They first appear in aged patients and become more frequent as the disease progresses. Atypical antipsychotics (such as clozapine, olanzapine, quetiapine, risperidone, and pimavanserin) are used in the handling of these mental episodes (Ke *et al.*, 2022).

Abnormality of impulse control has been reported to happen as over-expression of the past tendency or as a fresh occurrence in Parkinsonism patients, leading to shopping, obsessive gambling, sexual exercise, involvement in betting, and many other addictive behaviours (Perez-Lloret & Rascol, 2010; Ahlskog, 2011; Ke *et al.*, 2022; Choi & Horner, 2023). The percentage of a population that is affected with impulse control disorder varies within

15–25% in patients on dopamine agonist therapy as these cases are often under-reported by patients, their families and caregivers, and at most times unrecognised by health care professionals. The occurrence of this disorder has been associated with dopamine receptors (D2 or D3) stimulation in the mesocorticolimbic system and is not limited to any specific factor and could take place at any time post-treatment initiation. Although impulse control disorders are not dose-dependent, dose reduction can provide relief in some groups of patients. However, certain groups of patients may require the complete withdrawal of the offending drug for full recovery. The development of this disorder in persons requires the presence of certain risk factors, some of which include impulsive personality/behaviour, a history of drug use/abuse, as well as a familial history of gambling problems (Ahlskog, 2011; Bancos *et al.*, 2014; Ke *et al.*, 2022).

Other less frequently occurring adverse effects include congestion of nasal cavity, headache, high libido, pulmonary infiltrates, fibrosis (pleural and retroperitoneal), as well as erythromelalgia associated with arthralgia.

Erythromelalgia refers to reddish, tender, painful, swollen feet and hands; symptoms are completely cleared on withdrawal of the medication (Ravn *et al.*, 2017). In rare cases, patients might report an increased likelihood to sleep at times when they are not supposed to, especially in individuals who have been placed on pramipexole or ropinirole and might require to stop the drug intake (Choi & Horner, 2023).

13.4.3.8 Contraindications

Dopamine agonists' medications are inadvisable in several disease conditions, some of which include active peptic ulcers, history of psychotic diseases, and recent infarction of the myocardium. Dopamine agonists that are gotten from ergot should not be prescribed to patients with peripheral vascular disease (Choi & Horner, 2023).

13.4.4 Monoamine Oxidase Inhibitors (MAOIs)

MAOIs are a class of drugs that are known to reversibly or irreversibly inhibit the activities of the monoamine oxidase enzymes within the body. These enzymes are known to break down the different types of substances that transmit nerve impulses in synapse (neurotransmitters) (Hauser *et al.*, 2017; Tan *et al.*, 2022). Two different types of monoamine oxidase enzymes are known—monoamine oxidase A (MAOA) and monoamine oxidase B (MAOB) (Alborghetti & Nicoletti, 2019; Tan *et al.*, 2022). The MAOB is mostly distributed in the liver, brain, platelets, and selectively metabolizes dopamine, while the monoamine oxidase A (MAOA) enzyme is primarily found in the gastrointestinal region, placenta, and liver, and it metabolizes norepinephrine and serotonin in addition to dopamine (Lee & Yankee, 2021; Tan et al., 2022).

Rasagiline, Safinamide, and Selegiline are the three inhibitors of MAOB used for the management of Parkinsonism. All three drugs within this group have been reported to be potent in alleviating motor fluctuations, especially in patients who have been on dopamine therapy for a long time; and they are also known to improve the motor signs of Parkinsonism (Rizek *et al.*, 2016; Hauser *et al.*, 2017; Lee & Yankee, 2021; Church, 2021; Tan *et al.*, 2022).

Selegiline is a selective irreversible inhibitor of MAOB that reduces the breakdown of dopamine. As a result, it prolongs and enhances levodopa's antiparkinsonian impact allowing for a reduction in therapeutic doses, thereby reducing mild on-off or wearing-off phenomena (Gerlach *et al.*, 1996; Alborghetti & Nicoletti, 2019; Lee & Yankee, 2021). However, higher doses of Selegiline have been reported to also inhibit the activities of MAOA enzyme allowing for use in the treatment of depression. In most clinical cases, Selegiline is used as add-on therapy in patients who show a decline or fluctuation in response to levodopa therapy and has been reported to show little therapeutic effect when used alone (Cereda *et al.*, 2017; Lee & Yankee, 2021; Pringsheim *et al.*, 2021).

Selegiline is administered two times daily at a standard dose of 5 mg (with breakfast and lunch), as late dosing has been reported to result in difficulty in sleeping (insomnia) in many patients. There is a need to gradually taper down levodopa doses by 10–30% depending on patients' tolerance 2–3 days after commencing Selegiline therapy. Peak plasma concentration of 0.9–2.7 ng/mL following conventional 5 mg dosing occurs between 15–40 minutes following oral administration, with an onset of action of 1 hour (Moore & Saadabadi, 2023).

Adverse effects such as dyskinesia, whirling sensation in the head and tendency to fall, abdominal pains, sleeplessness, xerostomia, and headaches are common. Hypersensitivity reactions have been reported in patients on concomitant use with opioids and other derivatives. Patients have also reported increased somnolence while engaging in activities of daily living without prior warnings (Shulman *et al.*, 2013; Moore & Saadabadi, 2023).

Rasagiline, an MAOB inhibitor, acts by inhibiting the depletion of dopamine in the striatal region of the brain, thereby decreasing symptomatic movement impairment of Parkinsonism (Lecht *et al.*, 2007; Chen & Swope, 2013). Rasagiline was reported to be more effective in prevention of Parkinsonism caused by MPTP; thus, it is employed for early therapy of cases with moderate symptoms (Mandel *et al.*, 2005; Lecht *et al.*, 2007; Chen & Swope, 2013). However, in more severe cases of the disease and patients with response fluctuation, rasagiline is prescribed as an adjunct drug at a measured concentration of 1/2–1 mg daily, where it has been documented to prolong the healing impacts of carbidopa-levodopa therapy. However, a standard dosage of 1 mg daily is used in mono-therapy, especially in mild cases (Chen & Swope, 2013; Rizek *et al.*, 2016).

Safinamide is the third MAOB inhibitor approved as add-on therapy for patients already on levodopa-carbidopa therapy experiencing off episodes and is not effective as monotherapy (Blair & Dhillon, 2017; Alborghetti & Nicoletti, 2019; Tsuboi *et al.*, 2022). The exact way through which safinamide exerts its healing effect is not clear yet, but it is proposed to inhibit MAOB activity by impeding dopamine's breakdown, resulting in elevated levels within the brain and corresponding increased dopaminergic activity (Alborghetti & Nicoletti, 2019; Tsuboi *et al.*, 2022). Therapy is initiated using 50 mg orally once daily (qday) at the same time each day; doses are titrated upwards after 2 weeks till 100 mg qday is attained; doses greater than 100 mg qday have been reported to show no added therapeutic effect (Blair & Dhillon, 2017; Alborghetti & Nicoletti, 2019; Tsuboi *et al.*, 2022). Peak plasma concentration occurs 2–3 hours after administration of dose, with steady-state concentration achieved after 3–5 days. Discontinuation of therapy requires the gradual taper down of the dose to 50 mg qday for 1 week before stopping (Alborghetti & Nicoletti, 2019). Dyskinesia, hypertension, nausea, sleeplessness, and anxiety are just a few of the negative effects of MAOB inhibitors (Blair & Dhillon, 2017; Alborghetti & Nicoletti, 2019; Tsuboi *et al.*, 2022).

Side effects of levodopa, such as mental problems, dyskinesias, a feeling of vomiting, and sleeplessness, have been reported to be increased on addition of MAOIs to Parkinsonism therapy; in some cases, hypertension may be precipitated or aggravated (Riederer & Laux, 2011; Rizek *et al.*, 2016; Tsuboi *et al.*, 2022; Sub Laban & Saadabadi, 2023). Caution should be employed in an individual with safinamide hypersensitivity history, chronic liver damage, and who is on other MAOIs and opioids. In addition, patients on any of the MAOB inhibitors should not be prescribed propoxyphene, methadone, tramadol, meperidine, St. John's wort, or cyclobenzaprine (Rizek *et al.*, 2016; Thomas *et al.*, 2015; Sub Laban & Saadabadi, 2023).

The co-administration of MAOIs with the antitussive dextromethorphan cough syrup has been documented to lead to mental breakdown or psychosis episodes; hence, patients should be advised to avoid the use of all over-the-counter cold preparations and report any other medication taken to their prescribing physicians. Co-administration of MAOBIs and other classes of MAOIs is contraindicated. Also, the appearance of serotonin syndrome necessitates the use of caution in patients already on serotonin reuptake blockers or tricyclic antidepressants (Thomas *et al.*, 2015; Edinoff *et al.*, 2022; Sub Laban & Saadabadi, 2023).

The co-administration of a nonselective blocker of monoamine oxidase and levodopa should be discouraged, as this could result in crises of hypertension, especially in known hypertensive patients due to the peripheral accumulation of norepinephrine (Riederer & Laux, 2011; Thomas *et al.*, 2015; Edinoff *et al.*, 2022; Sub Laban & Saadabadi, 2023).

13.4.5 Catechol-*O*-Methyltransferase (COMT) Inhibitors

COMT inhibitors are a new class of medications used in the treatment of Parkinsonism and act by inhibiting the catechol-*o*-methyltransferase enzyme, thereby preventing breakdown of levodopa at the periphery, allowing larger concentrations to cross the BBB (Rivest *et al.*, 1999; Kalinderi *et al.*, 2011; Berger *et al.*, 2022). Alternative pathways for levodopa metabolism (the catechol-O-methyltransferase (COMT) pathway) are stimulated on inhibition of the dopa decarboxylase enzyme; this leads to an increase in the amount of 3-O-methyldopa (3-OMD) in the plasma and sub-optimal healing response to levodopa therapy. The poor levodopa healing response due to elevated levels of 3-OMD is a result of selective competition for a non-passive carrier process responsible for its movement through the mucous membrane of the intestine as well as the BBB (Jenner *et al.*, 2021; Berger *et al.*, 2022; Tuunainen *at al.*, 2023). The COMT inhibitors are indicated as adjuncts to levodopa-carbidopa therapy to alleviate off episodes, allowing for smoother therapeutic response, prolonged on times, and reduction in total daily doses of levodopa (Rivest *et al.*, 1999; Jenner *et al.*, 2021; Berger *et al.*, 2022). Table 13.5 shows COMT inhibitors with their dosage regimens for managing PD.

Tolcapone and **Entacapone** are selective peripheral-acting COMT inhibitors, while **Opicapone** is a novel peripheral-acting selective and reversible COMT inhibitor, which decreases the rate of conversion of levodopa to 3-OMD, thereby prolonging half-life and relative bioavailability of levodopa, thereby reducing motor fluctuations (Jenner *et al.*, 2021; Berger *et al.*, 2022). Although both members of this drug class are widely available, the shortcomings of the various COMT inhibitors have made it necessary for the development of a novel COMT inhibitor free of these adverse effects (Rivest *et al.*, 1999; Berger *et al.*, 2022).

Entacapone, a reversible peripheral COMT inhibitor, is used as adjunct therapy at a dose of 200 mg per oral with each dose of levodopa-carbidopa therapy, with the greatest approved daily dose of 1600 mg. Entacapone is used only by combining it with levodopa-carbidopa therapy; it is not to be used in persons on non-selective MAO inhibitors as the co-administration might inhibit the majority of the pathways responsible for catecholamine metabolism (Salamon *et al.*, 2022).

Tolcapone, a reversible COMT inhibitor with both central and peripheral impacts, is administered at a standard dose of 0.1 g 8 hourly (three times in a day) but might require a doubling of doses in some patients (Truong, 2009; Berger *et al.*, 2022). Tolcapone should be withdrawn if it fails to show the desired added therapeutic effect after an increase of the dose after three weeks.

The negative effects of the COMT inhibitors, which include dyskinesias, nausea, and confusion, could be partially as a result of the resultant increase in levodopa exposure (Truong, 2009; Berger *et al.*, 2022). Also, the severity and occurrence of adverse effects can be reduced and/or avoided when the daily administered dose of levodopa is reduced by 30% within the first two days of starting COMT inhibitor therapy (Spencer & Benfield, 1996; Salamon *et al.*, 2022). Other adverse effects reported by patients include diarrhea, low blood pressure caused by being in an upright posture, pain in the abdomen, difficulty in sleeping, and urine discolouration (Truong, 2009; Salamon *et al.*, 2022; Berger *et al.*, 2022).

Tolcapone is contraindicated in persons who have problems with hepatic function test (HFT) results, as its administration has been associated with elevated levels of hepatic enzymes and uncommonly with acute liver failure leading to death. However, there is a need for routine monitoring of liver function tests in patients on Tolcapone therapy every 2–4 weeks during the first 6 months of treatment and from time to time throughout therapy. Tolcapone should be withdrawn if hepatic damage becomes evident (Olanow & Watkins, 2007; Salamon *et al.*, 2022; Berger *et al.*, 2022).

Opicapone is a novel once-daily peripheral-acting selective and reversible Catechol-O-methyltransferase (COMT) inhibitor, with documented favourable adverse effects. Opicapone is adjunct therapy in patients on carbidopa/levodopa combination experiencing off periods, with doses as low as 25–50 mg producing a consistent increase in on times and a corresponding reduction in off times (Ferreira *et al.*, 2019; Jenner *et al.*, 2021; Berger *et al.*, 2022).

13.4.6 Dopamine Facilitators—Amantadine

Amantadine is a dopamine facilitator belonging to a group of medications called adamantanes. It prevents or fights viral infections with moderate dopaminergic function (Chang & Ramphul, 2022). This drug was accidentally discovered to have relatively weak anti-Parkinsonism properties; hence, it is indicated as combination therapy with levodopa-carbidopa, especially in patients with declining response to levodopa-carbidopa therapy (Hubsher *et al.*, 2012; Chang & Ramphul, 2022). Although the therapeutic effect of this drug is not as great as that of levodopa and its impacts are for a short duration (few weeks of medication), its effect on the motor symptoms of Parkinsonism (bradykinesia, stiffness, and tremors) is beneficial (Crosby *et al.*, 2003: Kalinderi *et al.*, 2011).

Amantadine has been proposed to synergistically augment the function of the dopaminergic system by enhancing the production, release, and or re-uptake of dopamine; however, the exact mechanism is unclear (Bailey & Stone, 1975; Chang & Ramphul, 2022).

Amantadine increases the release of catecholamine from peripheral reserves and counteracts the actions of adenosine A2A receptors, thereby limiting D2 receptor functions. It is also a non-strong and non-competitive antagonist of the N-methyl-D-aspartate (NMDA) type glutamate receptor, suggesting an antidyskinetic effect; hence, it is beneficial in inhibiting this adverse effect where levodopa-carbidopa therapy is used and also reduces iatrogenic dyskinesias in a person with severe Parkinson's (Crosby *et al.*, 2003; Chang & Ramphul, 2022).

Amantadine therapy is initiated at a dose of 100 mg once daily, which may be titrated upwards till an acceptable measured concentration of 100 mg twice or thrice (orally) in a day is achieved. This drug has wide absorption from the gastrointestinal tract after oral administration, with the greatest amount in the plasma occurring 1–4 hours post oral administration. The mean plasma half-life ranges from 10–14 hours, with much of the medication in the urine unchanged (Chang & Ramphul, 2022).

13.4.6.1 Adverse Effects

The side effect of Amantadine is very small, which serves as the major advantage of the drug when used for Parkinson's treatment. The undesirable central nervous system adverse effects of Amantadine therapy include depression, restlessness, loss of sleep, irritability, anxiety, excitement, confusion, and hallucinations, all of which are reversed on discontinuation of therapy.

Other adverse reactions to Amantadine include anorexia, nausea, constipation, dry mouth (gastrointestinal adverse effects), headache, heart failure, postural hypotension, and urinary retention (Bailey & Stone, 1975; Nikolaus *et al.*, 2019; Chang & Ramphul, 2022). Administration of doses greater than the standard daily recommended dose has been reported to result in acute toxic psychosis and convulsion in patients on Amantadine.

Livedo reticularis, a skin problem identified by a fixed, deep bluish-red, reticulated pattern on the skin is a less common adverse effect of Amantadine therapy and is reversible within a month of the medication stoppage (Quaresma *et al.*, 2014). The use of this drug should be with caution, and dosage adjustment is necessary in patients with liver and renal impairment, a history of seizures, or heart diseases (Isaacson *et al.*, 2018; Dragašević-Mišković *et al.*, 2019). Amantadine therapy should not be abruptly discontinued as this may result in neuroleptic malignant syndrome-like symptoms, which are characterised by tachycardia, high fever, muscle rigidity, and altered mental status. Doses should be tapered down by 50% for at least one week before the discontinuation of therapy (Crosby *et al.*, 2003: Chang & Ramphul, 2022).

13.5 DRUGS ACTING ON THE CHOLINERGIC SYSTEM

13.5.1 Acetylcholine-Blocking/Anticholinergic Drugs—Trihexyphenidyl (Benzhexol), Procyclidine, and Biperidine

The anticholinergic drugs, also referred to as acetylcholine-blocking agents, are the earliest group of medications used for treating Parkinsonism and date as far back as the early 1900s (Katzenschlager *et al.*, 2002; Kawabata & Katsuno, 2021). This class of drugs acts by blocking the activity of acetylcholine in the brain, thereby affecting movement. Anticholinergics are effective in the management of tremors, rigidity, and dystonia connected with wearing off or peak dose effects with small or no impacts on other symptoms of the disease (Brocks, 1999; Lee & Yankee, 2021). Table 13.6 shows the common anti-cholinergic drugs used in managing Parkinsonism with their dosage regimens.

Trihexyphenidyl (Benzhexol) is a centrally acting muscarinic antagonist utilised for treating stiffness, tremors, spasms, and poor muscle coordination often seen in people with Parkinsonism and drug-induced extrapyramidal disorders (Kawabata & Katsuno, 2021). The precise way or mechanism through which Trihexyphenidyl exerts its pharmacological impact is yet to be confirmed, but it is believed to act by directly inhibiting efferent impulses within the parasympathetic nervous system (Brocks, 1999: Jilani *et al.*, 2023).

Its dosage varies according to individual patient, but treatment usually begins with the administration of the lowest possible dose orally, which is gradually increased to the maximum effective tolerated dose by the patient. Doses greater than 10 mg daily are usually administered in four equal doses (with each meal and at bedtime) (Robottom & Reich, 2011). Abrupt withdrawal of Trihexyphenidyl in patients should be avoided, as this could result in the exacerbation of existing symptoms and or neuroleptic malignant syndrome.

Trihexyphenidyl's side effects are dependent on dosage, decreasing with time as the body tolerates the drug, and include mydriasis, headaches, dizziness, drowsiness, vertigo, problems with sweating, xerostomia, pain in the abdomen, nausea, constipation, and retention of urine.

Procyclidine, a synthetic cholinergic-blocking agent, acts by blocking the acetylcholine's excitatory effect at the muscarinic receptors; therefore, it is effective in the management of spasms, stiffness, tremors, as well as poor muscle coordination associated with Parkinsonism (Namali, 2007). Procyclidine is also indicated for the management of pseudo-parkinsonism, acute dystonic reactions, akathisia, and the control of extrapyramidal symptoms caused by neuroleptic drugs.

Procyclidine in Parkinsonism therapy is initiated at 2.5 mg twice daily, increased at 2.5–5.0 mg per day at an interval of 2–3 days till optimum clinical responses are achieved, and an effective maintenance dose of 15–30 mg daily. The common adverse effects of Procyclidine include drowsiness, dizziness, lightheadedness, constipation, flushing, dry mouth, and blurred vision amongst others, with reducing incidence and severity as the body develops tolerance to the drug (Namali, 2007).

Biperidine, a muscarinic receptor antagonist, is used as monotherapy or in combination with drugs like levodopa to manage Parkinsonism (Kostelnik *et al.*, 2017). Biperidine acts to improve muscle control and reduce stiffness, thereby improving the symptoms of the disease in patients (Brocks, 1999; Kostelnik *et al.*, 2017). Its side effects are connected to its anticholinergic effects.

13.5.2 Antihistaminics—Orphenadrine

Orphenadrine, an anticholinergic agent of the ethanolamine antihistamine class, has been used to treat Parkinsonism since the 1920s, although present-day use has been superseded by the dopamine agonist (Katzenschlager *et al.*, 2002; Gjerden *et al.*, 2009; Lee & Yankee, 2021). Orphenadrine is reported to bring back the physiological balance between dopaminergic and cholinergic neurotransmitters; hence, it is used as a skeletal muscle relaxant to relieve the muscle pains, discomforts, and rigidity associated with Parkinsonism (Katzenschlager *et al.*, 2002; Lee & Yankee, 2021).

Most of the time, Orphenadrine is used as an add-on therapy to the rest for the management and relief of discomforts related to excruciating musculoskeletal issues, initiated at a dose of 0.15 g in a day in divided doses and gradually increased by 0.05 g every 2 to 3 days till optimum therapeutic benefit is achieved. Adverse effects of Orphenadrine include dry mouth, somnolence, dizziness, vomiting, and abdominal pains amongst others (Katzenschlager *et al.*, 2002; Gjerden *et al.*, 2009; Lee & Yankee, 2021). Table 13.1 shows different classes of drugs for the management of Parkinsonism. It highlights the mechanism of actions and side effects. Examples are shown in Tables 13.2–13.6.

13.6 LIMITATIONS OF PHARMACOTHERAPY

Several guidelines for the management of Parkinsonism are currently available globally, and they contain information needed to guide the management of the disease with the overall goal of enhancing care and the patients' QoL (AAN, 2021). As a result of the increase in extent and severity of the disorder, patients on treatment will gradually experience motor complication episodes where the beneficial effects of the drug progressively become shorter (wearing off), episodes where the beneficial therapeutic effects of the drug become unpredictable (on/off episodes), episodes of no beneficial therapeutic effects for periods in time (freezing), and episodes of increased involuntary movements (dyskinesias). Treatment of Parkinsonism to date remains primarily pharmacological and involves the use of agents to replace depleted dopamine or mimic the activity of endogenous dopamine. However, as the disease progresses, there is a need to adjust the patient's medication regimen to balance the desired therapeutic effects with the undesired adverse effects of the drug, which often requires the adjustments of dosages, dosing regimens, and the use of adjuvant therapy.

The currently available anti-parkinsonism agents when used as monotherapy or together with other adjuvant anti-parkinsonism drugs have failed to confer therapeutic efficacy for the management of Parkinsonism on a long-term basis as the majority of the features of the disease are resistant to the drugs; combination therapies are often associated with severe adverse effects with little or no improvement in the patient's QoL and life expectancy. Hence, the major limitations of pharmacotherapy are seen at the terminal phases of the disorder and include the inability of the agents to improve many of the disabling movement and non-movement features of the disorder, associated troubling adverse effects of the different classes of agents used in the management of the disease, and the inability of the agents to halt the progress of the disease (Lertxundi *et al.*, 2016).

13.7 ADVANCES IN PHARMACOTHERAPY OF PARKINSON'S DISEASE

Parkinsonism treatment remains symptomatic, and dopamine remains the gold standard. There are several problems

TABLE 13.1
Summary of the Different Classes of Drugs Used in the Management of Parkinsonism

Drug Class	Examples	Mechanism of Action	Uses	Side Effects
Dopamine precursor	Levodopa	Passes the BBBC, dopaminergic neurons convert it to dopamine	Resolves symptoms of posture, speech, facial expressions, mood, and handwriting Improves hypokinesia and rigidity	Nausea Vomiting Postural hypotension
Peripheral decarboxylase inhibitors	Benserazide/ Carbidopa	Prevents the peripheral metabolism of L-DOPA by the decarboxylase enzyme	Given in combination with L-DOPA to increase its plasma half-life	Nausea Vomiting
Dopamine agonist	Bromocriptine Ropinirole Pramipexole Pergolide	Bromocriptine is an ergot derivative with powerful agonist activity of D2 receptors Pergolide has agonist activity on D1 and D2 receptors; it acts directly on the substantia nigra Ropinirole and Pramipexole are non-ergoline derivatives with selective agonist activity of D3 receptors	Used to improve parkinsonism symptoms	Nausea Dizziness Hallucination Postural Hypotension
MOA –B inhibitors	Selegiline Rasagiline	Selegiline is a selective irreversible inhibitor of MOA-B Co-administration with L-DOPA to enhance its action and hinder the effect of wearing-off	Prevents the increase in extent and severity of the disorder	Selegiline is metabolized to amphetamine causing agitation and sleep loss
COMT inhibitors	Tolcapone Entacapone	Tolcapone acts by prolonging and enhancing the activity of L-DOPA. Also prevents the peripheral decarboxylation by COMT It acts within the periphery	Co-administered with L-DOPA in persons with wearing-off effect	General side effects are: Nausea Vomiting Drug specific side effects are: Tolcapone – Diarrhoea Entacapone- yellow-orange urine
Dopamine facilitator	Amantadine	Aids the presynaptic production and release of brain's dopamine	Used to provide symptomatic benefits to patients	Insomnia Dizziness Confusion
Anticholinergics	Centrally acting anticholinergics - Trihexyphenidyl (Benzhexol) Procyclidine Biperidine Antihistamine-Orphenadrine	Decreases the cholinergic activity within the brain	Used in the early treatment of Parkinsonism Effective against tremors Used to reduce the severity of akinesia, rigidity, and control sialorrhea	Drowsiness, dizziness, lightheadedness, sleepiness, constipation, flushing, dry mouth, and blurred vision

TABLE 13.2
Examples of Levodopa Combinations Used to Treat Parkinsonism and Their Monitoring Parameters

Drug	Dosing	Monitoring Parameter(s)
Carbidopa/levodopa (10/100, 25/100 and 25/250 mg tablets)	Dosage: Patients are started on 25/100 mg TID; doses are titrated weekly based on individual patient's response	Routine monitoring of blood pressure, pulse changes in mental status and individualized clinical responses are necessary. Patients should be warned of the possible changes in the colour of urine to dark or blackish-brown colours.
Carbidopa/levodopa sustained-release tablet (25/100 and 50/200 mg tablets)	Dosage: Patients are started on 50/200 mg BID	
Carbidopa/levodopa ER tablets (23.75/95, 36.25/145, 48.75/195 and 61.25/245 mg tablets)	Dosage: Patients are placed on TID dosing; however, this may be increased in advanced cases to five times daily	
Carbidopa/levodopa/entacapone (carbidopa/levodopa/COMTI) (12.5/50/200, 18.75/75/200, 25/100/200, 31.25/125/200, 37.5/150/200, and 50/200/200 mg tablets)	Dosage: Patients are started on the lowest drug combination and titrated up depending on the patient's response	

Notes: BID = twice daily; COMTI = catechol-O-methyltransferase inhibitor; ER = extended release; TID = three times daily.

TABLE 13.3
Examples of Dopamine Agonists Commonly Employed in the Management of Parkinsonism and Their Monitoring Parameters

Drug	Dosing	Monitoring Parameter(s)
Bromocriptine mesylate (2.5mg tablets and 5 mg capsules)	• Dosage: Treatment is started at 1.25 mg BID ($^1/_2$ tablet) with meals, and may be titrated up every 2 to 4 weeks by 2.5 mg per day with meals. Doses should not be more than 100 mg/day (safety is in question) *** A pregnancy test (PT) should be conducted every month during amenorrhea and in delayed menstrual cycles	• Routine monitoring of blood pressure, pulmonary function, liver function, BUN/Cr levels, weight, and heart rates is necessary for patients on therapy. Patients should be warned of the possible decrease in daytime alertness; hence should avoid driving or using heavy machinery.
Ropinirole (2, 4, 8, and 12 mg tablets)	Dosage: Treatment is started at doses of 2 mg QD for 1 to 2 weeks; then titrated weekly till a maximum daily dose of 24 mg is attained *** Patients should be monitored for orthostatic hypotension and skin reactions during dose titration	
Pramipexole (0.375, 0.75, 1.5, 2.25, 3.0, 3.75, 4.5 mg ER tablets)	ER Dosage: starting dose: 0.375 mg QD; titrate weekly to a maximum of 4.5 mg QD *** Baseline creatinine levels, skin reactions and signs of orthostatic hypotension should be monitored	
• Rotigotine • (Transdermal patch of different strength: 1, 2, 3, 4, 6 and 8 mg/24 hours)	• Dosage: Dosing is started by using 2 mg/24 hours or 4 mg/24 hours transdermal patch, and may be titrated up at weekly intervals to a maximum of 6 mg/24 hours for the initial dose of 2 mg/24 hours and 8 mg/24 for the initial dose of 4 mg/24 hours are attained. • ***Patients should be advised to rotate the site on the skin used and to only apply patches to healthy skin.	
• Apomorphine • (30 mg/3 mL glass cartridge)	• Dosage: Treatment is initiated at a dose of 2 mg (0.2 mL); titrated to the maximum dose of 6 mg (0.6 mL) under strict medical supervision. • ***Starting doses need to be decreased in persons having problem with kidney functions. • *** Concomitant antiemetic therapy should be initiated 3 days before the first dose of Apomorphine therapy and continued for at least 2 months during therapy.	

Notes: BID = twice daily; ER = extended release; QD = once daily.

TABLE 13.4
Examples of MAOB Inhibitors Used in the Treatment of Parkinson's Disease and Their Monitoring Parameters

Drug	Dosing	Monitoring Parameter(s)
Selegiline (5mg tablets and capsules, 1.25 mg ODT, Transdermal patch of different strength: 6, 9, and 12 mg/24 hours)	Dosage: Treatment is begun at 5 mg one time daily with breakfast for 2–4 weeks; then increased to 10mg daily in equally divided doses with breakfast and lunch When using ODT, treatment is initiated at 1.25 mg once daily with breakfast for a minimum of 6 weeks, the dose can then be increased to 2.5 mg once daily based patient's response and tolerability Treatment with the transdermal patch is initiated at a dose of 6 mg/24hrs and can be increased by 3 mg/24hrs at intervals of 2 weeks based on tolerability and response of the patient till the optimum daily dose of 12 mg/24hrs is achieved	• Routine monitoring of blood pressure, signs of serotonin syndrome and periodic skin examinations for patients on transdermal patches Patient's BUN/creatinine clearance levels should be routinely monitored
Rasagiline (0.5 and 1mg tablets)	Dosage: Treatment in mono-therapy and adjunct without levodopa is initiated at a dose of 1mg QD However when taken to complement levodopa, therapy treatment is initiated at 0.5 mg QD, which can be increased to 1mg QD (in such cases the dose of levodopa co-administered should be reduced to avoid the occurrence of possible adverse effects)	Patients should be monitored for signs of new or worsening mental health, blood pressure fluctuations, and signs of Parkinsonism Patients should be warned of the possible increased somnolence and decrease in daytime alertness Also, there is a need for routine skin examinations to check for the presence of melanoma
Safinamide (50 and 100mg tablets)	Dosage: Treatment is initiated at a dose of 50 mg QD for two weeks in patients with off episodes; after which it can be increased to 100 mg QD dependent upon the patient's response and tolerability	Patients should be monitored for signs of blood pressure fluctuations and signs/symptoms of serotonin syndrome Patients should be warned of the possible changes in vision when on Safinamide

Notes: ODT = oral disintegrating tablet; QD = once daily;

TABLE 13.5
Examples of COMT Inhibitors Used in the Treatment of Parkinson's Disease and Their Monitoring Parameters

Drug	Dosing	Monitoring Parameter(s)
Tolcapone (100 mg tablets)	Dosage: Treatment is initiated at 100 mg orally TDS; and can be increased till a maximum daily dose of 600 mg/day is achieved administered in 3 equal divided doses (i.e. 200 mg TDS)	AST and ALT levels need to be assessed at treatment baseline; then every 2–4 weeks till 6 months of treatment • Routine monitoring of blood pressure, mental status evaluation and skin examinations are recommended in patients on Tolcapone The drug should be changed if no clinical benefit is achieved within 3 weeks of initiation of therapy Also, the drug should be discontinued in patients whose AST or ALT levels are extremely elevated
Opicapone (50 mg hard capsules)	Dosage: is used in the management of off episodes at doses of 50 mg once daily at bedtime without a meal and at least one hour before and an hour after the administration of levodopa Doses can be reduced to 25 mg at night in patients with hepatic dysfunction	• Routine monitoring of blood pressure (baseline and during clinical treatment); mental alertness; and liver and renal function (baseline and during treatment) are important in patients on Opicapone Patients should be warned of chances of increased daytime somnolence and worsening of pre-existing sleep disorders
Entacapone (200 mg tablets)	Dosage: Treatment is initiated at a dose of 200 mg Entacapone with each levodopa+carbidopa dose; the dose can be increased to the optimal daily dose of 1600 mg dependent upon the tolerance and response of the patients	Patients should be monitored regularly for blood pressure, heart rate and rhythm changes

Notes: TDS = three times daily; AST = Aspartate aminotransferase; ALT = Alanine aminotransferase;

TABLE 13.6
Examples of Anti-Cholinergic Agents Commonly Employed in the Management of Parkinsonism and Their Monitoring Parameters

Drug	Dosing	Monitoring Parameter(s)
Triphexyphendyl (5 mg tablets)	Dosage: Treatment began at a dose of 1 mg once every day and gradually increased to a maximum daily dose of 15 mg divided into three or four equal doses dependent upon the patient's response and tolerability	Routine monitoring of liver, kidney and cardiac functions is necessary Also, blood pressure monitoring is indicated in known hypertensive patients
Procyclidine (5 mg tablets)	Dosage: Treatment is initiated at a dose of 2.5 mg TDS; gradually increased by 2.5 mg per day at dosing intervals of 2–3 days dependent upon the patient's response and tolerance till the maximum daily dose of Procyclidine is attained	
Orphenadrine (100 mg tablets)	Dosage: Treatment is initiated at a dose of 50 mg TDS	Routine monitoring of blood, urine and liver function tests is necessary as safety during prolonged use is yet to be established
Biperiden (2 mg tablets)	Dosage: Treatment is started at 2 mg/day divided equally and administered TDS or QID; the dose could be increased gradually till the optimum daily dose of 16mg is achieved	Routine monitoring of blood pressure (standing and supine), BUN/creatinine levels, LF, IOP and periodic skin examinations are necessary for patients on Biperiden therapy
Benztropine (0.5, 1 and 2 mg tablets)	Dosage: Treatment is initiated at the lowest dose of 0.5 mg, and gradually increased by 0.5 mg till the maximum daily dose of 6 mg is achieved over a period of 5–6 days	Patients should be routinely monitored during the commencement of therapy for signs of exacerbation of mental health

Notes: TDS= three times daily; QID= four times daily; LFT= liver function, IOP= intraocular pressure.

with current treatment regimens for Parkinsonism. These include their inability to halt disease progression and improve the disabling features of the disease as well as their associated side effects. Current advances in pharmacotherapy aim at improving the patient's quality of life (QoL) by extending the duration of on time and reducing motor fluctuations; designing a novel drug delivery system (continuous drug delivery), drug repurposing, and modified drug formulations have made these goals possible (Mao *et al.*, 2020; Stoker & Barker, 2020).

One such promising medication is IPX203, a formulation containing intermittent-release (IR) carbidopa + levodopa and extended-release (ER) levodopa granules. According to the AAN (2023), IPX203 showed significant improvement in efficacy and on time as compared to intermittent-release (IR) carbidopa + levodopa therapy; hence, it is an option in patients with persisting motor fluctuations (Hauser *et al.*, 2023).

Another such promise is the use of novel delivery systems—KOLs ABBV-951 and NeuroDerm's (ND0612)—to provide a continuous supply of levodopa subcutaneously in advanced stages of the disease, therefore improving the delivery of dopamine to the brain (Doggrell, 2023).

Drugs currently undergoing repurposing are azathioprine (an immunosuppressant) and sargramostim (a T regulatory cell), medications that reduce neuro-inflammation to manage motor dysfunction in Parkinson's disease. Other promising drug candidates are the dopamine stabilizer (Pridopidine), D1 selective agonist, and allosteric modulators, etc. These newer agents are being screened for selective D1 receptor affinity, antioxidant activity, stability following oral administration, and their effects on α-synuclein aggregation and will be available soon (Poewe *et al.*, 2017; Athauda *et al.*, 2017; Stoker & Barker, 2020; Mao *et al.*, 2020).

However, the real hope in treatment lies in regenerative medicine or cell-based regenerative therapy, which is aimed at regenerating degenerate dopamine neurons, thereby halting the progression of the disease (Stoker & Barker, 2020). The use of regenerative therapy will not only be limited by cost but also by the non-availability of fetal tissues required for the process, and hence might not be readily available.

13.8 CONCLUSION

As of now Parkinson's disease remains incurable, as none of the available drug preparations are able to stop/halt or delay the progression of the disease. However, early detection and initiation of therapy based on individualised patient needs has been reported to improve the patients' quality of life, making living with the disease condition more bearable for several years.

REFERENCES

Ahlskog, J.E. (2011). Pathological behaviours provoked by dopamine agonist therapy of Parkinson's disease. Physiology & Behavior, 104(1):168–172.

Alborghetti, M., & Nicoletti, F. (2019). Different generations of type-B monoamine oxidase inhibitors in Parkinson's disease: From bench to bedside. Current Neuropharmacology, 1:861–873.

Alonso Cánovas, A., Luquin Piudo, R., García Ruiz-Espiga, P., Burguera, J.A., Campos Arillo, V., Castro, A., Linazasoro, G., López Del Val, J., Vela, L., & Martínez Castrillo, J.C. (2014). Dopaminergic agonists in Parkinson's disease. Neurología (English Edition), 29(4):230–241 (doi: 10.1016/j.nrleng.2011.04.010).

American Academy of Neurology (AAN). (2021). AAN issues guidelines for early Parkinson's disease. EurekAlert. News Release. November 15, 2021, Accessed August 17, 2023 (www.eurekalert.org/news-release/934998).

Armstrong, M.J., & Okun, M.S. (2020). Diagnosis and treatment of Parkinson disease: A review. *JAMA,* 323(6):548–560 (doi: 10.1001/jama.2019.22360).

Arora, A., & Fletcher, P. (2013). Parkinsonism hyperpyrexia syndrome is caused by abrupt withdrawal of ropinirole. British Journal of Hospital Medicine, 74:69.

Athauda, D., Maclagan, K., Skene, S.S., Bajwa-Joseph, M., Letchford, D., Chowdhury, K., Hibbert, S., Budnik, N., Zampedri, L., Dickson, J., Li, Y., Aviles-Olmos, I., Warner, T.T., Limousin, P., Lees, A.J., Greig, N.H., Tebbs, S., & Foltynie, T. (2017). Exenatide once weekly versus placebo in Parkinson's disease: A randomised, double-blind, placebo-controlled trial. Lancet, 390(10103):1664–1675.

Bailey, E.V., & Stone, T.W. (1975). The mechanism of action of amantadine in parkinsonism: A review. Archives Internationales de Pharmacodynamie et de Thérapie, 216(2):246–262.

Bancos, I., Nannenga, M.R., Bostwick, J.M., Silber, M.H., Erickson, D., & Nippoldt, T.B. (2014). Impulse control disorders in patients with dopamine agonist-treated prolactinomas and nonfunctioning pituitary adenomas: A case-control study. Clinical Endocrinology, 80(6):863–868.

Beckers, M., Bloem, B.R., & Verbeek, M.M. (2022). Mechanisms of peripheral levodopa resistance in Parkinson's disease. npj Parkinson's Disease, 8:56 (doi: 10.1038/s41531-022-00321-y).

Berger, A.A., Winnick, A., Izygon, J., Jacob, B.M., Kaye, J.S., Kaye, R.J., Neuchatel, E.E., Kaye, A.M., Alpaugh, E.S., Cornett, E.M., Han, A.H., & Kaye, A.D. (2022). Opicapone, a novel catechol-O-methyl transferase inhibitor, for treatment of Parkinson's disease "off" episodes. Health Psychology Research, 10(3):36074 (doi: 10.52965/001c.36074)

Blair, H.A., & Dhillon, S. (2017). Safinamide: A review in Parkinson's disease. CNS Drugs, 31:169–176 (doi: 10.1007/s40263-017-0408-1).

Bloem, B.R., Okun, M.S., & Klim, C. (2021). Parkinson's disease. The Lancet, 397(10291):2284–2303 (doi: 10.1016/S0140-6736(21)00218).

Borovac, J.A. (2016). Side effects of a dopamine agonist therapy for Parkinson's disease: A mini-review of clinical pharmacology. Yale Journal of Biology and Medicine, 89(1):37–47.

Brocks, D.R. (1999). Anticholinergic drugs used in Parkinson's disease: An overlooked class of drugs from a pharmacokinetic perspective. Journal of Pharmacy and Pharmaceutical Sciences, 2(2):39–46.

Brooks, D.J. (2000). Dopamine agonists: Their role in the treatment of Parkinson's disease. Journal of Neurology, Neurosurgery & Psychiatry, 68:685–689.

Cacabelos, R. (2017). Parkinson's disease: From pathogenesis to pharmacogenomics. International Journal of Molecular Science, 18(3).

Calabresi, P., Di Filippo, M., Ghiglieri, V., Tambasco, N., & Picconi, B. (2010). Levodopa-induced dyskinesias in patients with Parkinson's disease: Filling the bench-to-bedside gap. The Lancet Neurology, 9:1106–1117.

Carbone, F., Djamshidian, A., Seppi, K., & Poewe, W. (2019). Apomorphine for Parkinson's disease: Efficacy and safety of current and new formulations. CNS Drugs, 33:905–918 (doi: 10.1007/s40263-019-00661-z).

Cereda, E., Cilia, R., Canesi, M., Tesei, S., Mariani, C.B., Zecchinelli, A.L., & Pezzoli, G. (2017). Efficacy of rasagiline and selegiline in Parkinson's disease: A head-to-head 3-year retrospective case-control study. Journal of Neurology, 264:1254–1263.

Chagraoui, A., Boulain, M., Juvin, L., Anouar, Y., Barrière, G., & Deurwaerdère, P.D. (2020). L-DOPA in Parkinson's disease: Looking at the "false" neurotransmitters and their meaning. International Journal of Molecular Science, 21:294 (doi: 10.3390/ijms21010294).

Chang, C., & Ramphul, K. (2022). Amantadine. In: StatPearls. Treasure Island, FL: StatPearls Publishing.

Chen, J.J., & Swope, D.M. (2012). Pharmacotherapy of Parkinson's disease. Pharmacotherapy, 27(12p2):161S–173S (doi: 10.1592/phco.27.12part2.161S).

Chen, J.J., & Swope, D.M. (2013). Clinical pharmacology of rasagiline: A novel, second-generation propargylamine for the treatment of Parkinson's disease. Journal of Clinical Pharmacology, 45(8):875–894 (doi: 10.1177/0091270005277935).

Chen, J.J., & Swope, D.M. (2014). Parkinson's disease. In: DiPiro, J.T., Talbert, R.L., Yee, G.C., et al., editors. Pharmacotherapy: A Pathophysiologic Approach. 9th ed. New York, NY: McGraw-Hill.

Choi, J., & Horner, K.A. (2023). Dopamine agonists. [Updated June 26, 2023]. In: StatPearls [Internet]. Treasure Island, FL: StatPearls Publishing, January (www.ncbi.nlm.nih.gov/books/NBK551686/).

Church, F.C. (2021). Treatment options for motor and non-motor symptoms of Parkinson's disease. Biomolecules, 11(4):612 (doi:10.3390/biom11040612).

Connolly, B.S., & Lang, A.E. (2014). Pharmacological treatment of Parkinson's disease: A review. JAMA, 311:1670–1683.

Crosby, N.J., Deane, K., & Clarke, C.E. (2003). Amantadine in Parkinson's disease. Cochrane Database of Systematic Reviews, 2003(1): CD003468 (doi: 10.1002/14651858.CD003468).

Doggrell, S.A. (2023). Continuous subcutaneous levodopa-carbidopa for the treatment of advanced Parkinson's disease: Is it an improvement on other delivery? Expert opinion drug delivery, 10:1–11 (doi: 10.1080/17425247.2023.2253146).

Dorseya, E.R., Shererb, T., Okunc, M.S., & Bloemd, B.R. (2018). The emerging evidence of the Parkinson's pandemic. Journal of Parkinson's Disease 8:S3–S8 (doi:10.3233/JPD-181474).

Dragašević-Mišković, N., Petrović, I., Stanković, I., & Kostić, V.S. (2019). Chemical management of levodopa-induced dyskinesia in Parkinson's disease patients. Expert Opinion Pharmacotherapy, 20(2):219–230.

Duvoisin, R.C. (1992). A brief history of parkinsonism. Neurologic Clinics, 10(2):301–316 (doi: 10.1016/S0733-8619(18)30211-1).

Edinoff, A.N., Swinford, C.R., Odisho, A.S., Burroughs, C.R., Stark, C.W., Raslan, W.A., Cornett, E.M., Kaye, A.M., & Kaye, A.D. (2022). Clinically relevant drug interactions with monoamine oxidase inhibitors. Health Psychology Research, 10(4):39576 (doi: 10.52965/001c.39576).

Ferreire, J.J., Lees, A., Roacha, J.-F., Poewe, W., Rascol, O., & Soares-de-Silva, P. (2019). Long-term efficacy of opicapone in fluctuating Parkinson's disease patients: A pooled analysis of data from two phase 3 clinical trials and their open-label extensions. European Journal of Neurology, 26(7):953–960 (doi: 10.1111/ene.13914).

Frampton, J.E. (2019). Rotigotine transdermal patch: A review in Parkinson's disease. CNS Drugs, 33(7):707–718 (doi: 10.1007/s40263-019-00646-y).

Gancher, S.T., Nutt, J.G., & Woodward, W.R. (1991). Absorption of apomorphine by various routes in parkinsonism. Movement Disorder, 6(3):212–216.

Gerlach, M., Youdim, M.B.H., & Riederer, P. (1996). Pharmacology of selegiline. Neurology, 47(6 Suppl 3):137S–145S (doi: 10.1212/WNL.47.6_Suppl_3.137S).

Gjerden, P., Bramness, J.G., & Slørdal, L. (2009). The use and potential abuse of anticholinergic antiparkinson drugs in Norway: A pharmacoepidemiological study. British Journal of Clinical Pharmacology, 67(2):228–233 (doi: 10.1111/j.1365-2125.2008.03342.x).

Güneş, M., & Karavana, S.Y. (2021). Non-oral drug delivery in Parkinson's disease: Current applications and future. Turkey Journal of Pharmaceutical Sciences, 19(3):343–352 (doi: 10.4274/tjps.galenos.2021.95226).

Hauser, R.A. (2009). Levodopa: Past, present, and future. European Neurology, 62(1):1–8 (doi: 10.1159/000215875).

Hauser, R.A., Espay, A.J., Ellenbogen, A.L., Fernandez, H.H., Isaacson, S.H., LeWitt, P.A., et al. (2023). IPX203 vs immediate-release carbidopa-levodopa for the treatment of motor fluctuations in Parkinson disease: The RISE-PD randomized clinical trial. JAMA Neurology, 80(10):1062–1069 (doi:10.1001/jamaneurol.2023.2679).

Hauser, R.A., Li, R., Pérez, A., Ren, X., Weintraub, D., Elm, J., Goudreau, J.L., Morgan, J.C., Fang, J.Y., Aminoff, M.J., Christine, C.W., Dhall, R., Umeh, C.C., Boyd, J.T., Stover, N., Leehey, M., Zweig, R.M., Nicholas, A.P., Bodis-Wollner, I., Willis, A., Kieburtz, K., & Tilley, B.C. (2017). Longer duration of MAO-B inhibitor exposure is associated with less clinical decline in Parkinson's disease: An analysis of NET-PD LS1. Journal of Parkinson's Disease, 7:117–127.

Hornykiewicz, O. (2010). A brief history of levodopa. Journal of Neurology, 257(Suppl 2):249–252 (doi: 10.1007/s00415-010-5741-y).

Huot, P., Johnston, T.H., Koprich, J.B., Fox, S.H., Brotchie, J.M., & Sibley, D.R. (2013). The pharmacology of L-DOPA-induced dyskinesia in Parkinson's disease. Pharmacological Reviews, 65(1):171–222 (doi: 10.1124/pr.111.005678).

Hubsher, G., Haider, M., & Okun M.S. (2012). Amantadine: The journey from fighting flu to treating Parkinson disease. Neurology, 78(14):1096–1099 (doi: 10.1212/WNL.0b013e31824e8f0d).

Isaacson, S.H., Fahn, S., Pahwa, R., Tanner, C.M., Espay, A.J., Trenkwalder, C., Adler, C.H., Patni, R., & Johnson, R. (2018). Parkinson's patients with dyskinesia switched from immediate release amantadine to open-label ADS-5102. Movement Disorder Clinical Practice, 5(2):183–190.

Jenner, P., Rocha, J., Ferreira, J.J., Rascol, O., & Soares-da-Silva, P. (2021). Redefining the strategy for the use of COMT inhibitors in Parkinson's disease: The role of opicapone. Expert Review of Neurotherapeutics, 21(9):1019–1033 (doi: 10.1080/14737175.2021.1968298).

Jilani, T.N., Sabir, S., & Sharma, S. (2023). Trihexyphenidyl. [Updated February 17, 2023]. In: StatPearls [Internet]. Treasure Island, FL: StatPearls Publishing, January (www.ncbi.nlm.nih.gov/books/NBK519488/).

Jost, W.H., & Angersbach, D. (2005). Ropinirole, a non-ergoline dopamine agonist. CNS Drug Review, 11(3):253–272 (doi: 10.1111/j.1527-3458.2005.tb00046.x).

Kalinderi, K., Fidani, L., Katsarou, Z., & Bostantjopoulou, S. (2011). Pharmacological treatment and the prospect of pharmacogenetics in Parkinson's disease. The International Journal of Clinical Practice, 65(12):1289–1294 (doi: 10.1111/j.1742-1241.2011.02793.x).

Katzenschlager, R., Poewe, W., Rascol, O., Trenkwalder, C., Deuschl, G., Chaudhuri, K.R., Henriksen, T., van Laar, T., Spivey, K., Vel, S., Staines, H., & Lees, A. (2018). Apomorphine subcutaneous infusion in patients with Parkinson's disease with persistent motor fluctuations (TOLEDO): A multicentre, double-blind, randomised, placebo-controlled trial. Lancet Neurology, 17(9):749–759.

Katzenschlager, R., Sampaio, C., Costa, J., & Lees, A. (2002). Anticholinergics for symptomatic management of Parkinson's disease. Cochrane Database System Review, 2002(2):CD003735 (doi: 10.1002/14651858).

Kawabata, K., & Katsuno, M. (2021). Trihexyphenidyl, biperiden, and other anticholinergics in the treatment of Parkinson's disease. In: Riederer, P., Laux, G., Nagatsu, T., Le, W., & Riederer, C., editors. NeuroPsychopharmacotherapy. Cham: Springer (doi: 10.1007/978-3-319-56015-1_221-1).

Ke, X., Wang, L., Chen, M., Liu, S., Yu, N., Duan, L., Gong, F., & Zhu, H. (2022). The side effects of dopamine receptor agonist drugs in Chinese prolactinoma patients: A cross-sectional study. BMC Endocrinology Disorder, 22:97 (doi: 10.1186/s12902-022-01009-3).

Keener, A.M., & Bordelon, Y.M. (2016). Parkinsonism. Seminar in Neurology, 36(4):330–334 (doi: 10.1055/s-0036-1585097).

Kostelnik, A., Cegan, A., & Pohanka, M. (2017). Anti-Parkinson drug biperiden inhibits enzyme acetylcholinesterase. BioMed Research International, 2017:2532764 (doi: 10.1155/2017/2532764).

Lecht, S., Haroutiunian, S., Hoffman, A., & Lazarovici, P. (2007). Rasagiline – a novel MAO B inhibitor in Parkinson's disease therapy. Therapeutics and Clinical Risk Management, 3(3):467–474 (doi: 10.2147/term.s12160420).

Lee, T.K., & Yankee, E.L. (2021). A review on Parkinson's disease treatment. Neuroimmunology and Neuroinflammation, 8:222–244 (doi: 10.20517/2347-8659.2020.58).

Lertxundi, U., Hernández, R., Domingo-Echaburu, S., Peral-Aguirregoitia, J., & Medrano, J. (2016). Pharmacotherapeutic challenges in Parkinson's disease inpatients. InTech (doi: 10.5772/62561).

LeWitt, P.A. (2014). Levodopa therapy for Parkinson's disease: Pharmacokinetics and Pharmacodynamics. Movement Disorders, 64–72 (doi: 10.1002/mds.26082).

Li, T., Zou, S., Zhang, Z., Liu, M., & Liang, Z. (2022). Efficacy of pramipexole on quality of life in patients with Parkinson's disease: A systematic review and meta-analysis. BMC Neurology, 22:320 (doi: 10.1186/s12883-022-02830-y).

Lieberman, A., Goldstein, M., Leibowitz, M., Neophytides, A., Kupersmith, M., Pact, V., & Kleinberg, D. (1981). Neurology, 31(6):675 (doi: 10.1212/WNL.31.6.675).

Mandel, S., Weinreb, O., Amit, T., & Youdim, M.B.H. (2005). Mechanism of neuroprotective action of the anti-Parkinson drug rasagiline and its derivatives. Brain Research Reviews, 48(2):379–387 (doi: 10.1016/j.brainresrev.2004.12.027).

Mao, Q., Qin, Wz., Zhang, A., & Ye, N. (2020). Recent advances in dopaminergic strategies for the treatment of Parkinson's disease. Acta Pharmacologica Sinica, 41:471–482 (doi: 10.1038/s41401-020-0365-y).

Markham, C.H., & Diamond, S.G. (1989). Pergolide in the treatment of Parkinson's disease. In: Calne, D.B., editor. Drugs for the Treatment of Parkinson's Disease. Handbook of Experimental Pharmacology, vol. 88. Berlin, Heidelberg: Springer (doi: 10.1007/978-3-642-73899-9_18).

Moore, J.J., & Saadabadi, A. (2023). Selegiline. [Updated August 14, 2023]. In: StatPearls [Internet]. Treasure Island, FL: StatPearls Publishing, January (www.ncbi.nlm.nih.gov/books/NBK526094/).

Mysiris, D.S., Vavougios, G.D., Karamichali, E., Papoutsopoulou, S., Stavrou, V.T., Papayianni, E., Boutlas, S., Mavridis, T., Foka, P., Zarogiannis, S.G., Gourgoulianis, K, & Xiromerisiou, G. (2022). Post-COVID-19 parkinsonism and Parkinson's disease pathogenesis: The exosomal cargo hypothesis. International Journal of Molecular Science, 23:9739 (doi: 10.3390/ijms23179739).

Namali, C. (2007). Procyclidine. In: Enna, S.J., & Bylund, D.B., editors. xPharm: The Comprehensive Pharmacology Reference. Elsevier, pages 1–4 (doi: 10.1016/B978-008055232-3.62469-9).

Nikolaus, S., Wittsack, H.J., Beu, M., Antke, C., Hautzel, H., Wickrath, F., Müller-Lutz, A., De Souza Silva, M.A., Huston, J.P., Antoch, G., & Müller, H.W. (2019). Amantadine enhances nigrostriatal and mesolimbic dopamine function in the rat brain in relation to motor and exploratory activity. Pharmacology Biochemical Behaviour, 179:156–170.

Olanow, C.W., & Watkins, P. (2007). Tolcapone: An efficacy and safety review. Clinical Neuropharmacology, 30(5):287–294 (doi: 10.1097/wnf.0b013e318038d2b6)

Pagliaro, L.A., & Pagliaro, A.M. (2020). Levodopa [L-Dopa]. In: PNDR: Psychologists' Neuropsychotropic Drug Reference. Routledge, pages 185–189.

Parkinson, J. (2002). An essay on the shaking palsy. Journal of Neuropsychiatry Clinical Neuroscience, 14:223–223.

Patel, T., & Chang, F. (2014). Parkinson's disease guidelines for pharmacists. Canadian Pharmacist Journal, 147(3):161–170 (doi: 10.1177/1715163514529740).

Perez-Lloret, S., & Rascol, O. (2010). Dopamine receptor agonists for the treatment of early or advanced Parkinson's disease. CNS Drugs, 24:941–968 (doi: 10.2165/11537810-000000000-00000).

Poewe, W., Seppi, K., Tanner, C., Halliday, G.M., Brundin, P., Volkmann, J., Schrag, A.E., & Lang, A.E. (2017). Parkinson disease. Nature Reviews: Disease Primers, 3:17013 (doi: 10.1038/nrdp.2017.13).

Pringsheim, T., Day, G.S., Smith, D.B., Rae-Grant, A., Licking, N., Armstrong, M.J., de Bie, R.M.A., Roze, E., Miyasaki, J.M., Hauser, R.A., Espay, A.J., Martello, J.P., Gurwell, J.A., Billinghurst, L., Sullivan, K., Fitts, M.S., Cothros, N., Hall, D.A., Rafferty, M., Hagerbrant, L., Hastings, T., O'Brien, M.D., Silsbee, H., Gronseth, G., & Lang, A.E.; Guideline Subcommittee of the AAN. (2021). Dopaminergic therapy for motor symptoms in early Parkinson disease practice guideline summary: A report of the AAN Guideline Subcommittee. Neurology, 97(20):942–957.

Quaresma, M.V., Gomes-Dias, A.C., Serruya, A., Vendramini, D.L., Braga, L., & Buçard-Mota, A. (2014). Amantadine-induced livedo reticularis—Case report. Anais Brasileiros de Dermatologia, 90(5):745–747.

Ravn, A.H., Thyssen, J.P., & Egeberg, A. (2017). Skin disorders in Parkinson's disease: Potential biomarkers and risk factors. Clinical, Cosmetic and Investigational Dermatology, 10:87–92 (doi: 10.2147/CCID.S130319).

Riederer, P., & Laux, G. (2011). MAO-inhibitors in Parkinson's disease. Experimental Neurobiology, 20(1):1–17 (doi: 10.5607/en.2011.20.1.1).

Rivest, J., Barclay, C., & Suchowersky, O. (1999). COMT inhibitors in Parkinson's disease. Canadian Journal of Neurological Sciences, 26(S2):S34–S38.

Rizek, P., Kumar, N., & Jog, M.S. (2016). An update on the diagnosis and treatment of Parkinson's disease. Canadian Medical Association Journal, 188(16):1157–1165 (doi: 10.1503/cmaj.151179).

Robottom, B.J., & Reich, S.G. (2011). Exposure to high dosage trihexyphenidyl during pregnancy for treatment of generalized dystonia: Case report and literature review. Neurologist, 17(6):340–341.

Salamon, A., Zádori, D., Szpisjak, L., Klivényi, P., & Vécsei, L. (2022). What is the impact of catechol-O-methyltransferase (COMT) on Parkinson's disease treatment? Expert Opinion on Pharmacotherapy, 23(10):1123–1128 (doi: 10.1080/14656566.2022.2060738).

Scheller, D., Ullmer, C., Berkels, R., Gwarek, M., & Lübbert, H. (2009). The in vitro receptor profile of rotigotine: A new agent for the treatment of Parkinson's disease. Naunyn Schmiedebergs Arch Pharmacology, 379(1):73–86 (doi: 10.1007/s00210-008-0341-4).

Shill, H.A., & Stacy, M. (2009). Update on ropinirole in the treatment of Parkinson's disease. Neuropsychiatry Disease Treatment, 5:33–36.

Shulman, K.I., Herrmann, N., & Walker, S.E. (2013). Current place of monoamine oxidase inhibitors in the treatment of depression. CNS Drugs, (10):789–797.

Skorvanek, M., & Bhatia, K.P. (2017). The skin and Parkinson's disease: Review of clinical, diagnostic, and therapeutic issues. Movement Disorder Clinical Practice, 4(1):21–31.

Spencer, C.M., & Benfield, P. (1996). Tolcapone. CNS Drugs, 5:475–481 (doi: 10.2165/00023210-199605060-00006).

Stoker, T.B., & Barker, R.A. (2020). Recent developments in the treatment of Parkinson's disease. F1000Research, 9:862 (doi: 10.12688/f1000research.25634.1).

Sub Laban, T., & Saadabadi, A. (2023). Monoamine oxidase inhibitors (MAOI) [Updated July 17, 2023]. In: StatPearls [Internet]. Treasure Island, FL: StatPearls Publishing, January (www.ncbi.nlm.nih.gov/books/NBK539848/).

Sveinbjornsdottir, S. (2016). The clinical symptoms of Parkinson's disease. Journal of Neurochemistry, 139(1):318–324 (doi: 10.1111/jnc.13691)

Tambasco, N., Romoli, M., & Calabresi, P. (2018). Levodopa in Parkinson's disease: Current status and future developments. Current Neuropharmacology, 16(8):1229–1250 (doi: 10.2174/1570159X15666170510143821).

Tan, Y., Jenner, P., & Chena, S. (2022). Monoamine oxidase-B inhibitors for the treatment of Parkinson's disease: Past, present, and future. Journal of Parkinson's Disease, 12:477–493 (doi: 10.3233/JPD-212976).

Thomas, S.J., Shin, M., McInnis, M.G., & Bostwick, J.R. (2015). Combination therapy with monoamine oxidase inhibitors and other antidepressants or stimulants: Strategies for the management of treatment-resistant depression. Pharmacotherapy, 35(4):433–449.

Tran, T., Brophy, J.M., Suissa, S., & Renoux, C. (2015). Risks of cardiac valve regurgitation and heart failure associated with ergot- and non-ergot-derived dopamine agonist use in patients with Parkinson's disease: A systematic review of observational studies. CNS Drugs, 29(12):985–998.

Tran, T.N., Vo, T.N.N., Frei, K., & Truong, D.D. (2018). Levodopa-induced dyskinesia: Clinical features, incidence, and risk factors. Journal of Neural Transmission, 125:1109–1117 (doi: 10.1007/s00702-018-1900-6).

Truong, D.D. (2009). Tolcapone: Review of its pharmacology and use as adjunctive therapy in patients with Parkinson's disease. Clinical Interventions in Aging, 4:109–113 (doi: 10.2147/cia.s3787).

Tsuboi, T., Satake, Y., Hiraga, K., Yokoi, K., Hattori, M., Suzuki, M., Hara, K., Ramirez-Zamora, A., Okun, M.S., & Katsuno, M. (2022). Effects of MAO-B inhibitors on non-motor symptoms and quality of life in Parkinson's disease: A systematic review. npj Parkinson's Disease, 8:75 (doi: 10.1038/s41531-022-00339-2).

Tuunainen, J., Sjöstedt, N., Vahteristo, M., Ellmen, J., Kuoppamaki, M., Rouru, J., & Yliperttula, M. (2023). Effect of carbidopa dose on levodopa pharmacokinetics with and without catechol-*O*-methyltransferase inhibition in healthy subjects. European Journal of Drug Metabolism and Pharmacokinetics, 48:23–34 (doi: 10.1007/s13318-022-00800-w).

van Camp, G., Flamez, A., Cosyns, B., Weiytjen, C., Muyldermas, L., De Sutter, J., Santens, P., Decoodt, P., Moerman, C., & Schoors, D. (2004). Treatment of Parkinson's disease with pergolide and relation to restrictive valvular heart disease. The Lancet, 363(9416):1179–1183.

van Rumund, A., Pavelka, L., Esselink, R.A.J., Geurtz, B.P.M., Wevers, R.A., Mollenhauer, B., Kruger, R., Bloem, B.R., & Verbeek, M.M. (2021). Peripheral decarboxylase inhibitors paradoxically induce aromatic L-amino acid decarboxylase. npj Parkinson's Disease, 7:29 (doi: 10.1038/s41531-021-00172-z).

Vázquez-Vélez, G.E., & Zoghbi, H.Y. (2021). Parkinson's disease genetics and pathophysiology. Annual Review of Neuroscience, 44(1):87–108.

Via, M.A., Chandra, H., Araki, T., Potenza, M.V., & Skamagas, M. (2010). Bromocriptine approved as the first medication to target dopamine activity to improve glycemic control in patients with type 2 diabetes. Diabetes, Metabolic Syndrome and Obesity, 3:43–48 (doi: 10.2147/dmsott.s9575).

Weiener, C.P., & Buhimschi, C. (2009). Bromocriptine. In: Weiner, C.P., & Buhimschi, C., editors. Drugs for Pregnant and Lactating Women. 2nd ed. W.B. Saunders, pages 841–965 (doi: 10.1016/B978-1-4160-4013-2.00015-6).

Whitfield, A.C., Moore, B.T., & Daniels, R.N. (2014). Classics in the chemical neuroscience: Levodopa. ACS Chemical Neuroscience, 5:1192–1197.

Wood, M., Dubois, V., Scheller, D., & Gillard, M. (2015). Rotigotine is a potent agonist at dopamine D1 receptors as well as at dopamine D2 and D3 receptors. British Journal of Pharmacology, 172(4):1124–1135 (doi:10.1111/bph.12988).

Zanardi, A.P.J., da Silva, E.S., Costa, R.R., Passos-Monteiro, E., Dos Santos, I.O., Kruel, L.F.M., & Peyré-Tartaruga, L.A. (2021). Gait parameters of Parkinson's disease compared with healthy controls: A systematic review and meta-analysis. Science Report, 11:752 (doi: 10.1038/s41598-020-80768-2).

14 Induction and Validation of Zebrafish Model of Parkinson's Disease

Gisha Sivan, Rajesh Pamanji, and Joseph Selvin

14.1 INTRODUCTION

Zebrafish is one of the best alternate versatile models to understand the mechanistic pathways of most of the diseases and validate new disease management strategies for combating human diseases. Zebrafish compared to the human reference genome, 70% of the human genes exhibit similarity and 84% of human genes associated with the disease have zebrafish orthologue (Howe et al., 2013). Zebrafish behaviour and pathophysiology are easily quantifiable to humans (Razali et al., 2021). From the first studies on homozygous development of zebrafish by Streisinger and his colleagues in the 1980s (Streisinger et al., 1981), zebrafish research has evolved from a developmental study model organism to being used in various disciplines of biomedical sciences including behavioural studies, neurodegeneration, neurodevelopment, toxicology, and drug delivery thanks to the advancements in molecular tools and techniques in zebrafish husbandry. Recent developments in CRISPR and next-generation sequencing (NGS) technology have opened the role of zebrafish in disease modelling, providing new diagnostic and therapeutic strategies. The zebrafish model is advantageous over mammalian models, in expense, maintenance, assessment duration, and visualization of internal organ development.

The integration of the genetic and functional studies helps in understanding the underlying basic pathways in which each gene is involved and how genes can be altered for different disease models. The easy incorporation of genetic mutants and visualization of the genetic analysis in zebrafish renders mechanistic understanding of the molecular basis of neural development. The cognitive processes exhibited by zebrafish, including learning, memory, fear, anxiety, perception, social skills, and sleep patterns (Oliveira, 2013), are well translated to those exhibited by humans, showing that zebrafish are not only related to the genotypic and molecular mechanisms but they also provide relatable phenotypic information (Razali et al., 2021).

14.2 BENEFITS OF USING THE ZEBRAFISH MODEL

Zebrafish has several characteristics that make it a suitable model for disease study. Zebrafish being a vertebrate share close characteristics with humans. The transparent chorion helps in the visualization of neurodevelopmental stages in zebrafish. The high fecundity of 200–250 eggs per mating, the rapid development of external structure within 24 hpf and well-described internal organs like heart, liver, and kidney formation within 96 hpf, a sequenced characterized genome, and the ease of generating the required number of tissue-specific transgenic animals are characteristic features that make zebrafish suitable for disease models (Basnet et al., 2019). Zebrafish embryos are in vivo models with transparent chorion that reflect development using adequate photo-imaging tools. The minimal quantity (50 μl) of media/water required makes it comparable to the *in vitro* model. A significant number of publications have reported adult zebrafish as the model organism of neurodegeneration studies (Shenoy et al., 2022). The optic clarity of the embryo for visualization of the phenotypic changes of the underlying genetic expressions is one of the salient characteristics that makes zebrafish advantageous over other animal models.

14.3 CHALLENGES OF USING THE ZEBRAFISH MODEL

The zebrafish serves as a translational animal model for comprehending human pathogenesis and its potential therapeutic properties (Chia et al., 2022). However, it is essential to note that the zebrafish has its own limitations. Being an aquatic, poikilothermic vertebrate, the zebrafish primarily relies on gills for respiration. Despite having genetic similarities and body planning conservation, there exist differences in brain anatomy between zebrafish and humans (Cassar et al., 2020). Nevertheless, this does not diminish the prominence of the zebrafish as a model organism for neurological studies related to neurotransmitters, transporters, and receptors, among others (Kalueff et al., 2014). The zebrafish's susceptibility to diseases differs from humans due to having a simpler immune system, making it less compatible with complex immune system studies (Trede et al., 2004). Moreover, differences in enzyme and transporter expression and activity are observed in zebrafish when compared to humans.

14.4 ZEBRAFISH IN PARKINSON'S

PD is a disease that takes decades to develop in humans and does not occur in other animals (Barnhill et al., 2020).

DOI: 10.1201/9781003402893-16

Zebrafish have been shown to possess several orthologues of PD like pink1, dj-1, parkin, and lrrk2 (Xi et al., 2011), and share a high homology to all PD-related proteins. The zebrafish brain contains 100,000 (Randlett et al., 2015) neurons compared to 86 billion neurons in humans, making it a simple, manageable model for neurodegenerative studies. The transparency of the developing zebrafish embryos helps visualization of the development of the nervous system, in real time. The relatively simple nervous system coupled with molecular and optical techniques makes genetic expression studies highly translatable to the human nervous system. Zebrafish is considered a suitable neurodegenerative model as the gene/protein expressions are easily accessible, it has a neuroanatomical structure similar to humans, and the short life cycle leads to adulthood and aging with gene expressions showing visible phenotypic changes. The mapping of the dopaminergic nervous system was carried out based on extensive studies using the nervous system (Tay et al., 2011).

In zebrafish, the nervous system is simple and structurally similar to a mammalian brain (Shenoy, 2022). The morphogenetic movement involves eversion in zebrafish and inversion in the mammalian brain, which leads to the absence of the hippocampus, amygdala, and substantia nigra in zebrafish. The neural development in zebrafish embryos starts with the central nervous system forming at 6 hours post-fertilization (hpf) during the process of gastrulation. Around 10 hpf the forebrain, midbrain, and hindbrain start formation, with distinct boundaries and well defined by 24 hpf, laying the foundation for further differentiation into the adult brain. The forebrain region is further divided into telencephalon-containing regions like the pallium (dorsal telencephalon), subpallium (ventral telencephalon), and olfactory bulbs and diencephalon, hypothalamus, and retina (Rocha et al., 2020). The forebrain in zebrafish is concerned with receiving and processing sensory information and the lateral pallium plays a crucial role in spatial learning in zebrafish, and the medial pallium is responsible for avoidance learning.

The zebrafish midbrain is formed of tegmentum, optic tectum, torus semicircularis, torus longitudinalis, and concerned with vision and hearing. The zebrafish midbrain is found between the forebrain and hindbrain. The hindbrain in zebrafish comprises the medulla oblongata, pons, and cerebellum. The midbrain, medulla oblongata, and pons together form the brainstem that regulates the respiratory, circulatory, and wakefulness in fish. The cerebellum is concerned with emotional learning and cognition (Costa et al., 2021).

Zebrafish exhibit PD-related symptoms like anxiety-related behaviors, and changes in movement can be analyzed by zebrafish swimming behaviour and locomotion (Razali et al., 2022). Zebrafish exhibit parallels with the human dopaminergic system, a central component in PD. For instance, the zebrafish ventral telencephalon is considered homologous to the mammalian striatum, a brain region critical for motor and cognitive functions. It is also connected to the substantia nigra through dopaminergic projections (Du et al., 2016). Zebrafish, particularly during their larval stages, offer a robust model for investigating PD due to these advantages.

Larval zebrafish have emerged as a remarkably advantageous model for studying PD phenotypes. This is primarily attributed to their extensively researched nervous system during development, which exhibits surprising homology with several aspects of the human nervous system. Of particular significance is the dopaminergic system, which displays remarkable similarities between humans and zebrafish, making the latter a valuable asset for PD research. Furthermore, the zebrafish ventral telencephalon has been shown to be analogous to the mammalian striatum, an essential brain region responsible for motor and cognitive behaviors, and is connected to the substantia nigra via dopaminergic projections.

14.5 GENE-BASED MODELS

In zebrafish various genomic manipulation methods are employed to detect fish with mutation studies (Woods and Schier, 2008). Morpholino oligonucleotides (MOs) represents the most widely adopted anti-sense knockdown tools to accelerate gene discovery to verify mutant phenotypes in the zebrafish (Li et al., 2021). Morpholinos synthesized artificially are antisense molecules comprised of 25 nucleotides designed to impede the translation initiation complex of messenger RNA (mRNA) sequences by binding. This innovative technology has been utilized to ascertain the role of specific genes by temporarily inhibiting them, particularly during the developmental process (Bill et al., 2009). Morpholinos exhibit exceptional precision and efficiency in disrupting genes or mRNA function. Morpholino oligonucleotides (MOs) provide an economical and expeditious approach to evaluate gene function, even for genes with existing mutations. The preliminary screening is rapid and can engender focal points for subsequent work, contingent upon the acquisition of the mutant line. MOs can be deployed to sift through a small set of genes of interest or to screen a gene for susceptible exons, knockout of the most promising candidate. The main disadvantage of using morpholinos is that their transient effect lasts only up to 5 days. Furthermore, these oligonucleotides are typically introduced into the embryo and exert their influence on every cell in the embryo, compromising normal development if the disrupted protein is required ubiquitously, which renders them unsuitable for the study of gene function during later developmental stages. Morpholinos are oligonucleotides that have been modified and are capable of binding to specific target mRNA. This binding action results in the inhibition of gene expression. Moreover, morpholinos are capable of inducing abnormal intron inclusion or skipping, thereby causing alterations in protein production. These morpholinos are commonly injected into zebrafish embryos between the one-cell and four-cell stages. Although the dose, properties of the target transcript, and protein kinetics may alter their potency, their effects can extend for up to 5 days. Researchers frequently co-inject synthetic target mRNA and use controls like phenotypic rescue to verify their specificity and efficacy. Furthermore, the delivery of null mutant mRNA can be used as a separate technique to assess the outcomes of acute gene knockdown. When evaluating the effects of morpholino administration in zebrafish, it is essential to use caution and maintain stringent controls for efficacy and specificity, as per usual procedure.

When a specific target segment of DNA is targeted by restriction enzymes, zinc-finger nucleases (ZFNs) constitute a viable alternative molecular technique for somatic and germline gene disruption (Ekker, 2008). ZFNs target certain nucleotide sequences with a high degree of precision; however, it is possible for mutations to develop at the repair site. Because normal diploid genetics requires several generations and is expensive, the current limits of zinc finger technology result in a minimum time lag of roughly one year before it is possible to study the loss of function tissue (Bill et al., 2009). Numerous genes have undergone mutations as a result of conventional methods for causing random mutagenesis, such as the use of chemicals, retroviruses, or transposons. These techniques, however, have fallen short of providing the accuracy needed to purposefully design particular mutations. In contrast, new research has demonstrated a ground-breaking method for facilitating targeted mutations in zebrafish genes using zinc-finger nucleases (ZFNs), composite molecules made up of a DNA-binding zinc-finger domain and the FokI restriction endonuclease. ZFNs work by causing specific double-strand breaks in the DNA. By altering the zinc finger composition of a ZFN, each finger is uniquely designed to connect with a particular triplet of DNA base pairs, resulting in the specificity of this DNA cleavage. By combining three or four zinc fingers, precise binding to 9- or 12-bp motifs, respectively, is achieved. The actual double-strand breaks are effected when two ZFNs attach to the target DNA, causing their nuclease domains to come into proximity. It is important to note that these nuclease domains are only active in the form of a dimer, and they cleave the DNA between the bound ZFNs.

In order to generate mutations via nonhomologous end joining, researchers utilize a method whereby mRNAs encoding two ZFNs are introduced via injection into fertilized zebrafish eggs. The activity of the ZFNs is then assessed through the utilization of techniques such as polymerase chain reaction (PCR) and phenotypic screening in both the injected fish and their offspring. Notably, it has been observed that a substantial proportion of the injected fish (30–50%) pass on the mutations induced by ZFNs to their progeny, and a significant fraction of these progeny (18% in one study and 7% in the other) exhibit mutant characteristics. This breakthrough method allows for the precise disruption of any gene of interest, thus paving the way for more intricate genetic manipulations within the zebrafish genome.

The implementation of TILLING (targeting induced local lesions in genomes), which is a reverse genetic approach, greatly facilitated the process of identifying genes of interest. The methodology for TILLING in zebrafish involved the initial examination of genomic DNA from a vast collection of ENU-mutagenized zebrafish in search of rare mutations in target genes. This examination was conducted either by utilizing Cell on fish pools or by resequencing individual fish that were desired for analysis. Subsequently, the mutations were obtained by breeding the singular carrier that had been identified. The key components for a successful TILLING project include an effective screening technique, a well-mutagenized fish library, and a nearly flawless ability to retrieve valuable mutations once they have been identified (Moens et al., 2008). CRISPR-Cas9 can be used to create targeted gene knockouts and gene knock-in in zebrafish where guide RNAs (gRNAs) are introduced to target specific genes and introduce Cas9 protein, which cuts the DNA at the targeted site, or insert specific DNA sequence into the zebrafish genome allowing precise gene editing (Medishetti et al., 2022). This technique is invaluable for studying gene function and understanding the genetic basis of diseases and developing zebrafish lines with specific genetic mutations to understand their roles in developmental processes.

14.5.1 Synuclein

The zebrafish does not possess a homologous counterpart of the human SNCA gene, responsible for encoding the α-synuclein protein, which serves as a representative model for synucleinopathies. Nonetheless, the zebrafish does exhibit the expression of three synuclein genes, namely SNCB, SNCG1, and SNCG2, which encode for β-, γ1-, and γ2-synuclein proteins, respectively. It is crucial to mention that γ1 shares functional similarities with α-synuclein in humans (Barnhill et al., 2020). The advancement of genetic technology has enabled the creation of zebrafish transgenic models that encompass the human wild-type α-synuclein protein within their living system. In this specific model, the incorporation of human α-synuclein fused with enhanced green fluorescent protein (GFP) was executed in a transient manner. This fusion protein facilitates the monitoring of the subcellular distribution and behavior of α-synuclein over a period of time. In order to regulate the expression of this modified α-synuclein, the zebrafish neuroD promoter was utilized, which specifically directs expression within a distinct group of neurons. In order to comprehend the activity of α-synuclein within motor neurons, an analysis known as fluorescence recovery after photobleaching (FRAP) was conducted.

The punctate appearance of α-synuclein-GFP, situated within the motor axons, was readily observable, with minimal labeling detected in the axonal segments located between varicosities. This observation presents a striking contrast to the unfettered GFP, thus suggesting a significant accumulation of α-synuclein at the presynaptic terminals (Weston et al., 2021). Several studies have employed these innovative zebrafish models, in which the expression of either the human wild-type (wt) or A53T mutant α-SYN is facilitated, thereby mimicking essential characteristics of synucleinopathies encompassing genetic mutations. Furthermore, the cloning and characterization of three synucleins from zebrafish, namely sncga, sncgb, and sncb, were conducted. The sequences of these zebrafish synucleins bear resemblance to their human protein counterparts.

The examination of their spatial and temporal expression patterns during development was conducted using whole-mount in situ hybridization on staged embryos, employing the 3' UTR sequence for the synthesis of anti-sense riboprobe. It was observed that sncgb was exclusively expressed in the notochord, whereas sncga and sncb exhibited strong expression in the nervous system. This observation provides a foundation for future experiments aiming to evaluate their functions in normal physiology as well as their involvement in pathophysiology (Sun and Gitler, 2008). Initially, it was discovered that zebrafish Sncb shares 70% identity and 82% similarity with human SNCB. The detection of the zebrafish synuclein genes were mainly concentrated in the trigeminal placode, subsequently expanding to various regions of the brain. Meanwhile, sncga was detected in spinal cord and hindbrain neurons from 26 hpf zebrafish embryo and later in the brain and retina. Sncgb showed early notochord expression but became undetectable after 2 days post fertilization (dpf) and was also weakly expressed in adults. On the other hand, sncb and sncga were prominent in the brain and eyes and were co-expressed in catecholaminergic neurons (Sun and Gitler, 2008).

Effective elimination of toxic α-SYN is a crucial factor in preserving cellular balance and survival. However, the comprehension of this mechanism is restricted by the absence of appropriate in vivo models for studying α-SYN degradation. Recent advancements in imaging methods have facilitated the evaluation of tau protein clearance in living zebrafish neurons. One pioneering approach involved the fusion of human SNCA with the photoconvertible protein Dendra2, which enabled the investigation of α-SYN clearance kinetics, both for the wild-type (wt) and A53T mutant forms, in living organisms (Lopez et al., 2022). Syndecan-4 (Syn4), a heparin sulfate proteoglycan (HSPG), is a critical contributor to the formation of focal adhesions and interacts with various growth factors, regulating processes such as neural induction and cell migration during early zebrafish development. The expression of Syn4 is dynamic and widespread during these stages. Downregulation of Syn4 leads to an increase in the expression of neural cell marker genes. Thus, Syn4 is a key regulator of CaP axon formation during zebrafish embryogenesis and also inhibits cellular proliferation during neurogenesis (Luo et al., 2016).

14.5.2 Parkin

The PARK2 gene encodes cytosolic ubiquitin-E3-ligase, the Parkin protein involved in proteasome-dependent degradation of proteins. The zebrafish Parkin protein is similar to human Parkin protein with 62% of homology and 78% functional similarity (Flinn et al., 2009). The α-synuclein-GFP, located in the motor axons, exhibited a punctate appearance that was discernible, albeit with minimal prominence. Throughout the developmental stages of zebrafish larvae and in adult zebrafish, the Parkin gene was observed to be ubiquitously expressed. The present study entailed a comparative analysis between the protein sequence of human Parkin and the genome of zebrafish. For RNA extraction, zebrafish embryos at different developmental stages and organs of adult zebrafish were utilized. Anti-sense oligonucleotides were specifically designed to target specific splice sites in the Parkin transcript of zebrafish. To examine the effects of toxin exposure using MPP+, embryos that were wild-type (uninjected), control MO-injected, and had Parkin knockdown were allowed to mature for 48 hpf prior to toxin exposure.

The α-synuclein-GFP exhibits a punctate appearance and is located in the motor axons. Additionally, a modified in situ hybridization protocol was employed to measure the quantity of dopaminergic neurons. This protocol involved the co-incubation of the anti-digoxigenin-POD antibody with a primary antibody against tyrosine hydroxylase (TH) (Flinn et al., 2009). Mutations in the PARK2 gene have been associated with the early onset of PD. This particular gene encodes the Parkin protein, and mutations in PARK2 are the primary cause of autosomal-recessive PD (Selvaraj and Piramanayagam, 2019).

Under normal circumstances, Parkin acts as a ligase, recognizing damaged proteins that arise from cellular insult or injury and facilitating their degradation through autophagy and breakdown mechanisms (Barnhill et al., 2020). The α-synuclein-GFP exhibits a punctate appearance and is situated in the motor axons. Additionally, the modified in situ Parkin also plays a significant role in mitophagy concerning PD, which involves the process of eliminating damaged mitochondria, typically through the PINK1/Parkin pathway (Deas et al., 2011). Knockdown of PARK2 genes resulted in a 20% reduction in dopaminergic neurons (DA), while no noticeable changes were observed in the swimming pattern of zebrafish (Flinn et al., 2009). Despite being a neuropathological hallmark of PD, it is generally believed that PARK2-related PD is not associated with Lewy body formation (Madsen et al., 2021).

14.5.3 PARL

The punctate manifestation of α-synuclein-GFP, positioned within the motor axons, is further enhanced by a modified in situ Parkin. The PARL gene, which is associated with presenilin, undergoes translation into a protease that localizes at the inner mitochondrial membrane (Lysyk et al., 2020). Within the zebrafish genome, two prologues, namely PARLA and PARLB, exhibit a 67% and 55% homology in amino acid sequence with humans, as unveiled by various studies. Noble et al. (2012) observed that the ablation of PARLA and PARLB in zebrafish through the application of morpholino mediation resulted in the loss of DA neurons. The in vivo loss of function was effectively achieved through the utilization of translocation blocking and splice blocking morpho oligonucleotide techniques.

The α-synuclein-GFP, which is located in the motor axons, exhibits a punctate appearance. Furthermore, the temporal and spatial mRNA expression of PARLA and PARLB was examined using RT-PCR. Despite not being detected maternally, PARLA mRNA was present throughout embryonic development from 4 hpf to at least 7 dpf. In adult zebrafish, PARLA was found to be highly expressed in the brain and muscle tissues, with lower expression observed in the liver and heart. While the PARLB gene expression was initially observed at a low level maternally, transcripts became more abundant from 4 hpf to at least 7 dpf, as well as in the brain of adult zebrafish.

The synteny analysis of genes situated in close proximity to PARL in both human and zebrafish. This analysis illustrated that PARL, along with several genes upstream and downstream of PARL on human chromosome 3, can be mapped to chromosome 2 in zebrafish. Additionally, the zebrafish equivalent of the CHRD gene, which is located on chromosome 3 in humans, namely chrd, was identified to coexist with PARLA on chromosome 15 of the zebrafish organism. The temporal and spatial mRNA expression of PARLA and PARLB were meticulously examined through the utilization of the reverse transcription polymerase chain reaction (RT-PCR) methodology.

The α-synuclein-GFP exhibits a punctate appearance and is located in the motor axons. Additionally, a modified iParla mRNA is noticeably absent maternally at 2 hours post-fertilization (hpf), but can be observed throughout embryonic development from 4 hpf to at least 7 days post-fertilization (dpf). In the adult zebrafish, PARLA demonstrates high levels of expression in the brain and muscle tissues, while displaying relatively lower levels of expression in the liver and heart. Conversely, the PARLB gene shows low transcription rates maternally, although transcripts appear to be more abundant from 4 hpf to at least 7 dpf, as well as in the brain of the adult zebrafish organism. The lack of PARLA and PARLB significantly impacts the patterning of dopaminergic neurons and embryonic survival.

To investigate this occurrence, translation blocking MOs were employed, specifically the PARLA MO and the PARLB MO, along with splice-blocking MOs, namely the PARLA e1i1 MO and the PARLB i3e4 MO. These MOs were administered separately or in various combinations into one-cell embryos. The resulting phenotypes were subsequently compared to those of embryos that were injected with mismatched MOs, a standard control MO, or were left un-injected. Furthermore, PARL DA neuron loss had an impact on locomotor behavior and impaired olfaction. The Ensembl database was utilized for the annotation of the zebrafish genome (GRCZ11) in order to obtain information on gene coding and transcript data, as well as for the design of sgRNA, which was accomplished via the use of the CHOPCHOP software. Gene-specific oligonucleotides containing the T7 promoter sequence, a 20 base pair (bp) target site followed by the PAM sequence, and an 80 bp constant oligonucleotide complementary region were annealed in order to generate the templates for sgRNA transcription. Three sgRNAs were synthesized and subsequently injected simultaneously in order to achieve multi-exon deletion. It was discovered that the deletion of 1.4 kilobases (kb) occurred in the region encompassing exon 1 to exon 2 (referred to as EX1_2del according to the nomenclature for the description of sequence variations in the genomic DNA sequence of PARLA gene by Merhi et al., 2021) when two out of the three simultaneously injected sgRNAs were utilized. PARL functions upstream of PINK1 in terms of mitochondrial health. The research conducted by Shamchuk et al. (2017) demonstrates that the loss-of-function of the PARL genes in zebrafish leads to perturbation of dopaminergic neurons and increased mortality, thereby emphasizing its influence in the development of Parkinson's disease. The presence of the PARL mRNA was absent maternally [at 2 h post-fertilization (hpf)], but was detectable during embryonic development from 4 hpf onwards. The absence of PARLA mRNA was conspicuously observed.

14.5.4 LRKK2

The reduction of LRRK2 in zebrafish not only results in neuronal degeneration but also induces developmental perturbations encompassing curvature abnormalities along the axis, ocular irregularities, as well as edema in the eye and optic vesicle. Moreover, we provide evidence that improper expression of LRRK2 can instigate the upregulation and accumulation of β-synuclein. Despite being the most prevalent genetic cause of PD, the role of mutations in the leucine-rich repeat kinase 2 (LRRK2) gene remains largely unexplored.

LRRK2 displays pleiotropic properties and has been implicated in both neurodegeneration and peripheral processes, such as kidney functions in rodents. Although conflicting evidence has been presented in recent zebrafish studies regarding the effects of removing the LRRK2 WD40 domain on dopaminergic neurons and locomotion, developmental abnormalities have been documented. The utilization of Morpholino to inhibit the zebrafish LRRK2 protein led to embryonic lethality and significant developmental anomalies, including growth retardation and neuronal loss. The result of this analysis led to the identification of a transcript measuring 9168 bp, which contains both start and stop codons. The size of this transcript is consistent with that of the zlrrk2 mRNA detected through Northern analysis. The zLRRK2 protein possesses all the functional domains required by the hLRRK2 protein. Following a brief period of reduction, the expression of zlrrk2 increased once again during the hatching and larval stages, extending up to at least 10 dpf. At both 24 hpf and 6 dpf, a robust expression of zLRRK2 was observed in the brain through whole mount in situ hybridization (WISH) analysis.

Furthermore, the expression of zLRRK2 in the brain is widespread. In mature fish that surpass three months in age, the expression of zLRRK2 persists. The presence of zLRRK2 mRNA was identified in various tissues including the brain, muscle, ovary, and gut through the implementation of both Northern blot and qRT-PCR analyses. However, Western blot analysis disclosed that the complete zLRRK2 protein was primarily detected in the brain. Upon the microinjection of morpholinos targeting the ATG start site into embryos, the expression of zLRRK2 protein was effectively eradicated as determined by Western blot analysis.

The reduction of zLRRK2 expression led to significant embryonic mortality, with approximately 90% of the 64 examined embryos perishing within 3 dpf. The surviving embryos experienced developmental delay, including slow growth, reduced brain size, and heart edema, when compared to their wild-type counterparts. To remove the WD40 domain, morpholinos were designed to specifically disrupt the splicing of the 45th exon of zLRRK2, thereby introducing a premature stop codon just before the aforementioned domain. The introduction of the splicing-blocking morpholinos into the embryos resulted in the generation of truncated zLRRK2 mRNA that lacked the WD40 domain (zLRRK2-ΔWD40).

The outcome of this analysis yielded a value of 2. This conclusion was corroborated by the assessment of RT-PCR and sequencing findings. As anticipated, zLRRK2-ΔWD40 morphants predominantly exhibited normal embryonic development, at least until 7 dpf, with no discernible morphological abnormalities, aside from a mild accumulation of blood between the swim bladder and yolk sac. Additionally, a Western blot analysis of the entire zLRRK2-ΔWD40 morphant fish lysate (3 dpf) was conducted (Sheng et al., 2010). Investigations were conducted employing LRRK2 translational and splice MOs to comprehend the impact of LRRK2 knockdown on vertebrate development, neurodegeneration, and the function of LRRK2. The findings showcased a diverse range of phenotypes resulting from LRRK2 knockdown, including neuron loss, defects in axis curvature, microphthalmia, coloboma, and edema in various regions of the embryo. All of these phenotypes were rectified through the simultaneous injection of human LRRK2 mRNA. Subsequent studies were undertaken to explore the impact of a decrease in total LRRK2 levels on the zebrafish nervous system. It was observed that LRRK2 morphants exhibited a significant decrease in GFP expression within the brain and spinal cord, indicative of neuron loss in the tail region.

LRRK2 was utilized in the research that were undertaken. An examination was carried out to determine LRRK2's osmoregulation in the kidney. Similar to this, a test was run to see if the LRRK2 morphant phenotype was connected to the endogenous LRRK2 expression patterns. To do this, HuC-GFP was injected into wt embryos, and then at 48 hpf, whole-mount immunofluorescence analysis was performed on the embryos. LRRK2 has been used in studies. The objective was to determine whether LRRK2 contributes to renal osmoregulation. At 48 hours after fertilization (hpf), observations were done to assess this, and it was discovered that the gene LRRK2 is expressed in a number of organs, including the telencephalon, diencephalon, midbrain, hindbrain, lens, heart, spinal cord, and embryonic kidney. Lens edema is considerably more common in LRRK2 morphants when LRRK2 is present in the lens. In addition, it has been noted that zebrafish LRRK2 suppression affects kidney development. The edema seen in LRRK2 morphants is the result of fluid retention brought on by the kidney's poor osmoregulation (Prabhudesai et al., 2016).

14.5.5 DJ1

To assess the loss-of-function mutations in the human PARK7 gene, which encodes DJ-1, as a rare cause of autosomal-recessive Parkinson's disease (ARPD). The zebrafish counterpart of DJ-1, known as zDJ-1, shares 83% sequence identity with its human counterpart and is expressed in the adult zebrafish brain, muscle, and gut. Additionally, zDJ-1 is also expressed in the embryo from 24 hours after fertilization onwards (Bai et al., 2006). By utilizing CRISPR-Cas9 technology, a zebrafish line with a deficiency in DJ-1 was generated by introducing a 5-base pair deletion into exon 1 of the zebrafish park7 gene.

Our investigation has revealed the occurrence of an early protein dysregulation pattern in the DJ-1 null brain, prior to the loss of tyrosine hydroxylase or phenotypic change. This dysregulation pattern was identified through the injection of two out of the three sgRNA simultaneously. A variety of mosaic mutations, including deletions ranging from 2 to 7 base pairs or silent point mutations, were detected at the F0 generation through targeted mutation. Successful mutagenesis was determined by the loss of the BsrBI restriction site, which coincided with the PAM site. At the F1 generation, a stable deletion of 5 base pairs was observed at residue 27, resulting in a frameshift and the premature termination of translation after 11 non-sense amino acids.

Prior to the loss of tyrosine hydroxylase or any phenotypic changes, our study has successfully detected early protein dysregulation. This particular line is now open for expansion. In the F2 generation, the identification of individual fish genotypes from caudal fin clips was achieved by utilizing DNA fingerprinting, specifically by exploiting the loss of the BsrBI site. The validity of the knockout was confirmed at the transcription level by generating a cDNA from the mRNA of larvae and amplifying the park7 gene. In comparison to the wild type, a faint but still distinguishable band for the park7−/− was observed, which may be attributed to the nonsense-mediated decay of mRNA.

It has been determined that two out of the three simultaneously injected sgRNAs were absent prior to the loss of tyrosine hydroxylase or any phenotypic alteration. Our investigation has successfully identified the occurrence of early protein dysregulation. Additionally, the sequencing of the cDNA has confirmed the presence of a premature stop codon at the transcriptional level and the lack of alternative splicing. Lastly, the utilization of a DJ-1 antibody for Western blotting of larval lysates has effectively demonstrated the complete absence of park7 protein expression in the knockout.

14.6 CHEMICAL-BASED MODELS

The investigation of Parkinson's disease in zebrafish necessitates the utilization of diverse chemical-centric models to instigate and scrutinize the disease's pathology. The zebrafish, owing to its genetic affinity to human beings and the simplistic nature of genetic manipulation, has emerged as a prominent model organism in the field of neurobiology research. Presented herein are a few prevalent chemical-based models for PD in zebrafish.

14.6.1 6-OHDA

6-hydroxydopamine (6-OHDA) is frequently employed as a neurotoxin in animal models to elicit PD-like symptoms. The selective impairment caused by 6-OHDA targets dopaminergic neurons, which constitute the primary neuronal population affected in PD. Given the similarity of zebrafish dopaminergic systems to those of humans, these organisms serve as a valuable model for investigating the degeneration of these neurons. In evaluating the behavior of adult zebrafish with 6-OHDA-induced lesions, notable decrements in both speed and distance traversed were observed (Vijayanathan et al., 2017; Hamza et al., 2021).

6-hydroxydopamine (6-OHDA) is a frequently utilized neurotoxin in animal models for the purpose of inducing symptoms reminiscent of PD. The specific target of 6-OHDA is dopaminergic neurons, which are the primary neurons affected in PD. Zebrafish, possessing a dopaminergic system similar to that of humans, serve as a valuable model for studying the role of particular peptides in neurodegenerative diseases. In a case study, 6-OHDA was administered to the zebrafish brain to cause parkinsonian symptoms, and a subsequent examination showed that nine out of the 118 peptides analyzed showed notable changes. In more detail, eight peptides showed a drop in expression whereas just one peptide showed an increase. Notably, the cytosolic and extracellular proteins involved in lipid metabolism and the dynamic cytoskeleton showed the most dramatic alterations (Fiametti et al., 2021). 6-OHDA is a neurotoxin that is normally administered by direct injection into the targeted brain region because it is unable to cross the blood-brain barrier.

When 6-OHDA is given to zebrafish, it causes dopamine (DA) neurons to die, which lowers the concentration of dopamine in the afflicted area. The study used 6-OHDA to create a PD model in adult zebrafish, examining the regenerative potential of dopaminergic neurons. Different doses of 6-OHDA were administered to damage DpN, resulting in over 85% elimination in various brain regions. All subjects survived treatment, and locomotor behavior showed a significant decrease in speed and distance travelled. It is noteworthy that the study discovered that the Dn DpN of the lesioned zebrafish fully recovered after 30 days of the lesion. Consequently, the research has effectively established a stable PD zebrafish model using a straightforward and reproducible approach through 6-OHDA (Vijayanathan et al., 2017).

14.6.2 MPTP

The chemical compound known as 1-Methyl-4-phenyl-1,2,3,6-tetrahydropyridine (MPTP) has been utilized to induce the degeneration of dopaminergic neurons and is commonly employed as a model for PD in laboratory animals, specifically zebrafish (Bashirzade et al., 2022). The administration of MPTP to adult zebrafish has demonstrated the occurrence of hypolocomotion and motor incoordination (Razali et al., 2022). Additionally, it has been observed that the application of high doses of MPTP results in the manifestation of distinctive computational patterns in the swimming behavior of zebrafish.

The collective findings of this study imply that the administration of MPTP in adult zebrafish serves as a late-stage experimental model for PD. This model exhibits distinct cognitive characteristics (Bashirzade et al., 2022). It is significant to remember that MPTP can have negative effects on both adult zebrafish and zebrafish embryos. Previous studies have demonstrated that MPTP decreases the number of dopaminergic cells in the zebrafish embryonic diencephalon by harming dopaminergic neurons. As a result, there is a reduction in locomotor activity, AChE levels, antioxidant enzyme activities, mortality and hatching rates, and the expression of bdnf, dj1, lrrk, and pink1 (Cansiz et al., 2022). The telencephalon and olfactory bulbs (OB) of Tg (dat:eGFP) fish have significantly fewer dopaminergic neurons, according to CVMIs of MPTP in adult transgenic zebrafish. MPTP can cause mitochondrial fragmentation in dopaminergic neurons, as shown by the analysis of mCherry and mitochondrial gene expression in zebrafish. Furthermore, it is known that the pink1/parkin pathway participates in mitophagy. It has been shown that this decline in dopaminergic neurons causes momentary changes in the sensory and motor behavioral traits of zebrafish. The Tg (dat:tom20 MLS:mCherry) zebrafish strain has also been used in this investigation to show a notable increase in the fraction of fragmented mitochondria in zebrafish that were given MPTP injections.

The findings are reinforced by the analysis of gene expression, which has suggested that the process accountable for the demise of dopaminergic cells is associated with mitophagy. This, in turn, results in heightened mitochondrial fission and a more fragmented structure of the mitochondria, thereby contributing to a more profound understanding of the role of chemically induced mitochondrial dysfunction as a contributing element in the degeneration of dopaminergic neurons and other disorders associated with Parkinson's disease (Kalyn and Ekker, 2021).

14.6.3 Rotenone

Rotenone, a neurotoxic ketone derived naturally from lancepod, is used as an herbicide, insecticide, and piscicide and is notable for its ability to elicit PD-like symptoms that closely resemble clinical manifestations. This is primarily achieved through direct penetration of the CNS, facilitated by its ability to readily traverse the BBB and accumulate within mitochondria (Ilie et al., 2022). As an animal model of PD, the pesticide rotenone is extensively used to produce PD-like symptoms in zebrafish. When exposed to rotenone, zebrafish exhibited motor impairment, anxiety-like behavior, and olfactory dysfunction. The presence of reduced dopamine levels in the rotenone group may account for these PD-like symptoms (Wang et al., 2017). A 30% decline in dopamine release, as well as compromised dopamine uptake, was observed in rotenone-treated zebrafish, resulting in behavioral deficits such as impaired motor skills, cognitive function, and reduced locomotor activity that mirror PD symptoms (Hettiarachchi et al., 2022).

14.6.4 Paraquat

Paraquat, a polar herbicide, has been found to induce oxidative damage in dopaminergic neurons, leading to behavioral defects in vivo and consequent onset of PD. Prolonged exposure to paraquat is known to increase the risk of PD by disrupting dopaminergic systems in humans. The toxicity of paraquat on zebrafish has been observed to result in behavioral impairments, abnormal neural activities, and reduced motor response in zebrafish larvae (Kim et al., 2022). Treatment of 18 hpf embryonic zebrafish with low-dose PQ treatment (0.04 ppm, lower than the accepted human daily exposure) resulted in the display of neurodegenerative phenotypes and motor deficits at various developmental stages. A decrease in the GSH levels, increase in lipid peroxidation, and apoptosis, respectively, were observed at various developmental stages, indicating that paraquat causes PD-like motor defects in later life by perturbing the cholinergic system due to oxidative stress (Nellore and Nandita, 2015). In vivo PD studies using the zebrafish model indicate that paraquat impairs mitochondrial function and dopaminergic signaling (Wang et al., 2018). The utilization of environmental neurotoxins in the study of PD in zebrafish has proven to be a valuable resource for elucidating the pathophysiological mechanism of this disease. However, despite its efficacy and benefits, there exist several limitations that must be acknowledged with regards to the utilization of the neurotoxin-induced zebrafish model.

14.7 CONCLUSION

The epoch of translational medicine has made a noteworthy contribution to scientific progress, as evidenced by the emergence of *Danio rerio* as an invaluable platform, particularly in the realm of neurodevelopmental research. Understanding the causes of neurodegenerative diseases including Parkinson's disease, Alzheimer's disease, and Huntington's disease has been made possible thanks in large part to zebrafish. It is more susceptible to molecular studies than primates and rodents because of the transparency of its embryos, the simplicity of its genetic modification, and its short life cycle. Zebrafish are currently widely used in biology, neuroscience, pharmacology, and toxicology research because they are an excellent model for examining neurobehavioral aspects that are pertinent to humans.

This chapter underlines the value of zebrafish as a practical model for evaluating novel medications developed to treat different neurological disorders, with a particular focus on PD. The chapter looked at current developments in zebrafish models' use for studying PD pathophysiology. In these models, the appropriate zebrafish genes' activity is decreased, or pathogenic genes linked to human neurodegenerative diseases are introduced. In particular types of neurons, these alterations result in noticeable morphological, physiological, and biochemical abnormalities that are also seen in other animal and human models. This emphasizes the functional similarity between the genes linked to zebrafish and human neurodegenerative disorders.

The chapter emphasizes the zebrafish's potential as an alternate model for examining the molecular causes of PD because of its distinct features. Although it is known that no animal model will be able to accurately replicate complicated human disorders like PD, using a variety of animal models can assist in getting around these limits. Zebrafish, a relatively recent model organism, offers a variety of models that look at many PD-related topics. Research focused on genes has revealed that zebrafish possess homologous genes to human PD genes that generally have similar functions. However, gene-based models that rely on morpholinos (MOs) occasionally yield conflicting results, raising concerns about their reliability. It is crucial to adhere to more stringent guidelines and considerations regarding off-target effects when utilizing MOs. On the other hand, emerging genomic editing approaches such as CRISPR hold the potential for more consistent gene knockouts/ins. Researchers must also consider gene duplication within the zebrafish genome, which may lead to promising research directions.

While acknowledging that the replication of complex human diseases like PD in animal models is far from perfect, the use of multiple animal models can help alleviate their inherent limitations. In this regard, the zebrafish, a relatively new model organism, offers a range of models that explore various aspects of PD. Although chemical methods can readily induce disease-like states in zebrafish, they may not fully replicate all aspects of the disease.

Nonetheless, several well-established chemically induced PD models provide stable foundations for study, while newer methods show potential but require further investigation. Interestingly, zebrafish models typically exhibit similar effects as mammalian models when exposed to the same chemicals, thus confirming the validity of zebrafish for this application. With the advancements in genomic editing technology as well as the utilization of advanced imaging techniques, the zebrafish is swiftly emerging as a highly favoured model organism. Possessing a diverse array of behavioural paradigms and the ability to conduct high-throughput screening, the zebrafish, as a model for diseases, is positioned to yield novel breakthroughs in the realm of neurodegenerative diseases and other related fields. While it is acknowledged that animal models inherently possess certain limitations in replicating intricate human diseases such as Parkinson's disease, the utilization of multiple animal models can effectively address and alleviate these limitations. In this aspect, the zebrafish offers a range of models that delve into various facets of Parkinson's disease.

14.8 ACKNOWLEDGEMENT

The author, Rajesh Pamanji, is thankful to the Vice Chancellor and Registrar, Pondicherry University, Pondicherry, India, for providing the facilities for ongoing research work and to UGC, Government of India, for providing financial assistance in the form of DSK postdoctoral fellowship BL/19-20/0302.

REFERENCES

Bai, Q., Mullett, S. J., Garver, J. A., Hinkle, D. A., & Burton, E. A. (2006). Zebrafish DJ-1 is evolutionarily conserved and expressed in dopaminergic neurons. Brain Research, 1113(1), 33–44. https://doi.org/10.1016/j.brainres.2006.07.057

Barnhill, L. M., Murata, H., & Bronstein, J. M. (2020). Studying the pathophysiology of Parkinson's disease using Zebrafish. Biomedicines, 8(7), 197. https://doi.org/10.3390/biomedicines8070197

Bashirzade, A. A. O., Cheresiz, S. V., Belova, A. S., Drobkov, A. V., Korotaeva, A. D., Azizi-Arani, S., Azimirad, A., Odle, E., Gild, E.-Y. V., Ardashov, O. V., Volcho, K. P., Bozhko, D. V., Myrov, V. O., Kolchanova, S. M., Polovian, A. I., Galumov, G. K., Salakhutdinov, N. F., Amstislavskaya, T. G., & Kalueff, A. V. (2022). MPTP-Treated Zebrafish recapitulate "late-stage" Parkinson's-like cognitive decline. Toxics, 10(2), 69. https://doi.org/10.3390/toxics10020069

Basnet, R. M., Zizioli, D., Taweedet, S., Finazzi, D., & Memo, M. (2019). Zebrafish larvae as a behavioral model in neuropharmacology. Biomedicines, 7, 23. https://doi.org/10.3390/biomedicines7010023

Bill, B. R., Petzold, A. M., Clark, K. J., Schimmenti, L. A., & Ekker, S. C. (2009). A primer for morpholino use in zebrafish. Zebrafish, 6(1), 69–77. https://doi.org/10.1089/zeb.2008.0555

Cansız, D., Ustundag, U. V., Unal, I., Alturfan, A. A., & Emekli-Alturfan, E. (2022). Morphine attenuates neurotoxic effects of MPTP in zebrafish embryos by regulating oxidant/antioxidant balance and acetylcholinesterase activity. Drug and Chemical Toxicology, 45(6), 2439–2447. https://doi.org/10.1080/01480545.2021.1957558

Cassar, S., Adatto, I., Freeman, J. L., Gamse, J. T., Iturria, I., Lawrence, C., Muriana, A., Peterson, R. T., Van Cruchten, S., & Zon, L. I. (2020). Use of Zebrafish in drug discovery toxicology. Chemical Research in Toxicology, 33(1), 95–118. https://doi.org/10.1021/acs.chemrestox.9b00335

Chia, K., Klingseisen, A., Sieger, D., & Priller, J. (2022). Zebrafish as a model organism for neurodegenerative disease. Frontiers in Molecular Neuroscience, 15, 940484. https://doi.org/10.3389/fnmol.2022.940484

Costa, K. C. M., Brigante, T. A. V., Fernandes, G. G., Scomparin, D. S., Scarante, F. F., de Oliveira, D. P., & Campos, A. C. (2021). Zebrafish as a translational model: An experimental alternative to study the mechanisms involved in anosmia and possible neurodegenerative aspects of COVID-19? eNeuro, 8(3). https://doi.org/10.1523/ENEURO.0027-21.2021

Deas, E., Wood, N. W., & Plun-Favreau, H. (2011). Mitophagy and Parkinson's disease: The PINK1-parkin link. Biochimica et Biophysica Acta, 1813(4), 623–633. https://doi.org/10.1016/j.bbamcr.2010.08.007

Du, Y., Guo, Q., Shan, M., Wu, Y., Huang, S., Zhao, H., Hong, H., Yang, M., Yang, X., Ren, L., Peng, J., Sun, J., Zhou, H., Li, S., & Su, B. (2016). Spatial and temporal distribution of dopaminergic neurons during development in Zebrafish. Frontiers in Neuroanatomy, 10, 115. https://doi.org/10.3389/fnana.2016.00115

Ekker, S. C. (2008). Zinc finger-based knockout punches for zebrafish genes. Zebrafish, 5(2), 121–123. https://doi.org/10.1089/zeb.2008.9988

Fiametti, L. O., Correa, C. N., & de Castro, L. M. (2021). Peptide profile of Zebrafish brain in a 6-OHDA-induced Parkinson model. Zebrafish, 18(1), 55–65. https://doi.org/10.1089/zeb.2020.1945

Flinn, L., Mortiboys, H., Volkmann, K., Köster, R. W., Ingham, P. W., & Bandmann, O. (2009). Complex I deficiency and dopaminergic neuronal cell loss in Parkin-deficient zebrafish (*Danio rerio*). Brain: A Journal of Neurology, 132(Pt 6), 1613–1623. https://doi.org/10.1093/brain/awp108

Hettiarachchi, P., Niyangoda, S. S., Jarosova, R., & Johnson, M. A. (2022). Dopamine release impairments accompany locomotor and cognitive deficiencies in rotenone-treated Parkinson's disease model Zebrafish. Chemical Research in Toxicology, 35(11), 1974–1982. https://doi.org/10.1021/acs.chemrestox.2c00150

Howe, K., Clark, M. D., Torroja, C. F., Torrance, J., Berthelot, C., Muffato, M., Collins, J. E., Humphray, S., McLaren, K., Matthews, L., McLaren, S., Sealy, I., Caccamo, M., Churcher, C., Scott, C., Barrett, J. C., Koch, R., Rauch, G.-J., White, S., . . . Stemple, D. L. (2013). The zebrafish reference genome sequence and its relationship to the human genome. Nature, 496(7446), 498–503. https://doi.org/10.1038/nature12111

Ilie, O.-D., Duta, R., Jijie, R., Nita, I.-B., Nicoara, M., Faggio, C., Dobrin, R., Mavroudis, I., Ciobica, A., & Doroftei, B. (2022). Assessing anti-social and aggressive behavior in a Zebrafish (*Danio rerio*) model of Parkinson's disease chronically exposed to rotenone. Brain Sciences, 12(7), 898. https://doi.org/10.3390/brainsci12070898

Kalueff, A. V., Stewart, A. M., & Gerlai, R. (2014). Zebrafish as an emerging model for studying complex brain disorders. Trends in Pharmacological Sciences, 35(2), 63–75. https://doi.org/10.1016/j.tips.2013.12.002

Kalyn, M., & Ekker, M. (2021). Cerebroventricular microinjections of MPTP on adult Zebrafish induces dopaminergic neuronal death, mitochondrial fragmentation, and sensorimotor impairments. Frontiers in Neuroscience, 15, 718244. https://doi.org/10.3389/fnins.2021.718244

Kim, S. S., Hwang, K.-S., Kan, H., Yang, J. Y., Son, Y., Shin, D.-S., Lee, B. H., Chae, C. H., & Bae, M. A. (2022). Neurotoxicological profiling of paraquat in Zebrafish model. Neurochemical Research, 47(8), 2294–2306. https://doi.org/10.1007/s11064-022-03615-y

Li, Y., Jia, Z., Zhang, S., & He, X. (2021). Progress in gene-editing technology of Zebrafish. Biomolecules, 11(9), 1300. https://doi.org/10.3390/biom11091300

Lopez, A., Gorb, A., Palha, N., Fleming, A., & Rubinsztein, D. C. (2022). A new Zebrafish model to measure neuronal α-synuclein clearance in vivo. Genes, 13(5), 868. https://doi.org/10.3390/genes13050868

Luo, N., Li, H., Xiang, B., Qiao, L., He, J., Ji, Y., Liu, Y., Li, S., Lu, R., Li, Y., Meng, W., Wu, Y., Xu, H., & Mo, X. (2016). Syndecan-4 modulates the proliferation of neural cells and the formation of CaP axons during zebrafish embryonic neurogenesis. Scientific Reports, 6, 25300. https://doi.org/10.1038/srep25300

Lysyk, L., Brassard, R., Touret, N., & Lemieux, M. J. (2020). PARL protease: A glimpse at intramembrane proteolysis in the inner mitochondrial membrane. Journal of Molecular Biology, 432(18), 5052–5062. doi: 10.1016/j.jmb.2020.04.006. Epub 2020 Apr 19. PMID: 32320686.

Madsen, D. A., Schmidt, S. I., Blaabjerg, M., & Meyer, M. (2021). Interaction between Parkin and α-synuclein in PARK2-mediated Parkinson's disease. Cells, 10(2), 283. https://doi.org/10.3390/cells10020283

Md Hamzah, N., Lim, S. M., Vijayanathan, Y., Lim, F. T., Abdul Majeed, A. B., Tan, M. P., & Ramasamy, K. (2021). Locomotor assessment of 6-hydroxydopamine-induced adult Zebrafish-based Parkinson's disease model. Journal of Visualized Experiments, 178. https://doi.org/10.3791/63355

Medishetti, R., Balamurugan, K., Yadavalli, K., Rani, R., Sevilimedu, A., Challa, A. K., Parsa, K., & Chatti, K. (2022). CRISPR-Cas9-induced gene knockout in zebrafish. STAR Protocols, 3(4), 101779. https://doi.org/10.1016/j.xpro.2022.101779

Merhi, R., Kalyn, M., Zhu-Pawlowsky, A., & Ekker, M. (2021). Loss of parla function results in inactivity, olfactory impairment, and dopamine neuron loss in Zebrafish. Biomedicines, 9(2), 205. https://doi.org/10.3390/biomedicines9020205

Moens, C. B., Donn, T. M., Wolf-Saxon, E. R., & Ma, T. P. (2008). Reverse genetics in zebrafish by TILLING. Briefings in Functional Genomics & Proteomics, 7(6), 454–459. https://doi.org/10.1093/bfgp/eln046

Nellore, J., & Nandita, P. (2015). Paraquat exposure induces behavioral deficits in larval zebrafish during the window of dopamine neurogenesis. Toxicology Reports, 2, 950–956. https://doi.org/10.1016/j.toxrep.2015.06.007

Noble, S., Ismail, A., Godoy, R., Xi, Y., & Ekker, M. (2012). Zebrafish Parla- and Parlb-deficiency affects dopaminergic neuron patterning and embryonic survival. Journal of Neurochemistry, 122(1), 196–207. https://doi.org/10.1111/j.1471-4159.2012.07758.x

Oliveira, R. F. (2013). Mind the fish: Zebrafish as a model in cognitive social neuroscience. Frontiers in Neural Circuits, 7. https://doi.org/10.3389/fncir.2013.00131

Prabhudesai, S., Bensabeur, F. Z., Abdullah, R., Basak, I., Baez, S., Alves, G., Holtzman, N. G., Larsen, J. P., & Møller, S. G. (2016). LRRK2 knockdown in zebrafish causes developmental defects, neuronal loss, and synuclein aggregation. Journal of Neuroscience Research, 94(8), 717–735. https://doi.org/10.1002/jnr.23754

Randlett, O., Wee, C. L., Naumann, E. A., Nnaemeka, O., Schoppik, D., Fitzgerald, J. E., Portugues, R., Lacoste, A. M. B., Riegler, C., Engert, F., & Schier, A. F. (2015). Whole-brain activity mapping onto a zebrafish brain atlas. Nature Methods, 12(11), 1039–1046. https://doi.org/10.1038/nmeth.3581

Razali, K., Mohd Nasir, M. H., Othman, N., Doolaanea, A. A., Kumar, J., Nabeel Ibrahim, W., & Mohamed, W. M. Y. (2022). Characterization of neurobehavioral pattern in a zebrafish 1-methyl-4-phenyl-1,2,3,6-tetrahydropyridine (MPTP)-induced model: A 96-hour behavioral study. PloS One, 17(10), e0274844. https://doi.org/10.1371/journal.pone.0274844

Razali, K., Othman, N., Mohd Nasir, M. H., Doolaanea, A. A., Kumar, J., Ibrahim, W. N., Mohamed Ibrahim, N., & Mohamed, W. M. Y. (2021). The promise of the zebrafish model for Parkinson's disease: Today's science and tomorrow's treatment. Frontiers in Genetics, 12, 655550. doi: 10.3389/fgene.2021.655550. PMID: 33936174; PMCID: PMC8082503.

Rocha, M., Singh, N., Ahsan, K., Beiriger, A., & Prince, V. E. (2020). Neural crest development: Insights from the zebrafish. Developmental Dynamics: An Official Publication of the American Association of Anatomists, 249(1), 88–111. https://doi.org/10.1002/dvdy.122

Selvaraj, S., & Piramanayagam, S. (2019). Impact of gene mutation in the development of Parkinson's disease. Genes & Diseases, 6(2), 120–128. https://doi.org/10.1016/j.gendis.2019.01.004

Shamchuk, A. L., Allison, W. T., & Tierney, K. B. (2017). The Importance of olfactory and motor endpoints for Zebrafish models of neurodegenerative disease. In Animal Models for the Study of Human Disease (pp. 525–554). Elsevier. https://doi.org/10.1016/B978-0-12-809468-6.00021-8

Sheng, D., Qu, D., Kwok, K. H. H., Ng, S. S., Lim, A. Y. M., Aw, S. S., Lee, C. W. H., Sung, W. K., Tan, E. K., Lufkin, T., Jesuthasan, S., Sinnakaruppan, M., & Liu, J. (2010). Deletion of the WD40 domain of LRRK2 in Zebrafish causes Parkinsonism-like loss of neurons and locomotive defect. PLoS Genetics, 6(4), e1000914. https://doi.org/10.1371/journal.pgen.1000914

Shenoy, A., Banerjee, M., Upadhya, A., Bagwe-Parab, S., & Kaur, G. (2022). The brilliance of the zebrafish model: Perception on behavior and Alzheimer's disease. Frontiers in Behavioral Neuroscience, 16, Article 861155. https://doi.org/10.3389/fnbeh.2022.861155

Streisinger, G., Walker, C., Dower, N., Knauber, D., & Singer, F. (1981). Production of clones of homozygous diploid Zebrafish (*Brachydanio rerio*). Nature, 291(5813), 293–296. https://doi.org/10.1038/291293a0

Sun, Z., & Gitler, A. D. (2008). Discovery and characterization of three novel synuclein genes in zebrafish. Developmental Dynamics: An Official Publication of the American Association of Anatomists, 237(9), 2490–2495. https://doi.org/10.1002/dvdy.21569

Tay, T. L., Ronneberger, O., Ryu, S., Nitschke, R., & Driever, W. (2011). Comprehensive catecholaminergic projectome analysis reveals single-neuron integration of zebrafish ascending and descending dopaminergic systems. Nature Communications, 2, 1171. doi: 10.1038/ncomms1171.

Trede, N. S., Langenau, D. M., Traver, D., Look, A. T., & Zon, L. I. (2004). The use of zebrafish to understand immunity. Immunity, 20(4), 367–379. https://doi.org/10.1016/s1074-7613(04)00084-6

Vijayanathan, Y., Lim, F. T., Lim, S. M., Long, C. M., Tan, M. P., Majeed, A. B. A., & Ramasamy, K. (2017). 6-OHDA-lesioned adult Zebrafish as a useful Parkinson's disease model for dopaminergic neuroregeneration. Neurotoxicity Research, 32(3), 496–508. https://doi.org/10.1007/s12640-017-9778-x

Wang X. H., Souders C. L., Zhao Y. H., & Martyniuk C. J. (2018). Paraquat affects mitochondrial bioenergetics, dopamine system expression, and locomotor activity in zebrafish (Danio rerio). Chemosphere 191, 106–117. 10.1016/j.chemosphere.2017.10.032

Wang, Y., Liu, W., Yang, J., Wang, F., Sima, Y., Zhong, Z.-M., Wang, H., Hu, L.-F., & Liu, C.-F. (2017). Parkinson's disease-like motor and non-motor symptoms in rotenone-treated zebrafish. Neurotoxicology, 58, 103–109. https://doi.org/10.1016/j.neuro.2016.11.006

Weston, L. J., Cook, Z. T., Stackhouse, T. L., Sal, M. K., Schultz, B. I., Tobias, Z. J. C., Osterberg, V. R., Brockway, N. L., Pizano, S., Glover, G., Weissman, T. A., & Unni, V. K. (2021). In vivo aggregation of presynaptic alpha-synuclein is not influenced by its phosphorylation at serine-129. Neurobiology of Disease, 152, 105291. https://doi.org/10.1016/j.nbd.2021.105291

Woods, I. G., & Schier, A. F. (2008). Targeted mutagenesis in zebrafish. Nature Biotechnology, 26(6), 650–651. https://doi.org/10.1038/nbt0608-650

Xi, Y., Noble, S., & Ekker, M. (2011). Modeling neurodegeneration in zebrafish. Current Neurology and Neuroscience Reports, 11(3), 274–282. https://doi.org/10.1007/s11910-011-0182-2

15 Zebrafish as a Model for a Novel Neuroprotective Natural Product

Hagar F. Forsan

Abbreviations

PD	Parkinson's disease
ROS	reactive oxygen species
SOD	superoxide dismutase
CAT	catalase
PSB	pinostrobin
TCM	traditional Chinese medicine
MPTP	methyl-4-phenyl-1,2,3,6-tetra hydropyridine
LCH	ligusticum chuanxiong hort
HPF	hours post fertilization
EGb	extract of ginkgo biloba
AChE	acetylcholinesterase
CPO	chlorpyrifos oxon
ADHD	attention deficit hyperactivity disorder
PAF	platelet activating factor
MCAO	middle cerebral artery occlusion
BBB	blood brain barrier
DN	dominant negative

15.1 INTRODUCTION

The species of zebrafish that is common in freshwater biomes is *Danio rerio*. Due to their distinctive traits, zebrafish are popular for scientific studies and in aquariums (Caro et al., 2016; Lawrence, 2007).

Although zebrafish has been used in studies since the 1950s (O'Connor, 2006; Vascotto et al., 1997), professor George Streisinger first looked into its potential as a genetics model in the early 1980s (Streisinger et al., 1981).

Zebrafish's size is tiny, and adults rarely grow longer than 3 to 4 cm, while larvae and embryos can fit in 96-well plates. It is housed in large numbers in a compact area according to small bodies. Zebrafish are prolific producers, and reproduction each week can result in hundreds of fertilized eggs. The big size enables the collection of thousands of embryos for the assessment of a range of bioactive substances. It is also transparent during the early development stages, which makes it possible to observe a variety of intricate real-time dynamic processes.

Almost 1,500 mutations were found in more than 400 new genes during two large-scale zebrafish screens of mutants conducted in USA and Germany (Nüsslein-Volhard, 2012). These genes are engaged in a variety of mechanisms that control development and organogenesis. The zebrafish's use in developmental genetics and ultimately many other fields was made possible by this development. It is used for embryogenesis, neuronal networks, regeneration, and disease research (Grunwald & Eisen, 2002; Nüsslein-Volhard, 2012). It is suitable for genetic engineering because their entire genomic code is published (Howe et al., 2013). Management changed as the model system has expanded. Using the latest system provided with filters and alarm systems, zebrafish facilities have undergone a dramatic transition. Zebrafish systems with recirculating aquaculture, originally created to screen for Tubingen mutations in 1993, have established a standard for industrial zebrafish enclosures and are the basis for modern breeding systems. It is well known that model organisms can be used to study embryogenesis, neural networks, regeneration, and disease (Brand et al., 1995)

Even though it has made significant advancements, zebrafish nutrition still trails significantly. In a zebrafish research group, issues related to a lack of nutritional data and standardization have been addressed (Watts et al., 2012). At concentrations from 0.6–4.6 g/kg, the highest minimum was 0.2 g/kg dry feed that four diets of commercial zebrafish were analyzed for iron content (Penglase et al., 2012).

These increased concentrations are potentially hazardous to fish. This long-term objective aims to ascertain the dietary requirements of natural and mutant zebrafish on a regular basis, and create a standard feed formulation with high-quality ingredients. New studies are now focusing on the nutrition of zebrafish.

Zebrafish are an excellent model for examining human health and disease benefits, as they and humans share many genetic, anatomical, and physiological similarities (Howe et al., 2013).

Zebrafish have been generally used as an effective in vivo organism in numerous areas of research related to all of these beneficial characteristics. Natural products like microalgae, omega 3 fatty acids, and probiotics are becoming more and more popular (Barros de Medeiros et al., 2022), as well as phenolic compounds (Gumus & Gharibzahedi, 2021).

Zebrafish have become well-known as an innovative tool for quickly and affordably evaluating the biological effects and safety of natural products, which is crucial in creating functional foods, nutritional pharmaceuticals, and natural cosmetics.

DOI: 10.1201/9781003402893-17

Parkinson's disease (PD) is caused by the progressive degradation of dopaminergic neurons in the pars compacta substantia nigra of the midbrain. Decreased dopamine levels, such as neuroendocrine abnormalities, and impaired secretion of melatonin is a common cause of PD (De Pablo-Fernández et al., 2017). The HPA axis is altered, which raises cortisol levels, and serum TNF-α levels are a clinical symptom of PD patients, suggesting that the NEI system enhances progression of PD.

A recent chapter focused primarily on the genetic side of modeling developmental, metabolic, and mental disorders in zebrafish (Choi et al., 2021). It also summarizes zebrafish aquatic neurotoxicity contaminants (Fitzgerald et al., 2021). It offers an overview of the development models of zebrafish and assessing usage of different bioactivities of natural products including neuroprotection.

15.2 ZEBRAFISH LABORATORY MANAGEMENT

There are essential elements of zebrafish routine care and management, including environmental requirements, feeding, breeding, rearing larvae, and natural product delivery methods in zebrafish models.

15.2.1 Environmental Requirements

Zebrafish are widely utilized in many fields since they are simple to grow and can withstand a variety of environmental conditions (Tsang et al., 2017). A zebrafish laboratory requires provision of an ideal living habitat management, although they perform well in terms of environmental tolerance.

Zebrafish require an environment with a pH of 7–8, a temperature of 28°C, and 10–14 hours of light-to-dark lighting.

15.2.2 Feeding

To address the nutritional needs of zebrafish in the laboratory, a balanced and complete diet is required. Most of the time, they are given commercially prepared diets with 300–400 micron-sized ingredients, live prawns in brine developed in a laboratory, or both feeding twice daily to adult zebrafish (Avdesh et al., 2012). Typically, zebrafish in experimental settings should be fed food equivalent to 35% of their body weight per day (Zhao et al., 2012). Avoid overfeeding because it might negatively affect the quality of water by raising nitrate levels and abruptly changing pH, which would then affect zebrafish viability and reproduction (Tsang et al., 2017).

15.2.3 Breeding

In order to successfully breed zebrafish, a variety of variables must be taken into account, including both biotic and abiotic environment characteristics (Kolb et al., 2018; Tsang et al., 2017). Number of breeders and their sex ratio, as well as the size of the breeding chamber, have an impact on mating behavior, in addition to breeders' size, age, and health, as well as the frequency with which they are used to spawning.

Diet, light cycle, temperature, and other variables can have an impact on spawning and egg production. Males can be distinguished from females by their thin bodies and vibrant colors, while females have a rounded body shape in adult zebrafish and often have a bigger underbelly (Avdesh et al., 2012).

Breeding arrangements with an additional male, for example, two males and one female or three males and two females, were once thought to increase efficiency of fertilization. There should be less use of males because aggressive male behaviors are counterproductive (Tsang et al., 2017).

Zebrafish can remain reproductive for most of their life cycle and can reach sexual maturity by two months of age (Dabrowski & Miller, 2018; Lawrence et al., 2012); however, if they are too young or too old, they are rarely used in laboratories as breeders (Tsang et al., 2017). Additionally, the frequency of spawning is a crucial factor in zebrafish breeding, with once a week being considered ideal (Kurtzman et al., 2010). Too little spawning can impair reproductive activity and cause egg reabsorption, whereas too much spawning can reduce egg quality and quantity (Tsang et al., 2017).

Zebrafish begin mating at the first light sign following a period of complete darkness, so the photoperiod matters greatly for zebrafish reproduction (Avdesh et al., 2012). After fertilization, the fertilized eggs can be gently extracted using a disposable pipette and placed in a Petri plate or other containers with a water or embryo medium. After being cleansed with deionized water to lower the risk of infection, the unfertilized eggs should be discarded right away.

15.2.4 Larval Rearing

Until the larvae hatch, eggs that are fertilized can be maintained at 28 ± 0.5°C for two to three days; lower temperatures will lead to slower embryo development.

The first two days after hatching, the larval zebrafish exhibit just a little amount of spontaneous activity. The yolk-sac stores are enough to satisfy their nutritional needs throughout this time without further feeding. The yolk quickly runs out at that point, and diets must be introduced. For immature larval zebrafish, live feeds, for example, paramecium and rotifers, or produced commercial diets with a 100-micron size are advised. Finally, 200 microns can progressively replace the food, and then 300–400 microns (Avdesh et al., 2012). Larval zebrafish diet must be offered more frequently than those for adults but with fewer amounts each time. Zebrafish mature in 2–4 months on average, but under particular rearing conditions (such as algal turbidity, illumination, water salinity, and food availability), the process can be sped up to 45 days (Hitchcock & Raymond, 2004).

15.2.5 Ways of Delivering Natural Compounds

The most typical method involves adding these substances directly into the water or medium used to develop zebrafish eggs, larvae, or adults (Lin et al., 2022).

15.2.6 Parameters of Environment and Breeding Practices

A unique zebrafish advantage is its high ability to withstand environmental conditions (Lawrence, 2007). These characteristics allow these animals to be housed and raised using a number of scientific procedures. As environmental and traditional practices are straightforward, there are only a few principles for fish managing and caring. Breeding fish that have long been free of visible disease, thrive well, and have sufficiently high survival levels makes them good animal models (Lawrence, 2016).

Adaptability has played a significant role in the zebrafish model's overall growth, but paradoxically, it has had minimum usage as an application in pharmaceutical production. Indeed, pharmaceutical corporations have adopted and routinely utilized various animal models (rats, mice, dogs, and rabbits) for commercially standardized preclinical toxicity research.

There are two issues with experimental variability in the zebrafish model. Reports on housing and environmental issues are inaccurate or lacking. Today, tools such as ARRIVE Guidelines (Kilkenny et al., 2010), Protocols (Teytelman et al., 2016), or Benchmark Toxicity Level (Hsieh et al., 2019) can be used to improve reporting, but standards for nutrition, health surveillance, and genetic maintenance are generally lacking. This results in a bigger and trickier challenge. Standards that are resolved can be created by deepening scientific knowledge of these components to the point where we can prescribe standards with assurance (Varga et al., 2018).

15.3 TOXICITY

In addition to the initial somite formation, neurons also develop obviously at the gas conditioning end. Though, the neuronal onset fate happens first, as in the case of certain specifications (Blader & Strähle, 2000); e.g., as a neuronal marker cyp26 initially means the blastocyst stage on the other hand rapidly reduces due to gastric formation. Bone expression of morphogenetic protein antagonists controls how neurons develop (Blader & Strähle, 2000). The central nervous system is distinguished from the neural plate, which was recognized as the cutaneous epithelium. After migrating solid rod formation, a neural tube will eventually become hollow, and cells will then differentiate to produce the brain and medullary lineages (Lewis & Eisen, 2003). The primary nerve cells distinguish after the stomach (Blader & Strähle, 2000). Innate fast muscle development occurs concurrently with primary motor neurons differentiating in response to Hedgehog signaling from the spinal cord over gastrulation.

Secondary motor neurons are suggested in response to main motor neurons' Delta/Notch signals and connect to the pectoralis major, slow and fast muscles (Lewis & Eisen, 2003). The microtubules of neurons contain a-1-tubulin (a1-T), which is a neuron-specific isoform of tubulin. Expression of a1-T begins from the moment the progenitors commit to the neuronal lineage and continues throughout axon formation. It is adversely regulated in mature neurons.

This protein is a promising candidate for a marker of early neurons for monitoring neuronal development. Tg (a1-Fish was used for T: GFP) is used to detect mutations that impair neuronal growth; however, it is still being used to screen for chemicals that are damaging to neurogenesis. Chemicals like malathion, atrazine, 2,4-dichlorophenoxyacetic acid (2,4-D), and dichlorodiphenyltrichloroethane decrease anterior brain cells, distort caudal axons, and lower the overall number of neuronal cells (Ton et al., 2006). Cadmium changes trajectories and suppresses motor neuron growth. In contrast, sodium benzoate increased projections of axonal.

Some chemicals, for example, triclosan and sodium benzoate, have in the synapse higher signal transducers, while others such as chromium and chlorpyrifos oxon (CPO) reduce or inhibit (Jacobson et al., 2010). Changed synaptic role probably affects neurodevelopment that may not be apparent until later in life.

15.3.1 The Evaluation of Neuroprotection of Zebrafish Microorganism

Zebrafish are utilized as models for a number of neurological conditions, including sensorineural hearing loss, nerve injury, seizure, attention deficit hyperactivity disorder (ADHD), neurodegeneration, anxiety, and depression (Table 15.1).

15.3.1.1 *Ceratonia Siliqua L.*

Carob, also known as *Ceratonia siliqua L.* (*C. siliqua L.*), is a member of the subfamily Caesalpinaceae of the Leguminosae family (Obeidat et al., 2012). In the Mediterranean basin, it is pervasive (Obeidat et al., 2012). Carob is a perennial tree that is a hermaphrodite in some forms (El Hajaji et al., 2010), dioecious (Obeidat et al., 2012), and evergreen. *C. siliqua L.* is a plant that can withstand drought and salinity that grows naturally in desert environments (Vekiari et al., 2011).

Ceratonia siliqua L. is a medicinal plant native to the Mediterranean region, has anti-diarrheal, anti-diabetic, acetylcholine-enhancing, antioxidant, and anti-atherosclerotic properties as well as potential anti-neurodegenerative properties. Gallic acid was the most abundant phenolic component. Furthermore, large levels of rutin were discovered also in the water extract (Abidar et al., 2020).

TABLE 15.1
Evaluation of Neuroprotection of Zebrafish Microorganism

Natural Products	Extract/Compounds	Models	Main Effects and Mechanisms	References
Ceratonia siliqua L.	Aqueous extract from leaves	6-OHDA zebrafish model	Enhancing cognitive function; exhibiting favorable antioxidant and anti-AChE actions	(Abidar et al., 2020)
	Acteoside	6-OHDA-induced neural damage in zebrafish model	Preventing the death of dopaminergic neurons and movement impairments; increasing antioxidant enzymes by turning on the Nrf2/ARE signaling pathway	(C. Li et al., 2018)
Eucommia ulmoides Oliver	Pinostrobin	MPTP-induced PD zebrafish	Alleviating behavioural deficits and reducing dopaminergic neuron loss	(C. Li et al., 2018)

Reactive oxygen species (ROS) particles have a high degree of reactivity that are produced during aerobic life's normal cellular metabolism and have a single or several unpaired electrons in a single atomic orbital (Bhat et al., 2015; Gülçin et al., 2010). Defining oxidative stress as an excessive generation of ROS coupled with a deficient antioxidative defence (Limón-Pacheco & Gonsebatt, 2009). Several studies demonstrated a link among oxidative stress and PD pathogenesis (Kamdem et al., 2013).

The brain is especially susceptible to attacks from free radicals that result in neurodegeneration. It uses more oxygen under normal physiological settings while also having lower levels of endogenous antioxidant molecules (Roberts et al., 2010). This is supported by numerous lines of evidence, including ROS high levels situated in specific brain parts (Chong et al., 2013). PD caused elevated free amounts of iron to accumulate in the brain (Gaki & Papavassiliou, 2014), 4 hydroxynonenal, and malondialdehyde, which are indicators of lipid peroxidation substantia nigra in PD patients (Bhat et al., 2015).

In the substantia nigra of PD patients, the endogenous antioxidant system, which includes superoxide dismutase (SOD), catalase (CAT), and glutathione peroxidase enzymes, is found at reduced levels (Giordano et al., 2014).

Phenolic compounds are L-phenylalanine produces a broad set of metabolites, including phenolic acids, flavonoids, lignans, hydroxycinnamic acids, and tannins (Almanasrah et al., 2015). According to El Hajaji et al. (2010), phenolic chemicals improve the flavor, color, and quality of food (El Hajaji et al., 2010). They exhibit activity of antioxidants (Oniszczuk & Podgórski, 2015), which suppresses ROS (Kukula-Koch et al., 2013), for example, hydroxyl radical (OH), hydrogen peroxide (H_2O_2), and superoxide anion (O_2) (Park & Jhon, 2010), and have demonstrated human health beneficial effects, highlighting their potential use in the pharmaceutical industry (Kasrati et al., 2014). Numerous illnesses are prevented by it, including cancer, cardiovascular disease, and neurological disorders. The *C. siliqua* 6-OHDA zebrafish PD model offers an alternative supply of bioactive molecules that might develop memory capabilities.

Pinostrobin (PSB) is a bioflavonoid found in food that has been demonstrated to have 380 protective effects in AD mice brought on by preventing oxidative damage and reducing mitochondria mediated neuronal death. The MPP+/38 PSB's neuroprotective properties in a PD model caused by MPTP were investigated, and this could be a regimen for the treatment and prevention of PD (Kang et al., 2005).

Activation of the Nrf2/ARE pathway, which contributes to PSB-mediated neuroprotection against MPP+ insults, PSB-induced Nrf2/ARE activation. PSB can protect SH-SY5 cells from MPTP490-induced dopaminergic neuronal death and motor behavior impairments in zebrafish in vivo (C. Li et al., 2018).

15.3.1.2 *Palythoa Caribaeorum*

Palythoa caribaeorum (class Anthozoa) is an animal of the class Zoantharan, which, along with other cnidarians, such as jellyfish, water hyacinth. and anemones, has a specialized structure in its tissues, cnidocytes, which provide many toxins to capture prey and deter predators. Cnidaria comprises some of the most toxic marine organisms. Cnidocytes, a special type of stinging cell found only in this branch of species, are capable of delivering potent venom mixtures for immobilizing prey to prevent or hunt to protect themselves (Jouiaei et al., 2015).

Disulfide-rich, cysteine stabilized peptides found in cnidarian venom operate as poisons via modifying the properties of ion channels involved in the formation of action potentials in skeletal muscle, heart, and neurons. These peptides range in length from 10 to 60 amino acids and are arranged in secondary structures; several disulfide links serve to stabilize, and a number of ion channels and neurotransmitter receptors have proven that they are very effective inhibitors and modulators based on the number of disulfide bonds and residues of amino acids (King, 2015). The four main structural classes of these toxins are type 1, which has 35–37 residues and three disulfide bridges; type 2, which has 58–59 radicals and three disulfide bridges; type 3, which has 41–42 residues and three disulfide bridges; and type 4, which has 28 radicals and two disulfide bridges. First-class examples are BgK of the sea anemone

Bunodosoma granulifera and ShK of the sun sea anemone *Stichodactyla helianthus*. *P. caribaeroum*'s whole transcriptome has undergone comprehensive sequencing, and a number of peptide sequences related to toxins have been discovered. Some of these sequences have been isolated for functional investigation (C. Li et al., 2018).

In a zebrafish organism for drug testing, a novel PcShK3 ShK-like peptide from *P. caribaeorum* (a species of Zoantharan of the subclass Hexacorallia, Cnidaria) may have cardioprotective and neuroprotective properties. Electrophysiological tests using Kv1 and KCa channel subtypes on the PcShK3 peptide will be necessary to further demonstrate that ShK-like peptides from *P. caribaeorum* that function as powerful inhibitors of ion channels of the ShK toxin family (C. Li et al., 2018).

Additionally, more insightful data will support hypotheses made to define the fundamental molecular mechanism and action of PcShK3 on Kv1 or KCa by fusing the electrophysiology of a channel of the potassium subgroup with measurement of the calcium content inside cells. It offered knowledge of prospective characterization of fresh zoantharian peptide sequences that resemble ShK and its derivatives. The peptide's intriguing cardioprotective and neuroprotective activity, which, in combination with structurally guided peptide cleavage, is particularly noteworthy, may benefit from drug development by zebrafish microorganisms (C. Li et al., 2018).

15.3.1.3 *Eucommia Ulmoides Oliver*

Oliver ulmoides, commonly referred to as Du-Zhong, Gutta-percha tree, Sixian, Sizhong, and Sijinshu, is a rare and protected plant species acquired from the Tertiary glacial migration (Voss, 1983).

The species grows well in strong light circumstances, and it grows well on loamy soil and sandy soil with a pH of 4.5–8.0. It grows at altitudes ranging from 300 to 2500 m, with an average rainfall of around 1000 mm.

Eucommia ulmoides Oliver can withstand high temperature 44°C and low temperature 40°C (Xing et al., 2019). More than 99% of the resources of *Eucommia ulmoides Oliver* are found in China, where 33 million people farm the plant. It is indigenous to the highlands of southern China and found throughout the country. The provinces of Sichuan, Guizhou, Hubei, Hunan, Shaanxi, Gansu, Henan, Anhui, and Jiangxi are included in its natural range. There have also been successful introductions in Jiangsu, Zhejiang, Guangxi, Yunnan, Shanxi, Shandong, Hebei, Beijing, Xinjiang, and other provinces and regions (C. Liu et al., 2020). There are small variations in the morphology of the shell, leaf, and seed in various locations. According to the old medical record, *Eucommia ulmoides Oliver* has been recognized and used for at least 2,000 years. *Eucommia ulmoides Oliver* can first be found recorded in *Shennong Herbs Classic*. In Leningrad in 1955, there was the first international conference on the pharmacology of *Eucommia ulmoides Oliver*. More research was done after that. Based on their structural similarity, the chemical components isolated from *Eucommia ulmoides Oliver* can be divided into flavonoids, iridoids, phenols, lignans, and sterols.

As a result, it is included in the traditional Chinese medicine registry as having effects on kidney and liver functions and strengthening muscles and bones (Xiao et al., 2023).

Furthermore, *Eucommia ulmoides Oliver* is used for hypolipidemic, antihypertensive, antioxidant, antidiabetic, anti-inflammatory, bone metabolic, neuroprotective, nephroprotective, hepatoprotective, anti-fatigue, sedative and hypnotic, anti-aging, anti-depressant, immunomodulatory, cognitive enhancement, uterine smooth muscle relaxation, and erectile function enhancement properties.

PD is the second most common neurological illness, and there are limited preventive or therapeutic measures to treat PD. *Eucommia urmoides Oliver* bark extract has been reported to have anti-PD effects. Five extracts from *Eucommia urmoides Oliver* leaves also have a therapeutic effect on subsequent PD using the zebrafish model.

Growing emphasis has been placed on naturally occurring polyphenols because of their potentially broad biological action and low toxicity (Seong Su Kang et al., 2017). Natural polyphenols, such as chlorogenic acid and curcumin, are beneficial to the brain (De Pablo-Fernández et al., 2017; Kang et al., 2017). The geniposide acid analogous to geniposide has been reported to have neuroprotective effects. Furthermore, autophagy-related gene expression was influenced by extract of *Eucommia ulmoides Oliver* therapy when compared to the MPTP group, implying that components of extract of *Eucommia ulmoides Oliver* may have interactions with these substances. In the molecular docking investigation, two iridoids (geniposidic and aucubin acid) and three phenolic acids (crypto chlorogenic acid, neochlorogenic acid, and chlorogenic acid) as main components of extract of *Eucommia ulmoides Oliver* were chosen as probable binding ligands. As a result, autophagy regulators Pink1, Ulk2, Beclin1, and Atg5 interacted with the majority of the substances examined. There is chemical and pharmacological evidence that *Eucommia ulmoides Oliver* extract has anti-PD action. EEuOL may treat PD-like symptoms by activating autophagy, and phenolic acids in extract of *Eucommia ulmoides Oliver* may play a role in this biological process, meaning that EEuOL could be used as a treatment appealing to PD (S. Zhang et al., 2020).

15.3.1.4 Tongtian Oral Liquid

Traditional Chinese Medicine (TCM) is used for treating PD around the world. Tongtian Oral Liquid (Tongtian Oral Liquid, Tongtian Koufuye) is one of these TCM patents and several herbs formula, consisting of 11 herbal components, with antioxidant, neuroprotective, and analgesic properties. 1-Methyl-4-phenyl-1,2,3,6-tetra hydropyridine (MPTP; Y. Zhang et al., 2014).

It is one of these patented TCM products that include antioxidant and neuroprotective herbal supplements. A combination of 11 herbal ingredients, including Tea, Fangfeng, Qianghuo, *Ligusticum chuanxiong*, Mint, Red Peony Root,

Licorice, *Angelica dahurica*, Asarum, Chrysanthemum, and Gastrodia make up Tongtian Oral Liquid.

All of these plants are said to relieve pain, dissipate gas, increase blood flow, and remove blood stasis. They are mostly used to cure migraines (Dongjie et al., 2022).

Ligusticum chuanxiong Hort (LCH), present in Tongtian Oral Liquid (Tongtian Koufuye), reduced stroke and inflammatory response, improved neurological behavior (Vickers, 2017), and rats with middle cerebral artery occlusion (MCAO) have lower levels of neuronal and oxidative stress death. It is a very commonly used herbal remedy for pain relief (Gu et al., 2018). LCH also has neuroprotective benefits through stimulating neurogenesis, brainstem cell proliferation, and the maintenance of mature neurons, reducing capillary permeability, increasing cerebral blood flow, reducing arterial resistance, and improving microcirculation (Gu et al., 2018; B. Yu et al., 2020).

Gastroenteritis Blume found in Tongtian Oral Liquid (Tongtian Koufuye) is also a good option for neurological illnesses such as PD. Gastrodin possesses anti-inflammatory, anticonvulsant, antioxidant, anxiolytic, sedative qualities, and analgesic as well as the potential to improve learning and memory (Dai et al., 2011; X.-L. Wang et al., 2014).

Gastrogin suppresses inflammatory enzymes and proinflammatory cytokines caused by the inhibition of NF-B signalling in microglial cells and the phosphorylation of her MAPK by lipopolysaccharide (Dai et al., 2011). In cellular PD models, gastrogin also protects dopamine neurons by counteracting oxidative stress via the p38 MAPK/Nrf2 pathway (Jiang et al., 2014).

Licorice root (*Glycyrrhiza glabra L.*) is an herbal medication utilized as a neuroprotection for preventing PD and stroke. If the function of mitochondria is affected by MPP+, rotenone, and 6-OHDA, DNA damage results from PARP1 activation. Moreover, AIMP2 (Parkin substrate) causes the activation of PARP1 and apoptosis of dopaminergic neurons. Liquiritigenin, an estrogen binding substance in PD patients, receptor beta prevents dopamine neuron death (H. Kim et al., 2017).

Neurons are safeguarded by Imperatorin, a bioactive component of *Angelica Dahuricae radix* through antiapoptotic activity (Tun & Kang, 2017). Yeoldahanso-tang, an herbal remedy containing *A. Dahuricae radix*, protects neurons in PD models by elevating autophagy (Bae et al., 2011). *Acorus calamus -asarone* inhibits PD and Alzheimer's disease via controlling autophagy and synaptogenesis, via the stress pathway of IRE1/XBP1-ER (Lall & Singh, 2020).

Peony suffruticosa Andrews root bark, commonly known as Peony, removes heat from the blood and is often used to stimulate the blood. It promotes blood circulation and relieves congestion, paeonol, and paeoniflorin (Z. Wang et al., 2017). The bioactive components contained in this herb have been shown to have effective neuro-related action in PD-like animals. Paeonol strengthens cells' reductive survival in neural cells following 6-hydroxydopamine poisoning formation of reactive oxygen species (ROS), and enhancement activity of superoxide Dismutase (SOD) and expression of Bcl-2 (Tang et al., 2021).

Paeoniflorin has been demonstrated to promote health in both in vivo and in vitro investigations. Dopaminergic neuron degeneration and behavioural deficiencies are modulatory toxicity of MPTP, glutamate, and 6-hydroxydopamine Bcl-2 family proteins, as well as neuroinflammation inhibition (Liu, Zhang et al., 2006; Zheng et al., 2017).

Chrysanthemum Indicum Lin (CI), Methanol extracts at doses of 1, 10, and 100 μg decreased ROS production in controlling the prevented PARP proteolysis, MPP+-induced SH-SY5Y cells, and Bax/Bcl-2 ratio; in addition, CI also dose-dependently inhibited COX-2 expression, prostaglandin E2 production, NF-κB p65 activation in BV-2 cells, and IκB-α degradation. CI may entail molecular mechanisms that prevent both the neuroinflammatory NF-B/IB- pathway and neuronal apoptosis (I.-S. Kim et al., 2011). Of the 11 herbs included, the ones listed above are thought to be particularly effective in reducing the tremors related to PD.

Toxins utilized to generate PD in animal models, such as MPTP, rotenone, fenazaquin, and pyridaben, result in defective mitochondrial transport complex activity, increased permeability of mitochondrial transition, and reduced mitochondrial motility. A lipid-soluble neurotoxin titled 1-methyl-4-phenyl-1,2,3,6-tetrahydropyridine targets DA cells and is utilized to kill DA neurons to create a model for animals of PD (Hawkins & Byrne, 2015).

After systemic injection, 1-methyl-4-phenyl-1,2,3,6-tetrahydropyridine passes through the blood brain barrier (BBB) and is oxidized to 1-methyl-4-phenyl-1,2,3,6-tetrahydropyridine + by MAO-B situated on non-dopaminergic cells' outer membrane of mitochondria (Meredith & Rademacher, 2011). The labile chemical 1-methyl-4-phenyl-1,2,3,6-tetrahydropyridine + is further oxidized to yield the active poisonous compound MPP+.

Compared to 1-methyl-4-phenyl-1,2,3,6-tetrahydropyridine, MPP+ is significantly less lipophilic, and thus cannot readily diffuse across the lipid bilayer of the cell membrane (Meredith & Rademacher, 2011), but dopamine transporters actively transfer it to dopaminergic neurons (Butler et al., 2015).

MPP+'s positive charge causes it to attach to and prevent complex I from being used by dopaminergic neurons' electron transport chain, which produces irreversible oxidative damage and increases ROS production by damaging the mitochondrial membrane, causing the mitochondrial membrane potential to collapse, causing irreversible oxidative damage, and increasing ROS production (Martinez & Greenamyre, 2012). 1-methyl-4-phenyl-1,2,3,6-tetrahydropyridine causes dopaminergic neurons loss, decreased levels of indopamine, noradrenaline, and serotonin (Vaz et al., 2018), and motor deficits and has been widely utilized to induce PD models in zebrafish.

Zebrafish are an appealing tool for assessing drug prospects for toxicity and safety risks, as well as for drug discovery. Unlike mammals, zebrafish are small, have high functional conservation, a short life cycle, high fecundity, share 85% genetic homology with humans, and rapid development (Kalueff et al., 2014).

Zebrafish larvae also have optical transparency, which enables in-depth, in vivo monitoring of cellular and physiological processes in real time. Numerous studies have been carried out on the zebrafish dopaminergic system, which is fully formed by 96 hours post fertilization (hpf) (Rink & Wullimann, 2002). Zebrafish are a good model for the generation of the phenotype of PD. In zebrafish larvae, DA neuron populations mimic those found in the human body's substantia nigra. In addition, zebrafish provide a unique vertebrate example for investigating movement disorders. Because of these characteristics, zebrafish are now used to test new medications and analyze their mechanisms of action in addition to serving as a PD model.

The purpose of this study was to determine the brain-protective effects of Tongtian Oral Liquid on MPTP-induced developmental damage, and dopaminergic neurons loss, movement disorders, also gene expression associated with the development of the dopaminergic nerve system. When exposed to the neurotoxic 1-methyl-4-phenyl-1,2,3,6-tetrahydropyridin, dopaminergic neurons become sensitive in zebrafish brain development.

The antioxidant and neuroprotective Tongtian Oral Liquid properties were verified by a reduction in the generation of ROS and the removal of free radicals, which lengthened dopaminergic neurons, boosted motor performance, raised GPx, catalase, SOD's antioxidant activity, lowered lipid peroxidation, and increased activity. Dopamine-related gene mRNA expression has been restored in th1, drd1, drd2, drd3, and drd4 in 1-methyl-4-phenyl-1,2,3,6-tetrahydropyridin-exposed zebrafish, which provides further evidence that dopaminergic signaling was restored. Taken together, all of these findings support Tongtian Oral Liquid's neuroprotective potential in 1-methyl-4-phenyl-1,2,3,6-tetrahydropyridin-induced neurotoxicity in a zebrafish example of PD rather than its botanical origin, because of its free radical removal and antioxidant properties. More study is needed to determine the precise mechanism of action of Tongtian Oral Liquid for neuroprotection in 1-methyl-4-phenyl-1,2,3,6-tetrahydropyridin, and the zebrafish model was used to develop experimental parkinsonism.

15.3.1.5 *Alpiniae Oxyphyllae Fructus*

Alpiniae oxyphyllae fructus are dry-ripening *Alpinia purpurata* fruit. It has been utilized in traditional medical Chinese medicine for a long time (Yang et al., 2020). *Alpiniae oxyphyllae fructus* (AOF) are frequently used in conventional medicine to alleviate sickness, frequent urination, diarrhea, and semen leaking (Li et al., 2022). In recent pharmacological research, AOFs have anti-inflammatory (Qi et al., 2019), antioxidant (Bian et al., 2013), anti-AD, and neuroprotective effects (H. Liu et al., 2020). *Alpiniae oxyphyllae fructus* contain various chemical constituents, for example, sesquiterpenes, flavonoids, essential oils, diarylheptanoids, steroids, and glycosides (F. Chen et al., 2013; Sõukand et al., 2015). Currently, the efficacy of *Alpiniae oxyphyllae fructus* in AD is remarkable, but their potential therapeutic mechanisms are unclear (Zhou et al., 2023).

An ethanol extract from *Alpinia oxyphylla fructus* was used to test its efficacy in 6-OHDA-induced PD in zebrafish larvae. We compared this extract with previously identified protocatechinic acid and chrysin and found it to be more effective in preventing and reversing 6-OHDA-induced loss of her-dominant negative (DN) and musculoskeletal disorders. The same extract regulated pro-inflammatory genes expression and signaling pathways phosphatidylinositol-3-kinase and Akt/protein kinase B (PI3K-AKT) in zebrafish. This extract was also tested in a pheochromocytoma cell line (PC12) in which lesions were induced by 6-OHDA (Z.-J. Zhang et al., 2012). However, *Alpinia purpurata* extract has not yet been used in clinical studies.

15.3.1.6 *Ginkgo Biloba*

Ginkgo biloba is a plant leaf that is gaining popularity because of its potent therapeutic qualities. It is widely used to treat collateral obstruction due to congestion, chest pain, and stroke hemiplegia. Since the 1960s, the chemical composition of *ginkgo biloba* has been studied in detail by scientists to deepen our understanding of its pharmacological effects and clinical applications. According to research, the primary active medicinal components of *ginkgo biloba* are organic acids, flavonoids, and terpene lactones. These substances enhance memory and cognitive performance while also increasing blood circulation, preventing blood clots, lowering blood lipids, and regulating the central nervous system. The majority of known pharmacological actions are linked to processes that function as antioxidants, scavenge free radicals, and regulate the production of excitatory amino acids by inhibiting platelet activating factor (PAF) (Guo-Xia et al., 2006). Recent research has revealed that inadequate insulin secretion and insulin resistance are two of the most common causes of diabetes. Extract of ginkgo biloba (EGb) regulates insulin secretion in both directions. It boosts secretion of insulin in those who have compromised function of pancreas, suppression reduces levels of leptin, and energizes the glucose transporter for insulin resistance improvement (Sun et al., 2018).

Ginkgo drops tablets contain an extract of *ginkgo biloba*. It was evaluated on 6-OHDA-induced lesions in wild-type AB and melanin allele mutant albino zebrafish larvae that were produced in China. *Ginkgo biloba* pills have shown some effects. They reversed her 6-OHDA-induced musculoskeletal zebrafish damage and DN loss prevention. The advantages of tablets for neuroprotection may be controlled by the Akt/GSK3-signalling pathway, according to in vitro researches (D. Yu et al., 2021). The ability of MPTP to cause Parkinson's disease in mice was also examined (D. Yu et al., 2021). In addition, Egb has been previously tested in transgenic A53T-α-synuclein mice (Kuang et al., 2018). *Ginkgo biloba* drops have shown interesting results in clinical trials for dementia (DeKosky et al., 2006) and coronary artery disease (DeKosky et al., 2006).

15.3.1.7 Seaweeds

In Asian countries such as China, algae were an important part of the diet (Gómez-Zorita et al., 2020). In some regions around the world, populations increased to levels where food production is no longer sufficient to feed the population (Leandro et al., 2020), and so in recent decades, there has been a decline in food production for both human consumption and industrial processing. However, there is a growing demand for alternatives (Peñalver et al., 2020). Algae are viewed as superior to animal and manufactured products and have a rich, sustainable supply of macro and micronutrients in the ocean (Peñalver et al., 2020). In addition, several algal compounds have been reported with potential use as materials in function food as well as health and dietary supplements. Bioactive compounds are bioactive and therefore have positive health effects (Leandro et al., 2020). Seaweed is a great source of sulfated polysaccharides (15–76% dry matter), proteins that contain all nine necessary amino acids (5–47% dry matter), and lipids (1–5% dry matter) in relation to macronutrients. For example, B. polyunsaturated fatty acids (Cotas et al., 2020; Peñalver et al., 2020). From a micronutrient perspective, seaweed is an excellent source of water-soluble vitamins (B1, C, B12), lipids and minerals such as calcium, magnesium, iron, copper, and iodine (7–36% of dry matter) (Cherry et al., 2019; Peñalver et al., 2020). Additionally algae are a secondary sources of metabolites such as polyphenols (Cotas et al., 2020; Gómez-Zorita et al., 2020; Peñalver et al., 2020).

Some of these substances, especially the latter, provide beneficial health effects such as antioxidant, antibacterial, anti-inflammatory, anticancer, diabetic, antihypertensive, antihyperlipidemic, and antiobesity effects (Lomartire & Gonçalves, 2022). Algae are taxonomically classified into three major groups: brown, red, and green algae (Bizzaro et al., 2022). Algae color is determined by pigments such as green color for chlorophyll, red color for phycobilin, and brown color for fucoxanthin algae (Peñalver et al., 2020). Each algae have a different composition depending on its species as well as other geographical and environmental considerations, such as algal competition, the strength of the substrate, exposure to ice and waves, salinity, wave energy, and light (André et al., 2020; S. Kim et al., 2022). Three algae that are frequently used in food and nutritional supplements were chosen for this study, and their biological activity and composition were compared to promote health and be linked to disease. In particular, oxidative stress was reported to have the ability to treat disease.

Fucoxanthin, a marine carotenoid derived from seaweed, was tested for 6-OHDA-induced cytotoxicity in zebrafish larvae and the PC12 cell line. Fucoxanthin protects against 6-OHDA-induced neurotoxicity in zebrafish, rescues deficits in locomotor activity, and protects against antioxidant enzymes (nicotinamide heme oxygenase-1, glutamate-cysteine ligase modification subunit, GCLC (catalytic)). Fucoxanthin also protected neurons from damage by targeting Keap1 (H. Liu et al., 2020). 11-Dehydrocinulariolide is a crystallized marine-derived compound isolated from soft coral cineraria. It shows interesting activity in wild-type AB tribes zebrafish larvae.

The ability to reverse musculoskeletal damage caused by 6-OHDA prevents neurodegeneration in zebrafish. Mitochondrial DJ-1 expression and cytosolic increased, as well as downstream PI3K/Akt, cyclic AMP-responsive element binding protein (p-CREB), nuclear factor erythroid 2-related factor 2, and heme oxygenase-1 (Nrf2/HO-1) pathway (Feng et al., 2016).

11-Dehydrocinuraliolide was also tested in 6-OHDA-induced PD in the SH-SY5Y cell line (W.-F. Chen et al., 2012).

A marine substance called 1-tosylpentan-3-one was discovered in the soft coral *Cladiella australis*. This was tested for ultimate efficacy in relation to PD using 6-OHDA as a lesion inducer in AB strain zebrafish larvae (in the same investigation, the SH-SY5Y cell line was also experimented with).

1-Tosylpentan-3-one showed beneficial effects by inhibiting the activation of both p38 mitogen-activated protein kinase (p38 MAPK) and caspase-3 in the SH-SY5Y cell line. It also enhanced Nrf2 and heme oxygenase-1 (HO-1) expression via the PI3K/Akt signaling pathway. This compound exhibited anti-apoptotic activity and reversed the hypomotility of zebrafish larvae (Kao et al., 2017).

Sterin B is a marine Isomaravarican triterpenoid isolated from the sponge jasper sterifera. AB strain zebrafish larvae were tested against her 6-OHDA-induced lesions. Sterettin B has been shown to inhibit cellular apoptosis and oxidative stress through regulation of the PI3K/Akt, MAPK, and Nrf2/HO-1cascades. Moreover, this sea connection restored motor deficits in zebrafish. It is worth noting that Sterettin B was tested in her HSY5Y cell line and shielded against cell deterioration (Feng et al., 2019).

An ethanol extract of *Ericaulon buergerianum* (Gujingcao), a Chinese marine aquatic plant, was tested on six OHDA-induced lesions in zebrafish larvae and the PC12 cell line. This extract protected against DN loss, rescued the defect in locomotor activity, and suppressed apoptosis occurring in PC12. In addition, it inhibited NO overproduction and downregulated iNOS overexpression (M. Wang et al., 2011).

15.4 BIOACTIVE COMPOUNDS

Previous research has demonstrated the wide range of tasks that peptides extracted from C-phycocyanin (C-PC) can perform, including antioxidant and anticancer properties. There is, however, limited study on C-PC peptides used for the neuroprotective impact against a PD model generated by MPTP. Twelve new peptides from C-PC were extracted, purified, and identified in this study. A zebrafish PD model was used to assess the anti-PD effect of the synthesized peptides. Therefore, the loss of dopamine neurons and cerebral arteries was dramatically reversed by three of these peptides (MAAAHR, MPQPPAK, and MTAAAR), which also significantly lessened the locomotor impairment in PD zebrafish.

Three new peptides may also enhance the level of ROS and protein carbonylation while blocking the MPTP-induced decline of the antioxidant enzymes SOD, CAT, and GSH-Px. They can also lessen acetylcholinesterase (AChE) activity and the apoptosis of some brain areas in zebrafish. The probable molecular basis of the anti-PD activities of peptides in the larvae was clarified by additional research. According to the findings, C-PC peptides have the ability to alter a number of genes linked to oxidative stress, autophagy, and apoptotic signaling pathways, consequently reducing the severity of PD symptoms. Overall, our findings point to the neuroprotective properties of three new peptides and offer insightful molecular explanations as well as a prospective therapeutic target for the treatment of Parkinson's disease (Xu et al., 2023).

Flavonoids are polyphenolic secondary metabolites with the ability to affect a variety of biological pathways, including those involved in neural plasticity, learning, and memory. Flavonoids have come to light as a result of breakthroughs in pharmacognosy, biotechnology, and biochemistry throughout the last century. Additionally, oxidative stress, neuroinflammation, and mitochondrial dysfunction are three biological processes that lead to the neurodegenerative diseases development to be affected by flavonoids. Thus, a number of flavonoids could be employed as adjuvants to stop and reverse neurodegenerative illnesses including PD and AD. The zebrafish is an intriguing model organism that may present fresh perspectives on how flavonoids can treat neurodegenerative illnesses.

Zebrafish are a useful organism to study human neurodegenerative diseases, and determining how flavonoids affect those disorders because of their high genome homology to humans of 70%, their brain organization that is largely similar to that of humans, in addition to comparable neurochemical and neuroanatomical mechanisms (Mhalhel et al., 2023).

15.5 CONCLUSIONS AND PERSPECTIVES

This chapter's main topic was on the employed zebrafish models for natural product evaluation. This work primarily summarizes and discusses the routine management and laboratory zebrafish care, zebrafish models development, and indication for evaluation of natural products. The use of zebrafish as a trustworthy tool offers a viable approach for natural compounds being screened and evaluated with possible benefits to human health in a timely and cost-effective manner. It is simple to apply a variety of zebrafish disease models, namely those to investigate the regulatory effects of natural items, and researchers may use dietary manipulation, chemical induction, and transgenic lines with fluorescent labeling.

The advancement of automated tools and analytical procedures has boosted zebrafish usage as a highly efficient screening instrument evaluation of natural substances with diverse bioactivities. This may aid in searching for new medicines and useful chemicals originating from natural sources.

This chapter's main emphasis was on the often used models of zebrafish to assess natural products. Specifically highlighted and discussed are the daily management and care of zebrafish in the laboratory, zebrafish development models and evaluation indicators, and the assessment of natural resources using models of zebrafish.

The neuroprotective properties of numerous natural compounds are outlined in this chapter, while just Parkinson's disease-related keywords were used. This means that natural remedies that have been successfully utilized to cure other illnesses or safeguard the body may also be effective against neurodegenerative diseases. Therefore, there may be uses for other natural compounds with comparable potency. For more effective treatments, standards for daily intake should be set since natural ingredients can be eaten as food in daily life.

This is a dependable and practical method for screening and analyzing natural compounds with potential beneficial effects on health in a timely and cost-effective manner. Natural products' regulatory effects can be easily investigated using a variety of zebrafish illness models, namely those that entail chemical induction, food change, or transgenic lines with fluorescent labelling.

It has also been extensively noted about automated instruments development, innovation, and analytical support techniques for zebrafish as a powerful instrument for high-throughput screening and assessment of natural compounds with a wide range of bioactivities. The bioaccessibility and bioavailability of natural compounds in zebrafish require more study. Additional study is required to determine the drawbacks of employing zebrafish microorganisms for every disease or health problem, as well as potential remedies. Advanced sensitivity and precise technology automated to fulfill the requirements of high-output assessment as well as a reasonably priced smart automated system for zebrafish rearing are still required. In general, zebrafish modelling is a useful method for assessing natural items, which can be crucial for the creation of functional meals, nutraceuticals, and cosmetics.

REFERENCES

Abidar, S., Boiangiu, R. S., Dumitru, G., Todirascu-Ciornea, E., Amakran, A., Cioanca, O., . . . Nhiri, M. (2020). The aqueous extract from ceratonia siliqua leaves protects against 6-hydroxydopamine in zebrafish: Understanding the underlying mechanism. *Antioxidants*, *9*(4), 304.

Almanasrah, M., Roseiro, L. B., Bogel-Lukasik, R., Carvalheiro, F., Brazinha, C., Crespo, J., . . . Duarte, L. C. (2015). Selective recovery of phenolic compounds and carbohydrates from carob kibbles using water-based extraction. *Industrial Crops and Products*, *70*, 443–450.

André, R., Guedes, L., Melo, R., Ascensão, L., Pacheco, R., Vaz, P. D., & Serralheiro, M. L. (2020). Effect of food preparations on in vitro bioactivities and chemical components of Fucus vesiculosus. *Foods*, *9*(7), 955.

Avdesh, A., Chen, M., Martin-Iverson, M. T., Mondal, A., Ong, D., Rainey-Smith, S., . . . Verdile, G. (2012). Regular care and maintenance of a zebrafish (*Danio rerio*) laboratory: An introduction. *Journal of Visualized Experiments, 9*(3), e4196.

Bae, N., Ahn, T., Chung, S., Oh, M. S., Ko, H., Oh, H., . . . Yang, H. O. (2011). The neuroprotective effect of modified Yeoldahanso-tang via autophagy enhancement in models of Parkinson's disease. *Journal of Ethnopharmacology*, *134*(2), 313–322.

Barros de Medeiros, V. P., da Costa, W. K. A., da Silva, R. T., Pimentel, T. C., & Magnani, M. (2022). Microalgae as source of functional ingredients in new-generation foods: Challenges, technological effects, biological activity, and regulatory issues. *Critical Reviews in Food Science and Nutrition*, *62*(18), 4929–4950.

Bhat, A. H., Dar, K. B., Anees, S., Zargar, M. A., Masood, A., Sofi, M. A., & Ganie, S. A. (2015). Oxidative stress, mitochondrial dysfunction and neurodegenerative diseases; a mechanistic insight. *Biomedicine & Pharmacotherapy*, *74*, 101–110.

Bian, Q.-Y., Wang, S.-Y., Xu, L.-J., Chan, C.-O., Mok, D. K., & Chen, S.-B. (2013). Two new antioxidant diarylheptanoids from the fruits of Alpinia oxyphylla. *Journal of Asian Natural Products Research*, *15*(10), 1094–1099.

Bizzaro, G., Vatland, A. K., & Pampanin, D. M. (2022). The one-health approach in seaweed food production. *Environment International*, *158*, 106948.

Blader, P., & Strähle, U. (2000). Zebrafish developmental genetics and central nervous system development. *Human Molecular Genetics*, *9*(6), 945–951.

Brand, M., Beuchle, D., Endres, F., Haffter, P., Hammerschmidt, M., Mullins, M., . . . Schwarz, K. (1995). Keeping and raising zebrafish (*Danio rerio*) in Tübingen. *Zebrafish Science Monitor*, *3*, 2–7.

Butler, B., Saha, K., Rana, T., Becker, J. P., Sambo, D., Davari, P., . . . Khoshbouei, H. (2015). Dopamine transporter activity is modulated by α-synuclein. *Journal of Biological Chemistry*, *290*(49), 29542–29554.

Caro, M., Iturria, I., Martinez-Santos, M., Pardo, M., Rainieri, S., Tueros, I., & Navarro, V. (2016). Zebrafish dives into food research: Effectiveness assessment of bioactive compounds. *Food & Function*, *7*(6), 2615–2623.

Chen, F., Li, H.-L., Li, Y.-H., Tan, Y.-F., & Zhang, J.-Q. (2013). Quantitative analysis of the major constituents in Chinese medicinal preparation SuoQuan formulae by ultra fast high performance liquid chromatography/quadrupole tandem mass spectrometry. *Chemistry Central Journal*, *7*, 1–9.

Chen, W.-F., Chakraborty, C., Sung, C.-S., Feng, C.-W., Jean, Y.-H., Lin, Y.-Y., . . . Su, T.-M. (2012). Neuroprotection by marine-derived compound, 11-dehydrosinulariolide, in an in vitro Parkinson's model: A promising candidate for the treatment of Parkinson's disease. *Naunyn-Schmiedeberg's Archives of Pharmacology*, *385*, 265–275.

Cherry, P., O'Hara, C., Magee, P. J., McSorley, E. M., & Allsopp, P. J. (2019). Risks and benefits of consuming edible seaweeds. *Nutrition Reviews*, *77*(5), 307–329.

Choi, T.-Y., Choi, T.-I., Lee, Y.-R., Choe, S.-K., & Kim, C.-H. (2021). Zebrafish as an animal model for biomedical research. *Experimental & Molecular Medicine*, *53*(3), 310–317.

Chong, C.-M., Zhou, Z.-Y., Razmovski-Naumovski, V., Cui, G.-Z., Zhang, L.-Q., Sa, F., . . . Lee, S. M.-Y. (2013). Danshensu protects against 6-hydroxydopamine-induced damage of PC12 cells in vitro and dopaminergic neurons in zebrafish. *Neuroscience Letters*, *543*, 121–125.

Cotas, J., Leandro, A., Monteiro, P., Pacheco, D., Figueirinha, A., Gonçalves, A. M., . . . Pereira, L. (2020). Seaweed phenolics: From extraction to applications. *Marine Drugs*, *18*(8), 384.

Dabrowski, K., & Miller, M. (2018). Contested paradigm in raising zebrafish (*Danio rerio*). *Zebrafish*, *15*(3), 295–309.

Dai, J.-N., Zong, Y., Zhong, L.-M., Li, Y.-M., Zhang, W., Bian, L.-G., . . . Lu, D. (2011). Gastrodin inhibits expression of inducible NO synthase, cyclooxygenase-2 and proinflammatory cytokines in cultured LPS-stimulated microglia via MAPK pathways. *PLoS One*, *6*(7), e21891.

De Pablo-Fernández, E., Breen, D. P., Bouloux, P. M., Barker, R. A., Foltynie, T., & Warner, T. T. (2017). Neuroendocrine abnormalities in Parkinson's disease. *Journal of Neurology, Neurosurgery & Psychiatry*, *88*(2), 176–185.

DeKosky, S. T., Fitzpatrick, A., Ives, D. G., Saxton, J., Williamson, J., Lopez, O. L., . . . Robbins, J. (2006). The ginkgo evaluation of memory (GEM) study: Design and baseline data of a randomized trial of ginkgo biloba extract in prevention of dementia. *Contemporary Clinical Trials*, *27*(3), 238–253.

Dongjie, S., Rajendran, R. S., Xia, Q., She, G., Tu, P., Zhang, Y., & Liu, K. (2022). Neuroprotective effects of Tongtian oral liquid, a traditional Chinese medicine in the Parkinson's disease-induced zebrafish model. *Biomedicine & Pharmacotherapy*, *148*, 112706.

El Hajaji, H., Lachkar, N., Alaoui, K., Cherrah, Y., Farah, A., Ennabili, A., . . . Lachkar, M. (2010). Antioxidant properties and total phenolic content of three varieties of carob tree leaves from Morocco. *Records of Natural Products*, *4*(4), 193.

Feng, C.-W., Chen, N.-F., Wen, Z.-H., Yang, W.-Y., Kuo, H.-M., Sung, P.-J., . . . Chen, W.-F. (2019). In vitro and in vivo neuroprotective effects of stellettin B through anti-apoptosis and the Nrf2/HO-1 pathway. *Marine Drugs*, *17*(6), 315.

Feng, C.-W., Hung, H.-C., Huang, S.-Y., Chen, C.-H., Chen, Y.-R., Chen, C.-Y., . . . Sheu, J.-H. (2016). Neuroprotective effect of the marine-derived compound 11-dehydrosinulariolide through DJ-1-related pathway in in vitro and in vivo models of Parkinson's disease. *Marine Drugs*, *14*(10), 187.

Fitzgerald, J. A., Könemann, S., Krümpelmann, L., Županič, A., & Vom Berg, C. (2021). Approaches to test the neurotoxicity of environmental contaminants in the zebrafish model: From behavior to molecular mechanisms. *Environmental Toxicology and Chemistry*, *40*(4), 989–1006.

Gaki, G. S., & Papavassiliou, A. G. (2014). Oxidative stress-induced signaling pathways implicated in the pathogenesis of Parkinson's disease. *Neuromolecular Medicine*, *16*, 217–230.

Giordano, S., Darley-Usmar, V., & Zhang, J. (2014). Autophagy as an essential cellular antioxidant pathway in neurodegenerative disease. *Redox Biology*, *2*, 82–90.

Gómez-Zorita, S., González-Arceo, M., Trepiana, J., Eseberri, I., Fernández-Quintela, A., Milton-Laskibar, I., . . . Portillo, M. P. (2020). Anti-obesity effects of macroalgae. *Nutrients*, *12*(8), 2378.

Grunwald, D. J., & Eisen, J. S. (2002). Headwaters of the zebrafish—emergence of a new model vertebrate. *Nature Reviews Genetics*, *3*(9), 717–724.

Gu, J., Su, S., Guo, J., Zhu, Y., Zhao, M., & Duan, J.-a. (2018). Anti-inflammatory and anti-apoptotic effects of the combination of Ligusticum chuanxiong and Radix Paeoniae against focal cerebral ischaemia via TLR4/MyD88/MAPK/NF-κB signalling pathway in MCAO rats. *Journal of Pharmacy and Pharmacology*, *70*(2), 268–277.

Gülçin, I., Bursal, E., Şehitoğlu, M. H., Bilsel, M., & Gören, A. C. (2010). Polyphenol contents and antioxidant activity of lyophilized aqueous extract of propolis from Erzurum, Turkey. *Food and Chemical Toxicology*, *48*(8–9), 2227–2238.

Gumus, C. E., & Gharibzahedi, S. M. T. (2021). Yogurts supplemented with lipid emulsions rich in omega-3 fatty acids: New insights into the fortification, microencapsulation, quality properties, and health-promoting effects. *Trends in Food Science & Technology*, *110*, 267–279.

Guo-Xia, W., Fu-Liang, C., & Jun, C. (2006). Progress in researches on the pharmaceutical mechanism and clinical application of Ginkgo Biloba extract on various kinds of diseases. *Chinese Journal of Integrative Medicine*, *12*(3), 234–239.

Hawkins, R. D., & Byrne, J. H. (2015). Associative learning in invertebrates. *Cold Spring Harbor Perspectives in Biology*, *7*(5), a021709.

Hitchcock, P. F., & Raymond, P. A. (2004). The teleost retina as a model for developmental and regeneration biology. *Zebrafish*, *1*(3), 257–271.

Howe, K., Clark, M. D., Torroja, C. F., Torrance, J., Berthelot, C., Muffato, M., . . . Matthews, L. (2013). The zebrafish reference genome sequence and its relationship to the human genome. *Nature*, *496*(7446), 498–503.

Hsieh, J.-H., Ryan, K., Sedykh, A., Lin, J.-A., Shapiro, A. J., Parham, F., & Behl, M. (2019). Application of benchmark concentration (BMC) analysis on zebrafish data: A new perspective for quantifying toxicity in alternative animal models. *Toxicological Sciences*, *167*(1), 92–104.

Jacobson, S. M., Birkholz, D. A., McNamara, M. L., Bharate, S. B., & George, K. M. (2010). Subacute developmental exposure of zebrafish to the organophosphate pesticide metabolite, chlorpyrifos-oxon, results in defects in Rohon-Beard sensory neuron development. *Aquatic Toxicology*, *100*(1), 101–111.

Jiang, G., Hu, Y., Liu, L., Cai, J., Peng, C., & Li, Q. (2014). Gastrodin protects against MPP+-induced oxidative stress by up regulates heme oxygenase-1 expression through p38 MAPK/Nrf2 pathway in human dopaminergic cells. *Neurochemistry International*, *75*, 79–88.

Jouiaei, M., Sunagar, K., Federman Gross, A., Scheib, H., Alewood, P. F., Moran, Y., & Fry, B. G. (2015). Evolution of an ancient venom: Recognition of a novel family of cnidarian toxins and the common evolutionary origin of sodium and potassium neurotoxins in sea anemone. *Molecular Biology and Evolution*, *32*(6), 1598–1610.

Kalueff, A. V., Stewart, A. M., & Gerlai, R. (2014). Zebrafish as an emerging model for studying complex brain disorders. *Trends in Pharmacological Sciences*, *35*(2), 63–75.

Kamdem, J. P., Olalekan, E. O., Hassan, W., Kade, I. J., Yetunde, O., Boligon, A. A., . . . Rocha, J. B. T. (2013). Trichilia catigua (Catuaba) bark extract exerts neuroprotection against oxidative stress induced by different neurotoxic agents in rat hippocampal slices. *Industrial Crops and Products*, *50*, 625–632.

Kang, S. S., Lee, J. Y., Choi, Y. K., Song, S. S., Kim, J. S., Jeon, S. J., . . . Han, B. H. (2005). Neuroprotective effects of naturally occurring biflavonoids. *Bioorganic & Medicinal Chemistry Letters*, *15*(15), 3588–3591.

Kang, S. S., Zhang, Z., Liu, X., Manfredsson, F. P., Benskey, M. J., Cao, X., . . . Ye, K. (2017). TrkB neurotrophic activities are blocked by α-synuclein, triggering dopaminergic cell death in Parkinson's disease. *Proceedings of the National Academy of Sciences*, *114*(40), 10773–10778.

Kao, C.-J., Chen, W.-F., Guo, B.-L., Feng, C.-W., Hung, H.-C., Yang, W.-Y., . . . Chen, N.-F. (2017). The 1-tosylpentan-3-one protects against 6-hydroxydopamine-induced neurotoxicity. *International Journal of Molecular Sciences*, *18*(5), 1096.

Kasrati, A., Jamali, C. A., Fadli, M., Bekkouche, K., Hassani, L., Wohlmuth, H., . . . Abbad, A. (2014). Antioxidative activity and synergistic effect of *Thymus saturejoides* Coss. Essential oils with cefixime against selected food-borne bacteria. *Industrial Crops and Products*, *61*, 338–344.

Kilkenny, C., Browne, W. J., Cuthill, I. C., Emerson, M., & Altman, D. G. (2010). Improving bioscience research reporting: The ARRIVE guidelines for reporting animal research. *Journal of Pharmacology and Pharmacotherapeutics*, *1*(2), 94–99.

Kim, H., Ham, S., Lee, J. Y., Jo, A., Lee, G. H., Lee, Y.-S., . . . Pletnikova, O. (2017). Estrogen receptor activation contributes to RNF146 expression and neuroprotection in Parkinson's disease models. *Oncotarget*, *8*(63), 106721.

Kim, I.-S., Ko, H.-M., Koppula, S., Kim, B.-W., & Choi, D.-K. (2011). Protective effect of Chrysanthemum indicum Linne against 1-methyl-4-phenylpridinium ion and lipopolysaccharide-induced cytotoxicity in cellular model of Parkinson's disease. *Food and Chemical Toxicology*, *49*(4), 963–973.

Kim, S., Choi, S. K., Van, S., Kim, S. T., Kang, Y. H., & Park, S. R. (2022). Geographic differentiation of morphological characteristics in the brown seaweed sargassum thunbergii along the korean coast: A response to local environmental conditions. *Journal of Marine Science and Engineering*, *10*(4), 549.

King, G. (2015). *Venoms to drugs: Venom as a source for the development of human therapeutics*. London: Royal Society of Chemistry.

Kolb, A., Hildebrandt, F., & Lawrence, C. (2018). Effects of diet and social housing on reproductive success in adult zebrafish, *Danio rerio*. *Zebrafish*, *15*(5), 445–453.

Kuang, S., Yang, L., Rao, Z., Zhong, Z., Li, J., Zhong, H., . . . Tang, X. (2018). Effects of ginkgo biloba extract on A53T α-synuclein transgenic mouse models of Parkinson's disease. *Canadian Journal of Neurological Sciences*, *45*(2), 182–187.

Kukula-Koch, W., Aligiannis, N., Halabalaki, M., Skaltsounis, A.-L., Glowniak, K., & Kalpoutzakis, E. (2013). Influence of extraction procedures on phenolic content and antioxidant activity of cretan barberry herb. *Food Chemistry*, *138*(1), 406–413.

Kurtzman, M. S., Craig, M. P., Grizzle, B. K., & Hove, J. R. (2010). Sexually segregated housing results in improved early larval survival in zebrafish. *Lab Animal*, *39*(6), 183–189.

Lall, S., & Singh, N. (2020). COVID-19: Unmasking the new face of education. *International Journal of Pharmaceutical Sciences*, *11*(SPL1), 48–53.

Lawrence, C. (2007). The husbandry of zebrafish (*Danio rerio*): A review. *Aquaculture*, *269*(1–4), 1–20.

Lawrence, C. (2016). New frontiers for zebrafish management. *Methods in Cell Biology*, *135*, 483–508. https://doi.org/10.1016/bs.mcb.2016.04.015.

Lawrence, C., Best, J., James, A., & Maloney, K. (2012). The effects of feeding frequency on growth and reproduction in zebrafish (*Danio rerio*). *Aquaculture*, *368*, 103–108.

Leandro, A., Pacheco, D., Cotas, J., Marques, J. C., Pereira, L., & Gonçalves, A. M. (2020). Seaweed's bioactive candidate compounds to food industry and global food security. *Life, 10*(8), 140.

Lewis, K. E., & Eisen, J. S. (2003). From cells to circuits: Development of the zebrafish spinal cord. *Progress in Neurobiology, 69*(6), 419–449.

Li, C., Tang, B., Feng, Y., Tang, F., Pui-Man Hoi, M., Su, Z., & Ming-Yuen Lee, S. (2018). Pinostrobin exerts neuroprotective actions in neurotoxin-induced Parkinson's disease models through Nrf2 induction. *Journal of Agricultural and Food Chemistry, 66*(31), 8307–8318.

Li, X., Zhang, Y., Wang, Y., Xu, J., Xin, P., Meng, Y., . . . Kuang, H. (2017). The mechanisms of traditional Chinese medicine underlying the prevention and treatment of Parkinson's disease. *Frontiers in Pharmacology, 8*, 634.

Limón-Pacheco, J., & Gonsebatt, M. E. (2009). The role of antioxidants and antioxidant-related enzymes in protective responses to environmentally induced oxidative stress. *Mutation Research/Genetic Toxicology and Environmental Mutagenesis, 674*(1–2), 137–147.

Lin, F.-J., Li, H., Wu, D.-T., Zhuang, Q.-G., Li, H.-B., Geng, F., & Gan, R.-Y. (2022). Recent development in zebrafish model for bioactivity and safety evaluation of natural products. *Critical Reviews in Food Science and Nutrition, 62*(31), 8646–8674.

Liu, C., Guo, F.-F., Xiao, J.-P., Wei, J.-Y., Tang, L.-Y., & Yang, H.-J. (2020). Research advances in chemical constituents and pharmacological activities of different parts of Eucommia ulmoides. *Zhongguo Zhong yao za zhi= Zhongguo Zhongyao Zazhi= China Journal of Chinese Materia Medica, 45*(3), 497–512.

Liu, H., Wang, X., Shi, Q., Li, L., Zhang, Q., Wu, Z.-L., . . . Wang, Y. (2020). Dimeric diarylheptanoids with neuroprotective activities from rhizomes of Alpinia officinarum. *ACS omega, 5*(17), 10167–10175.

Liu, H.-Q., Zhang, W.-Y., Luo, X.-T., Ye, Y., & Zhu, X.-Z. (2006). Paeoniflorin attenuates neuroinflammation and dopaminergic neurodegeneration in the MPTP model of Parkinson's disease by activation of adenosine A1 receptor. *British Journal of Pharmacology, 148*(3), 314.

Lomartire, S., & Gonçalves, A. M. (2022). An overview of potential seaweed-derived bioactive compounds for pharmaceutical applications. *Marine Drugs, 20*(2), 141.

Martinez, T. N., & Greenamyre, J. T. (2012). Toxin models of mitochondrial dysfunction in Parkinson's disease. *Antioxidants & Redox Signaling, 16*(9), 920–934.

Meredith, G. E., & Rademacher, D. J. (2011). MPTP mouse models of Parkinson's disease: An update. *Journal of Parkinson's Disease, 1*(1), 19–33.

Mhalhel, K., Sicari, M., Pansera, L., Chen, J., Levanti, M., Diotel, N., . . . Montalbano, G. (2023). Zebrafish: A model deciphering the impact of flavonoids on neurodegenerative disorders. *Cells, 12*(2), 252.

Nüsslein-Volhard, C. (2012). The zebrafish issue of development. *Development, 139*(22), 4099–4103.

Obeidat, B., Alrababah, M., Alhamad, M., Gharaibeh, M., & Ishmais, M. A. (2012). Effects of feeding carob pods (Ceratonia siliqua L.) on nursing performance of Awassi ewes and their lambs. *Small Ruminant Research, 105*(1–3), 9–15.

O'Connor, P. M. (2006). Postcranial pneumaticity: An evaluation of soft-tissue influences on the postcranial skeleton and the reconstruction of pulmonary anatomy in archosaurs. *Journal of Morphology, 267*(10), 1199–1226.

Oniszczuk, A., & Podgórski, R. (2015). Influence of different extraction methods on the quantification of selected flavonoids and phenolic acids from Tilia cordata inflorescence. *Industrial Crops and Products, 76*, 509–514.

Park, E.-J., & Jhon, D.-Y. (2010). The antioxidant, angiotensin converting enzyme inhibition activity, and phenolic compounds of bamboo shoot extracts. *LWT-Food Science and Technology, 43*(4), 655–659.

Peñalver, R., Lorenzo, J. M., Ros, G., Amarowicz, R., Pateiro, M., & Nieto, G. (2020). Seaweeds as a functional ingredient for a healthy diet. *Marine Drugs, 18*(6), 301.

Penglase, S., Moren, M., & Hamre, K. (2012). Standardize the diet for zebrafish model. *Nature, 491*(7424), 333–333.

Qi, Y., Cheng, X., Jing, H., Yan, T., Xiao, F., Wu, B., . . . Jia, Y. (2019). Effect of *Alpinia oxyphylla—Schisandra chinensis* herb pair on inflammation and apoptosis in Alzheimer's disease mice model. *Journal of Ethnopharmacology, 237*, 28–38.

Rink, E., & Wullimann, M. F. (2002). Development of the catecholaminergic system in the early zebrafish brain: An immunohistochemical study. *Developmental Brain Research, 137*(1), 89–100.

Roberts, R. A., Smith, R. A., Safe, S., Szabo, C., Tjalkens, R. B., & Robertson, F. M. (2010). Toxicological and pathophysiological roles of reactive oxygen and nitrogen species. *Toxicology, 276*(2), 85–94.

Sõukand, R., Pieroni, A., Biró, M., Dénes, A., Dogan, Y., Hajdari, A., . . . Nedelcheva, A. (2015). An ethnobotanical perspective on traditional fermented plant foods and beverages in Eastern Europe. *Journal of Ethnopharmacology, 170*, 284–296.

Streisinger, G., Walker, C., Dower, N., Knauber, D., & Singer, F. (1981). Production of clones of homozygous diploid zebra fish (*Brachydanio rerio*). *Nature, 291*(5813), 293–296.

Sun, M., Chai, L., Lu, F., Zhao, Y., Li, Q., Cui, B., . . . Liu, Y. (2018). Efficacy and safety of Ginkgo Biloba pills for coronary heart disease with impaired glucose regulation: Study protocol for a series of N-of-1 randomized, double-blind, placebo-controlled trials. *Evidence-Based Complementary and Alternative Medicine, 2018*.

Tang, H., Li, K., Zhang, S., Lan, H., Liang, L., Huang, C., & Li, T. (2021). Inhibitory effect of paeonol on apoptosis, oxidative stress, and inflammatory response in human umbilical vein endothelial cells induced by high glucose and palmitic acid induced through regulating SIRT1/FOXO3a/NF-κB pathway. *Journal of Interferon & Cytokine Research, 41*(3), 111–124.

Teytelman, L., Stoliartchouk, A., Kindler, L., & Hurwitz, B. L. (2016). Protocols. io: Virtual communities for protocol development and discussion. *PLoS Biology, 14*(8), e1002538.

Ton, C., Lin, Y., & Willett, C. (2006). Zebrafish as a model for developmental neurotoxicity testing. *Birth Defects Research Part A: Clinical and Molecular Teratology, 76*(7), 553–567.

Tsang, B., Zahid, H., Ansari, R., Lee, R. C.-Y., Partap, A., & Gerlai, R. (2017). Breeding zebrafish: A review of different methods and a discussion on standardization. *Zebrafish, 14*(6), 561–573.

Tun, T., & Kang, Y.-S. (2017). Imperatorin is transported through blood-brain barrier by carrier-mediated transporters. *Biomolecules & Therapeutics, 25*(4), 441.

Varga, Z. M., Ekker, S. C., & Lawrence, C. (2018). *Workshop report: Zebrafish and other fish models—description of extrinsic environmental factors for rigorous experiments and reproducible results.* Mary Ann Liebert, Inc.

Vascotto, S. G., Beckham, Y., & Kelly, G. M. (1997). The zebrafish's swim to fame as an experimental model in biology. *Biochemistry and Cell Biology, 75*(5), 479–485.

Vaz, R. L., Outeiro, T. F., & Ferreira, J. J. (2018). Zebrafish as an animal model for drug discovery in Parkinson's disease and other movement disorders: A systematic review. *Frontiers in Neurology*, *9*, 347.

Vekiari, S. A., Ouzounidou, G., Ozturk, M., & Görk, G. (2011). Variation of quality characteristics in Greek and Turkish carob pods during fruit development. *Procedia-Social and Behavioral Sciences*, *19*, 750–755.

Vickers, N. J. (2017). Animal communication: When i'm calling you, will you answer too? *Current Biology*, *27*(14), R713-R715.

Voss, E. G. (1983). An integrated system of classification of flowering plants—Cronquist A. *Economic Botany*, *37*, 498–498.

Wang, M., Zhang, Z., Cheang, L. C.-V., Lin, Z., & Lee, S. M.-Y. (2011). Eriocaulon buergerianum extract protects PC12 cells and neurons in zebrafish against 6-hydroxydopamine-induced damage. *Chinese Medicine*, *6*, 1–10.

Wang, X.-L., Xing, G.-H., Hong, B., Li, X.-M., Zou, Y., Zhang, X.-J., & Dong, M.-X. (2014). Gastrodin prevents motor deficits and oxidative stress in the MPTP mouse model of Parkinson's disease: Involvement of ERK1/2–Nrf2 signaling pathway. *Life Sciences*, *114*(2), 77–85.

Wang, Z., He, C., Peng, Y., Chen, F., & Xiao, P. (2017). Origins, phytochemistry, pharmacology, analytical methods and safety of cortex moutan (Paeonia suffruticosa Andrew): A systematic review. *Molecules*, *22*(6), 946.

Watts, S. A., Powell, M., & D'Abramo, L. R. (2012). Fundamental approaches to the study of zebrafi sh nutrition. *ILAR Journal*, *53*(2), 144–160.

Xiao, S., Li, D., Tang, Z., Wei, H., Zhang, Y., Yang, J., . . . Wang, W. (2023). Supplementary UV-B radiation effects on photosynthetic characteristics and important secondary metabolites in Eucommia ulmoides leaves. *International Journal of Molecular Sciences*, *24*(9), 8168.

Xing, Y.-F., He, D., Wang, Y., Zeng, W., Zhang, C., Lu, Y., . . . Xing, X.-H. (2019). Chemical constituents, biological functions and pharmacological effects for comprehensive utilization of Eucommia ulmoides Oliver. *Food Science and Human Wellness*, *8*(2), 177–188.

Xu, F.-H., Qiu, Y.-Z., Zhang, Y., Yang, F.-H., Ji, M.-M., Liu, K.-C., . . . Li, B. (2023). The molecular mechanism of three novel peptides from C-phycocyanin alleviates MPTP-induced Parkinson's disease-like pathology in zebrafish. *Food & Function*, *14*(13), 6157-6171.

Yang, X., Yang, Y., Chen, H., Xu, T., Li, C., Zhou, R., . . . Chen, Y. (2020). Extraction, isolation, immunoregulatory activity, and characterization of Alpiniae oxyphyllae fructus polysaccharides. *International Journal of Biological Macromolecules*, *155*, 927–937.

Yu, B., Ruan, M., Liang, T., & Yu, Y. (2020). Synergy between borneol and extract of Ligusticum chuanxiong hort against cortex and striatum ischemia. *International Journal of Pharmacology*, *16*(2), 104–119.

Yu, D., Zhang, P., Li, J., Liu, T., Zhang, Y., Wang, Q., . . . Fan, X. (2021). Neuroprotective effects of Ginkgo biloba dropping pills in Parkinson's disease. *Journal of Pharmaceutical Analysis*, *11*(2), 220–231.

Zhang, S., Yu, Z., Xia, J., Zhang, X., Liu, K., Sik, A., & Jin, M. (2020). Anti-Parkinson's disease activity of phenolic acids from Eucommia ulmoides oliver leaf extracts and their autophagy activation mechanism. *Food & Function*, *11*(2), 1425–1440.

Zhang, Y., Wang, Z.-Z., Sun, H.-M., Li, P., Li, Y.-F., & Chen, N.-H. (2014). Systematic review of traditional Chinese medicine for depression in Parkinson's disease. *The American Journal of Chinese Medicine*, *42*(05), 1035–1051.

Zhang, Z.-J., Cheang, L. C., Wang, M.-W., Li, G.-H., Chu, I. K., Lin, Z.-X., & Lee, S. M. (2012). Ethanolic extract of fructus Alpinia oxyphylla protects against 6-hydroxydopamine-induced damage of PC12 cells in vitro and dopaminergic neurons in zebrafish. *Cellular and Molecular Neurobiology*, *32*, 27–40.

Zhao, J., Liu, Y., Jiang, J., Wu, P., Chen, G., Jiang, W., . . . Feng, L. (2012). Effects of dietary isoleucine on growth, the digestion and absorption capacity and gene expression in hepatopancreas and intestine of juvenile Jian carp (Cyprinus carpio var. Jian). *Aquaculture*, *368*, 117–128.

Zheng, M., Liu, C., Fan, Y., Yan, P., Shi, D., & Zhang, Y. (2017). Neuroprotection by Paeoniflorin in the MPTP mouse model of Parkinson's disease. *Neuropharmacology*, *116*, 412–420.

Zhou, S., Liu, L., Zhang, Y., Zhang, Z., Li, H., Fan, F., . . . Zuo, L. (2023). Integrated untargeted and targeted metabolomics to reveal therapeutic effect and mechanism of Alpiniae oxyphyllae fructus on Alzheimer's disease in APP/PS1 mice. *Frontiers in Pharmacology*, *13*, 1104954.

16 Zebrafish Model for Studying Microbiome-Associated Parkinson's Disease

Vishnu Prasad Nair R U, Rajesh Pamanji, and Joseph Selvin

16.1 INTRODUCTION

16.1.1 Exploring Zebrafish in Parkinson's Disease and Neurobehavioral Research

Parkinson's disease (PD) is a significant neurodegenerative condition marked by symptoms such as tremors, muscle stiffness, reduced mobility, and walking difficulties (Sveinbjornsdottir, 2016). It is characterized by the deterioration of dopaminergic neurons in the substantia nigra, accompanied by the accumulation of α-synuclein and the development of Lewy bodies in the remaining neurons (Schneider & Alcalay, 2017).

Zebrafish serve as a highly effective vertebrate system for studying the genetic and other elements that impact the development of neurobehavioral traits (Granato et al., 1996). Because of its compact size, rapid growth, transparent nature, external fertilization and development, aptitude for chemical genetic screening, and the abundance of genomic information available, this organism is widely recognized and valued as a prominent research subject (Howe et al., 2013). The zebrafish brain shares a structural resemblance to that of mammals, encompassing regions like the forebrain, midbrain, and hindbrain, which also include the diencephalon and telencephalon. Furthermore, the principal functions of neurotransmitters in zebrafish closely parallel those observed in humans (Caramillo & Echevarria, 2017; Lieschke & Currie, 2007; Panula et al., 2006; Toledo-Ibarra et al., 2013). Zebrafish, like mammals, possess a blood-brain barrier, which acts as a protective mechanism governing the transfer of substances between the bloodstream and the brain. Permeability tests have indicated that the physiological characteristics of the blood-brain barrier in zebrafish are preserved and resemble those found in humans. (Cuoghi & Mola, 2007). Apart from their physiological advantages, zebrafish also display complex cognitive behaviors, including the ability to learn and remember associations. They have been extensively studied for their capacity to exhibit anxiety-related behaviors, which are well-documented and can be observed and measured in experimental settings (Lieschke & Currie, 2007).

16.1.2 Zebrafish as a Model for Microbiome Research

Zebrafish provide a valuable alternative model for microbiome-based research due to their ease of generating axenic animals (Milligan-Myhre et al., 2011; Pham et al., 2008). Studies have demonstrated that the gut microbiota plays an active role in several facets of the host's physiology and behavior. These encompass involvement in immune responses, facilitation of gut maturation, processing of nutrients, and even exerting an influence on social behavior. (Bates et al., 2006; Rawls et al., 2004; Soares et al., 2019; Sullivan & Kim, 2008). Zebrafish and humans share numerous resemblances in terms of intestinal gene function, cell structure, and molecular mechanisms. Comparative genomics investigations have unveiled that approximately 70% of human genes have a counterpart in the zebrafish genome. Moreover, nearly 80% of human genes linked to diseases can be traced to at least one analogous gene in the zebrafish genome. These parallels establish zebrafish as an invaluable model organism for delving into human diseases and the genetic factors that underlie them (Shehwana & Konu, 2019).

The gene expression patterns related to carbohydrate metabolism in the intestine are remarkably similar between fish, including zebrafish, and humans. Zebrafish possess counterparts for various apolipoproteins, enzymes involved in lipid metabolism, and transporters for metabolites found in the human gut. This preservation of genes associated with lipid metabolism underscores the promise of zebrafish as a model organism for investigating human intestinal physiology and disorders related to lipid metabolism (Quinlivan & Farber, 2017). Based on reports, it has been discovered that gut microbes can impact as many as 212 genes within the digestive tract. This conclusion was drawn through DNA microarray analysis, which compared the gene expression profiles of germ-free zebrafish at 6 days post-fertilization (dpf), which were raised without microorganisms, to conventionally raised zebrafish. Among these influenced genes, 59 are shared with the mouse intestine. These genes play pivotal roles in crucial functions such as nutrient metabolism, epithelial cell growth, and the innate immune response. This research underscores how gut

DOI: 10.1201/9781003402893-18

microbes can profoundly influence the expression of essential genes related to various aspects of intestinal physiology and immune function. These findings may have implications not only for zebrafish but also for mammalian models like mice (Rawls et al., 2004).

Gut bacteria play a substantial role in human health by engaging in communication with human cells and triggering immune and inflammatory responses. This interaction between gut bacteria and human cells can result in the regulation of the immune system and the initiation of inflammatory processes. The impact of gut bacteria on immune and inflammatory functions is progressively gaining recognition as a pivotal component of overall human well-being. This recognition carries implications for a wide range of conditions and diseases, encompassing autoimmune disorders and inflammatory bowel diseases (Murdoch & Rawls, 2019; Semova et al., 2012).

In the intestinal epithelium of zebrafish, macrophages and leukocytes play a role in responding to pathogens and foreign invasions by getting activated. However, distinguishing between different subpopulations of these cells in zebrafish is challenging due to the need for specific antibodies for fish. This limitation hampers our understanding of the precise subpopulations and their particular functions within the zebrafish intestinal immune response. Efforts are ongoing to develop and optimize antibodies and markers that can aid in better characterizing these cell subpopulations in zebrafish (Li et al., 2017; Novoa & Figueras, 2012; Torraca et al., 2014).

16.2 FACTORS THAT INFLUENCE THE GUT MICROBIOTA

Multiple investigations indicate that gut microbiota diversity is influenced by factors that are either external or internal to each species (Figure 16.1), and the gut microbiota within host organisms exhibit variations from one day to another as well as distinct individual differences (Breen et al., 2019; Butt & Volkoff, 2019; Patula et al., 2021; Yukgehnaish et al., 2020).

16.2.1 External Factors

Various external factors can commonly influence the makeup and establishment of gut microbiota in zebrafish (Cornuault et al., 2022; Marques et al., 2006; Robinson et al., 2018; Zhang et al., 2020b). The extensively studied external factors encompass aquatic environmental pollutants, dietary patterns, antibiotic usage, and pathogens. Given zebrafish's aquatic nature, their gut microbiota is closely connected to their surrounding environment. Notably, zebrafish gut microbiomes exhibit substantial shifts across diverse aquatic habitats. Furthermore, the composition of gut bacterial communities differs among domestically reared adult zebrafish populations within distinct laboratory aquaculture settings situated in separate geographical locations (Roeselers et al., 2011). Concurrently, the alpha and beta diversity of the zebrafish intestinal microbiome undergoes rapid alterations immediately upon transitioning to different housing systems (Breen et al., 2019).

Antibiotics are employed to safeguard against and manage illnesses stemming from bacterial infections, especially for addressing critical or life-threatening situations such as pneumonia and sepsis in human patients (De Waele & Dhaese, 2019; Kollef & Betthauser, 2019; Svensson et al., 2015b). However, due to their high water solubility, low bioavailability, and extensive misuse, diverse residues of antibiotics are commonly identified in aquatic ecosystems. This environmental contamination by trace amounts of antibiotics within water sources can lead to an imbalance in the gut microbiota of zebrafish, potentially compromising their resistance to pathogenic infections (Kayani et al., 2021; Zhou et al., 2018b). To provide an example, prolonged exposure to oxytetracycline, a broad-spectrum antibiotic, leads to alterations in the makeup of intestinal bacteria. Specifically, this results in an increased presence of Alphaproteobacteria and Actinobacteria in adult zebrafish (Almeida et al., 2019a, 2019b). However, according to Zhou and colleagues, the exposure to oxytetracycline and sulfamethoxazole diminishes the diversity of intestinal microbes while promoting the prevalence of certain pathogens, notably Aeromonas and Enterococcus, in adult zebrafish (Zhou et al., 2018a). Studies have also indicated that subjecting zebrafish to environmentally relevant levels of antibiotics during their early developmental stages leads to an increase in harmful bacteria, the emergence of antibiotic-resistant genes, and an improvement in the metabolic capabilities of the zebrafish gut microbiome (Kayani et al., 2021).

Additionally, the administration of antibiotics reduces the diversity of intestinal microbes in adult zebrafish (Ding & He, 2010; Pindling et al., 2018). In a separate investigation, the application of low-dose ciprofloxacin, a broad-spectrum antibiotic, within the aquatic environment significantly disrupts the gut microbial communities in larval zebrafish. This disruption is associated with increased bacterial aggregation and heightened intestinal expulsion (Schlomann et al., 2019).

The presence of pathogens in the host's environment can be an external factor that triggers changes in the gut microbiota of zebrafish. The presence of the waterborne bacteria *Aeromonas hydrophila* causes disturbance in the equilibrium of the gut by suppressing beneficial bacteria and promoting harmful ones. This results in the activation of innate immune responses within the zebrafish intestine (Yang et al., 2017). The intestinal pathogen *Vibrio cholerae* colonizes the zebrafish intestine, changing the gut's microbial makeup. Additionally, it triggers the activation of type six secretion systems (T6SS), which play a crucial role in its pathogenicity (Breen et al., 2019, 2021; Logan et al., 2018).

16.2.2 Internal Factors

The gut microbiota of zebrafish is influenced by various intrinsic factors such as genetics, age, health status, and sex (Yukgehnaish et al., 2020). Host factors are

FIGURE 16.1 Factors that influence the gut microbiota.

central in influencing the differences in intestinal bacteria observed between zebrafish and other species. When mouse intestinal microbiota is transplanted into zebrafish, the resulting microbial composition is influenced by the zebrafish's specific physiological, hormonal, and neurological processes. This underlines the significance of host-related factors in shaping the diversity of bacterial communities, which can also be observed in other species like mammals. Dietary changes impact host physiology and development by modifying the gut microbiota, a phenomenon seen in mammals and zebrafish (Emani et al., 2013; López Nadal et al., 2020).

The composition of microbes undergoes dynamic shifts as the host ages, and the influence of gut microbiota on health becomes especially relevant in older individuals (O'Toole & Jeffery, 2015). Significantly, a recent study highlights that the predominant factor influencing the microbial composition of zebrafish is the developmental stage (Xiao et al., 2021). Until 75dpf, the microbial community composition in zebrafish undergoes consistent changes throughout their development (Stephens et al., 2016). Cantas et al. (2012) reported that zebrafish juveniles exhibit a greater bacterial richness in their intestinal microbiota than they do in later developmental stages. Likewise, zebrafish's microbiomes showed changes in composition and diversity across their lifespan, evident through the decrease in α-diversity and the increase in β-diversity throughout their development (Stephens et al., 2016).

The importance of host health in preserving a harmonious microbial community within the constantly changing intestinal environment is well-established. Zebrafish and humans exhibit particular resemblances in their innate and adaptive immune systems. During the early stages of life, especially up to 4–6 weeks after fertilization, larval zebrafish rely solely on their innate immune system to defend against microbial invasions, as their adaptive immunity has not yet matured (Flores et al., 2020; Lam et al., 2004).

16.2.3 The Role of Microbe-Associated Molecular Patterns (MAMPs)/Pathogen-Associated Molecular Patterns (PAMPs)

Epithelial cells in the host's intestines have the ability to detect the presence of gut microbes by recognizing certain molecular patterns associated with the microbes. An example of this is when pattern recognition receptors (PRRs) recognize lipopolysaccharides (LPS). This interaction triggers downstream signalling pathways, leading to the activation of conserved signalling pathways. As a result, the expression of antimicrobial factors and cytokines is induced. This sensing and response mechanism helps the host's intestinal epithelial cells mount immune responses against microbial threats and maintain a balanced microbial environment in the gut (Negi et al., 2019).

16.2.4 Antimicrobial Peptides (AMPs)

Antimicrobial peptides (AMPs) are essential in the innate immune systems of diverse organisms, such as humans, animals, and plants. They serve as the primary defence mechanism against external threats (Kościuczuk et al., 2012; Starr et al., 2018). In zebrafish, intestinal AMPs expression has been observed. These include defensin beta-like peptides and hepcidin. AMPs play a crucial role in the innate immune response of the intestine by targeting and eliminating pathogens, helping maintain gut homeostasis, and protecting against infections. The presence of these AMPs in zebrafish highlights their importance in the defence mechanisms of the zebrafish intestinal tract (Brugman, 2016). It has been observed that in various neurodegenerative diseases, including Parkinson's disease (PD), the characteristic proteins associated with these conditions can exhibit antimicrobial properties. In the case of PD, alpha-synuclein, a protein closely linked to the disease, has been found to have antimicrobial activities in some in vitro studies. Specifically, alpha-synuclein has demonstrated antimicrobial effects against *Escherichia coli*, *Staphylococcus aureus*, and certain types of fungi. These findings suggest that there may be a connection between microbial pathogens and the pathological processes involved in neurodegenerative diseases, including PD. Further research is needed to fully understand the implications of this antimicrobial activity and its significance in the context of PD (Park et al., 2016).

16.2.5 Brain-Derived Neurotrophic Factor (BDNF)

Gut bacteria also impact the central nervous system's (CNS) development and function (Diaz Heijtz et al., 2011). Germ-free (GF) models, which lack gut microbiota, have been observed to exhibit learning and memory deficits, hyperactivity, and alterations in neurotrophic factors compared to conventional models. One of the key neurotrophic factors involved in human neuronal and synaptic development is brain-derived neurotrophic factor (BDNF) (De Felice et al., 2014). BDNF synthesis begins with the translation of its precursor protein, pre-pro-BDNF, in the rough endoplasmic reticulum (ER). Pre-pro-BDNF contains a signal peptide that guides it into the ER lumen. The signal peptide cleaves off within the ER through endoproteolytic cleavage, forming pro-BDNF. Pro-BDNF is subsequently transported through the secretory pathway to the Golgi apparatus. In the Golgi, additional endoproteolytic cleavage occurs, removing a segment from pro-BDNF to generate the mature form of BDNF. The mature BDNF is then packaged into secretory granules in post-synaptic dendrites or vesicles in pre-synaptic terminals for transportation. These secretory granules or vesicles containing mature BDNF are then transported along the neuronal processes to their destinations. In post-synaptic dendrites, they are released to act locally, while in pre-synaptic terminals, they are involved in synaptic transmission and modulation (Leal et al., 2015).

Interestingly, BDNF has been detected in the brain and different organs of zebrafish. This suggests that BDNF may play a crucial role in various biological processes beyond the central nervous system in zebrafish (De Felice et al., 2014). Studies conducted by Borrelli et al. (2016) found that when zebrafish gut is administrated with *Lactobacillus rhamnosus*, it significantly increases the expression of the BDNF gene; this was in correlation with the shoaling pattern. The group that received probiotics (*Lactobacillus rhamnosus*) exhibited greater uniformity and more consistency within the shoal. The microbiota has the capability to raise the levels of BDNF in the hippocampal gyrus of the brain. This increase has been associated with animal functions such as memory, learning, and exploratory behavior (Bercik et al., 2011). Recent research indicates that BDNF has a significant role in gastrointestinal function and an elevated BDNF level was observed in the colon of individuals with diarrhoea-predominant IBS who experienced abdominal pain (Wang et al., 2015).

The exact mechanisms by which gut microbes influence BDNF and its impact on neurodevelopment and behavior warrant further investigation.

16.3 SEROTONERGIC REGULATION

In vertebrates, the serotonergic system plays a role in regulating socio-emotional behavior (Shams et al., 2015). Zebrafish is frequently employed as a model organism in studies centered on behavior regulated by the CNS. One of the reasons for this is their natural shoaling behavior and active social interactions, which can be easily quantified and analyzed. Shoaling refers to the tendency of zebrafish to swim and group together in coordinated patterns, making them an excellent model for studying social behavior and social cognition. Researchers can gain insights into the underlying neural circuits and molecular mechanisms involved in CNS-regulated behaviors by analyzing the quantitative aspects of their social behavior. The zebrafish's amenability to high-throughput behavioral assays further enhances its utility in behavioral research (Gerlai, 2014). Zebrafish possess a complex serotonergic system that includes homologous genes of serotonin transporters, receptors, and tryptophan hydroxylase. These components are crucial for synthesising, transporting, and signalling serotonin, a neurotransmitter known to regulate various physiological processes, including mood, behavior, and gastrointestinal function. The presence of homologous genes in zebrafish suggests that they share similarities with humans regarding serotonergic regulation, making zebrafish a valuable model organism for studying serotonin-related processes and disorders (Panula et al., 2010). Furthermore, the impact of microbes on the serotonergic system could serve as a significant mediator of gut–brain signalling (Yano et al., 2015).

Tempeh, a traditional food product from Southeast Asia and particularly popular in Indonesia, is created by fermenting cooked and dehulled soybeans using Rhizopus fungi (Nout & Kiers, 2005). The study examined how various types of tempeh affect the expression of specific genes related to serotonin signalling and metabolism (htr1aa, mao, slc6a4a, tph1a, tph1b, and tph2) in both the gut and brain of zebrafish. In the brain, there was a noticeable increase in the expression of tph1b, tph2, and slc6a4a genes, indicating changes in the serotonergic system as a result of tempeh consumption. However, in the gut, the expression of these genes decreased significantly, except for slc6a4a. This implies that different types of tempeh can lead to tissue-specific modifications in the serotonergic system, underscoring the intricate interplay between diet, gut, and brain functions (Chen et al., 2021).

16.4 HUMAN GUT VS. ZEBRAFISH GUT

The difference in the types of microorganisms found in the intestines of zebrafish and humans is more significant compared to what is seen in mice. Fish harbour approximately

10^7 to 10^{11} bacteria per gram of intestinal content, whereas mice exhibit around 10^8, and humans have about 10^{14} bacteria per gram (Wang et al., 2018). While the molecular responses towards microbes are conserved between zebrafish and humans, there are distinct differences in the composition of gut microbes between these two organisms. In the human gut, the dominant phyla of microbes are Bacteroidetes and Firmicutes. However, in the gut of zebrafish, Proteobacteria is the dominant phylum.

Additionally, the intestinal microbiome of zebrafish comprises Actinobacteria, Firmicutes, Verrucomicrobia, Bacteroidetes, Fusobacteria, and Proteobacteria. Although Bacteroidetes and Firmicutes are also present in zebrafish, obligate anaerobic bacteria find it difficult to survive in the fish gut due to differences in the gut environment. These differences in gut microbial composition between zebrafish and humans emphasize the need to consider the specific microbial communities when studying the gut–brain axis and other aspects of host-microbe interactions in different organisms (de Abreu et al., 2019; Wang et al., 2018).

16.5 COLONIZE THE HUMAN GUT MICROBIOME INTO ZEBRAFISH

You are correct that the host's genetically encoded physiology environment and the food and living environment play critical roles in shaping the gut microbiota. The differences between humans and zebrafish in genetics, physiology, diet, and environmental factors make colonising the zebrafish gut with a human-like microbiome challenging. Multiple factors, including host genetics, immune system, diet, and environmental exposures, among others, influence the gut microbiota of an organism. Therefore, while it is possible to manipulate the zebrafish microbiome to some extent, replicating the exact composition and functionality of the human gut microbiome in zebrafish is not a straightforward task. Researchers must consider these inherent differences and carefully design studies to investigate host-microbe interactions in zebrafish while acknowledging the limitations imposed by the species-specific variations in microbiota (Lu et al., 2021b). Transplanting human gut microbes into the intestines of zebrafish holds significant promise for using zebrafish as a model for investigating research related to human gut microbes. The successful colonization of human gut microbes in the zebrafish intestine is influenced by both the host's genome-encoded gut environment and environmental factors. In a study conducted by Rawls et al. (2006), research has indicated that the gut determines its microbial components by taking into account variations in host genome-encoded gut environments, rather than solely depending on the existing local microbial community. This was demonstrated through reciprocal transplantation experiments involving germ-free zebrafish and mouse recipients. These discoveries imply that the zebrafish gut environment and its specific genetic elements play a pivotal role in shaping the composition and establishment of the microbial community. By gaining a deeper understanding of these interactions between the host and microbes and by developing effective transplantation strategies, zebrafish could evolve into a valuable tool for studying the dynamics of the human gut microbiome and their implications for health and disease (Rawls et al., 2006).

This reduction of oxygen creates an anaerobic environment that favours the subsequent implantation and growth of anaerobic bacteria. This sequential colonisation process is crucial for establishing healthy and diverse gut microbiota in humans. However, replicating this process in fish, such as zebrafish, can be challenging due to differences in the initial microbial colonisation patterns and the specific environmental conditions required to implant anaerobic bacteria successfully. Overcoming these challenges and achieving stable colonisation of human gut microbes in fish intestines remains an ongoing area of research and may require further optimisation of experimental approaches and conditions (Toh et al., 2013), whereas the intestine of zebrafish larvae is expected to have a higher oxygen concentration than the intestine of mammals (Rawls et al., 2006).

Certainly, achieving a consistent colonization of human gut microbes within the fish intestine remains a challenging task. The establishment of the microbiota in newborn humans begins with the introduction of aerobic microorganisms that consume the available oxygen in the intestine. Experiments were carried out using conventionally raised zebrafish larvae that already had their own established aerobic gut microbiota. The goal was to facilitate the colonisation of zebrafish larvae by less aerotolerant or anaerobic human microorganisms. The study revealed the presence of certain microbial species, including Enterobacteriaceae, Bacillus, Roseburia, Prevotella, Oscillospira, and unclassified genus from the family Ruminococcaceae, in the 3-day-old larvae. However, only Bacillus remained detectable until the seventh day. It's worth noting that sporulating bacteria such as *Clostridioides difficile* and *Bacillus clausii* displayed the highest persistence among the microorganisms investigated. These findings suggest that the larvae's developmental stage and bacteria's sporulation ability may influence their colonization capacity. The results highlight the importance of considering both the larval development stage and the bacterial strains' characteristics in facilitating the successful colonization of zebrafish's less aerotolerant or anaerobic human microorganisms. Further research is needed to better understand the factors that impact colonization ability and to optimise strategies for manipulating the zebrafish gut microbiota (Lu et al., 2021b; Valenzuela et al., 2018).

The studies conducted thus far have revealed that only a small portion of the human intestinal microbiota can successfully establish itself in the zebrafish digestive system, which is lower than what has been observed in experiments with mouse models. There are several potential reasons for this difference. Firstly, zebrafish inhabit an aquatic environment, and the conditions in their gut are notably distinct from those in mammalian systems. The unique physiology, anatomy, and behavior of zebrafish are likely to exert an influence on the composition of their microbial communities. Secondly, there is a considerable difference in habitat temperature between zebrafish (26–29°C) and mammals (35–37°C), and this temperature variance may impede the colonization and growth of human intestinal bacteria within the zebrafish gut (Uzbay, 2019).

The higher oxygen levels present in the zebrafish intestine may create a less favourable environment for the colonization of anaerobic human microorganisms. Additionally, the absence of specific food sources in these studies could be a limiting factor for certain microbes attempting to colonize the zebrafish gut. However, introducing food as a variable would add further complexity to the experiments. Furthermore, it's crucial to note that the human intestinal microbiota used in these studies was sourced from a single donor. It would be advantageous to analyse samples from multiple individuals because the composition of gut microbiota varies with developmental age and among individuals. Lastly, when investigating microbial characteristics that contribute to host adaptation, it's essential to consider the entire host-microbe-environment system. Understanding the comprehensive microbial colonization in zebrafish requires taking into account the complex interactions among the host, microbiota, and environmental factors (Uzbay, 2019).

16.6 DIGESTIVE SYSTEM DEVELOPMENT

Zebrafish serve as valuable models for studying how the gut microbiota affects the development of the human intestinal tract, including gene expression, cell proliferation, tissue differentiation, and associated processes. The embryos first come into contact with microorganisms outside after hatching (about 48 or 96 hours after fertilization), when they are still essentially in an axenic chorion. When a larva hatches, most organs interact with the microbiota (1). The digestive tracts of GF and common zebrafish were initially compared for gene expression. Researchers have found that the gut microbiota has influenced over 200 genes, with 59 of them showing consistent expression in the mouse intestine. These genes that are linked to the microbiota mainly play roles in immune response, nutrient absorption, renewal of epithelial cells, and processing of foreign substances (Rawls et al., 2004). It is known that the overall growth kinetics of the gut microbiome is affected by the spatial dispersion of the microbiome in relation to both its host and itself (Jemielita et al., 2014). Similar to humans, zebrafish establish a complex microbial community in their intestines that changes throughout developmental time and shows significant inter-individual variance (Roeselers et al., 2011; Lozupone et al., 2012; Stephens et al., 2016).

16.7 GUT MICROBIOME AND PD

Lewy bodies are intracytoplasmic inclusion bodies that appear eosinophilic and serve as distinctive intracellular pathological markers for PD and dementia (Beyer et al., 2009). Brain tissue, as well as samples taken from the distal oesophagus and gastric wall of PD patients, were collected. The analysis uncovered the existence of immunoreactive α-synuclein inclusion bodies, which are abnormal aggregations of the α-synuclein protein, in the brain and within the neurons of Auerbach's and Meissner's plexuses (Braak et al., 2006). Expanding on these discoveries, Braak and collaborators developed a theory suggesting that the pathological advancement of PD is instigated by an unidentified pathogenic agent that breaches the gastric epithelial barrier. This pathogen initiates the abnormal accumulation of α-synuclein within nerve cells situated in the submucosal plexus. After these initial occurrences, it is postulated that the pathogen propagates in a retrograde manner through axonal connections and travels transneuronally along the vagus nerve, eventually reaching subcortical brain nuclei (Braak et al., 2006). Numerous animal experiments have aimed to investigate and substantiate Braak's theory by introducing different forms of α-synuclein into the intestinal wall of rats. The results of these investigations illustrated that α-synuclein can indeed travel in a retrograde manner along the vagus nerve. This transport mechanism was observed to extend to particular areas within the brain, including the dorsal motor nucleus of the vagus located in the brainstem and the substantia nigra pars compacta, where a notable decline in dopaminergic neurons was noted (Holmqvist et al., 2014; S. Kim et al., 2019; Uemura et al., 2018; Van Den Berge et al., 2019).

The microbiome-gut–brain axis represents a bidirectional communication system, wherein the gut microbiome can influence the functioning and activity of the brain through its metabolic byproducts and its capacity to stimulate the enteric nervous system (ENS). Additionally, it can transmit neural signals directly to the brain through the vagus nerve (Cryan et al., 2019; Galland, 2014). Therefore, as evidenced by both animal studies and clinical research, the vagus nerve assumes a crucial role in the neural pathway associated with the microbiome-gut–brain axis. It influences gut permeability and mediates alterations in the central nervous system (CNS) signalling in response to changes in the gut microbiome (Bravo et al., 2011; Dinan & Cryan, 2017; Svensson et al., 2015a). The brain has the capacity to impact gut functions through the stress response, as demonstrated by the connection between stress-related psychiatric conditions such as anxiety and gastrointestinal disorders, including irritable bowel syndrome and inflammatory bowel disease (Cryan & Dinan, 2012; Reber, 2012).

The gastrointestinal microbiota (GM) continuously interacts with the enteric nervous system (ENS). These interactions occur as bacteria within the GM ferment dietary fibres, producing and releasing neuroactive metabolites (Thomas et al., 2012). Moreover, various components of bacterial cell walls, such as the polysaccharide coating, can either inhibit or stimulate neuronal transmission within the ENS (Figure 16.2) (Forsythe & Kunze, 2013).

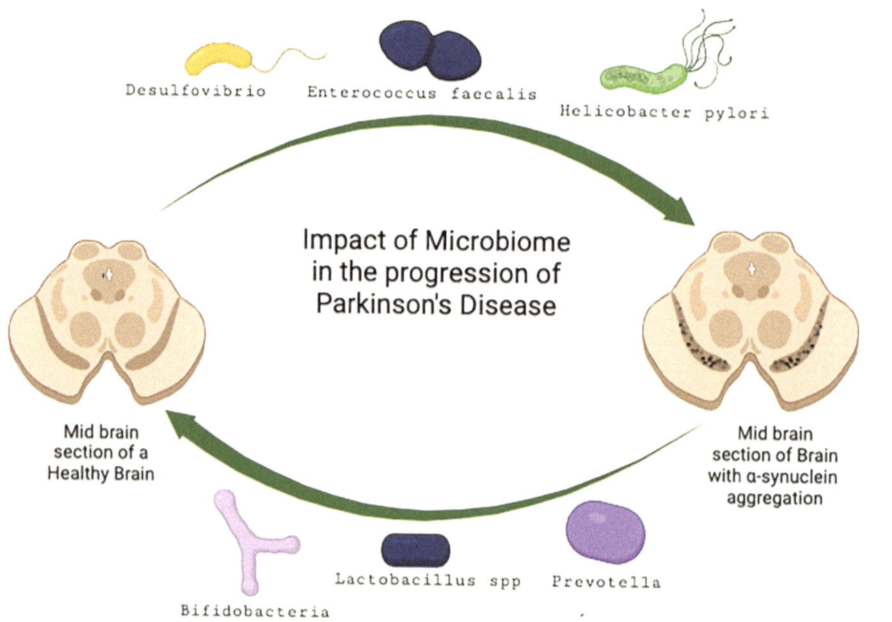

FIGURE 16.2 Impact of microbiome on Parkinson's disease progression.

Alterations in the composition of the gut microbiota lead to shifts in the levels of metabolic products produced by different genera of gut microbes, as well as neuroactive compounds and other bioactive substances whose production is influenced by changes in the gut microbiome. These alterations can potentially reduce the synthesis of endogenous neuroprotective substances through direct or indirect mechanisms (Elfil et al., 2020). Short-chain fatty acids (SCFAs) represent a significant metabolic product of the gut microbiome, and their production rate is subject to variation based on the specific composition of the gut microbiota (Baxter et al., 2019; Jang et al., 2019; Sanna et al., 2019). SCFAs can serve as an immediate energy source for the colonic mucosa, or they may enter the portal circulation and eventually influence the brain. This influence on the brain occurs through the elevation of factors such as glial cell line-derived neurotrophic factor and brain-derived neurotrophic factor (Ríos-Covián et al., 2016; Wu et al., 2008).

Newly found evidence indicates that the development of α-synucleinopathy, a disorder marked by the irregular accumulation of the α-synuclein protein, initiates in the ENS and then advances to the CNS in the early stages of the illness. This pattern coincides with the appearance of particular gastrointestinal symptoms (Lebouvier et al., 2009; Natale et al., 2011). Evidence gathered from a study involving a mouse model of PD, where α-synuclein is overexpressed, indicates that the microbiota plays a significant role in the progression of this neurodegenerative disorder. The colonization of GF mice with PD patients' faeces resulted in more physical impairments than colonization with healthy controls' faeces (Sampson et al., 2016). Additionally, individuals diagnosed with PD displayed an increased presence of bacteria that could potentially cause inflammation, as well as a decrease in bacteria that are known to have anti-inflammatory properties, as observed in samples from their feces and sigmoid mucosal biopsies. These findings correlate with the inflammation-related misfolding of α-synuclein and the pathological alterations observed in the CNS that are associated with PD (Keshavarzian et al., 2015a). The gut's bacterial composition plays a role in influencing PD, and further research has furnished substantial evidence for the disease's connection to gut dysbiosis. The severity of symptoms, including issues like postural instability and gait abnormalities, has been associated with shifts in the abundance of specific species within the Enterobacteriaceae family (Barichella et al., 2019; Scheperjans et al., 2015). Moreover, a reduction in the abundance of Lachnospiraceae bacteria is linked to the worsening of both motor and non-motor symptoms in individuals with PD (Unger et al., 2016).

16.8 THE GUT–BRAIN AXIS

It is now known that the gut microbiota significantly impacts zebrafish's neurochemistry and CNS. It's significant to note that microbial colonisation is necessary for the zebrafish nervous system's proper growth and physiological functioning. It has been discovered that sterile or antibiotic-treated zebrafish showed increased locomotor behavior or hyperactivity in this regard (Phelps et al., 2017). Additionally, therapy with the *Lactobacillus plantarum* strain in zebrafish with an intestinal disease reduced anxiety, depressive-like behavior and stress response (Davis et al., 2016). According to Borrelli et al., the probiotic *Lactobacillus rhamnosus* treatment group's gut microbial composition varied, with a notable rise in Firmicutes and a decline in Proteobacteria, including pathogens like Plesiomonas and Vibrio. In this regard, probiotic *L. rhamnosus* feeding considerably changed zebrafish's social and exploratory behavior and the expression levels of brain-derived neurotrophic factor, serotonin, and other endogenous neuroactive substances (Borrelli et al., 2016). According to research by Cuomo et al., feeding zebrafish larvae *L. rhamnosus* caused the DNA methylation for the Tph1A and BDNF promoter genes to be re-established in the gut and the brain. As a result, changes in the gut microbiota may impact the host's epigenetic environment and have long-term effects on particular gene regions (Cuomo et al., 2021). Gut microbiota may alter the physiology of the brain by controlling synaptogenesis, neurotransmitters, and neurotrophic factors (Sudo et al., 2004).

16.9 MICROBIOME AS AN INDICATOR FOR PD

A considerable proportion of individuals diagnosed with PD, as many as 80%, experience symptoms of constipation and delayed gastric emptying. Interestingly, these gastrointestinal problems often appear years prior to the development of the motor symptoms typically associated with PD (Mulak & Bonaz, 2015). The gut, which harbors a significant portion of our microbiome, serves as a connecting link between our bodies and the external environment. It also plays a pivotal role in the maturation and regulation of our immune system. Considering that the gastrointestinal (GI) microbiota has the capacity to impact synucleinopathies, there is evidence suggesting the participation of bacterial substances in influencing the aggregation of α-synuclein. Specifically, both lipopolysaccharide and various subunits of amyloid proteins from *E. coli* have been implicated in this process (Bhattacharyya et al., 2019; Chen et al., 2016; Evans et al., 2015; Kim et al., 2016).

16.10 INTERACTION BETWEEN HOST AND MICROBIOME

The interactions between the host and the microbiome are multifaceted, complex, and occur bidirectionally. It is widely acknowledged that gut bacteria have the ability to influence the host's immune response. For example, polysaccharides produced by *Bacteroides fragilis* and *Bifidobacterium breve* have been demonstrated to stimulate the production of regulatory T-cells (T regs) and enhance the secretion of the anti-inflammatory cytokine IL-10 (Blander et al., 2017; Donaldson et al., 2016). The broader consequences of gut inflammation and its potential connection to neurodegeneration have received considerable attention in recent times. Additionally, specific bacteria in the gut possess the ability to generate SCFAs. These SCFAs serve not only as a primary energy source for colonic epithelial cells but also have the capacity to promote the generation of regulatory T-cells (T regs) (Zuo & Ng, 2018). Butyrate, a specific type of SCFA, has been shown to have regulatory roles in the differentiation of dendritic cells (Qiang et al., 2017).

16.11 MICROBIOME FOR THE TREATMENT OF PD

Epidemiological research has linked various neuropsychiatric disorders, including anxiety, depression, autism spectrum disorder, and Parkinson's disease, to abnormal changes in the microbiota associated with the host (Kang et al., 2013; Keshavarzian et al., 2015b; Pirbaglou et al., 2016).

16.12 PROBIOTICS, PSYCHOBIOTICS, PREBIOTICS, SYNBIOTICS, AND POSTBIOTICS

Probiotics are living microorganisms that, when consumed in sufficient amounts, confer health benefits to the host. Psychobiotics, on the other hand, represent a distinct category of probiotics that specifically provide mental health advantages to individuals (Dinan et al., 2013). Psychobiotics are presently investigated as potential direct or supplementary treatments for psychiatric and neurodevelopmental disorders. There is also ongoing research exploring their potential application in neurodegenerative diseases. Consequently, psychobiotics have the potential to become innovative therapeutic choices in the clinical treatment of various brain-related conditions. Prebiotics, on the other hand, typically refer to non-digestible food components, often in the form of fibers, that have a positive influence on the host's health. They achieve this by selectively enhancing the growth and/or activity of particular genera of beneficial bacteria, such as Lactobacilli and Bifidobacteria (Klaenhammer et al., 2012). The term "synbiotics" is used to describe the combination of prebiotics and probiotics. Prebiotics are substances that promote the growth and activity of specific probiotic strains. On the other hand, postbiotics are functional bioactive compounds produced by live bacteria or released after bacterial cell breakdown. These can include cellular components like cell walls, secretions, and metabolites such as SCFAs. Similar to live probiotic cells, postbiotics also have beneficial effects on the host's health when consumed in sufficient quantities (Wegh et al., 2019).

The use of a probiotic blend comprising four strains—*Bifidobacterium bifidum*, *Lactobacillus reuteri*, *Lactobacillus acidophilus*, and *Lactobacillus fermentum*—has demonstrated a reduction in scores on the Movement Disorders Society–Unified Parkinson's Disease Rating Scale (MDS-UPDRS). This indicates that the probiotic mixture may benefit the symptoms associated with PD (Tamtaji et al., 2019). In a study conducted by Lu et al., it was shown that supplementation with *Lactobacillus plantarum* PS128 for a duration of 12 weeks, alongside regular antiparkinsonian medication, resulted in an improvement in the motor scores assessed using the UPDRS and the quality of life of patients with PD. This suggests that *L. plantarum* PS128 supplementation may have potential benefits in managing motor symptoms and enhancing the overall well-being of individuals with PD (Lu et al., 2021a). While the precise mechanisms through which probiotics exert their effects against PD remain not completely understood, they are believed to act through several pathways. These pathways may encompass the regulation of oxidative stress levels, control of inflammatory and anti-inflammatory processes, and potential influence on apoptosis (programmed cell death). Although additional research is necessary to clarify the exact mechanisms, these multiple avenues suggest that probiotics could play a multifaceted role in mitigating the onset and progression of PD (Zhu et al., 2022).

Research has demonstrated that *Lactobacillus plantarum* DP189 is an effective psychobiotic for reducing the aggravation of α-synuclein in mice with MPTP-induced PD. The beneficial effects of *L. plantarum* DP189 are attributed to its regulation of oxidative damage, inflammation, and the restoration of gut microbiota dysfunction. By modulating these factors, *L. plantarum* DP189 shows promise in mitigating α-synuclein-related pathology in PD (Wang et al., 2022). In a murine model of PD, a specific mixture of lactic acid bacteria was tested and found to have positive effects. This mixture included *Lactobacillus plantarum* CRL 2130, which is known for its ability to produce riboflavin, *Streptococcus thermophilus* CRL 807, an immunomodulatory strain, and *Streptococcus thermophilus* CRL 808, a producer of folate. The administration of this lactic acid bacteria mixture was associated with improvements in motor behavior and a reduction in neuroinflammation in the PD model. These findings suggest that the selected combination of lactic acid bacteria strains has the potential to positively impact motor symptoms and alleviate neuroinflammation associated with PD (Perez Visñuk et al., 2020). In the study conducted by Srivastav et al., it was discovered that a probiotic mixture comprising *Lactobacillus rhamnosus* GG (LGG), *Bifidobacterium animalis lactis*, and *Lactobacillus acidophilus* demonstrated a significant increase in butyrate levels. This increase in butyrate subsequently led to the protection and rescue of the nigral dopaminergic neurons from neurotoxicity induced by MPTP (1-methyl-4-phenyl-1,2,3,6-tetrahydropyridine) and rotenone. These findings suggest that the probiotic mixture's ability to modulate butyrate production may be crucial in mitigating neurotoxicity and preserving dopaminergic neurons in PD models (Srivastav et al., 2019).

16.13 MICROBIOMES PROMOTING PARKINSON'S DISEASE

16.13.1 *Helicobacter Pylori*

Research investigations have uncovered a significant prevalence of *Helicobacter pylori* infection in individuals with PD, suggesting a higher occurrence compared to the general population. *Helicobacter pylori* is typically located in the intestinal epithelium and is well-known as a common bacterial infection (Dobbs et al., 2000, 2012; Lee et al., 2008; Pierantozzi et al., 2006). *Helicobacter pylori* infection has been found to be strongly associated with the duodenum, which is the primary site for the absorption of levodopa. Levodopa is the most commonly prescribed medication for the symptomatic treatment of PD. When levodopa is ingested, it is absorbed primarily in the duodenum before being transported to the brain, where it is converted into dopamine to alleviate PD symptoms. Therefore, the presence of *Helicobacter pylori* infection in the duodenum can potentially affect the absorption and efficacy of levodopa, which could have implications for the management of PD (Hashim et al., 2014; Lyte, 2010; Sasahara et al., 1981; Zhong et al., 2022a).

16.13.2 *Enterococcus Faecalis*

Recent studies have confirmed that *Enterococcus faecalis*, a bacterium commonly found in the human gut, can decarboxylate levodopa and tyrosine. Significantly, *Enterococcus faecalis* and carbidopa, a medication commonly used in combination with levodopa for PD treatment, do not interfere with each other's decarboxylation effects on levodopa. This implies that in PD patients with a high abundance of *Enterococcus faecalis*, the peak concentration of levodopa in the serum may be reduced. Therefore, the presence of *Enterococcus faecalis* and its TyrDC (tyrosine decarboxylase) gene can potentially serve as predictors of interindividual variations in levodopa metabolism within the complex gut microbiome. Consequently, *Enterococcus faecalis* becomes one factor influencing levodopa's bioavailability. Inhibition of *Enterococcus faecalis* activity may represent a potential approach to enhancing the bioavailability of levodopa and holds promise for improving PD treatment outcomes (Fan et al., 2022; Maini Rekdal et al., 2019).

16.13.3 *Desulfovibrio*

Sulfate-reducing bacteria are a group of anaerobic bacteria that derive energy through dissimilatory sulfate reduction. This metabolic pathway involves the reduction of sulfate to sulfide. As a result of this metabolic activity, sulfate-reducing bacteria produce a significant amount of sulfide. Sulfide is a highly reactive compound affecting biological systems (Loubinoux et al., 2002). Dissimilatory sulfate-reducing bacteria (DSV) possesses two properties that are intricately linked to the pathogenesis of PD. Firstly, DSV has the ability to produce hydrogen sulfide (H_2S) through dissimilatory sulfate reduction, utilising sulfate as an electron acceptor during respiration. The production of H_2S by DSV is significant because H_2S has been implicated in various cellular processes and has neurotoxic effects. One of the mechanisms by which H_2S can contribute to PD pathogenesis is its ability to induce the release of mitochondrial cytochrome c into the cytoplasm. When released into the cytoplasm, the peroxidase activity of cytochrome c promotes the formation of α-synuclein free radicals. These free radicals, in turn, trigger the oligomerization of α-synuclein, a protein closely associated with PD pathology (Carbonero et al., 2012; Guo et al., 2015; Kumar et al., 2016). In addition to its role in promoting α-synuclein oligomerization, H_2S has been shown to interfere with iron metabolism by increasing the iron content in the cytoplasm. This disruption of iron homeostasis can affect PD pathogenesis. Increased levels of H_2S can lead to an elevation in cytoplasmic iron levels. Excess iron in the cytoplasm can trigger various processes contributing to α-synuclein pathology, including the induction of α-synuclein fibrosis. Iron has been found to promote the aggregation and fibrillation of α-synuclein, leading to the formation of insoluble protein aggregates, a PD hallmark. The interplay between H_2S-mediated alterations in iron metabolism and the induction of α-synuclein fibrosis represents another potential mechanism linking sulfate-reducing bacteria and PD (Cassanelli & Moulis, 2001; Golts et al., 2002; Hälldin & Land, 2008; Kostka et al., 2008; Uversky et al., 2001).

16.14 HOW CAN DIET AFFECT NEURODEGENERATION?

Diet significantly influences the microbial composition within the gut, consequently impacting various metabolic, hormonal, and neurological functions. The diversity of bacterial communities tends to fluctuate in response to dietary changes. This dietary impact extends to the physiological and developmental aspects of hosts, both directly and indirectly, by inducing alterations in the gut microbiota. This phenomenon is also observed in zebrafish, where diet-mediated changes affect host physiology through microbial modifications (Emani et al., 2013; Kashyap et al., 2013; Libao-Mercado et al., 2009).

16.14.1 Ketogenic Diet

The ketogenic diet (KD) is characterized by its low-carbohydrate, moderate protein, and high-fat composition. This diet induces a state known as ketosis, which occurs when the liver breaks down short-chain fatty acids (SCFAs), monounsaturated fatty acids (MUFAs), and polyunsaturated fatty acids (PUFAs) in the absence of sufficient carbohydrates and glycogen. This sets the KD apart from high-carbohydrate diets, as it promotes the production of molecules called ketone bodies, including acetoacetate and β-hydroxybutyrate (BHB). These ketone bodies serve as an alternative source of energy. They can be converted directly (ketosis) into acetyl-CoA, which enters the citric acid cycle (TCA cycle) to produce energy. Additionally, ketone bodies can influence various functions, including the regulation of protein activity and the modulation of gene expression, through a process known as β-hydroxybutyration, which helps reduce systemic inflammation (Zhang et al., 2020a).

The potential impact of ketosis on gut dysbiosis and whether long-term KD can alleviate early PD symptoms are not yet well understood. Short-term changes in the gut microbiome during the acute phase of KD are likely driven by the shift from carbohydrate-rich meals to a high-triglyceride diet, which affects the composition of intestinal microbes. However, in the long term, the production of ketone bodies seems to lead to distinct shifts in the gut microbiome, a phenomenon unique to high-fat diets. For instance, bacterial species like *A. muciniphila*, involved in mucin breakdown, and *Lactobacillus*, a type of probiotic bacteria, tend to increase in abundance due to the ketogenic diet. Conversely, species such as *Bifidobacterium*, *Desulfovibrio* (which produces sulfate), and *Turicibacter* tend to decrease in abundance in response to the ketogenic diet. These changes in gut microbiome composition caused by the KD may potentially influence the symptoms and progression of conditions like PD. However, more research is necessary to fully comprehend the relationship between ketosis, gut dysbiosis, and these neurological disorders (Ma et al., 2018).

16.14.2 Mediterranean Diet

The Mediterranean Diet (MD) has been linked to an increased presence of key bacteria such as *Eubacterium eligens*, *Faecalibacterium prausnitzii*, and *Bacteroides cellulosilyticus*. These bacteria are significant producers of SC (Wang et al., 2021). Certain nutrients in the MD are recognized for their specific functions. The MD stands out for its high fiber content, providing fermentable fibers to the gut that promote the growth of bacteria capable of producing SCFAs, such as *Prevotella* and *Lactobacillus*. On the other hand, the presence of these beneficial bacteria has been correlated with reduced levels of *Escherichia*, *Turicibacter*, and *Akkermansia*. Elevated levels of *Escherichia* and *Turicibacter* have been linked to increased gut inflammation, while a decrease in *Akkermansia* helps maintain the gut's mucosal barrier, reducing the risk of potential pathogenic infections (Hoffman et al., 2019). The type of dietary fat consumed has an impact on the composition of the gut microbiome. Consuming olive oil, which is rich in omega-6 polyunsaturated fatty acids (n6-PUFA), has been linked to an increase in the production of SCFAs (Millman et al., 2021). Omega-3 polyunsaturated fatty acids (n3-PUFA) derived from the consumption of fish have also garnered considerable interest due to their potential influence on gut microbiota. Notably, *Bifidobacterium* and *Lactobacillus* play a significant role in metabolizing n3-PUFA, fostering a favourable gut milieu. This process helps inhibit the proliferation of *Enterobacteria* and fosters beneficial effects. These fatty acids enhance the intestinal mucus barrier and diminish mild inflammatory reactions in the intestinal tract (Fu et al., 2021).

16.14.3 Plant-Based Diet

A plant-based diet primarily consists of fibers and complex carbohydrates, which are rich in antioxidants, vitamins A, C, and E, as well as phytochemicals. These diets typically have low levels of fat and simple sugars. Plant-based eating patterns can range from strict vegan diets that exclude all animal products and focus on vegetables, grains, nuts, and fruits, to less restrictive variations that eliminate meat, poultry, and seafood but may allow for dairy products and eggs. While the specific health benefits of plant-based diets are not fully understood, various cultural groups that follow different versions of these diets have a notably lower prevalence of PD (McCarty, 2001).

Among the various nutrients present in plant-based foods, soluble and insoluble dietary fibers are notably significant. While dietary fiber cannot be digested by humans, certain gut-related bacterial populations can use fiber as a source of nutrition. Increasing dietary fiber intake offers numerous metabolic and health benefits for overall human well-being, including promoting feelings of fullness and helping to prevent overeating. In terms of gut health, dietary fibers play a critical role in regulating gut pH and fostering the growth of beneficial probiotic bacteria such as *Bifidobacterium* and *Lactobacillus* in the intestinal tract. (So et al., 2018).

Large amounts of antioxidants are found in fruits and vegetables, especially in items like citrus, berries, and dried fruits. Elevated levels of natural antioxidants, such as glutathione and ascorbic acid (Vitamin C), have been linked to improved outcomes in children suffering from severe malnutrition. This is likely because antioxidants create a favourable environment for the growth of anaerobic organisms in an oxygen-rich setting. Antioxidants also directly counteract reactive oxygen species (ROS), which can help reduce inflammation both locally and systemically in conditions like PD and AD. Additionally, supplementing bacterial growth media with uric acid, ascorbate, and glutathione has increased the production of butyrate, a compound associated with reduced gut inflammation (Million et al., 2016, 2020). The positive effects of phytochemicals present in fruits, vegetables, grains, coffee beans, and tea leaves on gut dysbiosis are well-known. A recent study demonstrated these benefits by showing improvements in the gut microbiome through the consumption of phytochemicals. More specifically, individuals with PD who followed a selective vegetarian diet saw an increase in the presence of the beneficial strain *Ruminococcaceae*, which is linked to liver function and immune response. This suggests that incorporating phytochemical-rich foods into the diet can lead to positive changes in gut microbiota composition and potentially better health outcomes for PD patients (Hegelmaier et al., 2020; Liu et al., 2016).

Phytochemicals, which are frequently present in fruits, vegetables, grains, coffee beans, and tea leaves, offer advantages in combatting gut dysbiosis. A recent study with a randomized controlled trial design showed positive effects on the gut microbiome as a result of phytochemical consumption. Specifically, individuals with PD who followed a selective vegetarian diet experienced improvements, including an increased presence of *Ruminococcaceae*, a probiotic strain associated with liver metabolism and immune function. This suggests that consuming phytochemical-rich foods can lead to positive changes in the composition of the gut microbiota, potentially resulting in improved health outcomes for PD patients (Zhernakova et al., 2016). Nonetheless, the extent to which a plant-based diet's benefits stem from eliminating neurotoxic elements in a typical urban diet remains uncertain. Notably, diets rich in animal fat or cholesterol are linked to a notably elevated risk of PD, whereas foods prepared with plant-derived fats do not display the same association (McCarty, 2001).

16.15 FUTURE PERSPECTIVE

Crucially, during the early stages of development, zebrafish embryos are transparent and possess a straightforward vascular system. These characteristics can be utilised to create transgenic zebrafish strains containing fluorescent vascular or hematopoietic stem cells. By manipulating the gut microbiome and employing in vivo imaging, we can investigate how microbes impact the development of blood vessels and blood cell formation. Similarly, in studies related to the gut–brain connection, it is essential to create novel behavioral models for neurobehavioral disorders rooted in the intestinal microbiome. This approach allows us to explore the intricate relationship between gut bacteria, intestinal ailments, and brain function, ultimately advancing our understanding of the gut-microbiome-brain axis and potentially improving cognitive function in individuals with neurobehavioral disorders (Zhong et al., 2022b).

The gut microbiota plays a significant role in maintaining host well-being and influencing disease. There is growing evidence suggesting a link between gut dysbiosis and the onset and progression of PD. While the precise cause-and-effect relationships between the gut microbiota and PD are not fully understood, recent findings from studies using PD animal models support the idea that an imbalanced gut microbiota can exacerbate PD-related issues. Conversely, restoring a healthy gut microbiota may potentially delay or mitigate the onset of PD. This suggests that the gut microbiota could serve as both a diagnostic tool and a target for therapeutic interventions in PD (Zhu et al., 2022).

Nonetheless, it's important to note that the makeup of gut bacteria in zebrafish differs substantially from that in humans. Achieving a more human-like intestinal microbiota in zebrafish is both important and challenging. We still face the obstacle of fully understanding the shared molecular communication system between the host and gut bacteria, as much remains undiscovered (Lu et al., 2021a). In spite of this, zebrafish can prove highly valuable, particularly when we can establish disease or mutant models that exhibit similar traits to humans. By examining the roles of genes homologous to humans and the conserved molecular networks in zebrafish, we can gain insights into their interactions with gut microbes. This approach allows us to investigate the relationship between these genetic elements and the influence of gut bacteria in a meaningful way (Shehwana & Konu, 2019).

16.16 DISCUSSION

Zebrafish offer a multitude of compelling research possibilities. They share a significant overlap in genes and their functions with humans, and when exposed to various toxins, they often exhibit similar responses. This presents a remarkable chance to explore fundamental underlying mechanisms in a more accessible model system compared to traditional mammalian species. For example, zebrafish have revealed numerous orthologs of genes associated with PD, and there is evidence to support that these fish can offer valuable insights into both the regular functions of these genes and their involvement in the development of PD (Doyle & Croll, 2022).

The interactions between hosts and their microbiomes will undoubtedly remain a topic of keen interest across various research fields, particularly neurodegenerative diseases (Haikal et al., 2019). The gut microbiome (GM) significantly impacts the development and balance of various body systems, and alterations in its composition are believed to initiate early, even before the motor symptoms of PD become apparent. These changes may have connections to the underlying mechanisms of PD through various direct and indirect means (Elfil et al., 2020).

An imbalanced gut microbiota, which encompasses changes in microbial metabolites, could have significant implications in the development of PD through a range of mechanisms. These mechanisms include heightened intestinal permeability, inflammation in the gut and nervous system, irregular accumulation of alpha-synuclein fibrils, disrupted oxidative stress levels, and reduced production of neurotransmitters (Zhu et al., 2022). Research has shown that the microbiota plays a crucial role in regulating the host's serotonergic system and influencing stress responses both behaviorally and hormonally. This discovery has broader implications for our understanding of stress-related psychiatric and gastrointestinal disorders (Borrelli et al., 2016). The gut–brain axis concept is under extensive investigation, and numerous studies have validated the connection between changes in gut microbiota composition and specific clinical conditions (Ma et al., 2019). Improving our understanding of the consequences resulting from changes in the GM within theENS and the CNS can help uncover the link between GM, α-synucleinopathy, inflammation, neuroprotection, and neurotoxicity in individuals with PD (Elfil et al., 2020).

16.17 ACKNOWLEDGMENT

The author, Rajesh Pamanji, is thankful to the Vice Chancellor and Registrar, Pondicherry University, Pondicherry, India, for providing the facilities for ongoing research work and to UGC, Government of India, for providing financial assistance in the form of DSK post-doctoral fellowship BL/19-20/0302 and this work has also been supported by the Department of Biotechnology (DBT), Ministry of Science and Technology, Government of India for Vishnu Prasad Nair R U (DBT Reference: BT/IN/Indo-UK/AMR-Env/02/JS/2020-21). The Figures have been Created with BioRender.com

REFERENCES

Almeida, A. R., Alves, M., Domingues, I., & Henriques, I. (2019a). The impact of antibiotic exposure in water and zebrafish gut microbiomes: A 16S rRNA gene-based metagenomic analysis. Ecotoxicology and Environmental Safety, 186, 109771. https://doi.org/10.1016/j.ecoenv.2019.109771

Almeida, A. R., Tacão, M., Machado, A. L., Golovko, O., Zlabek, V., Domingues, I., & Henriques, I. (2019b). Long-term effects of oxytetracycline exposure in zebrafish: A multi-level perspective. Chemosphere, 222, 333–344. https://doi.org/10.1016/j.chemosphere.2019.01.147

Barichella, M., Severgnini, M., Cilia, R., Cassani, E., Bolliri, C., Caronni, S., Ferri, V., Cancello, R., Ceccarani, C., Faierman, S., Pinelli, G., De Bellis, G., Zecca, L., Cereda, E., Consolandi, C., & Pezzoli, G. (2019). Unraveling gut microbiota in Parkinson's disease and atypical parkinsonism. Movement Disorders: Official Journal of the Movement Disorder Society, 34(3), 396–405. https://doi.org/10.1002/mds.27581

Bates, J. M., Mittge, E., Kuhlman, J., Baden, K. N., Cheesman, S. E., & Guillemin, K. (2006). Distinct signals from the microbiota promote different aspects of zebrafish gut differentiation. Developmental Biology, 297(2), 374–386. https://doi.org/10.1016/j.ydbio.2006.05.006

Baxter, N. T., Schmidt, A. W., Venkataraman, A., Kim, K. S., Waldron, C., & Schmidt, T. M. (2019). Dynamics of human gut microbiota and short-chain fatty acids in response to dietary interventions with three fermentable fibers. MBio, 10(1), e02566–18. https://doi.org/10.1128/mBio.02566-18

Bercik, P., Denou, E., Collins, J., Jackson, W., Lu, J., Jury, J., Deng, Y., Blennerhassett, P., Macri, J., McCoy, K. D., Verdu, E. F., & Collins, S. M. (2011). The intestinal microbiota affect central levels of brain-derived neurotropic factor and behavior in mice. Gastroenterology, 141(2), 599–609, 609.e1–3. https://doi.org/10.1053/j.gastro.2011.04.052

Beyer, K., Domingo-Sàbat, M., & Ariza, A. (2009). Molecular pathology of Lewy body diseases. International Journal of Molecular Sciences, 10(3), 724–745. https://doi.org/10.3390/ijms10030724

Bhattacharyya, D., Mohite, G. M., Krishnamoorthy, J., Gayen, N., Mehra, S., Navalkar, A., Kotler, S. A., Ratha, B. N., Ghosh, A., Kumar, R., Garai, K., Mandal, A. K., Maji, S. K., & Bhunia, A. (2019). Lipopolysaccharide from gut microbiota modulates α-synuclein aggregation and alters its biological function. ACS Chemical Neuroscience, 10(5), 2229–2236. https://doi.org/10.1021/acschemneuro.8b00733

Blander, J. M., Longman, R. S., Iliev, I. D., Sonnenberg, G. F., & Artis, D. (2017). Regulation of inflammation by microbiota interactions with the host. Nature Immunology, 18(8), 851–860. https://doi.org/10.1038/ni.3780

Borrelli, L., Aceto, S., Agnisola, C., De Paolo, S., Dipineto, L., Stilling, R. M., Dinan, T. G., Cryan, J. F., Menna, L. F., & Fioretti, A. (2016). Probiotic modulation of the microbiota-gut–brain axis and behaviour in zebrafish. Scientific Reports, 6, 30046. https://doi.org/10.1038/srep30046

Braak, H., de Vos, R. A. I., Bohl, J., & Del Tredici, K. (2006). Gastric alpha-synuclein immunoreactive inclusions in Meissner's and Auerbach's plexuses in cases staged for Parkinson's disease-related brain pathology. Neuroscience Letters, 396(1), 67–72. https://doi.org/10.1016/j.neulet.2005.11.012

Bravo, J. A., Forsythe, P., Chew, M. V., Escaravage, E., Savignac, H. M., Dinan, T. G., Bienenstock, J., & Cryan, J. F. (2011). Ingestion of Lactobacillus strain regulates emotional behavior and central GABA receptor expression in a mouse via the vagus nerve. Proceedings of the National Academy of Sciences of the United States of America, 108(38), 16050–16055. https://doi.org/10.1073/pnas.1102999108

Breen, P., Winters, A. D., Nag, D., Ahmad, M. M., Theis, K. R., & Withey, J. H. (2019). Internal versus external pressures: Effect of housing systems on the zebrafish microbiome. Zebrafish, 16(4), 388–400. https://doi.org/10.1089/zeb.2018.1711

Breen, P., Winters, A. D., Theis, K. R., & Withey, J. H. (2021). The Vibrio cholerae type six secretion system is dispensable for colonization but affects pathogenesis and the structure of zebrafish intestinal microbiome. Infection and Immunity, 89(9), e0015121. https://doi.org/10.1128/IAI.00151-21

Brugman, S. (2016). The zebrafish as a model to study intestinal inflammation. Developmental and Comparative Immunology, 64, 82–92. https://doi.org/10.1016/j.dci.2016.02.020

Butt, R. L., & Volkoff, H. (2019). Gut microbiota and energy homeostasis in fish. Frontiers in Endocrinology, 10, 9. https://doi.org/10.3389/fendo.2019.00009

Cantas, L., Sørby, J. R. T., Aleström, P., & Sørum, H. (2012). Culturable gut microbiota diversity in zebrafish. Zebrafish, 9(1), 26–37. https://doi.org/10.1089/zeb.2011.0712

Caramillo, E. M., & Echevarria, D. J. (2017). Alzheimer's disease in the zebrafish: Where can we take it? Behavioural Pharmacology, 28(2 and 3), 179–186. https://doi.org/10.1097/FBP.0000000000000284

Carbonero, F., Benefiel, A. C., Alizadeh-Ghamsari, A. H., & Gaskins, H. R. (2012). Microbial pathways in colonic sulfur metabolism and links with health and disease. Frontiers in Physiology, 3, 448. https://doi.org/10.3389/fphys.2012.00448

Cassanelli, S., & Moulis, J. (2001). Sulfide is an efficient iron releasing agent for mammalian ferritins. Biochimica Et Biophysica Acta, 1547(1), 174–182. https://doi.org/10.1016/s0167-4838(01)00182-0

Chen, S. G., Stribinskis, V., Rane, M. J., Demuth, D. R., Gozal, E., Roberts, A. M., Jagadapillai, R., Liu, R., Choe, K., Shivakumar, B., Son, F., Jin, S., Kerber, R., Adame, A., Masliah, E., & Friedland, R. P. (2016). Exposure to the functional bacterial amyloid protein curli enhances alpha-synuclein aggregation in aged fischer 344 rats and caenorhabditis elegans. Scientific Reports, 6, 34477. https://doi.org/10.1038/srep34477

Chen, Y.-C., Tao, N.-L., Hu, S.-Y., Tsai, H.-Y., Liao, S.-C., Tsai, W.-L., & Hu, C.-Y. (2021). Effect of tempeh on gut microbiota and anti-stress activity in zebrafish. International Journal of Molecular Sciences, 22(23), 12660. https://doi.org/10.3390/ijms222312660

Cornuault, J. K., Byatt, G., Paquet, M.-E., De Koninck, P., & Moineau, S. (2022). Zebrafish: A big fish in the study of the gut microbiota. Current Opinion in Biotechnology, 73, 308–313. https://doi.org/10.1016/j.copbio.2021.09.007

Cryan, J. F., & Dinan, T. G. (2012). Mind-altering microorganisms: The impact of the gut microbiota on brain and behaviour. Nature Reviews Neuroscience, 13(10), 701–712. https://doi.org/10.1038/nrn3346

Cryan, J. F., O'Riordan, K. J., Cowan, C. S. M., Sandhu, K. V., Bastiaanssen, T. F. S., Boehme, M., Codagnone, M. G., Cussotto, S., Fulling, C., Golubeva, A. V., Guzzetta, K. E., Jaggar, M., Long-Smith, C. M., Lyte, J. M., Martin, J. A., Molinero-Perez, A., Moloney, G., Morelli, E., Morillas, E., . . . Dinan, T. G. (2019). The microbiota-gut–brain axis. Physiological Reviews, 99(4), 1877–2013. https://doi.org/10.1152/physrev.00018.2018

Cuoghi, B., & Mola, L. (2007). Microglia of teleosts: Facing a challenge in neurobiology. European Journal of Histochemistry, 51(4), 231–240.

Cuomo, M., Borrelli, L., Della Monica, R., Coretti, L., De Riso, G., D'Angelo Lancellotti di Durazzo, L., Fioretti, A., Lembo, F., Dinan, T. G., Cryan, J. F., Cocozza, S., & Chiariotti, L. (2021). DNA methylation profiles of Tph1A and BDNF in gut and brain of L. Rhamnosus-Treated Zebrafish. Biomolecules, 11(2), 142. https://doi.org/10.3390/biom11020142

Davis, D. J., Bryda, E. C., Gillespie, C. H., & Ericsson, A. C. (2016). Microbial modulation of behavior and stress responses in zebrafish larvae. Behavioural Brain Research, 311, 219–227. https://doi.org/10.1016/j.bbr.2016.05.040

de Abreu, M. S., Giacomini, A. C. V. V., Sysoev, M., Demin, K. A., Alekseeva, P. A., Spagnoli, S. T., & Kalueff, A. V. (2019). Modeling gut–brain interactions in zebrafish. Brain Research Bulletin, 148, 55–62. https://doi.org/10.1016/j.brainresbull.2019.03.003

De Felice, E., Porreca, I., Alleva, E., De Girolamo, P., Ambrosino, C., Ciriaco, E., Germanà, A., & Sordino, P. (2014). Localization of BDNF expression in the developing brain of zebrafish. Journal of Anatomy, 224(5), 564–574. https://doi.org/10.1111/joa.12168

De Waele, J. J., & Dhaese, S. (2019). Antibiotic stewardship in sepsis management: Toward a balanced use of antibiotics for the severely ill patient. Expert Review of Anti-Infective Therapy, 17(2), 89–97. https://doi.org/10.1080/14787210.2019.1568239

Diaz Heijtz, R., Wang, S., Anuar, F., Qian, Y., Björkholm, B., Samuelsson, A., Hibberd, M. L., Forssberg, H., & Pettersson, S. (2011). Normal gut microbiota modulates brain development and behavior. Proceedings of the National Academy of Sciences of the United States of America, 108(7), 3047–3052. https://doi.org/10.1073/pnas.1010529108

Dinan, T. G., & Cryan, J. F. (2017). The microbiome-gut–brain axis in health and disease. Gastroenterology Clinics of North America, 46(1), 77–89. https://doi.org/10.1016/j.gtc.2016.09.007

Dinan, T. G., Stanton, C., & Cryan, J. F. (2013). Psychobiotics: A novel class of psychotropic. Biological Psychiatry, 74(10), 720–726. https://doi.org/10.1016/j.biopsych.2013.05.001

Ding, C., & He, J. (2010). Effect of antibiotics in the environment on microbial populations. Applied Microbiology and Biotechnology, 87(3), 925–941. https://doi.org/10.1007/s00253-010-2649-5

Dobbs, R. J., Charlett, A., Dobbs, S. M., Weller, C., Ibrahim, M. A. A., Iguodala, O., Smee, C., Plant, J. M., Lawson, A. J., Taylor, D., & Bjarnason, I. (2012). Leukocyte-subset counts in idiopathic parkinsonism provide clues to a pathogenic pathway involving small intestinal bacterial overgrowth. A surveillance study. Gut Pathogens, 4(1), 12. https://doi.org/10.1186/1757-4749-4-12

Dobbs, R. J., Charlett, A., Dobbs, S. M., Weller, C., & Peterson, D. W. (2000). Parkinsonism: differential age-trend in helicobacter pylori antibody. Alimentary Pharmacology & Therapeutics, 14(9), 1199–1205. https://doi.org/10.1046/j.1365-2036.2000.00815.x

Donaldson, G. P., Lee, S. M., & Mazmanian, S. K. (2016). Gut biogeography of the bacterial microbiota. Nature Reviews. Microbiology, 14(1), 20–32. https://doi.org/10.1038/nrmicro3552

Doyle, J. M., & Croll, R. P. (2022). A Critical review of zebrafish models of Parkinson's disease. Frontiers in Pharmacology, 13, 835827. https://doi.org/10.3389/fphar.2022.835827

Elfil, M., Kamel, S., Kandil, M., Koo, B. B., & Schaefer, S. M. (2020). Implications of the gut microbiome in Parkinson's disease. Movement Disorders: Official Journal of the Movement Disorder Society, 35(6), 921–933. https://doi.org/10.1002/mds.28004

Emani, R., Asghar, M. N., Toivonen, R., Lauren, L., Söderström, M., Toivola, D. M., van Tol, E. a. F., & Hänninen, A. (2013). Casein hydrolysate diet controls intestinal T cell activation, free radical production and microbial colonisation in NOD mice. Diabetologia, 56(8), 1781–1791. https://doi.org/10.1007/s00125-013-2941-x

Evans, M. L., Chorell, E., Taylor, J. D., Åden, J., Götheson, A., Li, F., Koch, M., Sefer, L., Matthews, S. J., Wittung-Stafshede, P., Almqvist, F., & Chapman, M. R. (2015). The bacterial curli system possesses a potent and selective inhibitor of amyloid formation. Molecular Cell, 57(3), 445–455. https://doi.org/10.1016/j.molcel.2014.12.025

Fan, H., Sheng, S., & Zhang, F. (2022). New hope for Parkinson's disease treatment: Targeting gut microbiota. CNS Neuroscience & Therapeutics, 28(11), 1675–1688. https://doi.org/10.1111/cns.13916

Flores, E. M., Nguyen, A. T., Odem, M. A., Eisenhoffer, G. T., & Krachler, A. M. (2020). The zebrafish as a model for gastrointestinal tract-microbe interactions. Cellular Microbiology, 22(3), e13152. https://doi.org/10.1111/cmi.13152

Forsythe, P., & Kunze, W. A. (2013). Voices from within: Gut microbes and the CNS. Cellular and Molecular Life Sciences, 70(1), 55–69. https://doi.org/10.1007/s00018-012-1028-z

Fu, Y., Wang, Y., Gao, H., Li, D., Jiang, R., Ge, L., Tong, C., & Xu, K. (2021). Associations among dietary omega-3 polyunsaturated fatty acids, the gut microbiota, and intestinal immunity. Mediators of Inflammation, 2021, 8879227. https://doi.org/10.1155/2021/8879227

Galland, L. (2014). The gut microbiome and the brain. Journal of Medicinal Food, 17(12), 1261–1272. https://doi.org/10.1089/jmf.2014.7000

Gerlai, R. (2014). Social behavior of zebrafish: From synthetic images to biological mechanisms of shoaling. Journal of Neuroscience Methods, 234, 59–65. https://doi.org/10.1016/j.jneumeth.2014.04.028

Golts, N., Snyder, H., Frasier, M., Theisler, C., Choi, P., & Wolozin, B. (2002). Magnesium inhibits spontaneous and iron-induced aggregation of alpha-synuclein. The Journal of Biological Chemistry, 277(18), 16116–16123. https://doi.org/10.1074/jbc.M107866200

Granato, M., van Eeden, F. J., Schach, U., Trowe, T., Brand, M., Furutani-Seiki, M., Haffter, P., Hammerschmidt, M., Heisenberg, C. P., Jiang, Y. J., Kane, D. A., Kelsh, R. N., Mullins, M. C., Odenthal, J., & Nüsslein-Volhard, C. (1996). Genes controlling and mediating locomotion behavior of the zebrafish embryo and larva. Development (Cambridge, England), 123, 399–413. https://doi.org/10.1242/dev.123.1.399

Guo, Z., Li, C. S., Wang, C. M., Xie, Y. J., & Wang, A. L. (2015). CSE/H2S system protects mesenchymal stem cells from hypoxia and serum deprivation-induced apoptosis via mitochondrial injury, endoplasmic reticulum stress and PI3K/Akt activation pathways. Molecular Medicine Reports, 12(2), 2128–2134. https://doi.org/10.3892/mmr.2015.3651

Haikal, C., Chen, Q.-Q., & Li, J.-Y. (2019). Microbiome changes: An indicator of Parkinson's disease? Translational Neurodegeneration, 8(1), 38. https://doi.org/10.1186/s40035-019-0175-7

Hälldin, J., & Land, T. (2008). Sulfide increases labile iron pool in RD4 cells. Biometals: An International Journal on the Role of Metal Ions in Biology, Biochemistry, and Medicine, 21(2), 127–131. https://doi.org/10.1007/s10534-007-9099-2

Hashim, H., Azmin, S., Razlan, H., Yahya, N. W., Tan, H. J., Manaf, M. R. A., & Ibrahim, N. M. (2014). Eradication of Helicobacter pylori infection improves levodopa action, clinical symptoms and quality of life in patients with Parkinson's disease. PLoS One, 9(11), e112330. https://doi.org/10.1371/journal.pone.0112330

Hegelmaier, T., Lebbing, M., Duscha, A., Tomaske, L., Tönges, L., Holm, J. B., Bjørn Nielsen, H., Gatermann, S. G., Przuntek, H., & Haghikia, A. (2020). Interventional influence of the intestinal microbiome through dietary intervention and bowel cleansing might improve motor symptoms in Parkinson's disease. Cells, 9(2), 376. https://doi.org/10.3390/cells9020376

Hoffman, J. D., Yanckello, L. M., Chlipala, G., Hammond, T. C., McCulloch, S. D., Parikh, I., Sun, S., Morganti, J. M., Green, S. J., & Lin, A.-L. (2019). Dietary inulin alters the gut microbiome, enhances systemic metabolism and reduces neuroinflammation in an APOE4 mouse model. PLoS One, 14(8), e0221828. https://doi.org/10.1371/journal.pone.0221828

Holmqvist, S., Chutna, O., Bousset, L., Aldrin-Kirk, P., Li, W., Björklund, T., Wang, Z.-Y., Roybon, L., Melki, R., & Li, J.-Y. (2014). Direct evidence of Parkinson pathology spread from the gastrointestinal tract to the brain in rats. Acta Neuropathologica, 128(6), 805–820. https://doi.org/10.1007/s00401-014-1343-6

Howe, K., Clark, M. D., Torroja, C. F., Torrance, J., Berthelot, C., Muffato, M., Collins, J. E., Humphray, S., McLaren, K., Matthews, L., McLaren, S., Sealy, I., Caccamo, M., Churcher, C., Scott, C., Barrett, J. C., Koch, R., Rauch, G.-J., White, S., . . . Stemple, D. L. (2013). The zebrafish reference genome sequence and its relationship to the human genome. Nature, 496(7446), 498–503. https://doi.org/10.1038/nature12111

Jang, L.-G., Choi, G., Kim, S.-W., Kim, B.-Y., Lee, S., & Park, H. (2019). The combination of sport and sport-specific diet is associated with characteristics of gut microbiota: An observational study. Journal of the International Society of Sports Nutrition, 16(1), 21. https://doi.org/10.1186/s12970-019-0290-y

Jemielita, M., Taormina, M. J., Burns, A. R., Hampton, J. S., Rolig, A. S., Guillemin, K., & Parthasarathy, R. (2014). Spatial and temporal features of the growth of a bacterial species colonizing the zebrafish gut. MBio, 5(6), e01751–14. https://doi.org/10.1128/mBio.01751-14

Kang, D.-W., Park, J. G., Ilhan, Z. E., Wallstrom, G., Labaer, J., Adams, J. B., & Krajmalnik-Brown, R. (2013). Reduced incidence of Prevotella and other fermenters in intestinal microflora of autistic children. PLoS One, 8(7), e68322. https://doi.org/10.1371/journal.pone.0068322

Kashyap, P. C., Marcobal, A., Ursell, L. K., Larauche, M., Duboc, H., Earle, K. A., Sonnenburg, E. D., Ferreyra, J. A., Higginbottom, S. K., Million, M., Tache, Y., Pasricha, P. J., Knight, R., Farrugia, G., & Sonnenburg, J. L. (2013). Complex interactions among diet, gastrointestinal transit, and gut microbiota in humanized mice. Gastroenterology, 144(5), 967–977. https://doi.org/10.1053/j.gastro.2013.01.047

Kayani, M. U. R., Yu, K., Qiu, Y., Shen, Y., Gao, C., Feng, R., Zeng, X., Wang, W., Chen, L., & Su, H. L. (2021). Environmental concentrations of antibiotics alter the zebrafish gut microbiome structure and potential functions. Environmental Pollution (Barking, Essex: 1987), 278, 116760. https://doi.org/10.1016/j.envpol.2021.116760

Keshavarzian, A., Green, S. J., Engen, P. A., Voigt, R. M., Naqib, A., Forsyth, C. B., Mutlu, E., & Shannon, K. M. (2015a). Colonic bacterial composition in Parkinson's disease. Movement Disorders: Official Journal of the Movement Disorder Society, 30(10), 1351–1360. https://doi.org/10.1002/mds.26307

Kim, C., Lv, G., Lee, J. S., Jung, B. C., Masuda-Suzukake, M., Hong, C.-S., Valera, E., Lee, H.-J., Paik, S. R., Hasegawa, M., Masliah, E., Eliezer, D., & Lee, S.-J. (2016). Exposure to bacterial endotoxin generates a distinct strain of α-synuclein fibril. Scientific Reports, 6, 30891. https://doi.org/10.1038/srep30891

Kim, S., Kwon, S.-H., Kam, T.-I., Panicker, N., Karuppagounder, S. S., Lee, S., Lee, J. H., Kim, W. R., Kook, M., Foss, C. A., Shen, C., Lee, H., Kulkarni, S., Pasricha, P. J., Lee, G., Pomper, M. G., Dawson, V. L., Dawson, T. M., & Ko, H. S. (2019). Transneuronal propagation of pathologic α-synuclein from the gut to the brain models Parkinson's disease. Neuron, 103(4), 627–641.e7. https://doi.org/10.1016/j.neuron.2019.05.035

Klaenhammer, T. R., Kleerebezem, M., Kopp, M. V., & Rescigno, M. (2012). The impact of probiotics and prebiotics on the immune system. Nature Reviews. Immunology, 12(10), 728–734. https://doi.org/10.1038/nri3312

Kollef, M. H., & Betthauser, K. D. (2019). New antibiotics for community-acquired pneumonia. Current Opinion in Infectious Diseases, 32(2), 169–175. https://doi.org/10.1097/QCO.0000000000000526

Kościuczuk, E. M., Lisowski, P., Jarczak, J., Strzałkowska, N., Jóźwik, A., Horbańczuk, J., Krzyżewski, J., Zwierzchowski, L., & Bagnicka, E. (2012). Cathelicidins: Family of antimicrobial peptides. A review. Molecular Biology Reports, 39(12), 10957–10970. https://doi.org/10.1007/s11033-012-1997-x

Kostka, M., Högen, T., Danzer, K. M., Levin, J., Habeck, M., Wirth, A., Wagner, R., Glabe, C. G., Finger, S., Heinzelmann, U., Garidel, P., Duan, W., Ross, C. A., Kretzschmar, H., & Giese, A. (2008). Single particle characterization of iron-induced pore-forming alpha-synuclein oligomers. The Journal of Biological Chemistry, 283(16), 10992–11003. https://doi.org/10.1074/jbc.M709634200

Kumar, A., Ganini, D., & Mason, R. P. (2016). Role of cytochrome c in α-synuclein radical formation: Implications of α-synuclein in neuronal death in Maneb- and paraquat-induced model of Parkinson's disease. Molecular Neurodegeneration, 11(1), 70. https://doi.org/10.1186/s13024-016-0135-y

Lam, S. H., Chua, H. L., Gong, Z., Lam, T. J., & Sin, Y. M. (2004). Development and maturation of the immune system in zebrafish, *Danio rerio*: A gene expression profiling, in situ hybridization and immunological study. Developmental and Comparative Immunology, 28(1), 9–28. https://doi.org/10.1016/s0145-305x(03)00103-4

Leal, G., Afonso, P. M., Salazar, I. L., & Duarte, C. B. (2015). Regulation of hippocampal synaptic plasticity by BDNF. Brain Research, 1621, 82–101. https://doi.org/10.1016/j.brainres.2014.10.019

Lebouvier, T., Chaumette, T., Paillusson, S., Duyckaerts, C., Bruley des Varannes, S., Neunlist, M., & Derkinderen, P. (2009). The second brain and Parkinson's disease. The European Journal of Neuroscience, 30(5), 735–741. https://doi.org/10.1111/j.1460-9568.2009.06873.x

Lee, W. Y., Yoon, W. T., Shin, H. Y., Jeon, S. H., & Rhee, P.-L. (2008). Helicobacter pylori infection and motor fluctuations in patients with Parkinson's disease. Movement Disorders: Official Journal of the Movement Disorder Society, 23(12), 1696–1700. https://doi.org/10.1002/mds.22190

Li, Y., Li, Y., Cao, X., Jin, X., & Jin, T. (2017). Pattern recognition receptors in zebrafish provide functional and evolutionary insight into innate immune signaling pathways. Cellular & Molecular Immunology, 14(1), 80–89. https://doi.org/10.1038/cmi.2016.50

Libao-Mercado, A. J. O., Zhu, C. L., Cant, J. P., Lapierre, H., Thibault, J.-N., Sève, B., Fuller, M. F., & de Lange, C. F. M. (2009). Dietary and endogenous amino acids are the main contributors to microbial protein in the upper gut of normally nourished pigs. The Journal of Nutrition, 139(6), 1088–1094. https://doi.org/10.3945/jn.108.103267

Lieschke, G. J., & Currie, P. D. (2007). Animal models of human disease: Zebrafish swim into view. Nature Reviews Genetics, 8(5), 353–367. https://doi.org/10.1038/nrg2091

Liu, H.-X., Rocha, C. S., Dandekar, S., & Wan, Y.-J. Y. (2016). Functional analysis of the relationship between intestinal microbiota and the expression of hepatic genes and pathways during the course of liver regeneration. Journal of Hepatology, 64(3), 641–650. https://doi.org/10.1016/j.jhep.2015.09.022

Logan, S. L., Thomas, J., Yan, J., Baker, R. P., Shields, D. S., Xavier, J. B., Hammer, B. K., & Parthasarathy, R. (2018). The Vibrio cholerae type VI secretion system can modulate host intestinal mechanics to displace gut bacterial symbionts. Proceedings of the National Academy of Sciences of the United States of America, 115(16), E3779–E3787. https://doi.org/10.1073/pnas.1720133115

López Nadal, A., Ikeda-Ohtsubo, W., Sipkema, D., Peggs, D., McGurk, C., Forlenza, M., Wiegertjes, G. F., & Brugman, S. (2020). Feed, microbiota, and gut immunity: Using the zebrafish model to understand fish health. Frontiers in Immunology, 11, 114. https://doi.org/10.3389/fimmu.2020.00114

Loubinoux, J., Bronowicki, J.-P., Pereira, I. A. C., Mougenel, J.-L., & Faou, A. E. (2002). Sulfate-reducing bacteria in human feces and their association with inflammatory bowel diseases. FEMS Microbiology Ecology, 40(2), 107–112. https://doi.org/10.1111/j.1574-6941.2002.tb00942.x

Lozupone, C. A., Stombaugh, J. I., Gordon, J. I., Jansson, J. K., & Knight, R. (2012). Diversity, stability and resilience of the human gut microbiota. Nature, 489(7415), 220–230. https://doi.org/10.1038/nature11550

Lu, C.-S., Chang, H.-C., Weng, Y.-H., Chen, C.-C., Kuo, Y.-S., & Tsai, Y.-C. (2021a). The add-on effect of lactobacillus plantarum PS128 in patients with Parkinson's disease: A pilot study. Frontiers in Nutrition, 8, 650053. https://doi.org/10.3389/fnut.2021.650053

Lu, H., Li, P., Huang, X., Wang, C. H., Li, M., & Xu, Z. Z. (2021b). Zebrafish model for human gut microbiome-related studies: Advantages and limitations. Medicine in Microecology, 8, 100042. https://doi.org/10.1016/j.medmic.2021.100042

Lyte, M. (2010). Microbial endocrinology as a basis for improved L-DOPA bioavailability in Parkinson's patients treated for Helicobacter pylori. Medical Hypotheses, 74(5), 895–897. https://doi.org/10.1016/j.mehy.2009.11.001

Ma, D., Wang, A. C., Parikh, I., Green, S. J., Hoffman, J. D., Chlipala, G., Murphy, M. P., Sokola, B. S., Bauer, B., Hartz, A. M. S., & Lin, A.-L. (2018). Ketogenic diet enhances neurovascular function with altered gut microbiome in young healthy mice. Scientific Reports, 8(1), 6670. https://doi.org/10.1038/s41598-018-25190-5

Ma, Q., Xing, C., Long, W., Wang, H. Y., Liu, Q., & Wang, R.-F. (2019). Impact of microbiota on central nervous system and neurological diseases: The gut–brain axis. Journal of Neuroinflammation, 16(1), 53. https://doi.org/10.1186/s12974-019-1434-3

Maini Rekdal, V., Bess, E. N., Bisanz, J. E., Turnbaugh, P. J., & Balskus, E. P. (2019). Discovery and inhibition of an interspecies gut bacterial pathway for Levodopa metabolism. Science (New York, N.Y.), 364(6445), eaau6323. https://doi.org/10.1126/science.aau6323

Marques, A., Ollevier, F., Verstraete, W., Sorgeloos, P., & Bossier, P. (2006). Gnotobiotically grown aquatic animals: Opportunities to investigate host-microbe interactions. Journal of Applied Microbiology, 100(5), 903–918. https://doi.org/10.1111/j.1365-2672.2006.02961.x

McCarty, M. F. (2001). Does a vegan diet reduce risk for Parkinson's disease? Medical Hypotheses, 57(3), 318–323. https://doi.org/10.1054/mehy.2000.1321

Milligan-Myhre, K., Charette, J. R., Phennicie, R. T., Stephens, W. Z., Rawls, J. F., Guillemin, K., & Kim, C. H. (2011). Study of host-microbe interactions in zebrafish. Methods in Cell Biology, 105, 87–116. https://doi.org/10.1016/B978-0-12-381320-6.00004-7

Million, M., Armstrong, N., Khelaifia, S., Guilhot, E., Richez, M., Lagier, J.-C., Dubourg, G., Chabriere, E., & Raoult, D. (2020). The antioxidants glutathione, ascorbic acid and uric acid maintain butyrate production by human gut clostridia in the presence of oxygen in vitro. Scientific Reports, 10(1), 7705. https://doi.org/10.1038/s41598-020-64834-3

Million, M., Tidjani Alou, M., Khelaifia, S., Bachar, D., Lagier, J.-C., Dione, N., Brah, S., Hugon, P., Lombard, V., Armougom, F., Fromonot, J., Robert, C., Michelle, C., Diallo, A., Fabre, A., Guieu, R., Sokhna, C., Henrissat, B., Parola, P., & Raoult, D. (2016). Increased gut redox and depletion of anaerobic and methanogenic prokaryotes in severe acute malnutrition. Scientific Reports, 6, 26051. https://doi.org/10.1038/srep26051

Millman, J. F., Okamoto, S., Teruya, T., Uema, T., Ikematsu, S., Shimabukuro, M., & Masuzaki, H. (2021). Extra-virgin olive oil and the gut–brain axis: Influence on gut microbiota, mucosal immunity, and cardiometabolic and cognitive health. Nutrition Reviews, 79(12), 1362–1374. https://doi.org/10.1093/nutrit/nuaa148

Mulak, A., & Bonaz, B. (2015). Brain-gut-microbiota axis in Parkinson's disease. World Journal of Gastroenterology, 21(37), 10609–10620. https://doi.org/10.3748/wjg.v21.i37.10609

Murdoch, C. C., & Rawls, J. F. (2019). Commensal microbiota regulate vertebrate innate immunity-insights from the zebrafish. Frontiers in Immunology, 10, 2100. https://doi.org/10.3389/fimmu.2019.02100

Natale, G., Pasquali, L., Paparelli, A., & Fornai, F. (2011). Parallel manifestations of neuropathologies in the enteric and central nervous systems. Neurogastroenterology and Motility: The Official Journal of the European Gastrointestinal Motility Society, 23(12), 1056–1065. https://doi.org/10.1111/j.1365-2982.2011.01794.x

Negi, S., Das, D. K., Pahari, S., Nadeem, S., & Agrewala, J. N. (2019). Potential role of gut microbiota in induction and regulation of innate immune memory. Frontiers in Immunology, 10, 2441. https://doi.org/10.3389/fimmu.2019.02441

Nout, M. J. R., & Kiers, J. L. (2005). Tempe fermentation, innovation and functionality: Update into the third millenium. Journal of Applied Microbiology, 98(4), 789–805. https://doi.org/10.1111/j.1365-2672.2004.02471.x

Novoa, B., & Figueras, A. (2012). Zebrafish: Model for the study of inflammation and the innate immune response to infectious diseases. Advances in Experimental Medicine and Biology, 946, 253–275. https://doi.org/10.1007/978-1-4614-0106-3_15

O'Toole, P. W., & Jeffery, I. B. (2015). Gut microbiota and aging. Science (New York, N.Y.), 350(6265), 1214–1215. https://doi.org/10.1126/science.aac8469

Panula, P., Chen, Y.-C., Priyadarshini, M., Kudo, H., Semenova, S., Sundvik, M., & Sallinen, V. (2010). The comparative neuroanatomy and neurochemistry of zebrafish CNS systems of relevance to human neuropsychiatric diseases. Neurobiology of Disease, 40(1), 46–57. https://doi.org/10.1016/j.nbd.2010.05.010

Panula, P., Sallinen, V., Sundvik, M., Kolehmainen, J., Torkko, V., Tiittula, A., Moshnyakov, M., & Podlasz, P. (2006). Modulatory neurotransmitter systems and behavior: Towards zebrafish models of neurodegenerative diseases. Zebrafish, 3(2), 235–247. https://doi.org/10.1089/zeb.2006.3.235

Park, S.-C., Moon, J. C., Shin, S. Y., Son, H., Jung, Y. J., Kim, N.-H., Kim, Y.-M., Jang, M.-K., & Lee, J. R. (2016). Functional characterization of alpha-synuclein protein with antimicrobial activity. Biochemical and Biophysical Research Communications, 478(2), 924–928. https://doi.org/10.1016/j.bbrc.2016.08.052

Patula, S., Wojno, M., Pinnell, L. J., Oliaro, F., Cabay, C., Molinari, G. S., & Kwasek, K. (2021). Nutritional programming with dietary soybean meal and its effect on gut microbiota in zebrafish (*Danio rerio*). Zebrafish, 18(2), 125–138. https://doi.org/10.1089/zeb.2020.1952

Perez Visñuk, D., Savoy de Giori, G., LeBlanc, J. G., & de Moreno de LeBlanc, A. (2020). Neuroprotective effects associated with immune modulation by selected lactic acid bacteria in a Parkinson's disease model. Nutrition (Burbank, Los Angeles County, Calif.), 79–80, 110995. https://doi.org/10.1016/j.nut.2020.110995

Pham, L. N., Kanther, M., Semova, I., & Rawls, J. F. (2008). Methods for generating and colonizing gnotobiotic zebrafish. Nature Protocols, 3(12), 1862–1875. https://doi.org/10.1038/nprot.2008.186

Phelps, D., Brinkman, N. E., Keely, S. P., Anneken, E. M., Catron, T. R., Betancourt, D., Wood, C. E., Espenschied, S. T., Rawls, J. F., & Tal, T. (2017). Microbial colonization is required for normal neurobehavioral development in zebrafish. Scientific Reports, 7(1), 11244. https://doi.org/10.1038/s41598-017-10517-5

Pierantozzi, M., Pietroiusti, A., Brusa, L., Galati, S., Stefani, A., Lunardi, G., Fedele, E., Sancesario, G., Bernardi, G., Bergamaschi, A., Magrini, A., Stanzione, P., & Galante, A. (2006). Helicobacter pylori eradication and l-dopa absorption in patients with PD and motor fluctuations. Neurology, 66(12), 1824–1829. https://doi.org/10.1212/01.wnl.0000221672.01272.ba

Pindling, S., Azulai, D., Zheng, B., Dahan, D., & Perron, G. G. (2018). Dysbiosis and early mortality in zebrafish larvae exposed to subclinical concentrations of streptomycin. FEMS Microbiology Letters, 365(18), fny188. https://doi.org/10.1093/femsle/fny188

Pirbaglou, M., Katz, J., de Souza, R. J., Stearns, J. C., Motamed, M., & Ritvo, P. (2016). Probiotic supplementation can positively affect anxiety and depressive symptoms: A systematic review of randomized controlled trials. Nutrition Research (New York, N.Y.), 36(9), 889–898. https://doi.org/10.1016/j.nutres.2016.06.009

Qiang, Y., Xu, J., Yan, C., Jin, H., Xiao, T., Yan, N., Zhou, L., An, H., Zhou, X., Shao, Q., & Xia, S. (2017). Butyrate and retinoic acid imprint mucosal-like dendritic cell development synergistically from bone marrow cells. Clinical and Experimental Immunology, 189(3), 290–297. https://doi.org/10.1111/cei.12990

Quinlivan, V. H., & Farber, S. A. (2017). Lipid Uptake, Metabolism, and transport in the larval zebrafish. Frontiers in Endocrinology, 8, 319. https://doi.org/10.3389/fendo.2017.00319

Rawls, J. F., Mahowald, M. A., Ley, R. E., & Gordon, J. I. (2006). Reciprocal gut microbiota transplants from zebrafish and mice to germ-free recipients reveal host habitat selection. Cell, 127(2), 423–433. https://doi.org/10.1016/j.cell.2006.08.043

Rawls, J. F., Samuel, B. S., & Gordon, J. I. (2004). Gnotobiotic zebrafish reveal evolutionarily conserved responses to the gut microbiota. Proceedings of the National Academy of Sciences of the United States of America, 101(13), 4596–4601. https://doi.org/10.1073/pnas.0400706101

Reber, S. O. (2012). Stress and animal models of inflammatory bowel disease—An update on the role of the hypothalamo-pituitary-adrenal axis. Psychoneuroendocrinology, 37(1), 1–19. https://doi.org/10.1016/j.psyneuen.2011.05.014

Ríos-Covián, D., Ruas-Madiedo, P., Margolles, A., Gueimonde, M., de Los Reyes-Gavilán, C. G., & Salazar, N. (2016). Intestinal short chain fatty acids and their link with diet and human health. Frontiers in Microbiology, 7, 185. https://doi.org/10.3389/fmicb.2016.00185

Robinson, C. D., Klein, H. S., Murphy, K. D., Parthasarathy, R., Guillemin, K., & Bohannan, B. J. M. (2018). Experimental bacterial adaptation to the zebrafish gut reveals a primary role for immigration. PLoS Biology, 16(12), e2006893. https://doi.org/10.1371/journal.pbio.2006893

Roeselers, G., Mittge, E. K., Stephens, W. Z., Parichy, D. M., Cavanaugh, C. M., Guillemin, K., & Rawls, J. F. (2011). Evidence for a core gut microbiota in the zebrafish. The ISME Journal, 5(10), 1595–1608. https://doi.org/10.1038/ismej.2011.38

Sampson, T. R., Debelius, J. W., Thron, T., Janssen, S., Shastri, G. G., Ilhan, Z. E., Challis, C., Schretter, C. E., Rocha, S., Gradinaru, V., Chesselet, M.-F., Keshavarzian, A., Shannon, K. M., Krajmalnik-Brown, R., Wittung-Stafshede, P., Knight, R., & Mazmanian, S. K. (2016). Gut microbiota regulate motor deficits and neuroinflammation in a model of Parkinson's disease. Cell, 167(6), 1469–1480.e12. https://doi.org/10.1016/j.cell.2016.11.018

Sanna, S., van Zuydam, N. R., Mahajan, A., Kurilshikov, A., Vich Vila, A., Võsa, U., Mujagic, Z., Masclee, A. A. M., Jonkers, D. M. A. E., Oosting, M., Joosten, L. A. B., Netea, M. G., Franke, L., Zhernakova, A., Fu, J., Wijmenga, C., & McCarthy, M. I. (2019). Causal relationships among the gut microbiome, short-chain fatty acids and metabolic diseases. Nature Genetics, 51(4), 600–605. https://doi.org/10.1038/s41588-019-0350-x

Sasahara, K., Nitanai, T., Habara, T., Morioka, T., & Nakajima, E. (1981). Dosage form design for improvement of bioavailability of levodopa V: Absorption and metabolism of levodopa in intestinal segments of dogs. Journal of Pharmaceutical Sciences, 70(10), 1157–1160. https://doi.org/10.1002/jps.2600701018

Scheperjans, F., Aho, V., Pereira, P. A. B., Koskinen, K., Paulin, L., Pekkonen, E., Haapaniemi, E., Kaakkola, S., Eerola-Rautio, J., Pohja, M., Kinnunen, E., Murros, K., & Auvinen, P. (2015). Gut microbiota are related to Parkinson's disease and clinical phenotype. Movement Disorders: Official Journal of the Movement Disorder Society, 30(3), 350–358. https://doi.org/10.1002/mds.26069

Schlomann, B. H., Wiles, T. J., Wall, E. S., Guillemin, K., & Parthasarathy, R. (2019). Sublethal antibiotics collapse gut bacterial populations by enhancing aggregation and expulsion. Proceedings of the National Academy of Sciences of the United States of America, 116(43), 21392–21400. https://doi.org/10.1073/pnas.1907567116

Schneider, S. A., & Alcalay, R. N. (2017). Neuropathology of genetic synucleinopathies with parkinsonism: Review of the literature. Movement Disorders: Official Journal of the Movement Disorder Society, 32(11), 1504–1523. https://doi.org/10.1002/mds.27193

Semova, I., Carten, J. D., Stombaugh, J., Mackey, L. C., Knight, R., Farber, S. A., & Rawls, J. F. (2012). Microbiota regulate intestinal absorption and metabolism of fatty acids in the zebrafish. Cell Host & Microbe, 12(3), 277–288. https://doi.org/10.1016/j.chom.2012.08.003

Shams, S., Chatterjee, D., & Gerlai, R. (2015). Chronic social isolation affects thigmotaxis and whole-brain serotonin levels in adult zebrafish. Behavioural Brain Research, 292, 283–287. https://doi.org/10.1016/j.bbr.2015.05.061

Shehwana, H., & Konu, O. (2019). Comparative transcriptomics between zebrafish and mammals: A roadmap for discovery of conserved and uinique signaling pathways in physiology and disease. Frontiers in Cell and Developmental Biology, 7, 5. https://doi.org/10.3389/fcell.2019.00005

So, D., Whelan, K., Rossi, M., Morrison, M., Holtmann, G., Kelly, J. T., Shanahan, E. R., Staudacher, H. M., & Campbell, K. L. (2018). Dietary fiber intervention on gut microbiota composition in healthy adults: A systematic review and meta-analysis. The American Journal of Clinical Nutrition, 107(6), 965–983. https://doi.org/10.1093/ajcn/nqy041

Soares, M. C., Cable, J., Lima-Maximino, M. G., Maximino, C., & Xavier, R. (2019). Using fish models to investigate the links between microbiome and social behaviour: The next step for translational microbiome research? Fish and Fisheries, faf.12366. https://doi.org/10.1111/faf.12366

Srivastav, S., Neupane, S., Bhurtel, S., Katila, N., Maharjan, S., Choi, H., Hong, J. T., & Choi, D.-Y. (2019). Probiotics mixture increases butyrate, and subsequently rescues the nigral dopaminergic neurons from MPTP and rotenone-induced neurotoxicity. The Journal of Nutritional Biochemistry, 69, 73–86. https://doi.org/10.1016/j.jnutbio.2019.03.021

Starr, C. G., Maderdrut, J. L., He, J., Coy, D. H., & Wimley, W. C. (2018). Pituitary adenylate cyclase-activating polypeptide is a potent broad-spectrum antimicrobial peptide: Structure-activity relationships. Peptides, 104, 35–40. https://doi.org/10.1016/j.peptides.2018.04.006

Stephens, W. Z., Burns, A. R., Stagaman, K., Wong, S., Rawls, J. F., Guillemin, K., & Bohannan, B. J. M. (2016). The composition of the zebrafish intestinal microbial community varies across development. The ISME Journal, 10(3), 644–654. https://doi.org/10.1038/ismej.2015.140

Sudo, N., Chida, Y., Aiba, Y., Sonoda, J., Oyama, N., Yu, X.-N., Kubo, C., & Koga, Y. (2004). Postnatal microbial colonization programs the hypothalamic-pituitary-adrenal system for stress response in mice. The Journal of Physiology, 558(Pt 1), 263–275. https://doi.org/10.1113/jphysiol.2004.063388

Sullivan, C., & Kim, C. H. (2008). Zebrafish as a model for infectious disease and immune function. Fish & Shellfish Immunology, 25(4), 341–350. https://doi.org/10.1016/j.fsi.2008.05.005

Sveinbjornsdottir, S. (2016). The clinical symptoms of Parkinson's disease. Journal of Neurochemistry, 139(Suppl 1), 318–324. https://doi.org/10.1111/jnc.13691

Svensson, E., Horváth-Puhó, E., Thomsen, R. W., Djurhuus, J. C., Pedersen, L., Borghammer, P., & Sørensen, H. T. (2015a). Vagotomy and subsequent risk of Parkinson's disease. Annals of Neurology, 78(4), 522–529. https://doi.org/10.1002/ana.24448

Svensson, J. F., Patkova, B., Almström, M., Naji, H., Hall, N. J., Eaton, S., Pierro, A., & Wester, T. (2015b). Nonoperative treatment with antibiotics versus surgery for acute nonperforated appendicitis in children: A pilot randomized controlled trial. Annals of Surgery, 261(1), 67–71. https://doi.org/10.1097/SLA.0000000000000835

Tamtaji, O. R., Taghizadeh, M., Daneshvar Kakhaki, R., Kouchaki, E., Bahmani, F., Borzabadi, S., Oryan, S., Mafi, A., & Asemi, Z. (2019). Clinical and metabolic response to probiotic administration in people with Parkinson's disease: A randomized, double-blind, placebo-controlled trial. Clinical Nutrition (Edinburgh, Scotland), 38(3), 1031–1035. https://doi.org/10.1016/j.clnu.2018.05.018

Thomas, R. H., Meeking, M. M., Mepham, J. R., Tichenoff, L., Possmayer, F., Liu, S., & MacFabe, D. F. (2012). The enteric bacterial metabolite propionic acid alters brain and plasma phospholipid molecular species: Further development of a rodent model of autism spectrum disorders. Journal of Neuroinflammation, 9, 153. https://doi.org/10.1186/1742-2094-9-153

Toh, M. C., Goodyear, M., Daigneault, M., Allen-Vercoe, E., & Van Raay, T. J. (2013). Colonizing the embryonic zebrafish gut with anaerobic bacteria derived from the human gastrointestinal tract. Zebrafish, 10(2), 194–198. https://doi.org/10.1089/zeb.2012.0814

Toledo-Ibarra, G. A., Rojas-Mayorquín, A. E., & Girón-Pérez, M. I. (2013). Influence of the cholinergic system on the immune response of teleost fishes: Potential model in biomedical research. Clinical and Developmental Immunology, 2013, 1–9. https://doi.org/10.1155/2013/536534

Torraca, V., Masud, S., Spaink, H. P., & Meijer, A. H. (2014). Macrophage-pathogen interactions in infectious diseases: New therapeutic insights from the zebrafish host model. Disease Models & Mechanisms, 7(7), 785–797. https://doi.org/10.1242/dmm.015594

Uemura, N., Yagi, H., Uemura, M. T., Hatanaka, Y., Yamakado, H., & Takahashi, R. (2018). Inoculation of α-synuclein preformed fibrils into the mouse gastrointestinal tract induces Lewy body-like aggregates in the brainstem via the vagus nerve. Molecular Neurodegeneration, 13(1), 21. https://doi.org/10.1186/s13024-018-0257-5

Unger, M. M., Spiegel, J., Dillmann, K.-U., Grundmann, D., Philippeit, H., Bürmann, J., Faßbender, K., Schwiertz, A., & Schäfer, K.-H. (2016). Short chain fatty acids and gut microbiota differ between patients with Parkinson's disease and age-matched controls. Parkinsonism & Related Disorders, 32, 66–72. https://doi.org/10.1016/j.parkreldis.2016.08.019

Uversky, V. N., Li, J., & Fink, A. L. (2001). Metal-triggered structural transformations, aggregation, and fibrillation of human alpha-synuclein. A possible molecular NK between Parkinson's disease and heavy metal exposure. The Journal of Biological Chemistry, 276(47), 44284–44296. https://doi.org/10.1074/jbc.M105343200

Uzbay, T. (2019). Germ-free animal experiments in the gut microbiota studies. Current Opinion in Pharmacology, 49, 6–10. https://doi.org/10.1016/j.coph.2019.03.016

Valenzuela, M.-J., Caruffo, M., Herrera, Y., Medina, D. A., Coronado, M., Feijóo, C. G., Muñoz, S., Garrido, D., Troncoso, M., Figueroa, G., Toro, M., Reyes-Jara, A., Magne, F., & Navarrete, P. (2018). Evaluating the capacity of human gut microorganisms to colonize the zebrafish larvae (*Danio rerio*). Frontiers in Microbiology, 9, 1032. https://doi.org/10.3389/fmicb.2018.01032

Van Den Berge, N., Ferreira, N., Gram, H., Mikkelsen, T. W., Alstrup, A. K. O., Casadei, N., Tsung-Pin, P., Riess, O., Nyengaard, J. R., Tamgüney, G., Jensen, P. H., & Borghammer, P. (2019). Evidence for bidirectional and trans-synaptic parasympathetic and sympathetic propagation of alpha-synuclein in rats. Acta Neuropathologica, 138(4), 535–550. https://doi.org/10.1007/s00401-019-02040-w

Wang, A. R., Ran, C., Ringø, E., & Zhou, Z. G. (2018). Progress in fish gastrointestinal microbiota research. Reviews in Aquaculture, 10(3), 626–640. https://doi.org/10.1111/raq.12191

Wang, D. D., Nguyen, L. H., Li, Y., Yan, Y., Ma, W., Rinott, E., Ivey, K. L., Shai, I., Willett, W. C., Hu, F. B., Rimm, E. B., Stampfer, M. J., Chan, A. T., & Huttenhower, C. (2021). The gut microbiome modulates the protective association between a Mediterranean diet and cardiometabolic disease risk. Nature Medicine, 27(2), 333–343. https://doi.org/10.1038/s41591-020-01223-3

Wang, L., Zhao, Z., Zhao, L., Zhao, Y., Yang, G., Wang, C., Gao, L., Niu, C., & Li, S. (2022). Lactobacillus plantarum DP189 reduces α-SYN aggravation in MPTP-induced Parkinson's disease mice via regulating oxidative damage, inflammation, and gut microbiota disorder. Journal of Agricultural and Food Chemistry, 70(4), 1163–1173. https://doi.org/10.1021/acs.jafc.1c07711

Wang, P., Chen, F.-X., Du, C., Li, C.-Q., Yu, Y.-B., Zuo, X.-L., & Li, Y.-Q. (2015). Increased production of BDNF in colonic epithelial cells induced by fecal supernatants from diarrheic IBS patients. Scientific Reports, 5, 10121. https://doi.org/10.1038/srep10121

Wegh, C. A. M., Geerlings, S. Y., Knol, J., Roeselers, G., & Belzer, C. (2019). Postbiotics and their potential applications in early life nutrition and beyond. International Journal of Molecular Sciences, 20(19), 4673. https://doi.org/10.3390/ijms20194673

Wu, X., Chen, P. S., Dallas, S., Wilson, B., Block, M. L., Wang, C.-C., Kinyamu, H., Lu, N., Gao, X., Leng, Y., Chuang, D.-M., Zhang, W., Lu, R. B., & Hong, J.-S. (2008). Histone deacetylase inhibitors up-regulate astrocyte GDNF and BDNF gene transcription and protect dopaminergic neurons. The International Journal of Neuropsychopharmacology, 11(8), 1123–1134. https://doi.org/10.1017/S1461145708009024

Xiao, F., Zhu, W., Yu, Y., He, Z., Wu, B., Wang, C., Shu, L., Li, X., Yin, H., Wang, J., Juneau, P., Zheng, X., Wu, Y., Li, J., Chen, X., Hou, D., Huang, Z., He, J., Xu, G., . . . Yan, Q. (2021). Host development overwhelms environmental dispersal in governing the ecological succession of zebrafish gut microbiota. Npj Biofilms and Microbiomes, 7(1), 5. https://doi.org/10.1038/s41522-020-00176-2

Yang, H.-T., Zou, S.-S., Zhai, L.-J., Wang, Y., Zhang, F.-M., An, L.-G., & Yang, G.-W. (2017). Pathogen invasion changes the intestinal microbiota composition and induces innate immune responses in the zebrafish intestine. Fish & Shellfish Immunology, 71, 35–42. https://doi.org/10.1016/j.fsi.2017.09.075

Yano, J. M., Yu, K., Donaldson, G. P., Shastri, G. G., Ann, P., Ma, L., Nagler, C. R., Ismagilov, R. F., Mazmanian, S. K., & Hsiao, E. Y. (2015). Indigenous bacteria from the gut microbiota regulate host serotonin biosynthesis. Cell, 161(2), 264–276. https://doi.org/10.1016/j.cell.2015.02.047

Yukgehnaish, K., Kumar, P., Sivachandran, P., Marimuthu, K., Arshad, A., Paray, B. A., & Arockiaraj, J. (2020). Gut microbiota metagenomics in aquaculture: Factors influencing gut microbiome and its physiological role in fish. Reviews in Aquaculture, raq.12416. https://doi.org/10.1111/raq.12416

Zhang, H., Tang, K., Ma, J., Zhou, L., Liu, J., Zeng, L., Zhu, L., Xu, P., Chen, J., Wei, K., Liang, X., Lv, J., Xie, J., Liu, Y., Wan, Y., & Huang, B. (2020a). Ketogenesis-generated β-hydroxybutyrate is an epigenetic regulator of CD8+ T-cell memory development. Nature Cell Biology, 22(1), 18–25. https://doi.org/10.1038/s41556-019-0440-0

Zhang, M., Shan, C., Tan, F., Limbu, S. M., Chen, L., & Du, Z.-Y. (2020b). Gnotobiotic models: Powerful tools for deeply understanding intestinal microbiota-host interactions in aquaculture. Aquaculture, 517, 734800. https://doi.org/10.1016/j.aquaculture.2019.734800

Zhernakova, A., Kurilshikov, A., Bonder, M. J., Tigchelaar, E. F., Schirmer, M., Vatanen, T., Mujagic, Z., Vila, A. V., Falony, G., Vieira-Silva, S., Wang, J., Imhann, F., Brandsma, E., Jankipersadsing, S. A., Joossens, M., Cenit, M. C., Deelen, P., Swertz, M. A., LifeLines cohort study, . . . Fu, J. (2016). Population-based metagenomics analysis reveals markers for gut microbiome composition and diversity. Science (New York, N.Y.), 352(6285), 565–569. https://doi.org/10.1126/science.aad3369

Zhong, R., Chen, Q., Zhang, X., Li, M., & Lin, W. (2022a). Helicobacter pylori infection is associated with a poor response to levodopa in patients with Parkinson's disease: A systematic review and meta-analysis. Journal of Neurology, 269(2), 703–711. https://doi.org/10.1007/s00415-021-10473-1

Zhong, X., Li, J., Lu, F., Zhang, J., & Guo, L. (2022b). Application of zebrafish in the study of the gut microbiome. Animal Models and Experimental Medicine, 5(4), 323–336. https://doi.org/10.1002/ame2.12227

Zhou, L., Limbu, S. M., Qiao, F., Du, Z.-Y., & Zhang, M. (2018a). Influence of long-term feeding antibiotics on the gut health of zebrafish. Zebrafish, 15(4), 340–348. https://doi.org/10.1089/zeb.2017.1526

Zhou, L., Limbu, S. M., Shen, M., Zhai, W., Qiao, F., He, A., Du, Z.-Y., & Zhang, M. (2018b). Environmental concentrations of antibiotics impair zebrafish gut health. Environmental Pollution (Barking, Essex: 1987), 235, 245–254. https://doi.org/10.1016/j.envpol.2017.12.073

Zhu, M., Liu, X., Ye, Y., Yan, X., Cheng, Y., Zhao, L., Chen, F., & Ling, Z. (2022). Gut microbiota: A novel therapeutic target for Parkinson's disease. Frontiers in Immunology, 13, 937555. https://doi.org/10.3389/fimmu.2022.937555

Zuo, T., & Ng, S. C. (2018). The gut microbiota in the pathogenesis and therapeutics of inflammatory bowel disease. Frontiers in Microbiology, 9, 2247. https://doi.org/10.3389/fmicb.2018.02247

17 Transgenic Zebrafish Reporter Lines for in vivo Studying of Parkinson's Disease

Nicola Facchinello

Diving into the intricate realm of neurodegenerative research, the fusion of transgenic fluorescent zebrafish lines and the study of Parkinson's disease (PD) emerges as a captivating synergy. A wealth of recent investigations, as illuminated in (Chia et al., 2022) and (Doyle & Croll, 2022), showcases the pivotal role played by these genetically modified zebrafish models in unraveling the complexities of PD.

The utilization of transgenic fluorescent zebrafish provides a unique platform for researchers to dissect the genetic intricacies underlying Parkinson's pathology, as evidenced by studies such as (Lopez et al., 2022) and (Khuansuwan et al., 2019). These investigations delve into the specific genes and molecular mechanisms that contribute to the onset and progression of PD, offering invaluable insights into the genetic landscape of this neurodegenerative disorder.

Beyond mere observation, the utility of transgenic fluorescent zebrafish lines extends to therapeutic avenues, a theme explored in Choe et al. (2021). These models not only unravel the genetic underpinnings but also serve as testing grounds for potential interventions and drug discoveries, fostering optimism for targeted treatments for PD.

The transgenic zebrafish line *Tg(elavl3:Hsa.H2B-GCaMP6s)*, spotlighted in references (Panier et al., 2013; Vladimirov et al., 2014), emerges as a crucial asset in PD research. Engineered to express the calcium indicator GCaMP6s under the *elavl3* promoter, this line allows real-time imaging of neuronal activity, providing a unique lens into the dynamics of neurons implicated in PD. *Elavl3*, an mRNA binding protein, is expressed in almost all post-mitotic or freshly differentiated neurons. It is found in the neural tube and some brain regions early in zebrafish embryos during the somite formation process. By visualizing calcium fluctuations, researchers gain insights into aberrant neuronal signaling, offering a nuanced understanding of PD pathophysiology. *Tg(elavl3:Hsa.H2B-GCaMP6s)* serves as a powerful tool to dissect the intricate neural changes associated with PD, contributing to advancements in therapeutic strategies for this neurodegenerative disorder.

The use of various fluorescent proteins like Kaede, eGFP, and mCherry enhances the versatility of the *elavl3* promoter, enabling researchers to discern specific neuronal populations in combination with other transgenic lines (Aizawa et al., 2005; Faucherre & López-Schier, 2011; Sato et al., 2006).

Engineered to express eGFP-Map1lc3b under the elavl3 promoter, this line facilitates in vivo monitoring of autophagic activity in post-mitotic neurons for Parkinson's disease research: *Tg(elavl3:eGFP:map1lc3b)*. In combination with the zebrafish model's optical clarity, rapid development, and conserved brain structures, it becomes a powerful tool for studying neurodegenerative processes. This transgenic line enables screening for autophagy-modulating compounds, offering a platform to assess the impact of autophagy inducers and inhibitors (Khuansuwan et al., 2019). The line also serves as a robust in vivo vertebrate model to investigate the effects of FDA-approved drugs, such as isradipine and nilotinib, on autophagy, contributing to potential therapeutic avenues for PD.

The transgenic zebrafish line *Tg(UAS:Dendra2-Hsa.SNCA, myl7:eGFP)* (Lopez et al., 2022) plays a pivotal role in PD research. Engineered to express human alpha-synuclein (Hsa.SNCA) tagged with Dendra2, coupled with eGFP expression in the myocardium under *myl7* promoter, this line enables the visualization of alpha-synuclein pathology in cardiac and neural tissues. This model allows researchers to study the impact of alpha-synuclein on both cardiac and neural systems, providing insights into the multifaceted aspects of PD.

Additional transgenic zebrafish lines, such as *TgBAC(gad1b:GAL4-VP16,myl7:mCherry)*, *TgBAC(slc17a6b:Gal4FF)*, and *Tg(slc6a3:eGFP)*, enable selective marking of GABAergic, glutamatergic, and dopaminergic neurons, respectively. These tools illuminate distinct neural populations, providing a nuanced understanding of their roles in PD pathology (Filosa et al., 2016; Satou et al., 2013; Xi et al., 2011).

Moreover the laboratory of Masahiko Hibi at Nagoya University developed a transgenic line, *Tg(aldoca:gap43-mCherry)*, that expresses a membrane-tagged mCherry protein specifically in Purkinje cells (Takeuchi, 2015). Remarkably, this expression persists into adulthood, allowing sustained visibility in the adult cerebellum.

Two key transgenic zebrafish lines, *Tg(mpeg1:eGFP)* (Ellett et al., 2011) and (Tg(Xla.Eef1a1:mlseGFP) (Jung Kim et al., 2008), play pivotal roles in PD research. *Tg(mpeg1:eGFP)* provides real-time visualization of macrophages and microglia, unraveling their involvement in inflammation and neurodegeneration. Meanwhile, *Tg(Xla.Eef1a1:mlseGFP)* offers a unique lens into mitochondrial dynamics within motor neurons, elucidating the critical interplay between mitochondrial function and

DOI: 10.1201/9781003402893-19

neuronal health in the context of Parkinson's. These transgenic zebrafish lines serve as invaluable tools, advancing our understanding of the cellular intricacies underlying PD pathology (Table 17.1).

The transgenic zebrafish line *Tg(mbp:eGFP-CAAX)*, explored in the study (Jung et al., 2010), emerges as a crucial tool in unraveling PD's myelin alterations. This line, expressing eGFP-CAAX under the control of the myelin

TABLE 17.1
Zebrafish Reporter Lines Targeting the Brain

Line	Transgenic ID	Promoter	Cells Tagged	Features	Ref.
Tg(olig2:eGFP)^{vu12}	ZDB-ALT-041129-8	Oligodendrocyte lineage transcription factor 2	Oligodendrocytes, Radial glia, Motor neurons	eGFP expression	(Park et al., 2007)
Tg(mbp:eGFP-CAAX)^{ue2} *Tg(1mbp:eGFP)*^{ue1} *Tg(mbpa:GAL4-VP16)*^{km5}	ZDB-ALT-111118-32 ZDB-ALT-120103-1 ZDB-ALT-150803-1	Myelin basic protein a	Myelin sheaths, Oligodendrocytes, and Schwann cells	eGFP and GAL4 expression	(Almeida et al., 2021; Chung et al., 2013; Jung et al., 2010)
Tg(gfap:GFP)^{mi2001}	ZDB-ALT-060623-4	Glial fibrillary acidic protein	Radial glias in brain, retina, and peripheral spinal nerves	GFP expression	(Bernardos & Raymond, 2006)
Tg(-4.9sox10:eGFP)^{ba2}	ZDB-ALT-050913-4	SRY-box transcription factor 10	Neural crest derived cells and oligodendrocytes lineage cells	eGFP expression	(Malatesta & Götz, 2013)
Tg(pou4f1-hsp70l:GFP)^{rw0110b}	ZDB-ALT-050317-18	POU class 4 homeobox 1	Medial division of the dorsal habeula	GFP expression	(Aizawa et al., 2005)
*Tg(aldoca:GAP-mCherry)*nub5	ZDB-ALT-150414-1	Aldolase *Ca*	Purkinje cells	mcherry	(Takeuchi 2015)
TgBAC(aoc1:GAL4-VP16)^{rw0148a}	ZDB-ALT-150312-5	Amine oxidase copper containing 1	Ventral habenula	GAL4 expression	(Amo et al., 2014)
Tg(elavl3:eGFP)^{knu3} *Tg(elavl3:Kaede)*^{rw0130a} *Tg2(elavl3:GAL4)*^{zf409}	ZDB-ALT-060301-2 ZDB-ALT-060619-4 ZDB-ALT-130919-2	ELAV-like neuron-specific RNA-binding protein 3	Differentiated neurons (motor and sensory)	Enhanced GFP, Kaede, or GAL4 expression	(Aizawa et al., 2005; Faucherre & López-Schier, 2011; Sato et al., 2006)
Tg(elavl3:GCaMP3)^{ens100} *Tg(elavl3:Hsa.H2B-GCaMP6s)*^{jf5}	ZDB-ALT-131122-1 ZDB-ALT-141023-2	ELAV-like neuron-specific RNA-binding protein 3	Differentiated neurons (motor and sensory)	GCaMP6s expression	(Panier et al., 2013; Vladimirov et al., 2014)
Tg(slc6a3:eGFP)^{ot8}	ZDB-ALT-111206-10	Solute carrier family 6 member 3	Dopaminergic neurons	Enhanced GFP expression	(Xi et al., 2011)
TgBAC(slc17a6b:Gal4FF)^{nns20}	ZDB-ALT-131127-1	Solute carrier family 17 member 6b	Glutamatergic Neurons	GAL4 expression	(Satou et al., 2013)
Tg(UAS:Brb1.0L)^{s1997t} *Tg(hsp70l:Cre)*^{zdf13}	ZDB-ALT-130412-2 ZDB-ALT-080222-1	Heat-shock-inducible promoter hsp70l		dTomato, mCerulean, and eYFP expression	(Le et al., 2007; Pan et al., 2011)
TgBAC(gad1b:GAL4-VP16,myl7:mCherry)^{mpn155}	ZDB-ALT-160726-5	Glutamate decarboxylase 1b	GABAergic neurons	mcherry expression	(Filosa et al., 2016)
Tg (UAS:Dendra2-Hsa.SNCA, myl7:eGFP)^{cu49}	ZDB-ALT-220328-8	alpha-Synuclein	aggregation of α-synuclein (α-SYN)	Enhanced GFP expression	(Lopez et al., 2022)
Tg(mpeg1:eGFP)^{gl22} *Tg(mpeg1:mCherry)*^{gl23}	ZDB-ALT-120117-1 ZDB-ALT-120117-2	Macrophage expressed 1	Macrophages	Enhanced GFP and mcherry expression	(Ellett et al., 2011; Fang et al., 2011)
Tg(Xla.Eef1a1:mlseGFP)^{cms1}	ZDB-ALT-090309-2	Cytochrome c oxidase subunit 8A	Mitochondria	Enhanced GFP expression	(Facchinello et al., 2021; Jung Kim et al., 2008)
Tg(elavl3:eGFP:map1lc3)^{la50016}	ZDB-ALT-191122-9	Microtubule-associated protein 1 light chain 3 beta	Autophagic activity in neurons	Enhanced GFP expression	(Khuansuwan et al., 2019)
Tg(-2.5neurod1:eGFP)^{ia50}	ZDB-ALT-150626-1	Neurogenic differentiation 1	Proneural marker	Enhanced GFP expression	(Ronneberger et al., 2012)

basic protein (*mbp*) promoter, allows real-time visualization and assessment of myelin content. By observing myelin dynamics in vivo, researchers gain insights into the potential role of myelin alterations in Parkinson's pathology. The *Tg(mbp:eGFP-CAAX)* zebrafish line serves as a valuable model for investigating the intricate interplay between myelin integrity and neurodegenerative processes, contributing to our understanding of PD etiology.

The brainbow system in transgenic zebrafish (Le et al., 2007; Pan et al., 2011) offers a revolutionary approach in PD research. This genetic labeling technique allows distinct neurons to be uniquely color-coded, providing an intricate map of neural circuits affected in PD. By visualizing individual neurons with different colors, researchers gain unparalleled insights into the complex connectivity and potential alterations in the Parkinsonian brain. The brainbow system serves as a transformative tool, shedding light on the spatial and functional dynamics of neurons in PD, fostering a deeper understanding of the disease's underlying neural intricacies and paving the way for targeted therapeutic strategies.

In the intricate world of neurodegenerative research, the convergence of zebrafish as a model organism and responsive element-based reporter lines has opened a gateway to unprecedented insights into the signaling pathways implicated in PD (Chia et al., 2022; Choe et al., 2021; Moro et al., 2013). Each signaling pathway reporter line becomes a protagonist, revealing its dynamic role in the expression within the brain and its potential implications for PD (Table 17.2).

The fluorescent Stat3-reporter zebrafish based on repeats of minimal responsive elements, as detailed in (Peron et al., 2020), provides a real-time glimpse into its expression dynamics within the brain. Stat3, with its intricate role in neuroinflammation and neuroprotection, emerges as a central player in the molecular cascade of PD. The role of Stat3 in zebrafish becomes a crucial chapter in understanding how this signaling pathway influences the inflammatory milieu associated with the neurodegenerative disorder.

Zebrafish reporter lines expressing Wnt, detailed in (Moro et al., 2013), allow the spatial and temporal dynamics visualization of Wnt signaling within the brain. The exploration of Wnt signaling, renowned for its diverse roles in development, provides the possibility to gain insights into the survival and function of neurons, particularly dopaminergic neurons, paving the way for potential therapeutic interventions.

Furthermore, as demonstrated in (Casari et al., 2014), zebrafish reporter lines expressing Transforming Growth Factor-beta (TGFβ) offer a platform for researchers to investigate its temporal and spatial dynamics inside the brain. At 4 dpf, eGFP is expressed in the hindbrain, diencephalon, and telencephalon of the transgenic line Tg(12xSBE: eGFP).

Sonic Hedgehog (Shh), a morphogen integral to neural development, was described (Mich et al., 2014). It has been shown that the Hedgehog (Hh)/Gli signaling pathway is both necessary and sufficient for normal hypothalamus proliferation rates, and that it positively influences hypothalamic neurogenesis in both larval and adult zebrafish.

TABLE 17.2
Zebrafish Reporter Lines Targeting in vivo Signaling Pathway Activities Involved in Brain Development

Line	Transgenic ID	Responsive element (RE)	Signaling Pathway	Features	Ref.
Tg(7xSRE-HSV.Ul23:eGFP)ia28	ZDB-ALT-210317–3	SRE	Stat3 signaling	Enhanced GFP expression	(Peron et al., 2020)
Tg(BRE-AAVmlp:dmKO2)^{mw40}	ZDB-ALT-110310–2	BRE	BMP signaling	Destabilized monomeric Kusabira-Orange (dmKO2)	(Collery & Link, 2011)
Tg(BMPRE:NLS-mCherry)^{ia17}	ZDB-ALT-130115–2	BRE	BMP signaling	Nuclear localization signal mCherry	(Casari et al., 2014)
Tg(7xTCF-Xla.Siam:GFP)^{ia4}	ZDB-ALT-110113–1	TCF	Wnt signaling	GFP expression	(Moro et al., 2012)
Tg(7xTCF-Xla.Siam:nlsmCherry)^{ia5}	ZDB-ALT-110113–2	TCF	Wnt signaling	Nuclear localization signal mCherry	(Moro et al., 2012)
Tg(12xSBE: eGFP)^{ia16}	ZDB-ALT-130212–12	SBE	TGFβ/Smad3	Enhanced GFP expression	(Casari et al., 2014)
Tg(6xCRE:eGFP)		CRE	CREB signaling	Enhanced GFP expression	(Giuliodori et al., 2018)
Tg(EPV.Tpi-Mmu.Hbb:eGFP)^{ia12}	ZDB-ALT-130115–3	NRE	Notch signaling	Enhanced GFP expression	(Corallo et al., 2013; Parsons et al., 2009)
Tg(12xgli-HSV.Ul23:GFP)^{ia11}	ZDB-ALT-120404–2	gli	Shh signaling	GFP expression	(Moro et al., 2013)
Tg(8xGliBS:mCherry-NLS-Odc1)^{st1002}	ZDB-ALT-141030–2	gli	Shh signaling	Nuclear-localized fluorescent protein with an ornithine decarboxylase-derived destabilizing peptide	(Mich et al., 2014)
Tg(9xGCRE-HSV.Ul23:eGFP)^{ia20}	ZDB-ALT-130123–1	GRE	Glucocorticoid	Enhanced GFP expression	(Benato et al., 2014)

Dopaminergic, serotonergic, and GABAergic neurons originate from a relatively highly proliferative precursor population known as Shh-responsive radial glia (Male et al., 2020). Male et al. demonstrated that transgenic reporter analyses unveiled significant heterogeneity in cell-cell signaling within the hypothalamic niche, displaying overlapping populations of radial glia expressing Shh-responsive, Notch-responsive, and Wnt-responsive markers. Their epistatic interaction signifies the ability of one pathway to modify or override another's effects, shaping neural development.

Zebrafish reporter lines expressing glucocorticoid receptors, as detailed in (Benato et al., 2014), regulate a variety of physiological activities, including neuroinflammation and neuronal survival. This line exhibits a considerable rise in eGFP fluorescence, which first appears in a broad pattern during the early stages of somitogenesis and then mostly concentrates 24 hours after fertilization in the brain and trunk muscles.

Furthermore, zebrafish reporter lines expressing Notch and bone morphogenetic proteins (BMPs) and cAMP response element-binding protein (CREB), as detailed in (Collery & Link, 2011; Parsons et al., 2009), unveil their potential impact on the maintenance and regeneration of neurons affected by the neurodegenerative disorder, offering novel dimensions to the understanding of PD etiology.

In the grand tapestry of signaling pathway reporter lines in zebrafish and PD, each protagonist—Stat3, Shh, Wnt, Notch, BMP, TGFβ, CREB, and glucocorticoids—adds a unique piece to the puzzle. The exploration of their spatial and temporal dynamics within the brain enriches our understanding of neurodevelopment and molecular pathway in the onset of PD. In conclusion, the combination of transgenic fluorescent zebrafish and PD research not only elucidates the genetic intricacies of this debilitating condition but also holds the promise of transformative therapeutic approaches.

REFERENCES

Aizawa, H., Bianco, I. H., Hamaoka, T., Miyashita, T., Uemura, O., Concha, M. L., Russell, C., Wilson, S. W., & Okamoto, H. (2005). Laterotopic representation of left-right information onto the dorso-ventral axis of a Zebrafish midbrain target nucleus. *Current Biology*, *15*(3), 238–243. https://doi.org/10.1016/j.cub.2005.01.014

Almeida, R. G., Williamson, J. M., Madden, M. E., Early, J. J., Voas, M. G., Talbot, W. S., Bianco, I. H., & Lyons, D. A. (2021). Myelination induces axonal hotspots of synaptic vesicle fusion that promote sheath growth. *Current Biology*, *31*(17), Articolo 17. https://doi.org/10.1016/j.cub.2021.06.036

Amo, R., Fredes, F., Kinoshita, M., Aoki, R., Aizawa, H., Agetsuma, M., Aoki, T., Shiraki, T., Kakinuma, H., Matsuda, M., Yamazaki, M., Takahoko, M., Tsuboi, T., Higashijima, S., Miyasaka, N., Koide, T., Yabuki, Y., Yoshihara, Y., Fukai, T., & Okamoto, H. (2014). The habenulo-raphe serotonergic circuit encodes an aversive expectation value essential for adaptive active avoidance of danger. *Neuron*, *84*(5), 1034–1048. https://doi.org/10.1016/j.neuron.2014.10.035

Benato, F., Colletti, E., Skobo, T., Moro, E., Colombo, L., Argenton, F., & Dalla Valle, L. (2014). A living biosensor model to dynamically trace glucocorticoid transcriptional activity during development and adult life in zebrafish. *Molecular and Cellular Endocrinology*, *392*(1–2), 60–72. https://doi.org/10.1016/j.mce.2014.04.015

Bernardos, R. L., & Raymond, P. A. (2006). GFAP transgenic zebrafish. *Gene Expression Patterns*, *6*(8), 1007–1013. https://doi.org/10.1016/j.modgep.2006.04.006

Casari, A., Schiavone, M., Facchinello, N., Vettori, A., Meyer, D., Tiso, N., Moro, E., & Argenton, F. (2014). A Smad3 transgenic reporter reveals TGF-beta control of zebrafish spinal cord development. *Developmental Biology*, *396*(1), 81–93. https://doi.org/10.1016/j.ydbio.2014.09.025

Chia, K., Klingseisen, A., Sieger, D., & Priller, J. (2022). Zebrafish as a model organism for neurodegenerative disease. *Frontiers in Molecular Neuroscience*, *15*, 940484. https://doi.org/10.3389/fnmol.2022.940484

Choe, C. P., Choi, S.-Y., Kee, Y., Kim, M. J., Kim, S.-H., Lee, Y., Park, H.-C., & Ro, H. (2021). Transgenic fluorescent zebrafish lines that have revolutionized biomedical research. *Laboratory Animal Research*, *37*(1), 26. https://doi.org/10.1186/s42826-021-00103-2

Chung, A.-Y., Kim, P.-S., Kim, S., Kim, E., Kim, D., Jeong, I., Kim, H.-K., Ryu, J.-H., Kim, C.-H., Choi, J., Seo, J.-H., & Park, H.-C. (2013). Generation of demyelination models by targeted ablation of oligodendrocytes in the zebrafish CNS. *Molecules and Cells*, *36*(1), 82–87. https://doi.org/10.1007/s10059-013-0087-9

Collery, R. F., & Link, B. A. (2011). Dynamic smad-mediated BMP signaling revealed through transgenic zebrafish. *Developmental Dynamics*, *240*(3), 712–722. https://doi.org/10.1002/dvdy.22567

Corallo, D., Schiavinato, A., Trapani, V., Moro, E., Argenton, F., & Bonaldo, P. (2013). Emilin3 is required for notochord sheath integrity and interacts with Scube2 to regulate notochord-derived Hedgehog signals. *Development*, *140*(22), 4594–4601. https://doi.org/10.1242/dev.094078

Doyle, J. M., & Croll, R. P. (2022). A critical review of zebrafish models of parkinson's disease. *Frontiers in Pharmacology*, *13*, 835827. https://doi.org/10.3389/fphar.2022.835827

Ellett, F., Pase, L., Hayman, J. W., Andrianopoulos, A., & Lieschke, G. J. (2011). Mpeg1 promoter transgenes direct macrophage-lineage expression in zebrafish. *Blood*, *117*(4), e49–e56. https://doi.org/10.1182/blood-2010-10-314120

Facchinello, N., Laquatra, C., Locatello, L., Beffagna, G., Brañas Casas, R., Fornetto, C., Dinarello, A., Martorano, L., Vettori, A., Risato, G., Celeghin, R., Meneghetti, G., Santoro, M. M., Delahodde, A., Vanzi, F., Rasola, A., Dalla Valle, L., Rasotto, M. B., Lodi, T., . . . Tiso, N. (2021). Efficient clofilium tosylate-mediated rescue of POLG-related disease phenotypes in zebrafish. *Cell Death & Disease*, *12*(1), 100. https://doi.org/10.1038/s41419-020-03359-z

Fang, L., Green, S. R., Baek, J. S., Lee, S.-H., Ellett, F., Deer, E., Lieschke, G. J., Witztum, J. L., Tsimikas, S., & Miller, Y. I. (2011). In vivo visualization and attenuation of oxidized lipid accumulation in hypercholesterolemic zebrafish. *Journal of Clinical Investigation*, *121*(12), 4861–4869. https://doi.org/10.1172/JCI57755

Faucherre, A., & López-Schier, H. (2011). Delaying Gal4-driven gene expression in the zebrafish with morpholinos and Gal80. *PLoS One*, *6*(1), e16587. https://doi.org/10.1371/journal.pone.0016587

Filosa, A., Barker, A. J., Dal Maschio, M., & Baier, H. (2016). Feeding state modulates behavioral choice and processing of prey Stimuli in the zebrafish tectum. *Neuron*, *90*(3), 596–608. https://doi.org/10.1016/j.neuron.2016.03.014

Giuliodori, A., Beffagna, G., Marchetto, G., Fornetto, C., Vanzi, F., Toppo, S., Facchinello, N., Santimaria, M., Vettori, A., Rizzo, S., Della Barbera, M., Pilichou, K., Argenton, F., Thiene, G., Tiso, N., & Basso, C. (2018). Loss of cardiac Wnt/β-catenin signalling in desmoplakin-deficient AC8 zebrafish models is rescuable by genetic and pharmacological intervention. *Cardiovascular Research*, *114*(8), 1082–1097. https://doi.org/10.1093/cvr/cvy057

Jung, S., Kim, S., Chung, A., Kim, H., So, J., Ryu, J., Park, H., & Kim, C. (2010). Visualization of myelination in GFP-transgenic zebrafish. *Developmental Dynamics*, *239*(2), 592–597. https://doi.org/10.1002/dvdy.22166

Jung Kim, M., Ho Kang, K., Kim, C.-H., & Choi, S.-Y. (2008). Real-time imaging of mitochondria in transgenic zebrafish expressing mitochondrially targeted GFP. *BioTechniques*, *45*(3), Articolo 3. https://doi.org/10.2144/000112909

Khuansuwan, S., Barnhill, L. M., Cheng, S., & Bronstein, J. M. (2019). A novel transgenic zebrafish line allows for in vivo quantification of autophagic activity in neurons. *Autophagy*, *15*(8), 1322–1332. https://doi.org/10.1080/15548627.2019.1580511

Le, X., Langenau, D. M., Keefe, M. D., Kutok, J. L., Neuberg, D. S., & Zon, L. I. (2007). Heat shock-inducible Cre/Lox approaches to induce diverse types of tumors and hyperplasia in transgenic zebrafish. *Proceedings of the National Academy of Sciences*, *104*(22), 9410–9415. https://doi.org/10.1073/pnas.0611302104

Lopez, A., Gorb, A., Palha, N., Fleming, A., & Rubinsztein, D. C. (2022). A new zebrafish model to measure neuronal α-synuclein clearance in vivo. *Genes*, *13*(5), 868. https://doi.org/10.3390/genes13050868

Malatesta, P., & Götz, M. (2013). Radial glia–from boring cables to stem cell stars. *Development*, *140*(3), 483–486. https://doi.org/10.1242/dev.085852

Male, I., Ozacar, A. T., Fagan, R. R., Loring, M. D., Shen, M.-C., Pace, V. A., Devine, C. A., Lawson, G. E., Lutservitz, A., & Karlstrom, R. O. (2020). Hedgehog signaling regulates neurogenesis in the larval and adult zebrafish hypothalamus. *Eneuro*, *7*(6). https://doi.org/10.1523/ENEURO.0226-20.2020

Mich, J. K., Payumo, A. Y., Rack, P. G., & Chen, J. K. (2014). In vivo imaging of hedgehog pathway activation with a nuclear fluorescent reporter. *PLoS One*, *9*(7), e103661. https://doi.org/10.1371/journal.pone.0103661

Moro, E., Ozhan-Kizil, G., Mongera, A., Beis, D., Wierzbicki, C., Young, R. M., Bournele, D., Domenichini, A., Valdivia, L. E., Lum, L., Chen, C., Amatruda, J. F., Tiso, N., Weidinger, G., & Argenton, F. (2012). In vivo Wnt signaling tracing through a transgenic biosensor fish reveals novel activity domains. *Developmental Biology*, *366*(2), 327–340. https://doi.org/10.1016/j.ydbio.2012.03.023

Moro, E., Vettori, A., Porazzi, P., Schiavone, M., Rampazzo, E., Casari, A., Ek, O., Facchinello, N., Astone, M., Zancan, I., Milanetto, M., Tiso, N., & Argenton, F. (2013). Generation and application of signaling pathway reporter lines in zebrafish. *Molecular Genetics and Genomics*, *288*(5–6), 231–242. https://doi.org/10.1007/s00438-013-0750-z

Pan, Y. A., Livet, J., Sanes, J. R., Lichtman, J. W., & Schier, A. F. (2011). Multicolor brainbow imaging in Zebrafish. *Cold Spring Harbor Protocols*, *2011*(1). https://doi.org/10.1101/pdb.prot5546

Panier, T., Romano, S. A., Olive, R., Pietri, T., Sumbre, G., Candelier, R., & Debrégeas, G. (2013). Fast functional imaging of multiple brain regions in intact zebrafish larvae using selective plane illumination microscopy. *Frontiers in Neural Circuits*, *7*. https://doi.org/10.3389/fncir.2013.00065

Park, H., Shin, J., Roberts, R. K., & Appel, B. (2007). An *olig2* reporter gene marks oligodendrocyte precursors in the postembryonic spinal cord of zebrafish. *Developmental Dynamics*, *236*(12), 3402–3407. https://doi.org/10.1002/dvdy.21365

Parsons, M. J., Pisharath, H., Yusuff, S., Moore, J. C., Siekmann, A. F., Lawson, N., & Leach, S. D. (2009). Notch-responsive cells initiate the secondary transition in larval zebrafish pancreas. *Mechanisms of Development*, *126*(10), 898–912. https://doi.org/10.1016/j.mod.2009.07.002

Peron, M., Dinarello, A., Meneghetti, G., Martorano, L., Facchinello, N., Vettori, A., Licciardello, G., Tiso, N., & Argenton, F. (2020). The stem-like STAT3-responsive cells of zebrafish intestine are WNT/β-catenin dependent. *Development*. https://doi.org/10.1242/dev.188987

Ronneberger, O., Liu, K., Rath, M., Rueβ, D., Mueller, T., Skibbe, H., Drayer, B., Schmidt, T., Filippi, A., Nitschke, R., Brox, T., Burkhardt, H., & Driever, W. (2012). ViBE-Z: A framework for 3D virtual colocalization analysis in zebrafish larval brains. *Nature Methods*, *9*(7), 735–742. https://doi.org/10.1038/nmeth.2076

Sato, T., Takahoko, M., & Okamoto, H. (2006). HuC:Kaede, a useful tool to label neural morphologies in networks in vivo. *Genesis*, *44*(3), 136–142. https://doi.org/10.1002/gene.20196

Satou, C., Kimura, Y., Hirata, H., Suster, M. L., Kawakami, K., & Higashijima, S. (2013). Transgenic tools to characterize neuronal properties of discrete populations of zebrafish neurons. *Development*, *140*(18), 3927–3931. https://doi.org/10.1242/dev.099531

Takeuchi, M., Matsuda, K., Yamaguchi, S., Asakawa, K., Miyasaka, N., Lal, P., Yoshihara, Y., Koga, A., Kawakami, K., Shimizu, T., & Hibi, M. (2015). Establishment of Gal4 transgenic zebrafish lines for analysis of development of cerebellar neural circuitry. *Developmental Biology*, *397*(1), 1–17.

Vladimirov, N., Mu, Y., Kawashima, T., Bennett, D. V., Yang, C.-T., Looger, L. L., Keller, P. J., Freeman, J., & Ahrens, M. B. (2014). Light-sheet functional imaging in fictively behaving zebrafish. *Nature Methods*, *11*(9), 883–884. https://doi.org/10.1038/nmeth.3040

Xi, Y., Yu, M., Godoy, R., Hatch, G., Poitras, L., & Ekker, M. (2011). Transgenic zebrafish expressing green fluorescent protein in dopaminergic neurons of the ventral diencephalon. *Developmental Dynamics*, *240*(11), 2539–2547. https://doi.org/10.1002/dvdy.22742

18 Interplay of Neurodegenerative Disorders in Zebrafish
Unlocking the Mystery

Sanobar Shariff, Burhan Kantawala, Qamar Sultana, Hasiba Karimi, Ider Oujamaa, and Vivek Sanker

18.1 INTRODUCTION TO ZEBRAFISH—NEURONAL BASIS

18.1.1 Background

Zebrafish, scientifically classified as *Danio rerio*, encompasses two distinct species thriving in diverse aquatic habitats. Members of the genus *Danio rerio* populate freshwater environments, while their counterparts in the genus *Pterois* reside in saltwater ecosystems. Noteworthy for researchers and enthusiasts alike, freshwater zebrafish exhibit distinct blue and silver longitudinal stripes, whereas their saltwater counterparts boast vibrant vertical stripes, complemented by colorful pectoral fins and venomous spikes (Zebra*fish \ textbar types & facts \textbar Britannica*, 2023). Originating from the tropical regions of Southeast Asia, zebrafish represent a pivotal species within the phylum Chordata, the class Telostei, and the Cyprinidae or Minnow family. Their habitats include streams, ponds, and canals, contributing to their ecological significance ("Why Use the Zebrafish in Research?," n.d.; "Zebrafish—Characteristics, Life Cycle, Development, Genome Research," n.d.). Remarkably versatile, zebrafish lack teeth and exhibit omnivorous tendencies. The reproductive process involves the female zebrafish releasing eggs into the aquatic environment, where they undergo external fertilization. ("Zebrafish—Characteristics, Life Cycle, Development, Genome Research," n.d.). Following rapid embryonic development, the eggs hatch within 2–3 days, marking the initiation of their life cycle. Zebrafish larvae possess transparent bodies, transitioning into adulthood with the development of characteristic stripes. An adult zebrafish typically measures between 2.5 to 4 centimeters in length, underscoring their compact yet intricate biology ("Why Use the Zebrafish in Research?," n.d.; "Zebrafish—Characteristics, Life Cycle, Development, Genome Research," n.d.). The intricacies of zebrafish behavior, reproductive patterns, and developmental milestones are central to comprehensive scientific exploration. Successful breeding hinges upon meticulous attention to environmental nuances. Ideal housing conditions encompass the implementation of mechanical, carbon, and biologic filters to maintain water quality. Further considerations include ultraviolet sterilization, regulated lighting, and darkness cycles that facilitate breeding during dawn. Maintaining precise water temperatures between 23–28°C and sustaining an optimal pH range of 6.2–7.5 are imperative elements of zebrafish husbandry ("Zebrafish—Characteristics, Life Cycle, Development, Genome Research," n.d.). In essence, the multifaceted nature of zebrafish, coupled with their unique environmental adaptations, underscores their significance as a species of profound interest for researchers, contributing invaluable insights to the scientific community.

18.1.2 Zebrafish in Research: A Keystone in Genetics

In the realm of genetic research, freshwater zebrafish stand as paramount subjects, renowned for their unique attributes and immense utility. Strains such as AB, Casper, Nadia, Wild Indian karyotype, Tubigen, and wild-caught zebrafish have become instrumental tools in scientific inquiry (Teame et al., 2019). Their extensive usage stems from their conserved cell types, genes, and neural circuits, all of which bear relevance to human diseases. This unique genetic resemblance has positioned zebrafish at the forefront of studies in biomedicine, biology, and neurophysiology (Zebra Fish—an Overview | ScienceDirect Topics, n.d.).

The applications of zebrafish in genetic research are multifaceted. They provide invaluable insights into developmental disorders, mental health conditions, communication pathways within the brain and other organs, as well as metabolic disorders (Choi et al., 2021). Their transparency during the larval stage enables researchers to visualize the intricate structures of the nervous system and its functionality through in vivo imaging. Manipulating and analyzing the zebrafish genome is not only possible but also pivotal in understanding genetic intricacies, rendering them a superior model when compared to other non-mammalian vertebrates. Zebrafish play a pivotal role in drug discovery, where their larval forms find applications in electrophysiological recordings and neurobehavioral studies. This unique capability bridges critical gaps in our understanding of the neuronal

DOI: 10.1201/9781003402893-20

basis of various physiological processes (Zebra Fish—an Overview | ScienceDirect Topics, n.d.). The presence of a human HuC protein analog in zebrafish proves invaluable in neurogenesis research, serving as a marker for neuronal precursor cells. Advancements in technology, specifically tissue clearing techniques, have enabled comprehensive neural network visualization throughout the zebrafish brain, shedding light on the complexities of neural interactions.

Zebrafish serve as vital tools in studying phenotypic characteristics in human developmental disorder disease models. The ability to create mutant zebrafish, achieved through precise genetic alterations including gene knock-ins and knockouts, has opened new avenues for exploring intricate biomedical models (Teame et al., 2019). This versatility positions zebrafish as indispensable assets in the continually evolving landscape of genetic research, offering profound contributions to our understanding of genetics and its implications for human health.

18.1.3 When Did the Idea First Come About?

Many researchers were making constant efforts to study development in flies, but at the University of Oregon, the zebrafish was first introduced as a model for vertebrate research in the field of development and genetics by George Streisinger in 1972 (Zebra Fish—an Overview \textbar ScienceDirect Topics, n.d.). In 1981, George Streisinger et al. used zebrafish as an animal model for genetic studies by creating homozygous zebrafish in large numbers (Streisinger et al., 1981). Early on in the years after that, the study on mutations utilizing ethylnitrosourea in zebrafish and the resulting unique phenotypes were initiated and characterized (Haffter et al., 1996; Kimmel, 1989). It gained recognition in the 1990s when two significant mutants were developed. One mutant was developed by Christiane Nusslein-Volhard and the other by Wolfgang Driever and Mark Fishman. After that in the 20th century, Keith Cheng cloned the gene for abnormal pigmentation using zebrafish to establish its role in the skin color of humans in the year 2005. In Leonard Zon's lab, from the chemical screen of zebrafish, a novel treatment for melanoma was identified in the year 2011 (*Danio rerio* (Zebrafish) as a Model Organism \textbar Biology II \textbar JoVE, n.d.). The "human disease" that was first identified through zebrafish was a defect of hemoglobin production in the blood associated with the mutant gene ALAS2. Many mutants show phenotypic traits similar to human disease, and therefore zebrafish could be used as a useful tool to explore knowledge on cardiovascular disorders, hematological disorders, and neurological diseases including Parkinson's disease, anxiety, and post-traumatic stress disorder, among others. The zebrafish has functional homologs for 70% of the genes implicated in human disease. It has thus served as a model organism for the various diverse human diseases by facilitating the study of the developmental biology of different systems (Rahman Khan & Sulaiman Alhewairini, 2019).

18.1.4 Developments Since the Past

Over the past two decades, zebrafish have seen a gradual growth in becoming one of the best models for research studies. The several superior aspects of zebrafish, including short generation time and the possibility of manipulation, have contributed to this growth. The developments in knowledge of the scope of zebrafish as a model for research continue to be explored and have given a boost to its widespread use. In 2013 Howe et al. reported that the entire genome sequencing of zebrafish is complete. In the same year, Gaj et al. reported the introduction of genome editing technology. It was found that humans and zebrafish share 70% of genes, and over 80% of genes that cause genetic disease in humans are found in the zebrafish. First developed as a model for developmental genetic studies, the zebrafish at present has found expansive applications in biomedical sciences, biology, toxicology, drug development, physiology, and disease modeling (Zebra Fish—an Overview \ textbar ScienceDirect Topics, n.d.).

Zebrafish, with their transparent external embryos, provide a unique opportunity to observe complete development, starting from fertilization. This transparency has catalyzed a substantial increase in research output, with annual publications in biomedical studies utilizing zebrafish rising from double to triple digits (Teame et al., 2019). The comprehensive genome map of zebrafish has paved the way for advanced gene targeting, employing state-of-the-art CRISPR/Cas-9 technology to overcome challenges in mutant creation (Choi et al., 2021). Since the 20th century, zebrafish have been integral to disease studies, particularly in metabolic disorders, intestinal diseases, liver ailments, and diet-related health conditions. Their popularity in these areas highlights their significance in unraveling complex disease mechanisms. Zebrafish, therefore, stand as a revolutionary model, offering profound insights into developmental processes and serving as a cornerstone in disease research.

18.2 ZEBRAFISH: A PRAGMATIC PRECLINICAL MODEL IN BIOMEDICAL RESEARCH

18.2.1 Reasons for Using Zebrafish in Preclinical Studies

In the realm of model organisms, zebrafish have emerged as an exceptionally attractive choice, meeting key criteria for effective laboratory use. The selection of an ideal model organism hinges on its ability to reproduce abundantly and maintain simplicity in laboratory conditions ("Model Organism – Definition, Types, Characteristics and FAQs," n.d.). Zebrafish, recognized for their pivotal role in developmental biology, have swiftly become a prominent model for preclinical studies focusing on human diseases. This prominence is underpinned by several compelling factors:

Visual Clarity: Zebrafish exhibit external embryonic development, providing researchers with a direct visualization window into embryogenesis and organ morphology. The transparent nature of zebrafish larvae enables precise imaging studies, offering unparalleled insights into the intricacies of development (Parng et al., 2002; Teame et al., 2019).

High Reproductive Capacity: Female zebrafish lay hundreds of eggs weekly, coupled with a remarkably short generation time of 2–3 months. This prolific reproductive capability ensures a consistent supply of subjects for experimentation, facilitating continuous research progress (Zebra Fish—an Overview | ScienceDirect Topics, n.d.).

Rapid Development: Zebrafish undergo swift embryonic development, completing major organ system formation within 48 hours post-fertilization. This accelerated development timeline allows for efficient observations and studies, expediting research outcomes (Teame et al., 2019).

In light of these attributes, zebrafish stand as a cost-effective and efficient choice for preclinical research studies. They serve as a safe model, particularly conducive for phenotype-based screening and neurological studies (Rosa et al., 2022). Notably, preclinical screening is indispensable for evaluating the translation of novel drug delivery processes, assessing crucial factors such as biodistribution, efficacy, and biotoxicity. Zebrafish offer a valuable alternative for studying nanoscale drug delivery systems, owing to their optical clarity, substantial sample sizes, and the ability to conduct organ-specific analyses (Lee et al., 2017). Zebrafish's unique attributes position them as a pragmatic and invaluable preclinical model in biomedical research. Their adaptability, reproductive efficiency, and suitability for advanced studies make them indispensable tools, significantly advancing our understanding of complex biological processes and fostering innovative drug development initiatives.

18.2.2 Areas Where Zebrafish Are Utilized in Preclinical Studies

Zebrafish, as a model organism, have found extensive applications in diverse preclinical research areas, unveiling critical insights across several domains:

1. **Developmental Processes**: Zebrafish have been instrumental in dissecting the intricacies of developmental processes, including the derivation and migration of cell types within the enteric nervous system. Real-time tracking of blood vessels has enabled in-depth studies of angiogenesis. Remarkably, zebrafish exhibit regenerative capabilities, a phenomenon attributed to specific stem cells, offering valuable avenues for research (Rahman Khan & Sulaiman Alhewairini, 2019).
2. **Human Disease Exploration**: Zebrafish have played a pivotal role in exploring diseases affecting multiple systems, such as the central nervous system (CNS) and cardiovascular system (CVS). The development of phenotypically similar mutants has significantly contributed to understanding disease mechanisms and potential therapeutic interventions (Rahman Khan & Sulaiman Alhewairini, 2019).
3. **Drug Development and Toxicology**: Zebrafish serve as a valuable platform for drug development and toxicological studies. Researchers can target specific pharmacologic processes of drugs on organs, assess cell and tissue-specific effects through mutants and gene editing, and explore chemical and drug toxicity. Transgenic zebrafish models enable screening for cardiotoxicity-inducing molecules, providing crucial insights into drug safety (Patton et al., 2021; Rahman Khan & Sulaiman Alhewairini, 2019).
4. **Cancer Studies**: Zebrafish have facilitated in-depth exploration of genetic, cellular, and molecular aspects of cancer. Particularly, the effects of anti-angiogenic agents have been studied extensively. Coupled with advanced microscopy and imaging techniques, zebrafish models provide valuable data on cancer progression and treatment outcomes (Rahman Khan & Sulaiman Alhewairini, 2019).
5. **Neuroscience Investigations**: Zebrafish serve as valuable models for studying neurobehavioral diseases. Mutations in zebrafish genes, such as the glucocorticoid receptor gene, enable the modeling and analysis of conditions like depression. Zebrafish models also offer unique insights into complex neurodegenerative processes due to their transparent nature, allowing researchers to decode intricate mechanisms of neurodegeneration (Kalueff et al., 2014; J. Wang & Cao, 2021).

Zebrafish's versatile utility across developmental, disease, drug-related, cancer, and neuroscience studies underscores its significance in preclinical research. Its unique attributes empower researchers to delve deeper into various biological phenomena, making zebrafish a cornerstone in advancing our understanding of human health and diseases.

18.3 COMPARISON OF ZEBRAFISH WITH OTHER MURINE MODELS AND THE HUMAN BRAIN

18.3.1 Superiority of Zebrafish over Other Murine Models

The utilization of murine rodent animal models has been a longstanding and fundamental practice in research. This classic approach aims to faithfully replicate symptoms and disease progression in humans, prioritizing the crucial need for accurate disease modeling and understanding. For decades, mice have been a leading experimental model for biomedical research due to reasons including good genetic features, diagnostic parameters that could be comparable to humans, and a standardized convention of the development. However, significantly conflicting factors including different dietary requirements, long generation time, and high cost of maintenance have compelled the need to explore alternatives with the potential to serve as an animal model that is closer to the ideal requirements (Choi et al., 2021). In the recent past, a group of researchers meticulously investigated embryonic development, delving into cell lineage analysis and the formation of organ systems in zebrafish. Their groundbreaking discoveries illuminated a profound revelation: the pivotal pathways orchestrating human development are remarkably conserved in zebrafish. As mentioned earlier, this conservation is bolstered by the presence of functional homologs for 70% of the genes responsible for human diseases. This substantiates the distinct advantage of zebrafish over traditional murine animal models, underscoring its unparalleled relevance in scientific exploration (Rahman Khan & Sulaiman Alhewairini, 2019). Zebrafish possesses a distinctive advantage rooted in its highly sequenced genome, a trait that sets it apart. This fully mapped genome can be conveniently manipulated, offering researchers unprecedented opportunities. Thousands of mutations in protein-coding genes have been engineered, showcasing the breadth of possibilities. Moreover, researchers have crafted tissue-specific transgenic zebrafish lines, allowing the study of human diseases through precise gene promoters. By strategically integrating and eliminating specific genes, tailored biomedical systems can be crafted. This manipulation enables exhaustive studies on gene expression. Recently, advancements in gene regulation have introduced spatiotemporal control through regulatory element coupling, enhancing the Tol-2-based transgenic model. These developments further bolster the zebrafish's superior position, solidifying its dominance in the realm of genomics (Choi et al., 2021; Howe et al., 2013; Teame et al., 2019). Besides these attributes, the rapid development, shorter generation time, high fecundity, transparency of embryos, completely developed organ systems, and the relatively affordable maintenance of a huge number of multiple available strains optimizes the use of zebrafish as a fitting and effective animal model when compared to other murine models (Teame et al., 2019).

18.3.2 The Brain Development of Zebrafish and Humans

The segmental structure of the brain was studied and the neurons were characterized in the vertebrate zebrafish for the first time by Grunwald and Eisen. The further text will highlight the similarities in the CNS of humans and zebrafish (Grunwald & Eisen, 2002). The vertebrate neuroanatomy of zebrafish was comprehensively explored and resulted in the realization that the structure of its brain has a resemblance to that of mammals, emphasized in Figure 18.1. The zebrafish displayed homology to certain structures including the amygdala, which is involved in emotion control and other areas related to behavior (Perathoner et al., 2016; Randlett et al., 2015). The basic domains of CNS structure were found to be similar in fish and mammals, and the neurotransmitters including glutamate, dopamine, serotonin, GABA, acetylcholine, histamine, and noradrenaline are found in long pathways and interneuron systems (Panula et al., 2006). Key neuroanatomical similarities also include those of spinal motor neurons and brainstem pathways. The spinal motor neurons of the zebrafish are similar to alpha motor neurons in humans, and there are different subtypes among motor neurons with a percentage present in specific locations of the spinal cord innovating limb/fin muscles and axial muscles. Also as seen in humans, the fibers in zebrafish that come through vestibular nuclei and reticular formation descend from the hindbrain and project to the level of the spinal cord, especially on the interneurons and some also on the spinal motor neurons (Babin et al., 2014). Keeping that information in perspective, zebrafish has emerged as a promising model for cellular and molecular biology research on the vertebrate brain, yielding useful output on neuropsychiatric and neurological human diseases including Alzheimer's disease, epilepsy, schizophrenia, depressive disorder, Parkinson's disease, and autism spectrum disorders. The consequences of acute as well as chronic hyperglycemia on the homeostasis of the brain and neurogenesis have also been described (Lucini et al., 2018).

18.4 NEURAL BASIS IN NEURODEGENERATIVE DISORDERS

According to various studies (Blader, 2000; Fulwiler & Gilbert, 1991; Howe et al., 2013; Postlethwait et al., 2000), the zebrafish brain displays functional and structural resemblances with the mammalian brain. It also exhibits corresponding neural pathways that contain the majority of important brain chemical pathways for signal transduction. Their neurotransmitter system is preserved from initial in development and possesses clear resemblances to mammalian systems, like dopaminergic cells in the olfactory bulb and hypothalamus-producing neurotransmitters such as serotonin (5-HT), dopamine (DA), histamine (HA), acetylcholine (ACh), GABA (Kaslin & Panula, 2001; Panula et al., 2006; Wasel & Freeman, 2020), and glutamate. It makes zebrafish a powerful and ideal model for studying the neurological basis of four human neurodegenerative disorders (PD, AD, ALS, and HD) and understanding their pathogenesis.

18.4.1 Parkinson's Disease (PD)

PD is one of the most prevalent neurodegenerative disorders worldwide. Dopaminergic neuron loss in the substantia nigra and intracellular α-synuclein buildup in neurons of different brain areas are two features of Parkinson's disease (Lewy pathology) (Poewe et al., 2017). Zebrafish models for PD are among the most well-established

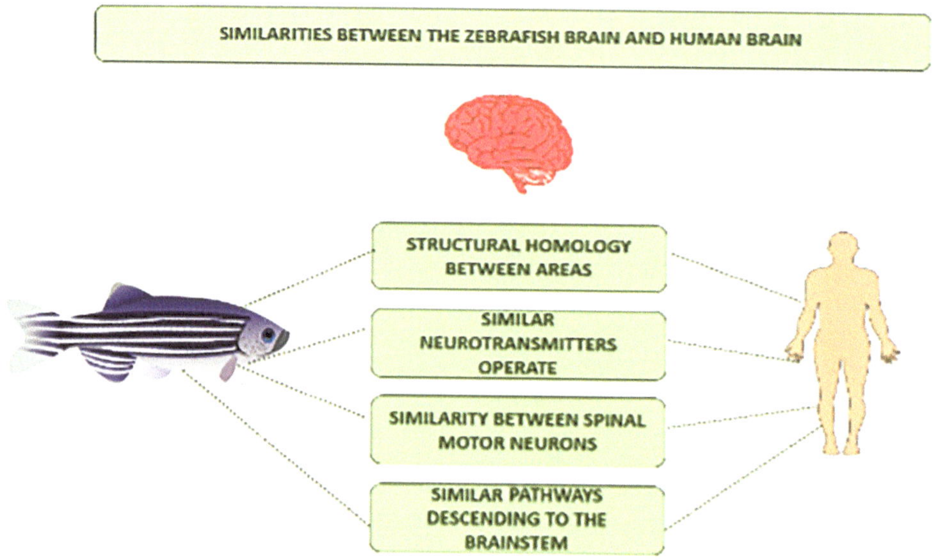

FIGURE 18.1 Zebrafish.

models (Razali et al., 2021). The development of several genetic, transgenic, and chemically induced models of PD has been made possible by the high preservation of genes related to PD and sensitivity to chemicals related to PD risk in zebrafish. The zebrafish model may be particularly helpful in investigating hypokinetic symptoms (bradykinesia) associated with PD, even if it cannot reproduce the entire disease spectrum.

18.4.1.1 Synucleins

Synucleins are a family of proteins that include α -, β -, and γ-synucleins. Although zebrafish lacks a homolog gene of the human α-synuclein gene, it expresses three isoforms of β- and γ-synucleins that are encoded for β-, γ1-, and γ2-synucleins, respectively. This is significant because human and zebrafish synucleins have substantial sequence similarities, with zebrafish γ1-synuclein being most comparable to human α-synuclein in terms of function (Sun & Gitler, 2008).

18.4.1.2 PTEN-Induced Kinase 1 (PINK1)

PINK1 is one of the main causes of autosomal-recessive earlier onset PD, and mutations in PINK1 result in decreased PINK1 protein translation. The PINK1 gene, which encodes the PINK1 protein, is essential for the regulation of oxidative stress and protection of neurons in case of mitochondrial dysfunction (Gonçalves & Morais, 2021). In another study, Anichtchik et al. (2008) reported substantial morphological anomalies and traits, including abnormalities of axonal tracts, dysfunction in escape reactions, and higher mortality, following PINK1 gene knockout by morpholino (MO) injection in a groundbreaking study that investigated PINK1 loss-of-function in zebrafish.

18.4.1.3 Parkin Gene Variants

These variants constitute the majority of autosomal-recessive mutations in early disease (Kitada et al., 2007). The Parkin gene codes for E3 ubiquitin ligase that is implicated in the degradation of proteasome pathway, and besides the other PD-related genes PINK1 and DJ-1, may potentially be involved in mitochondrial function (Flinn et al., 2009). At 3 days post fertilization (dpf), MO knockdown of Parkin resulted in a 20% reduction in the number of diencephalic dopaminergic neurons and an increase in the susceptibility of these neurons to the neurotoxin MPP+ (Flinn et al., 2009). Reduced mitochondrial complex I activity was observed as a result of the knockdown, as was previously observed in patients with Parkin mutations (Durcan & Fon, 2015). Parkin's ability to reduce proteotoxic stress, which results in cell death, suggests that it might also possess a protective role (Fett et al., 2010).

Several techniques have been used to induce neurotoxicity and clinical characteristics of PD in zebrafish chemically, in addition to the numerous genetic models of the disease (Chia et al., 2022).

18.4.1.4 MPTP

The infusion of the toxin MPTP is among the most popular models. Loss of dopaminergic neuronal cells and motor symptoms resembling PD have been observed in people after exposure to MPTP (Smeyne & Jackson-Lewis, 2005). Additionally, larval zebrafish were observed to be vulnerable to MPTP therapy, which specifically reduces diencephalic dopaminergic neurons and markedly decreases motor function (Bretaud et al., 2004). According to Hughes et al. (2020), bradykinesia is a motor disorder that manifests in PD patients with lower swimming speed and other aberrant swimming behaviors in the zebrafish PD model created using MPTP. In a research by McKinley et al. (2005), larvae treated with MPTP (5 Ag/mL) moved more slowly in reaction to touch than did control larvae, which reacted to touch by fleeing quickly. Although alive, most of the larvae that had been exposed to 10 μg/mL MPTP remained completely motionless and non-responsive to touch.

18.4.1.5 6-Hydroxydopamine (6-OHDA)

Another common neurotoxin is 6-hydroxydopamine (6-OHDA), which is an oxidative dopamine analog that kills dopaminergic and noradrenergic neurons by causing mitochondrial dysfunctions (Zeng et al., 2018). Intramuscular 6-OHDA injections resulted in locomotor abnormalities (O. Anichtchik et al., 2008). In a latest experiment, 6-OHDA was directly injected into the diencephalon of ZF, and the results showed that the swimming speed and distance travelled decreased, as well as the number of tytyrosineydroxylase (TH) immunoreactive neurons (Vijayanathan et al., 2017).

18.4.1.6 Rotenone

Rotenone is both an insectiscid and pescicid and a high-affinity inhibitor of mitochondrial ETC Complex I, which can also cross the blood-brain barrier (Sherer et al., 2003). When larval zebrafish were exposed to rotenone at LD50 (50 nM), the number of dopaminergic neurons in the anterior diencephalon decreased by 36%, and moderate locomotor abnormalities were observed (Kalyn et al., 2019). According to a study on environmental neurotoxins, high doses of rotenone killed adult zebrafish in a couple of days and caused pigmentation and eventual death in embryos (Bretaud et al., 2004).

18.4.1.7 Metronidazole

A 24-hour exposure to metronidazole caused a considerable decrease in dopaminergic neurons of the olfactory bulb and a reduction in the capacity of smell in adult transgenic zebrafish (Godoy et al., 2020).

18.4.2 Alzheimer's Disease

Alzheimer's disease is defined by extracellular amyloid (A) deposits derived from proteolytic amyloid precursor (APP) and neurofibrillary tangles (NFTs) composed of hyperphosphorylated tau protein aggregates (Frisoni et al., 2010).

18.4.2.1 Tau

The first tau model of zebrafish that briefly expressed human tau protein in neurons displayed hyperphosphorylation of tau, build-up of tau in neuronal cell bodies, and disruption of cytoskeletal that was strikingly similar to the NFTs found in human Alzheimer's disease (Tomasiewicz et al., 2002). The transcription of human tau (111) or mutant human tau (Paquet et al., 2009) was then expressed in the CNS using robust zebrafish transgenic tau lines, which produced noticeably greater tau expression levels and NFT-like tau accumulation. In vivo imaging showed rapid tau hyperphosphorylation and aggregation, and neuronal cell death (Paquet et al., 2009).

18.4.2.2 β-Amyloid

The major elements of γ-secretase complexes, which are necessary for intra-membrane cleavage of over 70 distinct substrates including APP protein, which generates the amyloid (Aβ) peptide, are the human Presenilin proteins (PSEN1 and PSEN2) (Lleo & Saura, 2011). The main component of a neuritic plaque used to diagnose Alzheimer's disease is aggregates of Aβ peptide (Fukumori et al., 2010). Higher quantities of Aβ have been linked to abnormal cerebrovascular branching and angiogenesis in the developing hindbrain, suggesting that Aβ has a significant role in the preservation of proper cerebrovascular function in zebrafish (Cameron et al., 2012). Interestingly, a different team found abnormalities in the cerebrovascular system of both APP and Aβ-deficient larvae. Cell death and aberrant vasculature development were observed in embryos treated with 2.5 M A1–40 (Donnini et al., 2010). Interestingly, administration of human Aβ1–42 peptide to embryos could cause these defects to disappear. These findings demonstrate the importance of Aβ in promoting cerebrovascular disease development.

18.4.2.3 Presinilin 1 and Its Homolog Presinilin2

The zebrafish PS1 gene (Psen1) is being widely expressed during embryogenes (Leimer et al., 1999). Zebrafish Psen1 cleaves AβPP to produce βA by interacting with other mammalian γ-secretase subunits to produce an active complex; this activity has been preserved throughout evolution (Leimer et al., 1999). Presenilin 2 (Psen2), the zebrafish Psen1 orthologue, and the human PS2 protein are quite similar in terms of sequence (Groth et al., 2002). The central nervous system (CNS), eye, and spinal column are the only areas of zebrafish that express large amounts of Psen2 protein. Psen2 expression is remarkably more abundant in melanocytes, which are produced by neural-crest cells and are essential for pigmentation. (Groth et al., 2002) Due to impaired Notch signalling with Psen1-dependent tyrosinase activity, the reduction of Psen1 or Psen2 results in melanocyte loss. Tyrosinase, an enzyme exclusive to melanocytes that is necessary for melanin formation, is encoded by the tyrosinase gene. Whole-mount in situ hybridization using a probe against tyrosinase has been used to identify melanocyte depletion (Camp & Lardelli, 2001).

Real-time imaging using stable transgenic zebrafish122 has allowed researchers to identify early pathogenic traits, including hyperphosphorylation and structural alterations of Tau within the first two days of embryonic development (Kozawa et al., 2016; Leung et al., 2013; Renninger & Orger, 2013) After five weeks of development, the larvae acquired significant neurodegeneration and displayed all pathological traits, including neurofibrillary tangles (Paquet et al., 2009).

18.4.3 Amyotrophic Lateral Sclerosis (ALS)

ALS is a deadly neurological condition defined by the gradual degradation of motor neurons in both the brain and spinal cord (Taylor et al., 2016).

The remarkable translucency of the larval zebrafish body makes it possible to directly observe certain proteins, cytoskeletons, and organelles in real time by probes tagged with fluorescence in a particular spinal motor neuron. Furthermore, the rapid development of zebrafish lines useful for the study of genes related to ALS has been made possible by their high genetic resemblance to humans and genetic amenability (Armstrong et al., 2016; Asakawa et al., 2020; Bercier et al., 2019; P. Bose et al., 2019; Bourefis et al., 2020; Da Costa et al., 2013; Hewamadduma et al., 2013; Kim et al., 2019; Lebedeva et al., 2017; Lissouba et al., 2018; Ohki et al., 2017; Ramesh et al., 2010; Sakowski et al., 2012; Schmid et al., 2013; Shaw et al., 2018). SOD1 (Da Costa et al., 2013), C9orf72 (Iyer et al., 2018), FUS (Morrice et al., 2020), and TARDBP (Kabashi et al., 2010) are the four genes most frequently linked to ALS and are highly conserved in zebrafish, with a sequence identity of over 70% (Hardiman et al., 2017). Stable mutant lines of zebrafish were created for all four of those genes, and they all successfully mimicked different ALS symptoms.

18.4.3.1 Superoxide Dismutase 1 (SOD1)

SOD1 mutations are a common and widely researched cause of familial ALS. Although research has been conducted to determine SOD1's function, it is still not known how mutant SOD1, which is widely expressed in the body, causes the selected loss of motor neurons. Currently, the pathology-causing factor in ALS is suggested to be a mutation in the gain of function. Zebrafish exhibited dramatically reduced axonal outgrowth, constrained branching, and poor swimming when human mutant SOD1 was overexpressed (Benedetti et al., 2016; Da Costa et al., 2013; Lemmens et al., 2007; Ramesh et al., 2010; Sakowski et al., 2012). In one of the earliest experiments, temporary activation of human G93A-SOD1 resulted in motor neuron loss, neuromuscular junction degeneration, and axonal pathology (Sakowski et al., 2012). Similarly, related to controls treated with wild-type SOD1 mRNA, zebrafish larvae infused with mutant human SOD1 mRNA revealed substantial axonopathy and diminished level of movements during a behavioral test (Robinson et al., 2019).

18.4.3.2 TARDBP

TARDBP produces the RNA/DNA-binding protein TDP43, which is involved in the metabolism and processing of RNA; in almost 90% of ALS patients, TARDBP mutations cause a buildup of mutated TDP43 in the inclusion bodies. In almost 90% of all ALS cases, TARDBP mutations cause a buildup of mutated TDP43 in inclusion bodies (Berning & Walker, 2019). Zebrafish transgenic with mutant human TARDBP display defects in motor axon development and branching as well as motor functions, which are hallmarks of ALS pathogenesis (P. Bose et al., 2019; Hewamadduma et al., 2013; Kabashi et al., 2010; Laird et al., 2010; Lissouba et al., 2018; Schmid et al., 2013). While testing the outcome of three distinct TARDBP mutants (G348C, A382T, and A315T) in patients with ALS, Kabashi et al. (2010) pioneered the finding that mutated TDP-43 translation in the zebrafish prompted to atypical axon growth, toxicity, and motor neuron defects, along with severe deficits in motor function.

18.4.3.3 C9orf72

The most prevalent mutation in familial and spontaneous ALS cases is the GGGGCC hexanucleotide repeat expansion in the C9orf72 gene (Balendra & Isaacs, 2018; Stepto et al., 2014). Additional insights into the pathophysiology of C9-ALS have been gained through the development of C9orf72 zebrafish models. However, it is significant to note that only humans and their nearest relatives, the chimpanzee, have a distinctive GGGGCC hexanucleotide repeat expansion in C9orf72 (Iyer et al., 2018). Future attempts to knock-in the repeats in zebrafish c9orf72 intron 1 using CRISPR genome editing techniques may be the most accurate method for modelling C9orf72-related ALS in zebrafish.

18.4.3.4 FUS

Mutation of the FUS gene results in severe and occasionally earlier onset dysfunction of motor neurons (Lattante et al., 2013). FUS binds to RNA and shares both structural and functional similarities with TDP-43. When FUS is mislocalized from the nucleus to the cytoplasm, it aggregates into inclusion bodies that resemble the inclusions formed by TDP-43 in motor neurons. According to certain theories, FUS may play a role in disease pathogenesis by either losing its role in the nucleus or gaining it after being mislocalized to the cytoplasm. Several zebrafish models of genetically altered FUS have been developed (Armstrong et al., 2016; Armstrong & Drapeau, 2013; Kabashi et al., 2011; Lebedeva et al., 2017). Antisense MOs were used in prior investigations to silence FUS, producing morphant ZF larvae with short motor axons, aberrant branching, and pronounced abnormal motor activity (Armstrong et al., 2016; Armstrong & Drapeau, 2013).

18.4.4 Huntington's Disease

HD is an inevitably fatal neurodegenerative condition caused by a mutant version of the huntington gene (HTT), which codes for an aberrant amplification of a nucleotide sequence coding glutamine (CAG) at amino terminus of the HTT protein (Graveland et al., 1985).

Diminished protein hemostasis of mHTT has a destructive impact that causes loss of neurons in the cortex and other brain regions as well as neuronal death of GABAergic medium spiny neurons in the striatum (Ehrlich, 2012; Eidelberg & Surmeier, 2011; Estrada-Sánchez & Rebec, 2013; Graveland et al., 1985).

18.4.4.1 Zebrafish Models for HD

The zebrafish HTT polypeptide encodes four glutamines (compared to around 35 glutamines in normal human HTT), and it shares approximately 70% similarity with the human HTT ortholog (Karlovich et al., 1998).

18.4.4.2 HTT Loss-of-Function

HTT is widely expressed in the zebrafish brain similar to humans and is essential for the development of pre-placodal and telencephalic progenitor cells (Lumsden et al., 2007). It has been suggested that the telencephalon of zebrafishes is anatomically equivalent to the striatum of mammals (Rink & Wullimann, 2002a). Henshall et al. (2009) examined how the absence of HTT in zebrafish affects several brain areas. The anterior-most portion of the neural plate was less properly formed when HTT mRNA translation was inhibited, as indicated by the decreased expression of genes that are typically expressed there (emx3, six1 dlx3b). Among other forebrain structures, pre-placodal cells and telencephalic precursors are induced by the anterior neural plate (Andoniadou & Martinez-Barbera, 2013; Schmid et al., 2013; Whitlock & Westerfield, 2000). HTT inhibition in zebrafish resulted in a 50% reduction in BDNF expression levels (Diekmann et al., 2009). High rates of neural cell apoptosis, small eyes and heads, expansion of the ventricles of the brain, and subsequent abnormalities of the lower jaw, including missing branchial arches, were all brought about by this effect (Diekmann et al., 2009).

18.4.4.3 HTT Gain-of-Function

The buildup of mHTT in neuron cells suggests a problem with protein quality control. To ensure appropriate protein folding and conformation, the C-terminus of the Hsc70-interacting protein (CHIP) functions as both a co-chaperone and ubiquitin ligase, connecting molecular chaperones with the ubiquitin protease system (McDonough & Patterson, 2003). The mHTT proteins in zebrafish were made more soluble and less likely to aggregate when CHIP was co-expressed with a pathogenic or a generic polyQ fragment of human HTT (Miller et al., 2005). In order to ensure appropriate protein folding and conformation, the C-terminus of the Hsc70-interacting protein (CHIP) functions as both a co-chaperone and a ubiquitin ligase, connecting molecular chaperones with ubiquitin-protease system (McDonough & Patterson, 2003).

18.5 EPIDEMIOLOGY AND CLINICAL CHARACTERISTICS

Parkinson's disease (PD) has features of cognitive, behavioral, and motor dysfunction secondary to a neurodegenerative process (Wasel & Freeman, 2020) of the dopaminergic (DA) neurons in the substantia nigra pars compacta (SNpc) part of the brain (DeMaagd & Philip, 2015). PD is the most common movement disorder (Nussbaum & Ellis, 2003) and the second most frequent neurodegenerative disorder in the elderly (Razali et al., 2021). Clinically, it manifests through a constellation of mainly motor but also non-motor features, including bradykinesia, postural instability, tremor, rigidity, olfactory dysfunction, depression, constipation, and REM sleep behavior disorder (RBD) (Pang et al., 2019; Rees et al., 2019).

The approximate economic burden of PD in the US is $51.9 billion in the year 2017 and is projected to be $79 billion in 2037 (Yang et al., 2020). The disease burden of non-motor symptoms is more and more being recognized as an integral part of the disease (Khoo et al., 2013). Furthermore, RBD (Ferini-Strambi & Zucconi, 2000) has been recognized to occur before the development of motor and cognitive symptoms of alpha-synucleinopathies (Postuma et al., 2009). In addition, other behavioral manifestations, such as depressive and anxiety symptoms, can predate diagnosis sometimes for years (Ahmad et al., 2023; Faivre et al., 2019; Shiba et al., 2000).

Unlike previously thought (Robbins & Cools, 2014), PD patients compared to matched controls have an overall three to six times greater risk of developing dementia, despite adjusting for confounders (Åström et al., 2022). Some authors propose a dual-syndrome hypothesis to explain the different nosological features seen in PD, with an early mild cognitive impairment (PD-MCI) phenotype appearing in nearly one-fifth of newly diagnosed and medication-naive PD cases and a late dementia phenotype (Aarsland et al., 2009). It is worth noting that this distinction is debated, as dementia complicates the course of most (83%) 20-year survivor PD patients, as reported by the Sydney Longitudinal Study (Hely et al., 2008). All of this justified the profiling of cognitive manifestations into a dopamine-related executive deficit originating in the fronto-striatal circuit and a more posterior syndrome associated with progression to PDD, pathological findings of tauopathies and non-dopamine related (Biundo et al., 2016).

18.6 ENVIRONMENTAL RISK FACTORS

Several environmental and genetic factors are known risk factors for the development of PD. Still, increased age is the primary risk factor (Reeve et al., 2014).

1-methyl-4-phenyl-1,2,3,6-tetrahydropyridine (MPTP) discovery during the 1980s as being a cause of a parkinsonism-like syndrome in drug users was the first proof of an environmental compound to induce PD (Langston et al., 1983). MPTP can pass across the blood-brain barrier (BBB) (Zeng et al., 2018) and is metabolized to MPP+, the active metabolite (Bajpai et al., 2013). MPP+ is further uptaked by dopamine transporter protein (DAT)-positive neurons, where it induces mitochondrial dysfunction through complex I blockade (Zawada et al., 2011). Other identified environmental risk factors have similar mechanisms. Oxidative stress, microglial-mediated inflammation, and mitochondrial dysfunction are described with 6-OHDA injection (Rodriguez-Pallares et al., 2007); they are mostly observed in experimental animal models as natural exposure is highly unlikely. Certain insecticides are historically recognized to increase the risk of PD in exposed populations, especially farmers (Tanner et al., 2011). Paraquat, a hydrophilic insecticide consequently transported in an active manner to the CNS, causes elevated production of ROS (X. H. Wang et al., 2018). On the other hand, rotenone crossing the BBB easily induces dopaminergic loss through mitochondrial dysfunction, oxidative stress (Ünal et al., 2019), microglial-associated neuroinflammation (Yurtsever et al., 2022), and can experimentally cause the aggregation of alpha-synuclein particles (Sherer et al., 2002). Although experimentally and epidemiologically, these agents are recognized to induce Parkinson's, they cannot explain but a small portion of cases (Pang et al., 2019).

Other environmental risk factors have been studied with varying evidence, such as the well-documented but controversial protective effects of smoking (Quik et al., 2012), through potentially neuroprotective and neurotrophic effects of nicotinic acetyl-choline receptors (nAChr) stimulation. On the other hand, uric acid levels, physical activity, ibuprofen, and calcium channel blockers are found to be protective while severe traumatic brain injury and beta blockers are regarded as risk factors (Ascherio & Schwarzschild, 2016; Bellou et al., 2016).

18.7 GENETIC RISK FACTORS

Besides the environmental factors, mutations to some of the genes also increase the vulnerability to early onset, familial, and/or spontaneous PD diagnoses. α-synuclein is encoded by the SNCA gene (PARK1). Missense mutations (Polymeropoulos et al., 1997) and multiplications (Singleton et al., 2003) cause the familial autosomal-dominant type of PD with high penetrance. Missense mutations potentially promote aggregation through loss of cell surface binding. On the other hand, multiplications increase gene expression and indicate that the wild type produced in high quantities is sufficient to induce PD (Pang et al., 2019). Furthermore, sporadic PD has been linked to certain normal variants (International Parkinson's Disease Genomics Consortium et al., 2017). Although the exact role of α-synuclein is unknown and thought to be not necessary for the viability of neurons, it is believed to be implicated in neurotransmission modulation (Sharma & Burré, 2023). Interestingly, many authors support the hypothesis of a prion-like CNS spread of α-synuclein aggregates in the brain (Goedert, 2015).

Leucin-rich repeat kinase 2 (LRRK2/PARK8) mutations are found in up to 10% of sporadic PD and 40% of familial forms (Hernandez et al., 2016), of which the G2019S mutation contributes up to 4% of familial PD (Spatola & Wider, 2014). It is the reduction of the GTPase activity (Steger et al., 2016) and the increase in activity of kinase (Greggio et al., 2006) causing mutations that are mostly implicated with PD. LRRK2 protein is implicated in dopaminergic neuronal loss possibly through different mechanisms, such as altering neuronal morphology and differentiation (A. Bose et al., 2022), increasing α-synuclein accumulation (Zhao, Keshiya, et al., 2020) and tau hyperphosphorylation (Ohta et al., 2015), dysfunctional endocytosis (Connor-Robson et al., 2019), mitochondrial impairment (Oun et al., 2022), and potentially neuroinflammation (Ahmadi Rastegar & Dzamko, 2020).

Other genes are also implicated in PD. For instance, VPS35 and CHCHD2 are implicated in autosomal-dominant forms of Parkinson's disease (AD-PD), while Parkin (PARK2), PINK1(PARK6), PARK7(DJ-1), ATP13A2 (PARK9), FBXO7 (PARK15), and PLA2G6 (PARK16) gene mutations lead to autosomal-recessive PD (AR-PD) (Billingsley et al., 2018; Helley et al., 2017). However, Mendelian inheritance accounts only for 5–10% of PD cases, and most patients are classified as having idiopathic PD.

18.8 PATHOLOGY

Dopaminergic neuronal loss in the SNpc and the presence of eosinophilic intracellular cytoplasmic inclusions of largely misfolded α-synuclein proteins are both considered the pathological hallmarks of PD (Baba et al., 1998; Lo Bianco et al., 2002). Aggregated α-synuclein particles in the axons and the dendrites form the Lewy neurites (LN), which are thought to occur earlier in the disease course (Koga et al., 2021), while the perikaryal equivalent are named the Lewy bodies (LB) (Braak et al., 2003).

It is well documented that the presence of LB and LN is not limited to the SNpc (Braak et al., 2003). Braak et al. observed the early presence of LBs in the olfactory structures, the dorsal IX/X motor nuclei, and the intermediate reticular zone (stage 1). Additional extension to the gigantocellular nucleus, caudal raphe nuclei, and coeruleus-subcoeruleus complex defines stage 2. These might explain early olfactory, some early autonomic, and behavioral symptoms that can predate the typical PD motor features (Prediger et al., 2012; Stanković et al., 2019; Vaswani et al., 2022). The destruction of the SN ensues at stage 3. It is worth noting that the ventral tegmental neurons remain relatively intact (Pang et al., 2019). Then thalamic and the anteromedial temporal mesocortex lesions ensue at stage 4, and neocortical involvement beginning at stage 5 and culminating at stage 6, potentially explaining the late course of PDD development.

In contrast, McKeith et al. proposed three stages, with a brainstem stage (Braak stages 1–3), a limbic stage (Braak stages 3–5), and a neocortical stage (stage 6) (Alafuzoff & Parkkinen, 2014; McKeith et al., 1996). It is worth noting that there is still much debate of the significance of such staging scores and the overall role of LBs in the etiopathogenesis of PD as well (Hindle, 2010; Zaccai et al., 2008).

18.9 NEURAL CIRCUITS

18.9.1 Neurotransmitters

Dopamine is the product of the rate-limiting step of hydroxylation of tyrosine (TH), followed by the decarboxylation of the dihydroxyphenylalanine (L-DOPA) (Yamamoto & Vernier, 2011). Dopamine is stored in synaptic vesicles with the help of vesicular monoamine transporter 2 (vMAT2) (Chaudhry et al., 2008). After release, DA is reuptaken by dopamine transporter (DAT) and then degraded intracellularly by either a catechol-O-methyl transferase (COMT) or monoamine oxidase (MAO) (Yamamoto & Vernier, 2011). The release of vesicles into the synaptic cleft is regulated by multiple factors. Neuronal depolarization is not the only controlling way to release dopamine. In fact, local Ca++ concentration is important to regulate DA release through a vesicular mechanism (Pucak & Grace, 1994). Besides the classical pathway, other minor synthesis pathways have been described, such as the Cyp2D pathway in rats or the tyrosinase pathway (Bromek et al., 2011; Sánchez-Ferrer et al., 1995).

PD symptoms were historically explained by neurodegeneration of the SNpc, especially the ventrolateral part (Fearnley & Lees, 1991). The replenishment of dopamine, initially through IV supplementation of its precursor L-DOPA, had quick miraculous effects on core motor symptoms, especially bradykinesia and rigidity (Hornykiewicz, 1974), which gave evidence of PD as a mainly dopamine-deficient state in the striatum. However, this view was challenged on the basis of many epidemiological, clinical, pathological, and experimental observations of extranigral neurodegeneration extent and dopamine-independent manifestations (Braak et al., 2003; Hely et al., 2005).

The other neurotransmitter systems also get affected in PD patients. Serotonin is mainly produced in the raphe nuclei region of the brainstem, which are impacted earlier according to Braak stage 2 by pathological changes (Braak et al., 2003), causing serotoninergic dysfunction documented in PET in vivo studies (Kerenyi et al., 2003; Kish et al., 2008; Politis et al., 2010), with less marked deficit and preferential reductions in the caudate (vs putamen) as opposed to dopamine. Some evidence supports that serotonergic imbalances are implicated in certain aspects of the disease, such as levodopa induced dyskinesias (Politis et al., 2014), tremor (Pasquini et al., 2018; Qamhawi et al., 2015), and to a lesser degree with mixed results depression (Boileau et al., 2008; Hesse et al., 2009), daytime sleepiness (Pavese et al., 2012), and possibly cognitive deficits (Frouni et al., 2022). The study of the importance of other receptors in PD yielded some new treatments for non-motor PD symptoms. Pimavanserin is a selective 5-HT_{2A} receptor inverse agonist and the only FDA-approved drug for psychosis caused by PD (Isaacson et al., 2020).

GABA plays an important role in the basal ganglia thalamocortical loop, as an inhibitory neurotransmitter. A couple of studies documented GABAergic imbalances and found it to correlate with axial rigidity (O'Gorman Tuura et al., 2018), overall disease severity (Nuland et al., 2020), and extent of dopamine depletion (Boccalaro et al., 2020).

Noradrenaline is mainly produced in the brainstem, specifically the locus coeruleus. This neurotransmitter has been implicated in many functions, such as cognitive processes, and also neural network reorganization (Sara, 2009). As early as Braak stage 2, lesions are observed in the gain-setting nuclei of the lower brainstem (Braak et al., 2003), raising the possibility of an adrenergic imbalance affecting projections to the entire forebrain, brainstem, and cerebellum. Postmortem studies documented adrenergic impairment. In one study comparing a PD parkinsonism brain with controls, a 50% decrease in NA levels in the cortex of PD cases was observed (Scatton et al., 1983). In vivo studies using functional imaging techniques implicated NA dysfunction in resting tremor (Nahimi et al., 2018), REM sleep behavior disorder, cognition, and autonomic denervation (Sommerauer et al., 2018).

The evidence for a histaminergic dysfunction was hypothesized as the hypothalamic tuberomammillary nucleus (TMN), which was observed to be heavily implicated by PD pathology during the disease process (Braak et al., 2003). Animal data suggested earlier endogenous histamine to accelerate dopaminergic loss via H1 receptor, and its blockade to have a protective role (Liu et al., 2007). However, a postmortem study failed to document any alteration in the mRNA of histidine decarboxylase, the enzyme required to produce histamine (Shan, Liu, et al., 2012), but the same group documented in another study SN and putamen changes of histamine metabolism strongly correlating with Shan, Bossers, et al. (2012). Interestingly, recent animal studies documented beneficial effects of blocking the H4 receptor in a rat rotenone model (Fang et al., 2021).

18.9.2 The Basal Ganglia-Thalamo-Cortical Loop

In contrast, PD circuit dysfunction, a set of abnormal activity patterns of the remaining neurons, has been emerging as a potential therapeutic target to treat many of the disease symptoms (McGregor & Nelson, 2019).

The BG circuit is at the heart of PD pathology. Early human studies inducing GPi lesions showed reductions in rigidity (Narabayashi, 1956), as well as the reversible silencing of the STN and GPi by DBS proving the relevance of the BG-thalamo-cortical loop in the pathogenesis of PD motor symptoms (Benazzouz et al., 1993; Pycroft et al., 2018). Different types of models were developed to explain healthy and abnormal functioning of the BG-thalamo-cortical circuits.

The parallel circuit model proposes a conserved segregated pathway of input and output. The ventral striatum receives input from the VTA and the limbic system. Located more caudally, the dorsomedial and the dorsolateral striatum integrate input from associative regions and the cortex, respectively. The last two channels integrate input from the SNc as well (Alexander & Crutcher, 1990; McGregor & Nelson, 2019; Moore & Bloom, 1978). Such a model can explain the predominance of motor symptoms following dopamine loss.

The classic model proposes two pathways opposing each other. The direct pathway projecting from the striatum to the GPi and the SNr via dMSNs decreases BG output to the thalamus and the cortex, facilitating movement. The indirect pathway through iMSNs projecting to the STN and the GPe increases the BG output and inhibits movements. This model predicts DA loss to cause excessive indirect pathway by suppressing GPe firing, increasing STN activity, and decreasing in the direct pathway by disinhibition of the GPi, with the overall net effect of suppressing the thalamus and the cortex, explaining rigidity and bradykinesia (Alexander & Crutcher, 1990; DeLong, 1990). Human, non-human primate, and rodent studies' findings using different approaches are mostly consistent with the classical model (Alcacer et al., 2017; Benazzouz et al., 1993; Filion & Tremblay, 1991; Narabayashi, 1956; Rees et al., 2019); however, some data paints a more complex picture. dMSNs can activate concurrently with iMSNs during action initiation (Cui et al., 2020). A model viewing the direct and indirect MSNs as complementary for selecting appropriate actions was first proposed by Mink (Mink & Thach, 1993). Suppression of competing actions through inhibition of the thalamus and the cortex by a GPi increased activity, which is activated through a cortical STN pathway. At the same time, a focal inhibition of GPi by dMSN firing is selective of the action needed (Schmidt et al., 2013).

While parameters such as activity pattern, bursting, synchronizations, and beta oscillations have been found to be associated with Parkinson's disease, their causal relationship to the disease is not yet well-established (Deffains et al., 2018; Hammond et al., 2007; Leblois et al., 2007; Singh et al., 2015; Starr et al., 2005).

18.9.3 Zebrafish as a Model for Studying Parkinson's Disease

Zebrafish (ZF) is a known and well-established model to study human disease, especially nervous system development, neurodegenerative, and pharmacological studies (Best, 2008; Vaz et al., 2018). Of the morbid genes listed in the OMIM database and of the human genes implicated in GWAS, 82% and 76% have ZF orthologues, respectively (Howe et al., 2013). In addition to other advantages, ZF emerged as a model to study PD, potentially replacing other classical animal models, such as rodents (Razali et al., 2021). Understanding the evolutionary preserved processes in nervous system functioning justifies the advantage of neurologic disorders study in multiple species (T. Gerlai, 2020).

The perception of sensory information and control of movement by the brain is a result of the continuous interaction among various nerve cells that are connected, often spanning multiple areas of the brain (Kunst et al., 2019). The structural and functional connectivity architecture is critical in understanding how behavioral, cognitive, and mental processes are shaped. Zebrafish is the first vertebrate to have a cellular-resolution neuronal circuit mapped into an online atlas (Kunst et al., 2019).

18.9.4 Neurodevelopmental Aspects of the ZD Dopaminergic System

DA is recognized as a common neurotransmitter to multiple systems in the brain (Yamamoto & Vernier, 2011), and its presence is conserved through evolution and can be found in Cnidarians (Carlberg & Anctil, 1993).

TH is used as an immunohistochemical marker for catecholaminergic neurons (Pickel et al., 1975). However, a paralogous gene named th2 was reported (Yamamoto et al., 2010), potentially explaining dopamine th1-negative neurons documented earlier in other fish species (Ekström et al., 1990). They share 73% amino acids similarity. Only the Th1 is considered the orthologue of human th, while th2 is thought to be lost in eutherian lineage (Yamamoto et al., 2010). Th1 mRNA is detected at 24 hpf and earlier than th2, and is expressed preferentially in the larval CNS and eyes as opposed to its paralogue found more in other organs such as the liver, heart, and kidneys (Chen et al., 2009).

The structure and size of the ZF brain differs from that of the human brain. However, certain areas in the teleost brain are remarkably comparable to some regions in the human brain, indicating similarities in the overall organization (Xi et al., 2011). Inferring by topology, the homologous striatum in ZF was historically complicated by differences in definitions and also by certain anatomic specificities of the ZF brain (Nieuwenhuys, 2009).

The telencephalon of ZF, as other members of the *Actinopterygii* class, is described as everted due to outward bending of the dorsal lateral parts, as opposed to most other vertebrates, where it's evaginated (Nieuwenhuys, 2009), with lateral ventricles lacking, giving place to two solid lobes. Most of the caudal surface is covered with an ependymal layer, as it is of a ventricular nature (Nieuwenhuys, 2009). The teleost telencephalon is regarded as the most complex in its class (Northcutt, 2008).

The posterior tuberculum is a large structure in the teleost diencephalon in comparison to tetrapods and is composed of two groups of nuclei. A rostral periventricular nucleus, the posterior tuberal nucleus, and the paraventricular organ all are situated between the ventral thalamus and hypothalamus. Caudally and laterally are a group of migrated nuclei, composed of the preglomerular nuclei, the tertiary gustatory nucleus, the nucleus subglomerulais, and the torus lateralis (Wullimann et al., 1996).

The hunt for the analogous system to the mammalian nigrostriatal pathway was challenging, as teleost fish dopaminergic and adrenergic cells were never found in the midbrain (Meek, 1994). Interestingly enough, all other brain regions contained catecholaminergic neurons (Rink & Wullimann, 2002b). From rostrally to caudally, they are the olfactory bulb, ventral telencephalon, preoptic area, ventral thalamus, posterior tuberculum, hypothalamus, pretectum, and rhomboencephalon.

The observation that amphibians have dopaminergic neurons with ascending projections spanning from the midbrain to the posterior tuberculum (Marın et al., 1998) suggests the dynamic nature of these neurons' location during evolution in different species. Rink and Wullimann (Rink & Wullimann, 2001) identified the posterior tuberculum to contain ascending dopaminergic neurons projecting to the ventral subpallium, the potential equivalent of the mesoencephalic pathways in mammals, and the larval diencephalic DA neurons were named DC1–DC7 according to their rostrocaudal location and morphology. Most of the dopaminergic neurons form by 96 hpf (Du et al., 2016), of which the OTP-dependent DC2 (the periventricular nucleus of the PT), DC4 (posterior tuberculum), DC5 (hypothalamic nucleus of the lateral recess), and DC6 (posterior tuberal nucleus extending to the caudal hypothalamus) (Löhr et al., 2009) have complex projections to different parts of the brain. DC2/4 were found to project to the subpalliam, DC5 to the LC, and all four clusters projected to the hindbrain (Tay et al., 2011). The hypothalamic DC3 (paraventricular organ) and DC7 (the hypothalamic nucleus of the posterior recess) clusters project locally and to adjacent regions as well as the preoptic, the hypothalamic anterior preoptic, and the DC1 (ventral thalamus) cell clusters. Only the OB and the amacrine retinal cells project exclusively locally.

The otp-dependent posterior tubercular dopaminergic neurons integrate the ascending dopamine system, the descending diencephalospinal system, and the endohypothalamic circuitry (Tay et al., 2011). These observations suggest that these neurons play a significant role in modulating both physiology and behavior, and they are comparable to the mammalian A11 group of dopaminergic neurons (Löhr et al., 2009). The development of DA neurons is dependent on certain transcription factors and important signaling pathways with varied effects on different points.

Mutant zebrafish with altered Shh and Fgf8 genes exhibited a late-forming absence of only pretectal and retinal amacrine cells, in comparison to a strong reduction in the DA masses, specifically the ventral telencephalon (Holzschuh et al., 2003). In contrast, early dkk1 overexpression actually increases DA neurons in the same region, probably by antagonizing the effects of the canonical WNT signaling pathway (Russek-Blum et al., 2008). In addition, the DeltaA/D Notch ligands, the orthologs of mammalian Delta1, control the early DA neurons' development and may be implicated in the maintenance of olig2 expressing precursor pool in the ventral telencephalon (Mahler et al., 2010).

Prethalamic DA neuron development require the Arx1 and Isl1 transcription factors (Filippi et al., 2012). Ascending projecting neurons numbers were reduced following lmx1b.1/lmx1b.2 knockdown. Furthermore, the Nurr1/nr4a2 conserved the role of specification of dopamine phenotype in the preoptic, retinal amacrine, and the pretectal DA neurons (only nr4a2a) (Filippi et al., 2007).

18.10 MODELING PD IN ZEBRAFISH

18.10.1 Genetic Models

Lewy bodies and the Lewy neurites are mostly composed of misfolded α-synuclein protein, and they're one of the pathological hallmarks of PD (Baba et al., 1998; Lo Bianco et al., 2002). ZF produces β-, γ1-, and γ2-synuclein proteins via SNCB, SNCG1, and SNCG2 genes expression, respectively, even though it lacks the SNCA gene that codes for α-synuclein (Toni & Cioni, 2015). Milanese et al. reported the γ1- and the β-synuclein morpholino oligonucleotides knockdown to cause severe hypokinesis and delayed differentiation of the DA neurons. Expression of the human α-synuclein protein reversed the abnormalities, which supports the conserved role in dopamine regulation and motor behavior (Milanese et al., 2012).

ZF transgenic models expressing the human α-synuclein protein were developed. Prabhudesai et al. studied the effects of CLR01, a potent inhibitor of amyloidogenic and α-synuclein protein aggregation. α-Syn-ZF embryos' neural apoptosis caused death between 48 and 72 hours. Toxicity was due to the ubiquitin proteasome system inhibition, and effects were prevented using the CLR01 in vivo (Prabhudesai et al., 2012). In addition, mitochondrial toxicity and ROS generation was also reported as a mechanism of the axonal dystrophy observed in a ZF peripheral nerve model expressing wild-type α-synuclein (O'Donnell et al., 2014). Weston et al. have recently developed a novel in vivo model in zebrafish to investigate the aggregation of human α-synuclein, by tagging it with enhanced green fluorescent protein (GFP). This approach enables real-time tracking of the changes in the precise subcellular localization of α-synuclein within neurons, providing a valuable tool for studying the dynamics of protein aggregation in vivo (Weston et al., 2021).

PARK2 gene expresses the Parkin protein, the most common cause of the autosomal-recessive type of PD (Ferreira & Massano, 2017). The ZF Parkin protein has 62% homology and 78% functional similarity to human orthologue (Flinn et al., 2009). A ZF model deficient in the Parkin protein developed reduced numbers of the PT ascending DA neurons and increased sensitivity to MPP+ (Flinn et al., 2009) through reduced activity of the mitochondrial complex I, with no abnormalities in swimming behavior. Parkin protein is implicated in protein degradation by autophagy and mitophagy through the PINK1/Parkin pathway (Ashrafi & Schwarz, 2013). In addition, the Parkin protein expression is associated with α-synuclein dephosphorylation protecting cells, probably through a maintenance of balance between phosphorylated and dephosphorylated forms (Khandelwal et al., 2010).

The PTEN-induced kinase 1 (Pink1) gene expresses the PINK1 protein, a frequent cause of autosomal-recessive PD with early onset. ZF PINK1 is 57.8% similar to the human protein (Sallinen et al., 2010). The ZF PINK1-deficient studies reported differences in severity of abnormalities observed. Anichtchik et al. first reported a severe developmental phenotype in a PINK1 knockdown ZF animal model, probably due to mitochondrial dysfunction and increase in ROS, with a potential implication of GSK3β (O. Anichtchik et al., 2008). Sallinen et al.'s animal model displayed a mild phenotype with increased vulnerability to the neurotoxin MPTP and loss of TH immunoreactivity in specific regions (Sallinen et al., 2010). Xi et al. reported relative sparing of ventral diencephalon DA neurons with perturbed ascending

PT projections, associated with locomotor impairment, especially decreased swimming behavior and response to tactile stimuli (Xi et al., 2010). The PINK1/Parkin pathway is an essential mitochondria quality control mechanism and leads to mitophagy (Agarwal & Muqit, 2022).

The presenilin-associated rhomboid-like (PARL) gene encodes a mitochondrial protease, whose physiological function is still unclear (Lysyk et al., 2020). ZF possesses two PARL paralogs, PARLA and PARLB, with 67% and 55% similarity to the human orthologue, respectively. Noble et al.'s knockdown model of both PARL genes was lethal, with one paralogue needed to ensure survival. In addition, severe mispatterning of DA neurons was observed only in PARLA knockdown (Noble et al., 2012). In another study using a CRISPR/Cas9-mediated mutagenesis to induce a PARLA loss of function in adult and larval ZF, specific loss of OB DA neurons with impaired locomotion was documented (Merhi et al., 2021).

The DJ-1 protein is encoded by the the PARK7 gene and is implicated in autosomal onset PD with early onset (Abou-Sleiman et al., 2004). Non-functional DJ-1 protein is associated with an increase in p53 target genes, which induces apoptosis (Kato et al., 2013) and potentially protects neurons by reducing BAX (Bretaud et al., 2006). A novel DJ-1 knockout ZF model using the CRISPR/Cas9 method showed abnormalities detected in early adult stages, with impaired protein functions implicating a wide range of functions, such as mitochondrial complex I dysfunction, redox regulation, mitophagy, and inflammation (Edson et al., 2019).

The leucine-rich repeat kinase 2 (LRRK2) mutations are a prevalent cause of the autosomal-dominant type of PD, first described in a Japanese family by Funayama et al. in 2002 (Funayama et al., 2002). ZF LRRK2 orthologue has 71% similar functionality with the human one (Sheng et al., 2010), and deletion of WD40 domain causes diencephalic neurodegeneration and movement impairment. ZF models of mutations in LRRK2 gene documented β-synuclein aggregation and neuronal cell mass reduction and developmental abnormalities (Prabhudesai et al., 2016). LRRK2 mutations also weakened antibacterial response and was associated paradoxically with hyperactivity in the study of Sheng et al. (2018).

18.10.2 Toxin-Induced Models

MPTP use in ZF can induce parkinsonism-like manifestations and has been studied at different stages of development (Bretaud et al., 2004; Kalyn & Ekker, 2021; Lam et al., 2005; Zhao, Han, et al., 2020). Toxicity from exposure is mainly induced by mitochondrial dysfunction through complex I blockade (Kalyn & Ekker, 2021; Zawada et al., 2011). The sensitivity of DA neurons varied by location, with sensitivity of diencephalic DA neurons (Wen et al., 2008), OB, and the periventricular pretectal nucleus (Kalyn & Ekker, 2021). Swim velocity (Bretaud et al., 2004; Kalyn & Ekker, 2021), distance traveled decrease (Kalyn & Ekker, 2021; Sallinen et al., 2009), touch stimulus abnormal reactions (Lam et al., 2005; McKinley et al., 2005), and reduced movement (Souza et al., 2011) were documented in larval ZF following MPTP exposure. In the adult ZF, only vascular microinjections elicited neuronal loss (Kalyn & Ekker, 2021), unlike intraperitoneal injection that only induced motor impairment with relative sparing of neuronal populations (O. V. Anichtchik et al., 2003; Bretaud et al., 2004).

6-OHDA uptaken by dopamine transporter causes neuronal cell loss through different mechanisms such as oxidative stress, microglial-mediated inflammation, and mitochondrial dysfunction (Rodriguez-Pallares et al., 2007). 6-OHDA can't pass the BBB, which might explain relative neuronal sparing in the Anichtchik et al. model, despite the reduction in total distance traveled, speed, and turning angle increase (O. V. Anichtchik et al., 2003). In another study with targeted intracerebroventricular injection, Vijayanathan et al. documented in addition to reduction in distance traveled and speed, important reductions in DA neurons in the OB, telencephalon, preoptic area, PT, and the hypothalamus (Vijayanathan et al., 2017). Feng et al. documented also the increase in anxious behavior indicated by progressive time spent in the bottom zone of the cuvette (Feng et al., 2014).

Rotenone's lipophilic nature permits the crossing of the BBB, inducing dopaminergic loss through mitochondrial dysfunction, oxidative stress (Ünal et al., 2019), microglial-associated neuroinflammation (Yurtsever et al., 2022), and can experimentally cause the aggregation of α-synuclein particles (Sherer et al., 2002). PD-like motor dysfunction was documented in animal models exposed to rotenone (Ünal et al., 2019; X. H. Wang et al., 2018), such as decreased swimming velocity and distance traveled. The light-dark box test abnormal reactions were reported to indicate anxiety and depression in an animal model exposed to rotenone (X. H. Wang et al., 2018).

Paraquat is a hydrophilic neurotoxin, thus transported actively into the brain. It has a close structure to MPTP (Vaccari et al., 2019). Exposed ZF animal models demonstrated motor impairments (Bortolotto et al., 2014; Joseph et al., 2020), anxiety (Nunes et al., 2017), aggression (Nunes et al., 2017), and memory problems (Bortolotto et al., 2014). Different mechanisms have been proposed, essentially impaired antioxidant systems (Joseph et al., 2020; Nunes et al., 2017), neuroinflammation, and energy production dysfunction (Joseph et al., 2020).

REFERENCES

Aarsland, D., Brønnick, K., Larsen, J. P., Tysnes, O. B., Alves, G., & Norwegian ParkWest Study Group. (2009). Cognitive impairment in incident, untreated Parkinson disease: The Norwegian ParkWest study. *Neurology*, *72*(13), 1121–1126. https://doi.org/10.1212/01.wnl.0000338632.00552.cb

Abou-Sleiman, P. M., Healy, D. G., & Wood, N. W. (2004). Causes of Parkinson's disease: Genetics of DJ-1. *Cell and Tissue Research*, *318*(1), 185–188. https://doi.org/10.1007/s00441-004-0922-6

Agarwal, S., & Muqit, M. M. K. (2022). PTEN-induced kinase 1 (PINK1) and Parkin: Unlocking a mitochondrial quality control pathway linked to Parkinson's disease. *Current Opinion in Neurobiology*, *72*, 111–119. https://doi.org/10.1016/j.conb.2021.09.005

Ahmad, M. H., Rizvi, M. A., Ali, M., & Mondal, A. C. (2023). Neurobiology of depression in Parkinson's disease: Insights into epidemiology, molecular mechanisms and treatment strategies. *Ageing Research Reviews*, *85*, 101840. https://doi.org/10.1016/j.arr.2022.101840

Ahmadi Rastegar, D., & Dzamko, N. (2020). Leucine rich repeat kinase 2 and innate immunity. *Frontiers in Neuroscience*, *14*, 193. https://doi.org/10.3389/fnins.2020.00193

Alafuzoff, I., & Parkkinen, L. (2014). Staged pathology in Parkinson's disease. *Parkinsonism & Related Disorders*, *20*, S57–S61. https://doi.org/10.1016/S1353-8020(13)70016-6

Alcacer, C., Andreoli, L., Sebastianutto, I., Jakobsson, J., Fieblinger, T., & Cenci, M. A. (2017). Chemogenetic stimulation of striatal projection neurons modulates responses to Parkinson's disease therapy. *Journal of Clinical Investigation*, *127*(2), 720–734. https://doi.org/10.1172/JCI90132

Alexander, G. E., & Crutcher, M. D. (1990). Functional architecture of basal ganglia circuits: Neural substrates of parallel processing. *Trends in Neurosciences*, *13*(7), 266–271. https://doi.org/10.1016/0166-2236(90)90107-L

Andoniadou, C. L., & Martinez-Barbera, J. P. (2013). Developmental mechanisms directing early anterior forebrain specification in vertebrates. *Cellular and Molecular Life Sciences*, *70*(20), 3739–3752. https://doi.org/10.1007/s00018-013-1269-5

Anichtchik, O., Diekmann, H., Fleming, A., Roach, A., Goldsmith, P., & Rubinsztein, D. C. (2008). Loss of PINK1 function affects development and results in neurodegeneration in Zebrafish. *The Journal of Neuroscience*, *28*(33), 8199–8207. https://doi.org/10.1523/JNEUROSCI.0979-08.2008

Anichtchik, O. V., Kaslin, J., Peitsaro, N., Scheinin, M., & Panula, P. (2003). Neurochemical and behavioural changes in zebrafish D*Danio rerio* after systemic administration of 6-hydroxydopamine and 1-methyl-4-phenyl-1,2,3,6-tetrahydropyridine: Effects of MPTP and 6-OHDA on zebrafish brain. *Journal of Neurochemistry*, *88*(2), 443–453. https://doi.org/10.1111/j.1471-4159.2004.02190.x

Armstrong, G. A. B., & Drapeau, P. (2013). Loss and gain of FUS function impair neuromuscular synaptic transmission in a genetic model of ALS. *Human Molecular Genetics*, *22*(21), 4282–4292. https://doi.org/10.1093/hmg/ddt278

Armstrong, G. A. B., Liao, M., You, Z., Lissouba, A., Chen, B. E., & Drapeau, P. (2016). Homology directed knockin of point mutations in the zebrafish tardbp and fus genes in ALS using the CRISPR/Cas9 system. *PLoS One*, *11*(3), e0150188. https://doi.org/10.1371/journal.pone.0150188

Asakawa, K., Handa, H., & Kawakami, K. (2020). Optogenetic modulation of TDP-43 oligomerization accelerates ALS-related pathologies in the spinal motor neurons. *Nature Communications*, *11*(1), 1004. https://doi.org/10.1038/s41467-020-14815-x

Ascherio, A., & Schwarzschild, M. A. (2016). The epidemiology of Parkinson's disease: Risk factors and prevention. *The Lancet Neurology*, *15*(12), 1257–1272. https://doi.org/10.1016/S1474-4422(16)30230-7

Ashrafi, G., & Schwarz, T. L. (2013). The pathways of mitophagy for quality control and clearance of mitochondria. *Cell Death & Differentiation*, *20*(1), 31–42. https://doi.org/10.1038/cdd.2012.81

Åström, D. O., Simonsen, J., Raket, L. L., Sgarbi, S., Hellsten, J., Hagell, P., Norlin, J. M., Kellerborg, K., Martinez-Martin, P., & Odin, P. (2022). High risk of developing dementia in Parkinson's disease: A Swedish registry-based study. *Scientific Reports*, *12*(1), 16759. https://doi.org/10.1038/s41598-022-21093-8

Baba, M., Nakajo, S., Tu, P. H., Tomita, T., Nakaya, K., Lee, V. M., Trojanowski, J. Q., & Iwatsubo, T. (1998). Aggregation of alpha-synuclcin in Lewy bodies of sporadic Parkinson's disease and dementia with Lewy bodies. *The American Journal of Pathology*, *152*(4), 879–884.

Babin, P. J., Goizet, C., & Raldúa, D. (2014). Zebrafish models of human motor neuron diseases: Advantages and limitations. *Progress in Neurobiology*, *118*, 36–58. https://doi.org/10.1016/j.pneurobio.2014.03.001

Bajpai, P., Sangar, M. C., Singh, S., Tang, W., Bansal, S., Chowdhury, G., Cheng, Q., Fang, J.-K., Martin, M. V., Guengerich, F. P., & Avadhani, N. G. (2013). Metabolism of 1-methyl-4-phenyl-1,2,3,6-tetrahydropyridine by mitochondrion-targeted cytochrome P450 2D6. *Journal of Biological Chemistry*, *288*(6), 4436–4451. https://doi.org/10.1074/jbc.M112.402123

Balendra, R., & Isaacs, A. M. (2018). C9orf72-mediated ALS and FTD: Multiple pathways to disease. *Nature Reviews Neurology*, *14*(9), 544–558. https://doi.org/10.1038/s41582-018-0047-2

Bellou, V., Belbasis, L., Tzoulaki, I., Evangelou, E., & Ioannidis, J. P. A. (2016). Environmental risk factors and Parkinson's disease: An umbrella review of meta-analyses. *Parkinsonism & Related Disorders*, *23*, 1–9. https://doi.org/10.1016/j.parkreldis.2015.12.008

Benazzouz, A., Gross, C., Féger, J., Boraud, T., & Bioulac, B. (1993). Reversal of rigidity and improvement in motor performance by subthalamic high-frequency stimulation in MPTP-treated monkeys. *European Journal of Neuroscience*, *5*(4), 382–389. https://doi.org/10.1111/j.1460-9568.1993.tb00505.x

Benedetti, L., Ghilardi, A., Rottoli, E., De Maglie, M., Prosperi, L., Perego, C., Baruscotti, M., Bucchi, A., Del Giacco, L., & Francolini, M. (2016). INaP selective inhibition reverts precocious inter-and motorneurons hyperexcitability in the Sod1-G93R zebrafish ALS model. *Scientific Reports*, *6*(1), 24515. https://doi.org/10.1038/srep24515

Bercier, V., Hubbard, J. M., Fidelin, K., Duroure, K., Auer, T. O., Revenu, C., Wyart, C., & Del Bene, F. (2019). Dynactin1 depletion leads to neuromuscular synapse instability and functional abnormalities. *Molecular Neurodegeneration*, *14*(1), 27. https://doi.org/10.1186/s13024-019-0327-3

Berning, B. A., & Walker, A. K. (2019). The pathobiology of TDP-43 C-terminal fragments in ALS and FTLD. *Frontiers in Neuroscience*, *13*, 335. https://doi.org/10.3389/fnins.2019.00335

Best, J. D., & Alderton, W. K. (2008). Zebrafish: An in vivo model for the study of neurological diseases. *Neuropsychiatric Disease and Treatment*, 4(3), 567–576. https://doi.org/10.2147/ndt.s2056

Billingsley, K. J., Bandres-Ciga, S., Saez-Atienzar, S., & Singleton, A. B. (2018). Genetic risk factors in Parkinson's disease. *Cell and Tissue Research*, *373*(1), 9–20. https://doi.org/10.1007/s00441-018-2817-y

Biundo, R., Weis, L., & Antonini, A. (2016). Cognitive decline in Parkinson's disease: The complex picture. *NPJ Parkinson's Disease*, *2*, 16018. https://doi.org/10.1038/npjparkd.2016.18

Blader, P. (2000). Zebrafish developmental genetics and central nervous system development. *Human Molecular Genetics*, *9*(6), 945–951. https://doi.org/10.1093/hmg/9.6.945

Boccalaro, I. L., Schwerdel, C., Cristiá-Lara, L., Fritschy, J., & Rubi, L. (2020). Dopamine depletion induces neuron-specific alterations of GABAergic transmission in the mouse striatum. *European Journal of Neuroscience*, *52*(5), 3353–3374. https://doi.org/10.1111/ejn.14886

Boileau, I., Warsh, J. J., Guttman, M., Saint-Cyr, J. A., McCluskey, T., Rusjan, P., Houle, S., Wilson, A. A., Meyer, J. H., & Kish, S. J. (2008). Elevated serotonin transporter binding in depressed patients with Parkinson's disease: A preliminary PET study with [^{11} C]DASB. *Movement Disorders*, *23*(12), 1776–1780. https://doi.org/10.1002/mds.22212

Bortolotto, J. W., Cognato, G. P., Christoff, R. R., Roesler, L. N., Leite, C. E., Kist, L. W., Bogo, M. R., Vianna, M. R., & Bonan, C. D. (2014). Long-term exposure to paraquat alters behavioral parameters and dopamine levels in adult zebrafish (*Danio rerio*). *Zebrafish*, *11*(2), 142–153. https://doi.org/10.1089/zeb.2013.0923

Bose, A., Petsko, G. A., & Studer, L. (2022). Induced pluripotent stem cells: A tool for modeling Parkinson's disease. *Trends in Neurosciences*, *45*(8), 608–620. https://doi.org/10.1016/j.tins.2022.05.001

Bose, P., Armstrong, G. A. B., & Drapeau, P. (2019). Neuromuscular junction abnormalities in a zebrafish loss-of-function model of TDP-43. *Journal of Neurophysiology*, *121*(1), 285–297. https://doi.org/10.1152/jn.00265.2018

Bourefis, A.-R., Campanari, M.-L., Buee-Scherrer, V., & Kabashi, E. (2020). Functional characterization of a FUS mutant zebrafish line as a novel genetic model for ALS. *Neurobiology of Disease*, *142*, 104935. https://doi.org/10.1016/j.nbd.2020.104935

Braak, H., Tredici, K. D., Rüb, U., de Vos, R. A. I., Jansen Steur, E. N. H., & Braak, E. (2003). Staging of brain pathology related to sporadic Parkinson's disease. *Neurobiology of Aging*, *24*(2), 197–211. https://doi.org/10.1016/S0197-4580(02)00065-9

Bretaud, S., Allen, C., Ingham, P. W., & Bandmann, O. (2006). P53-dependent neuronal cell death in a DJ-1-deficient zebrafish model of Parkinson's disease. *Journal of Neurochemistry*, *0*(0), 070209222715077-??? https://doi.org/10.1111/j.1471-4159.2006.04291.x

Bretaud, S., Lee, S., & Guo, S. (2004). Sensitivity of zebrafish to environmental toxins implicated in Parkinson's disease. *Neurotoxicology and Teratology*, *26*(6), 857–864. https://doi.org/10.1016/j.ntt.2004.06.014

Bromek, E., Haduch, A., Gołembiowska, K., & Daniel, W. A. (2011). Cytochrome P450 mediates dopamine formation in the brain in vivo: Dopamine formation by brain cytochrome P450 in vivo. *Journal of Neurochemistry*, *118*(5), 806–815. https://doi.org/10.1111/j.1471-4159.2011.07339.x

Cameron, D. J., Galvin, C., Alkam, T., Sidhu, H., Ellison, J., Luna, S., & Ethell, D. W. (2012). Alzheimer's-related peptide amyloid-β plays a conserved role in angiogenesis. *PLoS One*, *7*(7), e39598. https://doi.org/10.1371/journal.pone.0039598

Camp, E., & Lardelli, M. (2001). Tyrosinase gene expression in zebrafish embryos. *Development Genes and Evolution*, *211*(3), 150–153. https://doi.org/10.1007/s004270000125

Carlberg, M., & Anctil, M. (1993). Biogenic amines in coelenterates. *Comparative Biochemistry and Physiology Part C: Pharmacology, Toxicology and Endocrinology*, *106*(1), 1–9. https://doi.org/10.1016/0742-8413(93)90250-O

Chaudhry, F. A., Edwards, R. H., & Fonnum, F. (2008). Vesicular neurotransmitter transporters as targets for endogenous and exogenous toxic substances. *Annual Review of Pharmacology and Toxicology*, *48*(1), 277–301. https://doi.org/10.1146/annurev.pharmtox.46.120604.141146

Chen, Y.-C., Priyadarshini, M., & Panula, P. (2009). Complementary developmental expression of the two tyrosine hydroxylase transcripts in zebrafish. *Histochemistry and Cell Biology*, *132*(4), 375–381. https://doi.org/10.1007/s00418-009-0619-8

Chia, K., Klingseisen, A., Sieger, D., & Priller, J. (2022). Zebrafish as a model organism for neurodegenerative disease. *Frontiers in Molecular Neuroscience*, *15*, 940484. https://doi.org/10.3389/fnmol.2022.940484

Choi, T.-Y., Choi, T.-I., Lee, Y.-R., Choe, S.-K., & Kim, C.-H. (2021). Zebrafish as an animal model for biomedical research. *Experimental & Molecular Medicine*, *53*(3), 310–317. https://doi.org/10.1038/s12276-021-00571-5

Connor-Robson, N., Booth, H., Martin, J. G., Gao, B., Li, K., Doig, N., Vowles, J., Browne, C., Klinger, L., Juhasz, P., Klein, C., Cowley, S. A., Bolam, P., Hirst, W., & Wade-Martins, R. (2019). An integrated transcriptomics and proteomics analysis reveals functional endocytic dysregulation caused by mutations in LRRK2. *Neurobiology of Disease*, *127*, 512–526. https://doi.org/10.1016/j.nbd.2019.04.005

Cui, L.-Y., Chu, S.-F., & Chen, N.-H. (2020). The role of chemokines and chemokine receptors in multiple sclerosis. *International Immunopharmacology*, *83*, 106314. https://doi.org/10.1016/j.intimp.2020.106314

Da Costa, M. M. J., Allen, C. E., Higginbottom, A., Ramesh, T., Shaw, P. J., & McDermott, C. J. (2013). A new zebrafish model produced by TILLING of SOD1-related amyotrophic lateral sclerosis replicates key features of the disease and represents a tool for *in vivo* therapeutic screening. *Disease Models & Mechanisms*. https://doi.org/10.1242/dmm.012013sw

Danio rerio (Zebrafish) as a model organism\textbar biology II\textbar JoVE. (n.d.). Retrieved October 7, 2023, from www.jove.com/v/5128/danio-rerio-zebrafish-as-a-model-organism

Deffains, M., Iskhakova, L., Katabi, S., Israel, Z., & Bergman, H. (2018). Longer β oscillatory episodes reliably identify pathological subthalamic activity in Parkinsonism: Longer STN β episodes are a PD biomarker. *Movement Disorders*, *33*(10), 1609–1618. https://doi.org/10.1002/mds.27418

DeLong, M. R. (1990). Primate models of movement disorders of basal ganglia origin. *Trends in Neurosciences*, *13*(7), 281–285. https://doi.org/10.1016/0166-2236(90)90110-V

DeMaagd, G., & Philip, A. (2015). Parkinson's disease and its management: Part 1: Disease entity, risk factors, pathophysiology, clinical presentation, and diagnosis. *P & T: A Peer-Reviewed Journal for Formulary Management*, *40*(8), 504–532.

Diekmann, H., Anichtchik, O., Fleming, A., Futter, M., Goldsmith, P., Roach, A., & Rubinsztein, D. C. (2009). Decreased BDNF levels are a major contributor to the embryonic phenotype of huntingtin knockdown Zebrafish. *The Journal of Neuroscience*, *29*(5), 1343–1349. https://doi.org/10.1523/JNEUROSCI.6039-08.2009

Donnini, S., Solito, R., Cetti, E., Corti, F., Giachetti, A., Carra, S., Beltrame, M., Cotelli, F., & Ziche, M. (2010). Aß peptides accelerate the senescence of endothelial cells *in vitro* and *in vivo*, impairing angiogenesis. *The FASEB Journal*, *24*(7), 2385–2395. https://doi.org/10.1096/fj.09-146456

Du, Y., Guo, Q., Shan, M., Wu, Y., Huang, S., Zhao, H., Hong, H., Yang, M., Yang, X., Ren, L., Peng, J., Sun, J., Zhou, H., Li, S., & Su, B. (2016). Spatial and temporal distribution of dopaminergic neurons during development in Zebrafish. *Frontiers in Neuroanatomy*, *10*. https://doi.org/10.3389/fnana.2016.00115

Durcan, T. M., & Fon, E. A. (2015). The three 'P's of mitophagy: PARKIN, PINK1, and post-translational modifications. *Genes & Development*, *29*(10), 989–999. https://doi.org/10.1101/gad.262758.115

Edson, A. J., Hushagen, H. A., Frøyset, A. K., Elda, I., Khan, E. A., Di Stefano, A., & Fladmark, K. E. (2019). Dysregulation in the brain protein profile of zebrafish lacking the Parkinson's disease-related protein DJ-1. *Molecular Neurobiology*, *56*(12), 8306–8322. https://doi.org/10.1007/s12035-019-01667-w

Ehrlich, M. E. (2012). Huntington's disease and the striatal medium spiny neuron: Cell-autonomous and non-cell-autonomous mechanisms of disease. *Neurotherapeutics*, *9*(2), 270–284. https://doi.org/10.1007/s13311-012-0112-2

Eidelberg, D., & Surmeier, D. J. (2011). Brain networks in Huntington disease. *Journal of Clinical Investigation*, *121*(2), 484–492. https://doi.org/10.1172/JCI45646

Ekström, P., Honkanen, T., & Steinbusch, H. W. (1990). Distribution of dopamine-immunoreactive neuronal perikarya and fibres in the brain of a teleost, Gasterosteus aculeatus L. comparison with tyrosine hydroxylase- and dopamine-beta-hydroxylase-immunoreactive neurons. *Journal of Chemical Neuroanatomy*, *3*(4), 233–260.

Estrada-Sánchez, A. M., & Rebec, G. V. (2013). Role of cerebral cortex in the neuropathology of Huntington's disease. *Frontiers in Neural Circuits*, *7*. https://doi.org/10.3389/fncir.2013.00019

Faivre, F., Joshi, A., Bezard, E., & Barrot, M. (2019). The hidden side of Parkinson's disease: Studying pain, anxiety and depression in animal models. *Neuroscience & Biobehavioral Reviews*, *96*, 335–352. https://doi.org/10.1016/j.neubiorev.2018.10.004

Fang, Q., Xicoy, H., Shen, J., Luchetti, S., Dai, D., Zhou, P., Qi, X.-R., Martens, G. J. M., Huitinga, I., Swaab, D. F., Liu, C., & Shan, L. (2021). Histamine-4 receptor antagonist ameliorates Parkinson-like pathology in the striatum. *Brain, Behavior, and Immunity*, *92*, 127–138. https://doi.org/10.1016/j.bbi.2020.11.036

Fearnley, J. M., & Lees, A. J. (1991). Ageing and Parkinson's Disease: Substantia nigra regional selectivity. *Brain*, *114*(5), 2283–2301. https://doi.org/10.1093/brain/114.5.2283

Feng, C.-W., Wen, Z.-H., Huang, S.-Y., Hung, H.-C., Chen, C.-H., Yang, S.-N., Chen, N.-F., Wang, H.-M., Hsiao, C.-D., & Chen, W.-F. (2014). Effects of 6-hydroxydopamine exposure on motor activity and biochemical expression in zebrafish (*Danio rerio*) larvae. *Zebrafish*, *11*(3), 227–239. https://doi.org/10.1089/zeb.2013.0950

Ferini-Strambi, L., & Zucconi, M. (2000). REM sleep behavior disorder. *Clinical Neurophysiology*, *111*, S136–S140. https://doi.org/10.1016/S1388-2457(00)00414-4

Ferreira, M., & Massano, J. (2017). An updated review of Parkinson's disease genetics and clinicopathological correlations. *Acta Neurologica Scandinavica*, *135*(3), 273–284. https://doi.org/10.1111/ane.12616

Fett, M. E., Pilsl, A., Paquet, D., Van Bebber, F., Haass, C., Tatzelt, J., Schmid, B., & Winklhofer, K. F. (2010). Parkin is protective against proteotoxic stress in a transgenic zebrafish model. *PLoS One*, *5*(7), e11783. https://doi.org/10.1371/journal.pone.0011783

Filion, M., & Tremblay, L. (1991). Abnormal spontaneous activity of globus pallidus neurons in monkeys with MPTP-induced parkinsonism. *Brain Research*, *547*(1), 140–144. https://doi.org/10.1016/0006-8993(91)90585-J

Filippi, A., Dürr, K., Ryu, S., Willaredt, M., Holzschuh, J., & Driever, W. (2007). Expression and function of nr4a2, lmx1b, and pitx3in zebrafish dopaminergic and noradrenergic neuronal development. *BMC Developmental Biology*, *7*(1), 135. https://doi.org/10.1186/1471-213X-7-135

Filippi, A., Jainok, C., & Driever, W. (2012). Analysis of transcriptional codes for zebrafish dopaminergic neurons reveals essential functions of Arx and Isl1 in prethalamic dopaminergic neuron development. *Developmental Biology*, *369*(1), 133–149. https://doi.org/10.1016/j.ydbio.2012.06.010

Flinn, L., Mortiboys, H., Volkmann, K., Koster, R. W., Ingham, P. W., & Bandmann, O. (2009). Complex I deficiency and dopaminergic neuronal cell loss in parkin-deficient zebrafish (*Danio rerio*). *Brain*, *132*(6), 1613–1623. https://doi.org/10.1093/brain/awp108

Frisoni, G. B., Fox, N. C., Jack, C. R., Scheltens, P., & Thompson, P. M. (2010). The clinical use of structural MRI in Alzheimer disease. *Nature Reviews Neurology*, *6*(2), 67–77. https://doi.org/10.1038/nrneurol.2009.215

Frouni, I., Kwan, C., Belliveau, S., & Huot, P. (2022). Cognition and serotonin in Parkinson's disease. In *Progress in brain research* (Vol. 269, pp. 373–403). Elsevier. https://doi.org/10.1016/bs.pbr.2022.01.013

Fukumori, A., Fluhrer, R., Steiner, H., & Haass, C. (2010). Three-amino acid spacing of presenilin endoproteolysis suggests a general stepwise cleavage of-secretase-mediated intramembrane proteolysis. *Journal of Neuroscience*, *30*(23), 7853–7862. https://doi.org/10.1523/JNEUROSCI.1443-10.2010

Fulwiler, C., & Gilbert, W. (1991). Zebrafish embryology and neural development. *Current Opinion in Cell Biology*, *3*(6), 988–991. https://doi.org/10.1016/0955-0674(91)90118-I

Funayama, M., Hasegawa, K., Kowa, H., Saito, M., Tsuji, S., & Obata, F. (2002). A new locus for Parkinson's disease (PARK8) maps to chromosome 12p11.2-q13.1. *Annals of Neurology*, *51*(3), 296–301. https://doi.org/10.1002/ana.10113

Gerlai, R. T. (2020). Introduction: The Zebrafish in behavioral and neural genetics. In *Behavioral and neural genetics of zebrafish* (pp. xix–xxiii). Elsevier. https://doi.org/10.1016/B978-0-12-817528-6.02001-5

Godoy, R., Hua, K., Kalyn, M., Cusson, V.-M., Anisman, H., & Ekker, M. (2020). Dopaminergic neurons regenerate following chemogenetic ablation in the olfactory bulb of adult Zebrafish (*Danio rerio*). *Scientific Reports*, *10*(1), 12825. https://doi.org/10.1038/s41598-020-69734-0

Goedert, M. (2015). Alzheimer's and Parkinson's diseases: The prion concept in relation to assembled Aβ, tau, and α-synuclein. *Science*, *349*(6248), 1255555. https://doi.org/10.1126/science.1255555

Gonçalves, F. B., & Morais, V. A. (2021). PINK1: A bridge between mitochondria and Parkinson's disease. *Life*, *11*(5), 371. https://doi.org/10.3390/life11050371

Graveland, G. A., Williams, R. S., & DiFiglia, M. (1985). Evidence for degenerative and regenerative changes in neostriatal spiny neurons in Huntington's disease. *Science*, *227*(4688), 770–773. https://doi.org/10.1126/science.3155875

Greggio, E., Jain, S., Kingsbury, A., Bandopadhyay, R., Lewis, P., Kaganovich, A., van der Brug, M. P., Beilina, A., Blackinton, J., Thomas, K. J., Ahmad, R., Miller, D. W., Kesavapany, S., Singleton, A., Lees, A., Harvey, R. J., Harvey, K., & Cookson, M. R. (2006). Kinase activity is required for the toxic effects of mutant LRRK2/dardarin. *Neurobiology of Disease*, *23*(2), 329–341. https://doi.org/10.1016/j.nbd.2006.04.001

Groth, C., Nornes, S., McCarty, R., Tamme, R., & Lardelli, M. (2002). Identification of a second presenilin gene in zebrafish with similarity to the human Alzheimer's disease gene presenilin2. *Development Genes and Evolution*, *212*(10), 486–490. https://doi.org/10.1007/s00427-002-0269-5

Grunwald, D. J., & Eisen, J. S. (2002). Headwaters of the zebrafish—Emergence of a new model vertebrate. *Nature Reviews Genetics*, *3*(9), 717–724. https://doi.org/10.1038/nrg892

Haffter, P., Granato, M., Brand, M., Mullins, M. C., Hammerschmidt, M., Kane, D. A., Odenthal, J., J. M. Van Eeden, F., Jiang, Y.-J., Heisenberg, C.-P., Kelsh, R. N., Furutani-Seiki, M., Vogelsang, E., Beuchle, D., Schach, U., Fabian, C., & Nüsslein-Volhard, C. (1996). The identification of genes with unique and essential functions in the development of the zebrafish, *Danio rerio*. *Development*, *123*(1), 1–36. https://doi.org/10.1242/dev.123.1.1

Hammond, C., Bergman, H., & Brown, P. (2007). Pathological synchronization in Parkinson's disease: Networks, models and treatments. *Trends in Neurosciences*, *30*(7), 357–364. https://doi.org/10.1016/j.tins.2007.05.004

Hardiman, O., Al-Chalabi, A., Chio, A., Corr, E. M., Logroscino, G., Robberecht, W., Shaw, P. J., Simmons, Z., & Van Den Berg, L. H. (2017). Correction: Amyotrophic lateral sclerosis. *Nature Reviews Disease Primers*, *3*(1), 17085. https://doi.org/10.1038/nrdp.2017.85

Helley, M. P., Pinnell, J., Sportelli, C., & Tieu, K. (2017). Mitochondria: A common target for genetic mutations and environmental toxicants in Parkinson's disease. *Frontiers in Genetics*, *8*, 177. https://doi.org/10.3389/fgene.2017.00177

Hely, M. A., Morris, J. G. L., Reid, W. G. J., & Trafficante, R. (2005). Sydney multicenter study of Parkinson's disease: Non-L-dopa–responsive problems dominate at 15 years. *Movement Disorders*, *20*(2), 190–199. https://doi.org/10.1002/mds.20324

Hely, M. A., Reid, W. G. J., Adena, M. A., Halliday, G. M., & Morris, J. G. L. (2008). The Sydney multicenter study of Parkinson's disease: The inevitability of dementia at 20 years: Twenty year Sydney Parkinson's study. *Movement Disorders*, *23*(6), 837–844. https://doi.org/10.1002/mds.21956

Henshall, T. L., Tucker, B., Lumsden, A. L., Nornes, S., Lardelli, M. T., & Richards, R. I. (2009). Selective neuronal requirement for huntingtin in the developing zebrafish. *Human Molecular Genetics*, *18*(24), 4830–4842. https://doi.org/10.1093/hmg/ddp455

Hernandez, D. G., Reed, X., & Singleton, A. B. (2016). Genetics in Parkinson disease: Mendelian versus non-Mendelian inheritance. *Journal of Neurochemistry*, *139*, 59–74. https://doi.org/10.1111/jnc.13593

Hesse, S., Meyer, P. M., Strecker, K., Barthel, H., Wegner, F., Oehlwein, C., Isaias, I. U., Schwarz, J., & Sabri, O. (2009). Monoamine transporter availability in Parkinson's disease patients with or without depression. *European Journal of Nuclear Medicine and Molecular Imaging*, *36*(3), 428–435. https://doi.org/10.1007/s00259-008-0979-7

Hewamadduma, C. A. A., Grierson, A. J., Ma, T. P., Pan, L., Moens, C. B., Ingham, P. W., Ramesh, T., & Shaw, P. J. (2013). Tardbpl splicing rescues motor neuron and axonal development in a mutant tardbp zebrafish. *Human Molecular Genetics*, *22*(12), 2376–2386. https://doi.org/10.1093/hmg/ddt082

Hindle, J. V. (2010). Ageing, neurodegeneration and Parkinson's disease. *Age and Ageing*, *39*(2), 156–161. https://doi.org/10.1093/ageing/afp223

Holzschuh, J., Hauptmann, G., & Driever, W. (2003). Genetic Analysis of the roles of Hh, FGF8, and nodal signaling during catecholaminergic system development in the zebrafish brain. *The Journal of Neuroscience*, *23*(13), 5507–5519. https://doi.org/10.1523/JNEUROSCI.23-13-05507.2003

Hornykiewicz, O. (1974). The mechanisms of action of L-dopa in Parkinson's disease. *Life Sciences*, *15*(7), 1249–1259. https://doi.org/10.1016/0024-3205(74)90306-3

Howe, K., Clark, M. D., Torroja, C. F., Torrance, J., Berthelot, C., Muffato, M., Collins, J. E., Humphray, S., McLaren, K., Matthews, L., McLaren, S., Sealy, I., Caccamo, M., Churcher, C., Scott, C., Barrett, J. C., Koch, R., Rauch, G.-J., White, S., . . . Stemple, D. L. (2013). The zebrafish reference genome sequence and its relationship to the human genome. *Nature*, *496*(7446), 498–503. https://doi.org/10.1038/nature12111

Hughes, G. L., Lones, M. A., Bedder, M., Currie, P. D., Smith, S. L., & Pownall, M. E. (2020). Machine learning discriminates a movement disorder in a zebrafish model of Parkinson's disease. *Disease Models & Mechanisms*. https://doi.org/10.1242/dmm.045815

International Parkinson's Disease Genomics Consortium, 23 and Me Research Team, Chang, D., Nalls, M. A., Hallgrímsdóttir, I. B., Hunkapiller, J., van der Brug, M., Cai, F., Kerchner, G. A., Ayalon, G., Bingol, B., Sheng, M., Hinds, D., Behrens, T. W., Singleton, A. B., Bhangale, T. R., & Graham, R. R. (2017). A meta-analysis of genome-wide association studies identifies 17 new Parkinson's disease risk loci. *Nature Genetics*, *49*(10), 1511–1516. https://doi.org/10.1038/ng.3955

Isaacson, S. H., Coate, B., Norton, J., & Stankovic, S. (2020). Blinded SAPS-PD assessment after 10 weeks of pimavanserin treatment for Parkinson's disease psychosis. *Journal of Parkinson's Disease*, *10*(4), 1389–1396. https://doi.org/10.3233/JPD-202047

Iyer, S., Acharya, K. R., & Subramanian, V. (2018). A comparative bioinformatic analysis of *C9orf72*. *PeerJ*, *6*, e4391. https://doi.org/10.7717/peerj.4391

Joseph, T. P., Jagadeesan, N., Sai, L. Y., Lin, S. L., Sahu, S., & Schachner, M. (2020). Adhesion Molecule L1 Agonist Mimetics Protect Against the Pesticide Paraquat-Induced Locomotor Deficits and Biochemical Alterations in Zebrafish. *Frontiers in Neuroscience*, *14*, 458. https://doi.org/10.3389/fnins.2020.00458

Kabashi, E., Bercier, V., Lissouba, A., Liao, M., Brustein, E., Rouleau, G. A., & Drapeau, P. (2011). FUS and TARDBP but not SOD1 interact in genetic models of amyotrophic lateral sclerosis. *PLoS Genetics*, *7*(8), e1002214. https://doi.org/10.1371/journal.pgen.1002214

Kabashi, E., Lin, L., Tradewell, M. L., Dion, P. A., Bercier, V., Bourgouin, P., Rochefort, D., Bel Hadj, S., Durham, H. D., Velde, C. V., Rouleau, G. A., & Drapeau, P. (2010). Gain and loss of function of ALS-related mutations of TARDBP (TDP-43) cause motor deficits in vivo. *Human Molecular Genetics*, *19*(4), 671–683. https://doi.org/10.1093/hmg/ddp534

Kalueff, A. V., Stewart, A. M., & Gerlai, R. (2014). Zebrafish as an emerging model for studying complex brain disorders. *Trends in Pharmacological Sciences*, *35*(2), 63–75. https://doi.org/10.1016/j.tips.2013.12.002

Kalyn, M., & Ekker, M. (2021). Cerebroventricular microinjections of MPTP on adult zebrafish induces dopaminergic neuronal death, mitochondrial fragmentation, and sensorimotor impairments. *Frontiers in Neuroscience*, *15*, 718244. https://doi.org/10.3389/fnins.2021.718244

Kalyn, M., Hua, K., Mohd Noor, S., Wong, C. E. D., & Ekker, M. (2019). Comprehensive analysis of neurotoxin-induced ablation of dopaminergic neurons in zebrafish larvae. *Biomedicines*, *8*(1), 1. https://doi.org/10.3390/biomedicines8010001

Karlovich, C. A., John, R. M., Ramirez, L., Stainier, D. Y. R., & Myers, R. M. (1998). Characterization of the Huntington's disease (HD) gene homolog in the zebrafish *Danio rerio*. *Gene*,

217(1–2), 117–125. https://doi.org/10.1016/S0378-1119(98)00342-4

Kaslin, J., & Panula, P. (2001). Comparative anatomy of the histaminergic and other aminergic systems in zebrafish (*Danio rerio*). *The Journal of Comparative Neurology*, *440*(4), 342–377. https://doi.org/10.1002/cne.1390

Kato, I., Maita, H., Takahashi-Niki, K., Saito, Y., Noguchi, N., Iguchi-Ariga, S. M. M., & Ariga, H. (2013). Oxidized DJ-1 inhibits p53 by sequestering p53 from promoters in a DNA-binding affinity-dependent manner. *Molecular and Cellular Biology*, *33*(2), 340–359. https://doi.org/10.1128/MCB.01350-12

Kerenyi, L., Ricaurte, G. A., Schretlen, D. J., McCann, U., Varga, J., Mathews, W. B., Ravert, H. T., Dannals, R. F., Hilton, J., Wong, D. F., & Szabo, Z. (2003). Positron emission tomography of striatal serotonin transporters in Parkinson disease. *Archives of Neurology*, *60*(9). https://doi.org/10.1001/archneur.60.9.1223

Khandelwal, P. J., Dumanis, S. B., Feng, L. R., Maguire-Zeiss, K., Rebeck, G., Lashuel, H. A., & Moussa, C. E. (2010). Parkinson-related parkin reduces α-Synuclein phosphorylation in a gene transfer model. *Molecular Neurodegeneration*, *5*(1), 47. https://doi.org/10.1186/1750-1326-5-47

Khoo, T. K., Yarnall, A. J., Duncan, G. W., Coleman, S., O'Brien, J. T., Brooks, D. J., Barker, R. A., & Burn, D. J. (2013). The spectrum of nonmotor symptoms in early Parkinson disease. *Neurology*, *80*(3), 276–281. https://doi.org/10.1212/WNL.0b013e31827deb74

Kim, S., Chung, A., Na, J. E., Lee, S. J., Jeong, S. H., Kim, E., Sun, W., Rhyu, I. J., & Park, H. (2019). Myelin degeneration induced by mutant superoxide dismutase 1 accumulation promotes amyotrophic lateral sclerosis. *Glia*, *67*(10), 1910–1921. https://doi.org/10.1002/glia.23669

Kimmel, C. B. (1989). Genetics and early development of zebrafish. *Trends in Genetics*, *5*, 283–288. https://doi.org/10.1016/0168-9525(89)90103-0

Kish, S. J., Tong, J., Hornykiewicz, O., Rajput, A., Chang, L.-J., Guttman, M., & Furukawa, Y. (2008). Preferential loss of serotonin markers in caudate versus putamen in Parkinson's disease. *Brain: A Journal of Neurology*, *131*(Pt 1), 120–131. https://doi.org/10.1093/brain/awm239

Kitada, T., Pisani, A., Porter, D. R., Yamaguchi, H., Tscherter, A., Martella, G., Bonsi, P., Zhang, C., Pothos, E. N., & Shen, J. (2007). Impaired dopamine release and synaptic plasticity in the striatum of *PINK1* -deficient mice. *Proceedings of the National Academy of Sciences*, *104*(27), 11441–11446. https://doi.org/10.1073/pnas.0702717104

Koga, S., Sekiya, H., Kondru, N., Ross, O. A., & Dickson, D. W. (2021). Neuropathology and molecular diagnosis of Synucleinopathies. *Molecular Neurodegeneration*, *16*(1), 83. https://doi.org/10.1186/s13024-021-00501-z

Kozawa, S., Akanuma, T., Sato, T., Sato, Y. D., Ikeda, K., & Sato, T. N. (2016). Real-time prediction of cell division timing in developing zebrafish embryo. *Scientific Reports*, *6*(1), 32962. https://doi.org/10.1038/srep32962

Kunst, M., Laurell, E., Mokayes, N., Kramer, A., Kubo, F., Fernandes, A. M., Förster, D., Dal Maschio, M., & Baier, H. (2019). A cellular-resolution atlas of the larval zebrafish brain. *Neuron*, *103*(1), 21–38.e5. https://doi.org/10.1016/j.neuron.2019.04.034

Laird, A. S., Van Hoecke, A., De Muynck, L., Timmers, M., Van Den Bosch, L., Van Damme, P., & Robberecht, W. (2010). Progranulin is neurotrophic in vivo and protects against a mutant TDP-43 induced axonopathy. *PLoS One*, *5*(10), e13368. https://doi.org/10.1371/journal.pone.0013368

Lam, C. S., Korzh, V., & Strahle, U. (2005). Zebrafish embryos are susceptible to the dopaminergic neurotoxin MPTP. *European Journal of Neuroscience*, *21*(6), 1758–1762. https://doi.org/10.1111/j.1460-9568.2005.03988.x

Langston, J. W., Ballard, P., Tetrud, J. W., & Irwin, I. (1983). Chronic parkinsonism in humans due to a product of meperidine-analog synthesis. *Science*, *219*(4587), 979–980. https://doi.org/10.1126/science.6823561

Lattante, S., Rouleau, G. A., & Kabashi, E. (2013). *TARDBP* and *FUS* mutations associated with amyotrophic lateral sclerosis: Summary and update. *Human Mutation*, *34*(6), 812–826. https://doi.org/10.1002/humu.22319

Lebedeva, S., De Jesus Domingues, A. M., Butter, F., & Ketting, R. F. (2017). Characterization of genetic loss-of-function of Fus in zebrafish. *RNA Biology*, *14*(1), 29–35. https://doi.org/10.1080/15476286.2016.1256532

Leblois, A., Meissner, W., Bioulac, B., Gross, C. E., Hansel, D., & Boraud, T. (2007). Late emergence of synchronized oscillatory activity in the pallidum during progressive parkinsonism: Pallidal activity during progressive parkinsonism. *European Journal of Neuroscience*, *26*(6), 1701–1713. https://doi.org/10.1111/j.1460-9568.2007.05777.x

Lee, K. Y., Jang, G. H., Byun, C. H., Jeun, M., Searson, P. C., & Lee, K. H. (2017). Zebrafish models for functional and toxicological screening of nanoscale drug delivery systems: Promoting preclinical applications. *Bioscience Reports*, *37*(3), BSR20170199. https://doi.org/10.1042/BSR20170199

Leimer, U., Lun, K., Romig, H., Walter, J., Grünberg, J., Brand, M., & Haass, C. (1999). Zebrafish (*Danio rerio*) presenilin promotes aberrant amyloid β-peptide production and requires a critical aspartate residue for its function in amyloidogenesis. *Biochemistry*, *38*(41), 13602–13609. https://doi.org/10.1021/bi991453n

Lemmens, R., Van Hoecke, A., Hersmus, N., Geelen, V., D'Hollander, I., Thijs, V., Van Den Bosch, L., Carmeliet, P., & Robberecht, W. (2007). Overexpression of mutant superoxide dismutase 1 causes a motor axonopathy in the zebrafish. *Human Molecular Genetics*, *16*(19), 2359–2365. https://doi.org/10.1093/hmg/ddm193

Leung, L. C., Wang, G. X., & Mourrain, P. (2013). Imaging zebrafish neural circuitry from whole brain to synapse. *Frontiers in Neural Circuits*, *7*. https://doi.org/10.3389/fncir.2013.00076

Lissouba, A., Liao, M., Kabashi, E., & Drapeau, P. (2018). Transcriptomic analysis of zebrafish TDP-43 transgenic lines. *Frontiers in Molecular Neuroscience*, *11*, 463. https://doi.org/10.3389/fnmol.2018.00463

Liu, C.-Q., Chen, Z., Liu, F.-X., Hu, D.-N., & Luo, J.-H. (2007). Involvement of brain endogenous histamine in the degeneration of dopaminergic neurons in 6-hydroxydopamine-lesioned rats. *Neuropharmacology*, *53*(7), 832–841. https://doi.org/10.1016/j.neuropharm.2007.08.014

Lleo, A., & A. Saura, C. (2011). γ-Secretase substrates and their implications for drug development in Alzheimer's disease. *Current Topics in Medicinal Chemistry*, *11*(12), 1513–1527. https://doi.org/10.2174/156802611795861004

Lo Bianco, C., Ridet, J.-L., Schneider, B. L., Déglon, N., & Aebischer, P. (2002). α-Synucleinopathy and selective dopaminergic neuron loss in a rat lentiviral-based model of Parkinson's disease. *Proceedings of the National Academy of Sciences*, *99*(16), 10813–10818. https://doi.org/10.1073/pnas.152339799

Löhr, H., Ryu, S., & Driever, W. (2009). Zebrafish diencephalic A11-related dopaminergic neurons share a conserved transcriptional network with neuroendocrine cell lineages.

Development, *136*(6), 1007–1017. https://doi.org/10.1242/dev.033878

Lucini, C., D'Angelo, L., Cacialli, P., Palladino, A., & De Girolamo, P. (2018). BDNF, brain, and regeneration: Insights from zebrafish. *International Journal of Molecular Sciences*, *19*(10), 3155. https://doi.org/10.3390/ijms19103155

Lumsden, A. L., Henshall, T. L., Dayan, S., Lardelli, M. T., & Richards, R. I. (2007). Huntingtin-deficient zebrafish exhibit defects in iron utilization and development. *Human Molecular Genetics*, *16*(16), 1905–1920. https://doi.org/10.1093/hmg/ddm138

Lysyk, L., Brassard, R., Touret, N., & Lemieux, M. J. (2020). PARL protease: A glimpse at intramembrane proteolysis in the inner mitochondrial membrane. *Journal of Molecular Biology*, *432*(18), 5052–5062. https://doi.org/10.1016/j.jmb.2020.04.006

Mahler, J., Filippi, A., & Driever, W. (2010). DeltaA/DeltaD regulate multiple and temporally distinct phases of notch signaling during dopaminergic neurogenesis in Zebrafish. *The Journal of Neuroscience*, *30*(49), 16621–16635. https://doi.org/10.1523/JNEUROSCI.4769-10.2010

Marın, O., Smeets, W. J. A. J., & González, A. (1998). Evolution of the basal ganglia in tetrapods: A new perspective based on recent studies in amphibians. *Trends in Neurosciences*, *21*(11), 487–494. https://doi.org/10.1016/S0166-2236(98)01297-1

McDonough, H., & Patterson, C. (2003). CHIP: A link between the chaperone and proteasome systems. *Cell Stress & Chaperones*, *8*(4), 303. https://doi.org/10.1379/1466-1268(2003)008<0303:CALBTC>2.0.CO;2

McGregor, M. M., & Nelson, A. B. (2019). Circuit mechanisms of Parkinson's disease. *Neuron*, *101*(6), 1042–1056. https://doi.org/10.1016/j.neuron.2019.03.004

McKeith, I. G., Galasko, D., Kosaka, K., Perry, E. K., Dickson, D. W., Hansen, L. A., Salmon, D. P., Lowe, J., Mirra, S. S., Byrne, E. J., Lennox, G., Quinn, N. P., Edwardson, J. A., Ince, P. G., Bergeron, C., Burns, A., Miller, B. L., Lovestone, S., Collerton, D., . . . Perry, R. H. (1996). Consensus guidelines for the clinical and pathologic diagnosis of dementia with Lewy bodies (DLB): Report of the consortium on DLB international workshop. *Neurology*, *47*(5), 1113–1124. https://doi.org/10.1212/WNL.47.5.1113

McKinley, E. T., Baranowski, T. C., Blavo, D. O., Cato, C., Doan, T. N., & Rubinstein, A. L. (2005). Neuroprotection of MPTP-induced toxicity in zebrafish dopaminergic neurons. *Molecular Brain Research*, *141*(2), 128–137. https://doi.org/10.1016/j.molbrainres.2005.08.014

Meek, J. (1994). Catecholamines in the brain of osteichthyes (bony fishes). In W. J. A. J. Smeets and A. Reiner (Eds.), *Phylogeny and development of catecholamine systems in the CNS of Vertebrates*. Cambridge University Press.

Merhi, R., Kalyn, M., Zhu-Pawlowsky, A., & Ekker, M. (2021). Loss of parla function results in inactivity, olfactory impairment, and dopamine neuron loss in Zebrafish. *Biomedicines*, *9*(2), 205. https://doi.org/10.3390/biomedicines9020205

Milanese, C., Sager, J. J., Bai, Q., Farrell, T. C., Cannon, J. R., Greenamyre, J. T., & Burton, E. A. (2012). Hypokinesia and reduced dopamine levels in zebrafish lacking β- and γ1-synucleins. *Journal of Biological Chemistry*, *287*(5), 2971–2983. https://doi.org/10.1074/jbc.M111.308312

Miller, V. M., Nelson, R. F., Gouvion, C. M., Williams, A., Rodriguez-Lebron, E., Harper, S. Q., Davidson, B. L., Rebagliati, M. R., & Paulson, H. L. (2005). CHIP suppresses polyglutamine aggregation and toxicity *in vitro* and *in vivo*. *The Journal of Neuroscience*, *25*(40), 9152–9161. https://doi.org/10.1523/JNEUROSCI.3001-05.2005

Mink, J. W., & Thach, W. T. (1993). Basal ganglia intrinsic circuits and their role in behavior. *Current Opinion in Neurobiology*, *3*(6), 950–957. https://doi.org/10.1016/0959-4388(93)90167-W

Model Organism–Definition, Types, Characteristics and FAQs. (n.d.). In *Vedantu*. Retrieved October 7, 2023, from www.vedantu.com/biology/model-organism

Moore, R. Y., & Bloom, F. E. (1978). Central catecholamine neuron systems: Anatomy and physiology of the dopamine systems. *Annual Review of Neuroscience*, *1*(1), 129–169. https://doi.org/10.1146/annurev.ne.01.030178.001021

Morrice, J. R., Gregory-Evans, C. Y., & Shaw, C. A. (2020). Investigating microglia during motor neuron degeneration using a zebrafish model. *Micron*, *133*, 102852. https://doi.org/10.1016/j.micron.2020.102852

Nahimi, A., Sommerauer, M., Kinnerup, M. B., Østergaard, K., Wintherdahl, M., Jacobsen, J., Schacht, A., Johnsen, B., Damholdt, M. F., Borghammer, P., & Gjedde, A. (2018). Noradrenergic deficits in Parkinson disease imaged with ^{11}C-MeNER. *Journal of Nuclear Medicine*, *59*(4), 659–664. https://doi.org/10.2967/jnumed.117.190975

Narabayashi, H. (1956). Procaine oil blocking of the globus pattidus. *Archives of Neurology and Psychiatry*, *75*(1), 36. https://doi.org/10.1001/archneurpsyc.1956.02330190052003

Nieuwenhuys, R. (2009). The forebrain of actinopterygians revisited. *Brain, Behavior and Evolution*, *73*(4), 229–252. https://doi.org/10.1159/000225622

Noble, S., Ismail, A., Godoy, R., Xi, Y., & Ekker, M. (2012). Zebrafish Parla- and Parlb-deficiency affects dopaminergic neuron patterning and embryonic survival: Zebrafish Parla- and Parlb-deficiency. *Journal of Neurochemistry*, *122*(1), 196–207. https://doi.org/10.1111/j.1471-4159.2012.07758.x

Northcutt, R. G. (2008). Forebrain evolution in bony fishes. *Brain Research Bulletin*, *75*(2–4), 191–205. https://doi.org/10.1016/j.brainresbull.2007.10.058

Nuland, A. J. M., Ouden, H. E. M., Zach, H., Dirkx, M. F. M., Asten, J. J. A., Scheenen, T. W. J., Toni, I., Cools, R., & Helmich, R. C. (2020). GABAergic changes in the thalamocortical circuit in Parkinson's disease. *Human Brain Mapping*, *41*(4), 1017–1029. https://doi.org/10.1002/hbm.24857

Nunes, M. E., Müller, T. E., Braga, M. M., Fontana, B. D., Quadros, V. A., Marins, A., Rodrigues, C., Menezes, C., Rosemberg, D. B., & Loro, V. L. (2017). Chronic treatment with paraquat induces brain injury, changes in antioxidant defenses system, and modulates behavioral functions in zebrafish. *Molecular Neurobiology*, *54*(6), 3925–3934. https://doi.org/10.1007/s12035-016-9919-x

Nussbaum, R. L., & Ellis, C. E. (2003). Alzheimer's disease and Parkinson's disease. *New England Journal of Medicine*, *348*(14), 1356–1364. https://doi.org/10.1056/NEJM2003ra020003

O'Donnell, K. C., Lulla, A., Stahl, M. C., Wheat, N. D., Bronstein, J. M., & Sagasti, A. (2014). Axon degeneration and PGC1α-mediated protection in a vertebrate model of α-synuclein toxicity. *Disease Models & Mechanisms*. https://doi.org/10.1242/dmm.013185

O'Gorman Tuura, R. L., Baumann, C. R., & Baumann-Vogel, H. (2018). Beyond dopamine: GABA, glutamate, and the axial symptoms of Parkinson disease. *Frontiers in Neurology*, *9*, 806. https://doi.org/10.3389/fneur.2018.00806

Ohki, Y., Wenninger-Weinzierl, A., Hruscha, A., Asakawa, K., Kawakami, K., Haass, C., Edbauer, D., & Schmid, B.

(2017). Glycine-alanine dipeptide repeat protein contributes to toxicity in a zebrafish model of C9orf72 associated neurodegeneration. *Molecular Neurodegeneration*, *12*(1), 6. https://doi.org/10.1186/s13024-016-0146-8

Ohta, E., Nihira, T., Uchino, A., Imaizumi, Y., Okada, Y., Akamatsu, W., Takahashi, K., Hayakawa, H., Nagai, M., Ohyama, M., Ryo, M., Ogino, M., Murayama, S., Takashima, A., Nishiyama, K., Mizuno, Y., Mochizuki, H., Obata, F., & Okano, H. (2015). I2020T mutant LRRK2 iPSC-derived neurons in the Sagamihara family exhibit increased Tau phosphorylation through the AKT/GSK-3β signaling pathway. *Human Molecular Genetics*, *24*(17), 4879–4900. https://doi.org/10.1093/hmg/ddv212

Oun, A., Sabogal-Guaqueta, A. M., Galuh, S., Alexander, A., Kortholt, A., & Dolga, A. M. (2022). The multifaceted role of LRRK2 in Parkinson's disease: From human iPSC to organoids. *Neurobiology of Disease*, *173*, 105837. https://doi.org/10.1016/j.nbd.2022.105837

Pang, S. Y.-Y., Ho, P. W.-L., Liu, H.-F., Leung, C.-T., Li, L., Chang, E. E. S., Ramsden, D. B., & Ho, S.-L. (2019). The interplay of aging, genetics and environmental factors in the pathogenesis of Parkinson's disease. *Translational Neurodegeneration*, *8*(1), 23. https://doi.org/10.1186/s40035-019-0165-9

Panula, P., Sallinen, V., Sundvik, M., Kolehmainen, J., Torkko, V., Tiittula, A., Moshnyakov, M., & Podlasz, P. (2006). Modulatory neurotransmitter systems and behavior: Towards zebrafish models of neurodegenerative diseases. *Zebrafish*, *3*(2), 235–247. https://doi.org/10.1089/zeb.2006.3.235

Paquet, D., Bhat, R., Sydow, A., Mandelkow, E.-M., Berg, S., Hellberg, S., Fälting, J., Distel, M., Köster, R. W., Schmid, B., & Haass, C. (2009). A zebrafish model of tauopathy allows in vivo imaging of neuronal cell death and drug evaluation. *Journal of Clinical Investigation*, *119*(5), 1382–1395. https://doi.org/10.1172/JCI37537

Parng, C., Seng, W. L., Semino, C., & McGrath, P. (2002). Zebrafish: A preclinical model for drug screening. *ASSAY and Drug Development Technologies*, *1*(1), 41–48. https://doi.org/10.1089/154065802761001293

Pasquini, J., Ceravolo, R., Qamhawi, Z., Lee, J.-Y., Deuschl, G., Brooks, D. J., Bonuccelli, U., & Pavese, N. (2018). Progression of tremor in early stages of Parkinson's disease: A clinical and neuroimaging study. *Brain*, *141*(3), 811–821. https://doi.org/10.1093/brain/awx376

Patton, E. E., Zon, L. I., & Langenau, D. M. (2021). Zebrafish disease models in drug discovery: From preclinical modelling to clinical trials. *Nature Reviews Drug Discovery*, *20*(8), 611–628. https://doi.org/10.1038/s41573-021-00210-8

Pavese, N., Metta, V., Simpson, B. S., Marphy, T. A., Ramlackhansingh, A., Chaudhuri, K. R., & Brooks, D. J. (2012). 1.060 Sleep regulatory centres dysfunction in Parkinson's disease patients with excessive daytime sleepiness. An in vivo pet study. *Parkinsonism & Related Disorders*, *18*, S24–S25. https://doi.org/10.1016/S1353-8020(11)70174-2

Perathoner, S., Cordero-Maldonado, M. L., & Crawford, A. D. (2016). Potential of zebrafish as a model for exploring the role of the amygdala in emotional memory and motivational behavior. *Journal of Neuroscience Research*, *94*(6), 445–462. https://doi.org/10.1002/jnr.23712

Pickel, V. M., Joh, T. H., & Reis, D. J. (1975). Immunohistochemical localization of tyrosine hydroxylase in brain by light and electron microscopy. *Brain Research*, *85*(2), 295–300. https://doi.org/10.1016/0006-8993(75)90084-0

Poewe, W., Seppi, K., Tanner, C. M., Halliday, G. M., Brundin, P., Volkmann, J., Schrag, A.-E., & Lang, A. E. (2017). Parkinson disease. *Nature Reviews Disease Primers*, *3*(1), 17013. https://doi.org/10.1038/nrdp.2017.13

Politis, M., Wu, K., Loane, C., Brooks, D. J., Kiferle, L., Turkheimer, F. E., Bain, P., Molloy, S., & Piccini, P. (2014). Serotonergic mechanisms responsible for levodopa-induced dyskinesias in Parkinson's disease patients. *Journal of Clinical Investigation*, *124*(3), 1340–1349. https://doi.org/10.1172/JCI71640

Politis, M., Wu, K., Loane, C., Kiferle, L., Molloy, S., Brooks, D. J., & Piccini, P. (2010). Staging of serotonergic dysfunction in Parkinson's disease: An in vivo 11C-DASB PET study. *Neurobiology of Disease*, *40*(1), 216–221. https://doi.org/10.1016/j.nbd.2010.05.028

Polymeropoulos, M. H., Lavedan, C., Leroy, E., Ide, S. E., Dehejia, A., Dutra, A., Pike, B., Root, H., Rubenstein, J., Boyer, R., Stenroos, E. S., Chandrasekharappa, S., Athanassiadou, A., Papapetropoulos, T., Johnson, W. G., Lazzarini, A. M., Duvoisin, R. C., Di Iorio, G., Golbe, L. I., & Nussbaum, R. L. (1997). Mutation in the alpha-synuclein gene identified in families with Parkinson's disease. *Science (New York, N.Y.)*, *276*(5321), 2045–2047. https://doi.org/10.1126/science.276.5321.2045

Postlethwait, J. H., Woods, I. G., Ngo-Hazelett, P., Yan, Y.-L., Kelly, P. D., Chu, F., Huang, H., Hill-Force, A., & Talbot, W. S. (2000). Zebrafish comparative genomics and the origins of vertebrate chromosomes. *Genome Research*, *10*(12), 1890–1902. https://doi.org/10.1101/gr.164800

Postuma, R. B., Gagnon, J.-F., Vendette, M., & Montplaisir, J. Y. (2009). Idiopathic REM sleep behavior disorder in the transition to degenerative disease. *Movement Disorders*, *24*(15), 2225–2232. https://doi.org/10.1002/mds.22757

Prabhudesai, S., Bensabeur, F. Z., Abdullah, R., Basak, I., Baez, S., Alves, G., Holtzman, N. G., Larsen, J. P., & Møller, S. G. (2016). LRRK2 knockdown in zebrafish causes developmental defects, neuronal loss, and synuclein aggregation: Multiple roles of LRRK2. *Journal of Neuroscience Research*, *94*(8), 717–735. https://doi.org/10.1002/jnr.23754

Prabhudesai, S., Sinha, S., Attar, A., Kotagiri, A., Fitzmaurice, A. G., Lakshmanan, R., Ivanova, M. I., Loo, J. A., Klärner, F.-G., Schrader, T., Stahl, M., Bitan, G., & Bronstein, J. M. (2012). A novel "molecular tweezer" inhibitor of α-synuclein neurotoxicity in vitro and in vivo. *Neurotherapeutics*, *9*(2), 464–476. https://doi.org/10.1007/s13311-012-0105-1

Prediger, R. D. S., Matheus, F. C., Schwarzbold, M. L., Lima, M. M. S., & Vital, M. A. B. F. (2012). Anxiety in Parkinson's disease: A critical review of experimental and clinical studies. *Neuropharmacology*, *62*(1), 115–124. https://doi.org/10.1016/j.neuropharm.2011.08.039

Pucak, M. L., & Grace, A. A. (1994). Regulation of substantia nigra dopamine neurons. *Critical Reviews in Neurobiology*, *9*(1), 67–89.

Pycroft, L., Stein, J., & Aziz, T. (2018). Deep brain stimulation: An overview of history, methods, and future developments. *Brain and Neuroscience Advances*, *2*, 239821281881601. https://doi.org/10.1177/2398212818816017

Qamhawi, Z., Towey, D., Shah, B., Pagano, G., Seibyl, J., Marek, K., Borghammer, P., Brooks, D. J., & Pavese, N. (2015). Clinical correlates of raphe serotonergic dysfunction in early Parkinson's disease. *Brain*, *138*(10), 2964–2973. https://doi.org/10.1093/brain/awv215

Quik, M., Perez, X. A., & Bordia, T. (2012). Nicotine as a potential neuroprotective agent for Parkinson's disease. *Movement Disorders: Official Journal of the Movement Disorder Society*, *27*(8), 947–957. https://doi.org/10.1002/mds.25028

Rahman Khan, F., & Sulaiman Alhewairini, S. (2019). Zebrafish (*Danio* rerio) as a model organism. In L. Streba, D. Ionut Gheonea, & M. Schenker (Eds.), *Current trends in cancer management*. IntechOpen. https://doi.org/10.5772/intechopen.81517

Ramesh, T., Lyon, A. N., Pineda, R. H., Wang, C., Janssen, P. M. L., Canan, B. D., Burghes, A. H. M., & Beattie, C. E. (2010). A genetic model of amyotrophic lateral sclerosis in zebrafish displays phenotypic hallmarks of motoneuron disease. *Disease Models & Mechanisms*, *3*(9–10), 652–662. https://doi.org/10.1242/dmm.005538

Randlett, O., Wee, C. L., Naumann, E. A., Nnaemeka, O., Schoppik, D., Fitzgerald, J. E., Portugues, R., Lacoste, A. M. B., Riegler, C., Engert, F., & Schier, A. F. (2015). Whole-brain activity mapping onto a zebrafish brain atlas. *Nature Methods*, *12*(11), 1039–1046. https://doi.org/10.1038/nmeth.3581

Razali, K., Othman, N., Mohd Nasir, M. H., Doolaanea, A. A., Kumar, J., Ibrahim, W. N., Mohamed Ibrahim, N., & Mohamed, W. M. Y. (2021). The Promise of the zebrafish model for Parkinson's disease: Today's science and tomorrow's treatment. *Frontiers in Genetics*, *12*, 655550. https://doi.org/10.3389/fgene.2021.655550

Rees, R. N., Noyce, A. J., & Schrag, A. (2019). The prodromes of Parkinson's disease. *European Journal of Neuroscience*, *49*(3), 320–327. https://doi.org/10.1111/ejn.14269

Reeve, A., Simcox, E., & Turnbull, D. (2014). Ageing and Parkinson's disease: Why is advancing age the biggest risk factor? *Ageing Research Reviews*, *14*, 19–30. https://doi.org/10.1016/j.arr.2014.01.004

Renninger, S. L., & Orger, M. B. (2013). Two-photon imaging of neural population activity in zebrafish. *Methods*, *62*(3), 255–267. https://doi.org/10.1016/j.ymeth.2013.05.016

Rink, E., & Wullimann, M. F. (2001). The teleostean (zebrafish) dopaminergic system ascending to the subpallium (striatum) is located in the basal diencephalon (posterior tuberculum). *Brain Research*, *889*(1–2), 316–330. https://doi.org/10.1016/S0006-8993(00)03174-7

Rink, E., & Wullimann, M. F. (2002a). Connections of the ventral telencephalon and tyrosine hydroxylase distribution in the zebrafish brain (*Danio rerio*) lead to identification of an ascending dopaminergic system in a teleost. *Brain Research Bulletin*, *57*(3–4), 385–387. https://doi.org/10.1016/S0361-9230(01)00696-7

Rink, E., & Wullimann, M. F. (2002b). Development of the catecholaminergic system in the early zebrafish brain: An immunohistochemical study. *Developmental Brain Research*, *137*(1), 89–100. https://doi.org/10.1016/S0165-3806(02)00354-1

Robbins, T. W., & Cools, R. (2014). Cognitive deficits in Parkinson's disease: A cognitive neuroscience perspective: Cognitive neuroscience of Parkinson's disease. *Movement Disorders*, *29*(5), 597–607. https://doi.org/10.1002/mds.25853

Robinson, K. J., Yuan, K. C., Don, E. K., Hogan, A. L., Winnick, C. G., Tym, M. C., Lucas, C. W., Shahheydari, H., Watchon, M., Blair, I. P., Atkin, J. D., Nicholson, G. A., Cole, N. J., & Laird, A. S. (2019). Motor neuron abnormalities correlate with impaired movement in zebrafish that express mutant superoxide dismutase 1. *Zebrafish*, *16*(1), 8–14. https://doi.org/10.1089/zeb.2018.1588

Rodriguez-Pallares, J., Parga, J. A., Muñoz, A., Rey, P., Guerra, M. J., & Labandeira-Garcia, J. L. (2007). Mechanism of 6-hydroxydopamine neurotoxicity: The role of NADPH oxidase and microglial activation in 6-hydroxydopamine-induced degeneration of dopaminergic neurons. *Journal of Neurochemistry*, *0*(0), 070615193023005-??? https://doi.org/10.1111/j.1471-4159.2007.04699.x

Rosa, J. G. S., Lima, C., & Lopes-Ferreira, M. (2022). Zebrafish larvae behavior models as a tool for drug screenings and pre-clinical trials: A review. *International Journal of Molecular Sciences*, *23*(12), 6647. https://doi.org/10.3390/ijms23126647

Russek-Blum, N., Gutnick, A., Nabel-Rosen, H., Blechman, J., Staudt, N., Dorsky, R. I., Houart, C., & Levkowitz, G. (2008). Dopaminergic neuronal cluster size is determined during early forebrain patterning. *Development*, *135*(20), 3401–3413. https://doi.org/10.1242/dev.024232

Sakowski, S. A., Lunn, J. S., Busta, A. S., Oh, S. S., Zamora-Berridi, G., Palmer, M., Rosenberg, A. A., Philip, S. G., Dowling, J. J., & Feldman, E. L. (2012). Neuromuscular effects of G93A-SOD1 expression in zebrafish. *Molecular Neurodegeneration*, *7*(1), 44. https://doi.org/10.1186/1750-1326-7-44

Sallinen, V., Kolehmainen, J., Priyadarshini, M., Toleikyte, G., Chen, Y.-C., & Panula, P. (2010). Dopaminergic cell damage and vulnerability to MPTP in Pink1 knockdown zebrafish. *Neurobiology of Disease*, *40*(1), 93–101. https://doi.org/10.1016/j.nbd.2010.06.001

Sallinen, V., Torkko, V., Sundvik, M., Reenilä, I., Khrustalyov, D., Kaslin, J., & Panula, P. (2009). MPTP and MPP+ target specific aminergic cell populations in larval zebrafish. *Journal of Neurochemistry*, *108*(3), 719–731. https://doi.org/10.1111/j.1471-4159.2008.05793.x

Sánchez-Ferrer, A., Rodríguez-López, J. N., García-Cánovas, F., & García-Carmona, F. (1995). Tyrosinase: A comprehensive review of its mechanism. *Biochimica et Biophysica Acta*, *1247*(1), 1–11. https://doi.org/10.1016/0167-4838(94)00204-t

Sara, S. J. (2009). The locus coeruleus and noradrenergic modulation of cognition. *Nature Reviews Neuroscience*, *10*(3), 211–223. https://doi.org/10.1038/nrn2573

Scatton, B., Javoy-Agid, F., Rouquier, L., Dubois, B., & Agid, Y. (1983). Reduction of cortical dopamine, noradrenaline, serotonin and their metabolites in Parkinson's disease. *Brain Research*, *275*(2), 321–328. https://doi.org/10.1016/0006-8993(83)90993-9

Schmid, B., Hruscha, A., Hogl, S., Banzhaf-Strathmann, J., Strecker, K., Van Der Zee, J., Teucke, M., Eimer, S., Hegermann, J., Kittelmann, M., Kremmer, E., Cruts, M., Solchenberger, B., Hasenkamp, L., Van Bebber, F., Van Broeckhoven, C., Edbauer, D., Lichtenthaler, S. F., & Haass, C. (2013). Loss of ALS-associated TDP-43 in zebrafish causes muscle degeneration, vascular dysfunction, and reduced motor neuron axon outgrowth. *Proceedings of the National Academy of Sciences*, *110*(13), 4986–4991. https://doi.org/10.1073/pnas.1218311110

Schmidt, R., Leventhal, D. K., Mallet, N., Chen, F., & Berke, J. D. (2013). Canceling actions involves a race between basal ganglia pathways. *Nature Neuroscience*, *16*(8), 1118–1124. https://doi.org/10.1038/nn.3456

Shan, L., Bossers, K., Luchetti, S., Balesar, R., Lethbridge, N., Chazot, P. L., Bao, A.-M., & Swaab, D. F. (2012). Alterations in the histaminergic system in the substantia nigra and striatum of Parkinson's patients: A postmortem study. *Neurobiology of Aging*, *33*(7), 1488.e1–1488.e13. https://doi.org/10.1016/j.neurobiolaging.2011.10.016

Shan, L., Liu, C.-Q., Balesar, R., Hofman, M. A., Bao, A.-M., & Swaab, D. F. (2012). Neuronal histamine production remains unaltered in Parkinson's disease despite the accumulation of Lewy bodies and Lewy neurites in the tuberomamillary nucleus. *Neurobiology of Aging*, *33*(7), 1343–1344. https://doi.org/10.1016/j.neurobiolaging.2011.01.004

Sharma, M., & Burré, J. (2023). α-Synuclein in synaptic function and dysfunction. *Trends in Neurosciences*, *46*(2), 153–166. https://doi.org/10.1016/j.tins.2022.11.007

Shaw, M. P., Higginbottom, A., McGown, A., Castelli, L. M., James, E., Hautbergue, G. M., Shaw, P. J., & Ramesh, T. M. (2018). Stable transgenic C9orf72 zebrafish model key aspects of the ALS/FTD phenotype and reveal novel pathological features. *Acta Neuropathologica Communications*, *6*(1), 125. https://doi.org/10.1186/s40478-018-0629-7

Sheng, D., Qu, D., Kwok, K. H. H., Ng, S. S., Lim, A. Y. M., Aw, S. S., Lee, C. W. H., Sung, W. K., Tan, E. K., Lufkin, T., Jesuthasan, S., Sinnakaruppan, M., & Liu, J. (2010). Deletion of the WD40 domain of LRRK2 in zebrafish causes parkinsonism-like loss of neurons and locomotive defect. *PLoS Genetics*, *6*(4), e1000914. https://doi.org/10.1371/journal.pgen.1000914

Sheng, D., See, K., Hu, X., Yu, D., Wang, Y., Liu, Q., Li, F., Lu, M., Zhao, J., & Liu, J. (2018). Disruption of LRRK2 in Zebrafish leads to hyperactivity and weakened antibacterial response. *Biochemical and Biophysical Research Communications*, *497*(4), 1104–1109. https://doi.org/10.1016/j.bbrc.2018.02.186

Sherer, T. B., Betarbet, R., Stout, A. K., Lund, S., Baptista, M., Panov, A. V., Cookson, M. R., & Greenamyre, J. T. (2002). An *in vitro* model of Parkinson's disease: Linking mitochondrial impairment to altered α-synuclein metabolism and oxidative damage. *The Journal of Neuroscience*, *22*(16), 7006–7015. https://doi.org/10.1523/JNEUROSCI.22-16-07006.2002

Sherer, T. B., Betarbet, R., Testa, C. M., Seo, B. B., Richardson, J. R., Kim, J. H., Miller, G. W., Yagi, T., Matsuno-Yagi, A., & Greenamyre, J. T. (2003). Mechanism of toxicity in rotenone models of Parkinson's disease. *The Journal of Neuroscience*, *23*(34), 10756–10764. https://doi.org/10.1523/JNEUROSCI.23-34-10756.2003

Shiba, M., Bower, J. H., Maraganore, D. M., McDonnell, S. K., Peterson, B. J., Ahlskog, J. E., Schaid, D. J., & Rocca, W. A. (2000). Anxiety disorders and depressive disorders preceding Parkinson's disease: A case-control study. *Movement Disorders*, *15*(4), 669–677. https://doi.org/10.1002/1531-8257(200007)15:4<669::AID-MDS1011>3.0.CO;2-5

Singh, A., Liang, L., Kaneoke, Y., Cao, X., & Papa, S. M. (2015). Dopamine regulates distinctively the activity patterns of striatal output neurons in advanced parkinsonian primates. *Journal of Neurophysiology*, *113*(5), 1533–1544. https://doi.org/10.1152/jn.00910.2014

Singleton, A. B., Farrer, M., Johnson, J., Singleton, A., Hague, S., Kachergus, J., Hulihan, M., Peuralinna, T., Dutra, A., Nussbaum, R., Lincoln, S., Crawley, A., Hanson, M., Maraganore, D., Adler, C., Cookson, M. R., Muenter, M., Baptista, M., Miller, D., . . . Gwinn-Hardy, K. (2003). Alpha-Synuclein locus triplication causes Parkinson's disease. *Science (New York, N.Y.)*, *302*(5646), 841. https://doi.org/10.1126/science.1090278

Smeyne, R. J., & Jackson-Lewis, V. (2005). The MPTP model of Parkinson's disease. *Molecular Brain Research*, *134*(1), 57–66. https://doi.org/10.1016/j.molbrainres.2004.09.017

Sommerauer, M., Fedorova, T. D., Hansen, A. K., Knudsen, K., Otto, M., Jeppesen, J., Frederiksen, Y., Blicher, J. U., Geday, J., Nahimi, A., Damholdt, M. F., Brooks, D. J., & Borghammer, P. (2018). Evaluation of the noradrenergic system in Parkinson's disease: An 11C-MeNER PET and neuromelanin MRI study. *Brain*, *141*(2), 496–504. https://doi.org/10.1093/brain/awx348

Souza, B. R., Romano-Silva, M. A., & Tropepe, V. (2011). Dopamine D_2 receptor activity modulates akt signaling and alters GABAergic neuron development and motor behavior in zebrafish larvae. *The Journal of Neuroscience*, *31*(14), 5512–5525. https://doi.org/10.1523/JNEUROSCI.5548-10.2011

Spatola, M., & Wider, C. (2014). Genetics of Parkinson's disease: The yield. *Parkinsonism & Related Disorders*, *20*, S35–S38. https://doi.org/10.1016/S1353-8020(13)70011-7

Stanković, I., Petrović, I., Pekmezović, T., Marković, V., Stojković, T., Dragašević-Mišković, N., Svetel, M., & Kostić, V. (2019). Longitudinal assessment of autonomic dysfunction in early Parkinson's disease. *Parkinsonism & Related Disorders*, *66*, 74–79. https://doi.org/10.1016/j.parkreldis.2019.07.008

Starr, P. A., Rau, G. M., Davis, V., Marks, W. J., Ostrem, J. L., Simmons, D., Lindsey, N., & Turner, R. S. (2005). Spontaneous pallidal neuronal activity in human dystonia: Comparison with Parkinson's disease and normal macaque. *Journal of Neurophysiology*, *93*(6), 3165–3176. https://doi.org/10.1152/jn.00971.2004

Steger, M., Tonelli, F., Ito, G., Davies, P., Trost, M., Vetter, M., Wachter, S., Lorentzen, E., Duddy, G., Wilson, S., Baptista, M. A., Fiske, B. K., Fell, M. J., Morrow, J. A., Reith, A. D., Alessi, D. R., & Mann, M. (2016). Phosphoproteomics reveals that Parkinson's disease kinase LRRK2 regulates a subset of Rab GTPases. *ELife*, *5*, e12813. https://doi.org/10.7554/eLife.12813

Stepto, A., Gallo, J.-M., Shaw, C. E., & Hirth, F. (2014). Modelling C9ORF72 hexanucleotide repeat expansion in amyotrophic lateral sclerosis and frontotemporal dementia. *Acta Neuropathologica*, *127*(3), 377–389. https://doi.org/10.1007/s00401-013-1235-1

Streisinger, G., Walker, C., Dower, N., Knauber, D., & Singer, F. (1981). Production of clones of homozygous diploid zebra fish (Brachy *Danio rerio*). *Nature*, *291*(5813), 293–296. https://doi.org/10.1038/291293a0

Sun, Z., & Gitler, A. D. (2008). Discovery and characterization of three novel synuclein genes in zebrafish. *Developmental Dynamics*, *237*(9), 2490–2495. https://doi.org/10.1002/dvdy.21569

Tanner, C. M., Kamel, F., Ross, G. W., Hoppin, J. A., Goldman, S. M., Korell, M., Marras, C., Bhudhikanok, G. S., Kasten, M., Chade, A. R., Comyns, K., Richards, M. B., Meng, C., Priestley, B., Fernandez, H. H., Cambi, F., Umbach, D. M., Blair, A., Sandler, D. P., & Langston, J. W. (2011). Rotenone, paraquat, and Parkinson's disease. *Environmental Health Perspectives*, *119*(6), 866–872. https://doi.org/10.1289/ehp.1002839

Tay, T. L., Ronneberger, O., Ryu, S., Nitschke, R., & Driever, W. (2011). Comprehensive catecholaminergic projectome analysis reveals single-neuron integration of zebrafish ascending and descending dopaminergic systems. *Nature Communications*, *2*(1), 171. https://doi.org/10.1038/ncomms1171

Taylor, J. P., Brown, R. H., & Cleveland, D. W. (2016). Decoding ALS: From genes to mechanism. *Nature*, *539*(7628), 197–206. https://doi.org/10.1038/nature20413

Teame, T., Zhang, Z., Ran, C., Zhang, H., Yang, Y., Ding, Q., Xie, M., Gao, C., Ye, Y., Duan, M., & Zhou, Z. (2019). The use of zebrafish (*Danio rerio*) as biomedical models. *Animal Frontiers*, *9*(3), 68–77. https://doi.org/10.1093/af/vfz020

Tomasiewicz, H. G., Flaherty, D. B., Soria, J. P., & Wood, J. G. (2002). Transgenic zebrafish model of neurodegeneration. *Journal of Neuroscience Research*, *70*(6), 734–745. https://doi.org/10.1002/jnr.10451

Toni, M., & Cioni, C. (2015). Fish synucleins: An update. *Marine Drugs*, *13*(11), 6665–6686. https://doi.org/10.3390/md13116665

Ünal, İ., Üstündağ, Ü. V., Ateş, P. S., Eğilmezer, G., Alturfan, A. A., Yiğitbaşı, T., & Emekli-Alturfan, E. (2019). Rotenone impairs oxidant/antioxidant balance both in brain and intestines in zebrafish. *The International Journal of Neuroscience*, *129*(4), 363–368. https://doi.org/10.1080/00207454.2018.1538141

Vaccari, C., El Dib, R., Gomaa, H., Lopes, L. C., & de Camargo, J. L. (2019). Paraquat and Parkinson's disease: A systematic review and meta-analysis of observational studies. *Journal of Toxicology and Environmental Health, Part B*, *22*(5–6), 172–202. https://doi.org/10.1080/10937404.2019.1659197

Vaswani, P. A., Morley, J. F., Jennings, D., Siderowf, A., Marek, K., Jennings, D., Marek, K., Seibyl, J., Siderowf, A., Stern, M., Russell, D., Sethi, K., Frank, S., Simuni, T., Hauser, R., Ravina, B., Richards, I., Liang, G., Adler, C., . . . Chung, K. (2022). Serial olfactory testing for the diagnosis of prodromal Parkinson's disease in the PARS study. *Parkinsonism & Related Disorders*, *104*, 15–20. https://doi.org/10.1016/j.parkreldis.2022.09.007

Vaz, R. L., Outeiro, T. F., & Ferreira, J. J. (2018). Zebrafish as an animal model for drug discovery in Parkinson's disease and other movement disorders: A systematic review. *Frontiers in Neurology*. https://doi.org/10.3389/fneur.2018.00347

Vijayanathan, Y., Lim, F. T., Lim, S. M., Long, C. M., Tan, M. P., Majeed, A. B. A., & Ramasamy, K. (2017). 6-OHDA-Lesioned adult zebrafish as a useful Parkinson's disease model for dopaminergic neuroregeneration. *Neurotoxicity Research*, *32*(3), 496–508. https://doi.org/10.1007/s12640-017-9778-x

Wang, J., & Cao, H. (2021). Zebrafish and medaka: Important animal models for human neurodegenerative diseases. *International Journal of Molecular Sciences*, *22*(19), 10766. https://doi.org/10.3390/ijms221910766

Wang, X. H., Souders, C. L., Zhao, Y. H., & Martyniuk, C. J. (2018). Paraquat affects mitochondrial bioenergetics, dopamine system expression, and locomotor activity in zebrafish (*Danio rerio*). *Chemosphere*, *191*, 106–117. https://doi.org/10.1016/j.chemosphere.2017.10.032

Wasel, O., & Freeman, J. L. (2020). Chemical and genetic zebrafish models to define mechanisms of and treatments for dopaminergic neurodegeneration. *International Journal of Molecular Sciences*, *21*(17), 5981. https://doi.org/10.3390/ijms21175981

Wen, L., Wei, W., Gu, W., Huang, P., Ren, X., Zhang, Z., Zhu, Z., Lin, S., & Zhang, B. (2008). Visualization of monoaminergic neurons and neurotoxicity of MPTP in live transgenic zebrafish. *Developmental Biology*, *314*(1), 84–92. https://doi.org/10.1016/j.ydbio.2007.11.012

Weston, L. J., Cook, Z. T., Stackhouse, T. L., Sal, M. K., Schultz, B. I., Tobias, Z. J. C., Osterberg, V. R., Brockway, N. L., Pizano, S., Glover, G., Weissman, T. A., & Unni, V. K. (2021). In vivo aggregation of presynaptic alpha-synuclein is not influenced by its phosphorylation at serine-129. *Neurobiology of Disease*, *152*, 105291. https://doi.org/10.1016/j.nbd.2021.105291

Whitlock, K. E., & Westerfield, M. (2000). The olfactory placodes of the zebrafish form by convergence of cellular fields at the edge of the neural plate. *Development*, *127*(17), 3645–3653. https://doi.org/10.1242/dev.127.17.3645

Why use the zebrafish in research? (n.d.). In *@yourgenome Science website*. Retrieved October 7, 2023, from www.yourgenome.org/facts/why-use-the-zebrafish-in-research/

Wullimann, M. F., Rupp, B., & Reichert, H. (1996). *Neuroanatomy of the zebrafish brain: A topological atlas* (Softcover reprint of the hardcover 1st edition). Birkhäuser Verlag.

Xi, Y., Noble, S., & Ekker, M. (2011). Modeling neurodegeneration in zebrafish. *Current Neurology and Neuroscience Reports*, *11*(3), 274–282. https://doi.org/10.1007/s11910-011-0182-2

Xi, Y., Ryan, J., Noble, S., Yu, M., Yilbas, A. E., & Ekker, M. (2010). Impaired dopaminergic neuron development and locomotor function in zebrafish with loss of *pink1* function. *European Journal of Neuroscience*, *31*(4), 623–633. https://doi.org/10.1111/j.1460-9568.2010.07091.x

Yamamoto, K., Ruuskanen, J. O., Wullimann, M. F., & Vernier, P. (2010). Two tyrosine hydroxylase genes in vertebrates. *Molecular and Cellular Neuroscience*, *43*(4), 394–402. https://doi.org/10.1016/j.mcn.2010.01.006

Yamamoto, K., & Vernier, P. (2011). The evolution of dopamine systems in chordates. *Frontiers in Neuroanatomy*, *5*. https://doi.org/10.3389/fnana.2011.00021

Yang, W., Hamilton, J. L., Kopil, C., Beck, J. C., Tanner, C. M., Albin, R. L., Ray Dorsey, E., Dahodwala, N., Cintina, I., Hogan, P., & Thompson, T. (2020). Current and projected future economic burden of Parkinson's disease in the U.S. *NPJ Parkinson's Disease*, *6*(1), 15. https://doi.org/10.1038/s41531-020-0117-1

Yurtsever, İ., Üstündağ, Ü. V., Ünal, İ., Ateş, P. S., & Emekli-Alturfan, E. (2022). Rifampicin decreases neuroinflammation to maintain mitochondrial function and calcium homeostasis in rotenone-treated zebrafish. *Drug and Chemical Toxicology*, *45*(4), 1544–1551. https://doi.org/10.1080/01480545.2020.1846549

Zaccai, J., Brayne, C., McKeith, I., Matthews, F., Ince, P. G., & On behalf of the MRC Cognitive Function, Ageing Neuropathology Study. (2008). Patterns and stages of α-synucleinopathy: Relevance in a population-based cohort. *Neurology*, *70*(13), 1042–1048. https://doi.org/10.1212/01.wnl.0000306697.48738.b6

Zawada, W. M., Banninger, G. P., Thornton, J., Marriott, B., Cantu, D., Rachubinski, A. L., Das, M., Griffin, W. S. T., & Jones, S. M. (2011). Generation of reactive oxygen species in 1-methyl-4-phenylpyridinium (MPP+) treated dopaminergic neurons occurs as an NADPH oxidase-dependent two-wave cascade. *Journal of Neuroinflammation*, *8*(1), 129. https://doi.org/10.1186/1742-2094-8-129

Zebra*fish \textbar types & facts \textbar Britannica*. (2023). www.britannica.com/animal/zebra-fish

Zebrafish—an overview \textbar sciencedirect topics. (n.d.). Retrieved October 7, 2023, from www.sciencedirect.com/topics/biochemistry-genetics-and-molecular-biology/zebra-fish

Zebrafish—Characteristics, Life Cycle, Development, Genome Research. (n.d.). In *Microscope Master*. Retrieved October 7, 2023, from www.microscopemaster.com/zebrafish.html

Zeng, X.-S., Geng, W.-S., & Jia, J.-J. (2018). Neurotoxin-induced animal models of Parkinson disease: Pathogenic mechanism and assessment. *ASN Neuro*, *10*, 175909141877743. https://doi.org/10.1177/1759091418777438

Zhao, Y., Han, Y., Wang, Z., Chen, T., Qian, H., He, J., Li, J., Han, B., & Wang, T. (2020). Rosmarinic acid protects against 1-methyl-4-phenyl-1,2,3,6-tetrahydropyridine-induced dopaminergic neurotoxicity in zebrafish embryos. *Toxicology in Vitro*, *65*, 104823. https://doi.org/10.1016/j.tiv.2020.104823

Zhao, Y., Keshiya, S., Perera, G., Schramko, L., Halliday, G. M., & Dzamko, N. (2020). LRRK2 kinase inhibitors reduce alpha-synuclein in human neuronal cell lines with the G2019S mutation. *Neurobiology of Disease*, *144*, 105049. https://doi.org/10.1016/j.nbd.2020.105049

19 Smart Capsule and Edible Bird's Nest

Charting New Frontiers in Parkinson's Disease Defense: An Experimental Protocol Perspective

Wael Mohamed

19.1 INTRODUCTION

19.1.1 Hypothesis

EBN has neuroprotective effects, especially when combined with Deprenyl against the MPTP-induced zebrafish model of Parkinson's disease.

19.1.2 General Objective

To investigate the neuroprotective properties of EBN injection alone and in combination with Deprenyl in the MPTP-induced zebrafish model of Parkinson's disease.

19.1.3 Specific Objective 1

To establish and validate a zebrafish model of MPTP-induced Parkinson's disease.

19.1.4 Specific Objective 2

To evaluate the neuroprotective properties of EBN injection alone or in combination with Deprenyl in the MPTP-induced zebrafish model of Parkinson's disease.

19.1.5 Specific Objective 3

To examine the histopathological changes occurring in the substantia nigra, striatum, and other possible affected brain regions before and after treatment.

19.1.6 Specific Objective 4

To examine the delivery of EBN and Deprenyl using nanotechnology to the brain.

19.2 LITERATURE REVIEW

Parkinson's disease (PD) is the second most common neurodegenerative disease after Alzheimer's disease (AD), and it is expected to become more prevalent by 2050. According to the most recent published statistics, the incidence and prevalence of PD fluctuate around 4.5 to 21.0 and 18 to 328 cases, respectively, per 100,000 individuals annually (approximately 10 million people globally) (Dorsey et al., 2018; Savica et al., 2018). Mechanistically speaking, PD is characterized by a significant and constant loss of dopaminergic neurons from the substantia nigra pars compacta (SNpc) (DeMaagd and Philip, 2015; Kouli et al., 2018). The SNpc, which is found in the midbrain, is important for motor control because it supplies dopamine neurotransmitters to the striatum via a projection known as the nigrostriatal dopaminergic pathway (James et al., 2022). In PD, when dopaminergic neuronal cells die, the flow of dopamine to the striatum is stopped. This makes it hard to move. Because of this, motor disorders such as bradykinesia, dyskinesia, and resting tremors are seen as the gold standard for diagnosing PD (Kouli et al., 2018). Even though there has been a lot of research on PD, there is still a lot of uncertainty about what causes the specific loss of nigral dopaminergic neurons. As no curative or preventive therapy is currently available, exploring its pathophysiology is crucial to improving treatment. As of now, about 20 genes with rare but highly penetrant variants have been linked to familial or single-gene forms of PD, which are mostly found in people with European ancestry (Blauwendraat et al., 2020).

Since our understanding of PD is still far from complete, new promising therapeutics must be explored to improve patients' quality of life and maybe slow down the neurodegenerative process. Until now, there has been no effective treatment for this disorder despite the use of the "gold standard" for PD treatment, L-DOPA, which has long-term complications and drug resistance (wearing off, dyskinesias, freezing episodes, and "on-and-off" fluctuations) and the incidence of PD increasing. Within five years of chronic L-DOPA treatment, 40–50% of patients will develop dyskinesias and motor fluctuations (Korczyn, 2004). After 10 years, that number increases to 70–80% of patients. Neuroprotection implies the use of different agents exhibiting various neuroprotective strategies to prevent brain degeneration and neuron loss. Therefore, there is an urgent need for extensive scientific investigation to identify potential therapeutic targets for prevention and effective

DOI: 10.1201/9781003402893-21

treatment of the disorder (Rizek et al., 2016). Unfortunately, the extent of neurodegeneration is not reduced by the available treatments for PD, necessitating the creation of newer drugs and therapeutic strategies (Leathem et al., 2022). Therefore, a viable approach in the quest for new antiparkinsonian medications would be to use appropriate medicines that reduce the production of radicals and/or alter the effectiveness of uncontrolled monoamine oxidase type B.

It has been proposed that the monoamine oxidase B (MAO-B) enzyme mediates the breakdown of monoamine neurotransmitters like dopamine. In reactive astrocytes in PD, MAO-B actually mediates the abnormal production of GABA and hydrogen peroxide (H_2O_2). While H_2O_2 worsens astrocytic reactivity and dopaminergic neuronal death, astrocyte GABA tonically lowers dopaminergic neuronal activity. Reversible MAO-B drugs have recently been shown to minimize reactive astrogliosis and increase dopaminergic neuronal activity in animals suffering from PD (Nam et al., 2022). MAO-B are often used in the treatment of typical PD symptoms. Early-stage PD has been successfully treated with MAO-B inhibitor monotherapy, while later stages of the disease have been successfully treated with MAO-B inhibitors as adjuvant medications. Patients' motor and non-motor symptoms may be significantly improved by MAO-B inhibitors, which can shorten the "off" period and may possibly stop or postpone the course of the illness (Tan et al., 2022).

1-methyl-4-phenyl-1,2,3,6 tetrahydropyridine (MPTP) is a dopaminergic neurotoxin discovered accidently in the early 1980s, when it was commonly used as a synthetic heroin among young drug abusers, as a result of the appearance of Parkinsonian symptoms among the victims (Forno et al., 1993). Animal models of MPTP-induced PD were therefore developed to meet the research demand of studying PD (Speciale, 2002). MPTP is metabolized by MAO-B enzyme in the neuralgia cells to 1-methyl-4-phenylpyridinium (MPP), which is taken up by the dopaminergic neurons, causing mitochondrial damage leading to oxidative stress in the neurons and subsequent death (Sarath et al., 2016). Adult zebrafish exhibited behavioral alterations, such as decreased locomotor activity, upon exposure to MPTP (Bretaud et al., 2004). Accordingly, zebrafish developed as a promising research tool due to its high fecundity, high genetic resemblance with human diseases, ease of maintenance, and relatively low cost (Bretaud et al., 2004). MPTP-induced PD represented a more consistent model of disease compared with other neurotoxins such as 6-hydoxydopamine and rotenone (Vaz et al., 2018).

The targeted drug delivery is a prominent and efficient technique among all other PD treatment. The successful rate of this treatment is mainly based on selection of drug carrier. The drug carrier should be environmentally friendly, have a definite structure, less toxicity, high solubility, and easily penetrate the blood-brain barrier (BBB) (Ahlawat et al., 2020). The PAMAM dendrimers are monodispersed nanocarriers and are used in different central nervous system (CNS) related diseases, such as Parkinson's, Alzheimer's, Huntington's disease, and brain tumors (Florendo et al., 2018). Due to its specific properties, we use this as a drug carrier. Milowska et al. investigated the inhibition effect of PAMAM G4 and carbosilane dendrimers against α-synuclein (Milowska et al., 2015). The Pramipexole dihydrochloride loaded chitosan nanoparticles against PD via nose-to-brain pathways studied by Raj et al. (2018). Similarly, Sorokina et al. found the effect on cationic pyridyl phenylene dendrimers against beta amyloid protein (2019). The in vitro release and ex vivo permeation analysis of rotigotine NEs and chitosan-coated rotigotine NEs shows the prominent activity against PD (Choudhury et al., 2019), but the lack of penetration of the molecule into BBB is limited. So, the development of new designed PAMAM dendrimers loaded nanoparticles is urgently needed. In the present study, the virtual screening analysis is performed to identify a potential drug molecule against PD targets. From this, we perform molecular docking and molecular dynamics simulation and QM/MM analysis to find the intermolecular interactions and stability of the drug molecules in the active site of PD target. The FGF-PAMAM dendrimer is synthesized, which acts as a drug carrier against PD. In which, the classical PD drugs, designed drug models, and EBN molecules are loaded; this setup is used as a drug nanocarrier bioconjugate. The efficiency of the drug nanocarrier bioconjugate is analyzed with the help of a zebrafish model.

19.3 EDIBLE BIRD'S NEST (EBN)

19.3.1 Introduction and Background

A swiftlet is a unique, small insectivorous bird using its specialized salivary gland secretion in building its nest. Swiftlet species mainly live in the southeast Asia region (Macron, 2005; Chua et al., 2016). Its nest, edible bird's nest (EBN), is a valuable precious natural food built from the salivary secretion of glutinous glycoprotein through sublingual glands, mainly by male swiftlets during their nesting and breeding season. Swiftlet's nests are produced by more than 24 different species all over the world; however, only a few of them are edible. The *Collocalia*, *Aerodramus*, and *Hydrochous* are the three main genuses of swiftlets that produce EBN (Wong, 2013). *Aerodramus fuciphagus* produces a white nest, and *Aerodramus maximus* produces a black nest; both are the two highly exploited species of *Aerodramus* and the most widely identified swiftlets in Malaysia. *Aerodramus fuciphagus* is the most valuable type; it contains about 70–90% mucous secretion, with small percentages of feathers and insects that feed on the nest (Wong, 2013; Chua et al., 2016).

The Chinese community has regarded EBN since the Tang dynasty in AD681 as a precious food and medicine known as "Caviar of the East" (Macron, 2005). EBN soup was prepared by double boiling with rock sugar and was only available for the emperor and wealthy people in the olden days. Traditionally, it was widely used by the Chinese for its nutritional value and health beneficial effect for a thousand years despite being regarded as a costly traditional medicine. They maintain the convention that EBN is a balanced food and beauty enhancer, effective for various diseases of respiratory and gastrointestinal systems. Additionally, it increases immunity and improves ageing skin. Studies also showed that EBN is a potential curative agent in cancer, stomach ulcers, asthma, and cough (Kong et al., 1987; Ma and Liu, 2012; Hobbs, 2004). Lately, EBN was showed to have an antiviral effect by inhibiting influenza infection (Haghani et al., 2016, 2017) and neuroprotective activity (Xie et al., 2018). EBN also benefits as an antioxidant, anti-inflammatory agent, and can improve bone strength (Matsukawa et al., 2011; Yida et al., 2014). Today the awareness towards EBN has been increased worldwide as this salivary secretion is valuable for its therapeutic effect and delicacy (Ma and Liu, 2012). With the aid of modern technology, EBN has been produced as one of the key ingredients in health supplementation food, drinks, and cosmetics products (Kong et al., 1987).

19.3.2 Active Compounds in Edible Bird's Nest

The composition of EBN is different depending on the source (Macron, 2005; Quek et al., 2018). The major biologically active compound in EBN is protein and carbohydrates that are important in assessing the efficacy of EBN. The average raw protein in EBN is about 60% of its weight. The protein is made up of 17 different types of amino acids that is essential for the cells' growth, regeneration, production of the brain neurotransmitter, antibodies production, and immunoglobulin that help the immune system (Macron, 2005; Chua et al., 2014). The variation in EBN amino acid composition is mainly due to the different collection locations and types of cave or EBN's human-made houses (Seow et al., 2016). One study showed that serine, threonine, aspartic acid, glutamic acid, proline, and valine are the most abundant amino acids in EBN (Babji et al., 2018). The aromatic amino acid (tyrosine and phenylalanine) is highly present in white EBN, which has antidepressant and pain-alleviating effects (Macron, 2005). It was found that EBN also contains a 77KDa protein, which has the same properties as the ovotransferrin in chicken egg. It is responsible for the allergic reactions in children that consume EBN (Macron, 2005). The unique combination of protein and carbohydrates in EBN provides remarkable health benefits to humans (Macron, 2005; Xin et al., 2014; Babji et al., 2018). In 2017, a report indicated that mucin-like protein is the main hallmark peptide of EBN, which may help for the classification of EBN via its color and collection places (Wong et al., 2017). The gastric fluid and acidic enzyme digestion of EBN causes the release of protein that enhances the nutritional value of EBN (Wong et al., 2018).

Carbohydrates are considered the second major component of EBN. According to Kathan and Weeks (1969), 28% of EBN is made up of carbohydrates, including 9% N-acetylneuraminic acid (sialic acid), galactosamine, glucosamine, and galactose N-acetylglucosamine and N-acetyl galactosamine. These compounds are known as glyconutrients that give the unique value of EBN (Wong et al., 2017, 2018; Ma and Liu, 2012; Chua et al., 2015; Norhayati et al., 2010).

The unique indicator that allows the grading of various EBN is its sialic acid, which has the pharmacological effects on human health (Quek et al., 2018). N-acetylneuraminic acid (Neu5Ac or NANA) is the main type of sialic acid found in EBN (Pozsgay et al., 1987; Guo ct al., 2006; Yida et al., 2015b; Zhao et al., 2016). It plays an essential role in brain development, neuronal outgrowth, and synaptic transmission. A diet rich in sialic acid helps in increasing brain cell activity level and improving cognitive functions (Wang, 2012, 2009; Khalida et al., 2019). The amount of sialic acid in EBN is higher than the sialic acid-rich foods such as human milk and chicken egg yolk (Tong et al., 2020; Quek et al., 2018). This high content explains the beneficial effects of EBN in brain development, prevention of influenza virus infection, immunity enhancement, cell proliferation, and neurological improvement (Aswir and Wan Nazaimoon, 2011; Hou et al., 2017; Wong et al., 2018; Xie et al., 2018).

19.3.3 Neuroprotective Effects of Edible Bird's Nest

EBN has been recorded as an antioxidant with the amino acids present in the EBN after hydrolysis. After gastrointestinal digestion, the antioxidant effect of EBN will be enhanced (Yew et al., 2014; Yida et al., 2014). Research carried out by Yida et al. reported that EBN has the ability to reduce the risk of hypercoagulation associated with cardiovascular disease (CVD). The results showed that the EBN-treated group can improve the lipid profile, lower the blood sugar level, and decrease total cholesterol by reducing oxidative stress compared to the control group (Yida et al., 2015b). In 2015, the same research group demonstrated the effect of EBN on high-fat diet (HFD)-induced oxidative stress in a rat model. The result showed EBN could reduce the oxidative stress and inflammation triggered by HFD via transcriptional control of hepatic antioxidant genes expression related to inflammation (Yida et al., 2015c). The results support the effectiveness of EBN in the prevention of inflammation and oxidative stress induced by obesity. In addition, a study done by Ghassem et al. suggested that the protein hydrolysate of EBN possesses antioxidant properties and can scavenge the free radicals (Ghassem et al., 2017). A similar study was reported on the improvement of the level of superoxide dismutase (SOD), estrogen, malondialdehyde, and lipid profile of the ovariectomized rats with 12 weeks of EBN supplementation in the diet (Hou et al., 2017). These findings highlight the value of EBN to prevent the cardiometabolic disease induced by estrogen deficiency. Hu et al. studied the anti-aging effect of EBN in the *Drosophila melanogaster* model. The study showed that EBN could decrease mortality rates and lipid peroxidation via increasing the antioxidant enzyme activity (Hu et al., 2016). Likewise, research carried out by Albishtue and colleagues to evaluate the effect of EBN supplementation on uterine function and embryo implantation rate proved that EBN enhances anti-oxidative activity and decreases oxidative stress level, which enhances embryo implantation (Albishtue et al., 2019).

19.4 METHODS

19.4.1 Animal and Treatment

The adult wild type of *Danio rerio* zebrafish will be housed in the common zebrafish tanks in the central research and animal facility in IIUM (CREAM), in which all groups will be maintained in standard conditions before starting the experiments. EBN will be provided by Glycofood sdn bhd, Balakong, Selangor. EBN solution is prepared by dissolving 1g EBN powder in 100 ml of distilled water (DW), and then heated for 45 minutes at 60°C; the EBN solution was cooled at room temperature and supplied intraperitoneally in two doses depending. EBN will be given in one of two doses 60 mg/kg and 120 mg/kg given daily intraperitoneally for 8 consecutive weeks in the afternoon (Albishtue et al., 2018; Hou et al., 2017; Ismaeil et al., 2021). Deprenyl (Selegiline) will be given intraperitoneally in a dose of 5 mg/kg once daily for 8 consecutive weeks (Abdanipour et al., 2023; Kondeva-Burdina et al., 2022).

19.4.2 Zebrafish Parkinson Disease Model Induction and Validation

The method of induction and validation is adapted from Sarath et al. (2016), in which approximately 90 adult zebrafish will be selected (3–4 cm long, males and females), and all will be intraperitoneally injected with MPTP in two doses of 50 μg of (10 μg/μl) MPTP with 24-hour interval between the doses (total dose is 100 μg). Validation of the disease model will be completed according to Sarath et al. (2016) using real-time QPCR and behavioral analysis. In the RT-QPCR assay, the zebrafish brain tissues will be removed after anesthetizing the fish with 0.1% tricaine solution. After rinsing the brain tissue with isotonic solution, the tissues will be used for protein and RNA extraction following the standard protocol. The target genes include Lrrk7, Park2, Park7, and SNCGB. The internal controls include Beta Actin and GAPDH. Comparison will be made between the MPTP-injected zebrafish and a control (no injection) group to estimate the relative expression of the selected genes. The behavioral analysis of the zebrafish will be completed in a transparent 3L tank using the Danio vision tracing system to measure parameters including speed of swimming, number of freeze bouts, freeze duration, number of crosses between upper and lower sections, and distance travelled. In both methods, the parameters will be expressed as mean ± SEM, and comparison between the two groups (MPTP injected and the control) will be evaluated with independent t-test using SPSS software.

19.4.3 Zebrafish Experimental Settings

The total 100 adult zebrafish that were injected with MPTP and were validated to have features of Parkinson's disease will be divided into eight groups (ten fishes each) as follows (see Figure 19.1):

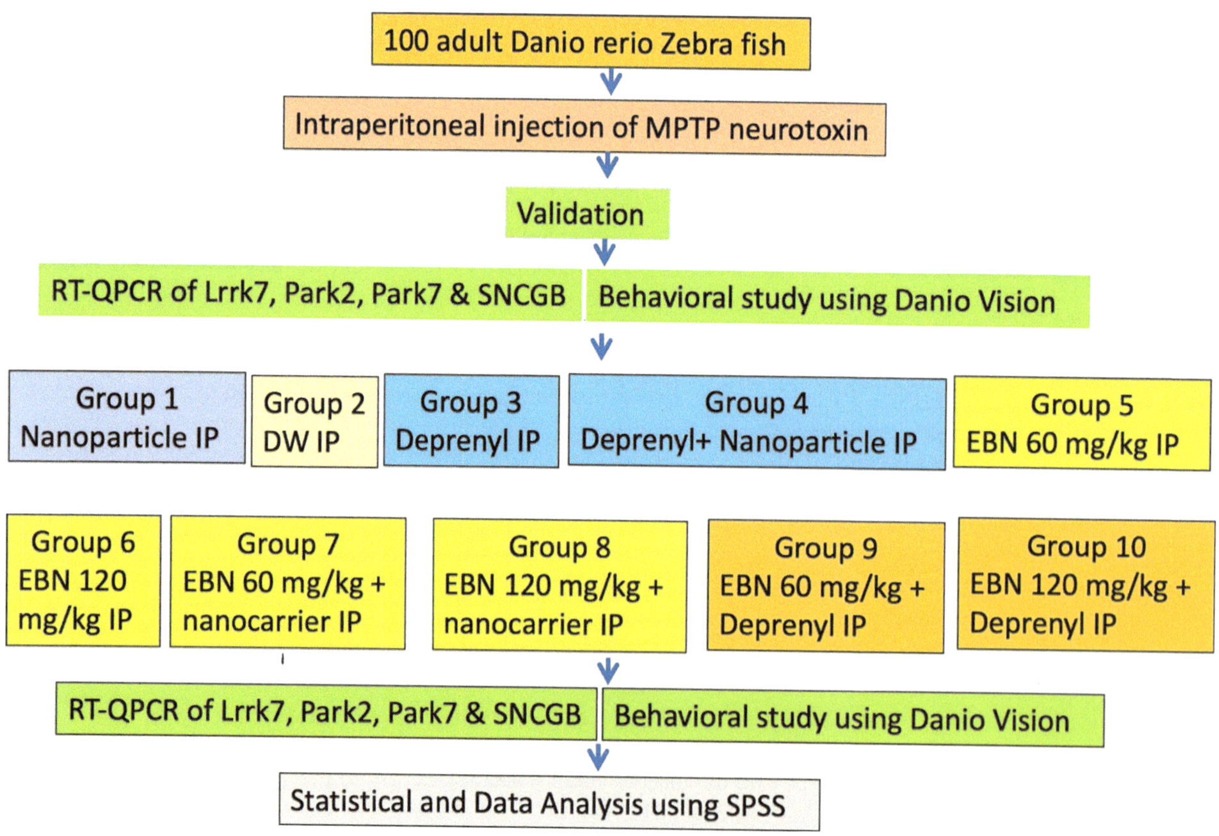

FIGURE 19.1 Flowchart of the experimental design.

Group 1: (sham control) will receive nanoparticles injection without treatment.
Group 2: (vehicle control) will receive a single intraperitoneal injection of DW.
Group 3: (positive control) will receive an intraperitoneal injection of Deprenyl (Selegiline) in a dose of 5 mg/kg once daily for 8 consecutive weeks (Abdanipour et al., 2023; Kondeva-Burdina et al., 2022).
Group 4: will receive an intraperitoneal injection of Deprenyl (Selegiline) in a dose of 5 mg/kg in a nanocarrier once daily for 8 consecutive weeks.
Group 5: will receive an intraperitoneal injection of EBN in 60 mg/kg twice daily for 8 consecutive weeks (Ismaeil et al., 2021).
Group 6: will receive an intraperitoneal injection of EBN in 120 mg/kg twice daily for 8 consecutive weeks (Ismaeil et al., 2021).
Group 7: will receive an intraperitoneal injection of EBN in 60 mg/kg in nanocarrier twice daily for 8 consecutive weeks.
Group 8: will receive an intraperitoneal injection of EBN in 120 mg/kg in nanocarrier twice daily for 8 consecutive weeks.
Group 9: will be treated for 8 consecutive weeks with an intraperitoneal injection of EBN in 60 mg/kg in nanocarrier twice daily and an intraperitoneal injection of Deprenyl (Selegiline) 5 mg/kg in a nanocarrier once daily.
Group 10: will be treated for 8 consecutive weeks with an intraperitoneal injection of EBN in 120 mg/kg in nanocarrier twice daily and an intraperitoneal injection of Deprenyl (Selegiline) 5 mg/kg in a nanocarrier once daily.

After ending the treatment, responses will be evaluated using gene expression analysis, western blotting, and behavioral studies as explained earlier using the same parameters.

19.4.4 Zebrafish Brain Tissue Specimens

As mentioned earlier, the brain tissues will be carefully dissected after anesthetizing the fish with 0.1% tricaine solution. After rinsing the brain tissue with isotonic solution, the tissues will be used for protein and RNA extraction following the standard protocol.

19.4.5 Gene Expression Analysis

The RNA extraction will be completed using TRI reagent (Sigma, USA). Then, cDNA formation and then RT- PCR using target genes in addition to two internal control genes. The amplicon sequence is listed in Table 19.1.

In the Western blotting, proteins will be extracted from the pooled brain tissue samples (10 brains) using RIPA lysis buffer with sonication for homogenization. Proteins will be quantified using bicinchoninic acid assay (BCA assay). The 50 μg of total protein will be resolved on the SDS-PAGE gel and then will be transferred to PVDF membrane by electroblotting. Membranes will be incubated with the primary antibody to SNCG (A and B) at respective optimum dilution for 1–2 hours at room temperature. After washing, the membrane will be incubated with their respective HRP tagged secondary antibody, followed by washing. Then, the membrane will be processed following the manufacturer's instructions for the chemiluminescent detection.

19.4.6 OMICS Analysis

We propose the inclusion of a transcriptomic study to unravel the potential mechanism of action of EBN and correlate the observed biological pathways with the phenotype. There are several reasons to consider incorporating transcriptomics into the study:

- Conducting a mechanistic study can address the "fundamental" scope of FRGS.
- By pinpointing specific protein markers through transcriptomics, we can avoid a trial-and-error approach in IHC.
- With the use of advanced bioinformatics, we can potentially identify new pathways regulated by EBN and correlate them with the phenotype to explain the observed results.

TABLE 19.1
Primer Sequences for RT-QPCR

Gene IDs	Forward Primer	Reverse Primer
Leucin-rich repeat kinase 2 Dardarin (Lrrk7)	ACTCGGATTAAGTTCCACCAGA	GAAAGAGGTGTTGAAGGACCAG
Parkinson protein 7 (DJ1) (Park 7)	ACAGACATCATGACTCCAGTGC	GACTAAAGCTGGGGTTGAAGAG
Parkin RBR E3 ubiquitin protein ligase (Park 2)	GGCATCACACCTTCTACAATGA	GTGGAGTCTACTGGTGTCTTC
Synuclein, gamma b (SNCGB)	CAGTGAGGGTTGATGGTCTGTA	TGATCACGTTACCATCCTTCTG
Actin	ACACGGAAATGATGAACCTCTT	CGTTCTCCAGTCCCTCTACTGT
GAPDH	TACGACCAGAAGCGTACAGAGA	GTGCAGGAGGCATTGCTTACA

19.4.7 Methodology for Most of the RNA Sequencing

RNA samples were analyzed for purity and processed using the Illumina—Stranded Total RNA Prep, Ligation with Ribo-Zero Plus (Illumina, CA, USA), as according to manufacturer's instructions. Samples were sequenced on the Illumina Novaseq 6000 system (Illumina, CA, USA). Libraries were normalized to 0.5 nM and pooled before loading into the NovaSeq sequencing platform. Demultiplexing of samples was performed using the Illumina DRAGEN Bio-IT Platform (Illumina) v3.9. Illumina DRAGEN includes an RNA-seq (splicing-aware) aligner, as well as RNA specific analysis components for gene expression quantification and gene fusion detection. Transcript quantification results were reported in the quant.sf text file. To perform differential gene expression (DEG) analysis, the output file was used as an input for differential gene expression using tools such as tximport and DESeq2.

19.5 PREPARATION OF NANOPARTICLES

19.5.1 Virtual High-Throughput Screening (VHTS)

The virtual screening analysis will be performed using the Autodock vina software package (Oleg et al., 2010). To perform the virtual screening, we require a pdbqt file format for both ligand and protein molecules. First, the receptor (α-Synuclein, MAO-B, COMT enzyme etc.) will be prepared using the Chimera software package (Pettersen et al., 2004). The ions, water molecules beyond the active site, cofactor, and extra chains will be removed from the receptor. Further, the hydrogen molecules and the charges will be added using the Chimera package, and it will be saved as a pdbqt file format. The grid will be generated for prepared receptor molecules to perform the virtual screening analysis. The ligand molecules will be downloaded from ZINC, PubChem, and DrugBank, which contains millions of natural and synthetic molecules (Teague and Joh, 2015; Wishart et al., 2018). The hydrogens, assign rotatable bonds, and gasteiger charges will be added using the Openbabel software package (O'Boyle et al., 2011); prepared ligand molecules are saved as a pdbqt format. The grid parameter defines the active site of receptor molecules; final poses are selected based on best binding energy values. The intermolecular interactions between the active site amino acid residues and the ligand molecules will be analyzed using a discovery studio visualizer (Biovia, 2020). The best conformer will be selected from the virtual screening analysis, and it will be used to perform molecular dynamics (MD) simulation and binding free energy analysis.

19.5.2 Molecular Dynamics Simulation Receptor and Ligand Preparation

The AMBER20 software package will be used to perform the molecular dynamics simulation analysis [8]. The receptor molecule will be prepared using the AMBER ff14SB force field, and the final topology and coordinates files of the receptor will be generated using a tleap module (Case et al., 2020). The generated topology files contain the molecular topology and force field parameters, and the coordinates file contains the information about the description of the atomic coordinates, initial velocities, and periodic box dimensions. The antechamber module available in the Ambertools20 package will be used to generate force field parameters of the ligand molecules with general AMBER (GAFF) force field. The TIP3P water molecules will be used to solvate the system with 10 Å each direction of the solute. Furthermore, the complex system will be minimized with steepest descent (500 steps) and conjugate gradient (1500 steps) methods to remove steric clashes. The minimized system will be annealed using NVT ensemble with 300 K up to 500 ps. Furthermore, the equilibration will be performed with constant temperature (300 K) and constant pressure (1 bar) using an NPT ensemble up to 1 ns. The final production phase will be performed up to 500 ns using a Langevin thermostat and Berendsen barostat with the 2 fs time step. The final trajectory files will be analyzed using the CPPTRAJ module available in the AMBERTOOLS20 package (Roe and Thomas, 2013).

19.5.3 MD Simulation Analysis

From the molecular dynamics (MD) simulation, the stability of the drug molecules will be determined; it contains four main steps: (1) energy minimization, (2) heating, (3) equilibration, and (4) production. The final production will be performed up to 500 ns to understand the stability of the protein-ligand complex using NPT ensemble with fixed temperature (300 K) and pressure (1 atm) and periodic boundary conditions. The root means square deviation (RMSD), root mean square fluctuations (RMSF), and radius of gyration (Rg) of the protein-ligand complex will be monitored separately for every 10 ns. The binding free energy and decomposition energy analysis will be calculated from the trajectory files.

19.5.4 Binding Free Energy Calculation

The binding free energies of the protein-ligand complex will be determined using MM/GBSA and MM/PBSA analysis. For the binding free energy calculation, 1,000 snapshots will be extracted from the MD production trajectories in which coordinates and data are saved 10 ns. The binding free energy energies of the complex will be calculated using the MMPBSA.py module (Miller et al., 2012).

19.5.5 Quantum Mechanics and Molecular Mechanics (QM/MM) Calculations

The enzymatic mechanism of the protein-ligand complex will be analyzed using hybrid QM/MM approaches. The two-layered ONIOM method encoded in the Gaussian16 program will be used (Chung et al., 2015) to calculate the enzymatic mechanism, in which the inner layer consists of the most critical parts involved in the enzymatic mechanism, and the rest of the system comprises the outer layer. The DFT method B3LYP will be used for the inner or model layer; it includes the key interacting residues of the receptor. The outer layer is treated using a standard molecular mechanic's force field – normally UFF. QM/MM methodology is one of the prominent techniques, and it calculates the enzymatic mechanism at a high accuracy level. The total energy of the system is calculated from the following equation:

$$E\,(\text{ONIOM, real}) = E\,(\text{high, model}) + E\,(\text{low, real}) - E\,(\text{low, model})$$

19.5.6 Synthesis of FGF-PAMAM Dendrimer

Fibroblast growth factor (FGF) is a polypeptide capable of crossing the blood brain-barrier (BBB) (Cuevas et al., 1998; Dordoe et al., 2021). When FGF is conjugated with polymers, the polymers are helped to cross the BBB (Paliwal et al., 2011; Lee et al., 2000). In the present proposal, FGF will be conjugated with 3G and/or 4G PAMAM dendrimers. The amine group of 3G and/or 4G PAMAM dendrimers will be conjugated with the carboxylate group of FGF (Shi et al., 1997) using the method stated by Paliwal et al. (2011). FGF is mixed with DCC and NHS in DMF. The ratio of FGF, DCC, and NHS are 1:1.6:1.6 as per recommended by Lee et al. (2000). Both DCC and NHS are needed slightly more to activate FGF and ensure conjugation reactions are complete. The mixture will then be left in 4°C temperature for 5 hours in a vacuum space. The precipitated dicyclohexylurea will then be filtered after 5 hours. The unreacted DCC will be removed by precipitation, followed by the purification of FGF-PAMAM by passing through a G-25 silica gel column. The synthesis of FGF-PAMAM will be confirmed by using an IR spectroscopic method. The NMR spectroscopy of the FGF-PAMAM dendrimer was carried out at 300 MHz after dissolving in CdCl3. A color reaction will be performed to confirm the degree of FGF sylation of the FGF-PAMAM dendrimer using the ninhydrin Kaiser test (Kumar et al., 2006) (Figure 19.2).

19.5.7 Drug Loading in Formulations

The known molar concentrations of the FGF-PAMAM dendrimer will be dissolved separately in methanol and mixed with the methanolic solution of the drug (approximately 10 mol). The mixed solutions will be incubated with slow magnetic stirring (50 rpm) using teflon beads for 24 hours. These solutions will be twice dialyzed in a cellulose dialysis bag (MWCO 1000 Da Sigma, Germany) against double-distilled water under sink conditions for 10 minutes to remove free drug from the formulations, which will be

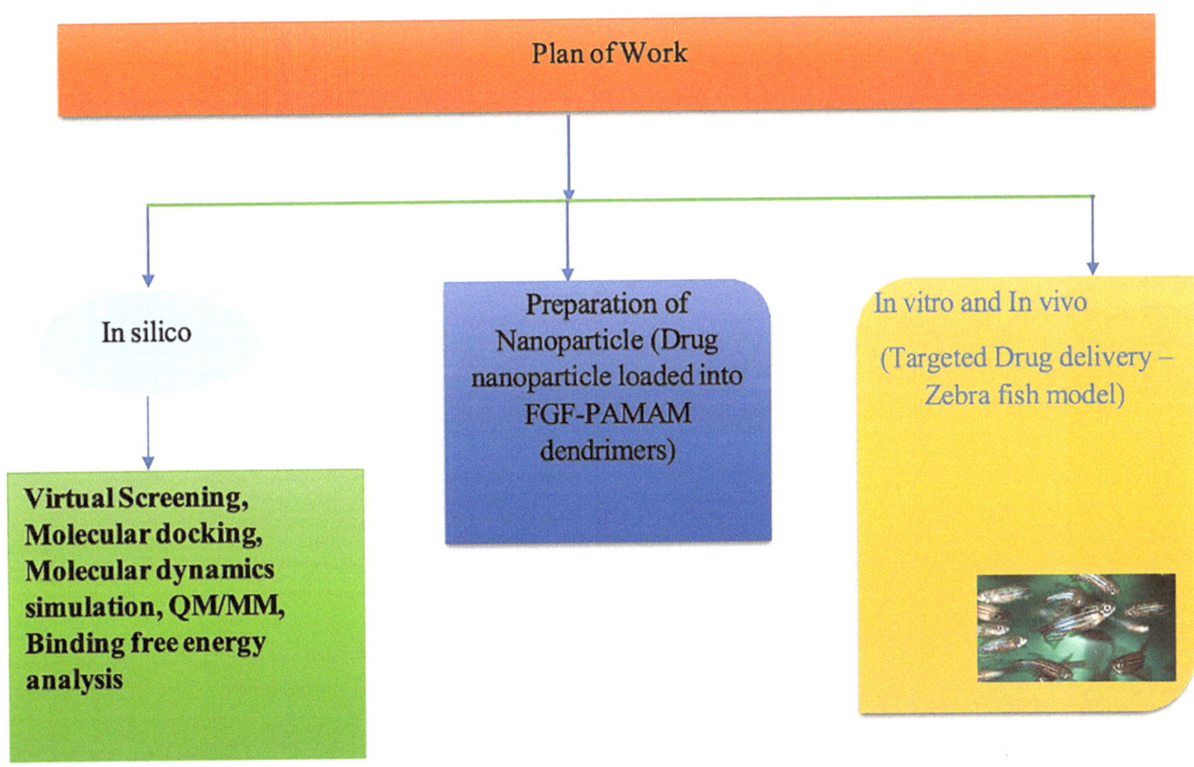

FIGURE 19.2 Plan of the work and work methodology.

estimated spectrophotometrically and/or using HPLC to determine the amount of drug loaded within the system. The dialyzed formulations will be lyophilized and used for further characterization (Kumar et al., 2007).

19.5.8 Morphology of the Drug-Loaded FGF-PAMAM Dendrimers

Transmission electron microscopy (TEM) will be performed to investigate the particle size and provide information on nanoparticle morphology. Prepared and dialyzed drug-loaded dendrimer formulations will be used for transmission electron microscopic studies. The TEM studies will be carried out using a 3 mm Forman (10.5% plastic powder in amyl acetate) coated copper grid (300 mesh) at 60 Kv using negative staining by 2% phosphotungstic acid (PTA) for drug-loaded FGF-PAMAM dendrimers at 150,000 times magnification at 50–60 Kv (Kumar, 2007).

19.5.9 Differential Scanning Calorimetry

Differential scanning calorimetry (DSC) will be performed to study the thermal stability and changes in crystallinity over a range of temperatures. The FGF-PAMAM dendrimer, drug, and FGF-drug FGF-PAMAM dendrimer particles will be studied by this method. A known mass of powder will be placed in an aluminium pan, and a lid will be crimped onto the pan. The pan will then be placed in the sample cell of a DSC module. The temperature of the DSC module will be equilibrated at 35°C and then increased at a rate of 10°C/min under a N2 gas purge until the material begins to degrade. The temperatures will be obtained for each peak in the resulting curve and provide indications of temperature stability and phase transitions (Lee et al., 2000).

19.6 STATISTICAL ANALYSIS

All of the data obtained from the experiment will be tabulated as mean ± SEM. The comparison between groups will made using one-way ANOVA with post hoc test analysis. A p-value of less than 0.05 is considered statically significant. The analysis will be completed using SPSS software version 17 (IBM, USA).

REFERENCES

Abdanipour, A., Mirzaei, M., Anarkooli, I. J., Mohammadi, P. (2023). Effect of selegiline as a monomine oxidase B inhibitor on the expression of neurotrophin mRNA levels in a contusion rat model of spinal cord injury. Neurological Research. 45(3):241–247. doi: 10.1080/01616412.2022.2129761.

Ahlawat, J., Gileydis, G. B., Shima, M. A., Melinda, A., Isabela, A., Jose, B., Ximena, C., Stephanie, C., Paulina, C., Vassti, E., Savana, E., Daniel, F., Carolina, L., Martin, M., Nabeel, A., Sergio, N., Karyme, O., Ximena, P., Brianna, P., Andrea, R., Raul, R., Veronica, R., Paola, R., Jessica, R., David, S., Graciela, U., Stephanie, V., Mahesh, N. (2020). Nanocarriers as potential drug delivery candidates for overcoming the blood–brain barrier: challenges and possibilities. ACS Omega. 5:12583–12595.

Albishtue, A. A., Yimer, N., Zakaria, M. Z. A., Haron, A. W., Yusoff, R., Assi, M. A., Almhanawi, B. H. (2018). Edible bird's nest impact on rats' uterine histomorphology, expressions of genes of growth factors and proliferating cell nuclear antigen, and oxidative stress level. Veterinary World. 11(1):71–79.

Albishtue, A. A., Yimer, N., Zakaria, M. Z. A., Haron, A. W., Babji, A. S., Abubakar, A. A., Almhanawi, B. H. (2019). Effects of EBN on embryo implantation, plasma concentrations of reproductive hormones, and uterine expressions of genes of PCNA, steroids, growth factors and their receptors in rats. Theriogenology. https://doi.org/10.1016/j.theriogenology.2018.12.026.

Aswir, A. R., Wan Nazaimoon, W. M. (2011). Effect of edible bird's nest on cell proliferation and tumor necrosis factor-alpha (TNF-α) release in vitro. International Food Research Journal. 18(3):1123–1127.

Babji, A. S., Etty Syarmila, I. K., Nur Aliah, D., Nurul Nadia, M., Hadi Akbar, D., Norrakiah, A. S., Ghassem, M., Najafian, L. and Salma, M. Y. (2018). Assessment on bioactive components of hydrolysed edible bird nest. International Food Research Journal. 25(5):1936–1941.

BIOVIA. Dassault Systems, BIOVIA Workbook, Release 2020; BIOVIA Pipeline Pilot, Release 2020, San Diego: Dassault Systèmes, 2020.

Blauwendraat, C., Nalls, M. A., Singleton, A. B. (2020). The genetic architecture of Parkinson's disease. Lancet Neurology. 19(2):170–178.

Bretaud, S., Lee, S., Guo, S. (2004). Sensitivity of zebrafish to environmental toxins implicated in Parkinson's disease. Neurotoxicology and Teratology. 26(6):857–864.

Case, D. A., et al. (2020). AMBER. San Francisco: University of California, 2020.

Choudhury, H., Nur Fadhilah, B. Z., Puteri Atdriann, B. T., Angeline, S. T., Manisha, P., Bappaditya, C., Nabil, A. A., Subrat, K. B., Prashant, K., Bapi, G., Shadab, M. (2019). Formulation development and evaluation of rotigotine mucoadhesive nanoemulsion for intranasal delivery. Journal of Drug Delivery Science and Technology. 54:101301.

Chua, K. H., Lee, T. H., Kamini, N., Nor Hamdan, M. Y., Lee, C. T., Eddie, T. T. T., Chau, L. S., Zukefl, S. N. (2016). A comprehensive review on edible bird nests and swiftlet farming. Journal of Integrative Medicine. 14(6):415–428.

Chua, Y. G., Bloodworth, B. C., Leong, L. P., Li, S. F. Y. (2014). Metabolite profiling of edible bird's nest using gas chromatography/mass spectrometry and liquid chromatography/mass spectrometry. Rapid Communications in Mass Spectrometry. 28(12):1387–1400. https://doi.org/10.1002/rcm.6914.

Chua, Y. G., Chan, S. H., Bloodworth, B. C., Li, S. F. Y., Leong, L. P. (2015). Identification of edible birds nest with amino acid and monosaccharide analysis. Journal of Agricultural and Food Chemistry. 63(1):279–289. https://doi.org/10.1021/jf503157n

Chung, L. W., Sameera, W. M. C., Ramozzi, R., Page, A. J., Hatanaka, M., Petrova, G. P., Harris, T. V., Li, X., Ke, Z., Liu, F., Li, H.-B., Ding, L., Morokuma, K. (2015). The ONIOM method and its applications. Chemical Reviews. 115:5678–5796.

Cuevas, P., Carceller, F., Muñoz-Willery, G.-G. (1998). Intravenous fibroblast growth factor penetrates the blood-brain barrier and protects hippocampal neurons against ischemia-reperfusion injury. Surgical Neurology. 49(1):77–84.

DeMaagd, G., Philip. (2015). A. Parkinson's disease and its management: part 1: Disease entity, risk factors, pathophysiology, clinical presentation, and diagnosis. Pharmacy and Therapeutics. 40 (8):504.

Dordoe, C., Chen, K., Huang, W., Chen, J., Hu, J., Wang, X., Lin, L. (2021). Roles of fibroblast growth factors and their therapeutic potential in treatment of ischemic stroke. Frontiers in Pharmacology. 12:671131.

Dorsey, E. R., Elbaz, A., Nichols, E., Abbasi, N., Abd-Allah, F., Abdelalim, A. (2018). Global, regional, and national burden of Parkinso's disease, 1990–2016: A systematic analysis for the Global Burden of Disease Study 2016. The Lancet Neurology. 17(11):939–953. https://doi.org/10.1016/S1474-4422 (18)30295-3.

Florendo, M., Alexander, F., Bhairavi, S., Ajit, S., Douglas, S., Gary, L. D., Julien, R. (2018). Use of polyamidoamine dendrimers in brain diseases. Molecules. 23(9):2238.

Forno, L., DeLanney, L., Irwin, I., Langston, J. (1993). Similarities and differences between MPTP-induced parkinsonsim and Parkinson's disease. Neuropathologic considerations. Advances in Neurology. 60:600–608.

Ghassem, M., Arihara, K., Mohammadi, S., Sani, N. A., Babji, A. S. (2017). Identification of two novel antioxidant peptides from edible bird's nest (Aerodramus fuciphagus) protein hydrolysates. Food and Function. 8(5):2046–2052. https://doi.org/10.1039/c6fo01615d.

Guo, C. T., Takahashi, T., Bukawa, W., Takahashi, N., Yagi, H., Kato, K., Hidari, K. I. P. J., Miyamoto, D., Suzuki, T., Suzuki, Y. (2006). Edible bird's nest extract inhibits influenza virus infection. Antiviral Research. 70(3):140–146. https://doi.org/10.1016/j.antiviral.2006.02.005.

Haghani, A., Mehrbod, P., Safi, N., Aminuddin, N. A., Bahadoran, A., Omar, A. R., Ideris, A. (2016). In vitro and in vivo mechanism of immunomodulatory and antiviral activity of Edible Bird's Nest (EBN) against influenza a virus (IAV) infection. Journal of Ethnopharmacology. 185:327–340. https://doi.org/10.1016/j.jep.2016.03.020.

Haghani, A., Mehrbod, P., Safi, N., Kadir F. A., Omar, A. R., Ideris, A. (2017). Edible bird's nest modulate intracellular molecular pathways of influenza A virus infected cells. BMC Complementary and Alternative Medicine. 17(1). https://doi.org/10.1186/s12906-016-1498-x.

Hobbs, J. J. (2004). Prolems in harvest of edible birds'nest in Sarwak and Sabah, Malaysia & Borneo. Biodiversity & Conservation. 13:2209–2226.

Hou, Z., He, P., Imam, M. U., Qi, J., Tang, S., Song, C., Ismail, M. (2017). Edible bird's nest prevents menopause-related memory and cognitive decline in rats via increased hippocampal sirtuin-1 expression. Oxidative Medicine and Cellular Longevity. 2017:7205082.

Hu, Q., Li, G., Yao, H., He, S., Li, H., Liu, S., Wu, Y., Lai, X. (2016). Edible bird's nest enhances antioxidant capacity and increases lifespan in Drosophila Melanogaster. Cellular and Molecular Biology (Noisyle-grand). 62(4):116–122.

Ismaeil, R., Affandi, K., Hui, C., Mohamed, W., Fadly, M. (2021). Neuroprotective effect of edible bird's nest on chronic cerebral hypoperfusion induced neurodegeneration in rats. Clinical Neurophysiology. 132(8):e121-e122, https://doi.org/10.1016/j.clinph.2021.02.299.

James, S., Reddy, V., Beato, M. Neuroanatomy, Substantia Nigra. Treasure Island, FL: StatPearls Publishing, 2022.

Kathan, R. H., Weeks, D. I. (1969). Structure studies of collocalia mucoid. I. Carbohydrate and amino acid composition. Archives of Biochemistry and Biophysics. 134(2), 572–576. https://doi.org/10.1016/0003-9861(69)90319-1.

Khalida, S. K. A., Rashed, A. A. Azizc, S. A., Hafandi, A. (2019). Effects of sialic acid from edible bird nest on cell viability associated with brain cognitive performance in mice. World Journal of Traditional Chinese Medicine. 5(4):214.

Kong, Y. C., Keung, W. M., Yip, T. T., Ko, K. M., Tsao, S. W., Ng, M. H. (1987). Evidence that epidermal growth factor is present in swiftlet's (Collocalia) nest. *Comparative* Biochemistry and Physiology–Part B: Biochemistry. 87(2):221–226. https://doi.org/10.1016/0305-0491(87)90133-7

Kondeva-Burdina, M., Mateev, E., Angelov, B., Tzankova, V., Georgieva, M. (2022). In silico evaluation and in vitro determination of neuroprotective and MAO-B inhibitory effects of pyrrole-based hydrazones: A therapeutic approach to Parkinson's disease. Molecules. 27(23):8485. doi: 10.3390/molecules27238485.

Korczyn, A. D. (2004). Drug treatment of Parkinson's disease. Dialogues in Clinical Neuroscience. 6(3):315–322. doi: 10.31887/DCNS.2004.6.3/akorczyn

Kouli, A., Torsney, K. M., Kuan, W. L. (2018). Parkinson's Disease: Etiology, Neuropathology, and Pathogenesis, 3–26. Exon Publications, 2018. https://doi.org/10.15586/codon-publications.parkinsonsdisease. 2018.ch1.

Kumar, P. V., Agashe, H., Dutta, T., Jain, N. K. (2007). PEGylated dendritic architecture for development of a prolonged drug delivery system for an antitubercular drug. Current Drug Delivery. 4(1):11–19.

Kumar, P. V., Asthana, A., Dutta, T., Jain, N. K. (2006). Intracellular macrophage uptake of rifampicin loaded mannosylated dendrimers. Journal of Drug Targeting. 14(8):546–556.

Leathem, A., Ortiz-Cerda, T., Dennis, J. M., Witting, P. K. (2022). Evidence for oxidative pathways in the pathogenesis of PD: Are antioxidants candidate drugs to ameliorate disease progression? International Journal of Molecular Sciences. 23(12):6923. doi: 10.3390/ijms23136923

Lee, Y., Kim, S. H., Byun, Y. (2000). Oral delivery of new heparin derivatives in rats. Pharmaceutical Research. 17(10):1259–1264.

Ma, F., Liu, D. (2012). Sketch of the edible bird's nest and its important bioactivities. Food Research International. 48:559–567.

Macron, M. F. (2005) Characterization of edible bird's nest the "caviar of East". Food Research International. 1125–1134.

Matsukawa, N., Matsumoto, B. W, Chiji, H., Nakayama, K., Hara, H., Tsukahara, T. (2011). Improvement of bone strength and dermal thickness due to dietary edible bird's nest extract in ovariectomized rats. Bioscience, Biotechnology, and Biochemistry. 75(3):590–592.

Miller, B. R. 3rd, McGee, T. D., Jr., Swails, J. M., Homeyer, N., Gohlke, H., Roitberg, A. E. (2012). MMPBSA.py: An efficient program for end-state free energy calculations. Journal of Chemical Theory and Computation. 8:3314–3321.

Milowska, K, Aleksandra, S., Marta, M., Rafael, G.-R., de la Mata, F. J., Teresa, G., Maria, B. (2015). Carbosilane dendrimers inhibit a-synuclein fibrillation and prevent cells from rotenone-induced damage. International Journal of Pharmaceutics. 484:268–275.

Nam, M.-H., Sa, M., Ju, Y. H., Park, M. G., Lee, C. J. (2022). Revisiting the role of astrocytic MAOB in Parkinson's disease. International Journal of Molecular Sciences. 23:4453. doi: 10.3390/ijms23084453.

Norhayati, M. K., Jr., Azman, O., Wan Nazaimoon, W. (2010). Preliminary study of the nutritional content of Malaysian edible bird's nest. Malaysian Journal of Nutrition. 16(3):389–396.

O'Boyle, N., Michael, B., Craig, A. J., Chris, M., Tim, V., Geoffrey, R. H. (2011). Open babel: An open chemical toolbox. Journal of Cheminformatics. 3:33, 1–14.

Oleg, T., Arthur, O. (2010). Autodock vina: Improving the speed and accuracy of docking with a new scoring function, efficient optimization, and multithreading. Journal of Computational Chemistry. 31:455–461. doi: 10.1002/jcc.21334

Paliwal, R., Paliwal, S. R., Agrawal, G. P., Vyas, S. P. (2011). Biomimetic solid lipid nanoparticles for oral bioavailability enhancement of low molecular weight heparin and its lipid conjugates: in vitro and in vivo evaluation. Molecular Pharmaceutics. 8(4):1314–1321.

Pettersen, E., Thomas, D. G., Conrad, C. H., Gregory, S. C., Daniel, M. G., Elaine, C. M., Thomas, E. F. (2004). UCSF chimera—a visualization system for exploratory research and analysis. Journal of Computational Chemistry. 25:1605–1612.

Pozsgay, V., Jennings, H., Kasper, D. L. (1987). 4,8-Anhyro-N acetylneuraminic acid. Isolation from edible bird's nest and structure deremination. European Journal of Biochemistry. 162(2):445–450. https://doi.org/10.1111/j.1432-1033.1987.tb10622.x.

Quek, M. C., Chin, N. L., Yusof, Y. A., Law, C. L., Tan, S. W. (2018). Characterization of edible bird's nest of different production, species and geographical origins using nutritional composition, physicochemical properties and antioxidant activities. Food Research International. 109:35–43. https://doi.org/10.1016/j.foodres.2018.03.078.

Raj, R., Sarika, W., Vinay, S., Ram, G. (2018). Pramipexole dihydrochloride loaded chitosan nanoparticles for nose to brain delivery: Development, characterization and in vivo anti-Parkinson activity. International Journal of Biological Macromolecules. 109:27–35.

Rizek, P., Kumar, N., Jog, M. (2016). An update on the diagnosis and treatment of Parkinson disease. Canadian Medical Association Journal. 188:1157–1165. doi: 10.1503/cmaj.151179

Roe, D. R., Thomas, E. C. 3rd. (2013). PTRAJ and CPPTRAJ: Software for processing and analysis of molecular dynamics trajectory data. Journal of Chemical Theory and Computation. 9:3084–3095.

Sarath, B. N., Murthy, C. L. N., Kakara, S., Sharma, R., Brahmendra Swamy, C. V., Idris, M. M. (2016). 1-Methyl-4-phenyl-1, 2, 3, 6-tetrahydropyridine induced Parkinson's disease in zebrafish. Proteomics. 16(9), 1407–1420.

Savica, R., Grossardt, B. R., Rocca, W. A., Bower, J. H. (2018). Parkinson disease with and without dementia: A prevalence study and future projections. Movement Disorders. 33:1–7. doi: 10.1002/mds.27277.

Seow, E. K., Ibrahim, B., Muhammad, S. A., Lee, L. H. Cheng, L.H. (2016). Differentiation between house and cave edible bird's nests by chemometric analysis of amino acid composition data. LWT Food Science and Technology. 65:428–435.

Shi, J., Friedman, S., Maciag, T. (1997). A carboxyl-terminal domain in fibroblast growth factor (FGF)-2 inhibits FGF-1 release in response to heat shock in vitro. Journal of Biological Chemistry. 272(2):1142–1147.

Sorokina, S. A., Stroylova, Y. Y., Tishina, S. A., Shifrina, Z. B., Muronetz, V. I. (2019). Promising anti-amyloid behaviour of cationic pyridylphenylene dendrimers: Role of structural features and mechanism of action. European Polymer Journal. 116:20–29.

Speciale, S. G. (2002). MPTP: Insights into parkinsonian neurodegeneration. Neurotoxicology and Teratology. 24(5):607–620.

Tan, Y. Y., Jenner, P., Chen, S. D. (2022). Monoamine oxidase-B inhibitors for the treatment of Parkinson's disease: Past, present, and future. Review. Journal of Parkinson's Disease. 12:477–493. doi: 10.3233/JPD-212976.

Teague, S., Joh, I. (2015). ZINC 15–Ligand discovery for everyone. Journal of Chemical Information and Modeling. 55:2324–2337.

Tong, S. R., Lee, T. H., Cheong, S. K., Lim, Y. M. (2020). Untargeted metabolite profiling on the water-soluble metabolites of edible bird's nest through liquid chromatography-mass spectrometry. Veterinary World, 13(2):304–316. https://doi.org/10.14202/vetworld.2020.304–316.

Vaz, R. L., Outeiro, T. F., Ferreira, J. J. (2018). Zebrafish as an animal model for drug discovery in Parkinson's disease and other movement disorders: A systematic review. Frontiers in Neurology. 9:347.

Wang, B. (2009). Sialic acid is an essential nutrient for brain development and cognition. In Annual Review of Nutrition. 29:177–222. https://doi.org/10.1146/annurev.nutr.28.061807.155515.

Wang, B. (2012). Molecular mechanism underlying sialic acid as an essential nutrient for brain development and cognition. Advances in Nutrition. 3(3). https://doi.org/10.3945/an.112.001875.

Wishart, D. S, Feunang, Y. D., Guo, A. C., Lo, E. J., Marcu, A., Grant, J. R., Sajed, T., Johnson, D., Li, C., Sayeeda, Z., Assempour, N., Iynkkaran, I., Liu, Y., Maciejewski, A., Gale, N., Wilson, A., Chin, L., Cummings, R., Le, D., Pon, A., Knox, C., Wilson, M. (2018). DrugBank 5.0: A major update to the DrugBank database for 2018. Nucleic Acids Research. 46:D1074–D1082.

Wong, C.-F., Chan, G. K.-L., Zhang, M.-L., Yao, P., Lin, H.-Q., Dong T. T.-X., Li, G., Lai, X.-P. Tsim, K. W.-K. (2017). Characterization of edible bird's nest by peptide fingerprinting with principal component analysis. Food Quality and Safety. 1:83–92.

Wong, R. S. Y. (2013). Edible bird's nest: Food or medicine? Chinese Journal of Integrative Medicine. 19(9):643–649. https://doi.org/10.1007/s11655-013-1563-y.

Wong, Z. C. F., Chan, G. K. L., Wu, K. Q. Y., Poon, K. K. M., Chen, Y., Dong, T. T. X., Tsim, K. W. K. (2018). Complete digestion of edible bird's nest releases free N-acetylneuraminic acid and small peptides: an efficient method to improve functional properties. Food and Function. 9(10):5139–5149. https://doi.org/10.1039/c8fo00991k.

Xie, Y., Zeng, H., Huang, Z., Xu, H., Fan, Q., Zhang, Y., Zheng, B. (2018). Effect of maternal administration of edible bird's nest on the learning and memory abilities of suckling offspring in mice. Neural Plasticity. https://doi.org/10.1155/2018/7697261

Xin, W. H., Kasim, Z. M., Babji, A. S., Ismail, N. H., Muhammad, N. N. Proximate analysis and amino acid composition in selected edible bird's nest. The 16th Food Innovation Asia Conference, Bangkok, Thailand. 2014.

Yew, M. Y., Koh, R. Y., Chye, S. M., Othman, I., Ng, K. Y. (2014). Edible bird's nest ameliorates oxidative stress-induced apoptosis in SH-SY5Y human neuroblastoma cells. BMC Complementary and Alternative Medicine. 14:391. https://doi.org/10.1186/1472-6882-14-391.

Yida, Z., Imam, U. U., Ismail, M. (2014). In vitro bioaccessibility and antioxidant properties of edible bird's nest following simulated human gastro-intestinal digestion. BMC Complementary and Alternative Medicine. 14(1). https://doi.org/10.1186/1472-6882-14-468.

Yida, Z., Imam, M. U., Ismail, M., Ismail, N., Hou, Z. (2015a). Edible bird's nest attenuates procoagulation effects of high-fat diet in rats. Drug Design, Development and Therapy. 9:3951–3959. https://doi.org/10.2147/DDDT.S87772.

Yida, Z., Imam, M. U., Ismail, M., Hou, Z., Abdullah, M. A., Ideris, A., Ismail, N. (2015b). Edible bird's nest attenuates high fat diet-induced oxidative stress and inflammation via regulation of hepatic antioxidant and inflammatory genes. BMC Complementary and Alternative Medicine. 15(1). https://doi.org/10.1186/s12906-015-0843-9.

Yida, Z., Imam, M. U., Ismail, M., Ooi, D. J., Sarega, N., Azmi, N. H., Ismail, N., Chan, K. W., Hou, Z., Yusuf, N. B. (2015c). Edible bird's nest prevents high fat diet-induced insulin resistance in rats. Journal of Diabetes Research. https://doi.org/10.1155/2015/760535.

Zhao, R., Li, G., Kong, X. J., Huang, X. Y., Li, W., Zeng, Y. Y., Lai, X. P. (2016). The improvement effects of edible bird's nest on proliferation and activation of B lymphocyte and its antagonistic effects on immunosuppression induced by cyclophosphamide. Drug Design, Development and Therapy. 10:371–381. https://doi.org/10.2147/DDDT.S88193.

20 Zebrafish Model for Drug Discovery and Screening

Sarah Hassan, Jabra Mustafa, Ahmad Hassan, and Wael Mohamed

20.1 INTRODUCTION

Over the past few decades, zebrafish (*Danio rerio*) have played a pivotal role in biomedical research, drug discovery, and drug screening. The zebrafish is a small, tropical fish ranging in size between 2–5 centimeters and naturally found in the rivers of South Asia (Patton et al., 2021). Zebrafish possess several unique characteristics that make it an excellent model organism for preclinical research, including small size, short life cycle, and inexpensive maintenance (Karaman et al., 2020). Additionally, because zebrafish are fertilized ex vivo, and a single breeding pair can produce hundreds of transparent embryos, it served as an ideal candidate for developmental and embryologic research in the 1930s when it was first introduced as a model organism. In the 1970s, zebrafish became an important animal model for genetic studies due to sharing 70% of their genome with humans, with more than 80% of disease proteins being conserved (Howe et al., 2013a). With the discovery of the CRISPR/Cas9 gene-editing technology, zebrafish models of human diseases have been developed, which have led to extraordinary advances in our understanding of disease pathogenesis and the discovery of novel treatments (Liu et al., 2019). Currently, zebrafish models have been developed for heart disease, liver disease, kidney disease, cancers, blood disorders, and behavioral disorders (Zhao et al., 2019; Goessling and Sadler, 2015; Outtandy et al., 2019; Cagan et al., 2019; Rissone and Burgess, 2018; Griffin et al., 2018). Zebrafish are one of the most commonly used model organisms in preclinical studies, along with mice, *C. elegans*, fruit flies, and yeast. Each organism possesses its own unique set of advantages and disadvantages, as summarized in Table 20.1, and careful selection of a specific animal model is a necessary step in outlining a successful preclinical study.

Because disease proteins and pathogenesis are often conserved between humans and zebrafish, drug molecules that interact with specific protein targets in humans are active in zebrafish models and display similar pharmacokinetic properties (MacRae and Peterson, 2015). This review outlines the history and current developments of drug discovery and screening of small molecules in zebrafish, as well as limitations of the zebrafish model.

20.2 OVERVIEW OF ZEBRAFISH AS A MODEL ORGANISM

When a new drug is discovered and being investigated, it is important to not only demonstrate its activity in vitro, but also to understand its mechanism of action within a living organism and its effects on diseased organs and the organism as a whole (Cagan et al., 2019). Phenotypic drug screening is a process in which potential drug compounds are delivered to whole living organisms in order to identify drugs that produce a desired phenotypic effect (Patton et al., 2021). In contrast, targeted drug screens involve focusing efforts on specific molecules that are known to be implicated in a disease pathogenesis. Drug discovery has traditionally relied heavily on phenotypic drug screening, but this approach has diminished in popularity with the emergence of targeted drug screens. This shift in approach is largely driven by a better understanding of the pathogenesis of common diseases. Phenotypic drug screening is still used, however, particularly in the search for first-in-class drugs.

In preclinical studies on zebrafish, embryos can actively absorb small compounds from the water, older zebrafish can absorb these compounds through the gills, or the compounds may be administered via oral galvage (Dang et al., 2016). The small size of the organism facilitates the phenotypic screening of a large quantity of embryos in the context of a whole, living vertebrate. Typically, the goal of high-throughput drug screens is to identify a drug compound that is highly selective towards a specific target. However, many drugs are able to bind multiple targets—a concept known as polypharmacology—and this characteristic is often incorporated into pharmaceutical designs in order to develop drugs that act on multiple biochemical pathways (Moya-Garcia et al., 2017). For instance, a recent study used behavioral assays to demonstrate that the drug haloperidol produces its desired antipsychotic effects through a multi-target mechanism of action (Bruni et al., 2016). In some instances, zebrafish models are better than mice at replicating phenotypic effects of drugs seen in humans. For example, the drug thalidomide, which was used in the 1950s as an anti-nausea medication in pregnant women, was found to result in congenital abnormalities (Grogan and Winston, 2023). Thalidomide was not found to result in any defects in the mouse model, but in the zebrafish model, the

DOI: 10.1201/9781003402893-22

TABLE 20.1
Advantages and Challenges of Common Model Organisms

Model Organism	Advantages	Challenges
Mice (*Mus musculus*)	• Genetic and physiological similarities to humans • Availability of transgenic and knockout models • Complex behaviors and organ systems	• Higher cost and resource-intensive • Longer gestation and lifespan compared to invertebrates • Ethical considerations in experimentation
Zebrafish (*Danio rerio*)	• Transparent embryos for easy observation • Rapid organ development; short generation time • High reproductive rate; large-scale screenings • Lack of ethics requirements during early developmental stages	• Limited representation of mammalian physiology • Limited behavioral complexity • Limited genetic homology with mammals
***C. elegans* (*Caenorhabditis elegans*)**	Simple, well-defined anatomy • Short lifespan and rapid generation time • Fully mapped and relatively small genome • Lack of ethics requirements	Lack of complex organ systems • Limited representation of mammalian physiology • Limited behavioral complexity
Fruit Fly (*Drosophila melanogaster*)	• Short generation time; large-scale genetic studies • Well-established genetic tools and resources • Share many biological processes with mammals • Lack of ethics requirements	• Limited representation of vertebrate physiology • Limited behavioral complexity
Yeast (*Saccharomyces cerevisiae*)	Rapid growth; easy genetic manipulation • Well-characterized cellular processes • Valuable for studying basic cellular functions • Lack of ethics requirements	• Lack of complex multicellular structures • Limited representation of higher organisms

drug induces similar morphological limb defects as is seen in humans (Patton et al., 2021).

20.3 DRUG DISCOVERY AND SCREENING ASSAYS IN ZEBRAFISH

The implementation of zebrafish into preclinical research experiments has provided numerous opportunities to accelerate the process of drug discovery and drug screening. Although these two terms are sometimes used interchangeably, they represent distinct stages of the drug development process. Drug discovery involves the initial stage of identifying novel compounds that possess therapeutic potential, while drug screening consists of testing a large number of known compounds for their therapeutic potential (Hughes et al., 2011). In drug screens, compounds with known therapeutic effects in a specific disease process are often shown to also be effective in separate disease processes. Two key approaches to drug discovery and screening and zebrafish are phenotype-based screens and high-throughput screens.

Phenotype-based screens laid the foundation for pharmaceutical drug discovery before molecular target-based screening became popularized (Zheng et al., 2013). Phenotype-based screens, however, are still very commonly used and have led to numerous landmark discoveries in medicine. For example, the discovery of penicillin and other antibiotics in bactericidal screens have largely been phenotypic in nature (Pina et al., 2009). In phenotype-based screens—often termed 'forward pharmacology'—the protein target and molecular mechanism of the drug can remain unknown even after establishing the drug's efficacy. In zebrafish, psychiatric and neurological disorders are often studied with behavioral phenotypic screens, in which a mutated zebrafish model with abnormal locomotion, feeding, or social behaviors is studied in the context of various compounds (Rosa et al., 2022). Cardiac phenotyping, which will be discussed further in a later section, is especially useful in zebrafish due to the similarity of their cardiac anatomy and pathways with humans (Miura and Yelon, 2011). Investigators have produced various zebrafish models of cardiac pathologies and can observe changes in heart rate, morphology, and contractility in response to different compounds. This is also a highly useful technique to discover previously unknown cardiotoxic effects of drugs. Finally, developmental phenotyping is especially useful in zebrafish due to the transparent nature of zebrafish embryos (Patton et al., 2021). This approach can be beneficial to determine if a drug possesses teratogenicity and can also provide insight into therapeutic interventions for developmental disorders.

High-throughput screens consist of systematically testing large numbers of molecules obtained from compound libraries for their biological activity (Lubin et al., 2021). The development of advanced automated techniques allow investigators to incorporate the use of robotics to not only handle zebrafish embryos and larvae in multi-well plates, but to also conduct rapid screens in conjunction with artificial intelligencedriven algorithms and advanced imaging techniques. Depending on the type of experiment being conducted, researchers can also combine advanced imaging technologies with fluorescent reporters, which involve genes or proteins that emit light upon activation (Abdelmoneim et al., 2020). The optically transparent embryos and larvae are placed into wells of microtiter plates and are combined with various genome-editing techniques to produce engineered zebrafish that can be screened automatically. Researchers are able to create genetically engineered zebrafish that express fluorescent proteins and allow for real-time visualization of drug responses and localization.

Several drugs that are currently being used to treat human diseases have their origins in zebrafish. Drug screening using zebrafish models provides several advantages over mouse models and in vitro cell cultures, providing more information regarding bioavailability, toxicity, and tissue selectivity (Dash and Patnaik, 2023). Several characteristics of zebrafish embryos make them ideal for screening a large number of compounds in a smaller time frame, including small size of the embryo, fast organogenesis rate, optic transparency, fully sequenced genome, more than 80% conservation of disease proteins when compared to humans, and ease with which gene-editing technology can be used. Recently, researchers have developed a whole-organism screening platform known as ARQiv-HTS, which allows for high-throughput in vivo drug discovery in zebrafish (White et al., 2016). An artificial intelligence–driven algorithm automatically detects zebrafish in brightfield images, identifies anatomical structures, divides the zebrafish into separate regions, and exclusively selects the correct orientation of the organism to display (Lubin et al., 2021). Not only does this method significantly reduce the time and cost of drug screening, but using a whole-organism approach provides the benefit of identifying toxic effects of medications relatively early in the investigative process.

20.4 ZEBRAFISH MODELS FOR SPECIFIC DISEASES

Small-molecule screening allows investigators to identify suitable targets in a particular model organism and select for compounds with certain characteristics, such as increased cell permeability, favorable pharmacokinetic and pharmacodynamic profiles, maximum effectiveness, and minimal toxicity (Dash et al., 2023). Several landmark discoveries in medicine have been produced via chemical screening assays, notably in the fields of heart disease, neurological disease, and cancer.

20.4.1 Cardiac Disease

Zebrafish have been used extensively to study various heart diseases, and high-throughput approaches have been utilized to examine characteristics of zebrafish cardiovascular physiology such as heart rate, contractility, and blood flow (Dash et al., 2023). Optical voltage mapping and Ca^{2+} imaging are examples of assays that can often be used in conjunction to provide further information (Dash et al., 2017; Jarvinen et al., 2018). Additionally, mutated zebrafish models of various cardiac diseases can be produced to aid in drug discovery. A study conducted in 2011 used the zebrafish *KCNH2* mutant *breakdance*—a model of congenital long QT (LQT) syndrome—to identify drugs that have the ability to provide functional rescue of LQT (Peal et al., 2011). LQT is a life-threatening disease caused by a genetic mutation that results in the prolongation of myocardial repolarization (Krahn et al., 2022). Discovery of new molecules to treat LQT has posed a challenge due to unsuitable animal models. For example, cardiac repolarization in mice and rats is substantially different from that in humans, and although rabbit models of LQT have been produced, rapid chemical screening is not feasible on larger mammals (Brunner et al., 2008). However, zebrafish retain key features of human cardiac repolarization, and successful drug-induced and genetic models of LQT have been produced (Dash et al., 2023). Namely, the zebrafish *breakdance* mutant is characterized by an I59S mutation in KCNH2—the protein associated with LQT type 2—and reliably produces the same 2:1 AV block seen in human pediatric patients with the disease (Peal et al., 2011). The investigators conducted a whole-organism chemical screening assay of 1,200 small molecules, which identified two compounds—flurandrenolide and 2-MMB—that rescued zebrafish with LQT. It was also demonstrated that flurandrenolide functions via the glucocorticoid receptor mediated pathway. The authors also found that dexamethasone, a canonically pure glucocorticoid, can rescue LQT as well.

20.4.2 Kidney Disease

The kidney is a complex organ, and nephrogenesis involves the development of the kidney through various stages, including the pronephros, mesonephros, and metanephros (Drummond and Davidson, 2010). Kidney disease involves several hallmark characteristics, such as a lessened glomerular filtration rate and increased excretion of urinary albumin (Kalantar-Zadeh et al., 2021). Although the gross structure and development of zebrafish varies from humans, the anatomy and function of individual nephrons is largely conserved among all vertebrates, which allows for the development of zebrafish models of kidney disease (Desgrange and Cereghini, 2015). For example, a zebrafish model of polycystic kidney disease (PKD), which is a disease characterized by the formation of renal cysts, has been developed in order to further study the molecular pathogenesis of the disease, which can aid in identifying new therapies (Swanhart et al., 2011). Autosomal-dominant polycystic kidney disease (ADPKD)—a subtype of PKD—is associated with mutations in two genes: *PKD1* (polycystin-1) and *PKD2* (polycystin-2) (Bergmann et al., 2018). Knockdown of *pkd2* function in zebrafish resulted in kidney cysts, hydrocephalus, and left-right asymmetry

(Bisgrove et al., 2005). A recent study conducted a chemical screen on zebrafish *pkd2* mutant embryos in order to identify molecules from a compound library that can rescue the mutant phenotype (Metzner et al., 2020). The investigators used dorsal tail curvature of the zebrafish as the assay readout and identified ALK5 (transforming growth factor beta receptor I) kinase inhibitors as strong suppressors of the *pkd2* tail phenotype.

20.4.3 Neurological Disease

Neurological diseases can be studied in zebrafish due to the evolutionary conservation of their nervous system with humans (Kalueff et al., 2014). A 2013 study identified the zebrafish mutant *grs357*, which produces a mutated glucocorticoid-receptor gene and results in glucocorticoid resistance and has shown to be a viable model of depression (Ziv et al., 2013). The *grs357* mutants become immobile when placed in an unfamiliar environment and demonstrate diminished exploratory behavior. However, the benzodiazepine diazepam and the antidepressant fluoxetine restore normal behavior. Interestingly, social interactions with other zebrafish also reversed the abnormal behavior. Parkinson's disease, a movement disorder characterized by the loss of dopaminergic neurons in the substantia nigra, has also been extensively studied in zebrafish. A 2005 study demonstrated that the neurotoxic compound MPTP (1-methyl-4-phenyl-1,2,3,6-tetrahydropyridine) induces the destruction of dopaminergic neurons in the zebrafish diencephalon and results in abnormal swimming and behavior (Lam et al., 2005). It was found that the MAO-B inhibitor deprenyl reverses this effect by catalyzing the conversion of MPTP to its metabolite MPP+. A separate study showed that the neurotoxin 6-hydroxydopamine resulted in a similar loss of dopaminergic neurons and compromised movement, but treatment with vitamin E, minocycline, or sinemet reversed these effects (Feng et al., 2014).

20.4.4 Cancer

Several characteristics of zebrafish make it a useful model to use in translational cancer research, including the transparency of zebrafish, the high rate of reproduction, and the ability to visualize tumors (Dash and Patnaik, 2023). The ability to conduct high-throughput drug screens on zebrafish also allows the animal model to be useful in the discovery of novel cancer treatments. A study conducted in 2014 utilized zebrafish in a small-molecule screen to identify compounds that demonstrate toxicity towards MYC-overexpressing thymocytes using a human T-cell acute lymphoblastic leukemia (T-ALL) cell line (Gutierrez et al., 2014). The study found that the antipsychotic perphenazine was found to induce apoptosis in T-ALL cells, and phenothiazines were thus identified as a new anti-cancer drug class. A separate study conducted in 2011 investigating melanoma used transgenic *mitf*-BRAF^{V600E}; p53$^{-/-}$ zebrafish embryos and found that dihydroorotate dehydrogenase inhibitors such as leflunomide led to a significant decrease in the growth of melanoma (White et al., 2011). These landmark studies establish zebrafish as a model organism that is crucial for drug discovery and screening, as well as investigating therapeutic applications that directly translate to clinical medicine.

20.5 ETHICAL CONSIDERATIONS

Zebrafish possess over 26,000 protein-coding genes, and nearly 71% of these genes have orthologs in humans (Howe et al., 2013b). Decades of preclinical studies have demonstrated striking similarities between zebrafish and humans on the molecular level, especially in the mechanism of action of different pharmaceuticals, as well as in enzymes and neurotransmitter receptors (de Abreu et al., 2019). This conservation between zebrafish and human physiology largely contributes to the appeal of zebrafish as a model organism. Zebrafish are particularly useful in research on pain, stress, and CNS drug discovery, which typically require careful ethical considerations prior to undertaking the study. For example, zebrafish respond to acute and chronic stress with an increase in cortisol, and studying their stress response has provided incredible insight into the molecular mechanisms of stress responses in humans (Song et al., 2018). Zebrafish are an indispensable tool in pain research and studying opioid pharmacology due to their evolutionarily conserved pain receptors (Demin et al., 2018).

Countries around the world have different sets of laws regulating the use of animals in research studies in order to promote animal welfare and engage in ethically sound research (de Abreu et al., 2019). It is imperative that the investigator, prior to conducting an experiment, perform a thorough analysis on the welfare of the animal and the anticipated benefit of the research experiment. In the United States, the Institutional Animal Care and Use Committee (IACUC)—which ensures the ethical and humane utilization of animals in research studies—is a federally mandated committee under the Animal Welfare Act (Everitt and Berridge, 2017). Investigators in the United States seeking to include adult or post-hatch larval zebrafish in their research protocol must first seek approval from the IACUC. Not every country mandates an IACUC, but most have some form of a regulatory body that oversees the use of animals in research protocols. Some countries, such as China, Japan, and Russia, do not cover zebrafish under their regulations (de Abreu et al., 2019).

Over 50 years ago, the principles of the 3Rs—replacement, reduction, and refinement—were introduced by Russell and Burch in their landmark paper, *The Principles of Humane Experimental Technique* (Russell and Burch, 1959). These principles guide regulatory bodies in their approach to promoting animal welfare in research. The principle of replacement involves using non-animal subjects in research experiments whenever possible (Tannenbaum and Bennett, 2015). For example, if a research hypothesis can be adequately tested using in vitro cell cultures or computerized

TABLE 20.2
The 3R Principles and Applications in Zebrafish Research

Principle	Definition	Application in Zebrafish Research
Replacement	Using alternatives to animal subjects in research protocols whenever possible.	• Utilizing non-animal models such as in vitro cell cultures or computerized simulations. • Using embryonic zebrafish or cell lines rather than adult zebrafish.
Reduction	Minimizing the number of animal subjects used to reach a meaningful conclusion.	• Using high-throughput screening methods and advanced imaging strategies to extract more data from each animal subject.
Refinement	Optimizing research protocols to enhance the welfare of animal subjects.	• Ensuring proper housing for zebrafish. • Refining anesthesia and euthanasia protocols. • Consideration of the psychological well-being of zebrafish.

simulations, then that would be preferable to using live animals. In zebrafish research, this typically consists of using embryonic zebrafish if possible, and using specific cell lines rather than adult zebrafish. The principle of reduction consists of minimizing the number of animal subjects used in order to arrive at a meaningful conclusion. Researchers often modify their research protocol in order to facilitate the use of fewer animal subjects and extract more data from each subject. In zebrafish research, studies can utilize high-throughput screening methods and various advanced imaging strategies in order to collect more data from each animal subject. The principle of refinement involves optimizing the welfare of the animal subjects in order to mitigate pain, stress, and overall harm. In zebrafish research, this involves providing the animals with proper housing, optimizing anesthesia or euthanasia protocols, and frequently taking into account the psychological well-being of the animal subjects. Table 20.2 provides a summary of the 3R principles and how they apply to research and zebrafish experimentation. Investigators conducting research on zebrafish are encouraged to take the 3Rs into account when designing their research protocols.

20.6 CHALLENGES OF THE ZEBRAFISH MODEL

Although screening in zebrafish offers a variety of advantages and has significantly accelerated the rate of drug discovery, investigators must be aware of several logistical challenges prior to conducting research with the zebrafish model. Primarily, robust research protocols and careful planning are required when conducting high-throughput screens in order to optimize data readouts (Gehrig et al., 2018). Typically, an investigator will be handling thousands of zebrafish at a time, and protocols must be established for animal husbandry, embryo culture, micromanipulation, and simple handling. Large-scale screens are often hindered by the optimization of current technology for in vitro assays, as the three-dimensional structure of zebrafish embryos leads to random orientation of the embryos within the microtiter plate wells (Peravali et al., 2011). This obscures the image of target structures, and further development of automated tools to examine zebrafish embryos in three dimensions is needed to maximize the potential of this model. Despite these challenges, investigators continue to successfully conduct research using the zebrafish model, and careful experimental design, robust research protocols, and improvements in technology enhance the utility of this model organism.

20.7 CONCLUSION AND FUTURE DIRECTIONS

Zebrafish have become an essential model organism for both drug screening and drug discovery. The conservation of disease-related genes and the ease with which genetic mutants can be produced to model human diseases have led to the discovery of several drugs that are currently undergoing clinical trials. The zebrafish model, however, possesses some limitations. Namely, studies on drug safety and toxicity are often complicated by the challenge of differentiating between lack of drug permeability and lack of toxicity. Antibody studies have also proven to be challenging due to the failure of anti-human antibodies to cross-react with proteins on the zebrafish cell surface. Tools available to study zebrafish are less developed than other animal models, and future work is needed to maximize the potential of the zebrafish model. Further application of AI techniques in automated phenotype-based high-throughput drug screening is anticipated to reduce time and labor costs. Overall, the zebrafish is an important animal model that has led to many discoveries, and further research will allow us to maximize its potential.

REFERENCES

Abdelmoneim A, Clark CL, Mukai M. Fluorescent reporter zebrafish line for estrogenic compound screening generated using a CRISPR/Cas9-mediated knock-in system. Toxicol Sci. 2020;173(2):336–346. doi:10.1093/toxsci/kfz224

Bergmann C, Guay-Woodford LM, Harris PC, Horie S, Peters DJM, Torres VE. Polycystic kidney disease. Nat Rev Dis Primers. 2018 Dec;4(1):50. doi:10.1038/s41572-018-0047-y

Bisgrove BW, Snarr BS, Emrazian A, Yost HJ. Polaris and Polycystin-2 in dorsal forerunner cells and Kupffer's vesicle are required for specification of the zebrafish left-right axis. Dev Biol. 2005 Nov;287(2):274–288.

Bruni G, Rennekamp AJ, Velenich A, McCarroll M, Gendelev L, Fertsch E, Taylor J, Lakhani P, Lensen D, Evron T, Lorello PJ, Huang XP, Kolczewski S, Carey G, Caldarone BJ, Prinssen E, Roth BL, Keiser MJ, Peterson RT, Kokel D. Zebrafish behavioral profiling identifies multitarget antipsychotic-like compounds. Nat Chem Biol. 2016 Jul;12(7):559–566. doi: 10.1038/nchembio.2097.

Brunner M, Peng X, Liu GX, Ren XQ, Ziv O, Choi BR, Mathur R, Hajjiri M, Odening KE, Steinberg E, Folco EJ, Pringa E, Centracchio J, Macharzina RR, Donahay T, Schofield L, Rana N, Kirk M, Mitchell GF, Poppas A, Zehender M, Koren G. Mechanisms of cardiac arrhythmias and sudden death in transgenic rabbits with long QT syndrome. J Clin Invest. 2008 Jun;118:2246–2259.

Cagan RL, Zon LI, White RM. Modeling cancer with flies and fish. Dev. Cell. 2019 May;49:317–324.

Dang M, Henderson RE, Garraway LA, Zon LI. Long-term drug administration in the adult zebrafish using oral gavage for cancer preclinical studies. Dis. Model. Mech. Jul 2016;9:811–820.

Dash SN, Narumanchi S, Pertunen S, Pavola J, Tikanen I, Lehtonen S. Sept7b is required for the subcellular organization of cardiomyocytes and cardiac function in zebrafish heart. Am. J. Physiol. 2017 May;312:H1085–H1095. doi: 10.1152/ajpheart.00394.2016

Dash SN, Patnaik L. Flight for fish in drug discovery: A review of zebrafish-based screening of molecules. Biol Lett. 2023 Aug;19(8):20220541. doi: 10.1098/rsbl.2022.0541.

de Abreu MS, Giacomini ACVV, Echevarria DJ, Kalueff AV. Legal aspects of zebrafish neuropharmacology and neurotoxicology research. Regul Toxicol Pharmacol. 2019 Feb;101:65–70. doi: 10.1016/j.yrtph.2018.11.007.

Demin KA, Meshalkina DA, Kysil EV, et al. Zebrafish models relevant to studying central opioid and endocannabinoid systems. Prog Neuropsychopharmacol Biol Psychiatry. 2018 Aug;86:301–312. doi:10.1016/j.pnpbp.2018.03.024

Desgrange A, Cereghini S. 2015. Nephron patterning: Lessons from *Xenopus*, zebrafish, and mouse studies. Cells. 2015 Sep;4:483–499. doi:10.3390/cells4030483

Drummond IA, Davidson AJ. Zebrafish kidney development. Methods Cell Biol. 2010;100:233–260. doi:10.1016/B978-0-12-384892-5.00009-8

Everitt JI, Berridge BR. The role of the IACUC in the design and conduct of animal experiments that contribute to translational success. ILAR J. 2017 Jul 1;58(1):129–134. doi: 10.1093/ilar/ilx003. PMID: 28838070.

Feng CW, Wen ZH, Huang SY, Hung HC, Chen CH, Yang SN, Chen NF, Wang HM, Hsiao CD, Chen WF. Effects of 6-hydroxydopamine exposure on motor activity and biochemical expression in zebrafish (*Danio rerio*) larvae. Zebrafish. 2014 Jun;11(3):227–239. doi: 10.1089/zeb.2013.0950.

Gehrig J, Pandey G, Westhoff JH. Zebrafish as a model for drug screening in genetic kidney diseases. Front Pediatr. 2018 Jun;6:183. doi:10.3389/fped.2018.00183

Goessling W, Sadler KC. Zebrafish: An important tool for liver disease research. Gastroenterology. 2015 Nov;149:1361–1377.

Griffin A, Hamling KR, Hong SN, Anvar M, Lee LP, Baraban SC. Preclinical animal models for dravet syndrome: Seizure phenotypes, comorbidities and drug screening. Front. Pharmacol. 2018 June;9:573.

Grogan DP, Winston NR. Thalidomide. [Updated 2023 Jun 12]. In: StatPearls [Internet]. Treasure Island, FL: StatPearls Publishing, 2023 Jan. Available from: www.ncbi.nlm.nih.gov/books/NBK557706/

Gutierrez A, Pan L, Groen RW, Baleydier F, Kentsis A, Marineau J, Grebliunaite R, Kozakewich E, Reed C, Pflumio F, Poglio S, Uzan B, Clemons P, VerPlank L, An F, Burbank J, Norton S, Tolliday N, Steen H, Weng AP, Yuan H, Bradner JE, Mitsiades C, Look AT, Aster JC. Phenothiazines induce PP2A-mediated apoptosis in T cell acute lymphoblastic leukemia. J Clin Invest. 2014 Feb;124(2):644–655. doi: 10.1172/JCI65093.

Howe K, et al. The zebrafish reference genome sequence and its relationship to the human genome. Nature. 2013a Apr 25;496:498–503.

Howe K, Clark MD, Torroja CF, et al. The zebrafish reference genome sequence and its relationship to the human genome. Nature. 2013b Apr 25;496(7446):498–503. doi: 10.1038/nature12111 [Erratum in: Nature. 2014 Jan 9;505(7482):248.

Hughes JP, Rees S, Kalindjian SB, Philpott KL. Principles of early drug discovery. Br J Pharmacol. 2011 Mar;162(6):1239–1249. doi: 10.1111/j.1476-5381.2010.01127.x.

Jarvinen J, Hasinen M, Dash SN, Vornanen M. 2018. Expression of calcium channel transcripts in zebrafish heart: Dominance of T-type channels. J. Expt. Biol. 2018 May;221:jeb179226.

Kalantar-Zadeh K, Jafar TH, Nitsch D, Neuen BL, Perkovic V. Chronic kidney disease. Lancet. 2021 Aug;398(10302):786–802. doi:10.1016/S0140-6736(21)00519-5

Kalueff AV, Stewart AM, Gerlai R. Zebrafish as an emerging model for studying complex brain disorders. Trends Pharmacol. Sci. 2014 Feb;35:63–75. doi: 10.1016/j.tips.2013.12.002

Karaman GE, Emekli-Alturfan E, Akyüz S. Zebrafish; an emerging model organism for studying toxicity and biocompatibility of dental materials. Cell Mol Biol (Noisy-le-grand). 2020 Dec 31;66(8):41–46.

Krahn AD, Laksman Z, Sy RW, Postema PG, Ackerman MJ, Wilde AAM, Han HC. Congenital long QT syndrome. JACC Clin Electrophysiol. 2022 May;8(5):687–706. doi: 10.1016/j.jacep.2022.02.017.

Lam CS, Korzh V, Strahle U. 2005. Zebrafish embryos are susceptible to the dopaminergic neurotoxin MPTP. Eur. J. Neurosci. 2005 Mar;21:1758–1762.

Liu K, Petree C, Requena T, Varshney P, Varshney GK. Expanding the CRISPR toolbox in zebrafish for studying development and disease. Front. Cell Dev. Biol. 2019 Mar;7:13.

Lubin A, Otterstrom J, Hoade Y, Bjedov I, Stead E, Whelan M, Gestri G, Paran Y, Payne E. A versatile, automated and high-throughput drug screening platform for zebrafish embryos. Biol Open. 2021 Sep;10(9):bio058513. doi:10.1242/bio.058513

MacRae CA, Peterson RT. Zebrafish as tools for drug discovery. Nat. Rev. Drug Discov. Sep 2015;14:721–731.

Metzner A, Griffiths JD, Streets AJ, Markham E, Philippou T, Van Eeden FJM, Ong, ACM. A high throughput zebrafish chemical screen reveals ALK5 and non-canonical androgen signalling as modulators of the pkd2$^{-/-}$ phenotype. Sci Rep. 2020 Jan;10(1):72. doi:10.1038/s41598-019-56995-7

Miura GI, Yelon D. A guide to analysis of cardiac phenotypes in the zebrafish embryo. Methods Cell Biol. 2011;101:161–180. doi: 10.1016/B978-0-12-387036-0.00007-4.

Moya-Garcia A, Adeyelu T, Kruger FA, Dawson NL, Lees JG, Overington JP, Orengo C, Ranea JAG. Structural and functional view of polypharmacology. Sci. Rep. Aug 2017;7:10102.

Outtandy P, Russell C, Kleta R, Bockenhauer D. Zebrafish as a model for kidney function and disease. Pediatr Nephrol. 2019 May;34(5):751–762. doi: 10.1007/s00467-018-3921-7.

Patton EE, Zon LI, Langenau DM. Zebrafish disease models in drug discovery: From preclinical modelling to clinical trials. Nat Rev Drug Discov. 2021 Aug;20(8):611–628. doi: 10.1038/s41573-021-00210-8.

Peal DS, Mills RW, Lynch SN, Mosley JM, Lim E, Ellinor PT, January CT, Peterson RT, Milan DJ. Novel chemical suppressors of long QT syndrome identified by an in vivo functional screen. Circulation. 2011 Jan;123(1):23–30. doi: 10.1161/CIRCULATIONAHA.110.003731.

Peravali R, Gehrig J, Giselbrecht S, Lutjohann DS, Hadzhiev Y, Muller F, Liebel U. Automated feature detection and imaging for high-resolution screening of zebrafish embryos. Biotechniques. 2011 May;50:319–324. doi: 10.2144/000113669

Pina AS, Hussain A, Roque ACA. An historical overview of drug discovery. Methods Mol Biol. 2009;572:3–12.

Rosa JGS, Lima C, Lopes-Ferreira M. Zebrafish larvae behavior models as a tool for drug screenings and pre-clinical trials: A review. Int J Mol Sci. 2022 Jun 14;23(12):6647. doi: 10.3390/ijms23126647.

Rissone A, Burgess SM. Rare genetic blood disease modeling in zebrafish. Front. Genet. 2018 Aug;9:348.

Russell WMS, Burch RL. (as reprinted 1992). The Principles of Humane Experimental Technique. Wheathampstead: Universities Federation for Animal Welfare, 1959.

Song C, Liu BP, Zhang YP, Peng Z, Wang J, Collier AD, Echevarria DJ, Savelieva KV, Lawrence RF, Rex CS, Meshalkina DA, Kalueff AV. Modeling consequences of prolonged strong unpredictable stress in zebrafish: Complex effects on behavior and physiology. Prog Neuropsychopharmacol Biol Psychiatry. 2018 Feb 2;81:384–394. doi: 10.1016/j.pnpbp.2017.08.021.

Swanhart LM, Cosentino CC, Diep CQ, Davidson AJ, de Caestecker M, Hukriede NA. Zebrafish kidney development: Basic science to translational research. Birth Defects Res C Embryo Today. 2011 Jun;93(2):141–156. doi:10.1002/bdrc.20209

Tannenbaum J, Bennett BT. Russell and Burch's 3Rs then and now: The need for clarity in definition and purpose. J Am Assoc Lab Anim Sci. 2015 Mar;54(2):120–32.

White DT, Eroglu AU, Wang G, Zhang L, Sengupta S, Ding D, Rajpurohit SK, Walker SL, Ji H, Qian J, Mumm JS. ARQiv-HTS, a versatile whole-organism screening platform enabling in vivo drug discovery at high-throughput rates. Nat Protoc. 2016 Dec;11(12):2432–2453. doi: 10.1038/nprot.2016.142.

White RM, Cech J, Ratanasirintrawoot S, Lin CY, Rahl PB, Burke CJ, Langdon E, Tomlinson ML, Mosher J, Kaufman C, Chen F, Long HK, Kramer M, Datta S, Neuberg D, Granter S, Young RA, Morrison S, Wheeler GN, Zon LI. DHODH modulates transcriptional elongation in the neural crest and melanoma. Nature. 2011 Mar 24;471(7339):518–522. doi: 10.1038/nature09882.

Zhao Y, Zhang K, Sips P, MacRae CA. Screening drugs for myocardial disease in vivo with zebrafish: An expert update. Expert Opin. Drug Discov. 2019 Apr;14:343–353.

Zheng W, Thorne N, McKew JC. Phenotypic screens as a renewed approach for drug discovery. Drug Discov Today. 2013 Nov;18(21–22):1067–1073. doi: 10.1016/j.drudis.2013.07.001.

Ziv L, Muto A, Schoonheim PJ, Meijsing SH, Strasser D, Ingraham HA, Schaaf MJ, Yamamoto KR, Baier H. An affective disorder in zebrafish with mutation of the glucocorticoid receptor. Mol Psychiatry. 2013 Jun;18(6):681–691. doi: 10.1038/mp.2012.64.

Index

Note: Page numbers in *italics* indicate a figure and page numbers in **bold** indicate a table on the corresponding page.